Sedges & Grasses of Eastern UTTAR PRADESH

Sedges & Grasses of Eastern *UTTAR PRADESH*

– Volume 1 –

by
Ajai Kumar Singh
Reader
Department of Botany
Udai Pratap Autonomous College
Varanasi - 221 002

2007
DAYA PUBLISHING HOUSE
Delhi - 110 035

ISBN 81-7035-463-3 (Set)
ISBN 81-7035-464-1 (Vol. 1)
ISBN 81-7035-465-X (Vol. 2)

Published by : **Daya Publishing House**
1123/74, Deva Ram Park
Tri Nagar, Delhi - 110 035
Phone: 27383999
Fax: (011) 23244987
e-mail : dayabooks@vsnl.com
website : www.dayabooks.com

Showroom : 4760-61/23, Ansari Road, Darya Ganj,
New Delhi - 110 002
Phone: 23245578, 23244987

Laser Typesetting : **Classic Computer Services**
Delhi - 110 035

Printed at : **Chawla Offset Printers**
Delhi - 110 052

PRINTED IN INDIA

DEDICATION

Dedicated to my family who went through the fatigues of seeing the book take shape; to my students who worked with me during the extensive and tiring field studies.

Preface

Most of the areas of reorganized Uttar Pradesh is cultivated. The region is under explored and very little is known about its plant resources. Some district floras viz., Flora of Agra, Flora of Allahabad and Flora Gorakhpurensis and checklists of some districts have been published but the detailed floristic account is not yet published and as such there is meagre information on plant resources of the state. The concept of endemism is also altered in many cases.

In Indian Plant Taxonomy what is needed more is elaborate description encompassing a complete range of variation with illustrations of all species. The vegetative characters neglected in the past, and others if any, should be studied to check whether any of them constitute more striking characters to enhance our understanding on flora. Keeping these facts in mind present work was started eight years back.

Though there have been some studies in eastern Uttar Pradesh in past, they are of a general nature. An in-depth study of limited groups of economic value has not been undertaken so far. The family Cyperaceae and Poaceae form an important and difficult group with the latter having considerable economic potential, and it was thought desirable to undertake a floristic survey of Eastern Uttar Pradesh for getting a through inventory of the members of these specialized group of flowering plants.

The present book is the outcome of about eight years study of author and his students. This book has been written without any research grants or any paid technical assistance. With the exception of author's own salary and the College infrastructure, this book has consumed no taxpayer's money. Total area was carefully explored, critical studies were done for identification, nomenclature and field notes. All specimens are deposited in the Herbarium, Department of Botany, Udai Pratap Autonomous College, Varanasi (India).

Sedges and Grasses of Eastern Uttar Pradesh deal with 320 species (86 species and 4 subspecies of sedges and 234 species and 2 varieties of Grasses). Thirty-two species of Sedges are new addition to the Upper Gangetic Plain, 10 species of Sedges; 3 genera and 23 species of Grasses are new to the Uttar Pradesh. Five species of Sedges are new to India. Species description, upto date name, field notes, synonyms and simple key for identification have been provided, line diagrams of 293 taxa are given. Appendices includes check-list of sedges and grasses of 7 districts of Eastern Uttar Pradesh, Upper Gangetic Plain, Uttar Pradesh and India

This book will be useful for the research scholars, teachers, students and persons, interested in weed and grassland management. In particular it will be more useful in Eastern Uttar Pradesh and also adjacent area of Bihar state.

I express my sincere thanks to the Principal, Udai Pratap Autonomous College, Varanasi for providing the facilities to complete the work. I also thanks to Daya Publishing House, Delhi for the publication of this book.

Ajai Kumar Singh

Contents

List of Illustrations

Contd...

Contd...

Contd...

Contd...

Contd...

Contd...

Contd...

Contd...

Chapter 1
Introduction

After a long period of antipathy, Biodiversity study has once again come to the centre-stage of biology, the world over. We now know, that the future quality of human life on this planet would very much depend upon the existing plant and animal diversity and are aware of the need for their documentation, conservation and for its better understanding. Unfortunately, this has come at a time when we are passing through an age of rapid loss of Biodiversity, which would have catastrophic results for all of us in future (Heywood, 1993).

Plants and plant products play a vital role in the material prosperity of any country, and therefore a thorough study of the occurrence, distribution, ecological association, phenological notes, local names, economic uses etc., of the natural vegetation assume a great importance. Vegetational and floristic studies have gained increased importance and relevance in recent years in view of the present need for a thorough, up to date assessment of the natural resources of our vast country. The need for conservation and balanced exploitation of the nation's plant wealth has also been keenly felt. In this context intensive exploration of limited areas for obtaining an inventory of the floristic elements present and identification of the potential sources of economic importance have become imperative. Initiating the work on Indian flora, Santapau (1958a), suggested that Universities should be encouraged and helped to produce the flora of University campuses, towns and neighbouring areas. Similar suggestions and recommendations were made at Summer School in Botany, held at Kodaikanal (1962) and by Johri and Doreswamy (1970) also. As a result, there has been a revival of interest among botanists in reinvesgating and modernizing the floristic studies in different regions of the Indian subcontinent. Since the attainment of Independence there has been an urge for better utilization of plant resources of the country. The late Prime Minister Sri Jawahar Lal Nehru appropriately wrote in his foreward to the first volume of Wealth of India (1948), "Nevertheless, India is wealthy and Wealth of India is there, but inspite of this wealth, the people are poor. The problem for us is to utilize this stored up wealth of the country in the soil and under the soil for the benefit of Indian humanity." His remarks are very significant because, though the country is very rich in its plant wealth, the information about it is rather really insufficient.

Venu (2004) suggested that in Indian plant taxonomy, what is needed more is elaborate description encompassing a complete range of variations with illustrations of all species. The vegetative characters, neglected in the past, and other if any, should be studied to check whether any of them constitute more striking characters to enhance our understanding of flora.

The opportunities for intensive plant exploration and introduction are now better, but there is an even-increasing destruction of the biota and forests leading to extinction and depletion of our plant resources. There is thus urgent need for understanding and accelerated survey of the plant resources of the country (Maheshwari, 1976).

Santapau (l.c.) suggested that, India now wants not only mere collection but critical observation on the spot with illustrative specinens also. This is possible only when limited areas explored intensively.

The flora of our country is very rich in many ways such as number of species, economic potential, endemism and not unoften in the uniqueness of biological characteristics. Only a few regions of the world of equal size have such diversity. Hooker (1907) has stated that the flora of British India is more than that of any other country of equal area in the Eastern hemisphere, if not the globe. Plant resources including forestry and agricultural, support the living of the large majority of our people. Further, the vast majority of our people live in villages, in relatively closer contact with nature and plant wealth. Yet, we are among the poor nations of the world. It has been stated that we are not utilizing our plant resources and vast land fully and properly (Jain, 1984; Pal, 1981; Vohra, 1980).

The convention on Biological Diversity ratified by India on 18th February, 1984 has focussed attention on the urgent need for floristic inventories and monographs, including generalised survey to assess conservation status and targeted survey to explore new genetic resources for the industry. This has necessiated the underlying need for generating taxonomic study (Particularly Taxonomic diversity of flora and fauna). A legally binding agreement, Convention on Biological Diversity (CBD), was adapted by the United Nations Conference on Environment and Development, held at Rio de Janeiro in June, 1992. The Biological Diversity Act (BDA). was formulated after India became signatory to the CBD. The Act was passed by the Parliament in December, 2002. The objectives of the Act are 'to provide for conservation of biological diversity, sustainable use of its components and equitable sharing of the benefits arising out of use of biological resources and for matter connected therewith or incidental there to.'

Taxonomic research in India is declining, despite its mega-diversity status. It is important to reprioritics the focus on issues like targeting the neglected taxonomic groups (like sedges and grasses), modern approaches in systematics and bioprospecting programmses for a better conservation strategy. These studies will provide a better future for biosystematics in India (Hariharan and Balaji, 2002).

In achieving success in sustainable development, economy and ecology, the primary activity in research protocol is taxonomy for authenticated product development and materials management. The implementation of knowledge-based development activities ought to be streamlined, in which working taxonomy will have a vital role. Participation of people being a vital component of development, the imperiative need is to evolve qualified human resources impowered to provide quality services with science and technology as the directive principles. Even high-level education in institutions need to be improved; this calls for new initiatives, of which taxonomy ought to be an area of focus, but is being neglected due to the ignorance about its importance in achieving the hopes and aspirations of the people in the emerging world order. Coming to environment-based activities, environment impact assessment has become an obligatory compoent of development. It is in the section of hydropower generation that taxonomy and biodiversity evaluation and management have become important. Taxonomy is equally significant in areas like agroforestry, mitigation of pollution, tourism development, social forestry needing an appropriate approach, among the other areas of socio-economic changes (Nair, 2004).

One of the foremost challenges faced by conservation biologists, in general and systematic biologists in particular is completing the inventory of the world's biodiversity and its components before they are last forever. During the last 300 years of their research systematic biologists have been able to identify, classify and catalogue only about 1.75 milion species, *i.e.* nearly 13% of the minimum estimated world total of 13.6 million species of plants, animals, fungi and microorganisms. The task ahead is enormous and systematic biologists have to play a pivotal role in expediently executing the mission of biodiversity inventorying and monitoring. The over riding importance of systematics and its creative role in fulfilling the emerging needs of science and society have never been realised so greatly as at this particular point of time. However, with the coming into force of CBD and launching-of new international initiatives like DIVERSITAS, Systematic Agenda 2000, and several national programmes in the line of Mexico's CONABI, Australia's ERIN, etc. the thrust now on the revival of alpha or classical taxonomy and its application, particularly in biodiversity inventorying and monitorying (Pushpangadan and Nair, 2001).

During the last part of the nineteenth century, Sir Joseph Dalton Hooker and his associates published a monumental work in seven volumes entilted, "The Flora of British India" (1872-97), which is the major and the first comprehensive work on the Indian Flora. In valedictory preface to the last volume of the Flora of British India, Sir Joseph Dalton Hooker expressed very rightly a statement that his flora would facilitate the preparation of, "local" Indian Floras. The necessity for such local floras to supplement the work of Sir J.D. Hooker *et al.*, which deals with several hundred species of plants described from a wide range of Indian subcontinent with diverse climate conditions, has been fully recognised by all concerned with the plant wealth of the country.

Since the publication of different State Floras of India during the first half of the twentieth century, a large number of addition of species and some genera have been described from different parts of India by various workers. Moreover, a considerable number of species and genera have extended their distribution in various parts of the country. Further, many exotic plants have become naturalized and are gradually spreading and aclimatising in different parts of India. The nomenclature of a large number of species and genera have also been changed according to the International Code of Botanical Nomenclature. Besides the publication of lists of valid names of plants as gathered from various taxonomic revisions, a large number of research papers are being published in different journals from time to time dealing with nomenclature of plants, floristic account of some smaller or larger areas, new distribution records, etc.. Further, several genera and families have also been revised in monographic studies by the recent taxonomists.

Most of the area of reorganised Uttar Pradesh is cultivated. The region is comparatively under-explored and very little is known about its plant resources. Some district floras *viz.*, Allahabad by Misra and Verma (1992), Agra by Dhakre and Sharma (1998), Gorakhpur by Srivastava (1976) and check-lists of some districts have been published but the detailed floristic account is not yet published and as such there is meagre information on plant resources of the state. According to Nair and Nair (1977), a review of floristic studies on the area during the period and after the publication of Duthie's Flora, reveals that still there is scope for further intensive exploration. Even the renewed interest in floristic studies observed in recent years has resulted in good knowledge of the flora of only a few more scattered areas. Rau (1969) has pointed out that the fertile plains of eastern Uttar Pradesh are of great botanical interest. The large tracts of cultivated lands, available here also form suitable breeding ground for many weeds. Though there have been some studies in these regions in the past, they are of a general nature. An in-depth study of limited groups of economic value has not been undertaken so far. The family Cyperaceae or sedges and Poaceae or grasses form an important group with the latter having considerable economic potential, and it was thought desirable to undertake a floristic survey of the Eastern Uttar Pradesh particularly the four districts of Varanasi Division (Chandauli, Ghazipur, Jaunpur & Varanasi districts) for getting a thorough inventory of the members of these specialised group of flowering plants. A fresh assessment in the field, as well as study in the laboratory and herbarium has thus become necessary and provides justification for the study. Though the present study cover mostly four districts but it may be considered for Eastern Uttar Pradesh because the climatic conditions of the other districts of Eastern Uttar Pradesh are similar to the area covered.

The family Cyperaceae and Poaceae belonging to the series Glumaceae which is marked by naked flowers in the axils of bracts or glumes arranged in spikelets. They differ from each other in the nature of stem, ligules and minor details of flowers, fruits and seeds. The Sedges have solid, 3-angled stem, 3-ranked leaves and the ligules are absent, while the Grasses have hollow and terete stem, 2-ranked leaves and mostly possess ligules.

Since sedge and grasses are of great economic importance, particularly in a predominantly agricultural and pastoral community of Eastern Uttar Pradesh, the present study would be advantageous to the local people of the study area. For proper utilization of raw materials in developing country like India, the need for basic economic surveys and assessments of the natural resources is often emphasized. To urdertake such surveys and investigation it is essential that the flora of the country should be well-known and only when all plants are correctly named, described and clasified, it would be possible to reconize useful or potential plants of economic value. The study of Sedges and Grasses has not attracted the desired attention in Eastern Uttar Pradesh, as these two groups are considered difficult for identification. Their spikelets being minute, the Sedges and Grasses requires careful dissection for correct determination and illustration. Being botanically "neglected group," sedges and grasses need a separate detailed taxonomic investigation. The present study has been carried out with the broad objective to present a comprehensive floristic accoount of the Sedges and Grasses of Eastern Uttar Pradesh, India.

Chapter 2

General Description and Climate

The present study area covers mostly four districts of Eastern Uttar Pradesh (Chandauli, Ghazipur, Jaunpur and Varanasi). (Plate 1).

GHAZIPUR

Ghazipur district (25°19' N to 25°54° N. Lat and 83° 4' E to 83° 58' E Long.) has an area of 3377 Km^2 form a part of the alluvial plain of the middle Ganga Valley, lying in the eastern U.P. At micro-level the district is actually a part of the, "Ganga-Ghaghra Doab east, "being chronically flood-prone area. It is bounded in the north-west by Azamgarh district, in the north-east and east by Ballia district, in the west by Jaunpur district and the south by Varanasi district, all in Uttar Pradesh and in south east by Bihar state, from which it is separated by the Karmanasa river.

The Ghazipur district lies in the alluvial plain of the Ganga. Kankar (Limestone conglomerate) is common and occurs in extensive strata throughout the upland area, at varying depths below the surface. Clay suitable for making bricks, is obtained every where. Saline earth is observed in all parts of the district, especially in rice tract between the Mangai and the Bhainasahi, where the effervescence known as, "reh" is also found. The river Ganga is the most important river of Ghazipur district. The Gomti, the Gangi, the Mangai, the Saryu and the Karmanasa are directly or indirectly its main affluents. The river Ganga enters the district with its north-easterly flow in Saidhpur tahsil at the confluence of the Ganga and Gomti.

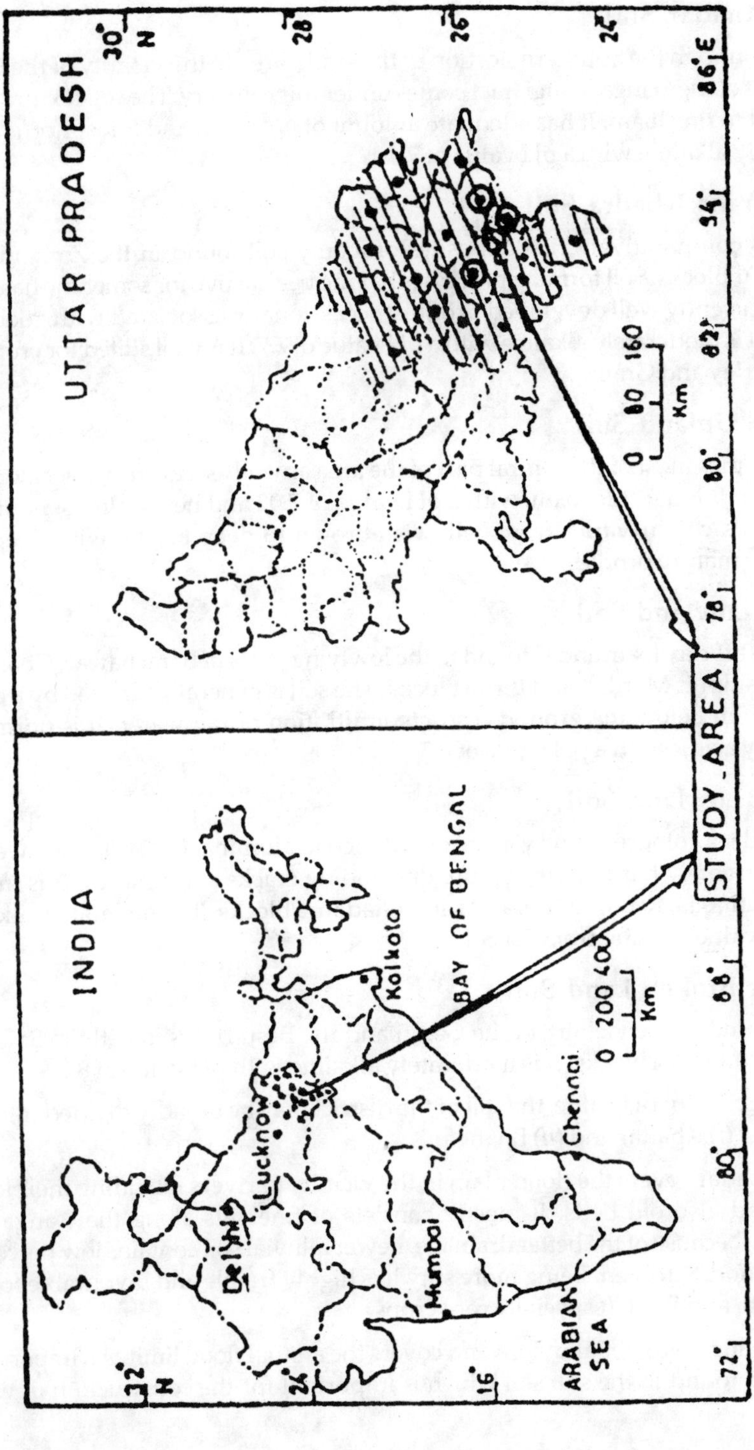

Plate 1: Map Showing Location of Study Area in India and Uttar Pradesh

The major soil types in Ghazipur district are as fallows:

Ganga-Khadar Soil

It is found in the Southern portion of the study area in the vicinity of the Ganga river. Most of the Ganga–gangi tract comes under this category. The soil texture varies from sand to fine loam. It has adequate amount of Nitrogen and Phosphorus and is moderately alkaline with a pH value of 7.4.

Ganga-Trans Khadra Soil

It has comparatively rich deposites of loamy soil found in the Zamania and Barachawar blocks. Soil forming processes having been active for sometime has given rise to sufficiently well developed profiles, occassionally associated with good lime reserves. It is moderately alkaline with a pH value of 7.7. It is well suited for crops like Wheat, Barley and Gram.

Northern Upland Soil

It occupies most of the central part of the area as well as a narrow tract along the Gangi river. It is a fertile loam, with a pH value of 7.02 and hence almost neutral in character. Its colour varies from yellowish-brown to deep brown with deep and sufficiently mature profiles.

Nothern Lowland Soil

It is a stiff clay loam and is found in the lowlying parts of Barachawar, Ghazipur, Deokali, Saidpur, Mardah and Birno blocks. The soil is generally affected by a depth of about 0.5m below the ground restricts infiltration of rainwater. It is normal to moderately acidic with a pH value of 6.7.

Southern Lowland Soil

It is a dark coloured clayey alluvial soil, rich in alumina. It is soft when wet but become extremely hard on drying up, developing cracks and fissures. It is mostly found in Zamania, Reotipur, Karanda and Bhadaura blocks. It is moderately alkaline with a pH value varying from 7.4-8.1.

Ravine or Broken Land Soil

It is found in the vicinity of the Gangi and the Besu rivers, mostly in Saidpur, Sadat and Manihari blocks. It is moderately alkaline with pH value of 8.2.

On the basis of pH value, the soil of the district may be broadly divided into two major types: (*i*) Khadar and (*ii*) Bhangar.

The khadar covers the flood plain in the vicinity of rivers including their lower reaches and also old beds. It mostly consists of fine silts along the Ganga and Karmanasa, because of the better drainage, newer alluvial soil contains low percetage of humous and Nitrogen. Being more silty it is highly friable and favourable for the Bhadai, Rabi and Zaid. (especially root crops).

The Bhanger is older alluvium and covers the annual flood limit. An impervious clay layer is found in the sub soil which is important for the construction of wells, most suitable for irrigation in this area. The Bhangar soil is often heavier with higher clay. It is sticky and usually not well drained, it is richer in lime content and kankar than the Khadar soil, it is more suitable to rice cultivation as it holds water for a long time.

The climate of the Ghazipur district conforms to the general pattern of subtropical monsoon climate. The climate is controlled by the temperature, rainfall, humidity, wind pressure etc.

Temperature

The record of 80 year's average show that the month of December to February very cold. The rapid increase in the temperature begins from March and thus the maximum mean diurnal range of temperature increase in latter months. The minimum temperatures 15.8°C and 24.9° C are recorded in the months of January and March respectively. The maximum temperature ranges from 30.3°C to 32.8°C in the months of May and June respectively. At the end of May or early June, the maximum temprature occasionally reaches over 40°C. The scorching wind locally known as 'Loo' flows in the area. With the advent of monsoon in the mid June, there is appreciable drop in the day temperature as compared to night.

Rainfall

Annual rainfall does not exhibit significant departures from the long term mean. For instance during 80 years periods of records the average annual rainfall is 95 cm in the district. Almost 90% of the total rainfall is received from the monsoon during the months froms June to September with maximum rainfall in the month of August. The monsoon weakens during the months of October. Only 5% of the annual rainfall occurs in the winter season from the cyclones.

Humidity

The months from July to September are very humid (about 80%). The humidity is lowest in the month of January (17%). The humidity decreases in after September (82.1%), till December (73.1%). Again it increases from January (17%) to April (38.7%).

JAUNPUR

Jaunpur district is situated in the north-west portion of Varanasi division. It is 90.2 Km in length from east to west and 85.3 Km in breadth from north to south. The total area of the district is about 401412.2 hectares. It is situated between the parallels of 25° 24' and 26° 12' north latitude and between 82° 7' and 83° 5' east longitude.

The land surface of the district is apparently a level tract with slight undulations caused by the rivers, which flow from north-west to south-east. The district is 81.74m high from the sea level. Gomti is the chief river which flows through the district. Gomti is fed in the district by several tributaries the chief being Pili and Sai.

River Pili which takes its origin in a string of 'Tals' in Sultanpur district joins Gomti at Dariaoganj. It has a very tortuous course and its banks are fringed with ravines which are not of great height. As a result during rains the river spreads to a considerable distance on either side.

The river Sai rises in Hardoi district and forms for 19.32 Km the boundary of Jaunpur and Mariahu tahsils. It falls in to Gomti near the village Rajapur. It resembles Gomti in following a very tortuous course and possessing a deep bed. The banks are broken by innumerable ravines and area often crowned with fine groves of *Mangifera indica* Linn. and other trees. The soil on either side is of sandy texture.

River Basuhi rises in Machhlishahar tahsil from Karnauli jhil and falls in to Varuna at the trijunction of Jaunpur, Varanasi and Mirzapur. River Varuna never enters the district but for a long distance forms the southern boundary.

Soil of the district consists mainly of loam and clay, the former prevailing of the higher levels especially in Jaunpur and Kirakat tahsils, as well as the portion of Khutahan lying to south of Gomti. Clay is confined to the depression and low-lying tracts in Machhlishahar tahsil and the greater part of Kirakat. Sand is not of common occurrense. On the whole, one-fourth of the cultivated area comprises pure clay which with sufficient moisture produces excellent paddy crop. It is generally known as 'Matiyar', 'Karail' clayey soil is black with a large amount of organic matter and is found in the beds of 'tals' and dried up ponds. It is extremely sticky and cohesive when wet but cracking into fissures when dry. The soil is used for the coarser varieties of late paddy. 'Balua' (sandy) soil is reserved for the cultivation of poor crops. Loam commonly called 'Dumat' comprises about half of the area and is excellent soil for various crops.

Jaunpur district falls within the region of monsoon climate. Due to good amount of rainfall during the rainy season and occasional showers during winter the district bears luxuriant vegetation. The hot summers and cold winters are characterized with strong westerly winds. On the basis of general meteorological conditions of the region the year may broadly be divided in to rainy, winter and summer seasons. Besides, there are two transition periods one between rainy and winter seasons and the another between winter and summers. A brief account of seasons is as follows:

Rainy Season

The rainy season commences from mid-June just after the dry and scorching summer months and lasts up to the middle of October. The rain fall received during these months is due to South-East monsoon arising from the Bay of Bengal. Of an average of total 1068.6mm annual rain fall, about 87.7% is received during this period. The wettest months of rainy season are July and August but some time even September. The mean minimum and maximum temperatures are 33.3°C and 25.0°C respectively. The relative humidity is very high during the rainy season, its mean value being 81.1% in morning and 61.4% in evening.

Winter Season

This lasts from middle of November to February. It is characterized by low temperature and low relative humidity. The easterly winds gradually change to westerly. January is supposed to be the coldest month of year. During this period the mean maximum and minimum temperature are 25.3°C and 11.1°C respectively. The winter rains, commonly accompanied by hail storms during the ending of January or the beginning of February, bring an average of 48.7 mm rainfall. The temperature starts rising from mid-February and by mid-march it becomes warmer.

Summer Season

The summer season starts from April and lasts upto the middle of June. Hot westerly winds, commonly known as 'Loo', are the characteristics of this period. The relative humidity is minimum, being 54.3% in morning and 27.8% in evening during this season. The air is much dry. The winds are strongest during June and dust-storms are also a common features.

VARANASI AND CHANDAULI

Varanasi and Chandauli districts are located in the eastern most part of Uttar Pradesh between 24° 43' N and 25° 35' N latitudes and between 82° 11' E and 83° 34' E longitudes. It covers on area of 5091 km^2. Varanasi district has an ancient past with rich cultural heritage and historical back ground. Varanasi the district head quarter is one of the oldest cities in the word. It is bounded by Bhadohi (Sant Ravidas Nager) district in the west, Jaunpur distrct in north and north-west, Ghazipur district in the north and north-east, Bhabhua and Baxur districts of Bihar in the east and Sonbhadra district in the South. The geology of the area consists of interbeded layers of sand, silt and clay and sometimes are associated with 'Kankar' in place. The Vindhyan section comprises of different types of Quartzite and Sandstone. The mineral products are bauxite, reh, Kankar and clay. The district of Chanduli and Varanasi may be broadaly divided in to two physiographic regions-the northern alluvial plain and the southern hilly region (now in Chanduli district). The Ganaga, Varuna, the Karmnasa, the Gomati, the Chandraprabha etc. are the main rivers traversing the districts of Varanasi and Chandauli. Soil of the area is generally alluviam. Mean daily maximum temperature is 47°C and Mean daily minimum temperature is about 9°C. The arable land comprises of 70% followed by forest 14.42%, Grazing land 0.6%, orchard plantation 1.55% (Mainly mango and guava), wasteland 22.6% and miscellaneous 11.18%. Chandraprabha is the only wild life sanctuary (now in Chandauli district).

The climate of Varanasi region is characterized by the monsoon which indicates a seasonal rhythm of weather. It is typical of the Upper Gangetic Plain as described by Dudgeon (1920) and Misra (1946b). All the elements of climate, *viz.*, temperature, rainfall, relative humidity and wind with their characteristic fluctuations mark out three distinct seasons, *viz.*, rainy, winter and summer; details of each are given below:

The Rainy Season

Varanasi is in the grip of monsoon usually by the 15th of June each year. The rainy season stretches well over three and half of month, commencing from the middle of June and extending up to the ending of September when 80-90% of the total annual rainfall is obtained.

The month of October show transition from the rainy to winter season. The characteristies features of the month are a widely fluctuations in day and night temperature, a clear sky and high insolation. The high insolation brings about rapid desiccation of wells and their vegetation. By the ends of this month a majority of the rainy season annuals dry out leaving behind a crop of mature seeds.

The Winter Season

The winter season extends for about three and a half months from November to the middle of February and is characterized by low temperature and dry weather. The month of March is a transition periods from the cold to the hot season. The month of March with a rise in temperature and its diurnal fluctuation with a longer photoperiod and high insolation induces vegetative and flowering growth in the perennial wall plants while winter annuals dry and decline.

The Summer Season

The summer season extends for about two and a half months from April to middle of the June, During the summer period hot winds locally called "Loo", continue to below from west and south-west till the advent of monsoon. These winds are diurnal which begin two to three hours before noon and blow through the heat of the day. They stop only towards the evening sometimes after sunset. They are intensely dry with very low relative humidity. Fitful dust storms blowing at high speed (locally called Andhi) are sometimes experienced.

Physiography

The study area is mainly composed of Gangetic alluvium the deposition of which commenced from the Pleistocene period after the final upheaval of the Himalayas and is still continuing. It consists of interbedded layers of sand, silt and clay which are associated by several places with kankar. Geologically the district is covered with thick alluvial deposits. The Maximum depth of the alluvial filling is about 1500-2000 m. The alluvial deposits with the extreme flood limit of the river may be classified in to two groups: (i) Newer alluvium (ii) Older alluvium.

The newer alluvium covers the flood plains in the vicinity replenished annually by new deposits. The older alluvium refers the upland tracts beyond the annual flood limit.

Drainage

The drainage lines affect the topography of the plain. The area can be divided in to draining systems.

Ganga

The most important physical feature of the area is the river Ganga. The Ganga touches the block at the Village Betabar, where it is joined by a small stream the Subbha nullah, which drains a small area. For a distance of about seven miles the river separates the district from Mirzapur district and runs roughly north-east wards across the district separating the study area on the left from village Sultanpur, Bheeti, Tengra and Salhupur on the right.

Varuna

Coming from the north west the river Varuna touches the district at the village Kotwa. By making a number of concave and convex bends the rivers separates the block from Harauva block. Near the village Chaittauni and Bendauli two Nalas have joined from the south of the river at right angal, which drain a large flow lying areas of Sarhari Lohata Nyaya Panchayats. From village BaraGaon the river leaves the block and runs in east direction for about thirty five miles till after shirting the north of the cantonment and passing through the civil station, it joins the Ganga at Sarai Mohan.

Soil

Soil is a dynamic physical composition of mineral and organic matter which acts as the natural medium that have capacity to grow and support the plants. The soil is a very important resource in the area under study, as the economy is mainly agro-based. The self nourished plant gets their essential nourishment from it, there fore in biotic pyramid, man lies on apex while soil makes base of pyramid. The study area is covered by alluvial soil varying from clay to sandy loam, deposited by river Ganga and Varuna. Alluviums are divided in Khadhar and Bhangar. The Khadhar tract is replenished annually by new deposits and contains deficient of humas and nitrogen, which may be made good by the uses of compost manures and fertilizers. This soil is suitable for Bhadai, Rabi and Zaid crops. The bhangar is the older alluvium and covers the upland tracts. It is under the process of denudation and contains patches of User land. It is richer in lime content and more suitable for rich cultivation provided there is adequate irrigation. The pH value of soil varies between 6.5 and 7.5. On the basis of minor variation in colour, texture, porosity, as well as the distance from the river the soil may be classified in to following type:

Sandy Loam

It occupies a narrow belt along the Varuna river and low lying tract over the sandy soil along the Ganga Khadar tracts. This soil is important to agricultural point of view because of better surface drainage, water holding capacity and soil fertility. This soil is much fertile for Rabi and Kharif.

Loamy Sand

It comprises the low lying tract along the Ganga are and has experienced along process of flooding which deposits flood sediments every year. This soil has very low productivity and is suitable for bajra or millets in the kharif and for Rabi crops like gram, pea and oilseeds.

Clayey Loam

It covers most of the bhangar tracts of the district. It has sufficient moisture retaining capacity and capable of producing paddy crops in wet season. A few patches of soil consisting alkalinity and salinity is found in this type of soil which is less fertile owing to white incrustation or 'Reh' on the ground surface.

Chapter 3

Methodology

The present work is based on the survey of Sedges and Grasses occurring in Eastern Uttar Pradesh, mostly in Chandauli, Ghazipur, Jaunpur and Varanasi districts during 1998-2005. The study area was surveyed on foot in their various parts and habitats at regular intervals with increasing frequency of visit during the monsoon and early winter seasons. The Sedges and Grasses have been collected in the flowering and the fruiting stages and were identified on the spot and the plants that could not be identified were brought to the Department of Botany, Udai Pratap Autonomous College, Varanasi for detailed studies and identified with the help of Floras (Duthie, 1903-29; Bor, 1940, 1960; Rao & Verma, 1982; Srivastava, 1976; SreeKumar & Nair, 1991; Mishra & Verma, 1992; Shukla, 1996; Moulik, 1997; Bhat & Nagendran, 2001; Prasad & Singh, 2002), different monographs and research articles published time to time. All the collected specimens were serially numbered. During the field studies; habit, habitat, height of plant, colour of flowers were also recorded. The herbarium sheets were prepared and completed by writing date, place of collection and remarks. The herbarium sheets were preserved and later checked up with the herbarium sheets of BSI (CC), Allahabad; NBRI, Lucknow and FRI, Dehradun. Phenological data and the physiognomic variations were also recorded during the field studies.

For the classification of the family Cyperaceae, the broad classification of Kern (1974) has been adopted with some modifications in the generic concept. For the family Poaceae, classification of Clayton & Renvoize (1986) has been followed with some modifications.

Chapter 4
Previous Work

Botanically, the study area remained under explored, and no comprehensive floristic account, of this region is available. Duthie's (1903-29) "Flora of the Upper Gangetic Plain and of the adjacent Siwalik and Sub-Himalayan tracts", Rau's (1969) "Flora of the upper Gangetic plain and of the adjacent Siwalik and Sub-Himalayan tracts-Check-List", Raizada's (1976)"Supplement to the Duthie's flora of the Upper Gangetic Plain and of the adjacent Siwalik and Sub-Himalayan tracts ", are recognized as the pioneer floristic works of this area. In these three most important publications grasses are not included. Kanjilal (1966) in his publication "A forest flora for the plains of Uttar Pradesh " has tried to revise the entire forest flora of undivided Uttar Pradesh. Only a very small part of the present study area (southern part of Chandauli district) have forests.

About 65 years ago an account of the grasses of undivided Uttar Pradesh has been published by Bor (1940) and after gap of 14 years of this valuable publication Grasses of undivided Uttar Pradesh has been reinvestigated by Raizada (1954) followed by Raizada, Bhardwaja and Jain (1961) and Raizada and Jain (1966).

Most of the area of Uttar Pradesh (reorganised) is cultivated. As compared to its hilly region (now in Uttaranchal State), the region is comparatively under-explored and very little is known about its plant resources. Some district flora *viz.* Gorakhpur by Srivastava (1976), Allahabad by Misra and Verma (1992), Agra by Dhakre and Sharma (1998) and check-lists of some districts have been published but the detailed floristic account is not yet published and as such there is meagre information on plant resources of the state. The concept of endemism is also altered in many cases. More recently Check-lists of Grasses by Uniyal etal., (1994), of Cyperaceae by Uniyal *et al.* (1997), of Dicotyledones by Khanna *et al.* (1999), and of Monocotyledons by Uniyal *et al.* (1999) of undivided Uttar Pradesh have been published, most of which are incomplete and erroneous. Very recently Check-list of Cyperaceae of reorganized Uttar Pradesh has been published by Singh and Srivastava (2004).

Only few contributions to the flora and plant resources of various parts of Varanasi division have been made during last 65 years (Mishra, 1944,1946a,1946b; Tripathi, 1964; Rao, 1968; Joshi and Reddi, 1969; Singh and Dixit, 1969; Shankar, 1970; Roy, 1971; Varshney, 1971; Singh and Singh, 1972; Gupta, 1979; Singh and Singh, 1979; Singh and Maheshwari, 1983,1985a, 1985b; Singh *et al.*, 1987; Sharma and Singh 1989 a, 1989b; Singh and Prakash, 1996 and Singh *et al.*, 2004). However these studies do not cover the aims and objectives of the present study.

Chapter 5

Systematic Treatments

(I) ARTIFICIAL KEY TO THE FAMILIES

1a. Stems trigonous, triquetrous, tetragonous, subtrigonous, terete, or even almost 3-winged, multyangular, or strongly compressed, solid, rarely spongy, not divided into nodes and internodes; leaves usually tristichous, polystichous, usually in basal clusteres, usually narrow, shorter than to much longer than the stem, glabrous; sheath usually reduced, closed, glabrous, colour varies from reddish, purplish, blackish, yellowish to greenish, mouth of sheath concave, convex, or truncate; ligule usually absent, rarely a fringe of hairs, or a membrane; inflorescence capitate, panicled, anthelate; glumes distichous, lemma, palea absent; perianths well developed, rarely reduced, hypogynous, bristly, or scaly; stamens usually 1-3; style solitary, hairy or glabrous, thickened towards the base; stigma 2 or 3 usually glabrous, rarely hairy; fruits usually a nut or achene. .... **1. Cyperaceae**

1b. Culms mostly cylindrical, hollow, slightly compressed, divided into nodes and internodes; leaves distichous, usually solitary at the nodes rarely in basal clusteres, much shorter than the culms, rarely equalling the culms, glabrous or hairy, usually broad; sheath well developed, flattened, glabrous or hairy, green, mouth of sheath not convex, concave, truncate; ligule usually membranous, or reduced to a fringe of hairs, rarely absent; inflorescence more often in spike, few or many-spiculate panicles, racemes or false racemes or in head of congested spikes; glumes rarely distichous; lemma and palea present; perianths reduced to minute lodicules and in between the stamens and pistil; stamens usually 3 or 6, rarely 2, 1 or many; styles usually 2, rarely 1 or 3, not thickened towards the base, glabrous; stigma usually 2, mostly plumose; fruits usually a caryopsis, rarely a berry, a drupe or achene.

.... **2. Poaceae**

(II) THE FAMILY CYPERACEAE

(*a*) Morphological Characters of the Family Cyperaceae

Cyperaceae members or the sedges are grass like or rush-like, mostly perennial but frequently annual (*Cyperus* spp, *Fimbristylis* spp. *Eleocharis* spp.) herbs (rarely pseudo-arboresent in the African genus *Microdrachoides*). They are usually tufted (caespitose) and erect, occasionally, species with long creeping rhizomes have a solitary or a series of single stems (*Carex* spp., *Cyperus* spp., *Scleria* spp).

The annuals have fibrous, usually shallow roots. The perennials have varied kinds of rhizomes and stolons. The rootstocks or rhizomes are clothed with scales which grade above into leaves.

The stem may be slender, almost capillary, weak and nodding, to very stout and erect or sometimes scandent (*Scleria* Berg.), and vary in height from a few millimeters (*Cyperus* L.) to more than four meters (*Scleria* Berg.); they are usually simple below the inflorescence, solid or sometimes hallow, and septate (*Eleocharis* R. Br.) trigonous with some what convex to flat faces or triquetrous with concave faces, or almost 3-winged, sometimes terete or subterete, quadrangular, multiangular or strongly compressed and ancipitous (*Fimbristylis* Vahl), smooth or scabrid.

The leaves are usually 3-ranked, sometimes distichous or polystychous, few to many, and usually clustered at the base, but sometimes one or few also occur upwards on the stem (*Rhynchospora* Vahl), sometimes all or the lower reduced to bladeless sheaths or apicula (*Eleocharis* R. Br.), rarely (*Scleria sumatrensis* Retz.) congested in groups and forming pseudo-whorls. In some of the *Carex* L. the leafy shoots are distinct from the lateral shoots. Generally the leaves are sessile, dorseventrally flattened, herbaceous or coriaceous, flat or plicate quite noticeable when young, linear or setaceous, much shorter than to far exceeding the stem, varying in width from less than a millimeter to centimeters, gradually acute- acuminate, revolute or involute margined, and smooth or antrorsely (sometimes retrorsely) scabrid margined, usually towards the tip, but frequently throughout, dull, pale to dark green or grayish-green (sometimes yellow), more or less 3-costate with the mid-costa raised on the lower surface and the laterals on the upper surface. The leaf-sheath are closed (rarely open), usually some what membranaceous, tightly or loosely enclosing the stem, varying in length from a few millimeters to several centimeters, of various shades of green, yellow, brown-red, purple, and black, usually fibrously splitting with it mouth concave, truncate or convex.

The inflorescence is always terminal though sometimes (*Scirpus* L.) it may appear to be lateral due to the erect posture of a stem-like bract; sometimes it is reduced to a solitary spikelet. Each peduncle being subtended by a foliceous, setaceous or glumaceous bract and enveloped at the base with in the prophyll.

The prophylls which are indicative of the beginning of a new branch are found distributed at various places on inflorescence, in different forms. It mat be a 2-keeled tubular structure enclosing the base of anthela ray in genera like *Cyperus* L. and *Scirpus* L. or a glume-like structure in the spikelets of the above genera (the second empty glume from the base). In *Lipocarpha* R. Br., it is like a membranaceous scale. The two basal of the hypogynous scales (glumeless) above the bract (usually with a stamen inflorescence in their axil) inflorescence *Mapania* Aubl. and *Hypolytrum* L. C. Rich., the utriculiform or ocreiform bract (cladoprophyll) and the sac-like structure (utricle) enclosing the pistil in *Carex* L., are also frequently interpretted to be prophyllar in origin. Prophylls have a significant role in the inertpretation of structure of the ultimate units of the inflorescence.

The spikelets are borne on the ultimale branches of the inflorescence and consist of rhachilla bearing glumes and flowers. Each spikelet is generally subtended by a prophyll, and below it, a bract. The spikelets are either bisexual or unisexual. In *Carex* L. the flowers are naked and unisexual and borne on a spike-like structure. The glumes exhibit a wide range of variations in their number, arrangement, shape, apex, nervation and coloration. These may be three to many in each spikelets, of which the lower one to several are empty and upper enclosing flowers, with the upper most few frequently absent. The fruit is indehiscent, one seeded, and is generally called a nut, though there is a tendency now in favour of calling it an achene. The former should have a hard fruit wall which is not found in Cyperaceae and the latter should have developed from a free carpel, whereas in Cyperaceae it develops from 2-3 fused carpels (Rao & Verma, 1981).

The Sedges though considerd a natural taxon, appears taxonomically intricate. The earlier view of its alliance with the Grasses is not based on sound argument and treated as highly superficial. On the contrary, the available data from various fields, including embryology, support the closer phyletic relationship of the Cyperaceae with the Juncaceae rather than with the Gramineae (Shah, 1967, 1972; Cronquist, 1968; Metcalf, 1971; Nijalingappa & Nagraj, 1980). In addition, the classification of the Cyperaceae on a natural basis has been a subject of much discussion and remains unsettled. The main difficulty about the classification lies with the obscure state of the morphology of the inflorescence, flower and its parts, and difficulties and differences of opinion among the cyprologists in their interpretations. With result, there is no satisfactory system for classifying the family.

The anther is 4-lobed, each lobe representing a microsporangium. Occasionally, anther becomes 3-lobed because of the fusion of 2 of its microsporangia (Shah, 1972). The microsporangial wall is 4-layered comprising epidermis, endothecium, middle layer and a glandular tapetum, and follows the Monocotyledonous type of development. However, in *Hypolytrum latifolium* it is 4–or– 5 layered due to periclinal division in some of the cells of the tapetum.

Unlike in many angiosperms where the microspore mother cells occur as a multilayered tissue, in the Cyperaceae they appear as a single circle surrounding the longitudinal axis of the microsporangium. At the onset of meiosis the microspore mother cells become wedge shaped with their narrower end being directed towards the center (adaxial) of the microsporangium. After meiosis, 4 microspore nuclei are formed in each microspore mother cell. They are located at the adaxial region of the cell showing decussate, tetrahedral, isobilateral or linear arrangement. Cytokinesis is of simultaneous type. The cell plates formed during cytokinesis are either complete or less complete without being reached to the periphery of the cell. In any case, the microspore tetrad consisting of 3 smaller adaxial abortive microspores with a scanty cytoplasm, and a larger, abaxial functional microspore having a dense cytoplasm. Another unique feature is that the individual microspores of a tetrad are not surrounded by a callose but the entire tetrad is invested by a callose. The first pollen mitosis starts simultaneously in all the 4 microspores but is completed only in the functional microspore and interrupted in the abortive microspores.

Certain deviations such as the functioning of more than one microspore (or microspore nucleus) in a tetrad and formation of double pollen grains reported in a few taxa represent the transitional stages in the derivation of the Cyperaceous type of pollen from the Juncaceous type.

In the majority of the members of the family Cyperaceae there is a single, basal, anatropous, bitegmic and crassinucellate ovule in each locule.

The development of endosperm in all the sedges is of the Nuclear type. Tiwari (1978) while enumerating the embryological characters of the family Cyperaceae states that the family is characterized by the endosperm lacking haustoria.

In most of the Cyperaceous taxa the development of embryo is Onagrad type (Johansen, 1950).

Despite uniform embryogeny there is a considerable variation in the morphology of the mature embryo. This led to the recognition of 6 different types of embryos based on the deposition of embryonal organs in the mature embryo (Veken, 1965). They are: *Carex* type, *Cyperus* type, *Scirpus* type, *Bulbostylis* type, *Fimbristylis* type, and *Schoenus* type. In all the types the position of cotyledon is terminal. Both the plumule and the radicle are sub-lateral in the *Schoenus* type whereas in the *Bulbostylis* type they are basal. In the *Carex* type the plumule is lateral and the radicle is basal. On the other hand, in the *Cyperus, Scirpus* and *Fimbristylis* types it is vice versa where the plumule is basal and the radicle is lateral. (PLATE 2A).

TYPES	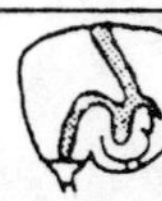 **_Bulbostylis_ type**	 **_Carex_ type**	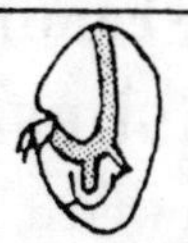 **_Cyperus_ type**	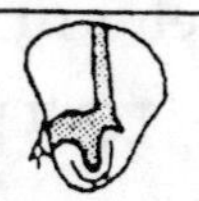 **_Fimbristylis_ type**	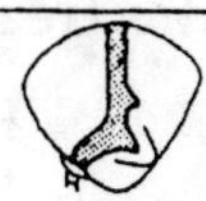 **_Schoenus_ type**	 **_Scirpus_ type**
SHAPE	Top-shaped	Top-shaped	Top-shaped	Ellipsoid	Mushroom-shaped	Broadly top shaped
PLUMULE	Lateral	Sub-lateral	Basal	Basal	Basal	Basal
RADICLE	Basal	Sub-lateral	Lateral	Lateral	Lateral	Basal
COTYLEDON	Terminal	Terminal	Terminal	Terminal	Terminal	Terminal

A

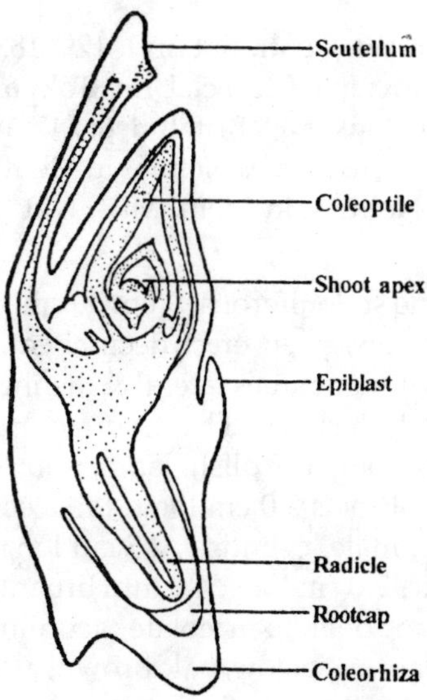

B

Plate 2: A: Embryo characteristics in Cyperaceae (based on data from Veken, 1965)
B: Mature embryo of _Triticum_ (median longitudinal section)

(*b*) Systematic Treatments of the Family Cyperaceae

KEY TO THE SUBFAMILIES

1a. Flowers bisexual; spikelets several flowered, if one or few flowered then without hypogynous bristles, empty glumes present at the base of spikelets, hypogynous bristles present or absent

.... **2. Cyperoideae**

1b. Flowers unisexual; spikelets not as above; emptyglumes absent at base of spikelets; hypogynous bristle absent

.............................**1. Caricoideae**

1. SUB FAMILY-CARIÇOIDEAE

KEY TO THE TRIBES

1a. Fruit an achene not bony, enclosed in a flask-shaped utricle

.... **1. Cariceae**

1b. Fruit a bony nut, not enclosed in a flask shaped utricle

.... **2. Sclerieae**

TRIBE-1. CARICEAE

Carex L.

Carex fedia Nees in Wt. Contr. Bot. Ind. 129. 1834; Kreczetowiez in Komar. Fl. USSR 3: 417. 1934; Koyama in J. Fac. Sci. Univ. Tokyo Bot. 8: 245. 1962. *C. wallichiana* Pres. ex Nees. Clarke in Hook. f. Fl. Brit. Ind. 6: 747. 1894; non spreng. 1826; Haines, Bot. Bih & Or. 956. 1924. *C.chaetogyne* Nelmes in Mem. Mus. Nat. Hist. Paris. n.s. B. 4: 170. 1955. *C. sacerdotis* Nelmes. in Mem. Mus. Nat. Hist. Paris n.s. B. 4: 172. 1955. (PLATE–3).

Perennials. Rhizome stoloniferous. Stem 50 cm long, 1.5 mm thick, triquetrous, scabrid above, smooth below, green, erect, decumbent, tufted, green, usually central & arising among the leaves, sometimes lateral & distinct from the leafy shoots. Leaves 53 cm long, 2.5 mm broad, clustered at base, surface rigid, margin scaberulous, linear, tricostate; sheath 2.0 cm long, purplish, ladder-fibrillosely splitting; ligule absent. Inflorescence racemose. Bracts 5.0 cm long, foliaceous, filiform. Spikelets 6, erect, upper 2-3 approximate, male, cylindric, 2.5 cm long, 5.0 mm broad, lower female rather distinct, cylindric, 2.0 cm long, 4.0 mm broad, peduncled. Female glume 6.0 mm long, 2.0 mm broad, ovate, lanceolate, acuminate, surface glabrous, margin scaberulous, at apex, glabrous below, pale brown, aristate, arista scabrid, 3- nerved; nerve brown, excurrent, central one thicker than lateral nerve.Utricles 4.0 mm long, 2.0 mm broad, shortly suberect, coriaceous, not inflated, ovoid- oblong, trigonous, abruptly beaked, several-nerved, greyish brown, villous, beakshort, deeply bifid. Stamens 3; anther 1.0 mm long, brown, oblong, glabrous; filament 0.3 mm long, uniformly flat, glabrous, shorter than anther. Pistil 2.5 mm long; ovary 0.5 mm long,

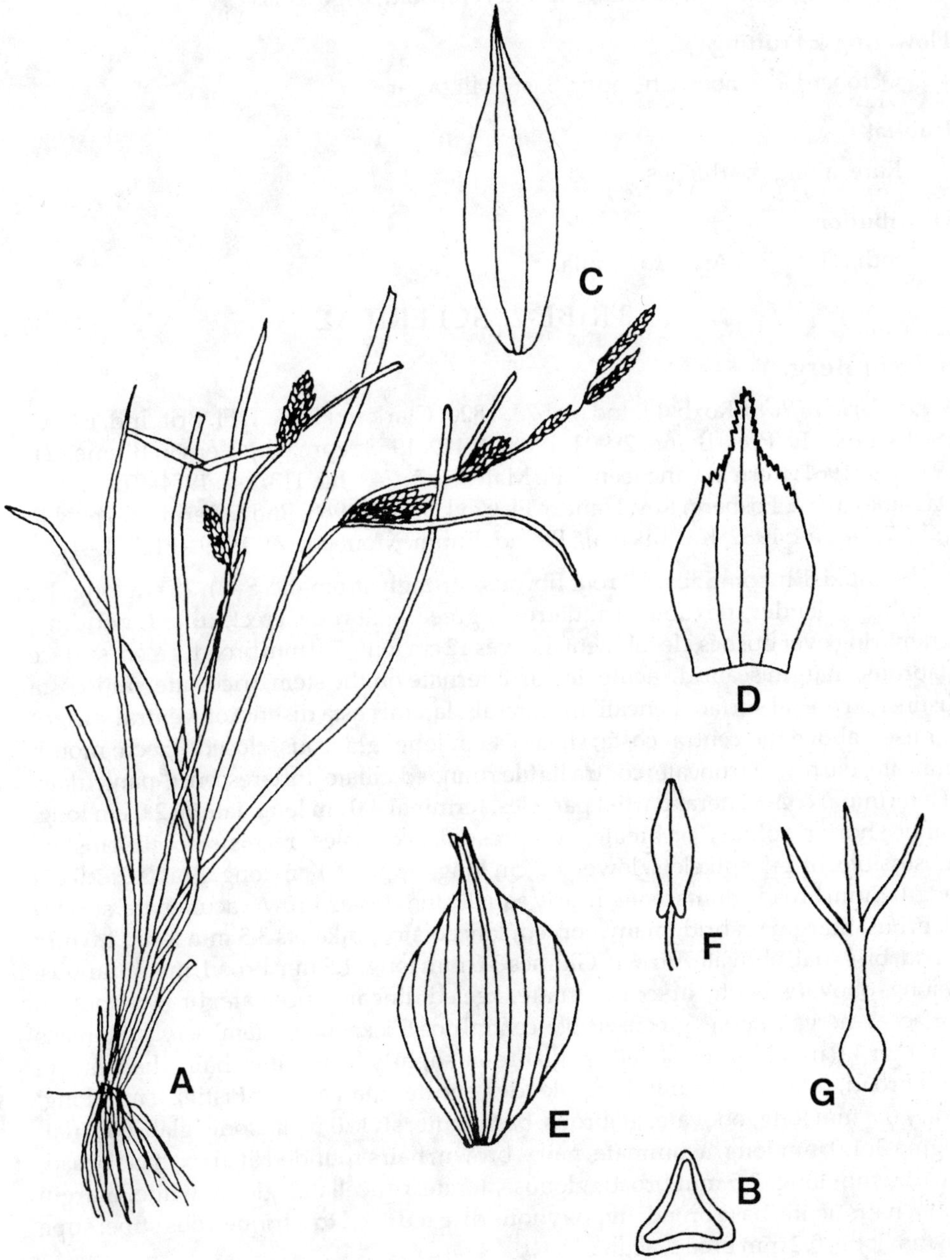

PLATE 3: *Carex fedia* Nees
A. Habit, B. T.S. of stem, C. Spikelet, D. Female glume, E. Utricle, F. Stamen, G. Pistil

glabrous; style 1 mm long, uniformly flat; stigma 3, 1.0 mm long, acuminate, glabrous. Nut 2.0 mm long, 1.0 mm broad, stipitate, ellipsoid, trigonous.

Flowering & Fruiting

October-December; Cholapur, Basti, Mirzapur.

Habitat

Rare, along the ditches.

Distribution

India, Tropical Asia, Australia.

TRIBE-2. SCLERIEAE

Scleria Berg.

Scleria biflora Roxb. Fl. Ind. 3: 573. 1820; Clark in Hook. f. Fl. Brit. Ind. 6: 687. 1893; Cooke, Fl. Pres. Bomb. 2: 904. 1908 (3: 419. 1958 repre.ed.); Kern in Blumea 11: 197. t. 7c. 1961; Kern in van Steenis, Fl. Males. 1, 7 : 743. fig. 113. a-g. 1974; Koyama in Dassanayake & Fosberg, Rev. Handb. Fl. Ceyl. 5: 362. 1985; Rao & Verma, Cyp. N.E. Ind. 57. t 42. a-c 1982; Karthik etal., Fl. Ind. Enum. Monocot. 71. 1989. (PLATE–4).

Annual. Rhizome absent; root fibrous, strongly aromatic. Stem 25 cm long, 1.0 mm thick, slender, flexuous, triquetrous, green, glabrous, erect, tufted, nodding, scrambring over bushes, decubment. Leaves 12 cm long, 7.0 mm broad, green, surface glabrous, margin scabrid, acute, linear, alternate on the stem, tricostate; mid-costa distinct on dorsal surface beneath the laterals, lateral costa distinct on ventral surface & raised above the central costa; sheath 4 cm long, glabrous, closed, green, mouth truncate, narrowly truncate; contra ligule rounded,ciliate. Inflorescence paniculate, of a terminal & 2-3 lateral partial panicles, terminal 4.0 cm long, lateral 2.0 cm long, rather shorter, solitary or binate, compressed peduncles; rachis smooth. clusters subsessible, of 2–5 spikelets, lower 1.2 cm long, upper 1.1 cm long. Bracts 2; 1.0 cm long, 0.5 mm broad, glumaceous, nearly equally the cluster, brown, acuminate, surface glabrous, margin scabrid, many nerved, lanceolate. Spikelets 3.5 mm long, 1.0 mm broad, bisexual, obovate, brown. Glumes 3.0 mm long, 1.5 mm broad, membranous, oblong–obovate, acute, apex tridentate, keeled, linear–lanceolate, brown spot on surface, 3–nerved; nerves green, visible, central one thicker than lateral nerves. Stamens 2; anther 1.0 mm long, crest yellow, glabrous, slightly ovate, apex hairy; hairs acute base broad; flilament 0.5 mm long, slender, shorter than anther. Pistil 2.0 mm long; ovary 0.5 mm long, obovate, glabrous, base acute; style 0.5 mm long, glabrous, flat; stigma 3; 1.0 mm long, acuminate, hairy, brown; hairs rounded at apex, base broad. Nut 2.0 mm long, 2.5 mm broad, globose, terete, cancellate, glossy white, margin hairy, hairs acute, base broad, hypogynous disc reduced to a triquetrous stipe; stipe 0.6 mm long, 0.2 mm broad, acute.

Flowering & Fruiting

August–September; Chakiya, Naugarh, Allahabad, Ballia, Padrauna.

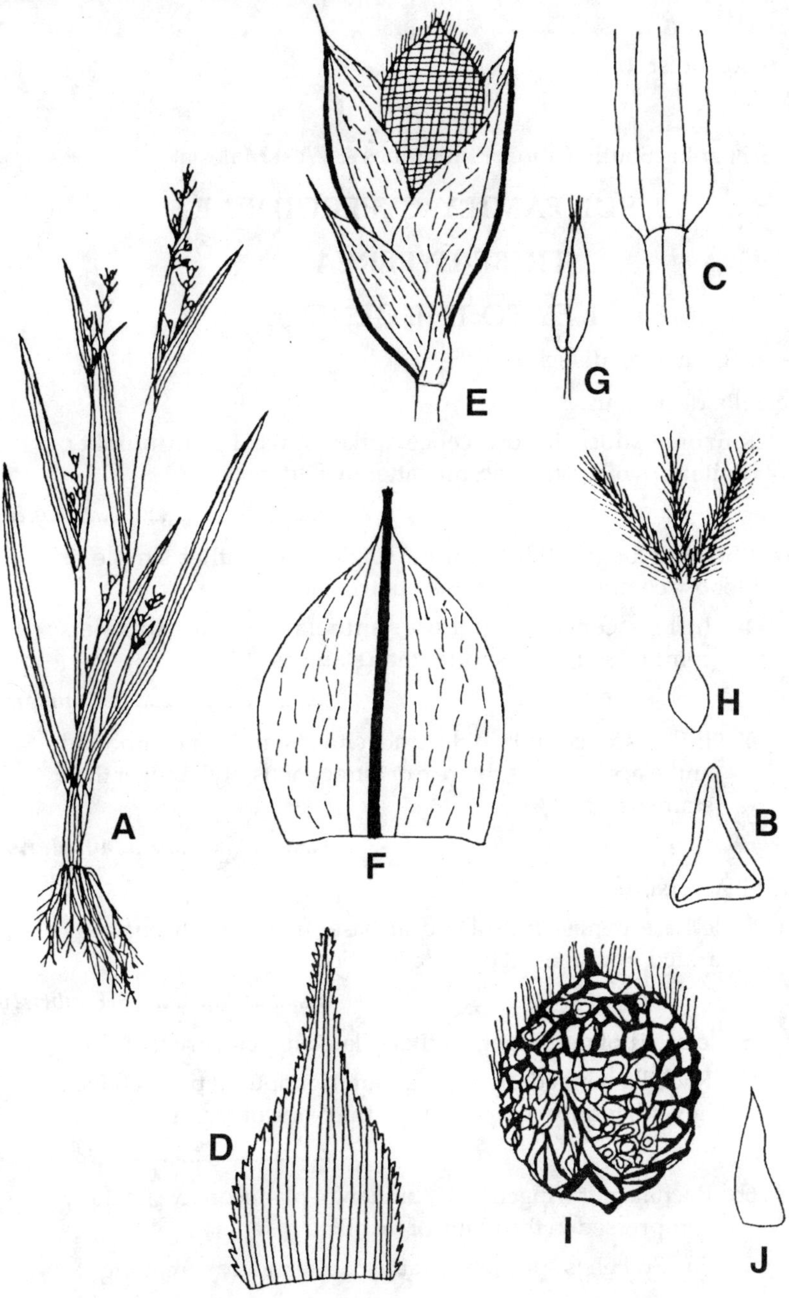

PLATE 4: ***Scleria biflora*** **Roxb.**
A. Habit, B. T.S. of stem, C. Leaf, D. Bract, E. Spikelet, F. Glume, G. Stamen, H. Pistil, I. Nut, J. Stipe

Habitat

Rare, along moist soil.

Distribution

India, Sri Lanka, South–China, Philippines & West Malesia.

2. SUB FAMILY–CYPEROIDEAE

TRIBE-CYPEREAE

KEY TO THE GENERA

1a. Flowering glumes all distichous:

2a. Rachilla deciduous:

3a. Rhizome short; inflorescence spike; spike 1–3 in number, capitate, ovoid, cylindric; nut latteraly flattened

.... 8. *Kyllinga*

3b. Rhizome long; inflorescence a single head, either simple or globose compound anthelate; nut trigonous:

4a. Inflorescence an globose, anthelate, spike; spikelets compressed; style shorter than stamens

.... 2. *Courtoisina*

4b. Inflorescence and single head, either simple or compound anthela; spikelets subterete or tetragonous; style longer than stamens

.... 10. *Mariscus*

2b. Rachilla persistent:

5a. Style base persistent, dilated at base, fimbriate throughout, separated from the nut by a constriction

.... 5. *Fimbristylis*

5b. Style base not persistent, neither dilated nor constricted:

6a. Rachilla winged; style trifid; nut trigonous, if planoconvex never laterally compressed in the plane of rachilla

.... 3. *Cyperus*

6b. Rachilla not winged; style bifid: Nut planoconvex, laterally compressed in the plane of rachilla:

7a. Spikelets spicately arranged, ultemately spreading

.... 7. *Juncellus*

7b. Spikelets digitately arranged, laterally flattened

.... 11. *Pycreus*

1b. Flowering glumes spirally arranged, rarely the basal one distichous:

8a. Style base dilated & constricted:

9a. Stem slender; leaves capillary, sheath with white hairs in the throats; inflorescence anthelate, capitate or rearely reduced to a single spikelet

.... 1. *Bulbostylis*

9b. Stem terete or triangular, transversely septate; leaves reduced to bladeless sheath, without hairs; inflorescence a single terminal, ebracteate spikelet

.... 4. *Eleocharis*

8b. Style base neither dilated nor constricted:

10a. Glumes awned:

11a. Glumes hairy; hypogynous bristles biseriate; outer bristles short reduced, inner bristles various in shaped, 3-nerved

.... 6. *Fuirena*

11b. Glumes glabrous; hypogynous bristles absent

.... 9. *Lipocarpha*

10b. Glumes not awned:

12a. Rhizome long; stem solid; nut trigonous or biconvex, often transversely wrinkled

.... 12. *Schoenoplectus*

12b. Rhizome short; stem semi–solid; nut elliptic, obovoid

.... 13. *Scirpus*

1. *Bulbostylis* Kunth.

KEY TO THE SPECIES

1a. Inflorescence anthelate with 2–many spikelets, sometimes reduced to a solitary spikelets

.... 2. *B. densa*

1b. Inflorescence capitate with 10–25 spikelets

.... 1. *B. barbata*

1. *Bulbostylis barbata*

Bulbostylis barbata (Rottb.) Clarke in Hook. f. Fl. Brit. Ind. 6: 651. 1893.Haines,Bot.Bih.& Or.923.1924; Singh, Bull.Bot.Surv.Ind.11: 19.1969; Ramaswami & Razi, Fl. Bangalore dist. 104. 1973; Kern in Fl. Male ser 1.7 (3): 539. 1974; Fischer in Gamble, Fl. Pres. Madr. 1662. 1931 (3): 1153. 1957 rep. ed.); Datar & Vartak in Journ. Univ. Poona Sci. Tech. 48: 24 t. 3. f. 1–6. 1976; Hooper in Saldanha & Nicolson, Fl. Hassan 659. 1976; Rao & Razi, Fl. Mysore 557. 1981; Rao & Verma, Cyp. N.E. Ind. 40 t. 21–21b. 1983; Sharma *et al.*, Fl. Karnataka 303. 1984; Koyama in Dassanayak &

Forsberg, Rev. Hand b. Fl. Ceyl. 5: 327. 1985; Singh, Fl. E. Karnataka 2: 627. 1988; Karthik *et al.*, Fl. Ind. Enum. Monocot. 33. 1989; Keshava Murthy & Yoganarasimhan, Fl. Coorg. 499. 1990. *Isolepis barbata* (Rottb.) R. Br. Prodr. Fl. Nov. Holl. 222. 1810; Nees in Wight. Contr. Bot. Ind. 109. 1834.*Scirpus barbatus* Rottb. Progr. 27. 1772 et in Desc. et. Incon. 52. t.17. f. 4. 1773. *Fimbristylis barbata* Benth. Fl. Austral. 7: 321. 1878. *Stenophyllous barbata* Cooke, Fl. Pres. Bomb. 2: 887. 1903 (3: 401. 1958 rep. ed.). (PLATE–5).

Annual. Rhizome absent; root fibrous. Stem15 cm long, 1.0 mm thick, brownish–green, very siender, striate,sulcate, glabrous, tufted. Leaves 13 cm long, 0.5 mm broad, one third as long as to subequalling the stem, acuminate, margin involute, capillary to very slender acute at apex, scabrous on lower surface towards the apex, several in basal clusters; sheath 1.0 cm long, 1.0 mm broad, stramineous to brown when dry, hairy at mouth; hairs falling with the age; hairs acute, base tubercled; ligule absent. Inflorescence capitate, 4.0 mm long, hemi–spherical. Bracts 3, 5.0 mm long, foliaceous, usually sub–equalling the capitula, dilated at base, usually scabrous on the lower surface, sometimes much exceeding upto 6.0 cm long. Spikelets 3.0 mm long, 1.5 mm broad, linear, lanceolate, polygonal, sub-acute, ovoid, many–flowered; rachilla stout, persistent, winged, zigzag. Flower bisexual, the terminal male or barren. Glume 1.5 mm long,1.0 mm broad, membranous, ovate, strongly keeled, laxly imbricating, mucronulate; mucro 0.2 mm long, glabrous, excurved, acute, brown, often glabrous usually hispidulous on outer surface, 3–nerved; nerve green, visible. Perianth absent. Stamens 1; anther 0.5 mm long, brown, linear–oblong, muticous, glabours; filament 1.5 mm long, slender, glabrous, slightly longer than anther, attached with pistil to the base. Pistil 3.0 mm long; ovary 1.0 mm long, obovate, glabrous, base acuminate; style 1.0 mm long, articulate with ovary, bulbiform at base, glabrous, style base persistent for a long time like a button on the top of the nut after rest of the style base fallen off, ultimately decaying; stigma 2,1.0 mm long, acute, brown, slightly fimbriate; fimbri acute base broad. Nut. 1.0 mm long, 0.5 mm broad, verruculose, transversely lineolate, stramineous to brown, surface cells minute, isodimetric, transversely oblong not very clear on the corrugate surface , triquetrous, broadly obovate, minutely stipitate, densely granular–puncticulate.

Flowering & Fruiting

June–October; Navalpur, Babatpur, Jaunpur, Lar, Sonbhadra, Ballia.

Habitat

In open to shady places, common in the paddy fields.

Distribution

India, widely distributed in the tropics & subtropics of Old world (South and East Asia, Tropical Africa, Australia).

2. *B. densa*

B. densa (Wall.) Hand–Mazz. in Karsten & Schenk, Vegetation sb. 20. 7: 16.1930; Ramaswami & Razi, Fl. Bangalore dist.105.1973; Hook. in Saldan a. & Nicolson, Fl. Hassan. 659. 1976; Datar & Vartak in Journ. Univ. Poona. Sci. Tenh. 48: 26. t. 4. f.1-

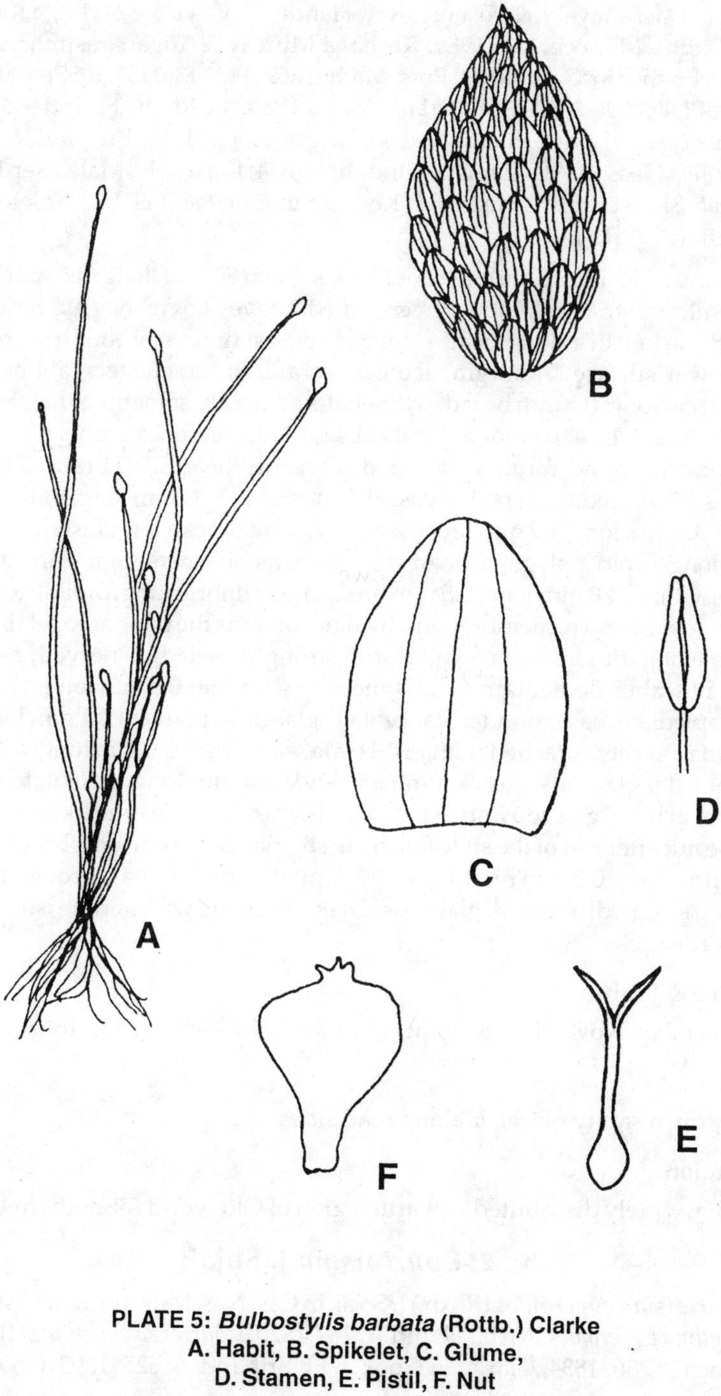

PLATE 5: ***Bulbostylis barbata*** **(Rottb.) Clarke**
A. Habit, B. Spikelet, C. Glume,
D. Stamen, E. Pistil, F. Nut

8.1976; Rao & Verma, Cyp. N.E. Ind. 40.1982; Sharma *et al.*, Fl. Karnataka 303. 1984; Koyama in Dassanayake & Fosberg, Rev. Handb. Fl. Ceyl. 5: 324. 1985; Karthik *et al.*, Fl. Ind. Enum. Monocot. 33. 1989; Keshava Murthy & Yoganarasimhan, Fl. Coorg 499. 1990; Fischer in Gamble, Fl. Pres. Madr. 1662. 1931 (3: 1153.1957repr.ed.); Cook, Aqu.Wet.Pl. Ind.96.f. 8.1996. *Scirpus densus* Wall. in Roxb. Fl. Ind. 1: 231. 1820. *Bulbostylis capillaris* var. *trifida* (Nees) Clark in Hook. f. Fl. Brit. Ind. 6: 652. 1893. *Isolepis trifida* Nees in Wt. Contr. Bot. Ind. 108. 1834; Kern in Fl. Males. ser 1.7 (3) : 538. 1974. *Fimbristylis capillaris* var. *trifida* Koyama in Jour. Fac. Sci. Uni. Tokyo sec. 3.8 (3) : 103. 1961. (PLATE–6).

Annual. Rhizome absent; root fibrous. Stem 7.0 cm long, 0.5 mm thick, very slender, sulcate, striate, glabrous, green, tufted. Leaves 4.5 cm long, 0.5 mm broad, one third to three fourth as long as the stem glabrous, at times scabrous towards the apex on the lower surface & margin, acute, several in basal clusters, almost capillary; sheath 1.0 cm long, 0.5 mm broad, white-hairs at mouth, sometimes hairs falling with age; hairs acute, base rounded; ligule absent. Inflorescence anthelate; anthela lax, simple or compound rarely contracted & nearly head-like. Bracts 3.0 mm long, setaceous, glumaceous, awned; awns, glabrous; rays 3, 1.0 cm long, filiform suberect. Spikelets 3.5 mm long, 1.2 mm broad, solitary, sometimes a few clustered, subacute at apex, oblong-ovoid, polygonal, many flowered; rachilla persistent, narrowly winged, glabrous. Glume 1.0 mm long, 0.5 mm broad, laxly imbricating, broadly ovate, acute-muticous, dark brown, membranous, hyaline towards the base, acropetally caducous from the elongating rachilla, cymbiform, strongly keeled, 1- nerved; nerves green prominent scaberulous. Stamens 2, sometimes 1; anther 0.5 mm long, yellow, linear, oblong, apiculate, base rounded, lanceolate, glabrous; filament 0.1 mm long, slender, shorter than anther, attached with Pistil to the base. Pistil 2.0 mm long; ovary 0.5 mm long, obovate, glabrous, base acuminate; style 1.0 mm long, articulate with ovary, bulbiform at base, glabrous, style base persistent for a long time like a button on the top of the nut after rest of the style fallen off; stigmas 2, 0.5 mm long, brown, subentire. Nut 1.0 mm long, 0.5 mm broad, shortly stipitate, smooth, pale brown, triquetrous, with obtuse angled, obovoid, glabrous, finely reticulate with minute isodimetric cells in many rows.

Flowering & Fruiting

September-November; Rajghat, Chandauli, Kashinagar, Jaunpur.

Habitat

Dry open sandy places & along road sides.

Distribution

India, widely distributed in warm region of Old world & Southern U.S.A.

2. *Courtoisina* J. Sojak

Courtoisina cyperoides (Roxb.) J. Sojak in Cas. Nas. Muz. (Prague) 148: 193 (1979) 1980. *Kyllina cyperoides* Roxb., Fl. Ind. 1: 187.1820. *Courtoisia cyperoides* (Roxb.) Nees in Linnaea 9: 286. 1834; Clarke in Hook. f. Fl. Brit. Ind. 6: 625. 1893; Cooke, Fl. Pres.

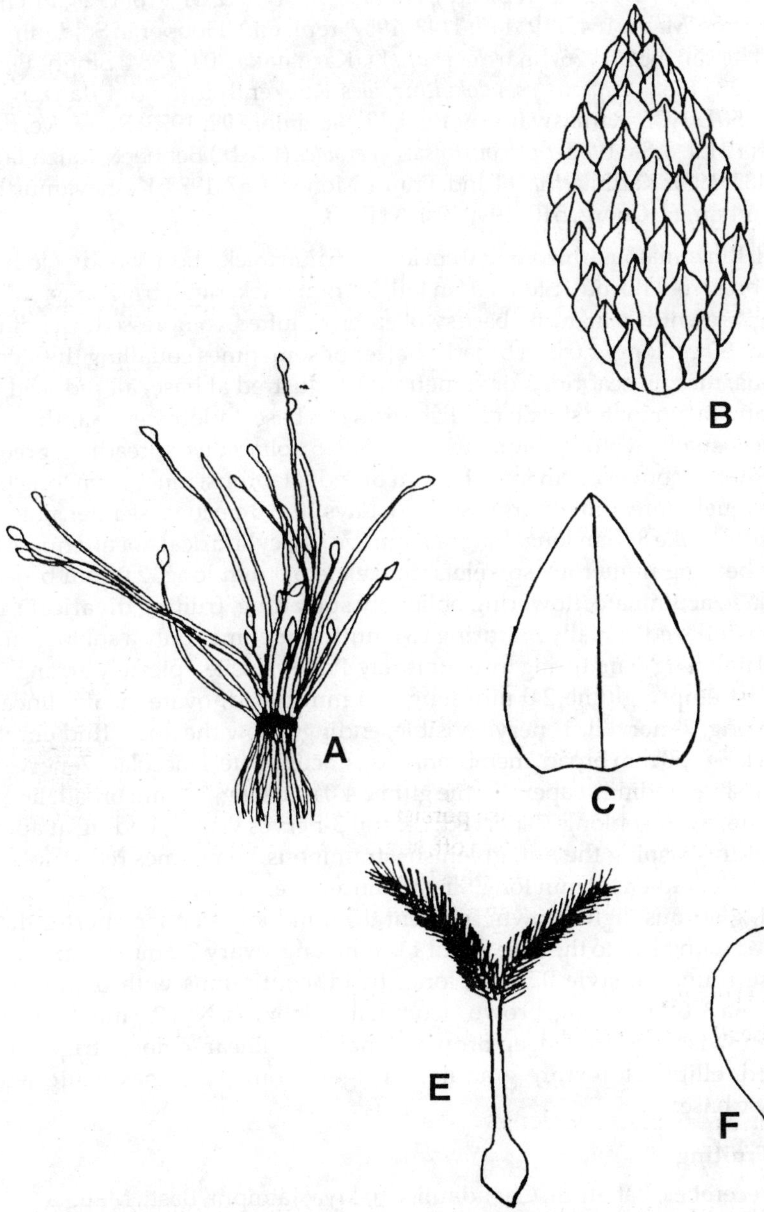

PLATE 6: ***Bulbostylis densa*** **(Wall.) Hand.-Mazz.**
A. Habit, B. Spikelet, C. Glume,
D. Stamen, E. Pistil, F. Nut

Bombay 2: 877. 1908 (3: 391. 1958. repr. ed); Haines, Bot. Bih. & Or. 910. 1924; Fischer in Gamble, Fl. Pres. Madr. 1645. 1931 (3: 1143. 1957. repr. ed.); Hooper in Saldanha & Nicolson, Fl. Hassan. 661. 1976; Sharma *et al.*, Fl. Karnataka 304. 1984; Singh, Fl. E. Karnataka 2: 628. 1988. *Cyperus pseudokyllingoides* Kuekenth in Engl. Pflanzenr. 4 (20), Heft 101 : 501. 1936; Ramaswamy & Razi, Fl. Bangalore 92. 1973; Rao & Verma, Cyp. N.E. India 21. f. 8-8a. 1982. *Indocourtoisia cyperoides* (Roxb.) Bennet & Raizada in Ind. For. 107: 432. 1981; Karthik *et al.*, Fl. Ind. Enum. Monocot. 57. 1989; Keshavamurthy & Yoganarasimhan, Fl. Coorg. 510. 1990. (PLATE–7).

Perennial. Rhizomatous; rhizome 3.0 cm long, 2.5 mm thick, short, woody, clothed with brown fibres; root fibrous. Stem 70 cm tall, 5.0 mm thick, rigid, trigonous, solid, green at young. brownish at maturity, base swollen, erect, tufted, compressed, asending, solitary. Leaves 50 cm long, 1.0 mm broad, shorter or sometimes equalling the stem, green, glabrous, margin scabrid, 3 or sometime 10, clustred at base, all reduced to bladeless sheaths, acuminate; sheath reddish-brown at base. Inflorescence anthelate; anthela simple, usually evolute, contracted. Bracts 3–5, foliaceous, spreading, green, arranged in whorles; outer 20 cm long,1.0 mm broad, glabrous; inner 3 cm long 0.5 mm broad, obliquely spreading, margin scabrid. Rays 10, 8.0 cm long, slender, patent, straight, smooth. Spike 8.0 cm long, long, 4.0 mm broad, cylindrical not attenuate at base, densely bearing numerous spikelets. Spikelete 6.0 mm long, 2.0 mm broad, linear-lanceolate, acuminate, flowering obliquely spreading, fruiting divaricate, or the lower even deflexed, usually maturing two nuts, sometimes only a solitary nut, sometimes 7.0 mm long & maturing three or rarely 4 nuts, sessile, spicately arranged, subcylindric; Ist empty glume 2.0 mm long, 1.0 mm broad, ovate, acute, linear-lanceolate, oblong, 7- nerved; 1- nerve, visible, ending below the apex. IInd empty glume 1.5 mm long, 1.0 mm broad, membranaceous, acuminate, lanceolate 7–nerved; central nerve visible, ending at apex. Fertile glume 4.0 mm long, 1.0 mm broad, linear lanceolate, acute, ovate–oblong, 7–nerved; central 3-nerves visible, ending at apex. All glumes tightly clasping the nut, greenish–straminous, sometimes red striolate, folded. Stamens 2; anther 1.0 mm long, linear–lanceolate, ovate, oblong, acuminat, base rounded, glabrous, light brown; filament 0.5 mm long, slender, shorter than anther, attached with pistil to the base. Pistil 1.0 mm long; ovary 0.2 mm long, ovate, glabrous, base rounded; style 0.5 mm long, trifid, continuous with ovary, flat, caducous; stigma 3, 0.4 mm long, brown, acuminate, glabrous. Nut 2.5 mm long, 0.5 mm broad, yellowish at young, deep brown at maturity, linear–oblong, trigonous, slightly curved, elliptical, texture granulate, ridges prominent, apex acute with remnant of style base.

Flowering & Fruiting

March–December; Babatpur, Chandauli, Chakiya, Jaunpur, Basti, Mau.

Habitat

Common in wastelands, cultivated fields, along road sides

Distribution

India, Tropical & sub tropical regions of Asia, Africa, Australia.

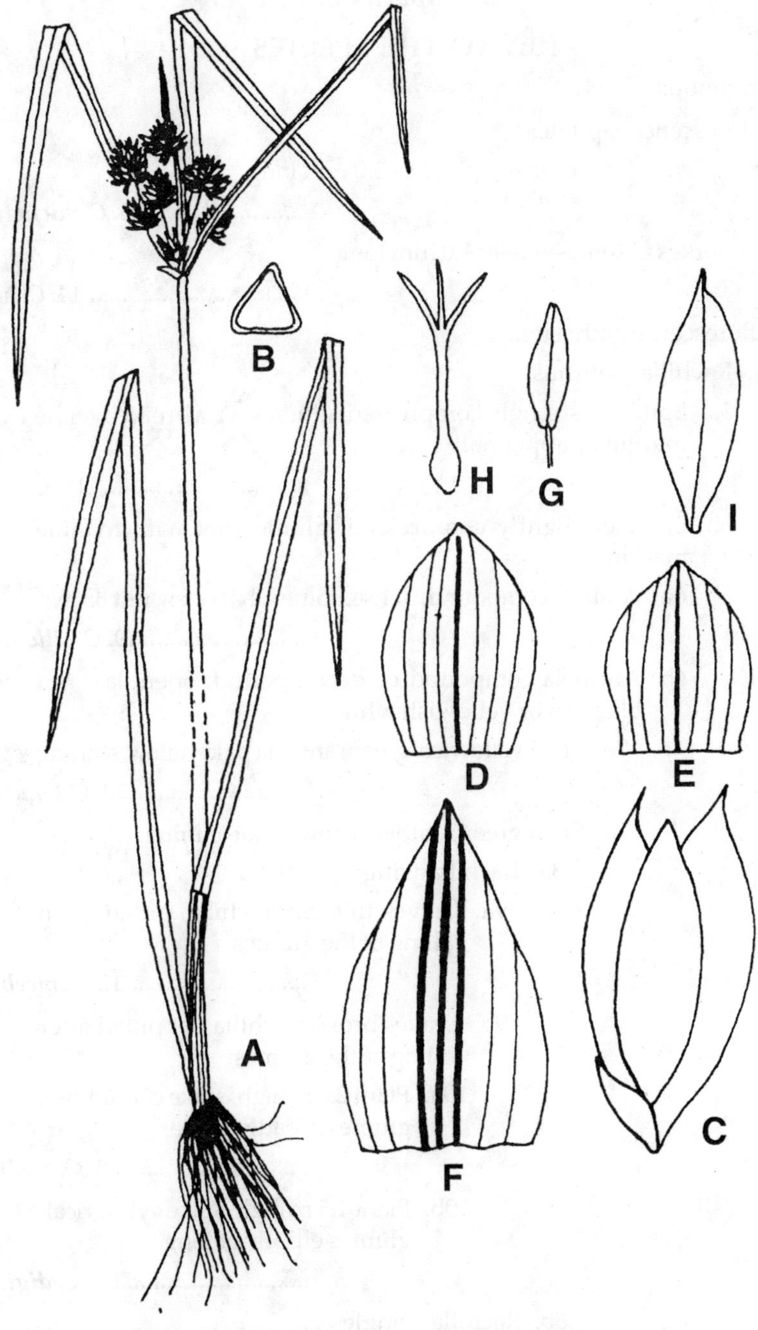

PLATE 7: ***Courtoisina cyperoides*** **(Roxb.) J. Sojak.**
A. Habit, B. T.S. of stem, C. Spikelet, D. Ist glume, E. IInd glume, F. IIIrd Glume, G. Stamen, H. Pistil, I. Nut

3. *Cyperus* L.

KEY TO THE SPECIES

1a. Stem trigonous:

2a. Inflorescence capitate:

3a. Spikes 14 in number; spikelets 3.0 mm long
.... **13. *C. michelianus***

3b. Spikes absent; spikelet 1.0 mm long
.... **14. *C. niveus***

2b. Inflorescence anthelate:

4a. Rachilla glabrous:

5a. Spikelets strongly compressed; glumes with broad hyaline margin in upper half
.... **15. *C. nutans***

5b. Spikelet slightly compressed; glumes not with hyaline margin:

6a. Anthela compound, dense, some what open, nut brown
.... **19. *C. sillentensis***

6b. Antthela compound or decompound; open, lax; nut blackish or yellowish white:

7a. Stem yellowish-green; anthela pale yellow, shining
.... **20. *C. tenuispica***

7b. Stem green; anthela brown, not shining:

8a. Rachilla winged:

9a. Leaves filiform; rachilla, pitted often falling of the glumes
.... **12. *C. meeboldii***

9b. Leaves broad; rachilla not pitted after falling of the glumes:

10a. Plant 2.0 m high; spike corymbose; glumes broadly ovate
.... **8. *C. exaltatus***

10b. Plant 1.5 m high; spike cylindrical; glumes elliptic oblong
.... **6. *C. digitatus***

8b. Rachilla wingless:

11a. Spikelets up to 22 together, linear
.... **4. *C.cuspidatus***

11b. Spikelets upto 13 together, digitate
.... **9. *C. halpan***

4b. Rachilla scabrid on angle or glabrous if glabrous with a red streak:

12a. Glumes loosely imbricating, yellow .... **2. *C. alulatus***

12b. Glumes tightly imbricating, reddish or deep brown .... **18. *C. rotundus***

1b. Stem triquetrous:

13a. Spikelets 3–8 mm long:

14a. Glumes with hyaline & with a reddish brown band on each side of the mid rib .... **5. *C. difformis***

14b. Glumes with cellular reticulate sides, yellow 1–2 striate, scarious on margin .... **10. *C. iria***

13b. Spikelets upto 1.0 cm long:

15a. Anther hairy .... **17. *C. platystylis***

15b. Anther glabrous:

16a. Leaves well developed:

17a. Rhizome absent; anthela simple lax .... **3. *C. compressus***

17b. Rhizome present, anthela compound or decumpound:

18a. Rhizome aromatic; spikelets clusters of 4–9 together; nut narrowly oboviod, trigonous .... **16. *C.pangorei***

18b. Rhizome not aromatic; spliketes solitary; nut ellipsode, triguetrous .... **11. *C. laxus*.**

16b. Leaves usually absent:

19a. Glumes elliptic, reddish-brown, margin slightly inrolled .... **7. *C. distans***

19b. Glumes orbicular, pale green, margin not inrolled .... **1. *C. alternifolius***

1. *Cyperus alternifolius*

Cyperus alternifolius L. Mant. 2: 28.1771; Herwig, Goodhon. Encycl. Hou. Pl.121.1984; Randhawa & Mukhopadhyay, Flori. Ind. 561. 1985. (PLATE–8).

PLATE 8: ***Cyperus alternifolius*** **L.**
A. Habit, B. Rhizome, C. T.S. of stem, D. Bract, E. Spikelet, F. Glume, G. Stamen, H. Pistil, I. Nut

Perennial. Rhizomatous; rhizome 10 cm long, blackish–brown, creeping, woody; root fibrous. Stem 1.5 m tall, 1.0 cm thick, green, glabrous, erect, triquetrous, densely tufted, ribbed. Leaves absent; sheath 22 cm long, 1.2 cm broad, deep green at upper half, dark brown at lower half, acuminate many nerved, membranous, glabrous, erect; ligule absent. Inflorescence anthelate; anthela 15 cm long, 10 cm broad, compound, lax, spreading. Bracts 25 cm long, 2.0 cm broad, green, suberect, arranged in whorls & forming an umbrella, numerous, acute, surface glabrous, margin scabrid, mid-vein distinct on lower surface, with 4–lateral veins on each side of mid vein; lateral veins distinct on upper surface; rays 8.0 cm long, erect, green, glabrous, slender, many. Spikelets 6.0 mm long 3.0 mm broad, ovate, oblong, linear, glabrous, sub-acute, 15– flowered; rachilla glabrous. Glume 2.0 mm long, 2.0 mm broad, membranous, glabrous on the both surface, ovate, orbicular, acute, laterally compressed, margin hyaline, white, scabrid at apex, glabrous below, 3–nerved; nerves deep green, glabrous, visiable, ending at apex. Stamens 3; anther 1.0 mm long, yellow, linear, oblong, glabrous; filament 1.5 mm long, slender, glabrous, longer than anther. Pistil 3.0 mm long; ovary 0.5 mm long, ovate, glabrous; style 1.0 mm long, slender, glabrous, solitary; stigma 3, 1.5 mm long, glabrous, white at young, turning into purple with age. Nut 1.0 mm long, 0.5 mm broad, brown, acuminate stipitate, trigonous, glabrous.

Flowering & Fruiting

January–March; BHU, Navalpur.

Habitat

Ornamental plant.

Distribution

India, Tropical Africa, Arabia, Asia, America..

2. *C. alulatus*

C. alulatus Kern in Reinwardtia 1 : 463–466.f. 1. 1952; Verma & Chandra in Rec. Bot. Surv. Ind. 21 (2): 236.1981. *C. iria* acute, non L. 1753; C.B. Clarke in III. Cyper. t. 14. 1909. *C. iria* var. *rectangularis* Kuekenth. in Engl., Planzenf. heft 101: 152. 1935; Sharma *et al.*, Fl. Karnataka 304. 1984; Singh, Fl. E. Karnataka 2: 630. 1988; Karthik *et al.*, Fl. Ind. Enum. Monocot. 43. 1989; Cook, Aqua.Wet. Pl. Ind.103.f.85.1996. *Cyperus rectangularis* Bennet in Indian For. 95: 692. 1969. (PLATE–9).

Annual. Rhizome absent; root fibrous. Stem 30 cm tall, 2.0 mm thick, trigonous, green, glabrous, eret, tufted, clothed with sheaths. Leaves 30 cm long, 2.5 mm broad, green, glabrous, acuminate, sheathing base arising from the lower parts of the stem; sheath. 1.0 cm long, 3.0 mm broad, loose, brown. Inflorescence anthelate; anthela 6.0 cm long, simple or compound, lax. Bracts 6.0 cm long 1.0 mm broad 2–7 in number, foliaceous obiliquely spreading; rays 8.0 cm long, 2–9 in number longer one often paniculately branched to 2–5 spikes slender, scabrite. Spikelets 5.0 mm long, 1.0 mm broad, ovate–oblong, linear, yellowish–green, spicately arranged, compressed, spreading at right angle, apex rounded, 15–flowered; rachilla wingless, with 1.0 mm long internodes, scabrid on angles. Glume 2.5 mm long, 1.5 mm broad, 1/3 imbricating, yellow, loosely arranged, broadly ovate, suborbicular, keeled, membranous,

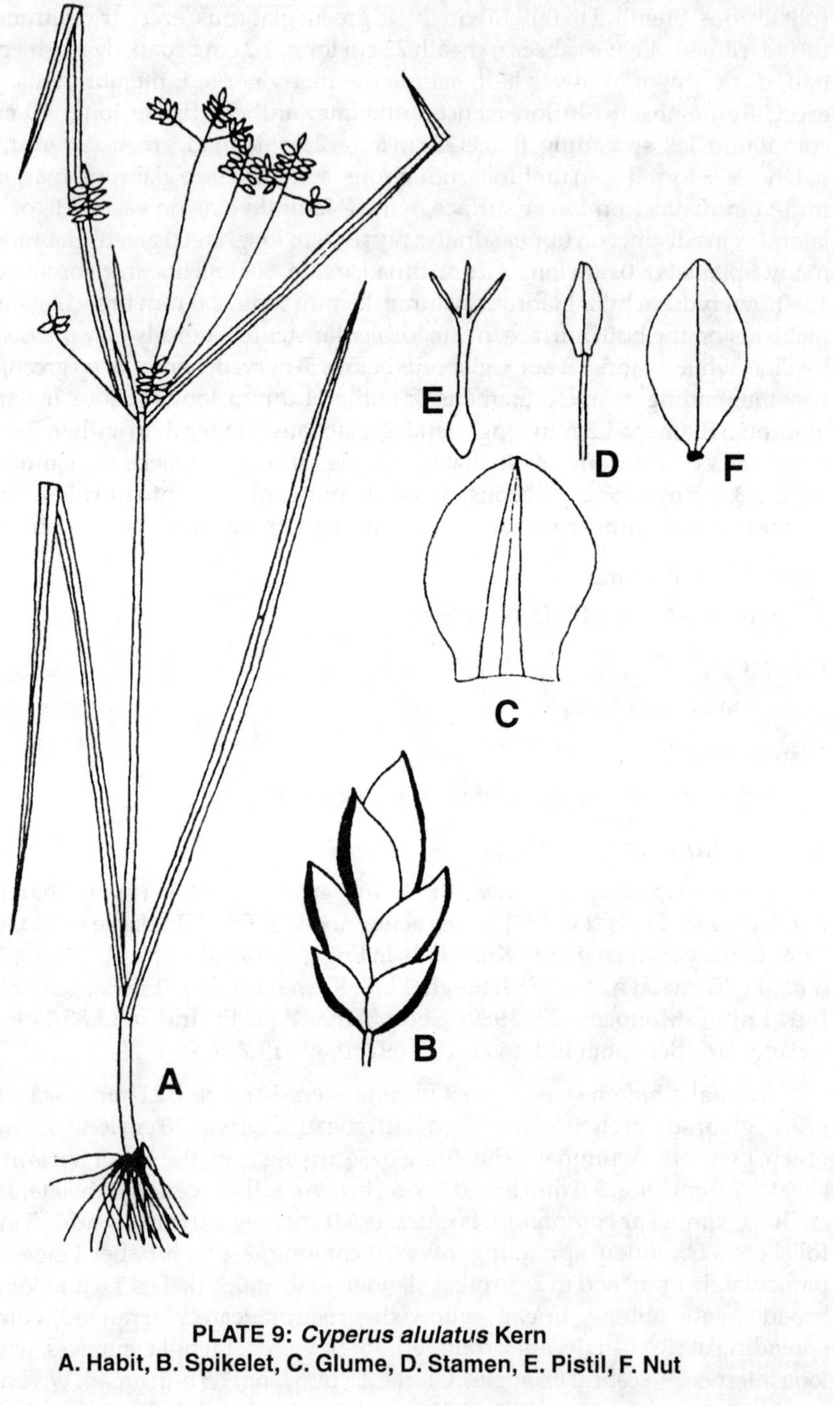

PLATE 9: ***Cyperus alulatus*** **Kern**
A. Habit, B. Spikelet, C. Glume, D. Stamen, E. Pistil, F. Nut

emarginate, muticous, mucronulate at apex, keeled, 3–nerved; nerves visible, serrate, winged towards top. Stamens 2; anther 0.5 mm long, yellowish–green, glabrous, acuminate, base rounded, oblong–ovate linear, lanceolate; filament 2.0 mm long, slender, longer than anther, attached with pistil to the base. Pistil 1.5 mm long; ovary 0.2 mm long, ovate, glabrous, base rounded; style 1.0 mm long, glabrous, flat; stigma 3, 0.3 mm long, acute, brown, glabrous. Nut 1.0 mm long, 0.5 mm broad, brown, obovoid, triquetrous, broadly stipitate with concave sides, mucronate at apex.

Flowering & Fruiting

July–December; Sarnath, Cholapur, Babatpur, Khalilabad, Sonbhadra, Jaunpur.

Habitat

Common in moist places, bunds of cultivated fields.

Distribution

India, Pakistan.

3. *C. compressus*

C. compressus L. Sp. Pl. 46. 1753; Clarke in Hook. f. Fl. Brit. Ind. 6: 605. 1893; Cooke, Fl. Pres. Bomb. 2: 866. 1908 (3: 379. 1958 repr. ed); Fischer in Gamble, Fl. Pres. Madr. 1640. 1931 (3: 1140. 1957 repr. ed.); Rajagopal, Proc. Nat.Acad. Sci. Ind. Sec. B. 35: 44. 1965; Kukenth in Engl. Pflanzenr. 4 (20). Heft 101: 156.1935. Ramaswamy & Razi, Fl. Bangalore 82.1973. Kern in van Steenis, Fl. Males. 1. 7: 617.1974; Hooper in Saldanha & Nicolson, Fl. Hassan 663. 1976; Rao & Razi, Fl. Mysore 558. 1981; Rao & Verma, Cyp. N.E. Ind. 15. 1982; Sharma *et al.*, Fl. Karnataka 305. 1984; Koyama in Dassanayak & Fosberg, Rev. Handb.Fl.Ceyl. 5: 194.1985; Singh, Fl. E. Karnataka 630. 1988; Karthik *et al.*, Fl. Ind. Enum. Monocot. 44. 1989; Misra & Verma, Fl. Allaha. 380. 1992; Cook, Aqu. Wet. Pl. Ind.106.f.92.1996. (PLATE–10).

Annual. Rhizome absent; root fibrous. Stem 40 cm long, 2.5 mm thick, green, glabrous, tufted, spreading, triquetrous, striate, erect. Leaves 150 cm long, 4.0 mm broad, sub-coriaceous, shorter than the stem, greyish green, flat, gradually narrowed at acuminate tip, margin scabrous towards the tip; sheath membranous, striate, pale brown to reddish brown toward base; ligule absent. Inflorescence anthelate; anthela simple, lax, open, sometimes reduced to a single sessile cluster. Involucral bracts 5, 30 cm long, 5.0 mm broad, foliaceous, spreading, unequal, slightly to much exceeding the anthela; rays 8, 15 cm long, unequal, slender, patent, diverging, sometimes absent in depauperated plants. Spikes 7, broadly ovoid, digitately–rays mostely bearing 12 spikelets on a short axis. Spikelets 3.5 cm long, 5.0 mm broad, spreading, strongly compressed, subacute, lanceolate, oblong, 40 flowered, flattened, greenish at young, turning straw–coloured at maturity; rachilla flexuous, persistent, with very thin caducous wings with 0.7 mm long internodes. Glume 5.0 mm long, 2.5 mm broad, ovate, acute, somewhat rigid, conspicuously multinerved, with green keel, pale green–yellowish, brown sides & silvery hyaline margins, mucro 0.8 mm long, straight to slightly curved. Stamens 3; anther 1.0 mm long, crest, ovate, reddish, linear, oblong, with slightly produced connective appendage; filament 3.5 mm long, slender, glabrous, longer than anther. Pistil 4.0 mm long; ovary 0.5 mm long, ovate, glabrous; style 2.5 mm long, tricleft, glabrous; stigma 2.0 mm long, glabrous. Nut 2.0 mm long, 1.0 mm

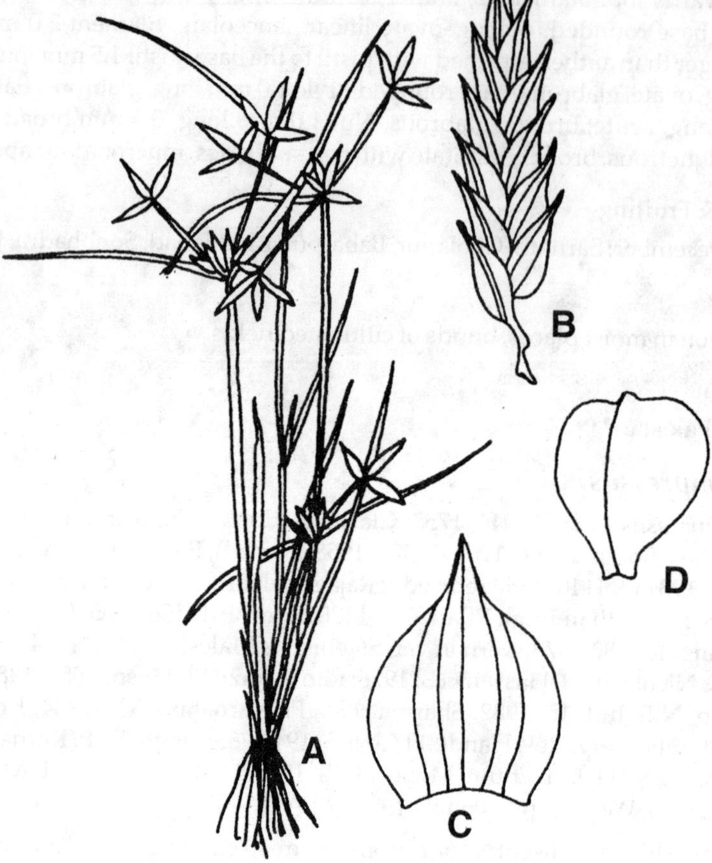

PLATE 10: *Cyperus compressus* **L.**
A.Habit, B. Spikelet, C. Glume, D. Nut

broad, triquetrous, stipitate, obovoid, apiculate, shiny, minutely punctate, deep blackish-brown, glossy.

Vernacular Name

Lavhala, Mothi.

Flowering & Fruiting

August–March; Raghat, Pindra, Ballia, Mau, Kushinagar,

Habitat

Common in agricultural fields, along railway tracks, near ponds & ditches.

Distribution

India, Pakistan.

4. *C. cuspidatus*

C. cuspidatus Kunth in H.B.K., Nov. Gen. et Sp. 1: 204.1815; Haines, Bot. Bih. & Or 895. 1924; Cook, Aqu.Wet. Pl. Ind. 108.f. 94.1996. *C. uncinatus* Clarke in Durand et Schinz, Consp. Fl. Afr. 5: 580. 1895 et in Journ. Linn. Soc. (Bot.) 36 : 219. 1903; C.B. Clarke in Hook. f.Fl. Brit. Ind. 6: 598. 1893; Kern in Fl. Males. ser. I. 7(3): 629. f. 59.1974. (PLATE–11)

Annual.Rhizome absent; root 3.0 cm long, blackish–brown, fibrous. Stem 15 cm long, 0.5 mm thick, trigonous, tufted, compressed, green, glabrous. Leaves 9.0 cm long, 1.0 mm thick, 4–6 in number, shorter than to longer than the stem, acuminate, green, glabrous, all reduced to bladeless sheaths, clustered at the base. Inflorescence. anthetate; anthela 2.0 cm long, 1.0 cm broad, simple or compound, contracted, evolute, green at young, browinish at maturity. Bracts 2–4, 3.0 cm long, foliaceous, exeeding the anthela; rays 1–3, 3.5 cm long, suberect, spreading, sometimes absent in depanperated plants. Spikelets 5.0 mm long, 1.3 mm broad, linear, upto 22 together, sides parallel, 10–26flowered, distichously arranged; rachilla zigzag, red–streaked, wingless. Glume 2.5 mm long, 1.0 mm broad, obliquely spreading, scarcely imbricating, ovate, aristate, fuscous–ferrugineous, membranous, acuminate, base rounded sometimes hyaline, keeled, 3–nerved; central nerves visible, lateral nerves faintly visible, aristate; arista 1.0 mm long, recurved. Stamens 1 rarely 2–3; anther 0.5 mm long, ovate, oblong, glabrous, acuminate, base rounded, crest reddish; filament 1.0 mm long, uniformly flat, longer than anther, attached with pistil to the base. Pistil 2.0 mm long; ovary 1.0 mm long, ovate, glabrous, base acuminate; style 0.5 mm long, deeply trifid, uniformly slender; stigma 0.5 mm long, acuminate, brown, glabrous. Nut 1.0 mm long, 0.5 mm broad, sessile, obovoid, trigonous, chest nut, puncticulate, apiculate.

Flowering & Fruiting

August–November; Harraua, Allahabad, Mirzapur, Basti.

Habitat

Fairly common & frequently found in moist localities.

Distribution

India, Srilanka, China, Australia, America.

5. *C. difformis*

C. difformis L. Cent. Pl. 2: 6 1756; C.B. Clarke in Hook. f. Fl. Brit. Ind. 9: 599. 1893; Haines, Bot. Bih. & Or. 893. 1924; Kern in Fl. Males ser. 1.7 (3): 629. 1974; Cooke, Fl. Pres. Bomb. 2: 862. 1908 (3: 376. 1958, repr. ed.); Fischer in Gamble, Fl.Pres.Madr.1640.1934 (3: 1139.1957 repr.ed.); Kukenth. in Engl., Pflanzenr. 4 (20) Heft 101: 237 f. 27 F–1.t.1.1936; Ramaswamy & Razi, Fl. Bangalore 84. 1973; Hooper in Saldanha & Nicolson, Fl. Hassan 664. 1976; Rao & Razi, Fl. Mysore 558.1981; Rao & Verma, Cyp. N.E. Ind. 12. 1982; Sharma *et al.*, Fl. Karnataka 305. 1984; Koyama in Dassanayake & Fosberg, Rev. Hand b. Fl. Ceylon 5: 206. 1985; Singh, Fl. E. Karnataka 631. 1988; Karthik. *et al.*, Fl. Ind. Enum.Monocot. 44 1989; Keshava Murthy & Yoganarasinhan, Fl. Coorg. 501. 1990. (PLATE–12).

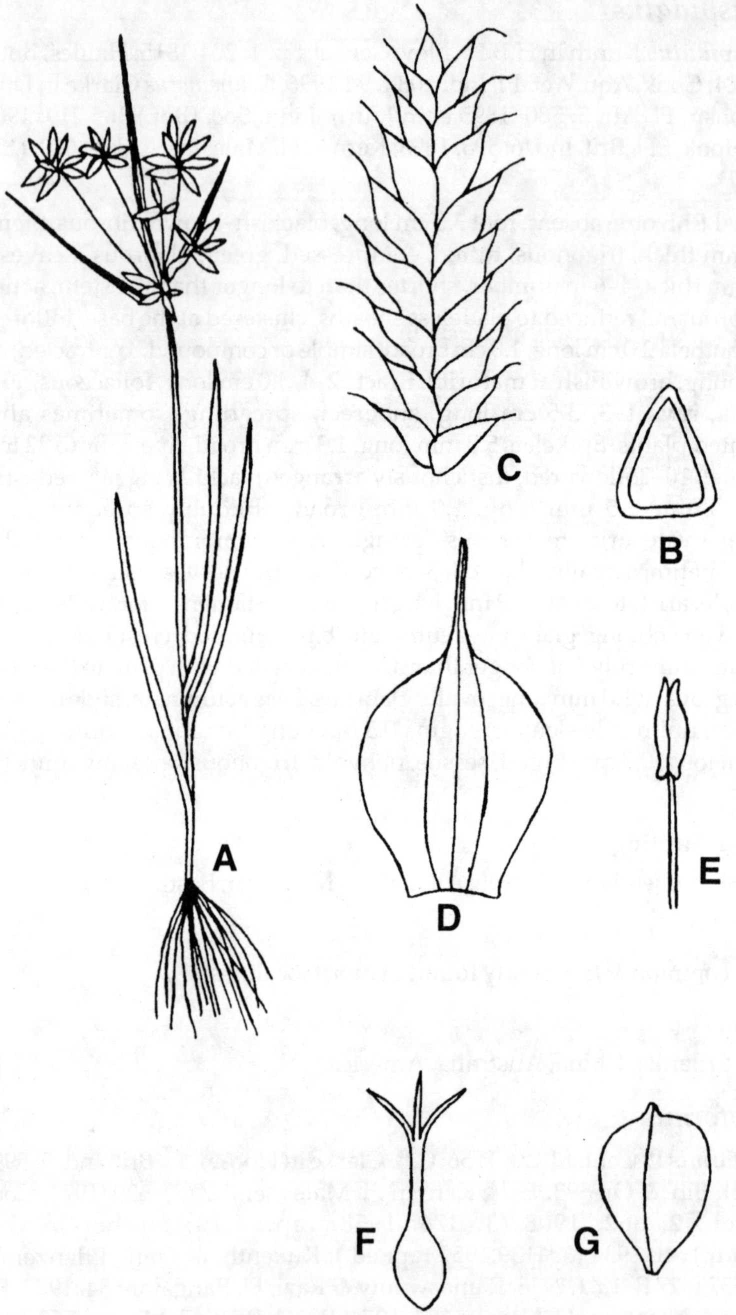

PLATE 11: ***Cyperus cuspidatus*** **Kunth**
A. Habit, B. T.S. of stem, C. Spikelet,
D. Glume, E. Stament, F. Pistil, G. Nut

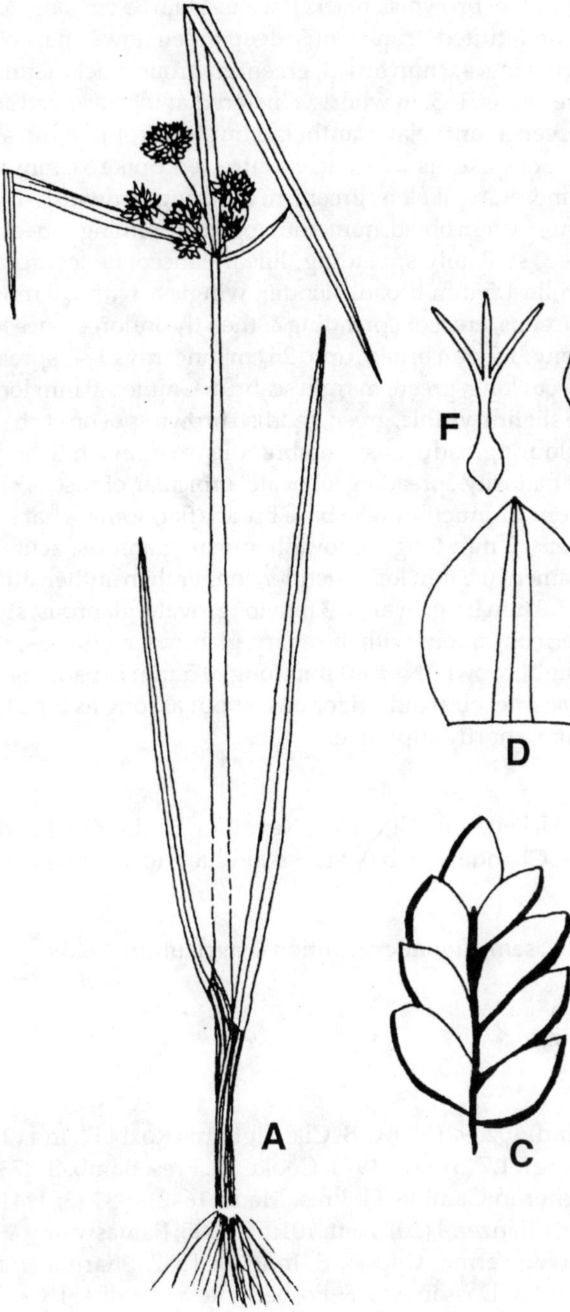

PLATE 12: *Cyperus difformis* L.
A. Habit, B. T.S. of stem, C. Spikelet, D. Glume, E. Stamen, F. Pistil, G. Nut

Annual. Rhizome absent; root brownish black, fibrous. Stem 35 cm long, 3.0 mm thick, erect, flacid, compressed, tufted, triquetrous, deep green, lower part of stem blackish brown. Leaves 20 cm long, 4.0 mm broad, green, glabrous, thick, acuminate, base broad, shorter than the stems, 1–3, in whorls, clustered at the base, reduced to bladeless sheaths. Inflorescence anthelate; anthela simple, evoluate, or almost congested in a globose cluster of spikelets, capetately contracted. Spike 6.0 mm across, dense, globose, bearing numerous spikelets, greenish, at young, brownish–black at maturity. Spikelet 2.0 mm long, 6.0 mm broad, numerous together, forming sub–globose. head, green, 12–36 flowered, stellately spreading, linear–lanceolate, compressed, oblong–ovate at apex; rachilla 1.5 mm broad, slender, wingless with 2.0 mm long internodes. Bracts 2–3, foliaceous, erect or spreading & then the inflorescence looks–like lateral, lowest 7.0 cm long, 1.0 mm broad, upto 25 cm long; rays 1–7, spreading, lowest 1.5 cm long, upto 5.0 cm long, green, margin scabrid. Glume 1.0 mm long, 0.5 mm broad, 3–nerved; nerve slightly visible, green, reddish brown spot on each side of midnerve, sometimes discolouring, caducous; membranous, margin whitish hyaline green at young, brownish at maturity, spreading, obovate–orbicular, obtuse or slightly emerginate at apex, muticous or mucronulate, base broad, flat, some what loosely arranged. Stamens 2; anther 0.2 mm long, yellowish–green, glabrous, acute, base rounded, linear, oblong; filament 0.5 mm long, slender, longer than anther, attached with pistil to the base. Pistil 1.0 mm long; ovary 0.3 mm long, ovate, glabrous; style 0.2 mm long, uniformly, slender, continuous with the ovary, glabrous, caducous; stigma 3; 0.5 mm long, glabrous, light brown. Nut 1.0 mm long, 0.5 mm broad, yellow at young, brown at maturity, sessile, obovoid, trigonous, about as long as the glumes, apiculate, elliposid, lenticular, shortly stipitatae.

Flowering & Fruiting

July–October; UPC, BHU, Babatpur, Cholapur, Navalpur, Sindhora, Chandmari, Taktakpur, Sarnath, Sarriya, Chandauli, Chakiya, Sakaldiha, Sadalpura, Jaunpur.

Habitat

Very common in aquatic, semi–aquatic condition & agriculture fields.

Distribution

India, Australia, America.

6. *C. digitatus*

C. digitatus Roxb. Fl. Ind. 1: 209. 1820; C.B. Cla. digitatus Roxb. Fl. Ind.. Ind. 6: 618. 1893; Kern in Fl. Males. ser. 1.7 (3): 601. 1974; Cooke, Fl. Pres. Bomb. 2: 873.1908 (3: 387.1958, repr. ed.); Fischer in Gamble, Fl. Pres. Madr. 1642. 1931 (3: 1141.1957 repr. ed.); Kukenth in Engler. Pflanzer. 4 (20), Heft. 101: 55. 1935; Ramaswamy & Razi, Fl. Bangalore 84. 1973; Rao & Verma, Cyp. N.E. Ind. 20: 1982; Sharma *et al.*, Fl. Karnataka 305. 1984; Koyama in Dassanayak & Fosberg, Rev. Handb. Fl. Ceylon 5: 170. 1985; Karthik *et al.*, Fl. Ind. Enum. Monocot. 44. 1989; Cook, Aqu.Wet. Pl. Ind.109.f.96 a-c. 1996. (PLATE–13).

Perennial. Rhizomatous; rhizome 2.0 cm long, 5.0 mm thick, woody creeping; root fibrous. Stem 50 cm tall, 9.0 mm thick, stout, trigonous above, triquetrous below,

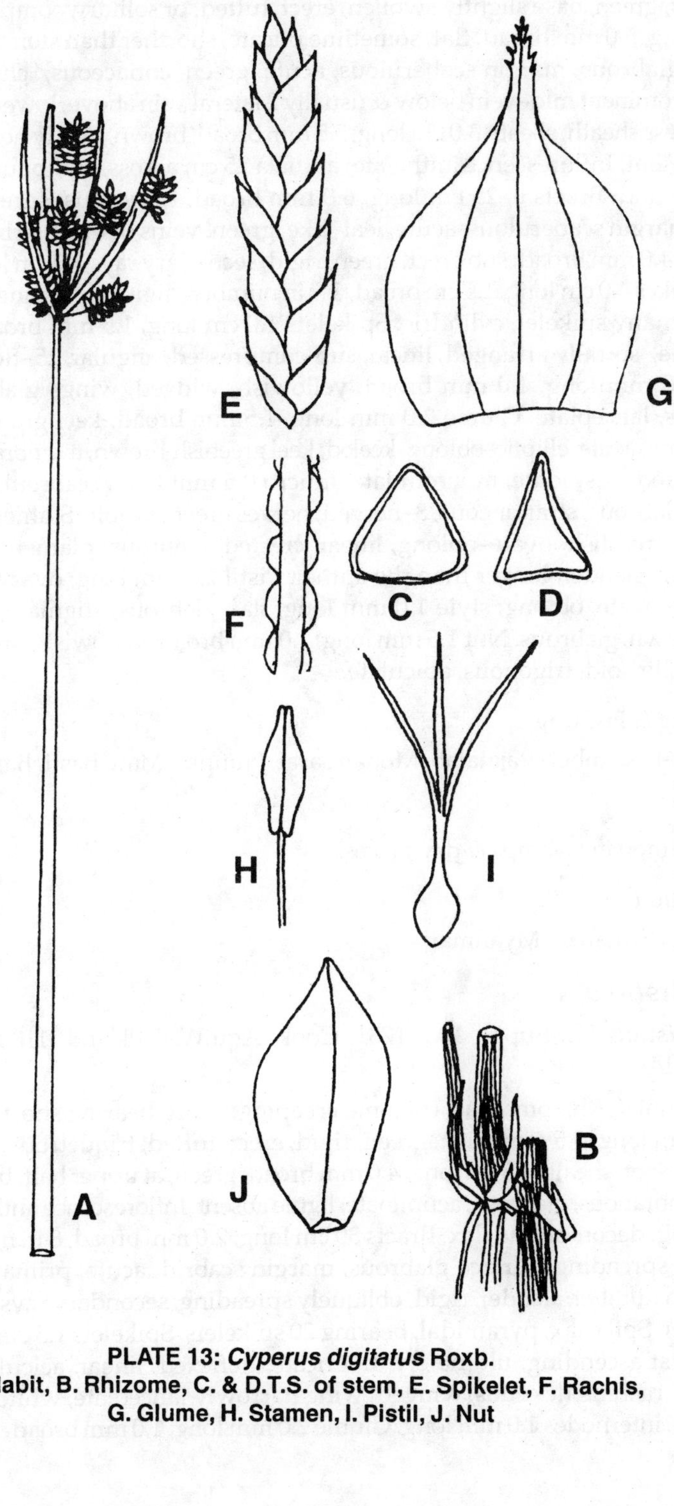

PLATE 13: ***Cyperus digitatus*** **Roxb.**
A. Habit, B. Rhizome, C. & D.T.S. of stem, E. Spikelet, F. Rachis, G. Glume, H. Stamen, I. Pistil, J. Nut

glabrous, green, base slightly swollen, erect, tufted, or solitary, compressed. Leaves 45 cm long, 5.0 mm broad, flat, sometime plicate, shorther than stem, veins distinct, surface glabrous, margin scaberulous, acute, green, coriaceous, clustered at base with a prominent mid–vein below & usually 2 lateral vein above; lower most reduced to bladeless sheath; sheath 3.0 cm long, 5.5 mm broad, brown, faintly nerved, glabrous; ligule absent. Inflorescence anthelate; anthela 25 cm across, compound, sometimes decompound. Bracts 6, 25 cm long, 6.0 mm broad, foliaceous, spreading, surface rough, margin scaberulous, acute, leaf–like, green, veins distinct; primary rays 6, 11 cm long, 4.0 mm broad, sub erect, green, rigid; secondary rays 3.0 cm long, rigid, sub erect. Spikes 4.0 cm long, 1.5 cm broad, 3-9 in number, digitately arranged, subloosely bearing many spikelet, cylindric. Spikelets 1.0 cm long, 1.5 mm broad, numerous, divericate, spirally arranged, linear, sub-compressed, angular, 25–flowered, acute; rachis 9.0 mm long, 1.0 mm broad, yellowish, winged, wings glabrous, hyaline caducous, lanceolate. Glume 2.0 mm long, 1.5 mm broad, keel green, short, half–imbricating, acute, elliptic–oblong, keeled; keel greenish-brown, membranous, laterally compressed, cuspidate, mucronulate; mucro 0.5 mm long, scabreulous, surface & margin glabrous, stramineous, 3–nerved; nerves green, visible. Stamens 2; anther 0.5 mm long, reddish, ovate–oblong, linear, crested, glabrous; filament 1.5 mm long, uniformly slender, longer than the anther. Pistil 2.5 mm long; ovary 0.5 mm long, glabrous, ovate, oblong; style 1.0 mm long, flat, glabrous; stigma 3; 1.0 mm long, acute, brown, glabrous. Nut 1.5 mm long, 1.0 mm broad, yellowish–brown, obovoid, oblong–ellipsoid, trigonous, apiculate.

Flowering & Fruiting

May-December; Rajatalab, Mohan sarai, Jaunpur, Mau, Basti, Ballia.

Habitat

Common in swamps & dry places.

Distribution

India, SriLanka, Myanmar.

7. *C. distans*

C. distans L.f. Suppl. 103. 1781; Cook, Aqu.Wet. Pl. Ind. 110.f.96 d-f. 1996. (PLATE–14).

Perennial. Rhizomatous; rhizome creeping, short, bearing short stout stolons. Stem 1.0 m long, 3.5 mm thick, green, rigid, erect, tufted, triquetrous, base swollen. Leaves absent; shealth 15 cm long, 4.0 mm broad, green, at upper half, brown at lower half, membranous, glabrous, acuminate; ligule absent. Inflorescence anthelate; anthela 25 cm long, decompound, lax. Bracts 30 cm long, 2.0 mm broad, 8 in number, green, suberect, spreading, surface glabrous, margin scabrid, acute; primary rays 20 cm long, 12 in number, slender, rigid, obliquely spreading; secondary rays 8.0 cm long, 6 in number. Spike lax, pyramidal, bearing 20 spikelets. Spikelets 1.5 cm long, 1.0 mm broad, first ascending, ultimately divericate, deflexed, linear, acicular, brown, 60 flowered; rachilla flexuous, winged; wing narrowly lanceolate, white hyaline soon caducous, internodes 1.0 mm long. Glume 2.0 mm long, 1.0 mm broad, membranous,

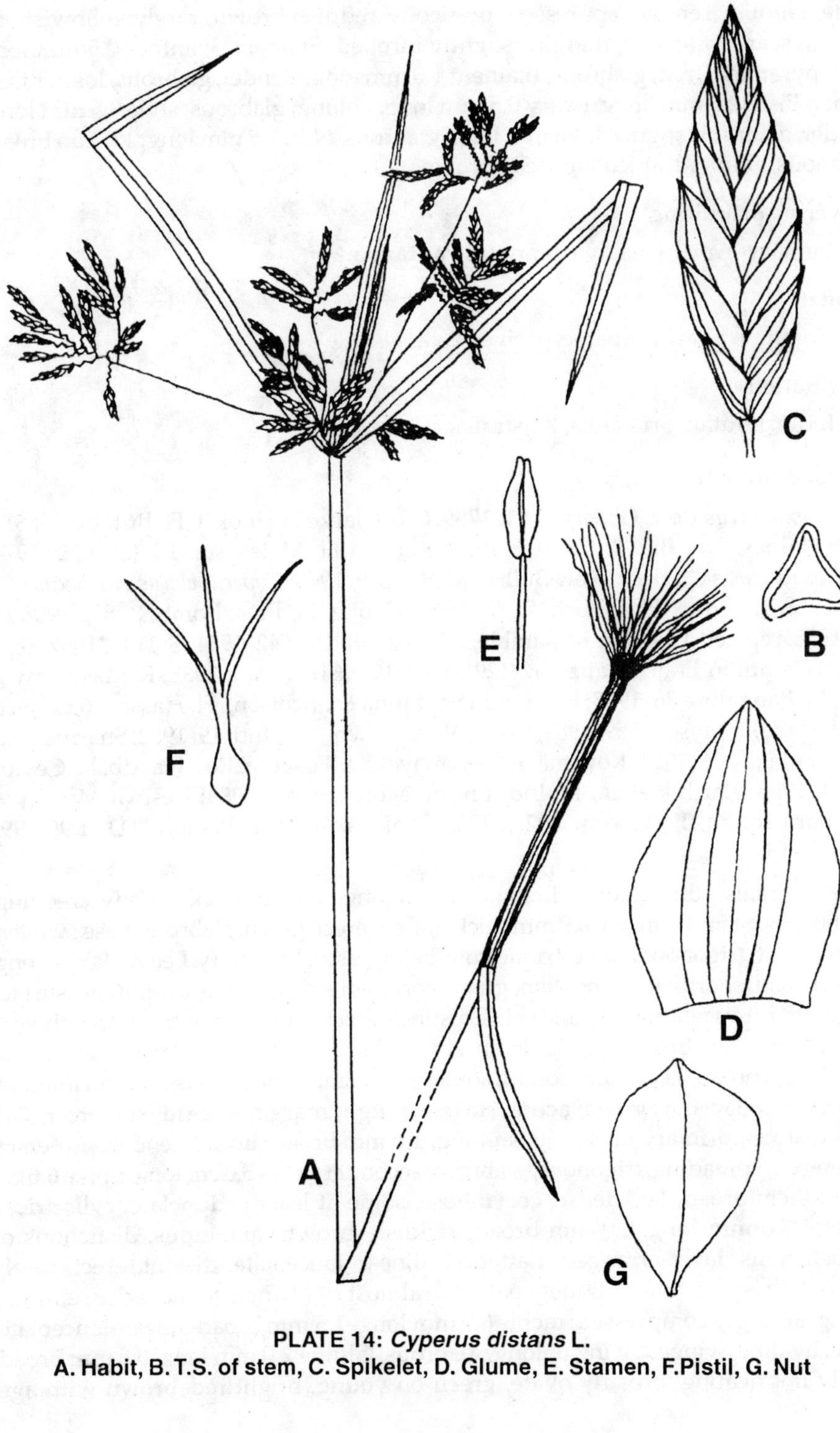

PLATE 14: ***Cyperus distans*** **L.**
A. Habit, B. T.S. of stem, C. Spikelet, D. Glume, E. Stamen, F. Pistil, G. Nut

ovate, elliptic, remote, appressed, muticous, reddish brown, rarely yellowish, 5–nerved, scarcely keeled, margins slightly inrolled. Stamens 3; anther 0.5 mm long crest, pyramidal, red, glabrous; filament 1.0 mm long, slender, glabrous, longer than anther. Pistil 2.0 mm long; ovary 0.5 mm long, oblong, glabrous; style 0.5 mm long, slender, glabrous; stigma 3, 1.0 mm long, glabrous. Nut. 1.5 mm long, 1.0 mm broad, trigonous, ellipioid, apiculate, deep brown.

Flowering & Fruiting

January–April; Chakiya, Jaunpur, Padrauna.

Habitat

Common along the banks of river & canals.

Distribution

India, Bhutan, Sri Lanka, Australia.

8. *C. exaltatus*

C. exaltatus Retz. Observ. 5: 11. 1789; C.B. Clarke in Hook. f. Fl. Brit. Ind. 6: 617. 1893; Haines, Bot. Bih. & Or. 902. 1924; Kern in Fl. Males. ser. 1.7 (3): 602. 1974. *Cyperus tokiensis* Clarke in Kew Bull. Add. Ser. 8: 9, 1908. *Cyperus elatus* var. *laxus* E.G. Camus in Lecomte, Fl. Indoch. 7: 70. 1912; Cooke, Fl. Pres. Bomb. 2: 872.1908 (3: 386.1958. repr. ed.); Fischer in Gamble, Fl. Pres. Madr. 1642. 1931 (3: 1141. 1957, repr. ed.); Kukenth in Engl. Pflanzenr. 4(20). Heft 101: 64, f. 9A–E 1935; Ramaswamy & Razi, Fl. Bangalore 86. 1973; Hooper in Saldanha & Nicolson, Fl. Hassan. 665. 1976; Rao & Razi, Fl. Mysore 559. 1981; Rao & Verma, Cyp. N.E. Ind. 19. 1982; Sharma *et al.*, Fl. Karnataka 306. 1984; Koyama in Dassanayake & Fosberg, Rev. Handb. Fl. Ceylon 5: 173. 1985; Karthik *et al.*, Fl. Ind. Enum. Monocot. 45. 1989; Keshva Murthy & Yoganarasinhan, Fl. Coorg. 502. 1990; Cook, Aqu. Wet. Pl. Ind. 111. f.99.1996 (PLATE–15).

Perennials. Rhizomatus; rhizome 1.0 cm long, 2.0 mm thick, woody, creeping; root fibrous. Stem 1.0 m tall, 6.0 mm thick, tufted, erect, green, glabrous, base swollen, stout, robust, trigonous above, triquetrous below, usually solitary. Leaves 90 cm long, 1.0 cm broad, shorter than the stem, green, coriaceous, flat, plicate acuminate, surface rough, margin scaberolous, midveins distinct, erect, clustered at base; sheath very reduced, spongy, brownish; ligule absent. Inflorescence anthelate; anthela 10 cm long, 5.0 cm broad, lax decompound, sometimes compound. Bracts 6, 70 cm long, 1.0 cm broad, foliaceous, green, acute, surface rough, margin scabrid, sub erect, flat, veins distinct; primary rays 7, 10 cm long, 1.0 mm broad, linear slender, sub-erect, compresed, spreading, trigonous, glabrous; secondary rays 5.0 cm long. Spike 6.0 cm long, 1.0 cm broad, lax, dense, corymbose, some at least pedunclate, cylindrical. Spiketet 7.0 mm long, 2.0 mm broad, reddish- brown, numerous, distichous or subdistichous, laxly arranged, flattened, linear–lanceolate, distant, erect, rarely horizontally spreading, obliquely patent to almost right angle to the rachis, elliptic–oblong, strongly compressed; rachis 6.0 mm long, 1.5 mm broad, linear–lanceolate, white, hyaline, winged; wing oblong, glabrous. Glume 2.0 mm long, 0.5 mm broad, closely imbricating, broadly ovate, green on young, bright red–brown with age,

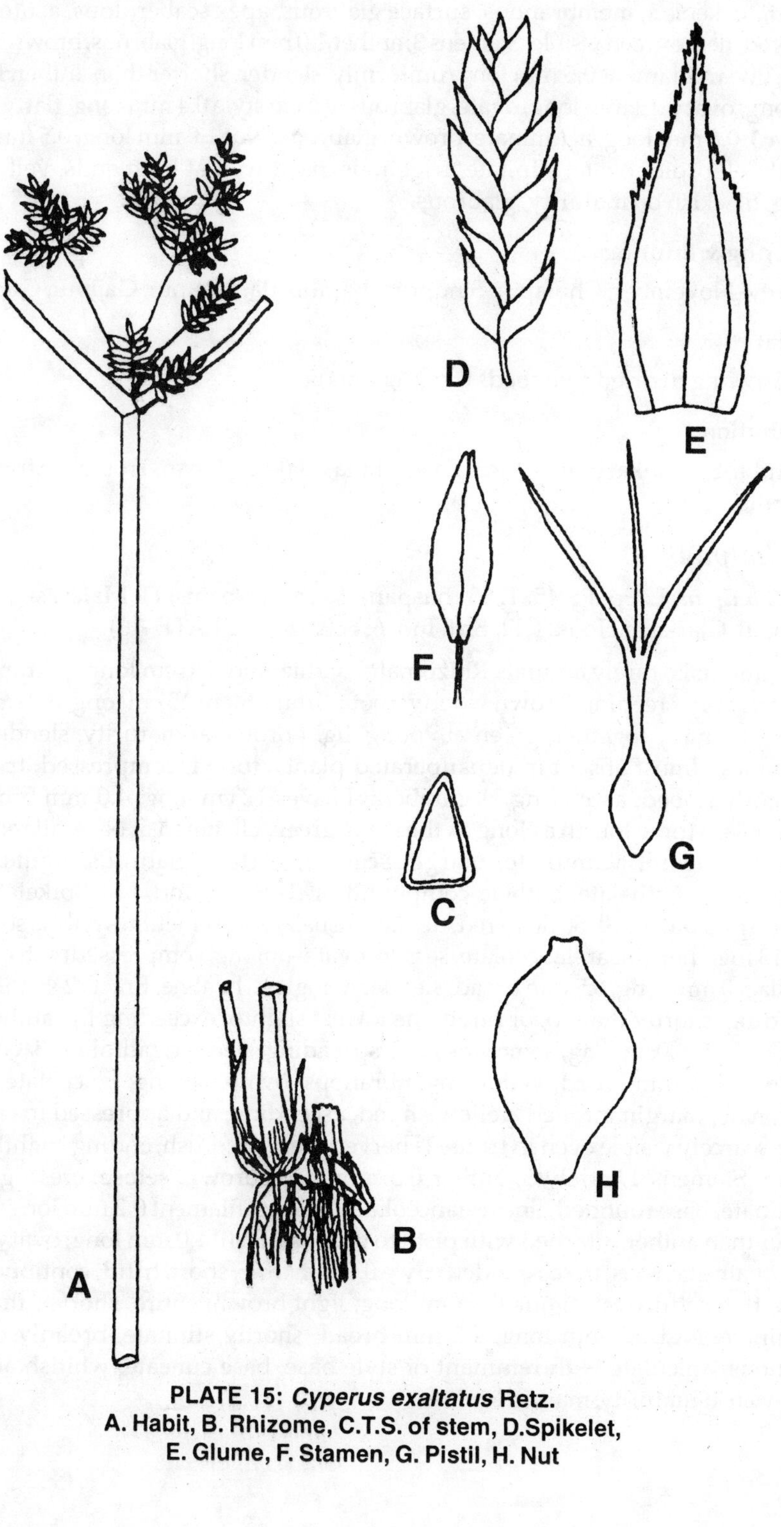

PLATE 15: *Cyperus exaltatus* Retz.
A. Habit, B. Rhizome, C.T.S. of stem, D.Spikelet,
E. Glume, F. Stamen, G. Pistil, H. Nut

cuspidate, keeled, membranous, surface glabrous, apex scaberulous, acute, glossy, 3–nerved; nerve green visible. Stamens 3; anther 1.0 mm long, glabrous, brown, oblong–ovate, linear; filament 0.5 mm long, uniformly slender, shorter than anther. Pistil 1.5 mm long; ovary 0.3 mm long, ovate, glabrous, green; style 0.4 mm long, flat, glabrous; stigma 3; 0.8 mm long, acuminate, brown, glabrous. Nut 1.0 mm long, 0.5 mm broad, broadly ellipsoid, ovate, stipitate, trigonous, narrowed at both ends, yellowish at young, blackish on maturity, glabrous.

Flowering & Fruiting

July–November; Cholapur, Sindhora, Barahikala, Jaunpur, Gajipur, Gorakhpur.

Habitat

Growing in dried river beds & along ponds.

Distribution

India, Sri Lanka, Indo–China, East–China, Malesia, Japan, Tropical Africa, South Australia.

9. *C. halpan*

C. halpan. L. Sp. Pl. 45. 1753 (haspan); Kern in Steenis, Fl. Males. ser. 1.7: 624. 1924; C.B. Clarke in Hook. f. Fl. Brit. Ind. 6: 660. 1893. (PLATE–16).

Perennials, rarely annuals.Rhizomatous; rhizomes 3.0 cm long, 3.0 mm broad, slender, short, creeping, brown woody; root fibrous. Stem 15 cm long, 3.0 mm thick, solitary to many together, green at young, light brown at maturity, slender, rigid, sometimes almost absent in depauperated plants, tufted, compressed, trigonous, frequently ribbed, ascending, decumbent. Leaves 12 cm long, 4.0 mm broad, 3–7, about half to three fourth as long as the stem, green, clustered at base, all reduced to bladeless sheath, acuminate, margin scabrid, surface glabrous; ligule absent. Inflorescence anthelate; anthela compound or decompound, lax. Spikelet 13 mm long, 1.5 mm broad, 10–50 flowered, digitate, usually 3–6, sometimes only 2, sometimes upto 13 together, linear–lanceolate, sessile, ovate–oblong, compressed, sub cylindric; rachilla 5.0 mm long, 0.5 mm broad, slender, wingless, hyaline. Bracts 2–3, foliaceous, spreading, shorter than to, or rarely the lowest slightly exceeding the anthela; rays usually 7–15, 8.0cm long, sometimes 1–2, spreading, slender; radiolii 1–3. Glume 1.0 mm long, 0.5 mm broad, yellow, membranous, ovate, oblong, lanceolate, keeled, mucronate, margin inrolled, yellowish red, ascending, and appressed to rachilla is not or scarcely visible even in fruits, 1- nerved; nerve reddish, ending slightly below the tip. Stamens 1, rarely 3; anther 0.5 mm long, brown, setose, crest, glabrous, acuminate, base rounded, linear–lanceolate, oblong; filament 0.2 mm long, slender, shorter than anther, attached with pistil to the base. Pistil 1.0 mm long; ovary 0.2 mm long, ovate glabrous; base rounded; style 0.5 mm long, short, trifid, continuous with ovary, flat, caducous; stigma 0.3 mm long, light brown, entire, shorter, than style, acuminate. Nut 0.5 mm long, 0.2 mm broad, shortly stipitate, broadly obovoid, trigonous, apiculate with remmant of style base, base cuneate, whitish at young; yellowish at maturity, margin keeled.

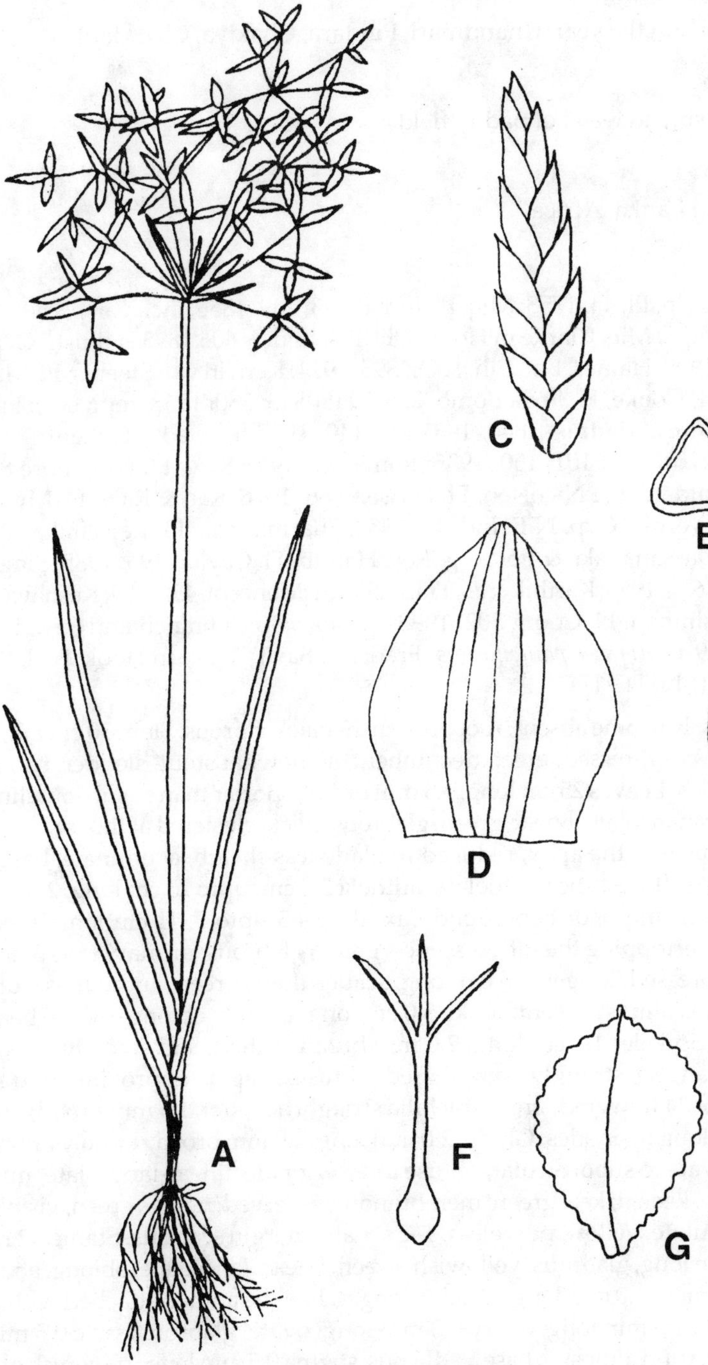

PLATE 16: ***Cyperus halpan*** **L.**
A. Habit, B. T.S. of stem, C. Spikelet, D. Glume, E. Stamen, F. Pistil, G. Nut

Flowering & Fruiting

Through out the year; Chandmari, Pindara, Chakiya, Chandauli.

Habitat

A wide spread weed of paddy fields.

Distribution

India, Sri Lanka, Africa.

10. *C. iria*

C. iria L. Sp. Pl. 45. 1753; Miq. Fl. Ind. Bat. 3: 269. 1856, incl. var. *parviflorus* Miq. et var. *diaphaniria* Miq, Clarke in Hook. f. Fl. Brit. Ind. 6: 606.1893 et illustr. Cyper. t. 14 f. 2, non f. l. 1909; Haines, Bot. Bih. & Or. 895. 1924; Kern in van Steenis, Fl. Males. ser. 1.7: 616. 1974, Cooke, Fl. Pres. Bomb. 2: 867. 1908 (3: 380. 1958, repr. ed.); Fischer in Gamble, Fl. Pres. Madras. 1640. 1931 (3: 1140. 1957 repr. ed.); Kukenth, in Engl., Pflanzenr 4 (20): Heft 101: 150. 1935; Ramaswamy & Razi, Fl. Bangalore 89. 1973; Hooper in Saldanha & Nicolson, Fl. Hassan 666. 1976; Rao & Razi, Fl. Mysore 559. 1981; Rao & Verma, Cyp. N.E. Ind. 15. 1982; Sharma *et al.*, Fl. Karnataka 306. 1984; Koyama in Dassanayake & Fosberg, Rev. Handb. Fl. Ceylon 196. 1985; Singh, Fl. E. Karnataka 2: 631. 1988; Karthik *et al.*, Fl.Ind. Enun. Monocot. 45. 1989; Keshava Murthy & Yoganarasimhan, Fl. Coorg 502. 1990. *C. paniciformis* Franch and Sav., Pl. Jap. 2: 103. 537. 1879. *C. iria* var. *paniciformis* (Franch & Sav.) Clarke in Hook. F. Fl. Brit. Ind. 6: 607. 1893. (PLATE-17).

Annuals. Rhizome absent; root brownish–black, fibrous. Stem 30 cm tall, 1.5 mm thick, tufted, compressed, erect, decumbent, somewhat stout, slender, triquetrous, green, glabrous. Leaves 25 cm long, 1.5 mm broad, shorter than to subequalling to the stem, flat or canaliculate, linear, green, glabrous, thick, clustered at the base, scaberous on margin towards the apex, reduced to blade less sheath, acuminate, base broad; ligule absent. Inflorescence anthelate; anthela 5.0 cm, upto 25 cm long, 2.0 mm upto 6.0 mm broad, simple or compound, lax. Bracts 3 upto 7, 11 cm long, foliaceous, spreading, overtopping the inflorescence, green, glabrous; primary rays 3–8 slender patent, compressed, longest 15 cm long, scaberulous, green, anthelules, subdense; secondary rays almost absent. Spike 5.0 cm long, usually oblong–ovoid, bearing 5–23 spikelets. Spikelet 1.3 cm long, 2.0 mm broad, patent, sub erect, linear–oblong, spicately arranged, strongly compressed, obtuse at apex, approximate to remote, pedicellate, 6–24 flowered, green; rachilla straight, slender, 1.5 mm broad, wingless, with 0.6 mm long intenodes. Glume 2.0 mm long, 1.0 mm broad, loosely imbricating, broadly obovate to suborbicular, obtuse to emerginate, linear-lanceolate, muticous, mucronutlate, keeled; keel green, membranous, 5-nerved; nerves green, visible; sides cellular, reticulate, yellow, nerveless, 1–2 striate, margin scarious. Stamen 2 rarely 3; anther 0.5 mm long, glabrous, yellowish–green, linear–lanceolate, oblong, acute, base rounded; filament 1.0 mm long, slender, longer than the anther, attached with pistil to the base. Pistil 1.5 mm long; ovary 0.5 mm long, ovate, glabrous; style 0.5 mm long, short, slender, trifid almost to base, caducous; stigma 0.5 mm long, brownish glabrous. Nut 1.5 mm long, 1.0 mm broad, broadly stipitate, obovoid–ellipsoid, triquetrous,

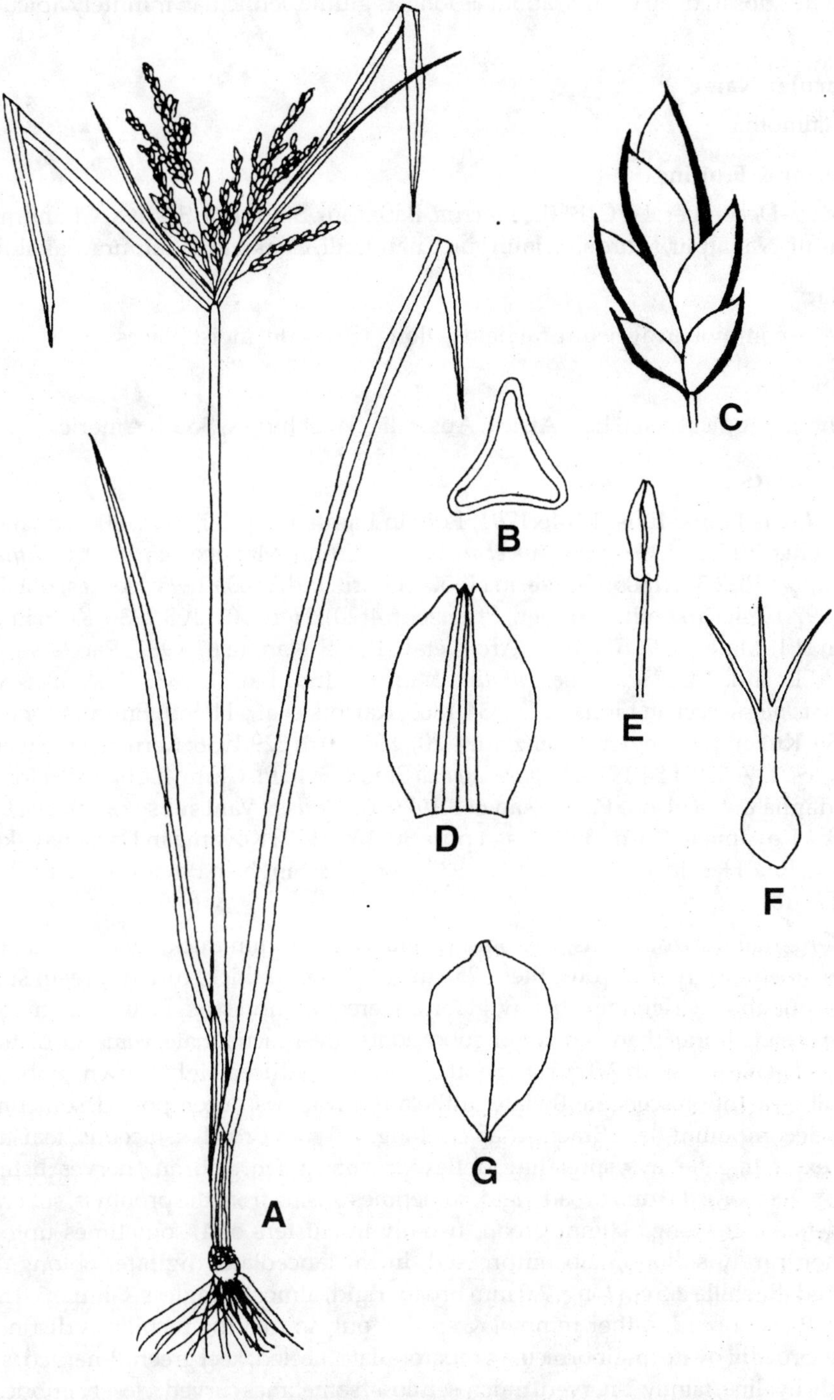

PLATE 17: ***Cyperus iria*** **L.**
A. Habit, B. T.S. of stem, C. Spikelet, D. Glume, E. Stamen, F. Pistil, G. Nut

apiculate, glossy, deep brown, about as long as glume, lenticular, minutely apiculate at apex.

Vernacular Name

Galmotha.

Flowering & Fruiting

May–December; UPC, BHU, Harrua, Babatpur, Sindhora, Sarriya, Chandmari; Cholapur, Navalpur, Murdaha, Jaunpur, Chandauli, Chakiya, Sadalpura, Sakaldiha.

Habitat

Very common in agricultural fields, along road side, moist places.

Distribution

India, Tropical Asia, East Africa, Australia, West Indies, South America.

11. *C. laxus*

C. laxus Lamk. Illus. 1: 146.1791; Poir. in Lamk. Encycl. 7: 260.1806; Raynal in Adansonia 2, 17: 277.1978; Karthik *et al.*, Fl. Ind. Enum. Monocot. 45.1989. *C. diffusus* Vahl Enum. Pl. 2.321.1805; Clarke, in Hook.f.Fl.Brit. Ind. 6: 603.1893; Haines, Bot. Bih. & Or. 897.1924; Kukenth in Engler., Pflanzenr. 4(20), Heft. 101: 108.1936; Kern in van Steenis, F1. Males. 1, 7: 619.1974; Arora etal., Bot. S. Kanara 61.1981; Rao & Verma, Cyp. N.E. Ind. 14.1982. *C. holophyllus* Miq. F1. Ind. Bat. 3: 283. *C. diffusus* var. *macrostachyus* Boeck in Linnaea 35: 534.1868; Karthik et at., Fl. Ind. Enum. Monocot. 46.1989; Kukenth in Engler., Pflanzenr, 4(20), Heft. 101: 229.1936; Kern in van Steenis, F1. Males. 1, 7: 619, f.54.1974; *C. pubesquama* Steud. Syn. Pl. Glum. 2: 20.1855; Hooper in Saldanha & Nicolson, Fl. Hassan 668.1976. *C. diffusus* Vahl subsp. *macrostachyus* (Boeck.) Koyama in Gard. Bull. Singa pore 30: 139.1977; Koyama in Dassanayake & Fosberg, Rev. Handb. F1. Ceyl. 5: 198. 1985; Prasad & Singh, in Phytotaxo. 1: 64.2001. (PLATE–18).

Perennial. Rhizomatous; rhizome 1.0 cm long, 3.0 mm thick, woody blackish-brown, creeping; root fibrous. Stem 25 cm tall, 2.0 mm thick, tufted, green, stout, triquetrous above, trigonours below, glabrous, erect, compressed. Leaves 22 cm long, 1.0mm broad, shorter than stem, rigid, subcaudate, green, acuminate, veins not distinct, clustered at base; sheath 2.0 cm long, 1.0 mm broad, reduced, light brown, glabrous; ligule absent. Inflorescence anthelate; anthela 10 cm across, decompound, sometimes supra-decompound, lax. Bracts 4, 25 cm long, 2.0 mm broad, setaceous, leaf-like, much exceeding the rays, spreading, deflexed, acute, green, glabrous, nerves distinct; rays 6, 5.0 cm long, 1.0 mm broad, rigid, sometimes arising from the prophyls, suberect. Spikelets 3.0 cm long, 3.0mm broad, usually in clusters of 4, sometimes upto 16 together, rarely solitary, subcompressed, linear-lanceolate, digitate, oblong, 20-flowered. Rachilla 2.5 cm long, 2.0 mm broad, rigid, almost wingless. Glume 3.0 mm long, 2.0 mm broad, rather remote, & spread out, so that the rachilla is distinctly visible, broadly ovate to suborbicular, mucronulate, keeled; keel green, 3-nerved; side reddish, hyaline, faintly 2-nerved, mucrosetulose, sometimes curved, closely imbricate, cuspidate, membranous. Stamens 3; anther 1.5 mm long, red, linear-oblong, glabrous;

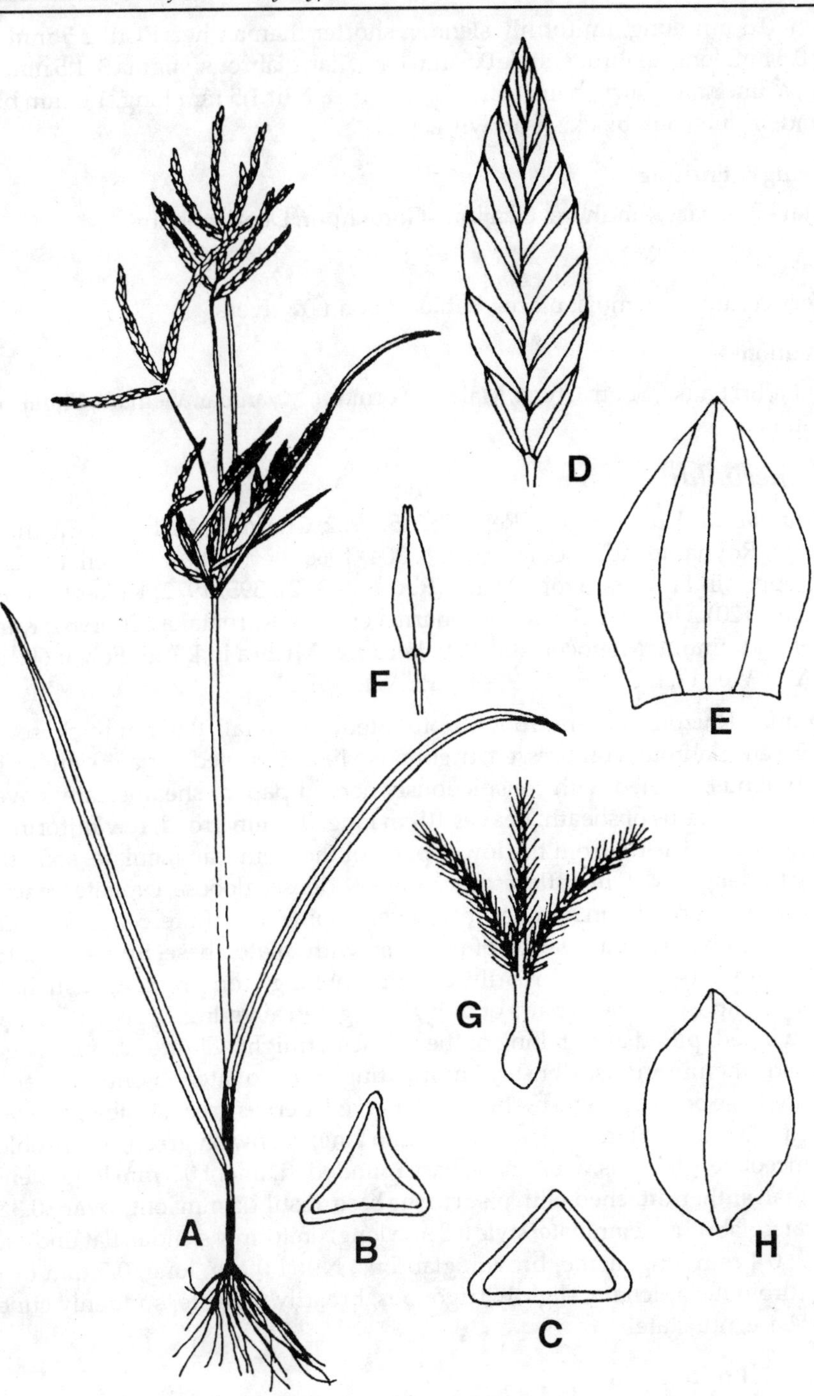

PLATE 18: ***Cyperus laxus*** **Lamk.**
A. Habit, B & C.T.S. of stem, D.Spikelet , E.Glume, F. Stamen, G. Pistil, H. Nut

filament 0.5 mm long, uniformly slender, shorter than anther. Pistil 2.5 mm long; ovary 0.5 mm long, glabrous; style 0.5 mm long, flat, glabrous; stigma 3, 1.5 mm long, brown, acuminate, hairy; hairs acute, base broad. Nut 1.5 mm long, 1.0 mm broad, ellipsoid, triquetrous, blackish-brown, acute.

Flowering & Fruiting

May-February; Sindhora, Ghajipur, Gorakhpur, Deoria, Bhatni.

Habitat

Very common on moist places, banks of ponds & rivers.

Distribution

India, Sri Lanka, South China, Malesia, Formosa, Myanmar, Vietnam, Phillippines, New Guinea.

12. *C. meeboldii*

C. meeboldii Kuk in Fedde, Rep. 18: 345. 1922; Cherm in Arch. Bot. Caem. 7 (4): 11, 1936; J. Roynal in Adansonia Ser. 2, 6: 304. 1966; Berhant in Fl. Sen. ed. 2: 375, 1967; Hepper in Fl. West. Trop. Africa (Rev. ed.) 3 (2): 395. 1972; Kukenth in Engl. Pflanzenr 4 (20). Heft 101: 309:. 1936; Sharma et al; Fl. Karnataka 306. 1984; Karthik *et al.,* Fl. Ind. Enum. Monocot. 40. 1989; Verma & Mishra in J. Ind. For. 4: 69. 1981; Cook, Aqu. Wet. Pl. Ind. 116.f.107.1996. (PLATE–19).

Annual. Rhizome absent; root fibrous. Stem 12 cm tall, 0.5 mm thick, several, tufted, green, glabrous, compressed, trigonous, slightly curved–striate, slender, base pseudobulbous, clothed with conspicuously nerved papery sheaths, later covered with fibrous remains of sheath. Leaves 10 cm long, 0.5 mm broad, few filiform with sheathing bases, arising from the lower parts of the stem, canaliculate, sometimes reaching the length of stem. Inflorescence solitary, dense, globose, capitate. Bracts 6.0 cm long, 0.5 mm broad, 3 in number, leafy, overtopping the inflorescence, foliaceous, spreading, often at if continuation of the stem, with dilated base. Spikelets 6.0 mm long, 0.5 mm broad,sessile, broadly elliptic, oblong, compressed, distichously arranged, composed of 5–10 pairs of thin glum, green at young, brown at maturity; rachilla winged, pitted after falling of the glumes, straight. Glume 2.0 mm long, 1.0 mm broad, membranous, densely imbricating, thin, ovate–oblong, lanceolate, acuminate, base broad, greenish, shining, 3–nerved; nerves green, visible, mid nerve produced as a mucro. Stamens 1; anther 1.0 mm long, yellowish-green, ovate–oblong, linear lanceolate, glabrous, acuminate, base rounded, filament 0.5 mm long, slender, shorter than anther, attached with pistil to the base. Pistil 1.5 mm long; ovary 0.3 mm long, ovate, glabrous, acuminate; style 0.8 mm long, uniformly slender, flat undivided stigma 2; 0.4 mm long, acute, brown, glabrous. Nut 1.0 mm long, 0.5 mm broad, brown, turbinate, apicuate, slightly trigonous, broadly obovate, suddenly cuneate towards base, punctate.

Flowering & Fruiting

January–April; Cholapur, Rajghat, Harraua, Jaunpur, Deoria, Basti.

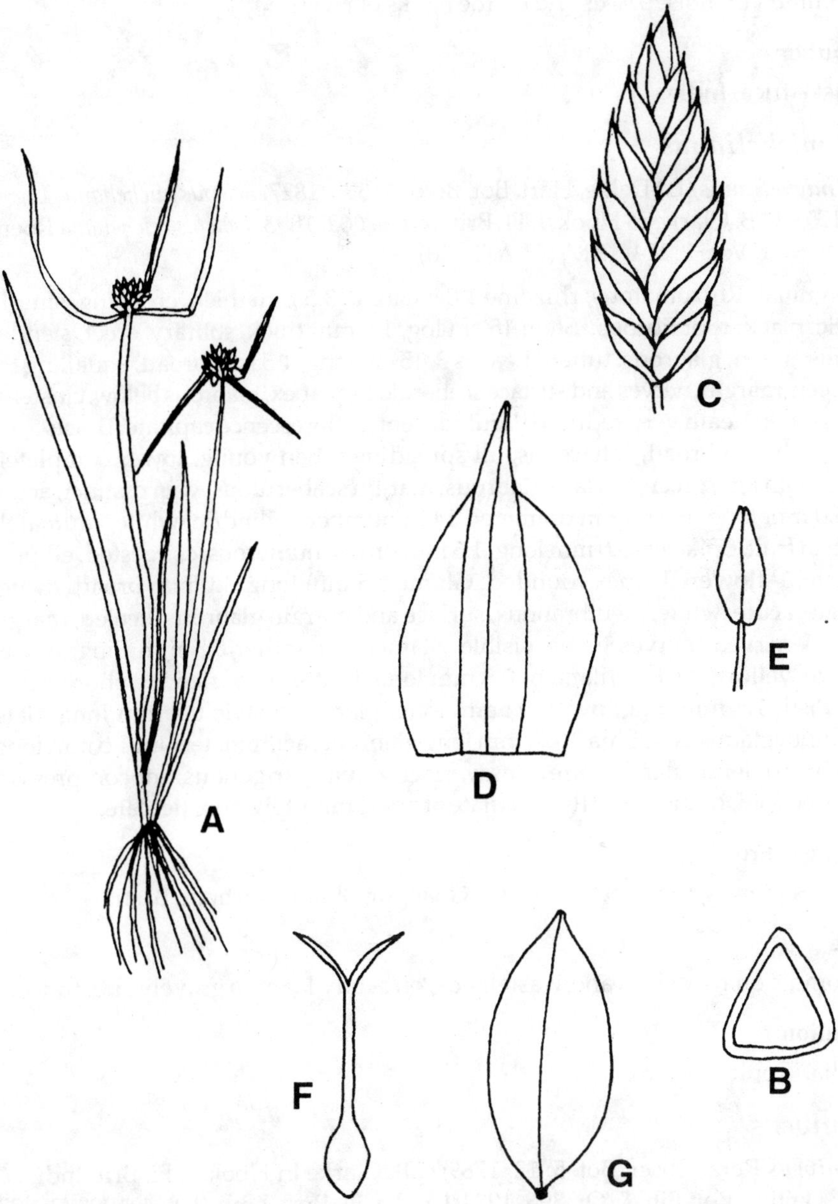

PLATE 19: ***Cyperus meeboldii*** **Kuk.**
A. Habit, B. T.S. of stem, C. Spikelet, D. Glume, E. Stemen, F. Pistil, G. Nut

Habitat

Common in moist piaces and on the banks of rivers.

Distribution

East Africa, India.

13. *C. michelianus*

C. michelianus (L.) Delile, Hort. Bot. Berol. 1: 303. 1827. *Scirpus michelianus* L., Sp. Pl. 52. 1753; C.B. Clarke in Hook. f. Fl. Brit. Ind. 6: 662. 1893. *Isolepis micheliana* Roem, & Schult. Syst. Veg. 2: 114. 1817. (PLATE–20).

Perennial. Rhizomatous; rhizome 1.0 cm long, 3.5 mm thick, creeping, simple, aromatic, black; root fibrous. Stem 15 cm log, 1.0 mm thick, solitary, erect, slender, trigonous, green, glabrous, tufted. Leaves 3, 15 cm long, 1.5 mm broad, equalling the stem, green, margin, nerves and surface scaberulous at apex, glabrous below, clustered at base, acute; sheath very reduced; ligule absent. Inflorescence capitate. Bracts 3, 6.0 cm long, 1.0 mm broad, foliaceous, all spreading when young, lower completely deflexed with age, green, surface glabrous, margin scaberulous, vein distinct, acute. Spike 5.0 mm long, 6.0 mm in diameter, 14 in number, cylindric, white, ultimately blackish in fruit. Spikelets 3.0 mm long, 1.5 mm broad, numerous, congested, elliptic–lanceolate, 1–flowered, apex rounded. Glume 2.5 mm long, 1.0 mm broad, ovate–lanceolate, acute, white, membranous, surface and margin glabrous, keeled, margin hyaline, 3–nerved; nerves green visible. Stamens 3; anther 0.5 mm long, ovate–lanceolate, yellow, keeled; filament 0.3 mm long, uniformaly slender, shorter than anther. Pistil 1.5 mm long; ovary 0.5mm long, glabrous; style 0.5 mm long, flate, deeply bifid, glabrous; stigma 3, 0.5 mm long glabrous, acuminate. Nut 1.5 mm long, 1.0 mm broad, lenticular, mature, glossy, black, equally trigonouș, subcompressed, oblong, obovoid, obtuse, shortly apiculate at apex, minutely puncticulate.

Flowering & Fruiting

July–September; Jaunpur, Chakiya, Ghajipur, Bhatni, Turtipar, Basti.

Habitat

Sporadic on roadsides and wastelands, on sandy loam to gravely substratum.

Distribution

India, Tropical Africa.

14. *C. niveus*

C. niveus Retz. Obser. Bot. 5: 12. 1789; C.B. Clarke in Hook. f. Fl. Brit. India. 6: 60.1893; Haines, Bot. Bih. & Or. 896. 1924; Cooke, Fl. Pres. omb. 2: 864. 1908 (3: 377. 1958. repr. ed.); Fischer in Gramble, Fl. Pres. Madr. 1640. 1931 (3: 1139. 1957 repr. ed.); Kukenth in Engl. Pflanzen 4 (20), Heft 101: 288. 1936; Kern in Blumea 10: 644. 1960 et in Reinwardtia 6: 60. 1961; Ramaswamy & Razi, Fl. Bangalore 90. 1973; Rao & Verma, Cyp. N.E. Ind. 14. 1982; Sharma *et al.*, Fl. Karnataka. 307. 1984; Karthik. *et al.*, Fl. Ind. Enum. Monocot. 46. 1989. (PLATE–21).

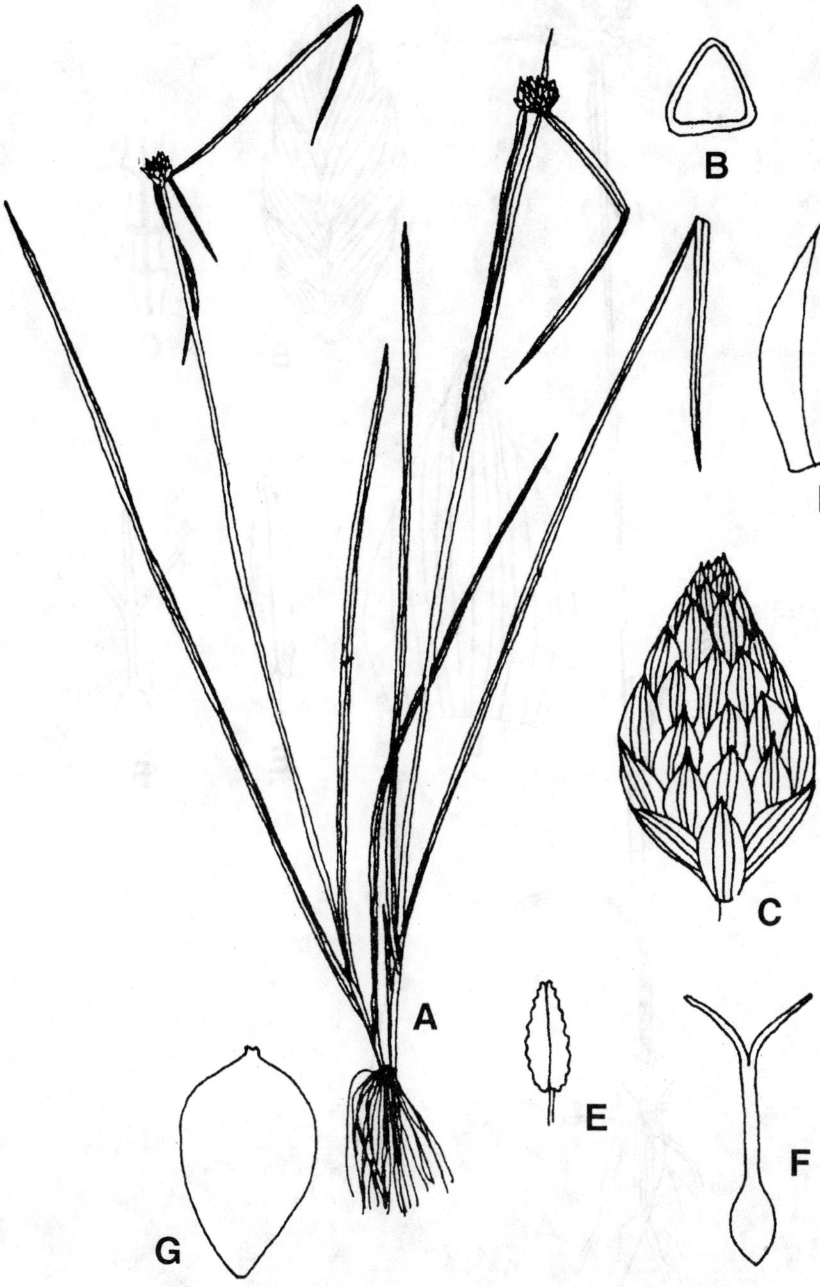

PLATE 20: *Cyperus michelianus* (L.) Delile
A. Habit, B. T.S. of stem, C. Spikelet, D. Glume, E. Stamen, F. Pistil, G. Nut

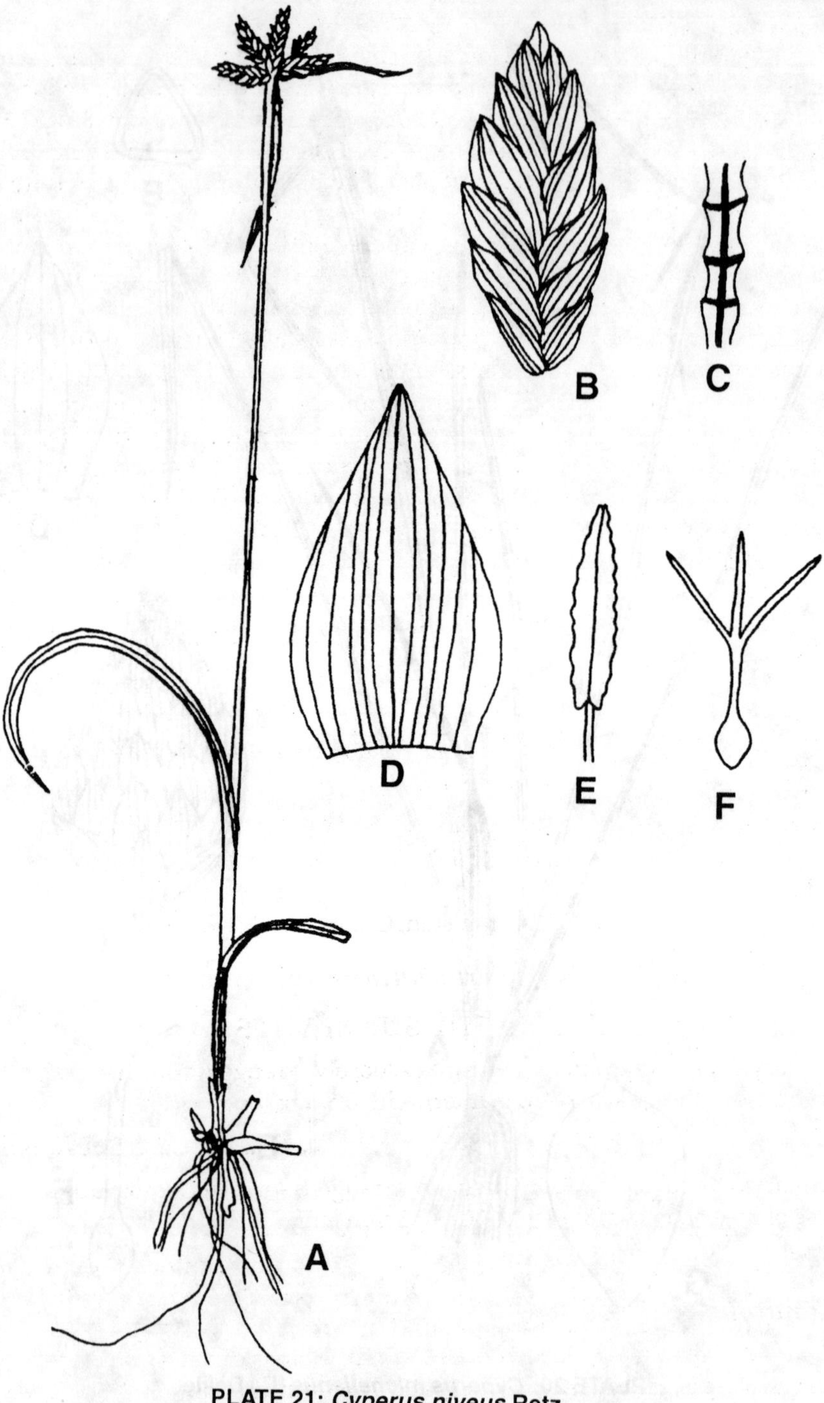

PLATE 21: *Cyperus niveus* Retz.
A. Habit, B. Spikelet, C. Rachis, D. Glume, E. Stamen, F. Pistil

Perennials. Rhizomatous; rhizome 1.0 cm long, 2.0 mm thick, creeping woody; root fibrous. Stem 15 cm long, 1.5 mm thick, usually in rows, slender trigonous, rigid, glabrous, green, erect, tufted, base swollen, compressed. Leaves 15 cm long, 1.5 mm broad, glaucous, green, surface glabrous, margin scaberulous, linear, acuminate, vein distinct on both surface, cluster at base, equalling the stem, erect; sheath very reduced; ligule absent. Inflorescence capitate, 1.0 cm long, 1.5 cm broad. Bracts 3, 8.0 cm long, 1.5 mm broad, green, spreading, deflexed, foliaceous, acute, surface glabrous margin scaberulous at apex, base glabrous, many nerved, dilated at base; rays absent. Spikelet 1.0 cm long, 3.0 mm broad, oblong–lanceolate, obtuse–elliptic, stellately spreading, flattened, sub acute, 15 flowered, sessile, white; rachis 8.0 mm long, 1.5 mm broad, wingless, keeled. Glume 4.0 mm long, 2.0 mm broad, acute & usually shortly sinuate at apex, chartaceous tightly imbricating, ovate–lanceolate, surfaces smooth, margin hyaline, whitish- stramineous at young; pale brown with age, acutely keeled towards apex; keel produced into a minute mucro, ll-nerved; central nerve slightly visible, lateral nerves faint. Stamens 3; anther 2.5 mm long, yellow, ovate, oblong–linear, exerted from the glume, margin keeled; filament 1.0 mm long, uniformly slender, shorter than anther. Pistil 3.0 mm long; ovary 0.5 mm long, obovate, glabrous, light green; style 1.0 mm long, flat, glabrous; stigma 3, 1.5 mm long, acuminate, brown. Nut 1.5 mm long, 1.0 mm broad, subsessile, obovoid, triquetrous, apiculate, deep–brown, densely puncticulate, obtuse.

Flowering & Fruiting

August–December; Chakiya, Chandauli, Jaunpur, Indara, Gorakhpur.

Habitat

Growing in wastelands & rock crevices.

Distribution

India, Myanmar, Pakistan, Afganistan, China, North Thailand, Asia.

15. *Cyperus nutans* Vahl.

KEY TO THE SUB SPECIES

1a. Primary rays erect to spreading; spikes loosely arranged; spikelets not upto base of the spikes; rachilla internodes 0.8 mm long

.... **2. Subsp. *nutans***

1b. Primary rays sub erect; spikes densely arranged; spikelets almost upto base of the spike; rachilla internodes 0.5 mm long

.... **1. Subsp. *eleusinoides***

1. *C. nutans*

C. nutans Vahl **subsp. *eleusinoides*** (Kunth) Haines, Bot. Bih. & Or. 898 1924; Kern in van Steenis, Fl. Males. ser. 1.7: 610. 1974; Koyama in Gard. Bull. Singapore 30: 136. 1977; Koyama in Dassanayake & Fosberg, Rev. Handb. Fl. Cey. 5: 191. 1985; Misra & Verma, Fl. Allah. 388. 1992; Hooper in Saldanha & Nicolson, Fl. Hassan 667.

1976; Rao & Razi, Fl. Mysore 560. 1981; Rao & Verma Cyp. N.E. Ind. 16. 1982; Sharma *et al.*, Fl. Karnataka 307. 1984. *Cyperus eleusinoides* Kunth, Enum. Pl. 2: 39. 1937; Clarke in Hook. f. Fl. Brit. Ind. 6: 608. 1893; Cooke, Fl. Bomb. 2: 868. 1908 (3: 382. 1958 repr. ed.); Fischer in Gamble, Fl. Pres. Madr. 1640. 1931 (3: 1140. 1957 repr. ed.); Kukenth in Engl., Pflanzenr 4 (20), Heft 101: 144. 1933; Ramaswamy & Razi, Fl. Bangalore 85. 1973. (PLATE–22A, 22B).

Perennial. Rhizomatous; rhizome cork- like, very short; root fibrous. Stem 120 cm long, 1.0 mm thick, green, glabrous, solitary, triquetrous below, triangular above, erect. Leaves 50 cm long, 1.5 cm broad, flat, sub-coriaceous, plicate, leathery, linear, glaucous, green above, whitish striate below, margin scabrous towards tip with very prominent mid–rib; sheath 30 cm long, membranous, purplish below, pale green above, basal one with short lamina; ligule absent. Inflorescence anthelate; anthela 20 cm long, less compound. Bracts 6 in number, 75 cm long, 8.0 mm wide, unequal, longer one overtopping the inflorescence; primary rays 10 in number, 30 cm long, sub-erect, very unequal, oblique, smooth; secondary rays 7 in number 6.0 cm long, often clothed in spikelets almost to the base. Spike 3.0 cm long, 1.0 cm wide, densely arranged, sessile, narrow, loose, bearing 24 spikelets. Spikelet 8.0 mm long, 2.5 mm broad, spicately arranged, compressed, dense, linear, oblong, 10 flowered, light brown; rachilla 0.2 mm wide, slightly flexuous, persistent, winged; wing whitish, hyaline, persistant, internodes 0.5 mm long. Glume 2.5 mm long, 1.0 mm wide, closely imbricate, obliquely erect to patulose, oblong, obtuse at apex, yellow brown, 3 nerved, mucro not prominant; keels greyish–green, margins of upper half white, broad & hyaline. Stamens 3; anther 1.0 mm long, crest, quadrate, red papillose, glabrous with a distinct & smooth connective appendage; filament 2.5 mm long, slender, glabrous, longer than anther. Pistil 3.5 mm long; ovary 0.5 mm long, ovate, obglabrous; style 2.0 mm long, tricleft, glabrous; stigma 3, 1.0 mm long, glabrous. Nut 1.5 mm long, 0.5 mm broad, triquetrous, oblong, obovoid, apiculate, brown.

Flowering & Fruiting

May–December, Jaunpur, Chandauli, Gorakhpur, Basti, Lar.

Habitat

Rare, along the canals, road sides & paddy fields.

Distribution

India, South–East Asia, Formosa, Tropical Africa, Tropical Australia.

2. *C. nutans*

C. nutans (Vahl) **subsp. *nutans*** Kern in Fl. Males. ser. 1. 7: 609. 1974; Koyama in Dassanayake & Fosberg, Handb. Fl. Ceyl. 5: 190. 1984; Misra & Verma, Fl. Allah. 388. 1992; Cook, Aqu.Wet. Pl. Ind. 117.f.108 c -e. 1996.*C. nutans* Vahl Enum. Pl. 2: 363. 1806; Clarke in Hook. f. Fl. Brit. Ind. 6: 607. 1893; Cooke, Fl. Pres. Bomb. 2: 868. 1908 (3: 381. 1958, repr. ed.); Fischer in Gamble, Fl. Pres. Madr. 1640. 1931 (3: 1140. 1957, repr. ed.); Kukenth. in Engl. Pflanzenr. 4 (20), Heft 101: 144. 1935; Rao & Razi, Fl. Mysore 560. 1981; Rao & Verma, Cyp. N.E. Ind. 16. 1982; Sharma *et al.*, Fl. Karnataka 307. 1984; Karthik *et al.*, Fl. Ind. Enum. Manocot. 46. 1989. (PLATE–22C, 22D, 22E).

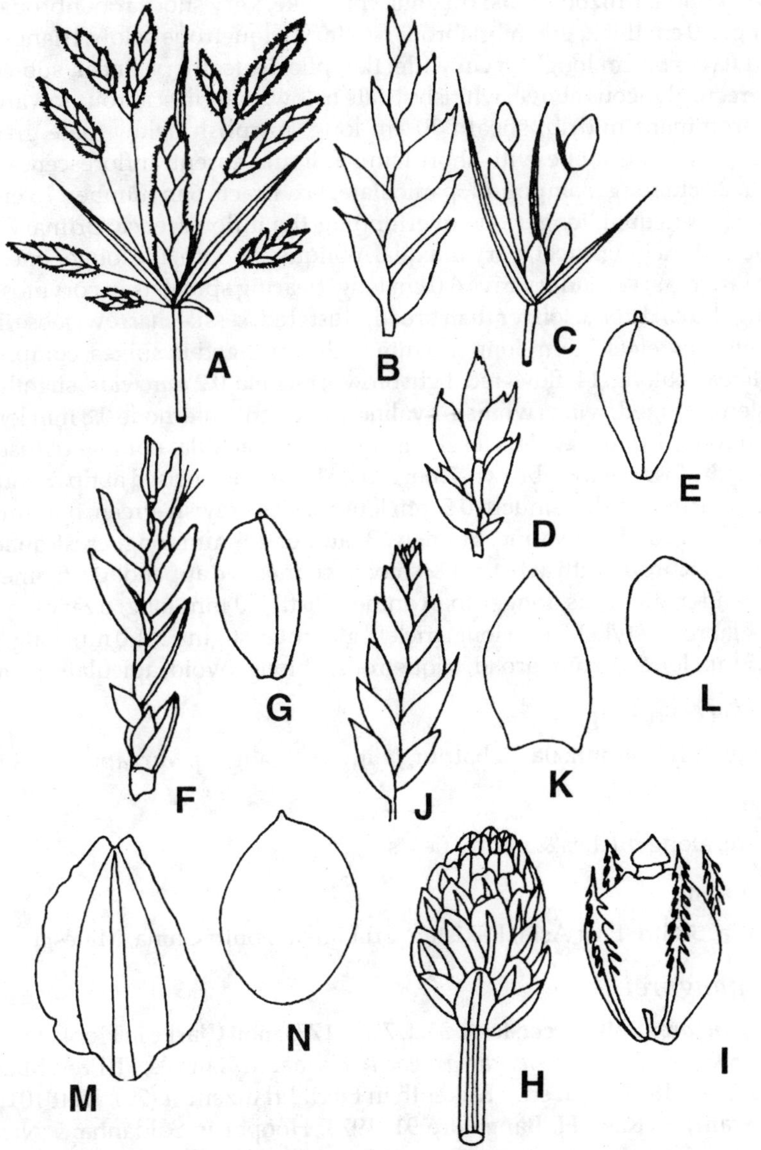

PLATE 22: ***Cyperus nutans*** **Vahl subsp.** ***eleusinoides*** **(Kunth) Haines**
A. Anthela, B. Spikelet
C. nutans **Vahl subsp.** ***nutans*** **Kern.–C. Anthela, D. Spikelet, E. Nut**
C. pangorei **Rottb.–F. Spikelet, G. Nut**
Eleocharis geniculata **(L.) Roem. & Schult.–H. Spikelet , I. Nut with bristle**
Juncellus alopecuroides **(Rottb.) C.B. Clarke–J. Spikelet, K. Glume, L.Nut**
J. laevigatus **(L.) C.B. Clarke–M. Glume, N. Nut**

Perennial. Rhizomatous; rhizome cork–like, very short; root fibrous. Stem 120 cm long, 1.0 cm thick, green, glabrous, solitary, triquetrous below, triangular above, erect. Leaves 50 cm long, 1.5 cm wide, flat, plicate, leathery, linear, sub-coriaceous, dark green, glaucous above, whitish striate below, margin scabrous towards tip with very prominent midrib; sheath 30 cm long, purplish below, pale green above, membranous, basal one with short lamina, ligule absent. Inflorescence anthelate; anthela 20 cm long, compound, paniculate, lax. Bracts 6 in number, 75 cm long, 8.0 mm wide, unequal, longer one overtopping the inflorescence; primary rays 10 in number, 30 cm long, erect, very unequal, oblique, smooth, the longer ones bearing 7 secondary rays; secondary rays 6.0 cm long, bearing spikelets in corymbs. Spike 4.0 cm long, 1.0 cm broad, longer than broad, clustered, sessile, narrow, loose, bearing 25 spikelets. Spikelets 1.5 cm long, 1.5 mm wide, arranged in spikes, compressed, sub erect, linear, oblong, 14–flowered, light brown; rachilla 0.2 mm wide, slightly flexuous, persistent, winged; wings whitish, hyaline, persistant; internode 0.8 mm long. Glume 2.5 mm long, 1.0 mm wide, closely appressed to rachilla, oblong, obtuse, oblique, erect to patulose, yellow–brown, 3 nerved on both side, incised at tip & mucronulate by the excurrent midrib, mucro 0.5 mm long; keeles greyish–green, margins of upper half, white, broad & hyaline. Stamens 3; anther 1.5 mm long, crest quadrate, red papillose, glabrous with a distinct & smooth connective appendage; filament 2.5 mm long, slender, glabrous, longer than anther. Pistil 3.0 mm long; ovary 0.5 mm long, ovate, glabrous; style 1.5 mm long, tricleft, glabrous; stigma 3, 1.0 mm long, glabrous. Nut 1.5 mm long, 0.5 mm broad, triquetrous, oblong, ovoid, apiculate, brown.

Flowering & Fruiting

May–July; Baranikala, Babatpur, Jaunpur, Shahganj, Mirzapur.

Habitat

Rare, along ditches & paddy fields.

Distribution

India, South–East Asia, Pakistan, Sri Lanka, South China, Malesia.

16. *C. pangorei*

C. pangorei Rottb. Decr. Icon. 31. t. 7. f 3. 1773; non Clarke in Hook. f. Fl. Brit. Ind. 6: 612. 1893 (as *C. corymbosus* var. *pangorei*); Fischer in Gamble, Fl. Pres. Madras 1641. 1931 (3: 1140. 1957, repr. ed.); Kukenth in Engl. Pflanzenr. 4 (20), Heft 101; 76. 1935; Ramaswamy & Razi. Fl. Bangalore 91. 1973; Hooper in Saldanha & Nicolson, Fl. Hassan 667.1976; Rao & Verma, Cyp. N.E.Ind.18.1982; Sharma *et al.*, Fl. Karnataka 307. 1984; Koyama in Dassanayak & Fosberg, Rev. Handb. Fl. Ceylon 5: 176. f. 9. 1985; Karthik *et al.*, Fl. Ind. Enum. Monocot. 46. 1989; Keshava Murthy & Yoganarasimhan, Fl. Coorg 503. 1990; Cook, Aqu. Wet. Pl. Ind. 118. f.110.1996. *C. tagetum* Roxb. Fl. Ind. 1: 208. 1820; C.B. Clarke in Hook. f. Fl. Brit. Ind. 6: 613. 1893; Cooke, Fl. Pres. Bomb. 2: 870. 1908 (3: 384. 1958 repr. ed.). (PLATE–22F, 22G).

Perennials. Rhizomatous; rhizome 10 cm long, 8.0 mm thick, creeping, stout, woody, clothed with brownish scale, strongly aromatic, bearing several stolons; stolon 1.0 cm long; roots some what thick, usually branched into blackish fibrous. Stem 15 cm upto 35 cm long, 5.0 mm thick, narrowed at apex, thick in middle, subcylindric below, trigonous above, green, glabrous, tufted, compressed, ascending, solitary or in row. Leaves 19 cm long, 5.0 mm broad, usually all reduced to bladeless subphyllous sheaths, sometimes the upper one bearing a blade surrounding the basal portion of stem, clustered at base, green, acuminate, coriaceous, thick, base broad, glabrous. Inflorescence anthelate; anthela compound or sometimes decumpound. Bracts 3–5, foliaceous, spreading, patent, forming an involucre; outer longest 15 cm long, 2.0 mm broad, glabrous; inner shortest 7.0 cm long, 1.5 mm broad, margin scabrid. Spikelet 3.0 cm long, 3.0 mm broad, clustered, 4–9 together, obliquely spreading, linear-lanceolate, ovate, subcylindric, subacute, sessite, spicately arranged, compressed, reddish brown–stramineous, 14–34 flowered; rachilla winged with 1.0 mm long internodes; wing lanceolate, white hyaline to sanguineous, caducous. Glume 4.5 mm long, 1.5 mm broad, spreading, loosely or closely imbricating, elliptic, oblong, linear-lanceolate, muticous, margin inrolled, membranous, keeled, sides hyaline, nerves, reddish, surface green, folded, 3–nerved on obtuse back; nerve visible, green, ending beyond the tip of glume. Stamens 3; anther 2.0 mm long, brown, crest, subquadrate, crenulate, glabrous, ascuminate, base rounded, linear–lanceolate, ovate, oblong; filament 2.5 mm long, slender. longer than anther. Pistil 2.5mm long; ovary 0.5 mm long, ovate, glabrous; style 1.0 mm long, deeply trifid,continuous to ovary, flat, caducous; stigma 1.0 mm long, exserted, brown, glabrous. Nut 1.5 mm long, 0.5 mm broad, narrowly oblong, obovoid, trigonous, scarely apiculate, blackish brown on maturity, yellowish brown on young.

Flowering & Fruiting

July–October; BHU, Banks of Varuna river, Jaunpur, Basti, Sonbhadra.

Habitat

Common along sandy soils, marshy area, along canals & rivers, rocky river beds & muddy river banks.

Distribution

India, Sri Lanka, Pakistan, Nepal, Myanmar.

17. *C. platystylis*

C. platystylis R. Br. Prodr. 214. 1810; C.B. Clarke in Hook. f. Fl. Brit. Ind. 6: 598. 1893; Haines, Bot. Bih. & Or. 893. 1924; Kern in Fl. Males. ser. 1.7 (3): 618. 1974; Fischer in Gamble, Fl. Pres. Madras 1639. 1931 (3: 1139. 1957, repr. ed.); Rao & Verma, Cyp. N.E. Ind. 12. 1982; Sharma *et al.*, Fl. Karnataka 307. 1984; Koyama in Dassanayake & Fosberg, Rev. Handb. Fl. Ceyon 5: 198, f. 12. 1985; Karthik *et al.*, Fl. Ind. Enum. Monocot 47. 1989; Cook, Aqu. Wet. Pl. Ind. 112.f.114.1996. *Cyperus pallidus* Nees (Linnaea 9: 284. 1835, nom and) et in Wight. Contr. Bot. Ind. 79.1834. *Anosporum pallidum* (Nees) Boeck in Linnaea 36: 412.1870. (PLATE–23).

Annual, aquatic herb. Rhizomatous; rhizome short, creeping; root thick, fibrous. Stem 1 m tall, 1.0 cm thick, triquetrous, green, smooth, scabrid at tip, erect, tufted, base slightly swollen, ascending, solitary, at times 2–3 together, solid. Leaves 110 cm long, 1 cm broad, 2–3 clustered at base, greeyish–green, flat, or chanelled, longer than the

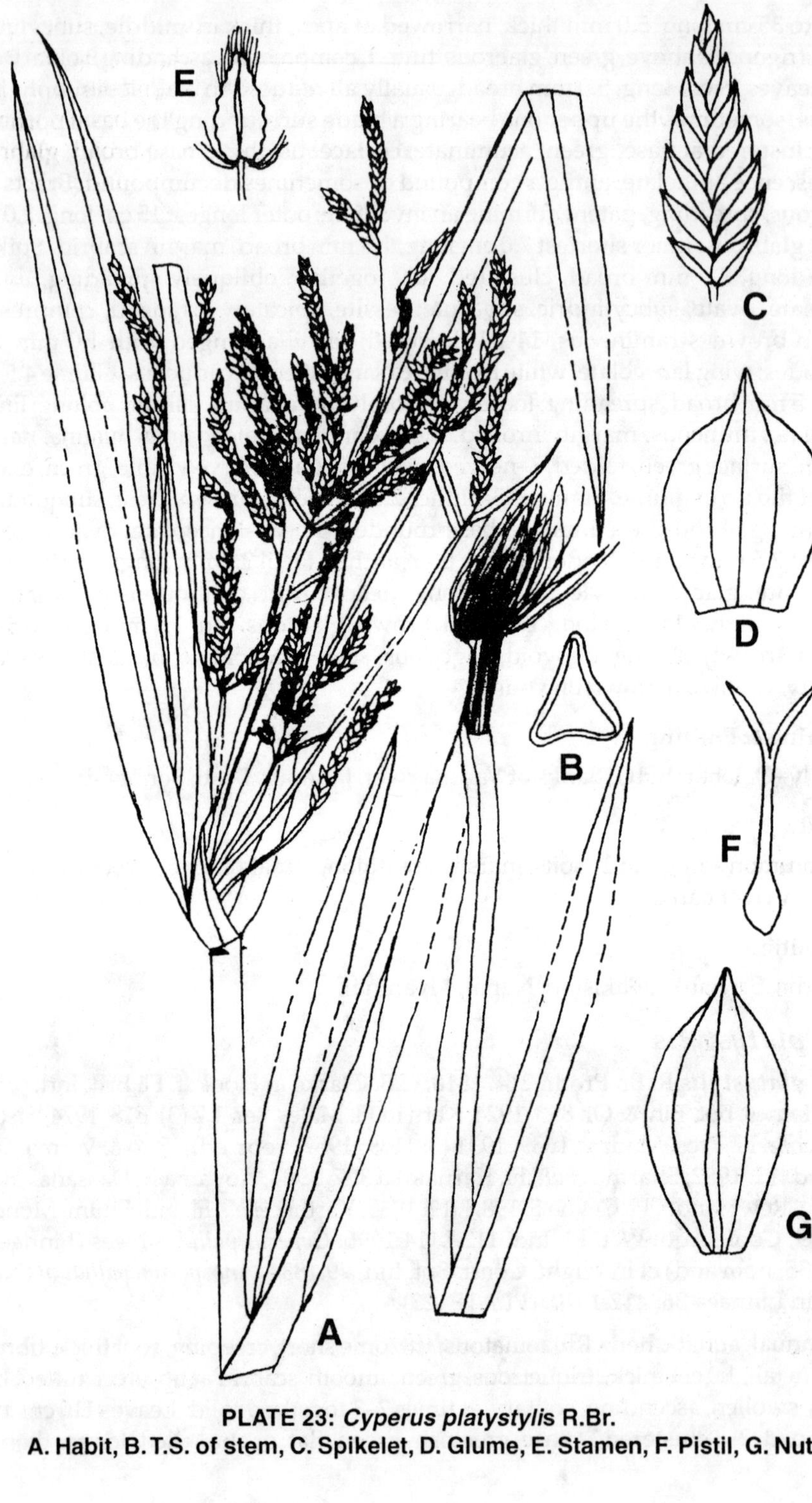

PLATE 23: ***Cyperus platystylis*** **R.Br.**
A. Habit, B. T.S. of stem, C. Spikelet, D. Glume, E. Stamen, F. Pistil, G. Nut

stem, acuminate, base broad, margin & midrib scabrid, coriaceous, transversely septate, nodulose, sheaths spongy, keeled, light brown, usually purplish shaded; ligule absent. Inflorescence anthelate; anthela decompound, sometimes compound, lax 10 cm long very dense to somewhat loose, depressed corymbose–semiglobose. Bracts 5–8 forming an involucer; outer whorls longest, 70 cm long, 1.0 cm broad, greeyish-green, glabrous, slightly spaced, overtopping the inflorescence, foliaceous, distant from each other, spreading to reflexed; inner whorls short, 10 cm long, 1.0 mm broad, diverging, margin scaberulous, same as outer whorls. Primary rays 10 cm long, 10–15 in number, spreading, smooth, secondary rays 4.0 cm long; bracteols 3.0 mm long, setaceous. Spikelet 1.0 cm long, 2.5 mm broad, stellately spreading, oblong, ovate, linear, lanceolate, numerous, congested, cluster of 3–8, 10–40 flowered, subsessile, digitately arranged, compressed, acute at apex, stramineous brown, shadded, rachilla wingless, glabrous, flat. Glume 2.0 mm long, 1.0 mm broad, broadly ovate, thick, membranous, folded, tightly imbricating, margin brownish, narrow, hyaline, cellular–reticulate, subobtuse & mucronate at apex, glossy, yellowish–brown, distichously arranged on the rachis, 3–nerved; nerves visible, brown ending beyond the tip of glume. Stamens 3; anther 1.0 mm long, crest, yellow, setose, linear–lanceolate, oblong, glabrous, keeled, acuminate, base rounded, apex and base hairy; hairs white, tip acute, base rounded; filament 2.5 mm long, slender, ferrugineous, longer than anther, attached with pistil to the base. Pistil 2.0 mm long; ovary 0.5 mm long, ovate, glabrous; style 1.0 mm long, deeply bifid, flattened, fimbriate, narrowed to base, continuous with the ovary, caducous; stigma 0.5 mm long, brown, glabrous, slender, apex acuminate. Nut 2.0 mm long, 0.5 mm broad, elliptic–ovate to ellopsoid, trigonous, dorsally compressed, ventrally concave with raised angles, apiculate the angles of mature nuts, yellowish, corky, thickened, glaucous–brownish.

Flowering & Fruiting

November–February; Rajghat, Barahikala, Jaunpur, Basti, Ballia.

Habitat

Floating on edges of the tanks or in marsh.

Distribution

India, Sri Lanka, Taiwan, Formosa, Malesia.

18. *Cyperus rotundus* L.

KEY TO THE SUBSPECIES

1a. Spikelets 2.0 mm broad; glumes reddish or deep brown, broadly ovate; nut obovoid-ellipsoid

......................**1. Sub sp. *rotundus***

1b. Spikelets 2.5 mm broad; glumes pale ferrugineous, deep yellow or reddish–yellow, ovate, lanceolate; nut broadly obovate

.... **2. Sub sp. *tuberosus***

1. *C. rotundus* subsp. *rotundus*

C. rotundus **subsp.** ***rotundus*** L. Sp. Pl. 45. 1753; C.B. Clarke in Hook. f. Fl. Brit. Ind. 6: 614. 1893; Haines, Bot. Bih. & Or. 903. 1924; Kern in Fl. Males. ser. 1.7 (3): 604. 1974; Cooke, Fl. Pres. Bomb. 2: 871. 1908 (3: 385. 1958 repr. ed.); Fischer in Gamble, Fl. Pres. Madr. 1641. 1931. (3: 1140. 1957 repr. ed.); Kukenth. in Engl., Pflanzenr. 4 (2), Heft 101: 107, f. 13. 1935; Satyanarayan & Shankarnarayan in Ann. Arid Zone 2: 146. 1964; Ramaswamy & Razi, Fl. Bangalore 93: 1973; Hooper in Saldanha & Nicolson, Fl. Hassan 669. 1976; Rao & Razi, Fl. Mysore 561; Rao & Verma Cyp. N.E. Ind. 19. 1982; Sharma *et al.*, Fl. Karnataka 307. 1984; Koyama in Dassanayake & Fosberg, Rev. Handb. Fl. Ceylon 5: 181. f. 10. 1985; Singh, Fl. E. Karnataka 2: 633. 1988; Karthik etal., Fl. Ind. Enum. Monocot. 47: 1989; Keshawamurthy & Yoganarasimhan, Fl. Coorg 503. 1990. (PLATE–24).

Perennial. Rhizomatous; rhizome creeping, stoloniferous, deeprooted; stolons 10 cm long, slender, ending with intermittent ellipsoid, obovoid, aromatic subglobose, often zonate, brownish, solitary or long chained tubers; root fibrous. Stem 90 cm tall, 3.0 mm broad, solitary or sometimes 2–3 together, erect, tufted, compressed, trigonous, glabrous, green, basal part light brown, sometimes tortuous, decumbent, solid. Leaves 67 cm long, 4.0 mm broad, several, usually shorter than, sometimes subequalling the stem, green, flat, linear, scabrid on the margins towards apex, acuminate, base broad, thick, clustered at base, reduced to bladeless sheath. Inflorescence anthelate; anthela usually simple, sometimes compound, lax. Bracts 2-4, 15 cm long, 1.5 mm broad, foliaceous, spreading, glabrous, green, acuminate; rays 3–in number, 6.0 cm long, 1.0 mm broad, spreading, green, slender, patent scabrid; secondary rays if present few. Spikelet 10 mm long, 2.0 mm broad, 3–10 together, linear–lanceolate, frequently curved, spicately arranged, suberect to spreading, acute at apex, 12–40 flowered, reddish–brown, subsessile; rachilla winged, with 0.8 mm long internodes, margin hyaline, glabrous, red streaked. Glume 4.0 mm long, 2.0 mm broad, membranous, obliquely erect, usually sanguineous, with narrow hyaline, margin tightly imbricating, broadly ovate, muticous, keeled, deep brown, acuminate, base broad, folded, distichous, 4–nerved; nerves excurrent, greenish–brown. Stamens 3; anther 2.0 mm long, red, crest–ovate, glabrous, linear, lanceolate, oblong, acuminate, base rounded; filament 3.0 mm long, slender, longer than anther, attached with pistil to the base. Pistil 2.0 mm long; ovary 0.5 mm long, ovate, glabrous; style 1.0 mm long, deeply trifid, long, exserted, caducous; stigma 0.5 mm long, brown, glabrous. Nut 2.0 mm long, 1.0 mm broad, oblong, obovoid, ellipsoid, trigonous, linear, blackish brown apiculate at apex.

Vernacular Name

Nagarmotha, Musta, Mustaka.

Flowering & Fruiting

May–September; UPC, BHU, Tarana, Pindra, Babatpur, Harrua, Lohata, Chaukhandi, Cholapur, Saranath, Jaunpur, Chandauli, Chakiya, Sadalpura, Sakaldiha, Basti, Padrauna, Indara, Mau, Mirzapur.

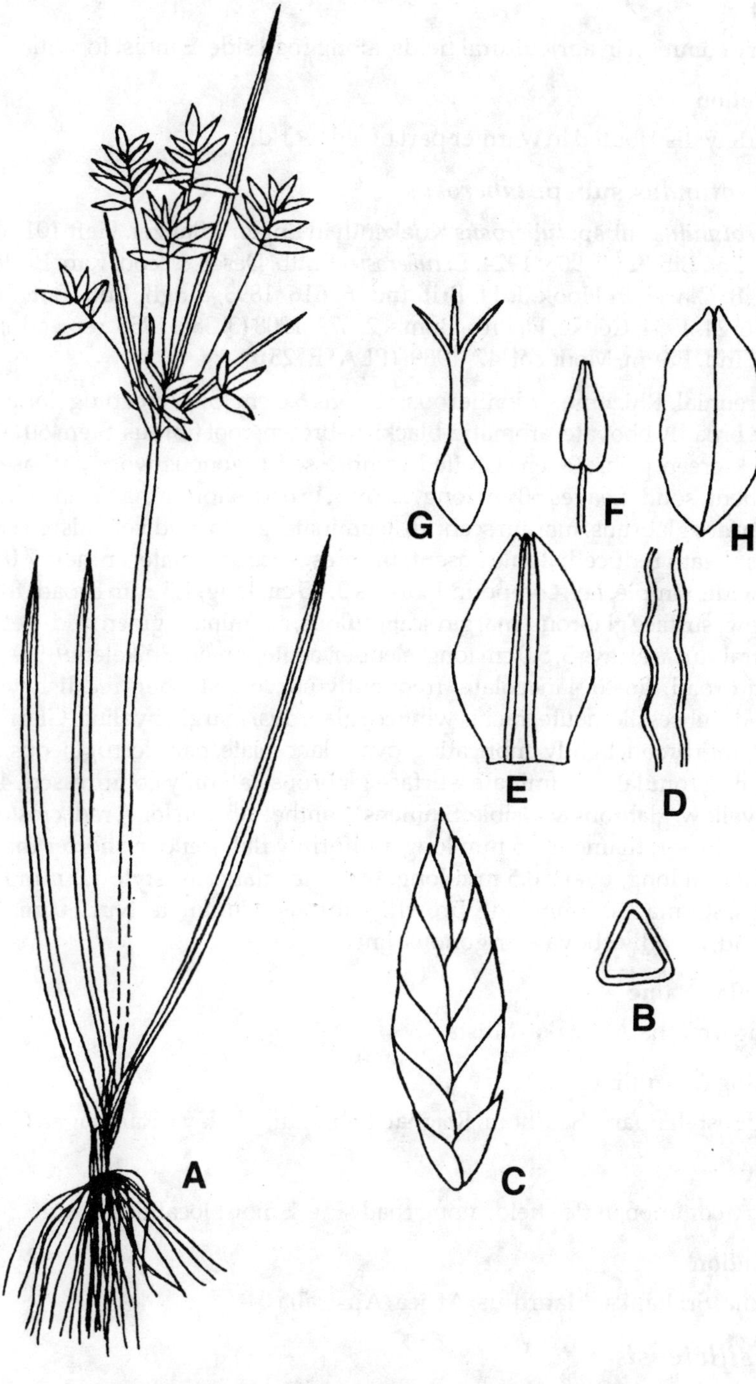

PLATE 24: ***Cyperus rotundus*** **subsp.** ***rotundus*** **L.**
A. Habit, B. T.S. of stem, C. Spikelet, D. Rachis, E. Glume, F. Stamen, G. Pistil, H. Nut

Habitat

Very common in agricultural fields, along road side & moist localities.

Distribution

Widely distributed in warmer part of old world.

2. *C. rotundus* subsp. *tuberosus*

C. rotundus* subsp. *tuberosus Kuekenth in Engler Pflanzer. Heft 101: 113. 1935; Haines, Bot. Bih. & Or. 903. 1924. *C. tuberosus* Rottb. Descr. & Icon. Rare Pl. 28. t. 7 f. 1. 1773; C.B. Clarke in Hook. f. Fl. Brit. Ind. 6: 616. 1893; Singh, Bull. Bot. Soc. Univ. Sangor 6: 24. 1954; Cooke, Fl. Pres. Bomb. 2: 872. 1908 (3: 385. 1958, repr. ed.); Karthik. *et al.*, Fl. Ind. Enum. Monocot. 47. 1989. (PLATE–25).

Perennial. Rhizome stoloniferous; stolons 6.0 cm long, creeping, long chained, slender, broadly obovate, aromatic, blackish–brown; root fibrous. Stem 60 cm tall, 2.0 mm thick, green, glabrous, erect, tufted, compressed, trigonous, woody at base, solitary, decumbent, solid. Leaves 50 cm long, 2.0 mm broad, shorter than stem, clustered at base, surface glabrous, margin scabrid, acuminate, green, mid vein distinct on dorsal surface; sheath reduced; ligule absent. Inflorescence anthelate; anthela 7.0 cm long, 4.0 cm wide, simple, lax, compound. Bracts 3, 15 cm long, 1.5 mm broad, foliaceous, spreading, surface glabrous, margin scaberulous, acuminate, green mid–vein distinct on ventral surface; rays 3, 5.0 cm long, slender, acute, green. Spikelets 9, 9.0 mm long, 2.5 mm broad, linear–lanceolate, frequently curved, slender, reddish–yellow, 20 flowered, subsessile, acute; rachis winged, glabrous, margin hyaline. Glume 4.0 mm long, 2.0 mm broad, tightly imbricating, ovate, lanceolate, pale–ferrugineous, reddish–yellow, mucronulate, acuminate, surface glabrous, laterally compressed, 4-nerved; nerves yellow, glabrous & visible. Stamens 3; anther 1.5 mm long, red, crested, ovate, glabrous, linear; filament 0.5 mm long, uniformly flat, slender, shorter than anther. Pistil 2.0 mm long; ovary 0.5 mm long, obovate, glabrous; style 1.0 mm long, flat, glabrous; stigma 3, 0.5 mm long, brown, glabrous, acuminate. Nut 2.0 mm long, 1.0 mm broad, broadly obovate, trigonous, linear, brown.

Vernacular Name

Nagarmotha, Mustaka, Musta.

Flowering & Fruiting

August–January; Sindhora, Rajghat, Babatpur, Chakiya, Ramnagar, Gorakhpur.

Habitat

Very common in rice field, along road side & moist localities.

Distribution

India, Sri Lanka, Mauritius, Africa, Australia.

19. *C. silletensis*

C. silletensis Nees in Wt. Contr. Bot. Ind. 79. 1834. (PLATE–26).

Annuals. Rhizome absent; root fibrous. Stem 10 cm long, 1.0 mm thick, slender, erect, trigonous, tufted, compressed, glabrous, green. Leaves 3, 10 cm long, 1.0 mm

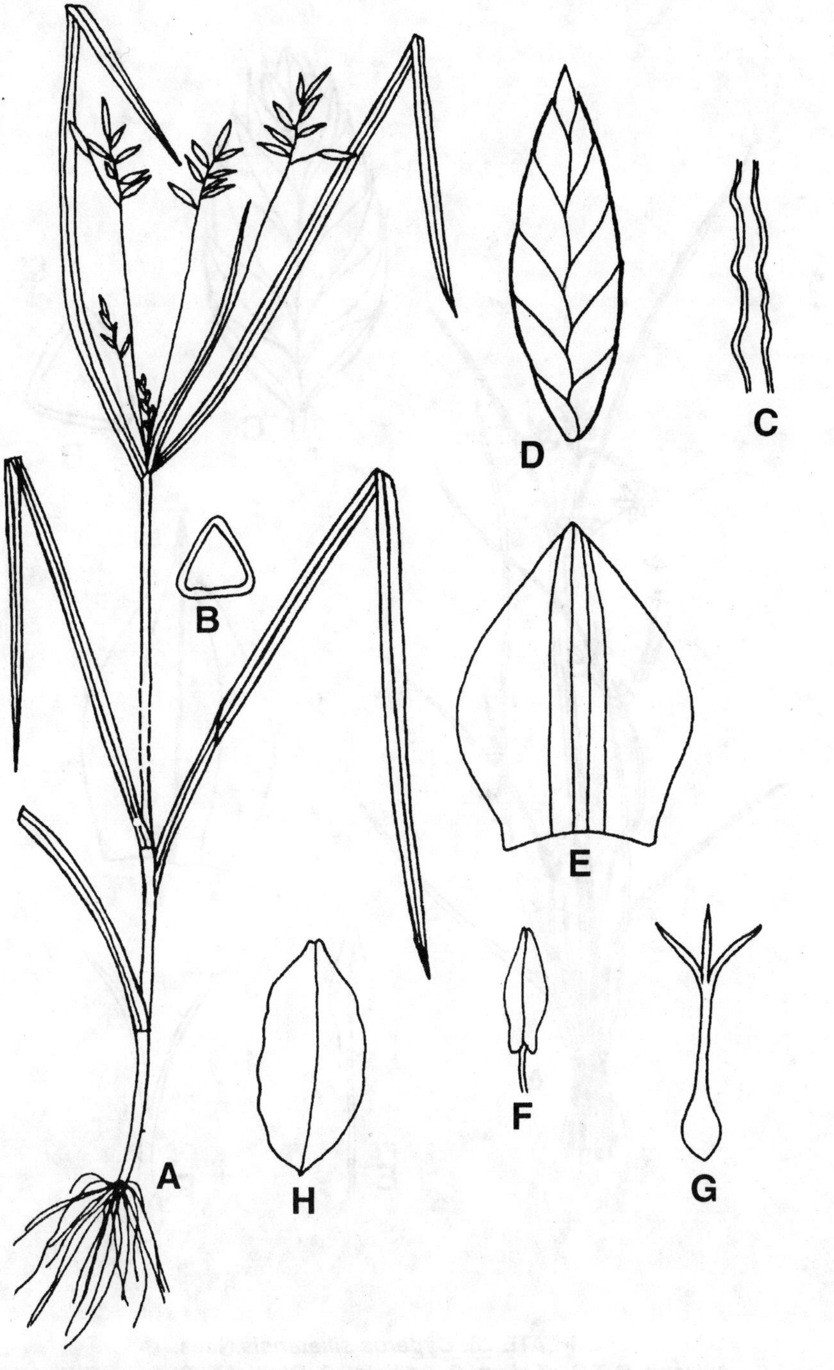

PLATE 25: *Cyperus rotundus* subsp. *tuberosus* (Rottb.) Kukenth.
A. Habit, B. T.S. of stem, C. Rachis, D. Spikelet, E. Glume, F. Stamen, G. Pistil, H. Nut

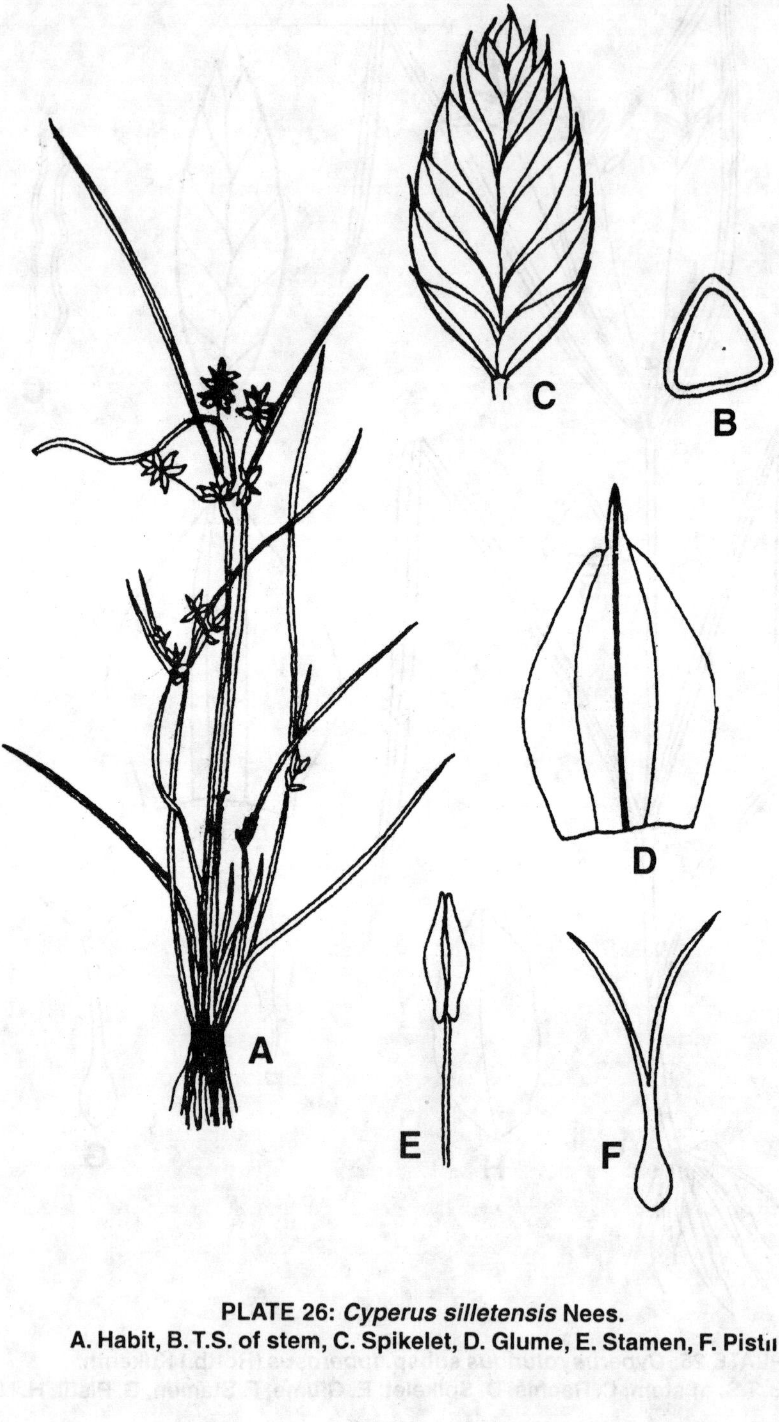

PLATE 26: ***Cyperus silletensis*** **Nees.**
A. Habit, B. T.S. of stem, C. Spikelet, D. Glume, E. Stamen, F. Pistil

thick, green, acute, equalling the stem, surface glabrous, margin scaberulous, veins distinct, clustered at base; lower most reduced to blade less sheath; sheatlı 3.0 cm long, 1.0 mm broad, greyish- green, glabrous; ligule absent. Inflorescence anthelate; anthela 4.0 cm across, compound, dense, somewhat open. Bracts 4, 8.0 cm long, 1.0 mm broad, leaf–like, green, surface & margin scaberulous, foliaceous, spreading, acute, vein distinct; rays 4, 2.5 cm long, sub-erect, rigid. Spikelet 4.0 mm long, 1.5 mm broad, 14 in cluster, lanceolate, ovate, compressed, acute, greyish- green; rachis 3.0 mm long, 1.0 mm broad, wingless, glabrous, reddish–brown. Glume 1.5 mm long, 1.0 mm broad, sublaxly imbricating, with tips ultimately spread out, ovate–lanceolate, muticous, greenish, stramineus at young, reddish with age, keeled, acute, surface & margin glabrous, membranous, 3–nerved; nerves green, visible. Stamens 1; anther 0.5 mm long, brown, ovate–oblong, linear, apiculate, glabrous; filament 1.0 mm long, uniformly slender, glabrous, longer than anther. Pistil 1.5 mm long, ovary 0.2 mm long, obovate, glabrous; style 0.3 mm long, flat, glabrous, deeply bifid; stigma 2, 1.0 mm long, acuminate, brown, glabrous. Nut 1.0 mm long, 0.5mm broad, sub sessile, ellipsoid, trigonous, brownish, apiculate.

Flowering & Fruiting

October–February; Tarana, Ghazipur, Deoria, Basti.

Habitat

In orchards & along river bed.

Distribution

India, Tropical Africa, America, Sri Lanka.

20. *C. tenuispica*

C. tenuispica Steud. Syn. Pl. Cyper. 2: 11. 1854; Kern in Fl. Males. ser. 1.7 (3): 625. 1974. *C. flavidus* C.B. Clarke in Journ. Linn. Soc. Bot. 21: 122. t. 3 f. 25. 1884 et in Hook. f. Fl. Brit. Ind. 6: 600. 1893, non Retz., 1789; Saldanha & Nicolson, Fl. Hasan 670.1976; Cooke, Fl. Bomb. 862. 1908 (Repr. ed. 3: 375. 1958); Fischer in Gamble, Fl. Pres. Madras. 1640.1931 (3: 1139. 1957 repr. ed.); Kukenth. in Engl. Pflanzenr. 4 (20), Heft 101. 245 t. 28 A–D. 1936; Ramaswamy & Razi, Fl. Bangalore 95. 1973; Rao & Razi, Fl. Mysore 561. 1981; Rao. & Verma, Cyp. N.E. Ind. 13. 1982. Sharma *et al.*, Fl. Karnataka 308. 1984; Koyama in Dassanayake & Fosberg, Rev. Handb. Fl. Ceylon 5: 205. 1985; Singh, Fl. E. Karnataka 2: 633. 1988; Karthik *et al.*, Fl. Ind. Enum. Monocot. 48. 1989; Keshava Murthy & Yoganarasimhan, Fl. Coorg 504. 1990. (PLATE–27).

Annual. Rhizome absent; root 8.0 cm long, fibrous, reddish. Stem 30 cm long, 1.0 mm thick, tufted, trigonous, yellowish-green when young, blackish with age, glabrous, tufted, compressed, erect, short lived. Leaves 15 cm long, 2.0 mm broad, clustered at base, half of the stem, trigonous, linear, yellow–green, acuminate, surface glabrous, margin scaberulous at apex, mid–vein distinct on dorsal surface, erect; sheath 1.0 cm long, 2.0 mm broad, reduced, scarious, stramineous to reddish brown, surface & margin glabrous, lower ones bladeless; ligule absent. Inflorescence anthelate; anthela 6.0 cm long, compound or decompound, lax, yellow with age. Brats 3, 8.0 cm long, 1.0 mm broad, foliaceous, leaf–like, surface glabrous, margin scaberulous; primary rays

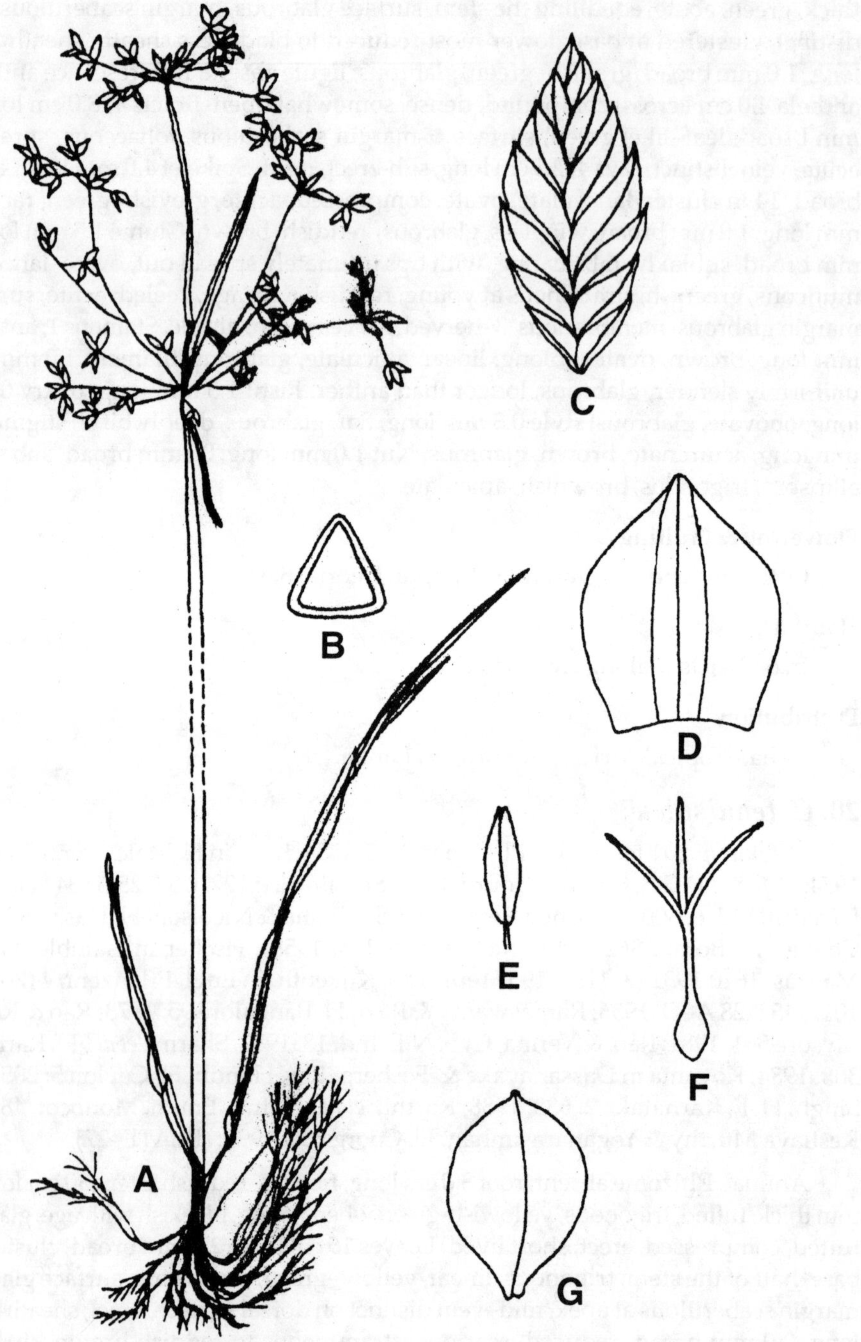

PLATE 27: ***Cyperus tenuispica*** **Steud.**
A. Habit, B. T.S. of stem C. Spikelet, D. Glume, E. Stamen, F. Pistil, G. Nut

5 in number, 8.0 cm long, sub erect, patent, slender, yellow; secondary rays 2.0 cm long, glabrous. Spikelet 5.0 mm long, 1.5 mm broad, 3-9 together, digitate, linear–lanceolate, acuminate, 18-flowered; rachis 4.0 mm long, 1.0 mm broad, wingless, glabrous with 0.5 mm long internodes. Glume 2.0 mm long, 1.0 mm broad, laxly imbricating, or ultimately widely spreading so that the rachilla is distinctly visible, yellowish–green, membranous, margin nerveless, hyline with purple striolate, ovate, oblong, obtuse, truncate, keeled, acute; apices usually excurved, lanceolate 3–nerved; nerves brownish green, visible on back view, faintly distinct on upper view. Stamens 1; anther 0.5 mm long, brown glabrous, linear–oblong, connective smooth; filament 0.1 mm long, uniformaly slender, shorter than anther. Pistil 1.5 mm long; ovary 0.3 mm long, obovate, oblong, glabrous; style 0.5 long, flat, glabrous; stigma 3, 0.7 mm long, acuminate, brown, glabrous. Nut 0.5 mm long, 0.2 mm broad, white when young, brown with age, broadly obovoid, subglobose, obscurely trigonous, obtuse, apiculate, subsessile, tuberculate, distinctly stipitate.

Flowering & Fruiting

July–November; Cholapur, Chunar, Chakiya, Chandauli, Ghazipur, Jaunpur.

Habitat

Growing on moist localities, banks of rivers.

Distribution

India, Sri Lanka, Chaina, Malesia, South Japan, Tropical Africa & Australia.

4. *Eleocharis* R.Br.

KEY TO THE SPECIES

1a. Plant 12 cm long; glumes notched; stamens one

.... **2.** ***E. anceps***

1b. Plant more than 17 cm long; glumes not as above; stamens 1–3:

2a. Spikelets 1.5 cm long, cylindric:

3a. Glumes obovate-oblong; hypogynous bristles longer than the nut

.... **4.** ***E. dulcis***

3b. Glumes ovate; hypogynous bristles as long as nut

.... **1.** ***E. acutangula***

2b. Spikelets less than 1.0 cm long, not cylindric:

4a. Glumes loosely imbricating, elliptic, oblong; hypogynous bristles white

.... **3.** ***E. atropurpurea***

4b. Glumes tightly inbricating, ovate, orbicular; hypogyhous bristles pink

.... **5.** ***E. geniculata***

1. *Eleocharis acutangula*

Eleocharis acutangula (Roxb.) Schult. in R. & S. Syst. Mant. 2: 91. 1824; Rama swamy & Razi, Fl. Bangalore 101. 1973; Kern in van Steenis, Fl. Males. 1, 7: 525. 1974; Hooper in Saldanha & Nicolson, Fl. Hassan 651. 1976; Rao & Razi, Fl. Mysore 561.1981; Rao & Verma, Cyp. N.E. Ind. 23.f. 9–9a. 1982; Sharma *et al.*, Fl. Karnataka 308. 1984; Koyama in Dassanayake & Fosberg, Rev. Handb. Fl. Ceylon 5: 256. 1985; Singh., Fl. E. Karnataka 634. 1988; Karthik *et al.*, Fl. Ind. Enum. Monocot. 48. 1989; Cook, Aqu. Wet. Pl. Ind. 127.f.121a-d.1996. *E. fistulosa* (Poir.) Lamk. in Spreng Jahrb Gewach, Heft 3: 78. 1820; Clarke in Hook. f. Fl. Brit. Ind. 6: 626. 1893; Schult in Roem. & Schult., Mant. 2 : 89. 1824; Cooke, Fl. Pres. Bomb 2: 888. 1908 (3: 402. 1958. repr. ed.); Fischer in Gamble, Fl. Pres. Madras 1648. 1931 (3: 1145. 1957 repr. ed.). *Scirpus acutangulus* Roxb. Fl. Ind. 1: 216.1820. *S.fistulosus* Poir. Encycl. 6: 749.1805; non Forssk.1775. (PLATE–28).

Perennials. Rhizomatous; rhizome 1.0 cm long, 2.0 mm thick, rooting at nodes. Stem 60 cm tall, 6.0 mm thick, stout, acute at apex, triquetrous, concave on the side, spongy, sepatate, brownish-green. Leaves reduced to few basal sheaths; sheaths 2–3, 8.0 cm long, 7.0 mm broad, loose, purplish, acute, oblique at mouth, soon disintegrating, membranous, tubular, truncate. Inflorescence solitary terminal spikelets. Bracts absent. Spikelets 3.5 cm long, 2.5 mm thick, terete, cylindrical, slightly thicker than the stem, acute, glaucous green–stramineous, 8–32 flowered; rachilla persistent. Glume 6.0 mm long, 4.0 mm broad, coriaceous, tightly imbricating, ovate, oblong, scarious margined, stramineus, red–brown dotted, several, striate, acute, obtuse at apex, spiral, caducous, mid-nerves visible brown with many faint nerves on both sides. Perianth–bristles 6, very unequal, 2.0 mm long, slightly broader at apex, brown, smallest half to two third as long as the nut, longest equalling to rather exceeding the beak of nut, retrorsely scabrid, acute, base rounded. Stamens 3, rarely 2; anther 2.0 mm long, deep brown, linear–lanceolate, ovate–oblong, muticous, glabrous, acuminate, base rounded, connective appendage very minute; filament 4.0 mm long, slender, hyaline, longer than anther, attched with pistil to the base. Pistil 4.0 mm long; ovary 1.0 mm long, obovate, glabrous, base acuminate; style 2.0 mm long, glabrous, dilated at the base, with a constriction above the nut, slender; stigma 3, 1.0 mm long, acute, brown, fimbriate; fimbri acute, base rounded. Nut 2.0 mm long, 1.0 mm broad, broadly obovoid, biconvex, slightly compressed, contricted at the apex into a short but distinct neck, glossy, brownish- stramineus, surface cells transversely oblong in 15–20 vertical rows on each face, style base ovoid–conical, 1.0 mm long, 1.0 mm broad, brown.

Flowering & Fruiting

June–October; Babatpur, Rajghat, Jaunpur, Gorakhpur, Padrauna

Habitat

In swamp, paddy fields, ponds, ditches, often half submerged.

Distribution

India, Nepal, Myanmar, Malesia, Japan, Formosa, Tropical Africa, Madagascar, West Indies, North Australia, South America.

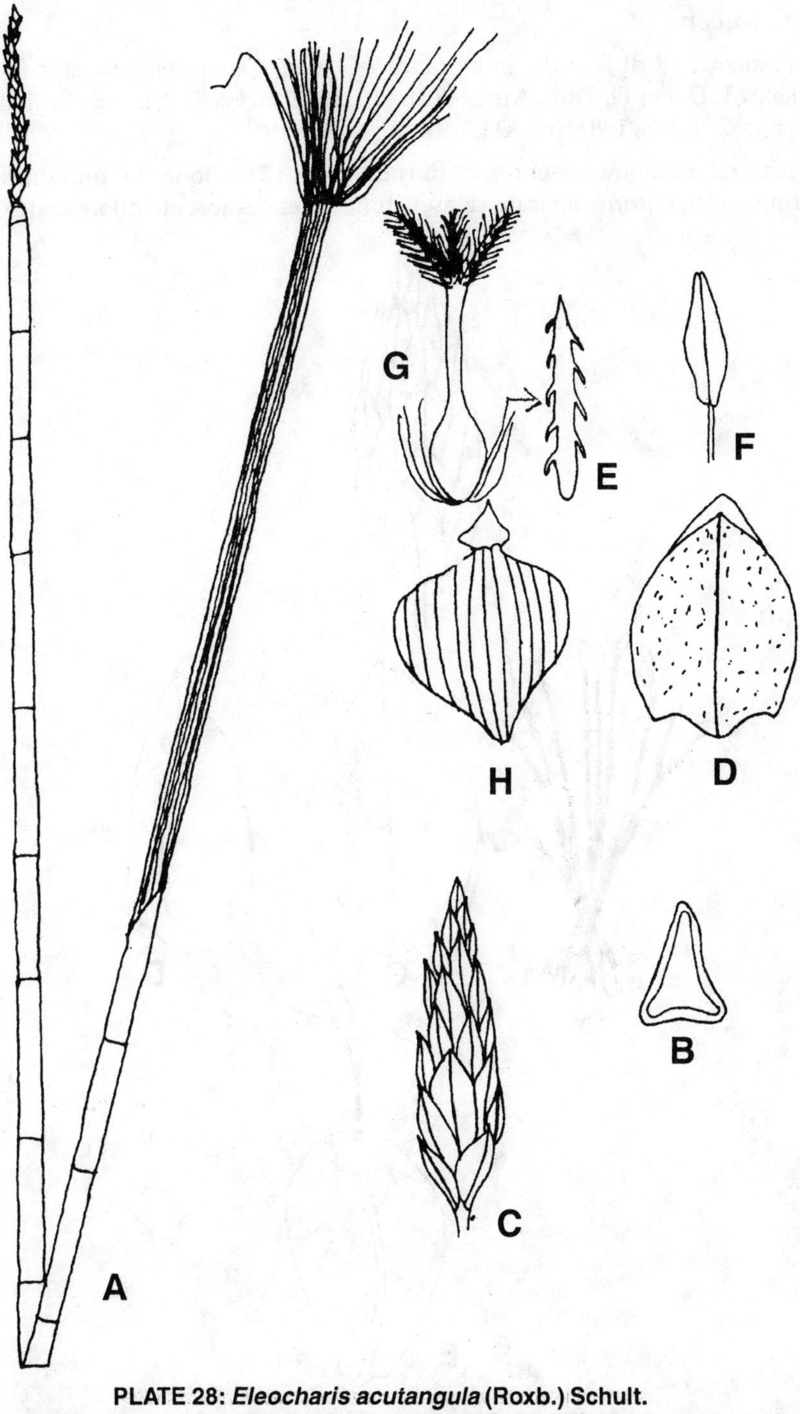

PLATE 28: ***Eleocharis acutangula*** **(Roxb.) Schult.**
A. Habit, B. T.S. of stem, C. Spikelet, D. Glume, E. Bristle, F. Stamen, G. Pistil, H. Nut

2. *E. anceps*

E. anceps (Ridl.) Clarke in Durand & Schinz. Consp. Fl. Africa. 5: 596. 1895; Clarke in T. Dyer., Fl. Trop. Africa 8(3): 410. 1902. *Heleocharis anceps* Ridl. in Trans. Linn. Soc. 2, 2: 148. 1884. (PLATE–29).

Annual. Rhizome absent; root fibrous. Stem 12 cm long, 1.0 mm thick, green, glabrous, erect, filiform, delicate, densely tufted. Leaves absent. Inflorescence consists

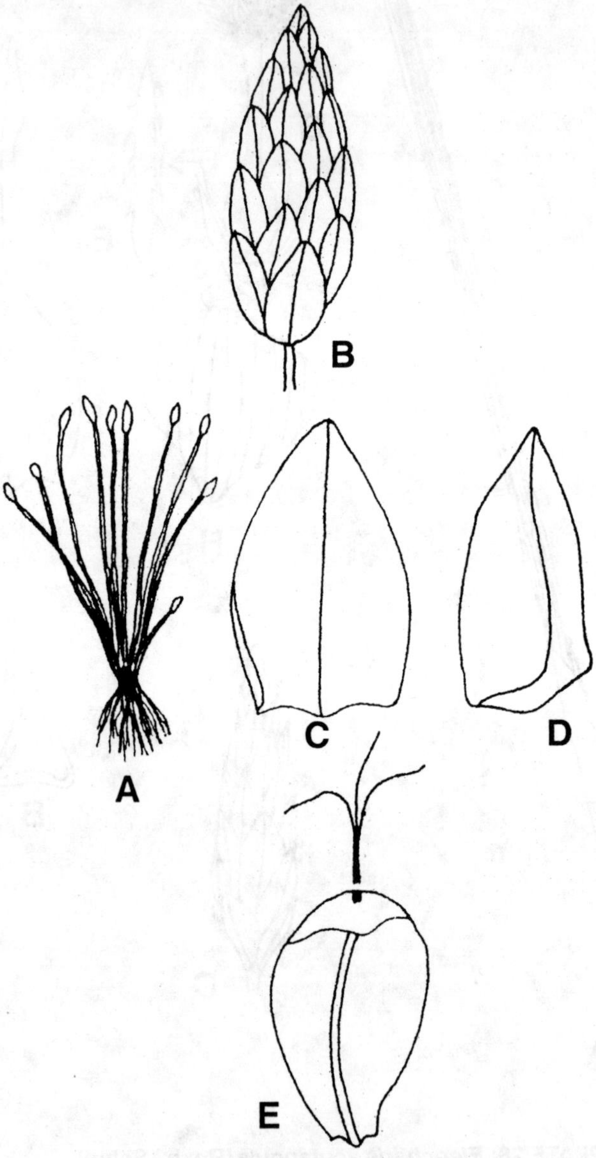

PLATE 29: *Eleocharis anceps* (Ridl.) Clark
A. Habit, B. Spikelet, C&D. Glume, E. Nut

of spikelet. Involucral bracts 2, 2.0 cm long, 1.0 mm broad, longer than the glumes. Spikelet 5.0 mm long, 1.0 mm broad, terminal, ovoid, oblong, greenish, tinged with red–brown, many flowered, solitary. Glume 5.0 mm long, 2.5 mm broad, membranous, surface glabrous, ovate, lanceolate, notched at apex, muticous, with brown, vertical & along the uninerved keel; keeled glabrous; nerves green, visible, glabrous, ending at apex. Stamens 1; anther 1.0 mm long, yellow, oblong, glabrous; filament 0.2 mm long, slender, glabrous, shorter than anther. Pistil 1.5 mm long; ovary 0.2 mm long, ovoid, glabrous; style 0.5 mm long, flat, glabrous; stigma 3, 0.8 mm long, brown, hairy; hairs acute, base broad. Nut 1.0 mm long, 0.5 mm broad, brown, tricostate, smooth; style base decurrent on the shoulders of the nuts.

Flowering & Fruiting

October–November; Barahikala, Ghazipur, Khalilabad

Habitat

Rare, in marshes of open grassland.

Distribution

India, Formosa, Maleshiya, Japan.

3. *E. atropurpurea*

E. atropurpurea (Retz.) J. Presal, Reliq. Haenk 1: 196. 1828; C.B. Clarke in Hook f. Fl. Brit. Ind. 6: 627. 1896; Bhattacharya, Bull. Bot. Surv. Ind. 6: 208. 1964; Panigrahi *et al.*, Bull. Bot. Surv. Ind. 11: 111. 1969; Cooke, Fl. Pres. Bombay, 2: 889. 1908 (3: 403. 1958 repr. ed.); Fischer in Gamble, Fl. Pres. Madr. 1648. 1931 (3: 1145. 1957 repr. ed.); Ramaswamy & Razi, Fl. Bangalore 101. 1973; Kern in van Steenis, Fl. Males. 1, 7: 536. 1974; Hooper in Saldanha & Nicolson, Fl. Hassan 671. 1976; Rao & Razi, Fl. Mysore 561. 1981; Rao & Verma, Cyp. N.E. Ind. 25. 1985; Sharma *et al.*, Fl. Karna. 308. 1984; Karthik *et al.*, Fl. Ind. Enum. Monocot. 48. 1989; Keshava Murthy & Yoganarasimhan, Fl. Coorg. 504. 1990; Rao & Verma, Cyp. N.E. Ind. 25. 1984; Misra & Verma, Fl. Allaha. 393. 1992; Cook, Aqu. Wet. Pl. Ind. 127. f. 121 c-i. 1996. *Scirpus atropurpurea* Retz. Obs. Bot. 5: 14. 1789. (PLATE–30).

Annual. Rhizome absent; root fibrous. Stem 27 cm long, 1.0 mm thick, green, glabrous, erect, tufted, slender. Leaves absent; sheath 2, 1.5 cm long, obliquely truncate at apex, membranous, purplish below, pale brownish above, glabrous. Inflorescence solitary, terminal spikelet. Bract absent. Spikelet 5.0 mm long, 2.5 mm broad, ovoid, oblong, lanceolate, subterete, brownish turning fuscous dark in fruit, acute, wider than the stem, many flowered. Glumes 1.5 mm long, 1.0 mm broad, membranous, elliptic, oblong, loosely imbricating, muticous, keeled; keels green, sides hyaline, nerveless with purplish bands. Perianth–bristles 7 in number, 1.0 mm long, slender, white, translucent, antrorsely scabrid, shorter than nut, sometimes much reduced or absent. Stamens 2; anther 0.5 mm long, yellow, oblong, glabrous; filament 1.0 mm long, slender, glabrous, longer than anther. Pistil 1.5 mm long; ovary 0.3 mm long, ovate glabrous; style 0.5 mm long, bicleft, base discoid, glabrous; stigma 2, 0.7 mm long, glabrous, acuminate. Nut 0.5 mm long, 0.3 mm broad, obovoid, biconvex, smooth, glossy, brownish–black, stipitate at base, truncate at apex with a greyish & flattened

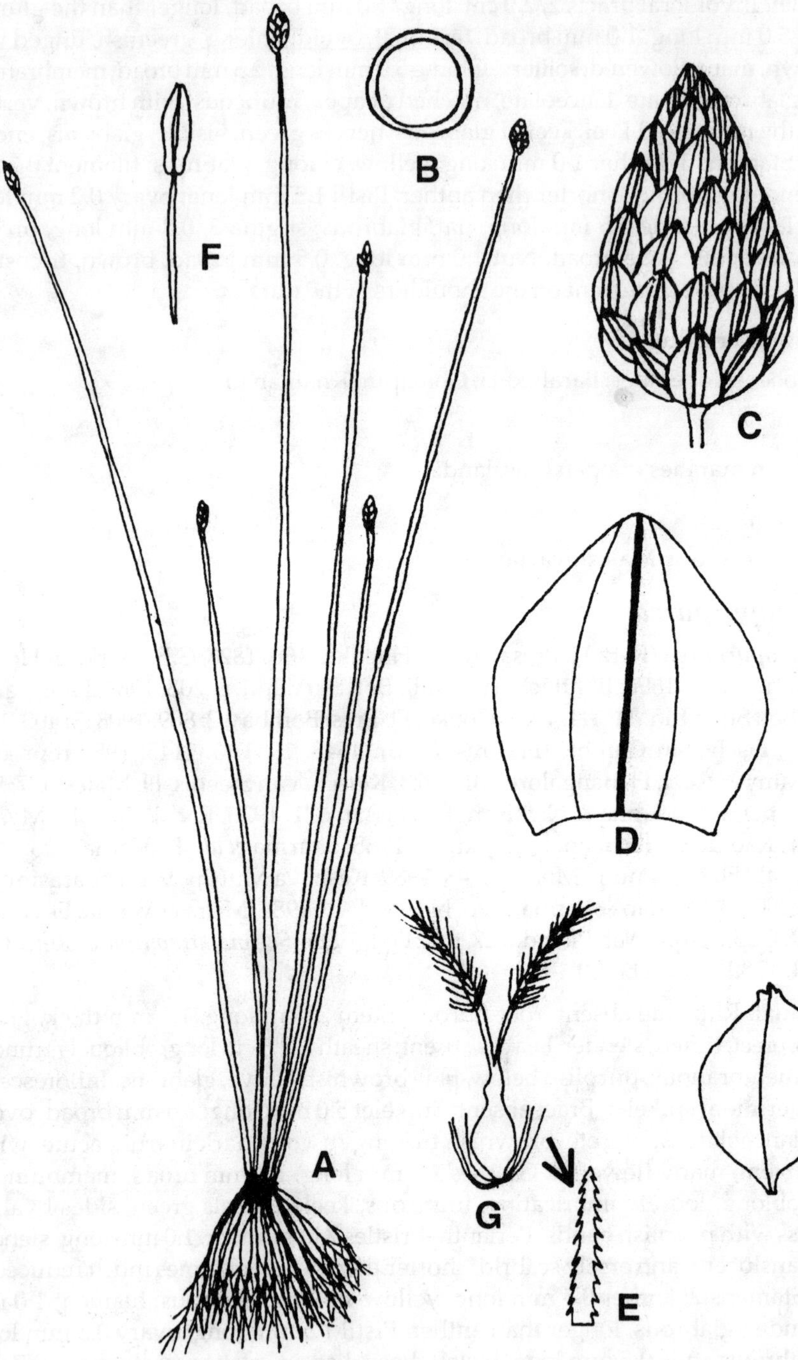

PLATE 30: ***Eleocharis atropurpurea*** **(Retz.), K.B. Presl**
A. Habit, B. T.S. of stem, C. Spikelet, D. Glume, E.Bristle, F. Stamen, G. Pistil, H. Nut

appendage, surface cells obscure, subquadrate; style base like a minute button, scarcely 0.5 mm long.

Flowering & Fruiting

August–December; Babatpur, Barahikala, Jaunpur, Basti, Bhatpar.

Habitat

Common in moist places.

Distribution

India, Pakistan, China, Malesia, Australia, Formusa, Japan, South Europe, America.

4. *E. dulcis*

E. dulqis (Burm. f.) Trin. ex Hensch., Vita Rumph. 186. 1833; Verma & Chandra in Rec. Bot. Surv. Ind. 21 (2): 249. 1981; Kern in van Steenis, Fl. Males. 1, 7: 529. 1974; Hooper in Saldanha & Nicolson, Fl. Hassan 672. 1976; Rao & Verma, Cyp. N.E. India 23. 1982; Sharma *et al.*, Fl. Karnataka 308. 1984; Koyama in Dassanayake & Fosberg, Rev. Handb. Fl. Ceylon 5: 259. 1985; Karthik *et al.*, Fl. Ind. Enum. Monocot. 49. 1989; Cook, Aqu. Wet. Pl. Ind. 128. f. 122 e-g. 1996. *Eleocharis plantaginea* (Retz.) Roem. & Schult. Syst. 2: 150. 1817; Miq. Fl. Ind. Bot. 3: 302. 1856; Clarke in Hook. f. Fl. Brit. Ind. 6: 625. 1893; Cooke, Fl. Pres. Bombay 2: 888. 1908 (3: 402. 1958, repr. ed.); Fischer in Gamble, Fl. Pres. Madras 1647. 1931 (3: 1145. 1957 repr. ed.). *Eleocharis equisetina* Presl. Rel. Haenk. 1: 195. 1828; Miq. Fl. Ind. Bot. 3: 302. 1856; Clarke in Hooke f. Fl. Brit. Ind. 6: 626. 1893. *Andropogon dulce* Burm. f. Fl. Ind. 219. 1768. *Scirpus tuberosus* Roxb. Fl. Ind. 1: 213. 1820. *S. tumidus* Roxb. Fl. Ind. 1: 215. 1820. *S. plantaginoides* Rottb. Descr. et. Icon. 45. t. 15. f. 2. 1773. 6: 626. 1893. (PLATE–31).

Perennials. Rhizomatous; rhizome stoloniferous, 4.0 cm long, 4.0 mm thick, blackish, sometime bearing tubers; root yellowish–brown, fibrous. Stem 50 cm long, 8.0 mm thick, glaucous, acute at apex, green, terete, transversely septate, spongy, erect, tufted, glabrous. Leaves absent; sheath 8.0 cm long, 9.0 mm broad, usually reddish, membranceous, faintly nerved, tubular, apiculate; ligule absent. Inflorescence a solitary terminal spikelet. Bracts absent. Spikelets 2.0 cm long, 6.0 mm broad, cylindric, terete, obtuse, acute, linear, lanceolate, slightly broader then stem; rachis 1.5 mm long, 4.0 mm broad, glabrous, perisitent. Glume 6.0 mm long, 4.0 mm broad, stramineous with red dots, obovate–oblong, linear–lanceolate, coriaceous, surface glabrous, margin scarious, purple, obtuse–acute at apex, winged; wing hyaline, wavy, densely imbricating, several nerved, central nerve visible, deep brown than lateral nerves. Perianth–bristles 8, subequal, hypogynous, slightly broader towards the base, reterorsely scabrid, yellowish; 6 well developed, 2.0 mm long, longer than nut, 2 small, 0.8 mm long, thread like. Stamens 3; anther 2.5 mm long, yellow, glabrous, ovate–oblong, linear–lanceolate, muticous, apex hairy, hairs acute, base broad; filament 1.0 mm long, uniformely flat, slender, shorter than anther. Pistil 6.5 mm long; ovary 1.0 mm long, obovate, green, glabrous; style 2.0 mm long, base dilated, glabrous, flat; stigma 3, 3.5 mm long, brown, acuminate, hairy; hairs apiculate, base broad. Nut 2.0 mm long, 1.0 mm broad, orbicular, obovoid, glossy, yellowish-brown, trigonous,

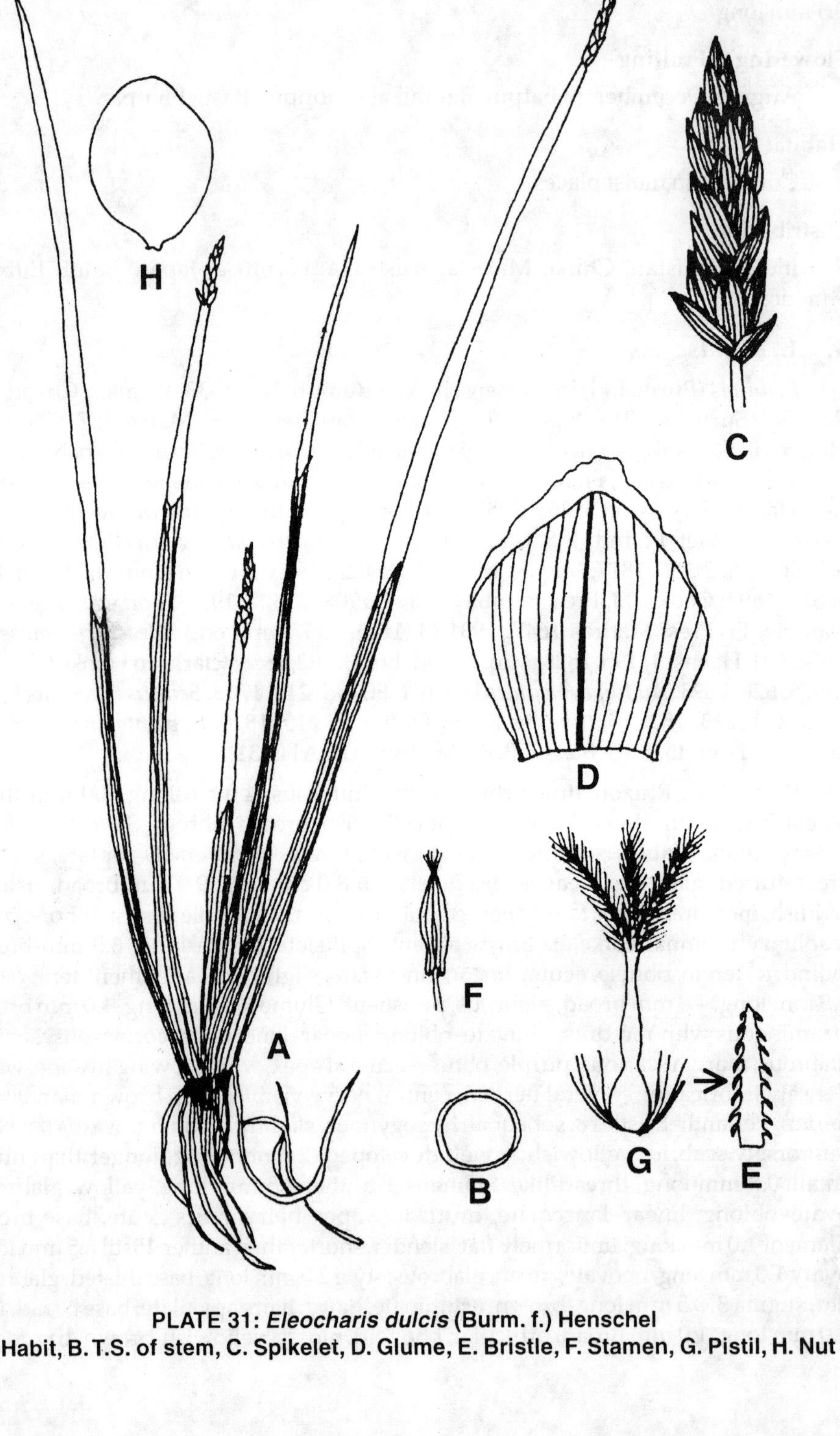

PLATE 31: *Eleocharis dulcis* (Burm. f.) Henschel
A. Habit, B. T.S. of stem, C. Spikelet, D. Glume, E. Bristle, F. Stamen, G. Pistil, H. Nut

prominently convex on the anticous face with longitudinal furrows on the convex faces, glabrous, truncate at top, obscurely reticulate, style base pyramidal, about half of the nut, trigonal, dark–brown, usually persistent.

Flowering & Fruiting

September–December; Barahikala, Gajipur, Saidpur, Jaunpur, Kushinagar, Sonbhadra.

Habitat

Common in ponds, rice fields & other marshy areas & muddy banks of canals.

Distribution

India, Sri–Lanka, Pakistan, China, Malesia, Formosa, Japan, Tropical West Africa, Madagascar, Australia.

5. *E. geniculata*

E. geniculata (L.) Roem. & Schult., Syst. Veg. 2: 150. 1817; Verma & Chandra in Rec. Bot. Surv. India 21 (2): 249. 1981; Ramaswamy & Razi. Fl. Banglore 102. 1973; Kern in van Steenis, Fl. Males 1, 7: 536. 1974; Hooper in Saldanha & Nicolson, Fl. Hassan 672. 1976; Rao & Razi, Fl. Mysore 562. 1981; Sharma *et al.*, Fl. Karnataka 308. 1984; Koyama in Dassanayake & Fosberg, Rev. Handb. Fl. Ceylon 5: 264. 1985; Singh, Fl. E. Karnataka 2: 634. 1988; Karthik *et al.*, Fl. Ind. Enum. Monocot. 49. 1989; Keshavamurthy & Yoganarasimhan, Fl. Coorg 505. 1990. *Scirpus geniculatus* L. Sp. Pl. 48. 1753. *Eleocharis capitata* R. Br. Prodr. 225. 1810; C.B. Clarke in Hook. f. Fl. Brit. Ind. 6: 627. 1893; Cooke, Fl. Pres. Bomb. 2: 889. 1908 (3: 404. 1958 repr. ed.); Fischer in Gamble, Fl. Pres. Madras 1648. 1931 (3: 1145. 1957 repr. ed.). (PLATE–22H, 22I).

Annual. Rhizomatous; rhizome 1.0 cm long, root fibrous, black. Stem 25 cm long, 1.5 mm thick, terete, rigid, tufted, erect or reflexed, angular–sulcate, green, glabrous. Leaves absent; sheath 3, 9.0 cm long, 2.0 mm broad, broad, membranous, tubular, truncate, oblique at mouth, acute–attenuate at apex, usually purplisn–brown towards base, faintly nerved; nerves several. Inflorescence solitary, termind spikelet. Bracts absent. Spikelet 3.5 cm long, 4.0 mm broad, subglobose, conical, ovoid, much broader then stem, obtuse at apex, brownish, terete; rachilla 2.0 cm long, 2.5 mm broad, glabrous, persistent. Glumes 4.5 mm long, 2.0 mm broad, elliptic–oblong, ovate–orbicular, acute, membranaceous, pale brown with red tinged, surface & margin glabrous, tightly imbricating, even during fruiting, apex winged; wing glabrous, 1–nerved, nerve green visible. Perianth–bristles 6; 2.0 mm long, hypogynous, retrosely scabrid, pink, usually longer than nut. Stamens 3; anther 2.0 mm long, oblong, linear–lanceolate, glabrous, connective appendage very minute, apiculate, crested, yellow; filament 0.5 mm long, uniformaly slender, shorter than anther. Pistil 3.5 mm long; ovary 1.0 mm long, obovate–oblong, glabrous, base apiculate; style 1.5 mm long, glabrous, dilated at base with a constriction above the nut, the nut base small, depressed, conical, divided half to three fourth of the length into 2 stigma; stigma 1.0 mm long, glabrous, brown, acuminate. Nut 2.0 mm long, 1.0 mm broad, shining, dark brown, blackish obovoid, biconvex, scarsely half as long the glume; style base persistent, minute, concial.

Flowering & Fruiting

August–January; Barahikala, Babatpur, Chakiya, Cholapur, Sakaldina, Basti, Lar.

Habitat

Very common in drying ponds & marshy places.

Distribution

Widely distributed throughout the hotter parts of old and new world, India.

5. *Fimbristylis* Vahl

KEY TO THE SPECIES

1a. Glumes aristate:

2a. Anthela simple, lax; spikelets terete .... **9. *F. dipsacea***

2b. Anthela compound, dense; spikelets polygonal .... **20. *F. squarrosa***

1b. Glumes nut aristate:

3a. Inflorescence a solitary spikelet:

4a. Leaves several; spikelets subterete in the upper half, compressed in lower half .... **18. *F. ovata***

4b. Leaves usually absent; spikelets not compressed in lower half:

5a. Plant 30 cm long, glume ovate, oblong, acute, or mucronulate .... **1. *F. acuminata***

5b. Plant 60 high; glumes spiral, obtuse .... **23. *F. tetragona***

3b. Inflorescence anthelate:

6a. Stems trigonous or triquetrous:

7a. Stems trigonous:

8a. Spikelets 4.0 mm long, ellipsoid, globose, terete .... **8. *F. dichotoma***

8b. Spikelets 5–10 mm long, ovate, oblong, polygonal:

9a. Spikelets greyish–green, subacute .... **4. *F. alboviridis***

9b. Spikelets brown, acute:

10a. Stems ribbed; leaf-sheath usually densely hairy .... **3. *F. aestivalis***

10b. Stems not ribbed; leaf-sheath glabrous:

11a. Plant 20 cm long; spikelets 4–20 clustered in a solitary head .... **5. *F. argentea***

11b. Plants upto 30 cm long; spikelets solitary:

12a. Ligule hairy; nut suborbicular .... **24. *F. tomentosa***

12b. Ligule membranous or absent; nut obovoid:

13a. Stigma recurved at maturity; nut trabeculate .... **6. *F. bisumbellata***

13b. Stigma not recurved at maturity; nut not trabeculate:

14a. Stem bearing 2–3 spikelets; nut biconvex .... **19. *F. schoenoides***

14b. Stem bearing 4– numerous spikelets; nut trigonous:

15a. Leaves with a triangular tip; anthela compound or supra decompound .... **7. *F. complanata***

15b. Leaves with out triangular tip; anthela simple .... **12. *F. ferruginea***

7b. Stem triquetrous:

16a. Leaves 7.0 cm long, all strongly curved; glumes subdistichous .... **13. *F. fimbristyloides***

16b. Leaves absent; glumes spiral .... **16. *F. nutans***

6b. Stems tetragonous or slender:

17a. Stem tetragonous:

18a. Leaves filiform; bracts setiform .. .. **22. *F. tenera***

18b. Leaves flat; bracts foliceous:

19a. Plants with many flowering leaf-less stem, and few sterile leafy stem .... **15. *F. miliacea***

19b. Plants only with flowering leafy stems:

20a. Plants 40 cm high, spikelets usually solitary; stamens 1–2 .... **14. *F. litoralis***

20b. Plants upto 60 cm high; spikelets usually in clusters of 2–5; stamens 3:

21a. Leaves 4.0 mm broad; anthela bearing 20–30 spikelets; nut pale stramineous, smooth .... **11. *F. falcata***

21b. Leaves 5.0 mm broad; anthela bearing 2 to many spikelets; nut brownish, verrucose .... **10. *F. eragrostis***

17b. Stems slender:

22a. Plants 10 cm high; nut obtusely trigonous .... **2. *F. adenolepis***

22b. Plants 35–40 cm high; nut not obtusely trigonous;

23a. Anthela compound, dense; spikelets 2.0 mm wide; glumes shortly murcronate .... **17. *F. oblonga***

23b. Anthela simple, open; spikelets 3.0 mm wide; glumes acute .... **21. *F. striolata***

1. *Fimbristylis acuminata*

Fimbristylis acuminata Vahl Enum. Pl. 2: 285. 1806; C.B. Clarke in Hook. f. Fl. Brit. Ind. 6: 631. 1893 et in I 11. Cyp. t. 40. f. 5–7. 1909; Kern in Fl. Males. ser. 1.7 (3): 588. 1974; Fischer in Gamble, Fl. Pres. Madras 1658. 1931 (3: 1150. 1957 repr. ed.); Hooper in Saldanha & Nicolson, Fl. Hassan 675. 1976; Rao and Razi, Fl. Mysore 562. 1981; Rao & Verma, Cyp. N.E. Ind. 29. 1982; Sharma *et al.*, Fl. Karnataka 308. 1984; Koyama in Dassanayake & Fosberg. Rev. Handb. Fl. Ceylon 5. 319. 1985; Karthik. et al. Fl. Ind. Enum. Monocot. 50. 1989; KeshavaMurthy & Yoganarasinhan, Fl. Coorg 506. 1990; Cook, Aqu. Wet. Pl. Ind. 135. f. 129. 1996. *F. setacea* acute. non. Benth, 1873; C.B. Clarke in Hook. f. Fl. Brit. Ind. 6: 632. 1893. (PLATE–32).

Perennial. Rhizomatous; rhizome 1.0 cm long; root fibrous. Stem 30 cm long, 1.5 mm thick, obtusely sulcate, quadrangular, green, tufted, erect, glabrous. Leaves absent, rarely reduced to obliquly truncate sheaths or with short triangular blade; ligule absent. Inflorescence of one terminal spikelet. Bracts 3.0 mm long, 2.0 mm broad, glume–like, triangular–ovate, keeled; keel slightly curved at apex towards inside. Rays absent. Spikelet 6.0 mm long, 2.0 mm broad, solitary, terminal, erect on the stem, ovoid–oblong, lanceolate, terete stramineus or brown tinged, acute. Glumes 4.5 mm long, 3.5 mm broad, broadly ovate, oblong, membranous, acute, slightly mucronulate, stramineous with brown tinged towards the middle portion, papery towards the

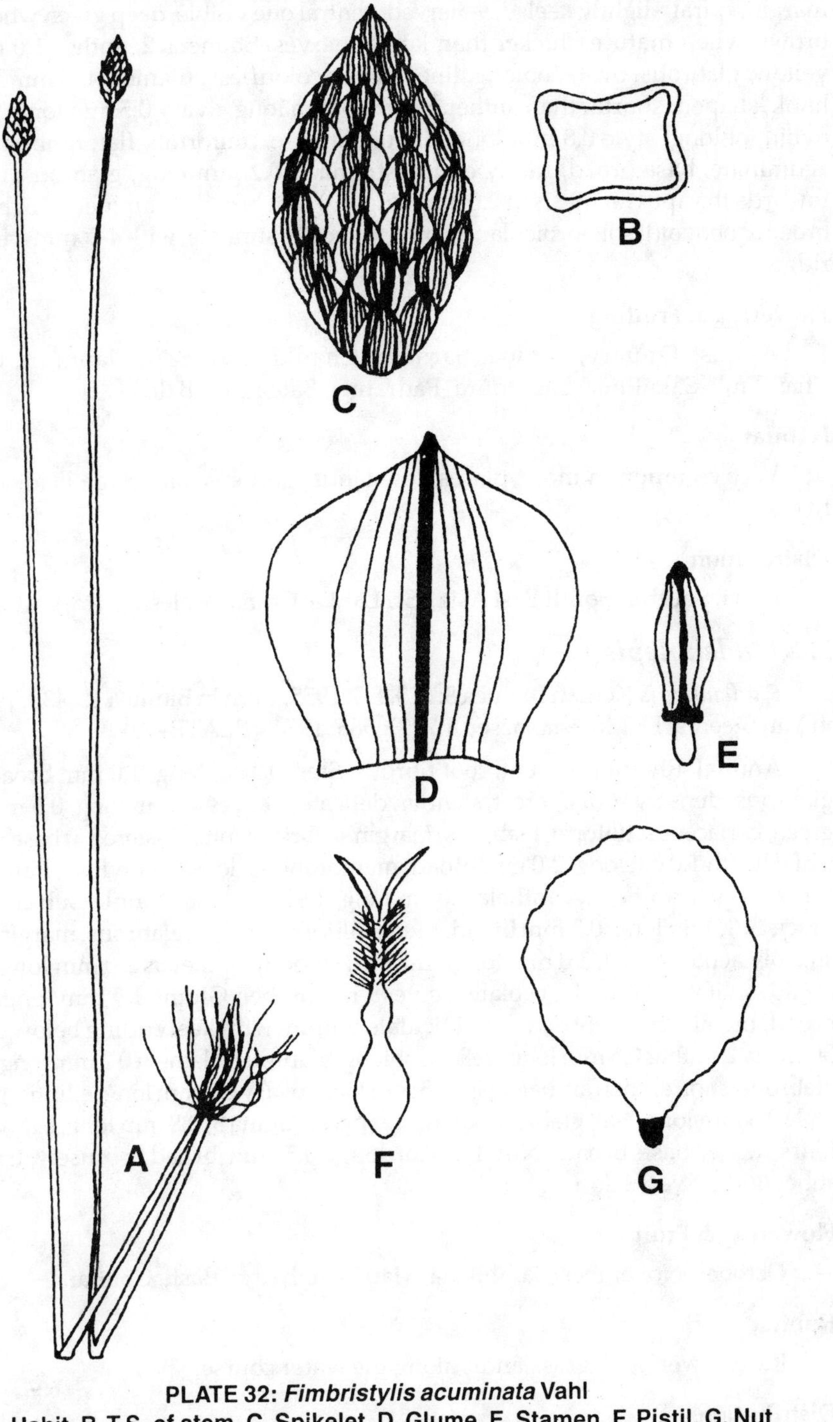

PLATE 32: *Fimbristylis acuminata* Vahl
A. Habit, B. T.S. of stem, C. Spikelet, D. Glume, E. Stamen, F. Pistil, G. Nut

margin, spiral, slightly keeled, 9–nerved; central one visible, deep green when young, brown when mature, thicker than lateral nerves. Stamens 2; anther 1.0 mm long, yellow, glabrous, ovate, oblong, linear, apex colourless; filament 0.5 mm long, flat, hook–shaped, shorter than anther. Pistil 1.5 mm long; ovary 0.5 mm long, glabrous, ovoid, oblong; style 0.8 mm long, dilated at base, uniformly flat, fimbriate; fimbri acuminate, base broad; rarely entire; stigma 2; 0.2 mm long, glabrous, narrowed towards the apex, acute. Nut 2.0 mm long, 1.0 mm broad, stramineous or brownish, broadly obovoid, sub–orbicular, biconvex, shortly stipitate, with 4–7 transverse wavy ridges.

Flowering & Fruiting

August January; Tarana, Sarnath, Chaukhandi, BHU, Jaunpur, Chakiya, Chandauli, Sakaldina, Sadalpura, Padrauna, Salempur, Ballia.

Habitat

Very common in moist places, low lands, gardens, along road side, banks of rivers.

Distribution

India, South & South East Asia, (Sri Lanka, China, Malesia), Tropical Australia.

2. *F. adenolepis*

F. adenolepis Kern in Blumea 8: 123. f. 3. 1955; Kern in Blumea 15: 433. 1967; Kern in Van Steenis, Fl. Malesiana. ser. 1.7 (3): 566. 1974. (PLATE–33).

Annual. Rhizome absent; root fibrous. Stem 10 cm long, 1.0 mm broad, green, glabrous, densely tufted, erect, slender, delicate. Leaves 4.0 cm long, 0.5 mm broad, green, coriaceous, filiform, glabrous, margin scabrid, acute, clustered at base, suberect; leaf–sheath 1.0 cm long, 2.0 mm broad, membranous, loose, brownish; ligule absent. Inflorescence anthelate; anthela 1.0 cm long, 1.0 cm broad, simple–sub compound. Bracts 3, 1.0 cm long, 0.5 mm broad, green, filiform, surface glabrous, margin scabrid, sharply acute; rays 2, 2.0 mm long, filiform, smooth. Spikelets 2.0 mm long, 1.0 mm broad, solitary, ovate, lanceolate, acute, 9 in number. Glume 1.5 mm long, 1.0 mm broad, membranous, brown, puncticulate within, with keel ending below the apex. Stamens 2; anther 0.5 mm long, yellow, oblong, glabrous; filament 0.2 mm long, slender, glabrous, shorter than anther. Pistil 1.5 mm long; ovary 0.2 mm long, elliptic, glabrous; style 0.5 mm long, flat, glabrous, dilated at base; stigma 3, 0.8 mm long, brown, hairy; hairs acute, base broad. Nut 1.0 mm long, 0.5 mm broad, obtusely trigonous, subcylindric, verruclose.

Flowering & Fruiting

October–November; Barahikala, Mau, Sonbhadra, Basti, Chunar.

Habitat

Rare in wet open grasslands, along the water course.

Distribution

India, Malesiya, SriLanka.

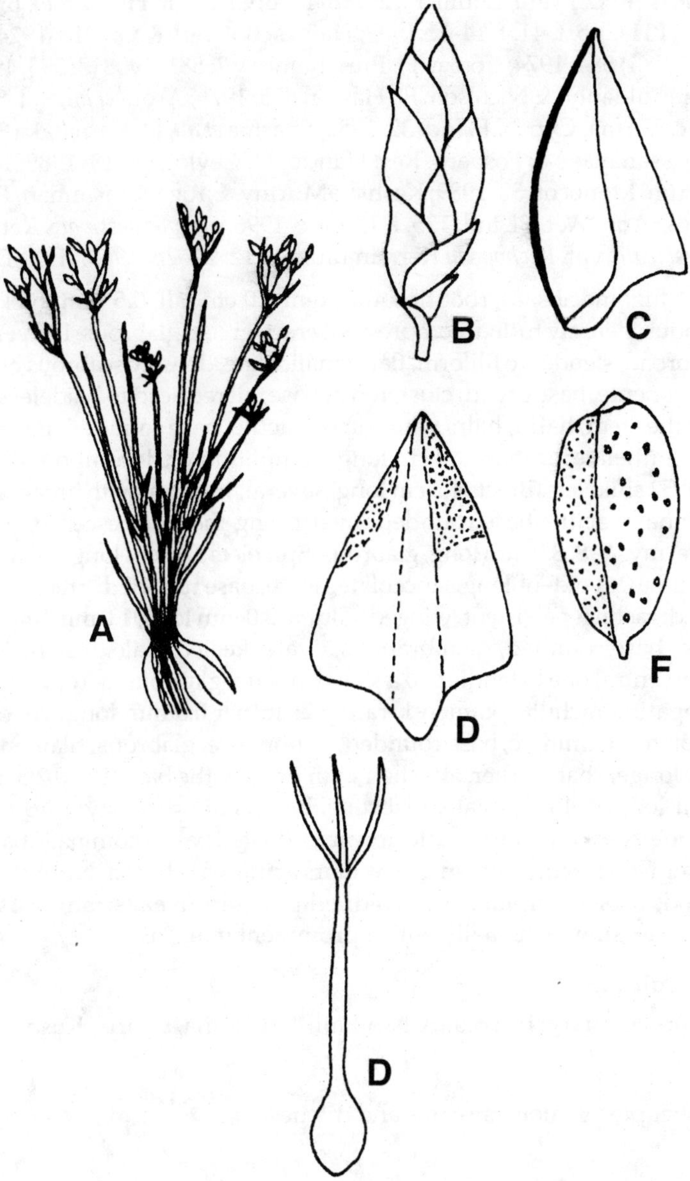

PLATE 33: *Fimbristylis adenolepis* Kern
A. Habit, B. Spikelet, C&D. Glume, E. Pistil, F. Nut

3. *F. aestivalis*

F. aestivalis (Retz.) Vahl, Enum. Pl. 2: 288. 1806; Clarke in Hook. f. Fl. Brit. Ind. 6: 637. 1893. et in I 11 Cyp. t. 41. f. 14–15 1909; Haines, Bot. Bih. & Or. 910. 1924; Kern in Fl. Males ser. 1.7 (3) 584. 1974; Cooke, Fl. Pres. Bombay 2: 881. 1908 (3: 395. 1958, repr. ed.); Hooper in Saldanha & Nicolson, Fl. Hassan 675. 1976; Arora *et al.*, Bot. S. Kanara 61. 1981; Rao & Verma, Cyp. N.E. Ind. 32. 1982; Sharma *et al.*, Fl. Karnataka 308. 1984; Koyama in Dassanayake & Fosberg, Rev. Handb. Fl. Ceylon 5: 313. 1985; Karthik *et al.*, Fl. Ind. Enum. Monocot. 50. 1989; KeshavaMurthy & Yoganarasimhan, Fl. Coorg. 506. 1990; Cook, Aqu. Wet. Pl. Ind. 135. f. 130. a-c. 1996. *Scirpus aestivalis* Retz. Obs. 4: 12. 1768. *S. aestivalis* var. *trichopoda* Kern in Blumea 12: 27–28. 1963. (PLATE–34).

Annual. Rhizome absent; root fibrous. Stem 5.0 cm tall, 0.5 mm thick slender, ribbed, trigonous, densely tufted, compressed, erect, green, glabrous. Leaves 3, 14 cm long, 1.0 mm broad, slender to filiform, flat–canaliculate, densely scabrous on margin, green, glabrous acute, base broad, clustered at base, all reduced to bladeless sheaths, subequalling the stem, hairy; hairs white, short, acute, base rounded; ligule absent. Inflorescence anthelate; anthela 2.0 cm long, compound or decompound–corymb, lax, bearing 8–52 spikelets. Bracts 1.5 cm long, several, filiform, with broad & usually hairy at base; hairs acute, base rounded; overtoping the inflorescence, with stout bristles at base; rays 5–8, 8.0 mm long, glabrous. Spikelets 3.0 mm long, 1.0 mm broad, solitary, polygonal, ovoid–oblong, lanceolate, acute, base rounded, greenish–brown, many flowered; rachilla persistent, winged. Glume 2.0 mm long, 1.0 mm broad, spiral, aristate, acute, base rounded, membranous, ovate, keeled, pale brown, 4–nerved; nerves visible, central one extending to a sharp mucuo, glabrous, acropetably falling from the elongating rachilla. Stamen 1, rarely 2; anther 0.5 mm long, yellow, ovate obong, lanceolate, acuminate, base rounded, gynophore, glabrous; filament 1.0 mm long, slender, longer than anther, attached with pistil to the base. Pistil 2.5 mm long; ovary 0.5 mm long, oblong, ovate, glabrous, base acuminate, style 1.5 mm long, uniformly slender, deciduous, fimbriate upwards, dilated base; acuminate, base broad; stigma 2; 0.5 mm long, acute, brown, hairy; hairs acute, base broad. Nut 1.0 mm long, 0.5 mm broad, shortly stipitate, obovoid, elliptic, biconvex, stramineus–brown, smooth, apiculate, shining, usually with a prominent margins.

Flowering & Fruiting

September–February; Harraua, Chandauli, Sakaldiha, Deoria, Kushinagar.

Habitat

In open wet places, along streams and ditches.

Distribution

India, Sri–Lanka, China, Indo–China, Thailand, Malesia, Japan, Amurland, Tropical Australia.

4. *F. albo-viridis*

F. albo-viridis Clarke in Hook. f. Fl. Brit. Ind. 6: 638. 1893; Kern in Fl. Males. ser. 1, 7 (3): 580. f. 46. 1974; Kern in Blumea 8: 140. 1995; Hooper in Saldanha & Nicolson, Fl. Hassan 675. 1976; Rao & Verma, Cyp. N.E. Ind. 35. 1982; Sharma *et al.*, F. Karnataka

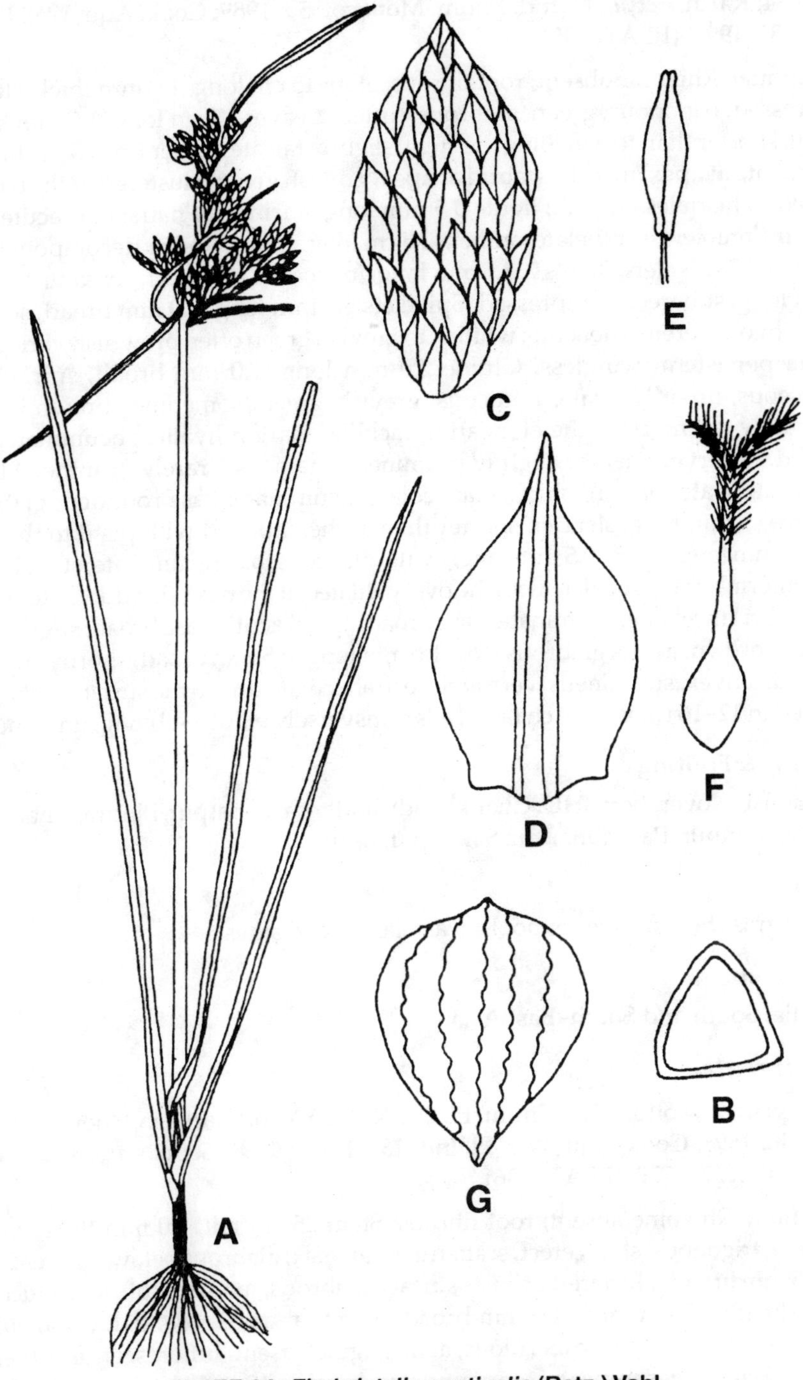

PLATE 34: ***Fimbristylis aestivalis*** **(Retz.) Vahl**
A. Habit, B. T.S. of stem, C. Spikelet, D. Glume, E. Stamen, F. Pistil, G. Nut

309. 1984; Karthik *et al.*, Fl. Ind. Enum. Monocot. 50. 1989; Cook, Aqu. Wet. Pl. Ind. 136. f. 131. 1996. (PLATE–35).

Annual. Rhizome absent; root fibrous. Stem 15 cm long, 1.0 mm thick, slender, compressed, trigonous, green, glabrous, tufted. Leaves 13 cm long, 1.5 mm broad, several, shorter than to equalling the stem, slightly falcate, rather firm, flat, obtuse to rather acute at apex, margin scabrid at apex, greyish green, clustered at the base, all reduced to bladeless sheath; ligule 0.5 mm long, a fringe of hairs; hair acute, base broad. Inflorescence anthelate; anthela 2.5 cm long, compound or decompound, lax, bearing 5–27 spikelets. Bracts 2, 3.5 mm long, foliaceous, suberect, spreading; rays 5, 12 mm long, suberect, compressed. Spikelets 5.0 mm long, 2.0 mm broad, solitary, oblong, ovoid, terete, subacute, whitish to greyish green often brownish variegated; rachilla persistent, wingless. Glume 2.0 mm long, 1.0 mm broad, spiral, sub–chartaceous, broadly ovate, muticous, greyish–green, sometimes brown, tinged, acropetably falling from the elongating rachilla, margin hyaline, acuminate, base rounded, 3–nerved; nerves slightly prominent. Stamens 1 rarely 2; anther 1.0 mm long, crest, ovate, brown, oblong–lanceolate, acuminate, base rounded, glabrous; filament 0.5 mm long, slender, shorter than anther, attched with pistil to the base. Pistil 2.5 mm long; ovary 0.5 mm long, ovate, glabrous; base acuminate; style 1.0 mm long, uniformly flat, articulate with the ovary, dilated at the base, deciduous, fimbriate; basal fimbri lanceolate, acuminate, base broad; apical acute, base broad; stigma 2; 1.0 mm long, brown, acute, glabrous. Nut 1.0 mm long, 0.5 mm broad, shortly stipitate, obvoid, biconvex, stramineus, verruculose, trabeculate due to the surface cells being arranged in 12–16 rows on each face, cells transversely elliptic–oblong, umbonulate.

Flowering & Fruiting

August–November; BHU, Chaukhandi Sindhora, Babatpur, Pindra, Chandauli, Chakiya, Jaunpur, Padrauna, Lar, Salempur, Basti.

Habitat

Dry grass fields, edges of pools, wasteland, road sides.

Distribution

India, South and South–East Asia.

5. *F. argentea*

F. argentea (Rottb.) Vahl Enum. Pl. 2: 294. 1805; C.B. Clarke in Hook. f. Fl. Brit. Ind. 6: 640. 1893; Cook, Aqu. Wet. Pl. Ind. 137. f. 133. 1996. *Scirpus argenteus* Rottb. Descr. Pl. Rar. 27. 1772. (PLATE–36).

Annual. Rhizome absent; root fibrous. Stem 25 cm tall, 2.0 mm thick, green, caespitose, trigonous, stout, erect, scaberulous at apex, glabrous below. Leaves 20 cm long, 1.5 mm broad, clustered at base, surface glabrous, margin scabrid, reduced to sheath; sheath 2.5 cm long, 4.0 mm broad, membranous, loose, flat, canaliculate, surface glabrous, margin scaberulous, acute; ligule absent. Inflorescence anthelate; anthela 3.0 cm long, simple or compound, lax, bearing 10 spikelet. Bracts 3, 8.0 mm long, 0.5 mm broad, acute, green, surface glabrous, margin scabrid, mid vein distinct

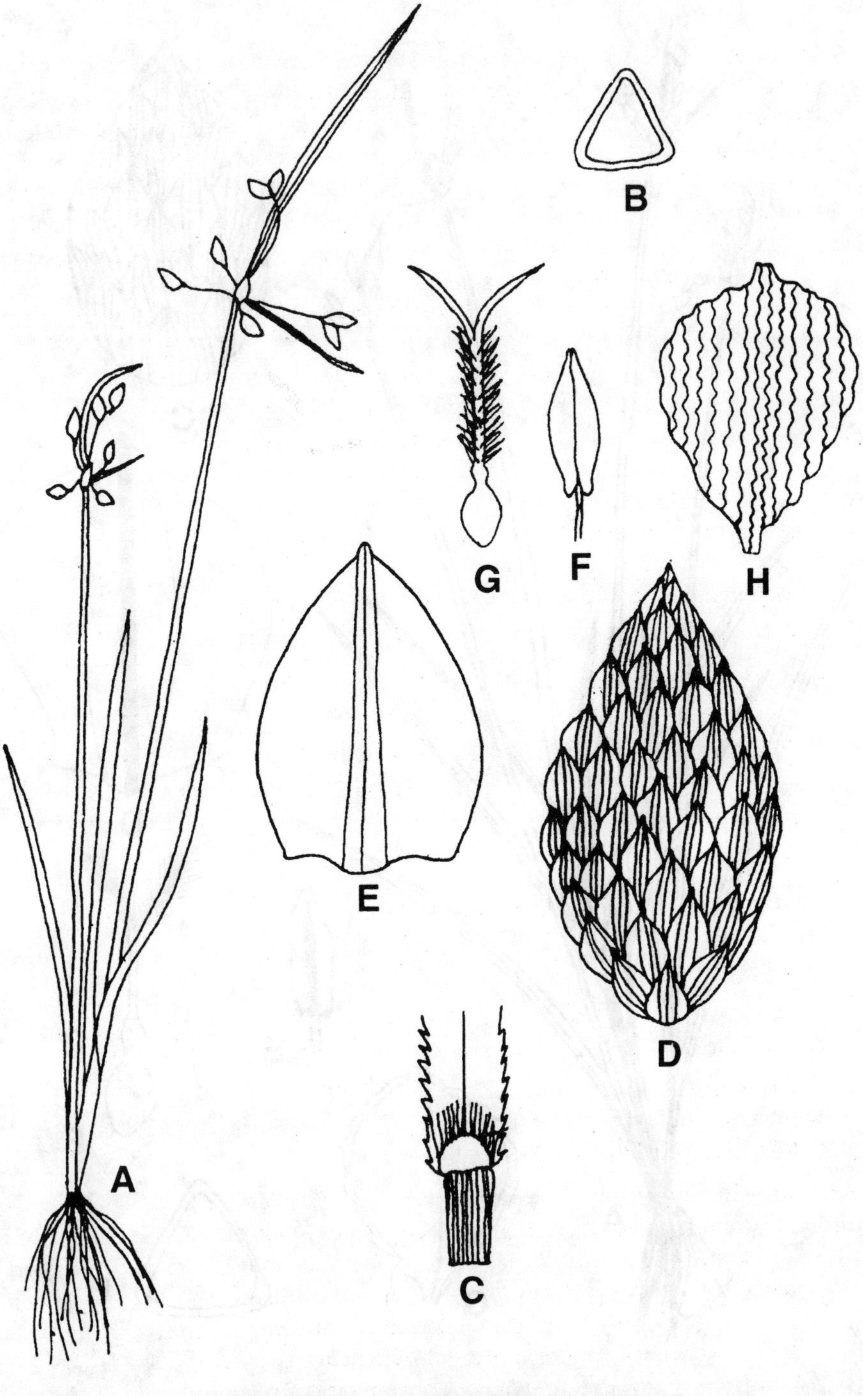

PLATE 35: ***Fimbristylis albo-viridis*** **Clarke**

A. Habit, B. T.S. of stem, C. Ligule, D. Spikelet, E. Glume, F. Stamen, G. Pistil, H. Nut

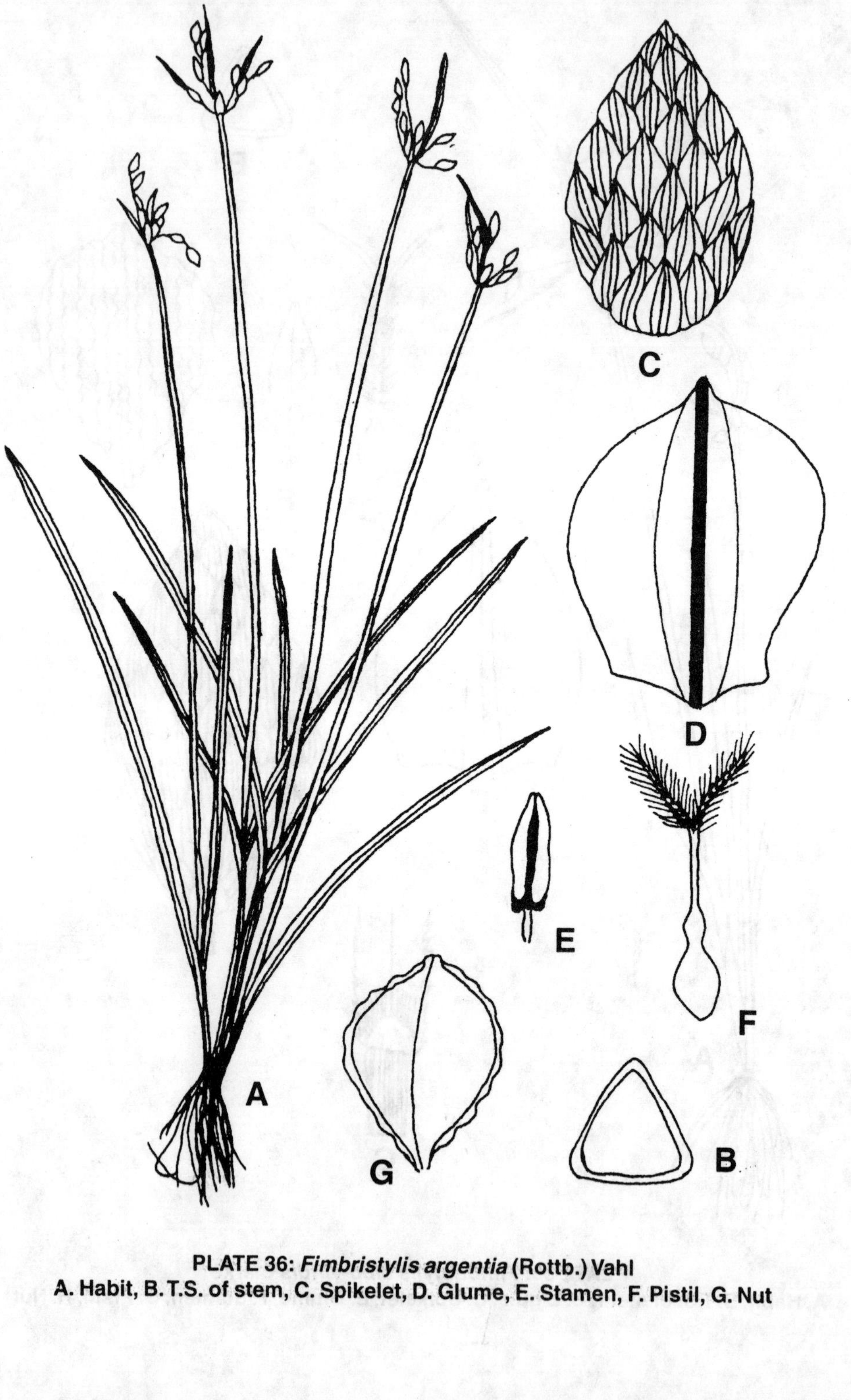

PLATE 36: *Fimbristylis argentia* (Rottb.) Vahl
A. Habit, B. T.S. of stem, C. Spikelet, D. Glume, E. Stamen, F. Pistil, G. Nut

on dorsal surface; rays 6; 1.0 cm long, ascendig, green, glabrous, acute. Spikelet 4.0 mm long, 1.5 mm broad, solitary, sub globose, terete, obtuse, numerous flowered; rachilla 1.0 mm broad, winged, wing glabrous. Glume 2.0 mm long, 2.0 mm broad, membranous, spiral, ovate, muticous, surface glabrous, margin scarious, green at young, brown with age, sometimes the upper glume sterile and much elongated, 3–nerved; nerves green, central one ending somewhat below the tip, & visible, thicker than lateral nerves. Stamens 1 rarely 2; anther 0.5 mm long, yellow, glabrous, oblong, linear, base colourless; filament 0.2 mm long, uniformly slender, glabrous, shorter than anther. Pistil 2.0 mm long; ovary, 0.5 mm long, oblong, glabrous; style 1.0 mm long, dilated at base, glabrous, flat; stigma 2; 0.5 mm long, acuminate, brown, hairy; hairs acuminate, base broad. Nut 1.0 mm long, 0.5 mm broad, subsessile, obovoid, trigonous, or somewhat lenticular, verruculose, apiculate, brown, surface cells in 12–15 vertical rows on each face; margin hyaline.

Flowering & Fruiting

June–September; UPC, BHU, Babatpur, Chakiya, Niyamatabad, Sultanpur.

Habitat

Common along moist habitats.

Distribution

India, Malaysia, Sri Lanka.

6. *F. bisumbellata*

F. bisumbellata (Forsk.) Bubani in Dodec. 30. 1850; Fischer in Gamble, Fl. Pres. Madras 1898. 1931 (repr. ed. 3: 1151. 1957); Ramaswamy & Razi, Fl. Bangalore 104. 1973; Kern in van Steenis, Fl. Males 1, 7: 579. 1974; Hooper in Saldanha & Nicolson. Fl. Hassan 676. 1976; Yoganarasimhan *et al.*, Fl. Chickmagalur 361. 1981; Rao & Razi, Fl. Mysore 562. 1981; Rao & Verma, Cyp. N.E. India 31. 1982; Sharma *et al.*, Fl. Karnataka 309. 1984; Koyama in Dassanayake & Fosberg. Rev. Handb. Fl. Ceylon 5: 312. 1985; Singh, Fl. E. Karnataka 635. 1988; Karthik *et al.*, Fl. Ind. Enum. Monocot. 51. 1989; KeshavaMurthy & Yoganarasimhan, Fl. Coorg. 507. 1990. *Scirpus bisumbellatus* Forsk. Fl. Aeg.-Arab. 1: 15. 1775. *Fimbristylis dichotoma* Vahl Enum. Pl. 2: 287. 1806; Clarke in Hook f. Fl. Brit. Ind. 6: 635; 1993; Kunth. Enum Pl. 2: 225. 1837; Cooke, Fl. Pres. Bombay 2: 880. 1908 (3: 394. 1958. repr. ed.). *Scirpus dichotomus* acut. non. L. 1753; Rottb., Descr. & Ic 57, t. 13, f. 1. 1786. *Scirpus palliscens* Roxb. Fl. Ind. 1: 229. 1820. (PLATE–37).

Annual. Rhizome absent; root fibrous. Stem 25 cm tall, 1.0 mm thick, slender, compressed, trigonous, erect, tufted, green, glabrous. Leaves 24 cm long, 1.5 mm thick, several, flat, falcate, linear, slightly shorter than to sometimes subequalling to stem, glabrous above, scabrous below, obtuse–abruptly acuminate at apex, pubesent beneath, green, sheath 3.0 mm long, margin hairy, membranous; hairs acute base broad; ligule 1.0 mm long, membranous, tip hairy; hairs acute, base broad. Inflorescence anthelate; anthela compound to supradecompound, lax to rather dense bearing few to numerous loose spikelets. Bracts 5, 8.0 cm long, overtopping the inflorencence, foliaceous, as long as to exceeding the anthela; rays 15, 3.0 cm long, slender, sub–erect, spreading. Spikelet 4.0 mm long, 2.0 mm broad, solitary, polygonal,

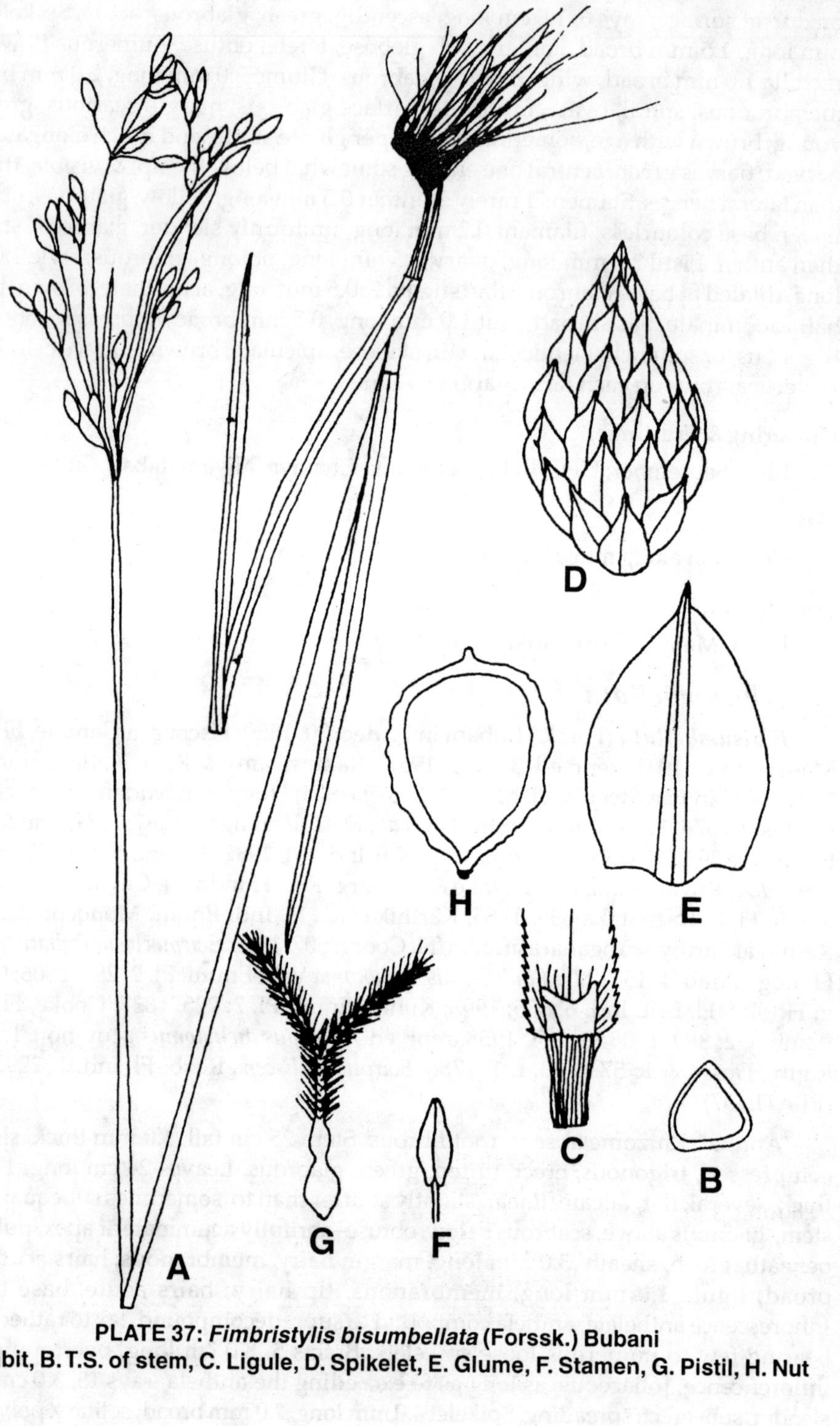

PLATE 37: *Fimbristylis bisumbellata* (Forssk.) Bubani
A. Habit, B. T.S. of stem, C. Ligule, D. Spikelet, E. Glume, F. Stamen, G. Pistil, H. Nut

narrowly oblong–ovoid, ellipsoid, angled due to strongly keeled glume, lanceolate; rachilla persistent, winged. Glume 1.5 mm long, 1.0 mm broad, spiral, membranous broadly ovate, sharply keeled, mucronate, yellowish brown, acuminate, base rounded, 3–nerved; nerves visible, margin purplish about half the length of glume, acropetally falling from elongating the rachilla. Stamens 2; anther 0.5 mm long, brown, gynophore, apiculate, base rounded, ovate–oblong, lanceolate, glabrous; filament 1.0 mm long, slender, longer than anther, attached with the pistil to the base. Pistil 2.0 mm long; ovary 0.3 mm long, oblong–acute, glabrous, base acuminate; style 0.7 mm long, slightly flattened, articulate with the ovary, dilated at the base, usually fimbriate upwards, fimbri acute, base broad; stigma 2, 1.0 mm long, acuminate, recurved in later stage, brown, hairy; hairs acuminate, base broad. Nut 1.0 mm long, 0.5 mm broad, shortly stipitate, biconvex, broadly obovoid, umbonulate, whitish–stramineus, smooth or verruculose, trabeculate by the surface cells being arranged in 5–9 vertical rows on each rows.

Flowering & Fruiting

February–July; Rameshwar, Lohata, Chaukhandi, Harrua, Babatpur, Chandauli, Jaunpur, Deoria, Basti, Azamgarh, Allahabad, Mirzapur.

Habitat

Frequent in moist sandy loam along road side and river banks.

Distribution

Widely distributed in the old world tropics, extending to the mediterannian region of Europe, India.

7. *F. complanata*

F. complanata (Retz.) Link., Hort. Berol 1: 292. 1827; C.B. clarke in Hook. f. Fl. Brit. Ind. 6: 646. 1893; Haines, Bot. Bih.& Or. 916. 1924; Kern in Fl. Males. ser. 1, 7 (3): 548. 1974; Cooke. Fl. Pres. Bombay 2: 885. 1908 (3: 400. 1958, repr. ed.); Fischer in Gamble, Fl. Pres. Madras, 1659. 1931 (3: 1151. 1957, repr. ed); Ramaswamy & Razi, Fl. Bangalore 105. 1973; Hooper in Saldanha & Nicolson Fl. Hassan 677. 1976; Rao & Razi, Fl. Mysore 562.1981; Rao & Verma, Cyp. N.E. India 37. 1982; Sharma *et al.*, Fl. Karnataka 309. 1984; Koyama in Dassanayake & Fosberg, Rev. Handb. Fl. Ceylon 5: 284. 1985; Karthik *et al.*, Fl. Ind. Enum. Monocot. 51. 1989; KeshavaMurthy & Yoganarasimhan, Fl. Coorg. 507. 1990; Cook, Aqu. Wet. Pl. Ind. 138. f. 136 a–c. 1996. *Scirpus complanatus* Retz. Obs. 5: 14. 1789. (PLATE–38).

Annual. Rhizome absent; root fibrous. Stem 75 cm long, 1.5 mm thick, strongly compressed, smooth, sometimes scabrid & winged upwards, trigonous towards base, flattened and ancipitous towards upper parts, erect, green, tufted. Leaves 13 cm long, 3.0 mm broad, shorter than to subequalling the stem, clustered at base, abruptly narrowed into a triangular tip, flat, margin thickened, densely scabrous towards apex, green, glabrous; sheath reduced; ligule 1.0 mm long, a fringe of hairs; hairs acute, base broad. Inflorescence anthelate; anthela 4.0 cm long, 1.0 cm broad, compound to supradecompound, lax, bearing 12–90 spikelets. Bracts 3.0 cm long, erect, foliaceous, shorter than inflorescence, acuminate at apex, densely scabrid on

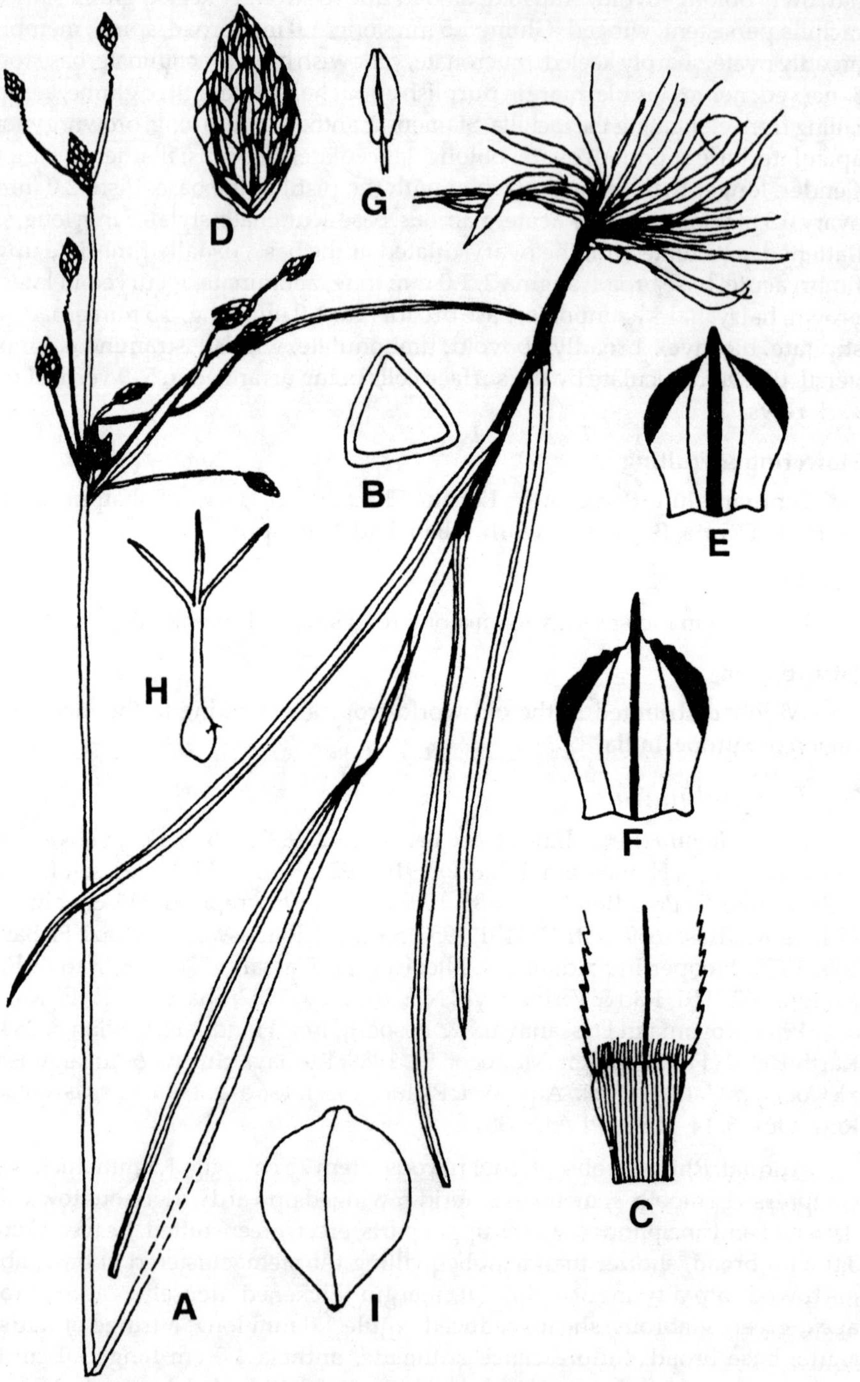

PLATE 38: *Fimbristylis complanata* (Retz.) Link.
A. Habit, B. T.S. of stem, C. Ligule, D. Spikelet, E & F. Glume, G. Stamen, H. Pistil, I. Nut

margin, rays 6; 4.0 cm long, flattened, spreading, compressed, sulcate, scabrous. Spikelet 5.0 mm long, 2.0 mm broad, solitary, ovoid, oblong–lanceolate, polygonal, brownish, obtuse; rachilla persistent, winged. Glume 2.0 mm long, 1.5 mm broad, spiral, acute, ovate, oblong, keeled, sides deep brown, membranous, spirally arranged, hyaline towards the margin, keel green, 3 nerved; nerves visible; mucronate, mucro 0.5 mm long, slightly curved; upper mucticous. Stamens 3; anther 1.0 mm long, yellow, acuminate, base rounded, glabrous, oblong, lanceolate, gynophore; filament 0.5 mm long, slender, longer than anther, attached with pistil to the base. Pistil 4.0 mm long; ovary 0.5 mm long, oblong, ovate, glabrous, base acuminate; style 2.0 mm long, glabrous, deciduous, triquetrous uniformly flat, pyramidally thickened at base; stigma 3; 1.5 mm long, glabrous. Nut 1.0 mm long, 0.5 mm broad, subsessile, broadly obovid, trigonous, whitish–stramineus, verrucullose, shortly stipitate and minutely umbonulate, epidermal cells very minute, in many rows.

Flowering & Fruiting

June–September; Babatpur, Basti, Deoria, Sultanpur, Sonbhadra.

Habitat

In dry to swampy wastelands & open grasslands.

Distribution

India, very common in South–East Asia.

8. *F. dichotoma*

F. dichotoma (L.) Vahl, Enum. Pl. 2: 287. 1806; Haines, Bot. Bih. & Or. 921. 1924; Fischer in Gamble, Fl. Pres. Madras 1898. 1936 (3: 1151. 1957 repr. ed.); Ramaswamy & Razi, Fl. Bangalore 106. 1973; Kern in van Steenis, Fl. Males 1.7: 575. 1974; Hooper in Saldanha & Nicolson, Fl. Hassan 677. 1976; Rao & Razi, Fl. Mysore 562. 1981; Rao & Verma Cyp. N.E. Ind. 31. 1982; Sharma *et al.*, Fl. Karnataka 309. 1984; Koyama in Dassanayake & Fosberg, Rev. Handb. Fl. Ceylon 5: 306. 1985; Singh, Fl. E. Karnataka 2: 636. 1988; Karthik *et al.*, Fl. Ind. Enum. Monocot. 51. 1989; Keshava Murthy & Yoganarasimhan, Fl. Coorg 507. 1990. *Scirpus dichotomus* L. Sp. Pl. 1: 50. 1753. *S. diphyllus* Retz., Obs. Bot. 5: 15. 1789. *Fimbristylis diphylla* (Retz.) Vahl, Enum. Pl. 2: 289. 1805; Clarke in Hook. f. Fl. Brit. Ind. 6: 636. 1884; Cooke, Fl. Pres. Bombay 2: 882. 1908 (3: 396. 1958, repr. ed.). *F. annua* var. *diphylla* Kukenth., Act. Hort. Got. 5: 109. 1929. *F. annua* var. *paucispiculata* Blatt. & Mc C. in J. Bomb. nat. Hist. Soc. 37: 544. 1934; Sharma et al. Fl. Karnataka 309. 1984. (PLATE–39).

Annual. Rhizomatous; rhizome 1.0 cm long, woody; root fibrous. Stem 30 cm tall, 1.5 mm thick, trigonous, striate, compressed below the anthella, tufted, erect, slender to stout, green, glabrous. Leaves 15 cm long, 3.0 mm broad, flat, linear, acute to obtuse at apex, sparsely scabrous–glabrous, pubescent beneath, sometimes glaucescent, clustered at the base, green; sheath 6.0 cm long, 2.0 mm broad, glabrous–densely puberculous; ligule 0.3 mm long, a fringe of hairs; hair acute, base rounded. Inflorescence anthelate; anthela 4.0 cm long, 2.0 cm broad, simple, compound to supredecompound–corymb, dense, lax, bearing few to numerous spikelets. Bracts 2–4, 3.0 cm long, shorter than anthela, sometime exceeding the anthela, foliaceous,

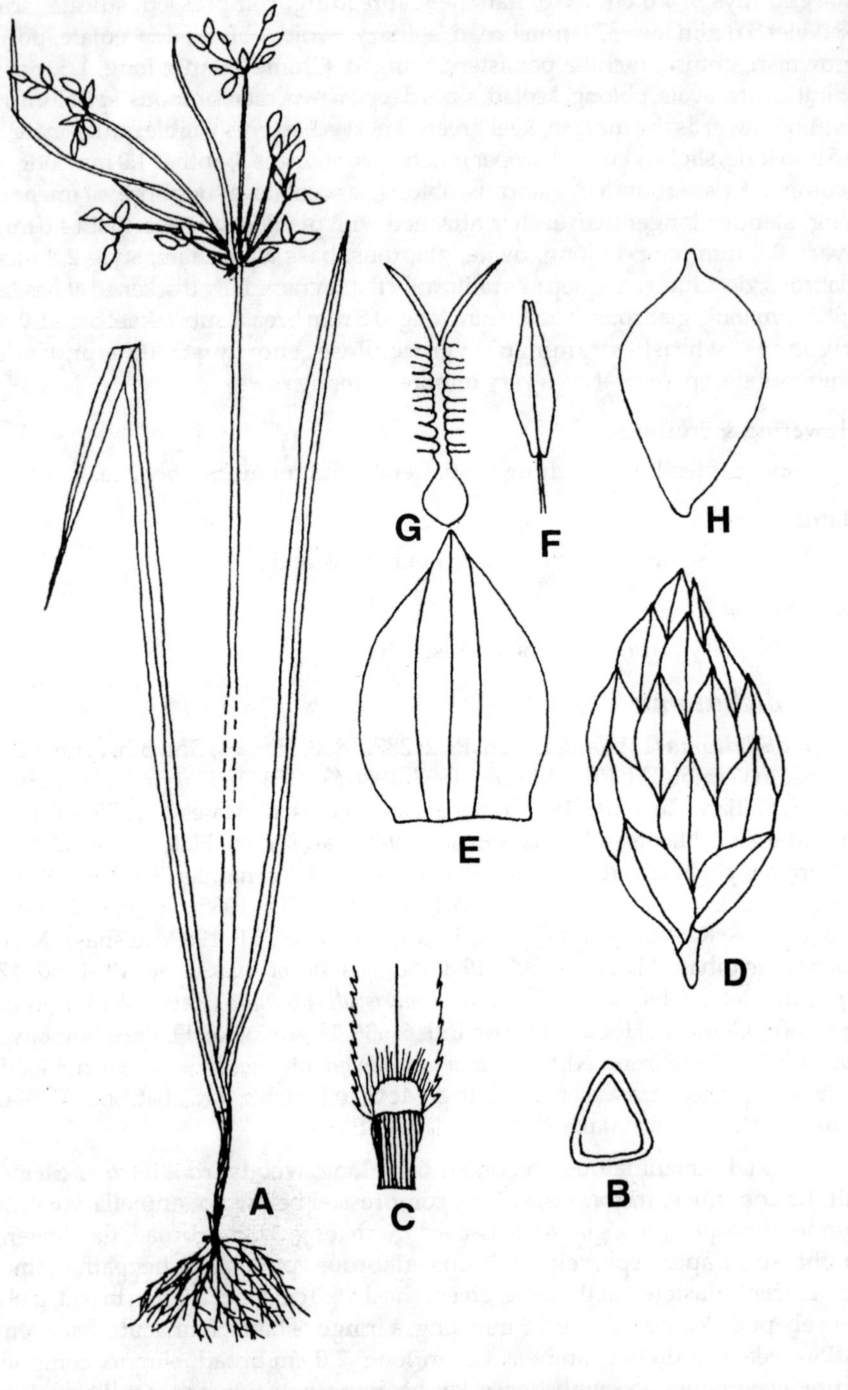

PLATE 39: *Fimbristylis dichotoma* (L.) Vahl
A. Habit, B. T.S. of stem, C. Ligule, D. Spikelet, E. Glume, F. Stamen, G. Pistil, H. Nut

pilose at base, pubsescent upwards. Rays 6.0 cm long angled, striate. Spikelet 4.0 mm long, 2.0 mm broad, ellipsoid–ovoid, brown, globose, terete, acute, many flowered; rachilla winged, persistent. Glume 1.5 mm long, 1.0 mm broad, chartaceous, spiral, ovate, oblong, acute, keeled; keel protuding above the apex as mucro, glossy, chest nut brown, 3–nerved; nerves prominent, acropetably falling from the elongating rachilla. Stamens 1, rarely 2, 3; anther 1.0 mm long, yellowish–green, oblong, ovate, lanceolate, acuminate, base rounded; filament 0.5 mm long, slender, shorter than anther, attached with pistil to the base. Pistil 4.5 mm long; ovary 1.0 mm long, ovate, oblong, glabrous stipitate; style 2.5 mm long, uniformly flat, deciduous, dilated at base, fimbriate throughout; fimbri acute, base broad; stigma 2, 1.0 mm long acute, glabrous, brown. Nut 1.0 mm long, 0.5 mm broad, shortly stipitate, obovoid, biconvex, apiculate, pale stramineus, ultimately brownish, smooth or sometime verruculose, trabeculate due to the surface cells being arranged in 5–10 vertical rows on each face.

Flowering & Fruiting

March–December; UPC, BHU, Tarana, Rajghat, Sindhora, Babatpur, Cholapur, Chandauli, Chakiya, Sadalpura, Sakaldiha, Jaunpur, Ballia, Mau, Mirzapur.

Habitat

Widely spread in open to shade, dry to swampy places.

Distribution

Very common in South–East Asia.

9. *F. dipsacea*

F. dipsacea (Rottb.) Clarke in Hook. f. Fl. Brit. Ind. 6: 635. 1893. *Scirpus dipsaceus* Rottb. Decr. et Incon. 56. t. 12. f.l. 1773. *Isolepis dipsacea* R. & S. Syst. 2: 119. 1817. (PLATE–40).

Annual. Rhizome absent; root fibrous. Stem 15 cm long, 1.0 mm thick, green glabrous, erect, densely tufted, slender. Leaves 18 cm long, 0.5 mm broad, green, glabrous, capillary, longer than the stem; sheath reduced; ligule absent. Inflorescence anthellate; anthela 4.0 cm long, simple, compound bearing several spikelets. Bracts 8.0 cm long, 0.5 mm broad, green, glabrous, longer than the anthela, foliaceous; ray absent. Spikelets 7.0 mm long, 3.0 mm broad, solitary, oblong, sub–globose, terete, squarrose. Glume 1.0 mm long, 1.0 mm broad, membranous, spiral, oblong, lanceolate, aristate, stramineus. Stamens 2; anther 0.5 mm long, yellow, apiculate; filament 0.2 mm long, slender, glabrous, shorter than anther. Pistil 1.5 mm long; ovary 0.3 mm long, glabrous; style 0.5 mm long, smooth with bulbous base; stigma 2, 0.7 mm long, glabrous, Nut 1.0 mm long, 0.5 mm broad, subsessile, linear, oblong, with almost parallel sides & two rows of glands when young, biconvex, faintly transversely lanceolate, brownish.

Flowring & Fruiting

July–October; Chakiya, Chandauli, Salempur, Ballia, Mardah.

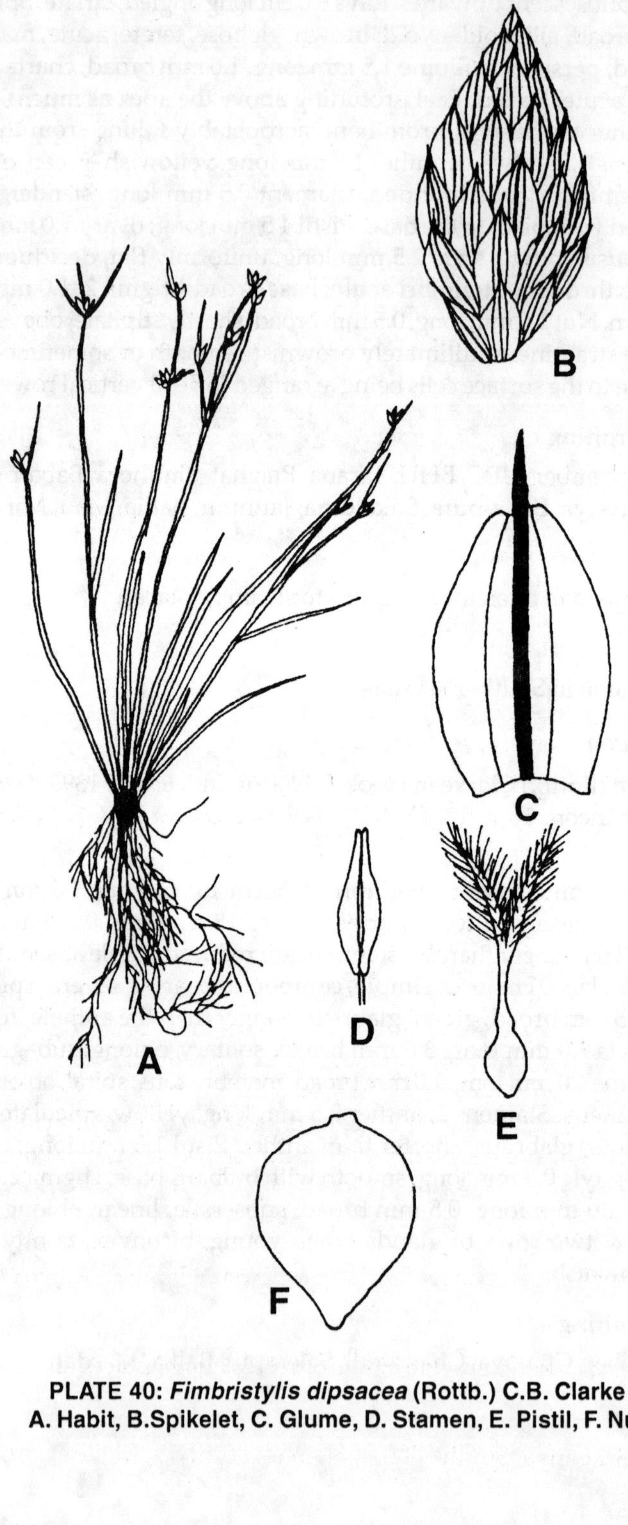

PLATE 40: *Fimbristylis dipsacea* (Rottb.) C.B. Clarke
A. Habit, B.Spikelet, C. Glume, D. Stamen, E. Pistil, F. Nut

Habitat

Common along the rocky places.

Distribution

India, Newzealand, WestIndies.

10. *F. eragrostis*

F. eragrostis (Nees). Hance in J. Linn. Soc. London (Bot.) 13: 132. 1873; Kern in Steenis, Fl. Males. ser. I. 7: 567. 1974; Rao & Verma, Cyp. N.E. Ind. 38. 1982; Koyama in Dassanayake & Fosberg, Rev. Handb. Fl. Ceylon 5: 275. 1985; Karthik *et al.*, Fl. Ind. Enum. Monocot. 52. 1989. *Abildgaardia eragrostis* Nees in Wt. Contr. Bot. Ind. 95. 1834. *F. nigrobrunnea* Thw. Enum. Pl. Zeyl. 434. 1864; C.B. Clarke in Hook. f. Fl. Brit. Ind. 6: 648. 1893; Fischer in Gamble, Fl. Pres. Madras 1659. 1931 (3: 1151. 1957, repr. ed.). *F. eragrostis* var. *nigrobrunnea* (Thw.) Hooper in Saldhana & Nicolson, Fl. Hasan 678. 1976; Sharma *et al.*, Fl. Karnataka 309. 1984. (PLATE–41).

Perennial. Rhizomatous; rhizome 2.0 cm long, creeping, woody, often clothed with old decayed leaf–sheaths; root fibrous. Stem 30 cm long, 3.0 mm wide, solitary at times, tufted, sulcate, tetragonous, scaberulous at tip, glabrous below, compressed & often ancipitous below the inflorescence, green, erect. Leaves 15 cm long, 1.5 mm broad, shorter than stem, flat, stiff, usually subfalcate, linear, clustered at base, numerous, green, margin scaberulous, tip triangular, obtuse, subcoriaceous; sheath 2.0 cm long, 4.0 mm broad, striate, dilated towards the base, glabrous; ligule absent. Inflorescence anthelate; anthela 2.5 cm long, 1.5 cm wide, compound, simple, subcapitate, rarely decompound, bearing 10 spikelets. Bracts 4, 2.0 cm long, 0.5 mm broad, foliaceous, acute, margin not spreading, dilated base, scabrid; rays 5, 5.0 cm long, suberect, green, slightly compressed, striate, smooth. Spikelets 7.0 mm long, 2.0 mm broad, clusteres of 2–4, rarely solitary, ellipsoid, ovate–oblong, lanceolate, compressed– subterete, dark brown, often contorted, acute; rachilla 5.0 mm long, 1.0 mm broad, persistent, winged. Glumes 2.5 mm long, 1.5 mm broad, lower distichous, subdistichous, or spiral, upper spiral, membranous, shining, keeled, brown, ovate, obtuse, mucronulate, glabrous, 1–nerved; nerves green, excurrent. Stamens 2; anther 1.5 mm long, yellow, linear, glabrous, accuminate, base rounded, slightly hairy at apex; hairs acute, base broad; filament 0.5 mm long, uniformly slender, glabrous, shorter than anther. Pistil 2.0 mm long; ovary 0.5 mm long, keeled, oblong; style 0.5 mm long, triquetrous, pyramidal, thickened at base, uniformly slender, entire; stigma 3, 1.0 mm long, brown, hairy; hairs acute, base broad. Nut 1.0 mm long, 0.5 mm broad, brown, obovoid, trigonous, shortly stipitate, verrucose, umbonulate, epidermal cells minute, isodiametric, 13 in vertical rows.

Flowering & Fruiting

April–September; Babatpur, Harrua, Sarnath, Jaunpur, Chandauli, Basti.

Habitat

Grassy hillsides, along the banks of rivers, moist places along road sides.

Distribution

India, Sri Lanka, South–China, Hainan, Formosa, Malesia, Tropical Australia.

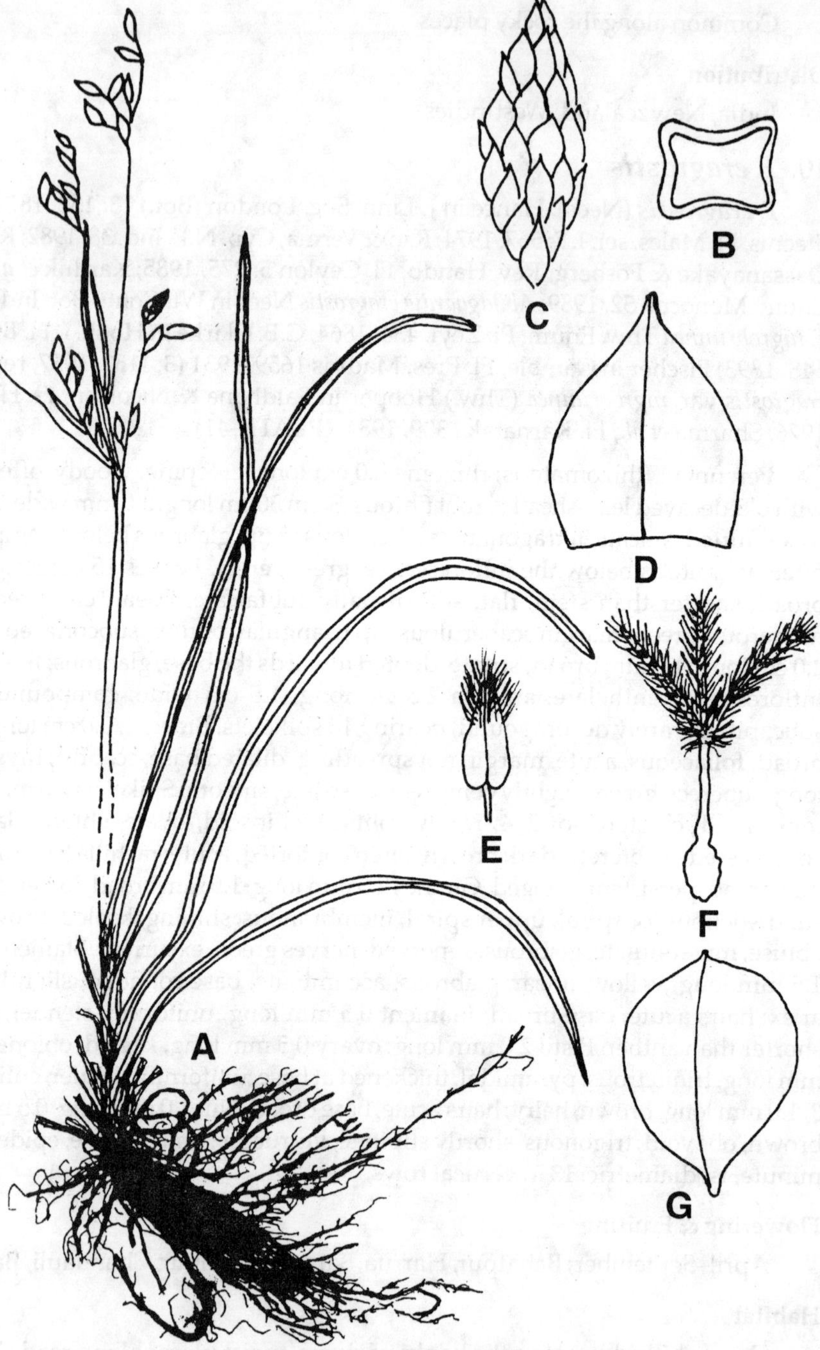

PLATE 41: ***Fimbristylis eragrostis*** **(Nees) Hance**
A. Habit, B. T.S. of stem, C. Spikelet, D. Glume, E. Stamen, F. Pistil, G. Nut

11. *F. falcata*

F. falcata (Vahl) Kunth. Enum. Pl. 2: 239. 1837; Verma & Chandra, Rec. Bot. Surv. Ind. 21 (2) : 255. 1981. *Scirpus falacatus* Vahl Enum. Pl. 2: 275. 1805. *Fimbristylis junciformis* Kunth. Enum. Pl. 2: 239. 1837; Clarke in Hook. f. Fl. Brit. Ind. 6: 647; 1893. (PLATE–42).

Perennial. Rhizomatous; rhizome 2.0 cm long, creeping; root fibrous. Stem 25 cm long, 1.5 mm thick, quadrangular, erect, compressed, tufted, green, glabrous. Leaves 15 cm long, 1.0 mm broad, 2 in number, green, glabrous, margin scabrid, clustered at base, reduced to blade less sheath, acute. Inflorescence anthelate; anthela 5.0 cm long, compound, decumpound or lax, bearing 25 spikelets. Bracts 3, 5.0 cm long, 1.0 mm broad, foliaceous, surface glabrous, margin scabrid; rays 1.5 cm long, slender. Spikelet 5.0 mm long, 1.0 mm broad, solitary, oblong–ovoid, polygonal, acute, base rounded, many–flowered; rachilla persistent, wingless, glabrous. Glume 1.5 mm long, 1.0 mm broad, spiral, broadly lanceolate, ovate, keeled, membranous, pale–brown mucronulate, 2–nerved; nerves prominent, aristate; arista straight, acropetabaly falling from elongating rachilla. Stamens 2; anther 1.0 mm long, crest–ovate, lanceolate, yellow, acuminate, base rounded, glabrous, gynophore; filament 0.5 mm long, slender, shorter thant anther, attached with pistil to the base. Pistil 2.0 mm long; ovary 0.5 mm long, ovate, glabrous; style 1.0 mm long, uniformaly slender, slightly flat, dilated at base, deciduous, fimbriate, fimbri acuminate, base broad; stigma 3; 0.5 mm long, acute, brown, glabrous. Nut 1.0 mm long, 0.5 mm broad, shortly stipitate, obovoid, biconvex, glossy, stramineus, smoot, blackish brown.

Flowering & Fruiting

May–July; Pindra, Jaunpur, Aurai, Chunar, Allahabad, Deoria.

Habitat

Common along moist places.

Distribution

India, Tropical Asia, Australia.

12. *F. ferruginea*

F. ferruginea (L.) Vahl Enum. Pl. 2: 291. 1805; C.B. Clarke in Hook. f. Fl. Brit. Ind. 6: 638. 1893. *Scirpus ferrugineus* L., Sp. Pl. 50. 1753. (PLATE–43).

Annuals. Rhizome absent; root fibrous black. Stem 50 cm long, 2.0 mm thick, trigonous, slenderly ribbed, green, glabrous. Leaves 23 cm long, 2.0 mm broad, several, green, surface glabrous, margin scaberulous, hairy; hair acute, base broad; mid–vein distinct on dorsal surface, clustered at base; sheath 2.0 cm long, 2.5 mm broad, membranous, brown, glabrous; ligule absent. Inflorescence anthelate; anthela 6.0 cm long, compound, rarely decompound, rarely simple in depapuerated plants, bearing 14 spikelets. Bracts 3, 15 cm long, 3.0 mm broad, foliaceous, green, margin scaberulous, surface glabrous, acuminate, midvein distinct on dorsal surface; rays 4; 4.0 cm long, 1.0 mm broad, rigid, sub–compressed, green, acute. Spikelet 5.0 mm long, 2.0 mm broad, usually in clusteres of 2–6 with a few solitary ones, rarely all solitary, ovate, linear–lanceolate; rachilla 3.5 mm long, 0.5 mm broad, wingless. Glume 3.5 mm long,

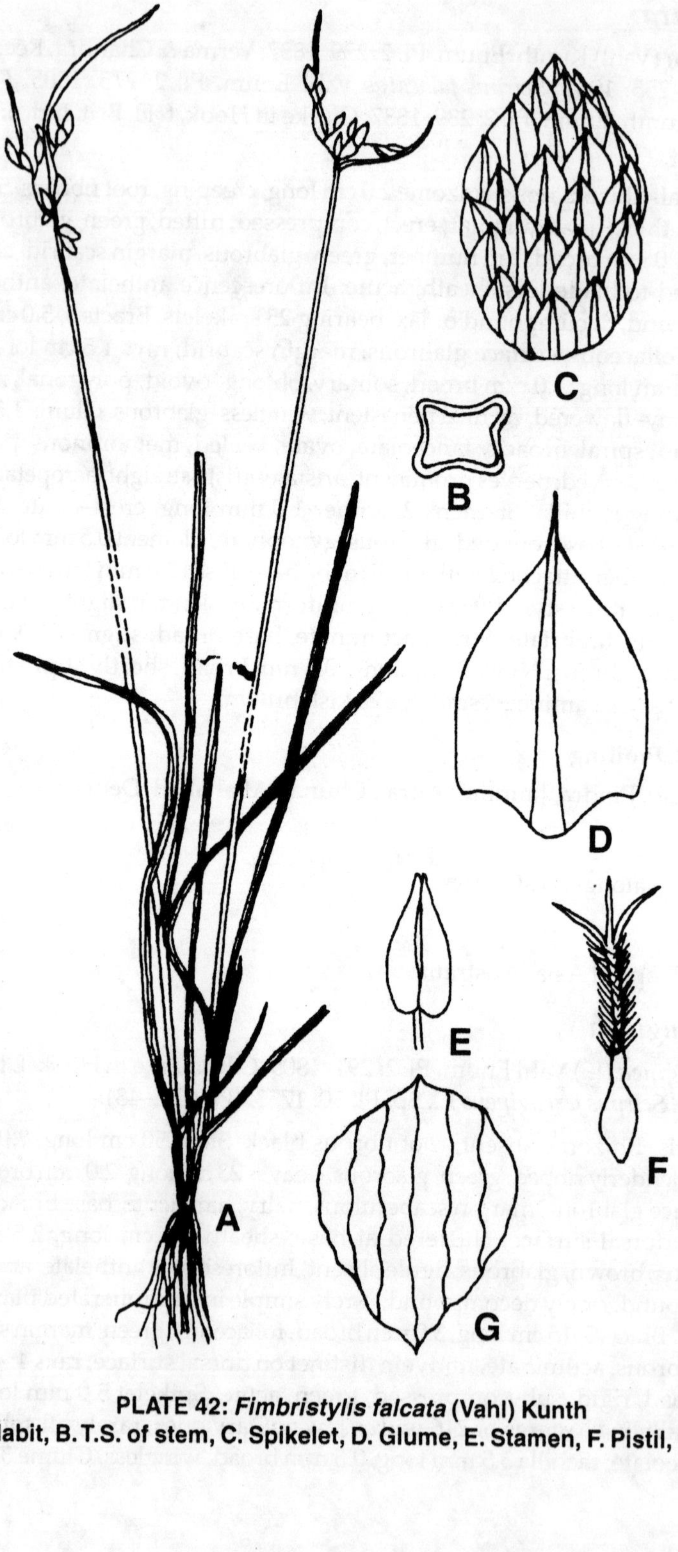

PLATE 42: *Fimbristylis falcata* (Vahl) Kunth
A. Habit, B. T.S. of stem, C. Spikelet, D. Glume, E. Stamen, F. Pistil, G. Nut

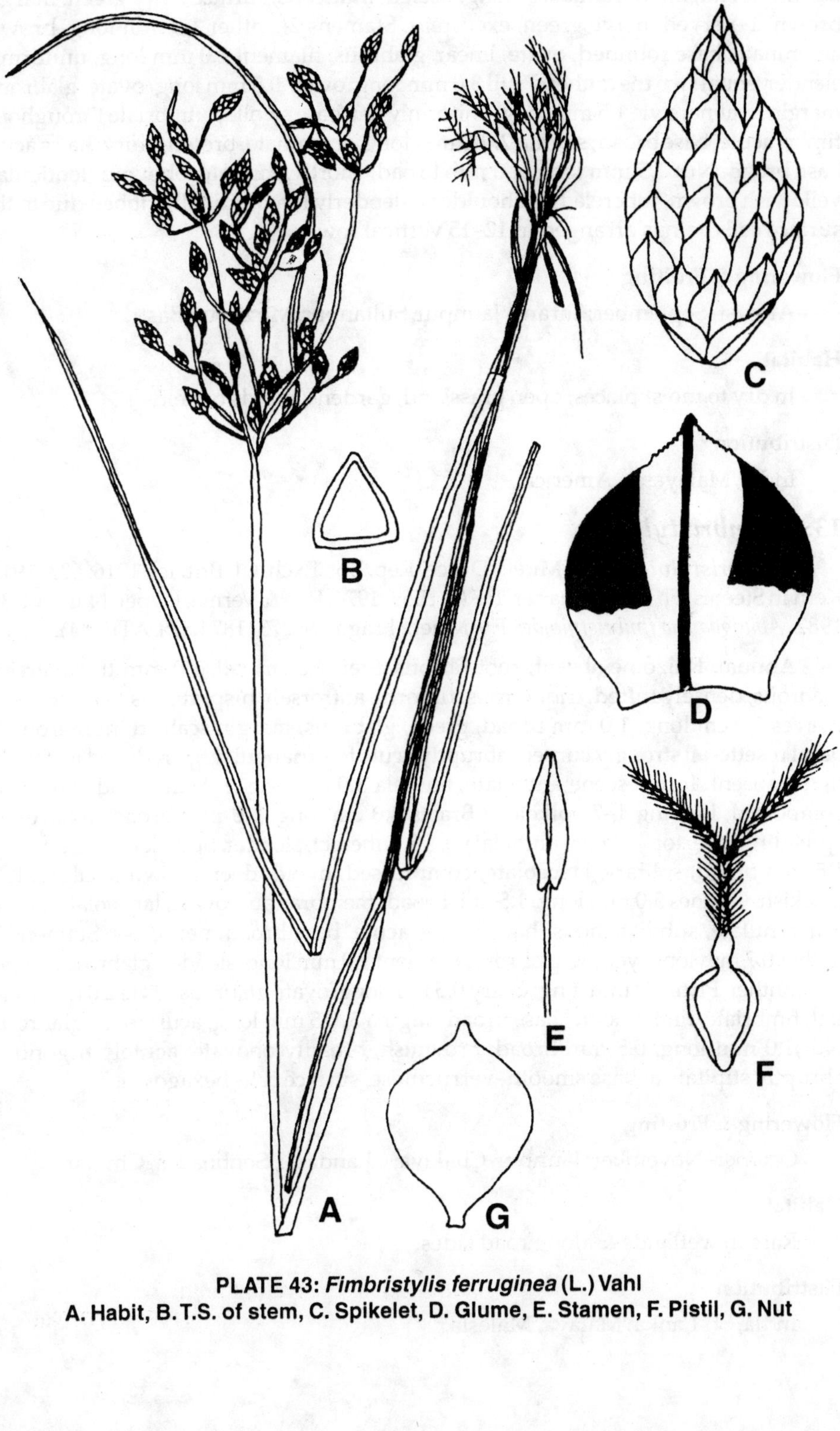

PLATE 43: *Fimbristylis ferruginea* (L.) Vahl
A. Habit, B. T.S. of stem, C. Spikelet, D. Glume, E. Stamen, F. Pistil, G. Nut

2.0 mm broad, membranous, oblong, keeled, muticous, surface rusty green, margin brown, 1–nerved, nerve green, excurrent. Stamens 2; anther 1.0 mm long, brown, acuminate, base rounded, ovate, linear, glabrous; filament 1.0 mm long, uniformly slender, equal two the anther. Pistil 3.0 mm long; ovary 0.5 mm long, ovate, glabrous, margin hyaline; style 1.5 mm long, uniformly flat, base swollen, fimbriate throughout; fimbri acute, base broad; stigma 2, 1.0 mm long, acuminate, brown, hairy; hairs acute base broad. Nut 1.5 mm long, 1.0 mm broad, shortly stipitate, obovoid, lenticular, yellowish brown, tubercled on shoulders, slenderly longitudinally ribbed due to the surface cells beings arranged in 12–15 vertical rows.

Flowering & Fruiting

August–September; Tarana, Jaunpur, Sultanpur, Mirzapur, Basti.

Habitat

In dry to moist places, open grassland, gardens, bunds of fields.

Distribution

India, Malayasia, America.

13. *F. fimbristyloides*

F. fimbristyloides (F.V. Muel.) Druce, Rep. Bot. Exch. Cl. Brit. Isl. 1916: 623. 1917; Kern in Steenis, Fl. Malesiana ser. 1: 7 (3) : 569. 1974; Rao & Verma, Cyper. N.E. Ind. 39. 1982. *Abildgaardia fimbristyloides* F.v. Muel., Fragm. 8: 273. 1874. (PLATE–44).

Annual. Rhizome absent; root fibrous. Stem 20 cm tall, 1.0 mm thick, green, glabrous, densely tufted, triquetrous, filiform, antrorsely hispidulous all over, erect. Leaves 5.0 cm long, 1.0 mm broad, green, glabrous, margin scabrid, numerous in basal rosette, all strongly curved, abruptly acute to a triangular tip, reduced to sheath; ligule absent. Inflorescene anthelate; anthela 3.0 cm long, 1.0 cm broad, simple to compound, bearing 1–7 spikelets. Bracts 4.0 cm long, 1.0 mm broad, foliaceaus, spreading, overtopping the anthela; rays 5, suberect, slender. Spikelets 3.0 mm long, 1.5 mm broad, solitary, lanceolate, compressed, ovoid, deep brown to ultimately blackish. Glumes 3.0 mm long, 1.5 mm broad, membranous, ovate, lanceolate, acute, mucronulate, subdistichous, hairy; hairs acute, base broad, nerveless. Stamens 3; anther 0.7 mm long, yellow, glabrous; filament 1.0 mm long, slender, glabrous, longer than anther. Pistil 3.0 mm long; ovary 0.5 mm long, ovate glabrous; style 2.0 mm long, flat, fimbriate; fimbri acute, base broad; stigma 3, 0.5 mm long, acuminate, glabrous. Nut 1.0 mm long, 0.5 mm broad, brownish, broadly obovate, acutely trigonous, abruptly stipitate at base smooth–verruculose, surface cells hexagonal.

Flowering & Fruiting

October–November; Jaunpur, Chakiya, Chandauli, Sonbhadra, Chunar.

Habitat

Rare in wetlands & along road sides.

Distribution

India, Sri Lanka, Malaya, Malesia.

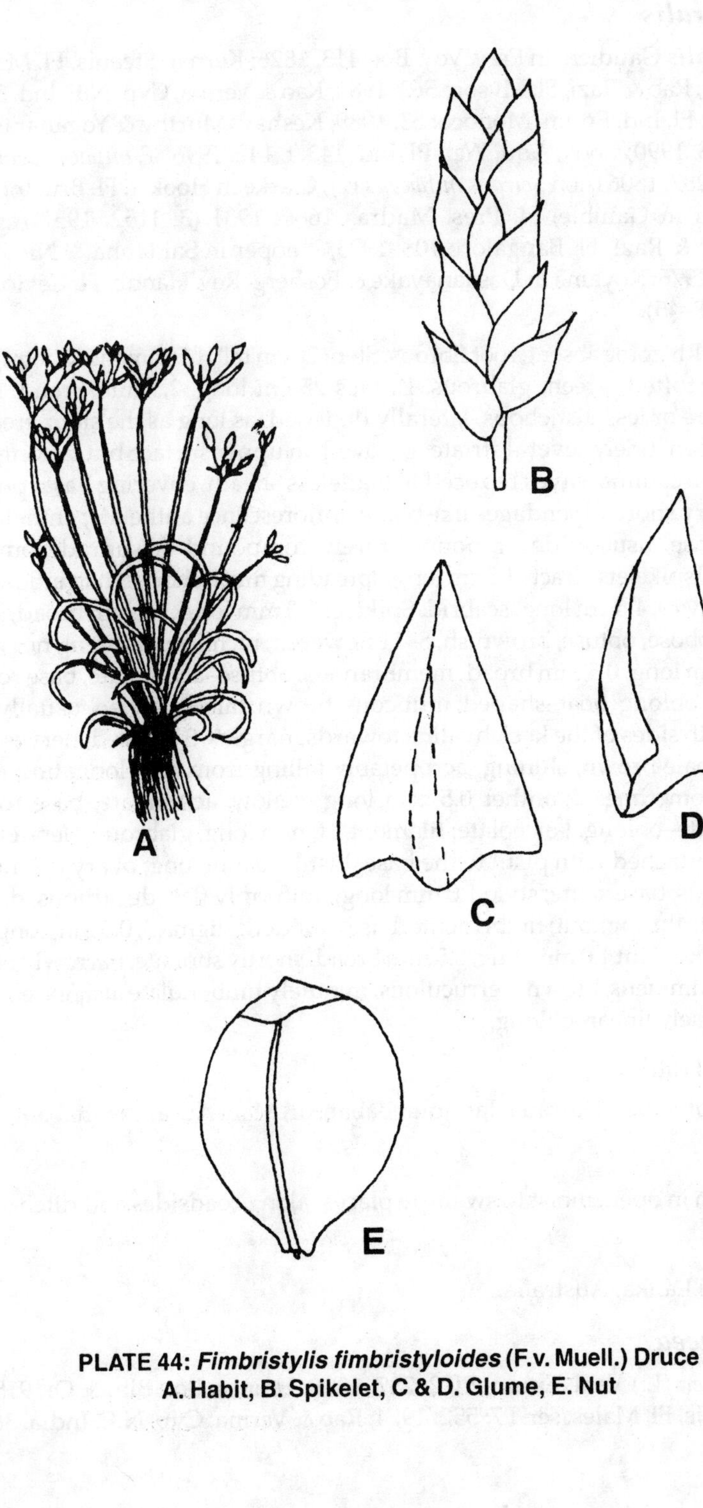

PLATE 44: *Fimbristylis fimbristyloides* (F.v. Muell.) Druce
A. Habit, B. Spikelet, C & D. Glume, E. Nut

14. *F. littoralis*

F. littoralis Gaudich. in Frey. Voy. Bot. 413. 1826; Kern in Steenis, Fl. Males. ser. 1.7: 551. 1974; Rao & Razi, Fl. Mysore 563. 1981; Rao & Verma, Cyp. N.E. Ind. 36. 1982; Karthik *et al.*, Fl. Ind. Enum. Monocot 53. 1989; KeshavaMurthy & Yoganarasimhan, Fl. Coorg 508. 1990; Cook, Aqu. Wet. Pl. Ind. 143. f. 142. 1996. *F. miliacea* sensu. Vahl Enum. Pl. 2: 287. 1806 (non *Scirpus miliaceus* L.); Clarke in Hook. f. Fl. Brit. Ind. 6: 644. 1893; Fischer in Gamble, Fl. Pres. Madras 1660. 1931 (3: 1152. 1957 repr. ed.); Ramaswamy & Razi, Fl. Bangalore 108. 1973; Hooper in Saldanha & Nicolson, Fl. Hassan 679. 1976; Koyama in Dasaanayake & Fosberg, Rev. Handb. Fl. Ceylon 5: 296. 1985. (PLATE–45).

Annual. Rhizome absent; root fibrous. Stem 20 cm tall, 1.5 mm thick, tetragonous, compressed, tufted, green, glabrous. Leaves 25 cm long, 2.5 mm broad, several, equitant, more or less distichous, laterally flattened, as long as the stem, frequently yellowish green, finely several striate, grooved on upper surface but lacking a mid–rib; upper two cauline leaves reduced to bladeless sheath, covering basal portion of stem with very short appendages like blade. Inflorescence anthelate; anthela 6.0 cm long lax, dense, usually decompound, rarely compound or supradecompound, bearing 12–30 spikelets. Bracts 1.5 cm long, spreading much shorter than inflorescence, foliaceous; rays 9; 4.0 cm long, scabrid. Spikelet 2.0 mm long, 1.0 mm broad, solitary, terete, subglobose, obtuse, brownish, 8–21 flowered; rachilla persistent, not winged. Glume 1.5 mm long, 0.5 mm broad, membranous, obtuse–acuminate, base rounded, spiral, ovate–oblong, boat–shaped, muticous, brown, faintly keeled, usually brown streak, on both sides of the keel, hyaline towards margins, 3–nerved; nerves visible, mid–nerve, pale brown, shining, acropetably falling from the elongating rachilla. Stamens 1, sometimes 2; anther 0.5 mm long, yellow, acuminate, base rounded, glabrous, ovate–oblong, lanceolate; filament 1.0 mm long, glabrous, slender longer than anther, attached with pistil to the base. Pistil 2.0 mm long; ovary 0.5 mm long, ovate, glabrous, base acute; style 1.0 mm long, uniformly flat, deciduous, dilated at base, fimbriate throughout; fimbri acute, base rounded; stigma 3, 0.5 mm long, acute, brown, glabrous. Nut 1.0 mm long, 0.5 mm broad, shortly stipitate, narrowly obovoid, trigonous, stramineus, brown, verruculous, minutely umbonulate at apex, epidermal cells transversely linear–oblong.

Flowering & Fruiting

June–September; Harraua, Jaunpur, Babatpur, Mau, Aurai, Allahabad.

Habitat

Common in open, moist to swampy places, along roadsides and ditches.

Distribution

India, Sri Lanka, Australia.

15. *F. miliacea*

F. miliacea (L.) Vahl. Enum. Pl. 2: 287. 1806; Haines, Bot. Bih. & Or. 915. 1924; Kern in Steenis, Fl. Males. ser. I 7: 552. 1974; Rao & Verma, Cyp. N.E. India. 36. 1982;

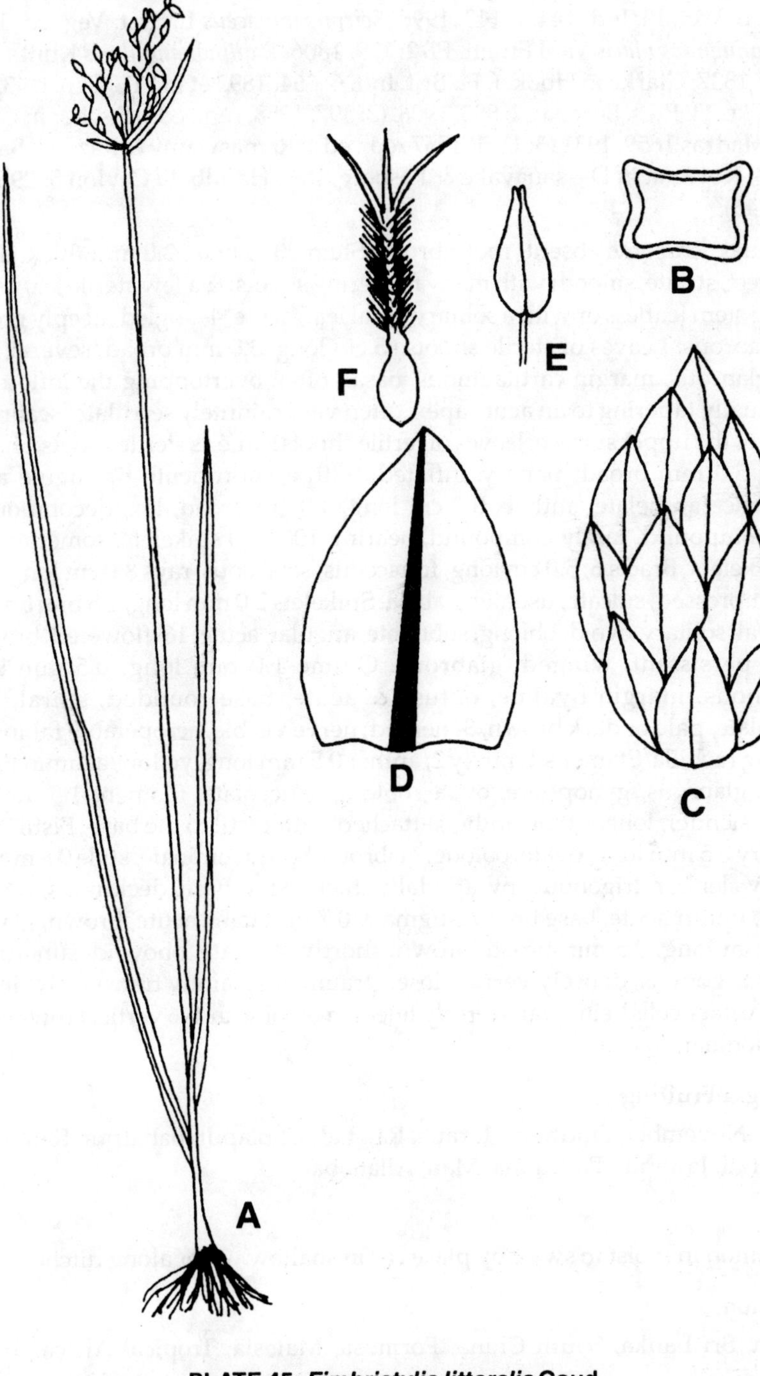

PLATE 45: *Fimbristylis littoralis* Gaud.
A. Habit, B. T.S. of stem, C. Spikelet, D. Glume, E. Stamen, F. Pistil

Sharma *et al.*, Fl. Karnataka 310. 1984; Karthik *et al.*, Fl. Ind. Enum. Monocot. 53. 1989; Cook, Aqu. Wet. Pl. Ind. 144. f. 143. 1996. *Scirpus miliaceus* L. Syst. Veg. ed. 10. 868. 1759. *S. quinquangularis* Vahl Enum. Pl. 2: 179. 1806. *F. quinquangularis* Kunth Enum. Pl. 2: 229. 1837; Clarke in Hook. f. Fl. Brit. Ind. 6: 664. 1893 et in I 11. Cyp. t. 43. f. 4–6. 1909; Cooke, Fl. Pres. Bombay 2: 883. 1908. (3: 397. 1958. repr. ed.); Fischer in Gamble, Fl. Pres. Madras 1659. 1931 (3: 1151. 1957 repr. ed.); Ramaswamy & Razi, Fl. Banglore 109. 1973; Koyama in Dassanayake & Fosberg, Rev. Handb. Fl Ceylon 5: 295. 1985. (PLATE–46).

Annual. Rhizome absent; root fibrous. Stem 90 cm tall, 2.0 mm thick, densely tufted, erect, striate, smooth with many flowering shoots, & a few sterile leafy shoots; flowering stem leafless or with a solitary basal leaf, acute, 4–angled, deeply grooved, green, glabrous. Leaves of sterile shoots 65 cm long, 3.0 mm broad, several, green, surface glabrous, margin cartilaginous, basal, often overtopping the inflorescene, flat, gradually tapering to an acute apex, often very minutely serrulate–scabrous on the veins of the upper surface; leaves of fertile shoots same as sterile shoots; sheath 10 cm long, 5.0 mm broad, papery, inflated, with a short acute tip; ligule absent. Inflorescence anthelate; anthela 8.0 cm long, 4.0 cm broad, lax, decompound or supradecompound, rarely compound, bearing 100–300 spikelets, sometimes only 10–15 spikelets. Bracts 6; 5.0 cm long, foliaceous, setaceous; rays 8.0 cm long, few to many compressed, sulcate, usually scabrid. Spikelets 2.0 mm long, 1.5 mm broad, all or nearly all solitary, ovoid, oblong, subterete, angular, acute, 16–flowered, brownish; rachilla, persistent, winged, glabrous. Glume 1.0 mm long, 0.5 mm broad, membranous, margin hyaline, obtuse & acute, base rounded, spiral, ovate, mucronulate, pale or dark brown, 3–nerved; nerve visible, acropetably falling from elongating rachilla. Stamens 1, rarely 2; anther 0.5 mm long, yellow, acuminate, base rounded, glabrous, gynophore, ovate, oblong, lanceolate; filament 1.5 mm long, glabrous, slender, longer than anther, attached with pistil to the base. Pistil 2.0 mm long; ovary 0.5 mm long, ovate, oblong, glabrous, base acuminate; style 0.8 mm long, uniformly slender, trigonous, pyramidally thickned at base, deciduous, fimbriate upwards; fimbri acute, base broad; stigma 3; 0.7 mm long, acute, brown, glabrous. Nut 1.0 mm long, 0.5 mm broad, brown, shortly stipitate, obovoid–suborbicular, obtuselly trigonous, densely verruculose, stramineus, faintly transversly lineolate with the surface cells being transversely linear in oblong, in 4–6 vertical rows on each face, umbonulate.

Flowering & Fruiting

May–November; Sindhora, Tarana, Rajghat, Cholapur, Babatpur, Rameshwar, Chaukhandi, Jaunpur, Padrauna, Mau, Allahabad.

Habitat

Common in moist to swampy places or in shallow water along ditches.

Distribution

India, Sri Lanka, South China, Formosa, Malesia, Tropical Africa, Tropical Australia.

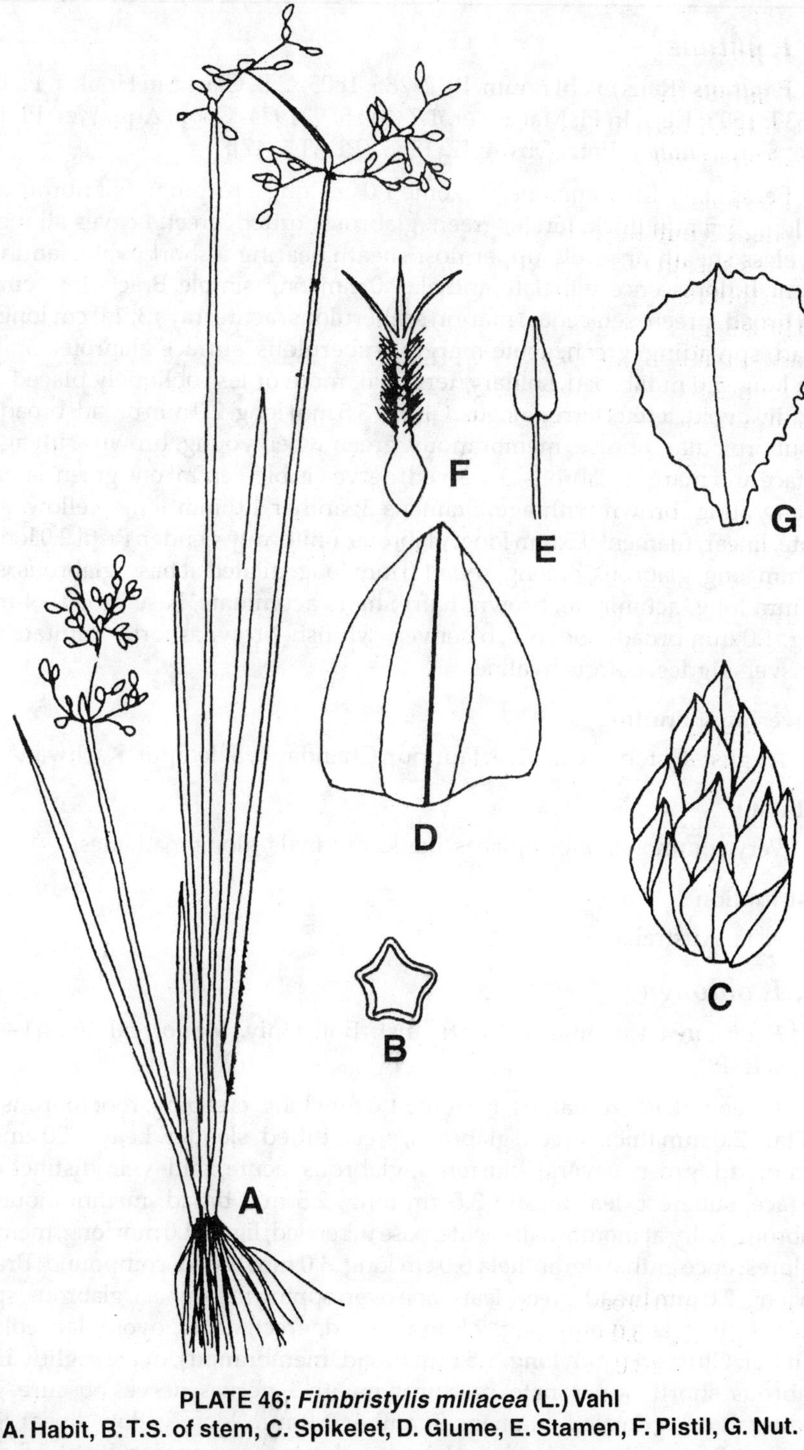

PLATE 46: ***Fimbristylis miliacea*** **(L.) Vahl**
A. Habit, B. T.S. of stem, C. Spikelet, D. Glume, E. Stamen, F. Pistil, G. Nut.

16. *F. nutans*

F. nutans (Retz.) Vahl Enum. Pl. 2: 285. 1805; C.B. Clarke in Hook. f. Fl. Brit. Ind. 6`: 632. 1893; Kern in Fl. Males. ser. 1.7 (3) : 589. 1974; Cook, Aqu. Wet. Pl. Ind. 144. 1996. *Scirpus nutans* Retz. Obs. 4: 12. 1786. (PLATE–47).

Perennials. Rhizomatous; rhizome 1.0 cm long, creeping; root fibrous. Stem 50 cm long, 1.5 mm thick, terete, green, glabrous, tufted, erect. Leaves all reduced to leaveless sheath or rarely uppermost sheath bearing a short ovate, lamina; ligule absent. Inflorescence anthelate; anthela 9.0 mm long, simple. Bracts 3, 3.0 cm long 1.0 mm broad, green, setaceous, margin scaberulous, acute; rays 3, 1.0 cm long 0.5 mm broad, spreading, green, acute margin scaberulous, surface glabrous. Spikelet 8.0 mm long, 3.0 mm broad, solitary, terminal, more or less obliquely placed on stem, broadly ovoid, terete, ferrugineous. Glume 3.5 mm long, 2.0 mm broad, broadly ovate, or suborbicular, obtuse, membranous, green when young, brown with age, acute, surface and margin glabrous, 3–nerved; nerve visible, central one green, lateral green when young, brown with age. Stamens 3; anther 1.0 mm long, yellow, glabrous, ovate, linear; filament 0.5 mm long, glabrous, uniformly slender. Pistil 2.0 long; ovary, 0.5 mm long, glabrous, oblong; style 1.0 mm long, dilated at base, glabrous; stigma 2; 0.5 mm long, acuminate, brown, hairy; hairs acuminate, base broad. Nut 1.5 mm long, 1.0 mm broad, obovoid, biconvex, whitish–brown, shortly stipitate with 3–5 transvers rigdes, margin hyaline.

Flowering & Fruiting

August–October; Chakiya, Jaunpur, Chandauli, Salempur, Kachwa, Aurai.

Habitat

Very common in moist places, banks of canals, along roadsides.

Distribution

Asia, Australia.

17. *F. oblonga*

F. oblonga Koyama in Contr. Inst. Bot. Univ. Monotreal. 70: 41–43. 1957. (PLATE–48).

Perennial. Rhizomatous; rhizome 1.0 mm long, creeping; root fobrous. Stem 40 cm tall, 2.0 mm thick, green, glabrous, erect, tufted, slender. Leaves 20 cm long, 2.5 mm broad, green, several, clustered, glabrous, acute, mid–vein distinct on lower surface, suberect, leaf–sheath 3.0 cm long, 2.5 mm broad, membranous, brown, glabrous, hairy at mouth; hairs acute, base tubercled; ligule 1.0 mm long, membranous. Inflorescence anthelate; anthela 6.0 cm long, 3.0 mm broad, compound. Bracts 3, 7.0 cm long, 2.0 mm broad, green, leafy, one overtopping the anthela, glabrous, spreading, acute. Spikelets 5.0 mm long, 2.5 mm broad, erect, acute, ovoid, lanceolate, 17 in number. Glume 3.0 mm long, 1.5 mm broad, membranous, ovate, tightly imbricate, glabrous, shortly mucronate, pale brown with 5–nerves; nerves obscure, glabrous, ending at apex, keeled. Stamens 3; anther 1.0 mm long, yellow, linear, glabrous; filament 0.5 mm long, slender, glabrous, shorter than anther. Pistil 3.5 mm long;

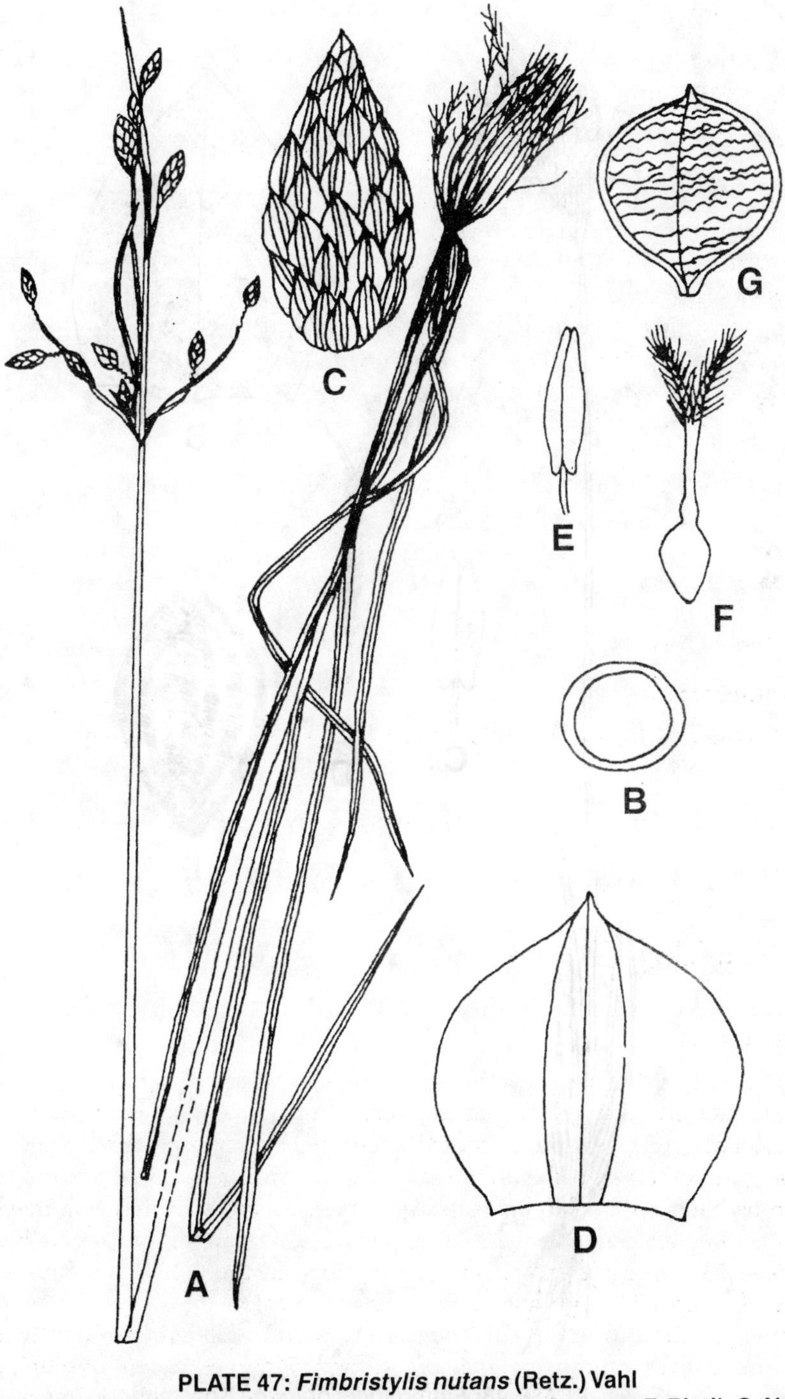

PLATE 47: ***Fimbristylis nutans*** **(Retz.) Vahl**
A. Habit, B. T.S. of stem, C. Spikelet, D. Glume, E. Stamen, F. Pistil, G. Nut

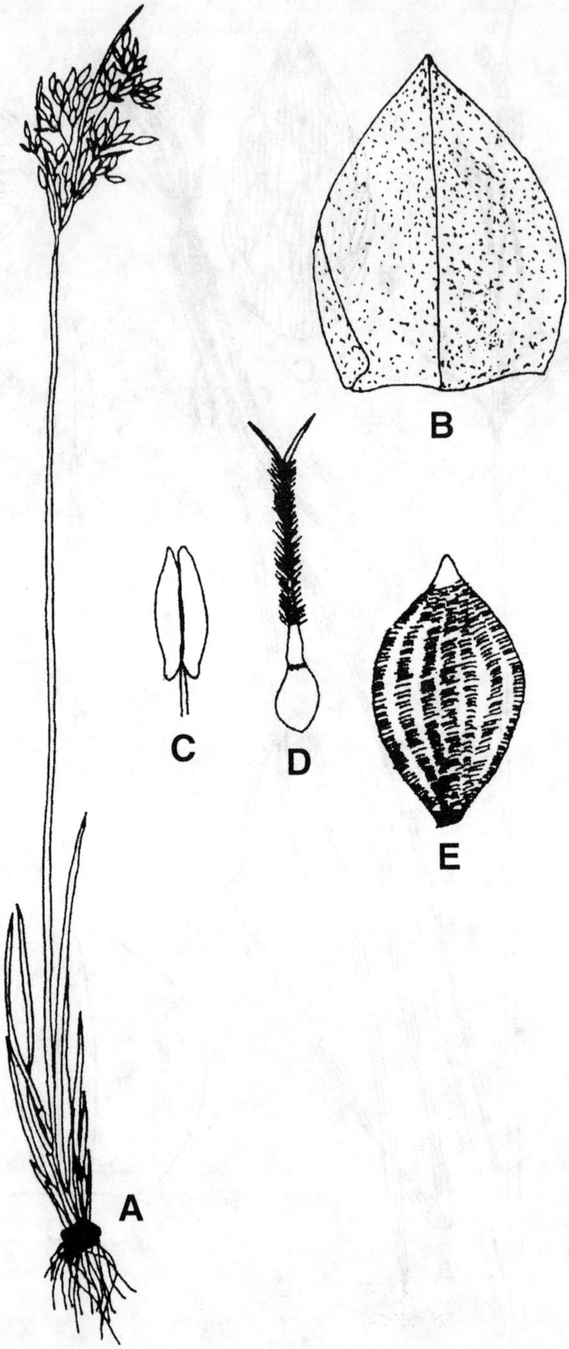

PLATE 48: *Fimbristylis oblonga* Koyama
A. Habit, B. Glume, C. Stamen, D. Pistil, E. Nut

ovary, 0.5 mm long ovate, glabrous; style 2.0 mm long, flat, dilated at base, much longer than nut, hairy & with distinctly hyaline margin; hairs acute, base broad; stigma 2, 1.0 mm long, brown, glabrous. Nut 1.0 mm long, 0.5 mm broad, brown, with 6–9 vertical ridges, not umbonulate, muticous with epidermal cells not impressed, shortly brown, stipitate.

Flowering & Fruiting

September–November; Ghazipur, Jaunpur, Kachwa, Sonbhadra, Kushinagar.

Habitat

Occasional on the banks of rivers, wet open grasslands.

Distribution

India, Sri Lanka, Malesia, Tropical Africa.

18. *F. ovata*

F. ovata (Burm f.) Kern in Blumea, 15: 126. 1967, Ramaswamy & Razi, Fl. Bangalore 108. 1973; Kern in van Steenis, Fl. Males. 1, 7: 565. 1974; Hooper in Saldanha & Nicolson, Fl. Hassan 679. 1976; Rao & Razi, Fl. Mysore 563. 1981; Arora *et al.*, Bot. S. Kanara 61. 1981; Rao & Verma, Cyp. N.E. Ind. 39. 1982; Sharma *et al.*, Fl. Karnataka 310. 1984; Koyama in Dassanayake & Fosberg, Rev. Handb. Fl. Ceylon 5: 273. 1985; Singh Fl. E. Karnataka 2: 637. 1988; Karthik *et al.*, Fl. Ind. Enum. Monocot. 54. 1989; KeshavaMurthy & Yoganarasimhan, Fl. Coorg 508. 1990; Cook, Aqu. Wet. Pl. Ind. 145. f. 144. a–c. 1996. *Carex ovata* Burm. f. Fl. Ind. 194. 1968. *Cyperus monostachyos* L. Mant. 2: 180. 1771. *Fimbristylis monostachyos* Hassk., Pl. Jav. Rar. 61. 1848 (*monostachyos*); Clarke in Hook. f. Fl. Brit. Ind. 6: 649. 1893; Cooke, Fl. Pres. Bombay 2: 885. 1908 (3: 399. 1958 repr. ed.); Fischer in Gamble, Fl. Pres. Madras 1660. 1931 (3: 1152. 1957 repr. ed.). (PLATE–49).

Perennials. Rhizomatous; rhizome 2.0 cm long, short, woody; root fibrous. Stem 30 cm tall, 1.5 mm thick, green, glabrous, sulcute, slender, suberect, slightly swollen at the base, fattened at the top, trigonous, sometime scabrid, tufted, compressed. Leaves 10 cm long, 1.0 mm broad, one–third to half as long as the stem, green, glabrous, clustered at the base, several, flat, canaliculate, with incurved margin, acuminate, often scaberous on the slightly thickened margins, reduced to blade less sheath; ligule absent. Inflorescence usually a solitary terminal spikelets, sometimes 3 peduncled spikelets. Bracts 2, 5.0 mm long, setaceous, scale–like, with the mid–nerve elongated into an awned; rays absent. Spikelet 6.0 mm long, 4.0 mm broad, compressed in lower half, subterete in upper half, ovate, lanceolate, acute, yellowish–stramineus, twisted; rachilla persistent winged, glabrous. Glumes 4.0 mm long, 3.5 mm broad, lower distichous, upper spiral, membranous, yellowish, broadly ovate–oblong, keeled, mucronate, acute, base rounded, acropetably falling from the rachilla, 3–nerved; nerve deeply excurrent, brownish spot between the nerves. Stamens 2 rarely 3; anther 1.5 mm long, brown, crest, rounded, linear–lanceolate, oblong, ovate, acute, gynophore, glabrous; filament 0.5 mm long, glabrous, uniformly slender, shorter than anther, attached with pistil to the base. Pistil 5.0 mm long, ovary 1.0 mm long, ovate, base acuminate; style 3.0 mm long, uniformly flat, deciduous, dilated at base, fimbriate

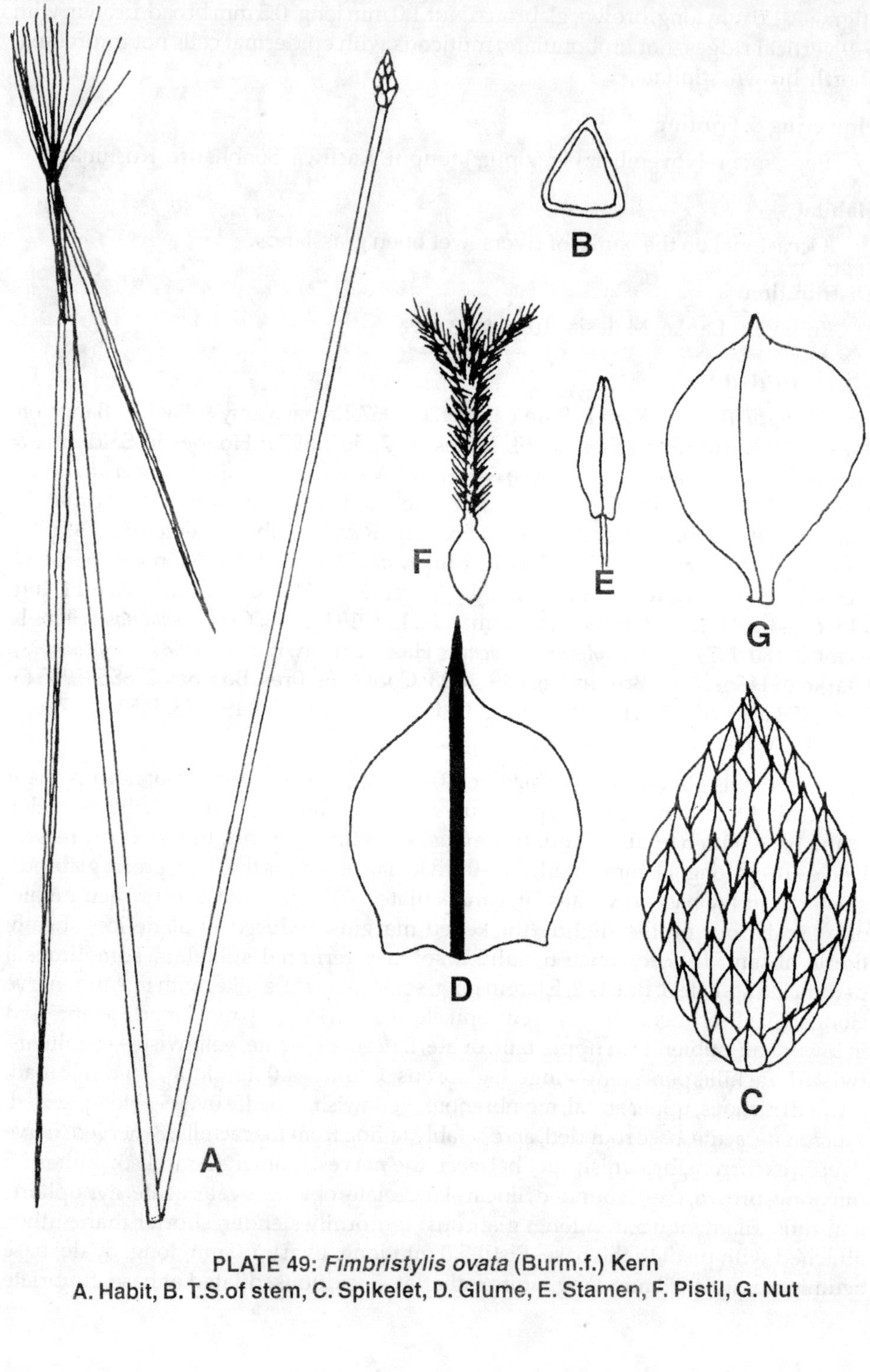

PLATE 49: *Fimbristylis ovata* (Burm.f.) Kern
A. Habit, B. T.S.of stem, C. Spikelet, D. Glume, E. Stamen, F. Pistil, G. Nut

throughout; fimbri acute base broad; stigma 3; 1.0 mm long, acute, brown, hairy; hairs acuminate, base rounded. Nut 1.5 mm long, 1.0 mm broad, stipitate, obovoid, trigonous, apiculate, white at young, brown tinged at maturity, trabeculate, almost truncate with umbonate apex, muricate except on the narrow base, epidermal cells isodiametric, in many vertical rows.

Flowering & Fruiting

July–December; Sindhora, Jaunpur, Chunar, Sultanpur, Ballia.

Habitat

Common in shady places along road sides, open moist places, bunds of cultivated fields.

Distribution

Indian, Bhutan, Combodia.

19. *F. schoenoides*

F. schoenoides (Retz.) Vahl, Enum. Pl. 2: 286. 1806; Clarke in Hook. f. Fl. Brit. Ind. 6: 634. 1893; Haines, Bot. Bih. & Or 918. 1924; Kern in Fl. Males. ser. 1.7 (3): 573. 1974; Cooke Fl. Pres. Bombay 2: 880. 1908 (3: 394. 1958 repr. ed.); Fischer in Gamble, Fl. Pres. Madras 1658. 1931 (3: 1150. 1957 repr. ed.); Ramaswamy & Razi, Fl. Bangalore 109. 1973; Hooper in Saldanha & Nicolson, Fl. Hassan 680. 1976; Rao & Razi, Fl. Mysore 563. 1981; Rao & Verma, Cyp. N.E. Ind. 30. 1982; Sharma *et al.*, Fl. Karnataka 310. 1984; Koyama in Dassanayake & Fosberg. Rev. Handb Fl. Ceylon 315. 1985; Singh, Fl. E. Karnataka 637. 1988; Karthik *et al.*, Fl. Ind. Enum. Monocot. 54. 1989. KeshavaMurthy & Yoganarasimhan Fl. Coorg 509. 1990. *Scirpus schoenoides* Retz., Obs. Bot. 5: 14. 1788. (PLATE–50).

Perennial. Rhizomatous; rhizome short, woody; root fibrous. Stem 35 cm tall, 1.0 mm thick, green, glabrous, slender, sulcate, suberect, compressed, densely tufted, trigonous. Leaves 30 cm long, 1.0 mm broad, filiform, canaliculate, shorter than to equalling the stem, obtuse, clustered at the base; sheath 3.0 cm long, glabrous; ligule 1.0 mm long, membranous, margin hairy; hairs acute, base broad. Inflorescence usually solitary, simple, terminal spikelet, or often bearing 2–3 peduncled spikelets. Bracts 1, 1.5 mm long, 1.0 mm broad foliaceous–glumaceous, triangular, apiculate, spinular–scabrous on the margins. Spikelets 5.0 mm long, 2.0 mm broad, broadly ovoid, oblong, terete, acutish; rachilla persistent, narrowly winged. Glume 2.0 mm long, 2.5 mm broad, spiral, broadly ovate, suborbicular, muticous, mucronulate, glossy, keeled, stramineous, red tinged, between nerves, membranous, sides 8 striate, 4–nerved; nerves faintly visible. Stamens 3; anther 1.0 mm long, ovate–oblong, yellow, acuminate, base rounded, glabrous; filament 0.5 mm long, glabrous, slender, shorter than to anther, attached with pistil to the base. Pistil 3.0 mm long; ovary 0.5 mm long, obovate, glabrous; style 1.5 mm long, slender, articulate with the ovary, dilated at the base, fimbriate upwards; fimbri acute, base broad, stigma 2, 1.0 mm long, acuminate, glabrous, brown. Nut 2.0 mm long, 1.0 mm broad broad, obovoid, biconvex, umbonulate, whithish at young, stramineous, on maturity, stipitate; stipe 0.3 mm long, lobed, epidermal cells isodiametric.

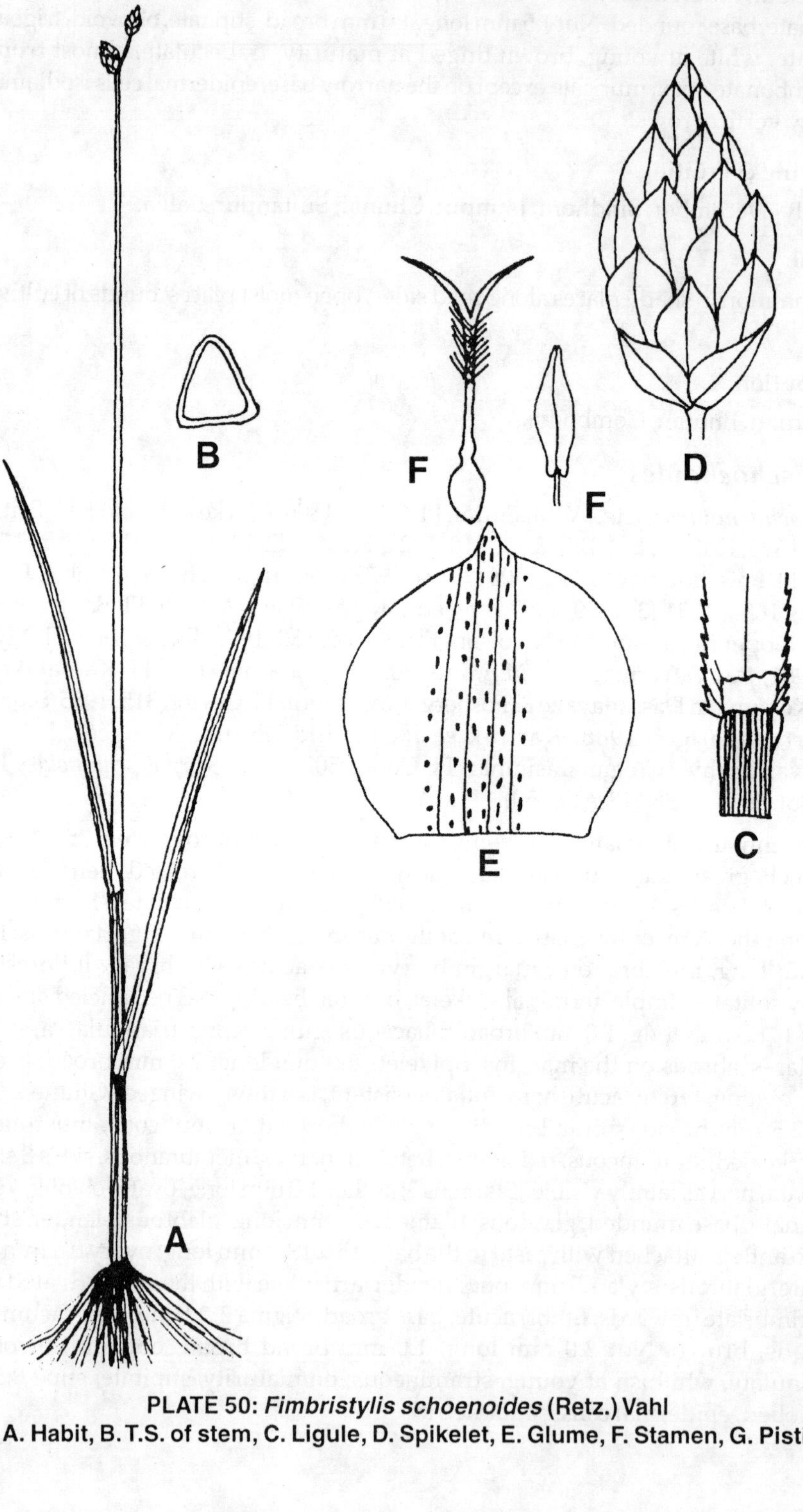

PLATE 50: *Fimbristylis schoenoides* (Retz.) Vahl
A. Habit, B. T.S. of stem, C. Ligule, D. Spikelet, E. Glume, F. Stamen, G. Pistil

Flowering & Fruiting

July–October; Babatpur, Basti, Jaunpur, Chandauli.

Habitat

Frequent in wastlands & paddy fields.

Distribution

India, Sri Lanka, China, Malesia, Thailand, Formosa, Tropical Australia.

20. *F. squarrosa*

F. squarrosa Vahl Enum. Pl. 2: 289. 1806; Cook, Aqu. Wet. Pl. Ind. 148. f. 149. 1996. *F. aestivalis* var. *squarrosa* Koyama in J. Fasc. Sci. Uni. Tokyo. 3. 8: 116. 1962. (PLATE–51).

Annual. Rhizome absent; root fibrous. Stem 10 cm long, 0.5 mm thick, green, glabrous, erect, tufted, trigonous. Leaves 10 cm long, 1.0 mm broad, green, surface glabrous, margin scabrid, acute, equalling the stem, several; sheath 4.0 cm long, 1.0 mm broad, membranous, brown, hairy; hairs acute, base broad; ligule absent. Inflorescence anthelate; anthela 6.0 cm long, compound, dense, bearing 19–spikelets. Bracts 4, 6.0 cm long, 1.0 mm broad, equalling the anthella, foliaceous, green, surface glabrous, margin scabrid, acute; rays 3, 5.0 cm long, slender, spreading. Spikelets 7.0 mm long. 2.0 mm broad, solitary, linear, oblong, polygonal, squarrose; rachilla narrowly winged. Glume 2.5 mm long, 1.5 mm broad, membranous, spiral, ovate, oblong, keeled, pale brown, 3–nerved, aristate; arista 1.0 mm long, excurved, glabrous. Stamens 1; anther 0.5 mm long, pale yellow, apiculate; filament 1.0 mm long, slender, glabrous, longer than anther. Pistil 1.5 mm long, ovary 0.2 mm long, ovate, glabrous; style 1.0 mm long, slender with long white pendent hairs from the base, covering half of the nut; hairs acute, base broad; stigma 2, 0.3 mm long, glabrous. Nut 1.0 mm long, 0.5 mm broad, pale brown, obovoid, shortly stipitate, biconvex, smooth.

Flowering & Fruiting

May–June; Sindhora, Basti, Sultanpur, Allahabad.

Habitat

In open marshy places, near ditches.

Distribution

India, Burma, Australia, Newzealand.

21. *F. striolata*

F. striolata Nepper in Kew Bull. 25 (3): 438. 1971. (PLATE–52).

Annual. Rhizome absent; root fibrous. Stem 35 cm tall, 2.0 mm thick, slender, glabrous, tufted. Leaves 15 cm long, 1.5 mm broad, green, flat, glabrous, coriaceous, acute, clustered at base, mid–vein distinct on lower surface; leaf–sheath 1.0 cm long, 2.5 mm broad, membranous, brown, glabrous; ligule 0.5 mm long, membranous. Inflorescence anthelate; anthela simple open. Bracts 2, 2.0 cm long, 1.0 mm broad,

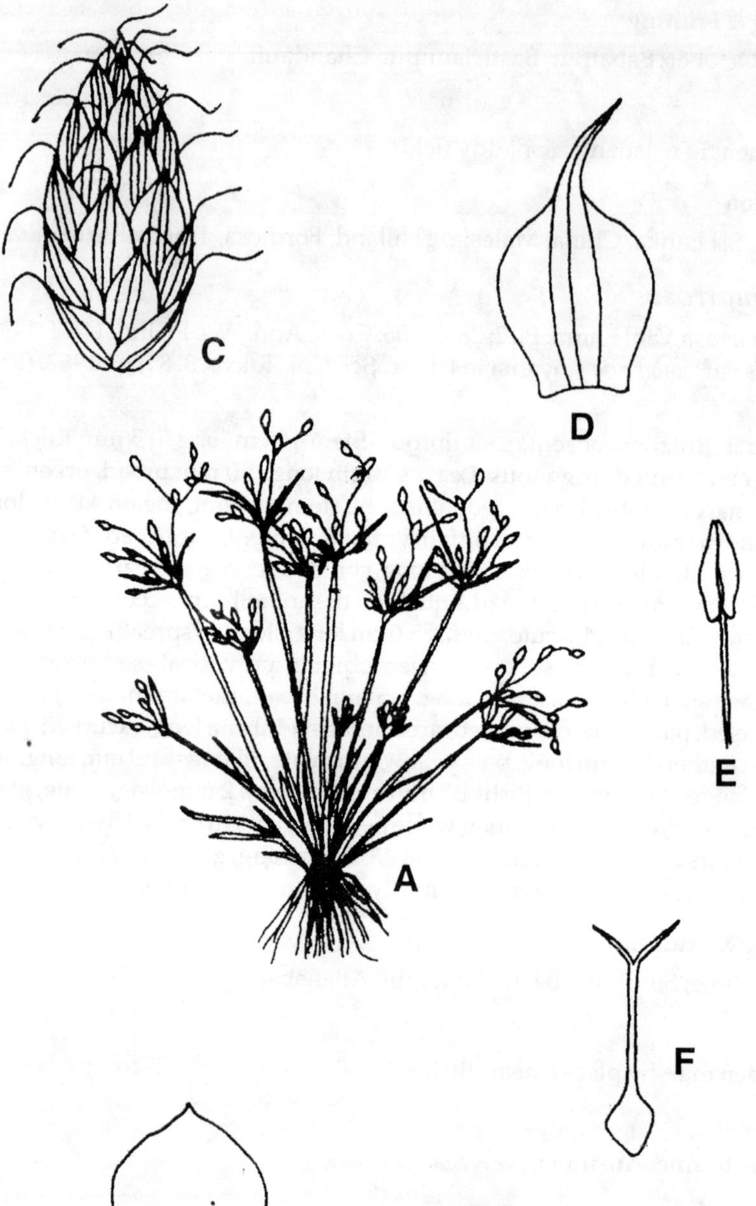

PLATE 51: ***Fimbristylis squarrosa*** **Vahl**

A. Habit, B. T.S. of stem, C. Spikelet, D. Glume, E. Stamen, F. Pistil, G. Nut

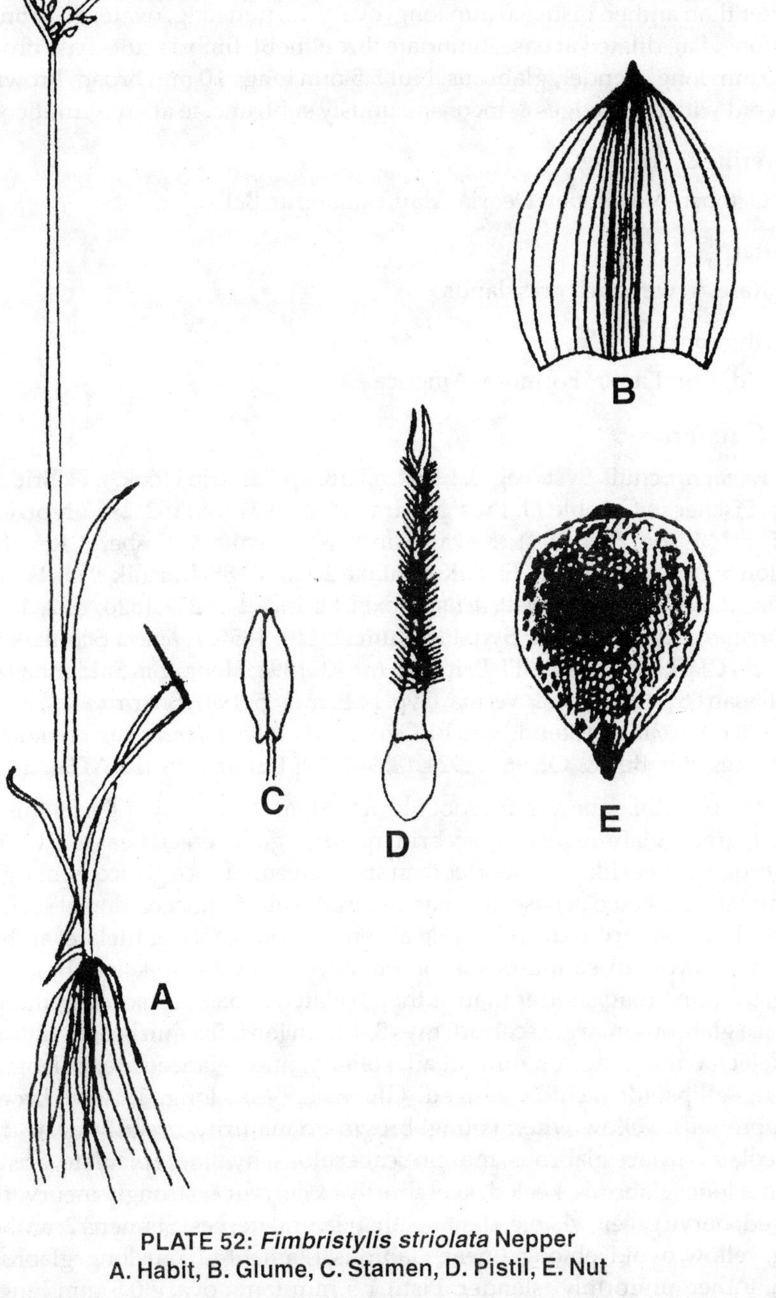

PLATE 52: ***Fimbristylis striolata*** **Nepper**
A. Habit, B. Glume, C. Stamen, D. Pistil, E. Nut

green, leafy, overtopping the anthela, acute, filiform. Spikelets 5.0 mm long, 3.0 mm wide, obtuse, 3–7 in number. Glume 5.0 mm long, 2.5 mm broad, membranous, ovate, lanceolate, acute, glabrous, brownish, with green keel & faintly striated sides. Stamens 3; anther 2.0 mm long, yellow, glabrous, linear; filament 1.0 mm long, slender glabrous, shorter than anther. Pistil 5.0 mm long; ovary 1.0 mm long, ovate, glabrous; style 3.0 mm long, flat, dilated at base, fimbriate throughout; fimbri acute, base broad; stigma 2, 1.0 mm long, slender, glabrous. Nut 5.5 mm long, 3.0 mm broad, brown, broadly obovoid with thick edges & inconspicuously subtruncate at apex, muticous.

Flowering & Fruiting

October–November; Deoria, Mau, Sultanpur, Ballia.

Habitat

Rare, in wet open grasslands.

Distribution

India, Sri Lanka, Formosa, America.

22. *F. tenera*

F. tenera Schult. Syst. Veg. 2. Mant. 57. 1824; Clarke in Hook. f. Fl. Brit. Ind. 6: 642. 1893; Fischer in Gamble Fl. Pres. Madras 1660. 1931 (3: 1152. 1957 repr. ed.); Rao & Razi, Fl. Mysore 563. 1981; Koyama in Dassanayake & Fosberg, Rev. Handb. Fl. Ceylon 5: 293. 1985; Singh, Fl. E. Karnataka 2: 637. 1988; Karthik *et al.*, Fl. Ind. Enum. Monocot. 55. 1989. *Scirpus tenellus* Roxb. Fl. Ind. 1,1: 227.1820; & 2, 1: 224. 1832. *Fimbristylis oxylepis* Steud. Syn. Pl. Glum. 2: 110. 1855. *F. tenera* Schult. var. *oxylepis* (Steud.) Clarke in Hook. f. Fl. Brit. Ind. 6: 642. 1894; Hooper in Saldanha & Nicolson Fl. Hassan 681. 1976; Rao & Verma, Cyp. N.E. Ind. 35. 1982; Sharma *et al.*, Fl. Karnataka 311. 1984. *F. monticola* Steud. Syn. Pl. Cyp. 2: 111. 1855. *F. tenera* var. *verrucaeum* Clarke ex Haines, Bot. Bih. & Or. 961. 1924. (3: 961. 1961. repr. ed.). (PLATE–53).

Annual. Rhizome absent; root fibrous. Stem 10 cm long, 1.0 mm thick, slender, tufted, green, glabrous, rarely scabrid, quadrangular, erect. Leaves 7.0 cm long, 0.5 mm broad, almost filiform, shorter than stem, green, surface glabrous, margin scabrid, acuminate, clustered at base, flat, narrow, mid–vein distinct on dorsal surface; sheath 3.0 cm long, papery, reduced; ligule absent. Inflorescence anthelate; anthela 1.0 cm long, 5.0 mm broad, simple or subcompound, bearing 7–9 spikelets. Bracts, 3, 1.5 mm long, 0.5 mm broad shorter than anthela, dilated at base, setaceous, filiform, green, surface glabrous, margin scabrid; rays 3, 1.5 cm long, 0.5 mm broad, filiform, green. Spikelet 4.0 mm long, 1.5 mm broad, solitary, linear–lanceolate, palebrown ovate, oblong–ellipsoid; rachilla winged. Glume 2.5 mm long, 1.0 mm broad, spiral, membranous, yellow when young, brown on maturity, broadly ovate, triangular, lanceolate, surface glabrous, margin scaberulous, hyaline, apiculate, aristate; arista 0.2 mm long, glabrous, keeled, keel shortly ex current & strongly incurved at base, 3 nerved; nerve green, visible, dentate, than lateral nerves. Stamens 2; anther 0.5 mm long, yellow, ovoid, oblong, linear, glabrous; filament 1.0 mm long, glabrous, longer than anther, uniformly slender. Pistil 1.5 mm long; ovary 0.5 mm long, oblong–obovoid, shortly stipitate, glabrous; style 0.5 mm long, dilated & angled at base,

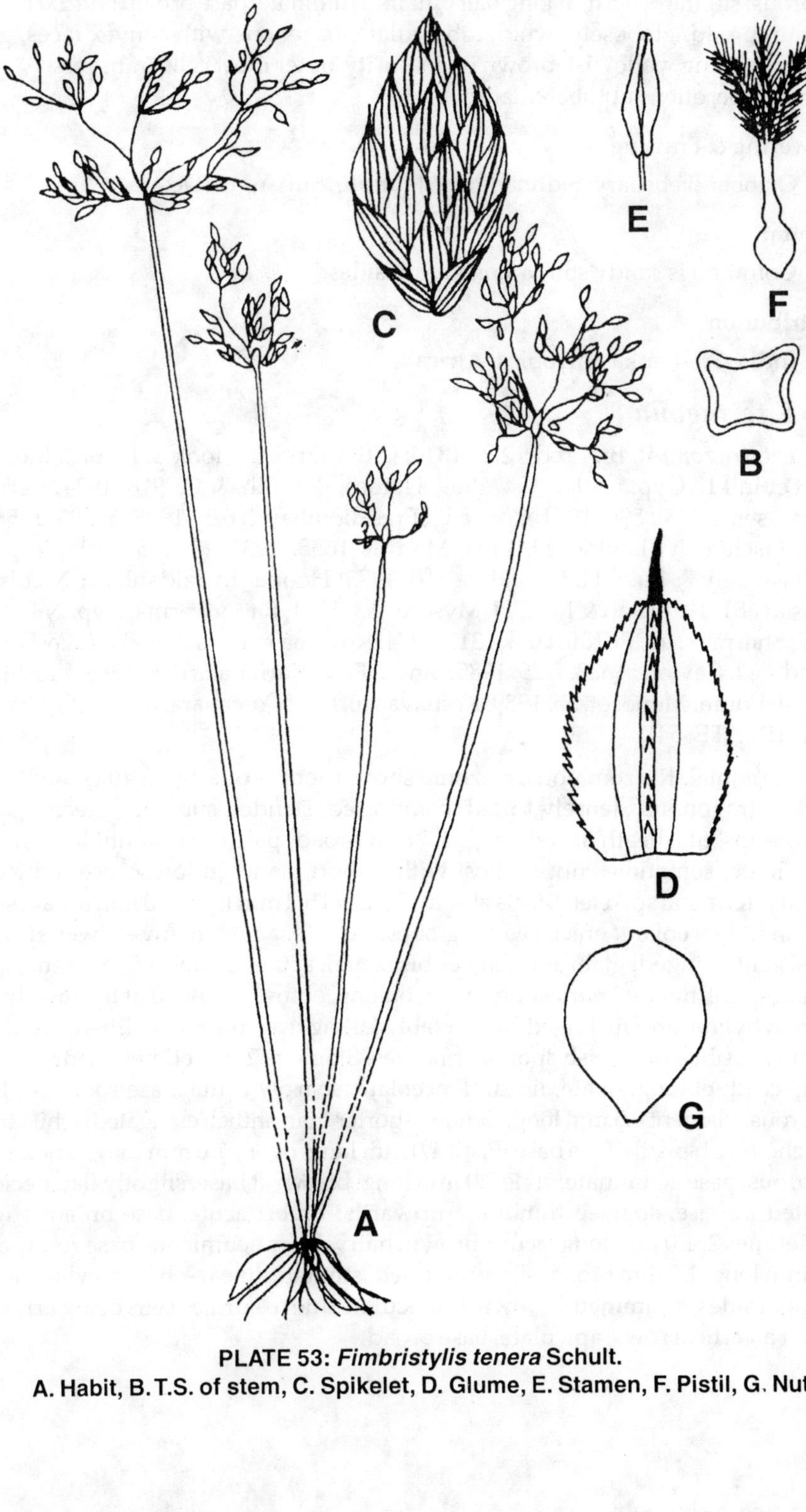

PLATE 53: *Fimbristylis tenera* Schult.
A. Habit, B. T.S. of stem, C. Spikelet, D. Glume, E. Stamen, F. Pistil, G. Nut

glabrous; stigma 3, 0.5 mm long hairy; hairs acuminate, base broad. Nut 1.0 mm long, 0.5 mm broad, globosely ovoid, tuberculate, trigonous with convex faces, greyish white at young, yellowish brown on maturity, outer most cells transversely oblong, verruculose, often subtuberculate.

Flowering & Fruiting

October–February; Sidhnathdari, Jaunpur, Sultanpur, Deoria.

Habitat

Common is sandy soil & along road–sides.

Distribution

India, Sri–Lanka & Tropical Africa.

23. *F. tetragona*

F. tetragona R. Br. Prodr. 226. 1810; C.B. Clarke in Hook. f. Fl. Brit. Ind. 6: 631. 1893 et in I Il. Cyp. t. 40. f. 1–4. 1909; Haines, Bot. Bih. & Or 918. 1924; Kern in Fl. Males. ser. 1.7 (3): 590. 1974; Cooke Fl. Pres. Bombay 2: 679. 1908 (3: 393. 1958. repr. ed.); Fischer in Gamble, Fl. Pres. Madras 1658. 1931. (3: 1150. 1957 repr. ed.); Ramaswamy & Razi, Fl. Bangalore 110. 1973; Hooper in Saldanha & Nicolson, Fl. Hassan 681. 1976; Rao & Razi, Fl. Mysore 563. 1981; Rao & Verma, Cyp. N.E. Ind. 29. 1982; Sharma *et al.*, Fl. Karnataka 311: 1984; Koyama in Dassanayake & Fosberg, Rev. Handb. Fl. Ceylon 5: 623. f. 25. 1985; Singh, Fl. E. Karnataka 638. 1988; Karthik *et al.*, Fl. Ind. Enum. Monocot. 55. 1989; KeshavaMurthy & Yoganarasimhan, Fl. Coorg 509. 1990. (PLATE–54).

Perennial. Rhizomatous; rhizome short; root fibrous. Stem 40 cm tall, 1.5 mm thick, tetragonous, densely tufted, compresed, slender, suberect, green, glabrous. Leaves absent; sheath 3.5 cm long, 1.5 mm broad, pale or chest nut brown, acute–acuminate, sometimes upper most with a short blade. Inflorescence reduced to a solitary, terminal spikelet. Bracts absent. Spikelet 5.0 mm long, 4.0 mm broad, solitary, terminal, lanceolate, erect, ovoid–globose, terete, acute, many–flowered; rachilla persistent, winged, glabrous. Flower bisexcual, acute. Glume 4.0 mm long, 3.5 mm broad, spiral, tightly imbricating, ovate–oblong, obtuse, rounded at tip, membranous, margin hyline, brown, tinged, acropetably falling from the rachilla, 3–nerved; nerves slightly visible, brownish spot near nerves. Stamens 2 sometimes 3; anther 1.5 mm long, crest, oblong, ovate, linear–lanceolate, sharply acute, base rounded, brown, glabrous; filament 0.5 mm long, slender, shorter than anther, elongated, while fruiting, attached with pistil to the base. Pistil 4.0 mm long; ovary 1.0 mm long, oblong, ovate, glabrous, base acuminate; style 2.0 mm long, brown at base, slightly flat, deciduous, dilated at base, sparsely fimbriate upwards; fimbri acute, base broad; stigma 3, sometimes 2, 1.0 mm long, acute, brown, hairy; hairs acuminate, base rounded. Nut 1.5 mm long, 1.0 mm broad, slightly curved, stipitate, linear–oblong, cylindrical with parallel sides, stramineus–brown, trabeculate due to surface cells being arranged in about a vertical rows, apiculate, base broad.

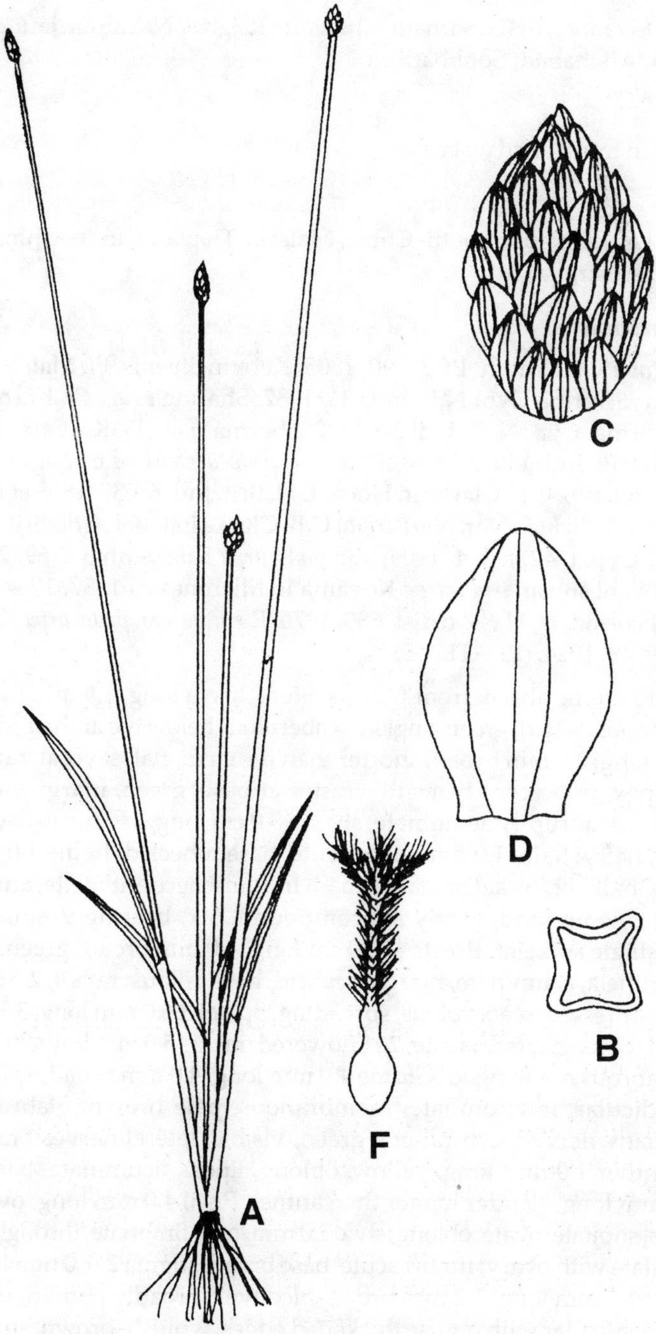

PLATE 54: ***Fimbristylis tetragona*** **R. Br.**
A. Habit, B. T.S. of stem, C. Spikelet, D. Glume, E. Stamen, F. Pistil

Flowering & Fruiting

August–December; BHU, Sarnath, Cholapur, Rajghat, Navalpur, Jaunpur, Deoria, Padrauna, Lar, Allahabad, Sonbhadhra.

Habitat

Common in moist shady places.

Distribution

India, Sri Lanka, Nepal, South–China, Malesia, Thailand, Indo–China, Formosa, tropical North Australia.

24. *F. tomentosa*

F. tomentosa Vahl Enum. Pl. 2: 290. 1805; Kern in Steenis, Fl. Males. ser. I. 7 (3): 576. 1974; Rao & Verma Cyp. N.E. Ind. 34. 1982; Sharma *et al.*, Fl. Karnataka. 310. 1984; Rao & Verma, Cyp. N. E. Ind. 34. 1982; Sharma *et al.*, Fl. Karnataka 310. 1984; Cook, Aqu. Wet. Pl. Ind. 149. f. 150 d–f. 1996. *F. podocarpa* Nees in Wight, Contr. Bot. India 98. 1834. pp. typ. C.B. Clarke in Hook. f. Fl. Brit. Ind. 6: 637. 1893 et in I 11. Cyp. t. 42. f. 5–6. 1909. *F. diphylla* var. *pluristriata* C.B. Clarke in Hook f. Fl. Brit. Ind. 6: 637. 1893 et in I, 11 Cyp. t. 42. f. 3–4. 1909; var. *podocarpa* Kuekenth. l.c. 69: 257. 1938. *F. dichotoma* (L.) Vahl subsp. *podocarpa* Koyama in Micronesica 1: 87. 1964; Hooper in Saldanha & Nicolson, Fl. Hassan dist. 680. 1976. *F. annua* var. *podocarpa* Kuekenth. in Bot. Jahrab. 59: 48. 1924. (PLATE–55).

Annual. Rhizome absent; root fibrous. Stem 30 cm long, 1.5 mm thick slender, nodding, trigonous, tufted, green, angled, scaberulous below the anthela, compressed. Leaves 23 cm long, 2.0 mm broad, shorter than the stem, flat, several, rarely falcate, glabrous at apex, pubescent beneath, cluster at base, green, margin scaberulous, mid–vein distinct, abruptly acuminate; sheath 3.0 cm long, 2.0 mm broad, glabrous, membranous, hairy; hairs 1.0 mm long, acute, base tubecled; ligule 1.0 mm long, a fringe of hairs; hair, white, acute, base broad. Inflorescnece anthelate; anthela 4.0 cm across, simple, compound, rarely subcompound, lax, bearing 9–spikelets, often reduced to a single spikelet. Bracts 4, 6.0 cm long, 1.0 mm broad, green, foliaceous, longer than anthela, acuminate, margin scabrid, base villous; rays 4, 2.5 cm long, 0.5 mm broad, compressed, scaberulous, spreading. Spikelet 6.0 mm long, 3.5 mm broad, solitary, ovoid–oblong, terete, acute, 20– flowered; rachis 5.0 mm long, 1.0 mm broad, persistent, glabrous not winged. Glume 4.0 mm long, 4.0 mm broad, spiral, broadly ovate, subordicular, mucronulate, membranous, pale brown, glabrous, keeled, acuminate, many nerved, central one green, visible, lateral nerves faintly visible. Stamens 2; anther 1.0 mm long, yellow, oblong, linear, acuminate, base rounded; filament 0.5 mm long, slender longer than anther. Pistil 4.0 mm long; ovary 1.0 mm long, glabrous, stipitate, ovate, oblong; style 2.0 mm long, fimbriate, throughout, dilated at base, articulate with ovary; fimbri acute, base broad; stigma 2, 1.0 mm long, brown, glabrous. Nut 1.5 mm long, 1.2 mm broad, biconvex, broadly obovid, umbonulate, subordicular, lenticular with obtuse thickened edges, whitish–brown, smooth, rarely verrucullose, trabeculate due to the surface cells being arranged in 16–24 vertical rows on each face, stipitate; stipe 0.5 mm long, obpyramidal, sparsely tubercled in the uper half towards the apex.

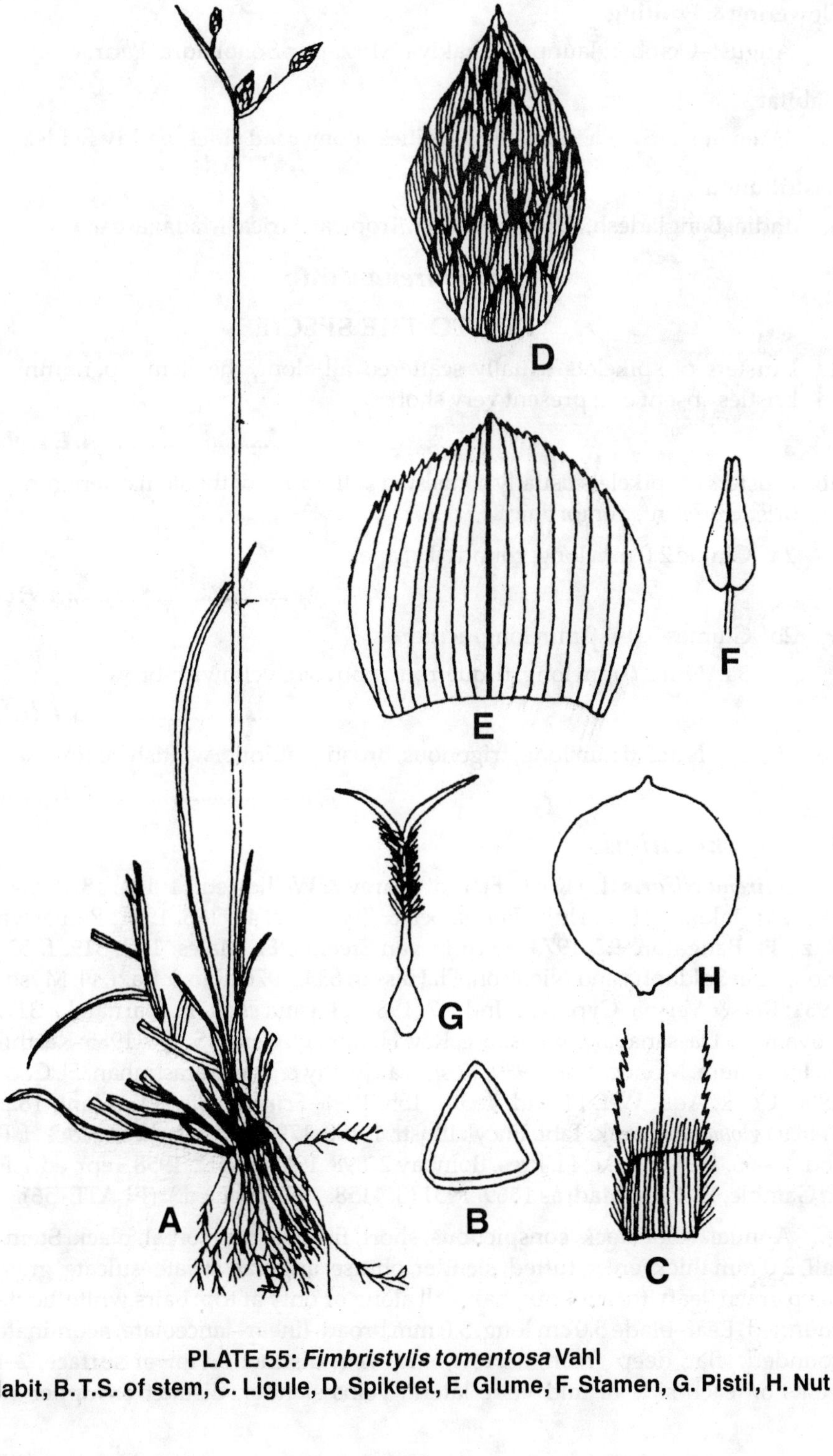

PLATE 55: *Fimbristylis tomentosa* Vahl
A. Habit, B. T.S. of stem, C. Ligule, D. Spikelet, E. Glume, F. Stamen, G. Pistil, H. Nut

Flowering & Fruiting

August–October; Jaunpur, Chakiya, Mirzapur, Sonbhadra, Padrauna.

Habitat

Open or shady places, moist localities, along road sides, paddy fields.

Distribution

India, Bangladesh, China, Malesia, Tropical Africa, Madagascar.

6. *Fuirena* Rottb.

KEY TO THE SPECIES

1a. Clusters of spikelets usually scattered all along the stems; perianth bristles absent or if present very short

.... **4.** ***F. umbellata***

1b. Clusters of spikelets usually congested at the apex of the stems; perianth bristle present, domorphic:

2a. Glume 2.0 mm long, rounded at apex

.... **1.** ***F. ciliaris***

2b. Glumes 3.0–4.0 mm long, acute:

3a. Nut 2.0 mm long, triquetrous, obovoid, yellowish-brown

.... **3.** ***F. trilobites***

3b. Nut 5.0 mm long, trigonous, broadly elliptic, whitish yellow

.... **2.** ***F. pubescens***

1. *Fuirena ciliaris*

Fuirena ciliaris (L.) Roxb. Fl. Ind. (Carey & Wallich ed.) 1: 184. 1820; non Nees, 1834; Vartak, in Journ. Univ. Poona Sci. & Tech. Sec. 40: 198. 1971; Ramaswamy & Razi, Fl. Bangalore 97. 1973; Kern in van Steenis, Fl. Males. 1. 7: 519. f. 32. 1974; Hooper in Saldanha and Nicolson, Fl. Hassan 683. 1976; Rao & Razi, Fl. Mysore 563. 1981; Rao & Verma, Cyp. N.E. Ind. 47. 1982; Sharma *et al.*, Fl. Karnataka 311. 1984; Koyama in Dassanayake & Fosberg, Rev. Handb. Fl. Ceyon 5: 150. 1985; Karthik *et al.*, Fl. Ind. Enum. Monocot. 56. 1989; KeshavaMurthy & Yoganarasimhan, Fl. Coorg 509. 1990; Cook, Aqu. Wet. Pl. Ind. 153. f. 155. 1996. *Scirpus ciliaris* L. Mant. 182. 1771. *Fuirena glomarate* Lamk. Tabl. Encyl. Illustr. 1: 150. 1791; C.B. Clarke in Hook. f. Fl. Brit. Ind. 6: 666. 1893; Cooke, Fl. Pres. Bombay 2: 898. 1908 (3: 412. 1958, repr. ed.); Fischer in Gamble, Fl. Pres. Madras 1669. 1931 (3: 1158. 1957 repr. ed.). (PLATE–56).

Annual. Root stock, conspicuous, short, fibrous, horizontal, black. Stem 15 cm tall, 2.0 mm thick, erect, tufted, slender, obtuse, angular, striate–sulcate, grass–like, deep green, leafy throughout, hairy all along or only at top; hairs white, acute, base rounded. Leaf–blade 5.0 cm long, 3.0 mm broad, linear–lanceolate, acuminate, base rounded, flat, deep green, narrow, mid–vein distinct on lower surface, 2–lateral veins on each side of mid vein; lateral veins slender, distinct on upper surface,

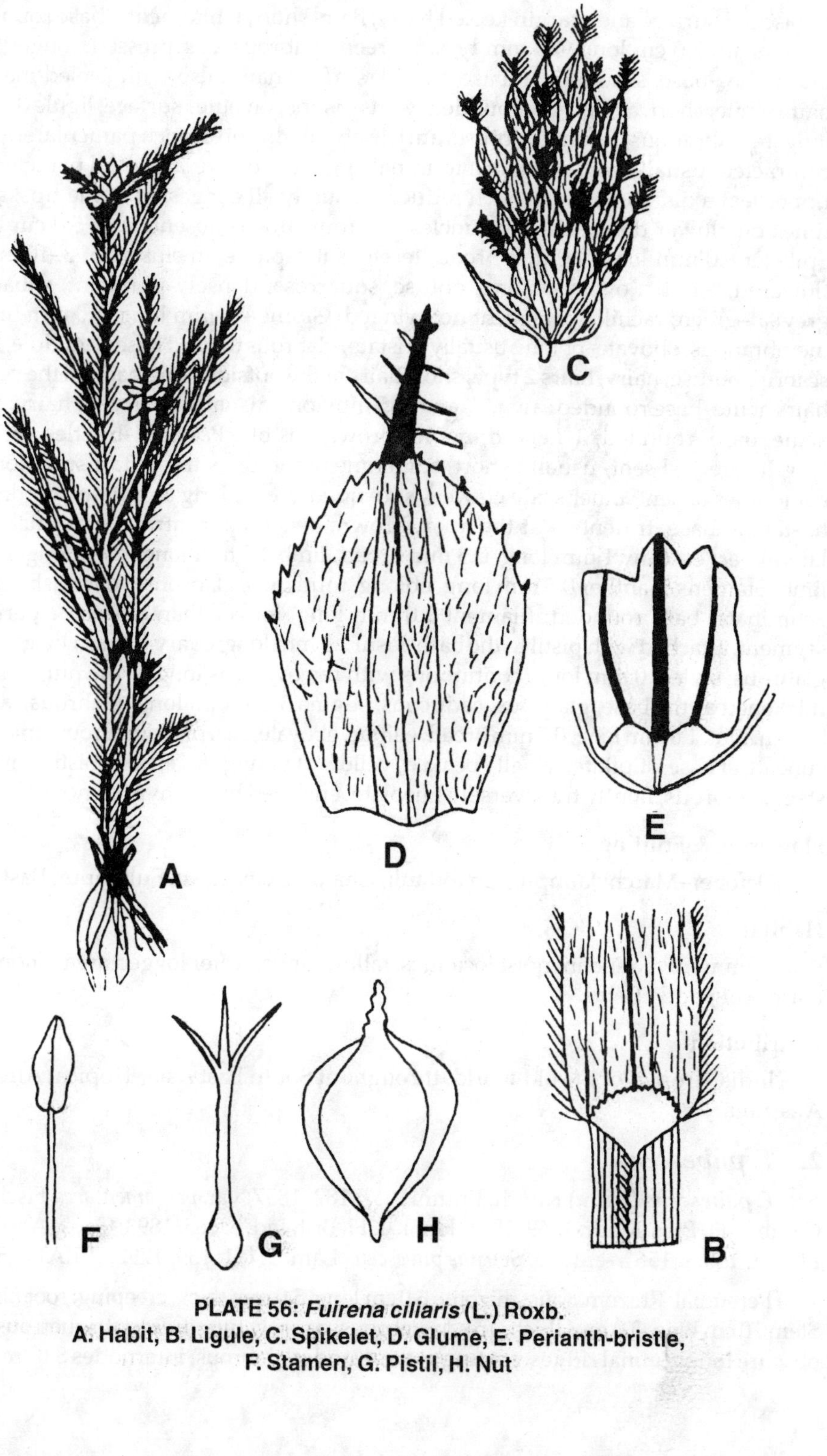

PLATE 56: ***Furirena ciliaris*** **(L.) Roxb.**
A. Habit, B. Ligule, C. Spikelet, D. Glume, E. Perianth-bristle, F. Stamen, G. Pistil, H. Nut

pubiscent, surface and margin keeled hairy; hairs short, white, acute, base rounded; leaf–sheath 7.0 cm long, 4.0 mm broad, green, glabrous, compressed, longer than internode, loose, sheathing at base, hispid, surface hairy also with keeled margin; hairs white, short, acute, base rounded, veins distinct on outer surface; ligule 0.2 mm long, membranous, lacerate, obtuse, auricle absent. Inflorescence paniculate, much contracted, usually consisting of a terminal, and one to several partial panicles in a upper leaf axils, sometimes much reduced, usually all congested at the tips, some times one lower down also; peduncles glabrous. Bracts absent; rays 5.0 cm long. Spikelet 6.0 mm long, 4.0 mm broad, terete, subcapitate, in cluster of 3–10, many flowered, elliptic, ovoid–oblong, obtuse, squarrose, densely pubscent, subacute, greyish–green; rachilla persistent not winged. Glume 2.0 mm long, 1.0 mm broad, membranous, obovate–oblong usually aristate, glabrous toward base to middle, apex scabrid, obtuse, hairy; hairs 2 type, short hairs at the outside, long hairs in the center; hairs acute, base rounded; awned; awn 0.5 mm long, usually recurved, hairy; hairs acute, base rounded; 3–nerved; nerves brown, visible. Perianth–bristles variable length rarely absent, usually short, sometimes as long as the nut, hispidulous or sometimes absent; inner scaly clawed, sub–quadrate, as long as the nut, cordate or hastate at base, tridentate at the apex, brown, 3–nerved; central nerve thick than lateral nerves; claw 1 mm long, 0.5 mm broad, limb 0.5 mm long, claw longer than limb. Stamens 3; anther 0.5 mm long, oblong, muticous, glabrous, yellowish–green, acuminate, base rounded; filament 1.0 mm long, longer than anther & perianth segment, attached with pistil to the base. Pistil 2.5 mm long; ovary 0.5 mm long, ovate, glabrous; style 1.0 mm long, continuous with the ovary, as long as the nut, scarcely thickened at the base, glabrous, caducous; stigma 3, 1.0 mm long, glabrous, acute, brown. Nut. 1.0 mm long, 0.5 mm broad, elliptic, obovate, sharply triquetrous, apiculate, cuneate at base, stipitate, usually beaked, yellow at young stage, brownish at mature stage, glabrous, faintly transversely lineolate, enclosed by the hypogynous scales.

Flowering & Fruiting

October–March; Jaunpur, Chandauli, Chakiya, Ghazipur, Sultanpur, Basti.

Habitat

Commonly found in moist localities, fallow lands, water logged areas, along the margin of paddy fields.

Distribution

India, Tropics of the old world, throughout South East Asia, Tropical Africa, & Australia.

2. *F. pubescens*

F. pubescens (Lam.) Kunth. Enum. Pl. 2: 182. 1837; var. *pergamentacea* Fischer in Gamble, Fl. Pres. Madr. 1669. 1931; Hook. f. Fl. Brit. Ind. 6: 665. 1893; Cook, Aqu. Wet. Pl. Ind. 153. f. 156 a–e. 1996. *Scirpus pubescens* Lam. 111. 1: 139. 1792. (PLATE–57).

Perennial. Rhizomatous; rhizome 6.0 cm long, 5.0 mm thick, creeping; root fibrous. Stem 70 cm long, 2.0 mm thick, green, glabrous, densely tufted, erect, triquetrous with obscure longitudinal ridges, rooting at base; nodes glabrous; internodes 8.0 cm long.

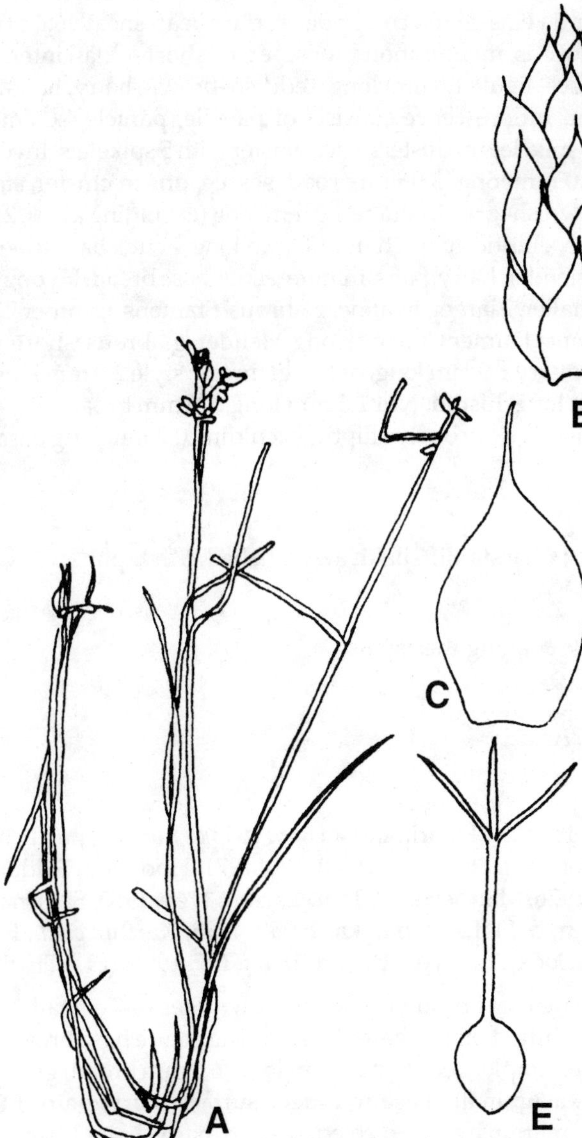

PLATE 57: ***Fuirena pubescens*** **(Lamk.) Kunth**
A. Habit, B. Spikelet, C & D. Glume, E. Pistil

Leaf–blade 20 cm long, 4.0 mm broad, green, glabrous, narrow, flat, linear–lanceolate, acuminate, base rounded, mid–vein distinct on lower surface, 2–lateral veins on each side of mid vein; lateral veins distinct on upper surface; leaf–sheath 6.0 cm long, 3.0 mm broad, brown, glabrous, membranous, loose, erect, shorter than internodes, veins distinct on both surface; ligule 1.0 mm long, reddish–brown, hairy; hairs acute, base broad; auride absent. Inflorescence consists of pancile; panicle 4.0 cm long, open, contracted, terminal spikelets in dusters, each cluster with 5 spikelets. Involucral bracts absent. Spikelets 6.0 mm long, 3.0 mm broad, sessile, one in cluster, stalked, ovate, many flowered, brownish–green. Glume 4.0 mm long (excluding awn), 2.0 mm broad, membranous, ovate, elliptic, hairy; hairs 1.0 mm long, acute, base tubercled; awned, own 0.5 mm long, slender, hairy, hairs minute, acute, base broad. Hypogynous bristle 3, 1.0 mm long, 1 smaller, 2 larger, hyaline, glabrous. Stamens 3; anther 2.0 mm long, yellow, linear, glabrous; filament 1.0 mm long, slender, glabrous, shorter than anther. Pistil 4.0 mm long; ovary 1.0 mm long, ovate, glabrous; style 2.0 mm long, flat, glabrous; stigma 3, 1.0 mm long, hispid. Nut 1.5 mm long, 1.0 mm broad, trigonous, with 3 distinct longitudinal ridges, broadly elliptic, exculding 0.2 mm long basal part of style, whitish–brown.

Flowering & Fruiting

September–December; Chandauli, Chakiya, Allahabad, Mirzapur.

Habitat

Rare in marshes, pools & along the canals.

Distribution

India, Europe, Africa, Asia.

3. *F. trilobites*

F. trilobites Clarke in Hook. f. Fl. Brit. Ind. 6: 666. 1893 & Illus. Cyper. t. 59. f. 5. 1909; Vartak in J. Univ. Poona Sci. & Tech. Sect. 40: 196. 1971; Hooper in Saldanha & Nicolson, Fl. Hassan 683. 1976; Matthews, Fl. Tam. Carn. 3: 1764. 1983, Sharma *et al.*, Fl. Karnataka 311. 1984; Singh, Fl. E. Karnataka. 2: 639. 1988; Karthik *et al.*, Fl. Ind. Enum. Monocot. 56. 1989; Cook, Aqu. Wet. Pl. Ind. 154. f. 157. 1996. (PLATE–58).

Annual. Rhizome absent; root fibrous, blackish–brown. Stem. 40 cm tall, 1.0 mm thick, slender, striate, erect, tufted, hairy: hairs 1.0 mm long, acute base broad, leafy throughout, rooting at base, simple. Leafblade 10 cm long, 6.0 mm broad, green, flat, narrow, linear, lanceolate, acuminate, base rounded, surface hairy; hairs 1.0 mm long, acute, base tubercled; margin hyaline, keeled, scabrid, hairy; hairs 1.0 mm long, acute, base broad, mid–vein distinct on lower surface, 3–lateral veins on each side of mid–vein; lateral veins distinct on upper surface; leaf–sheath 3.0 cm long, 3.0 mm broad, green, hairy; hairs 1.0 mm long, acute, base tubercled, margin hyaline, glabrous, erect, striate, loose; ligule 1.0 mm long. Inflorescence of solitary, axilary and terminal 3–9 spikelets in clusters, but inflorescence invariably arising from leaf axil. Bracts absent. Spikelets 7.0 mm long, 1.0 mm broad, ovoid, oblong, pale brown, squarrose, many flowered; rachilla winged, glabrous. Glumes 3.0 mm long, 1.5 mm broad, obovate, greyish–black, surface hairy between the nerves also with margin hairs 1.0

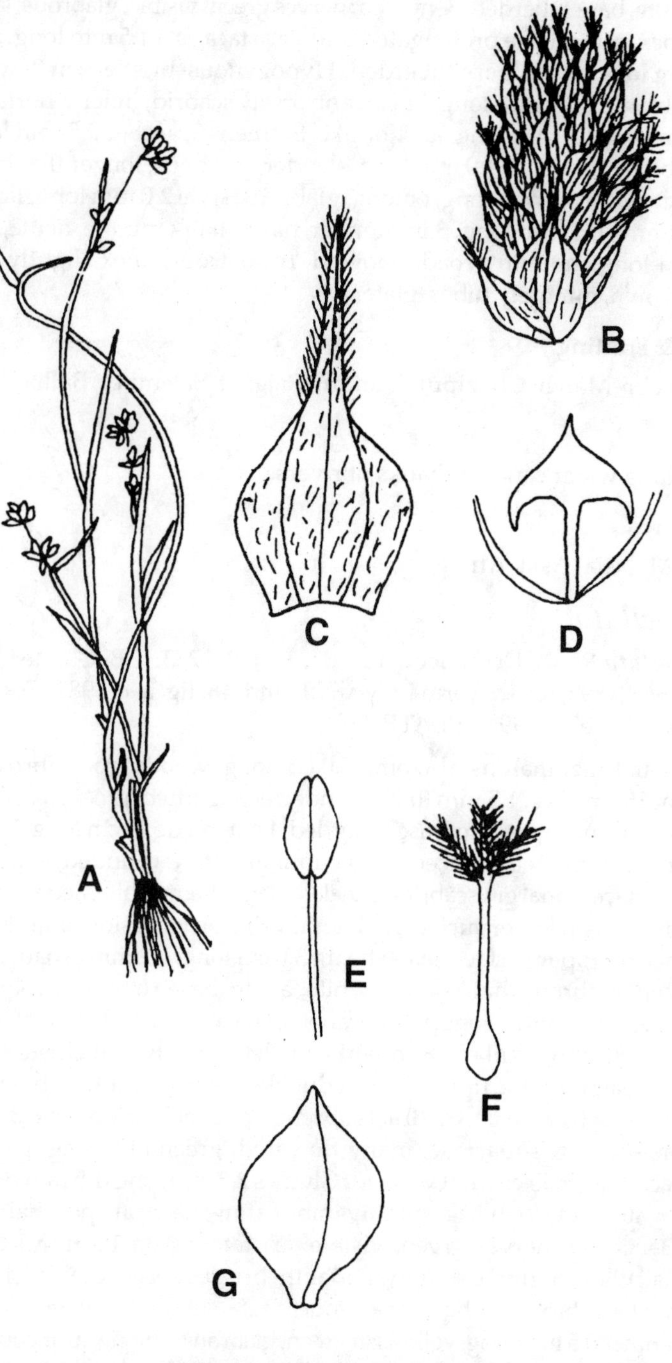

PLATE 58: ***Furirena trilobites*** **Clarke**
A. Habit, B. Spikelet, C. Glume, D. Perianth-bristle, E. Stamen, F. Pistil, G. Nut

mm long, acute, base tuberded, 3–nerved; nerves green, visible, glabrous, lateral nerves ending at apex, central one prolongated into a arista; arista 1.5 mm long, acute, hairy; hairs 1.0 mm long, acute, base tuberded. Hypogynous–bristle 6; in 2–whorls; outer whorls of 3 bristles, 1.0 mm long, linear, antrorsely scabrid, inner whorls of 3–scales, petaloid, laminae 1.5 mm long, anchor like. Stamens 3; anther 0.5 mm long, yellow, glabrous, oblong, filament 1.0 mm long, slender, glarbous, longer than anther. Pistil 3.0 mm long; ovary 0.5 mm long, oblong, glabrous; style 2.0 mm long, flat, undivide, glabrous; stigma 0.5 mm long, 3 in number, hairy; hairs minute, acute, base broad. Nut 2.0 mm long, 1.0 mm broad, obovoid, triquetrous, apex slightly protudring, yellowish brown, minutely tuberculate.

Flowering & Fruiting

Septemebr–March; Ghazipur, Jaunpur, Rajghat, Salempur, Ballia.

Habitat

Rare along water canals & banks of rivers.

Distribution

India, Malasia, Asia. Africa.

4. *F. umbellata*

F. umbellata Rottb. Descr. Icon. Pl. 70 t. 19. f. 3. 1773; C.B. Clarke in Hook. f. Fl. Brit. Ind. 6: 666. 1893; Rao & Verma. Cyp. N.E. Ind. 48. fig. a–c. 1982; Cook, Aqu. Wet. Pl. Ind. 154. f. 158 a–d. 1996. (PLATE–59).

Perennial. Rhizomatous; rhizome 1.0 cm long, woody; root fibrous, blackish–brown. Stem 15 cm long, 1.5 mm thick, slender, erect, tufted, green, leafy throughout, hairy; hairs 1.0 mm long, acute, base tuberded. Leaf–blade 10 cm long, 3.0 mm broad, green, narrow, flat, linear, lanceolate, acuminate, base rounded, narrow, densely pubescent, surface & margin scabrid & keeled, hairy; hairs white, acute, base rounded; mid–vein distinct on lower surface, 3–lateral veins on each side of mid–vein; lateral veins distinct on upper surface; leaf–sheath 3.0 cm long, 1.2 mm broad, loose striate–sulcate, green, glabrous, hairy; hairs white, acute, base rounded; veins distinct on outer surface; ligule 1.0 mm long, membranous, lancerate; auricle absent. Inflorescence paniculate, solitary or axillary, termianl spikelet, 4 rarely 9 in clusteres with loose peduncled cluster form the upper axis; peduncles densely scabrous but inflorescence invariably arising from leaf axil. Bracts absent. Spikelet 7.0 mm long, 3.0 mm broad, terete, ovoid–oblong, squarrose, many flowered, green at young, pale–brown on maturity; rachilla persistent, not winged. Glumes 2.5 mm long, 1.5 mm broad, oblong, obtuse, obovate, greyish–black, spirally imbricating, hairy at apex, hairs acute, base bulbous, 3–nerved; nerves green, visible; aristate; arista 1.5 mm long, glabrous recurved, hairy, lower glumes empty. Perianth–bristles absent or very short; perianth scales 1.0 mm long, brown, subquadrate, mucronate, subsessile or very shortly clawed. Stamens 3; anther 0.5 mm long, yellowish green, glabrous, oblong, muticous, acuminate, base rounded; filament 1.0 mm long, flat, longer than anther. Pistil 1.5 mm long; ovary 0.5 mm long, ovate, glabrous; style 0.5 mm long, continuous with ovary, slightly thickened at base, glabrous, caducous; stigma 3, 0.5 mm long, minutely hairy; hairs

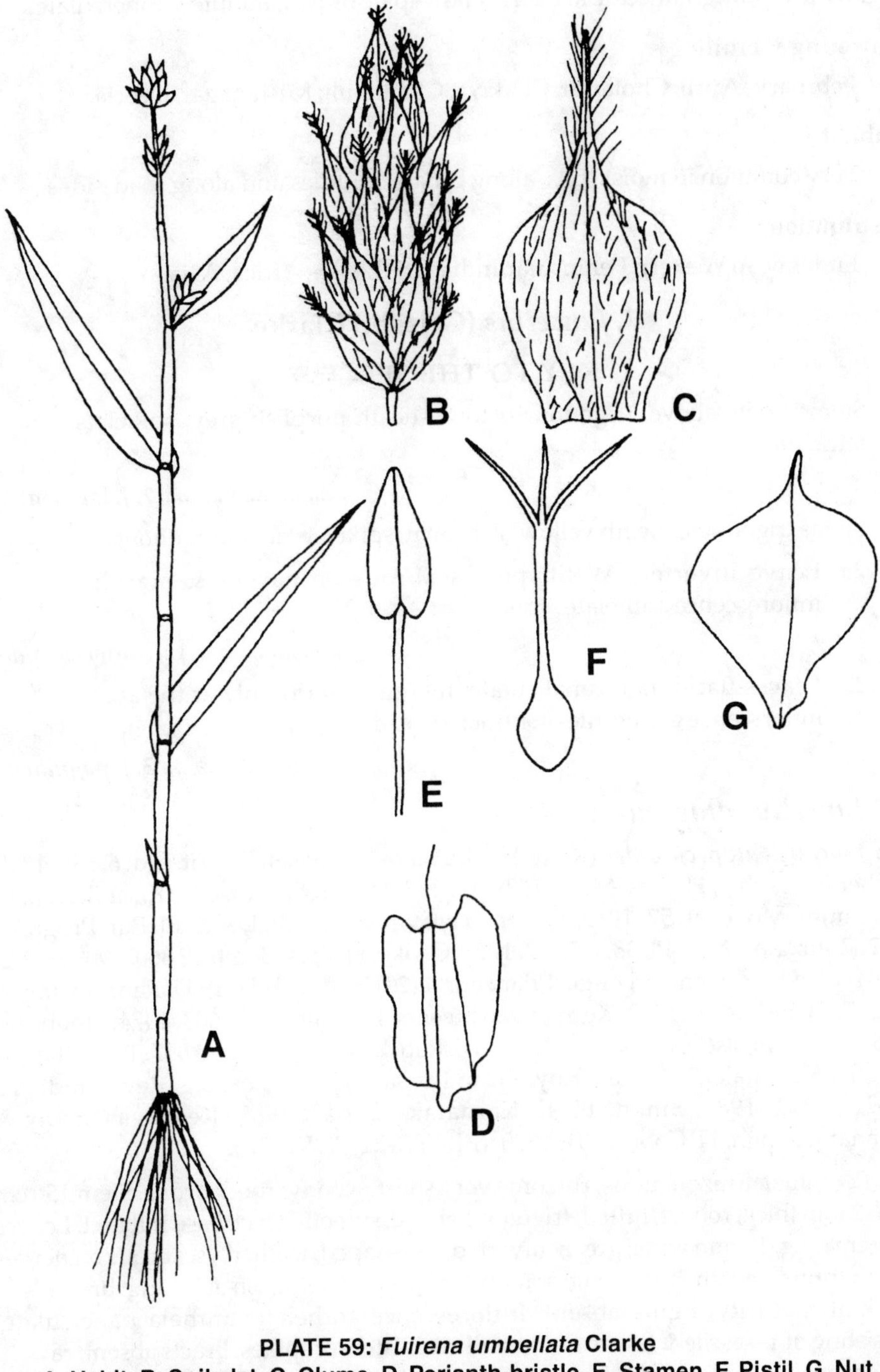

PLATE 59: *Fuirena umbellata* Clarke
A. Habit, B. Spikelet, C. Glume, D. Perianth-bristle, E. Stamen, F. Pistil, G. Nut

acuminate, base broad. Nut 1.4 mm long, 0.5 mm broad, obovoid, triquetrous, apex slightly protuding, cunceate at base, yellow–pale brown, minutely tuberculate.

Flowering & Fruiting

February–April; Cholapur, Chakiya, Chandauli, Kushinagar, Deoria.

Habitat

Very common in moist soils, along canals, puddles and along road-sides.

Distribution

Endemic to Western Peninsular India, Sri–Lanka, Africa, Asia.

7. *Juncellus* (Griseb.) Clarke

KEY TO THE SPECIES

1a. Stems terete above, trigonous below; sheath purplish grey; spikelets tetragonous

.... *2. J. laevigatus.*

1b. Stems trigonous; sheath yellowish brown; spikelets not tetragonous:

2a. Leave inverted, W–Shaped, scabrous on nerves & margin; inflorescence anthelate

.... **1.** ***J. alopecuroides***

2b. Leaves flacid, flat, canaliculate, margin scabrid only at the apex, inflorscence glomerules contracted head

.... **3.** ***J. pygmaeus.***

1. *Juncellus alopecuroides*

Juncellus alopecuroides (Rottb.) C.B. Clarke in Hook. f. Fl. Brit. Ind. 6: 595. 1893; Fischer in Gamble, Fl. Pres. Madr. 1629. 1931 (3: 1133. 1957 repr. ed.); Karthik *et al.*, Fl. Ind. Enum. Monocot. 57. 1989. *Cyperus alopecuroides* Rottb. Descr. Pl. Rar. Progr. 20. 1772; Rottb. Rov. Nov. Pl. 38. t. 8. f. 2. 1773; Cooke, Fl. Pres. Bomb. 2: 860. 1908 (3: 373. 1958 repr. ed.); Kukenth in Engl., Pflanzenr. 4 (20) Heft. 101: 71. 1935; Ramaswamy & Razi, Fl. Banglore 80. 1973; Kern in van Steenis, Fl. Males. 1, 7: 603. 1974; Hooper in Saldanha & Nicolson, Fl. Hassan 663. 1976; Rao & Razi, Fl. Mysore 557. 1981; Sharma *et al.*, Fl. Karnataka 304. 1984; Koyama in Dassanayake & Fosberg, Rev. Handb. Fl. Ceyl. 5: 213. 1985; Singh, Fl. E. Karnataka 2: 629. 1988; KeshavaMurthy & Yoganarasimhan, Fl. Coorg. 501. 1990. (PLATE–22J, 22K, 22L).

Perennial. Rhizomatous; rhizome very short, woody; root fibrous. Stem 150 cm tall, 12 mm thick, robust, tufted, trigonous above, smooth, erect, base inflated. Leaves 5, 70 cm long, 17 mm wide, green, inverted, W–shaped, leathery, scabrous on nerves and margins; sheath 25 cm long, somewhat spongy, yellowish at young, brownish–black at maturity; ligule absent. Inflorescence anthelate; anthela panciculate, consisting of 1–sessile & few to many stalked culsters of spikes; Bracts absent; rays 10 in number, 20 cm long, the larger ones sometimes with tertiary clusters of spikes.

Spikes 4.0 cm long, 15 mm in diameter, digitate, upright, cylindrical, oblong. Spikelets 6.0 mm long, 2.5 mm broad, spicate, eventually spreading, ovate, oblong, lanceolate, 30 flowered, pale golden at young, reddish–brown with age; rachilla peresistent, strongly flattened, winged; wings whitish, narrow, long persistant. Glumes 2.5 mm long, 1.5 mm broad, ovate, elliptic, obtuse, apiculate, minutely mucronate, 7 nerved, yellowish–golden with purple stripes, densely imbricate; keel absent, midrib excurrent in a short mucro. Stamens 2; anther 0.5 mm long, yellow, oblong, glabrous; filament 1.0 mm long, slender, glabrous, longer than anther. Pistil 2.0 mm long; ovary 0.5 mm long, ovate, glabrous; style 1.0 mm long, bicleft, glabrous; stigma 0.2 mm long, acuminate. Nut 1.0 mm long, 0.5 mm broad, dorsoventrally flattened, planoconvex, ellipsoid, golden yellow, minutely reticulate.

Flowering & Fruiting

July–March; Jaunpur, Ghazipur, Allahabad, Sultanpur, Azamgarh.

Habitat

Usually in standing water, in pools.

Distribution

India, Australia, Sri Lanka, Malesia, North & Tropical Africa, Madagascar, West Indies & Australia.

2. *J. laevigatus*

J. laevigatus (L.) C.B. Clarke in Hook. f. Fl. Brit. Ind. 6: 596. 1893; Fischer in Gamble, Fl. Pres. Madr. 1629. 1931 (3: 1133. 1957 repr. ed.) Karthik *et al.*, Fl. Ind. Enum. Monocot. 57. 1989. *Cyperus laevigatus* L. Mant. Alt. 179. 1771; Kern in Fl. Males. ser. 1. 7: 661. 1974; Misra & Verma, Fl. Allaha. 384. 1992; Cooke, Fl. Pres. Bomb. 2: 860. 1908 (3: 373. 1958 repr. ed.); Kukenth in Engl., Pflanzenr. 4 (20), Heft 101: 321. 1936; Hooper in Saldanha & Nicolson, Fl. Hassan 666. 1976; Rao & Razi, Fl. Mysore 560. 1981; Sharma *et al.*, Fl. Karnataka 306. 1984; Singh, Fl. E. Karnataka 2: 632. 1988. (PLATE–22M, 22N).

Perennial. Rhizomatous; rhizome 30 cm long, 5.0 mm thick, creeping, lightly brown–purplish black; becoming woody, root fibrous. Stem 35 cm long, 2.0 mm thick, green, glabrous, close together in rows, rarely solitary along the creeping rhizome, terete above, trigonous below. Leaves 2.5 cm long, 2.0 mm broad, inrolled, stem like, reduced to sheath; sheath grey at young, purplish brown with age; membranous. Bracts 2, 3.0 cm long, erect, stem like, overtopping the inflorescence; rays absent. Inflorescence pseudolateral, 1.5 cm in diameter, a single cluster of spikelet. Spikelet 10 mm long, 3.5 mm broad, sessile, linear, lanceolate, rarely curved, 30 flowered; rachilla straight, persistent, tetragonous, hardly winged. Glume 3.5 mm long, 2.0 mm broad, broadly elliptic, ovate, subapiculate, mucronate, obscurely 4–nerved, whitish–straw at young, brownish with age, with reddish dots, sides faintly striped, margin thin; keel distinct. Stamens 3; anther 1.0 mm long, yellow, linear, oblong; filament 1.5 mm long, slender, glabrous, longer than anther. Pistil 3.0 mm long; ovary 0.5 mm long, ovate, glabrous; style 1.5 mm long, bicleft, glabrous; stigma 1.0 mm long, slender, glabrous. Nut 2.0 mm long, 1.0 mm broad, planoconvex, obovoid, stipitate, apiculate, brown, surface smooth, but densely punctate with isodiametric surface cells.

Flowering & Fruiting

October–February; Chakiya, Chandauli, Ghazipur, Khalilabad, Allahabad.

Habitat

Rare, along moist sandy beds of rivers & canals.

Distribution

India, West Africa, Mediterranean Europe, South America, Africa, Australia.

3. *J. pygmaeus*

J. pygmaeus (Rottb.) Clarke in Hook. f. Fl. Brit. Ind. 6: 596. 1893; Fischer in Gamble, Fl. Pres. Madr. 1629. 1931 (3: 1133. 1957 repr. ed.); Karthik *et al.*, Fl. Ind. Enum. Monocot. 58. 1989. *Cyperus michelianus* (L.) Delile. subsp. *pygmaeus* (Rottb.) Aschers and Graebn., Synops. Mitteleur. Fl. 2(2): 273. 1903; Ramaswamy & Razi, Fl. Bangalore 90. 1973; Hooper in Saldhana & Nicolson, Fl. Hassan 667. 1976; Rao & Razi, Fl. Mysore 560. 1981; Sharma *et al.*, Fl. Karnataka 307. 1984; Singh, Fl. E. Karnataka 2: 632. 1988. *C. pygmaeus* Rottb. Descr. Incon. Rar. Pl. 20. t. 14. f. 4–5. 1773; Kern in Fl. Males ser. 1, 7 (3): 634. 1974; Cooke, Fl. Pres. Bomb. 2: 859. 1908 (3: 372. 1958 repr. ed.); Koyama in Dassanayake & Fosberg. Fl. Ceylon 5: 214. 1985; Panigrahi *et al.*, in Bull. Bot. Surv. Ind. 11: 111. 1969. (PLATE–60).

Annual. Rhizome absent; root fibrous. Stem 20 cm tall, 1.0 mm thick, green, glabrous, tufted, erect–patent, base brownish, often pulvinate, trigonous. Leaves 16 cm long, 2.0 mm broad, numerous, flat, canaliculate, flacid, linear, acuminate, surfce glabrous, margin scabrid at apex, clustered at base, reduced to bladeless sheath. Inflorescence various, distinctly terminal of a few glomerules contracted head. Bracts 2–6, 10 cm long, foliaceous, patent, overtopping the inflorescence, dilated at base. Spikelets 6.0 mm long, 1.5 mm broad, strongly compressed, numerous, congested, ovate–oblong, lanceolate, distichouly arranged, whitish green–stramineous; rachilla slightly flexous, curved, twisted, winged. Glume 3.0 mm long, 1.0 mm broad, membranous, brown streaks, oblong–lanceolate, acuminate, base flat, cuspidate, broadly hyaline on margin, yellowish, subtransluscent, mucronate; mucro 0.5 mm long, scabrid, 3–nerved; nerves green, visible. Stamens 2; anther 0.5 mm long, brown, acuminate, base rounded, glabrous, oblong, lanceolate; filament 1.0 mm long, uniformly flat, slender, longer than anther, attached with pistil to the base. Pistil 1.5 mm long; ovary 0.5 mm long, ovate, glabrous, base slightly rounded; style 0.5 mm long, deeply bifid, flat, glabrous; stigma 0.5 mm long, acute, glabrous, brown. Nut 1.0 mm long, 0.5 mm broad, orange–brown, ellipsoid, oblong, plano–convex, trigonous, minutely puncticulate.

Flowering & Fruiting

August–November; Chandauli, Ghazipur, Lar, Sultanpur, Gorakhpur.

Habitat

Along roadside ditches, river beds, dry mudy areas, near ponds.

Distribution

India, Widely distributed in South & South East Asia, Mediterranean region. Africa, Australia.

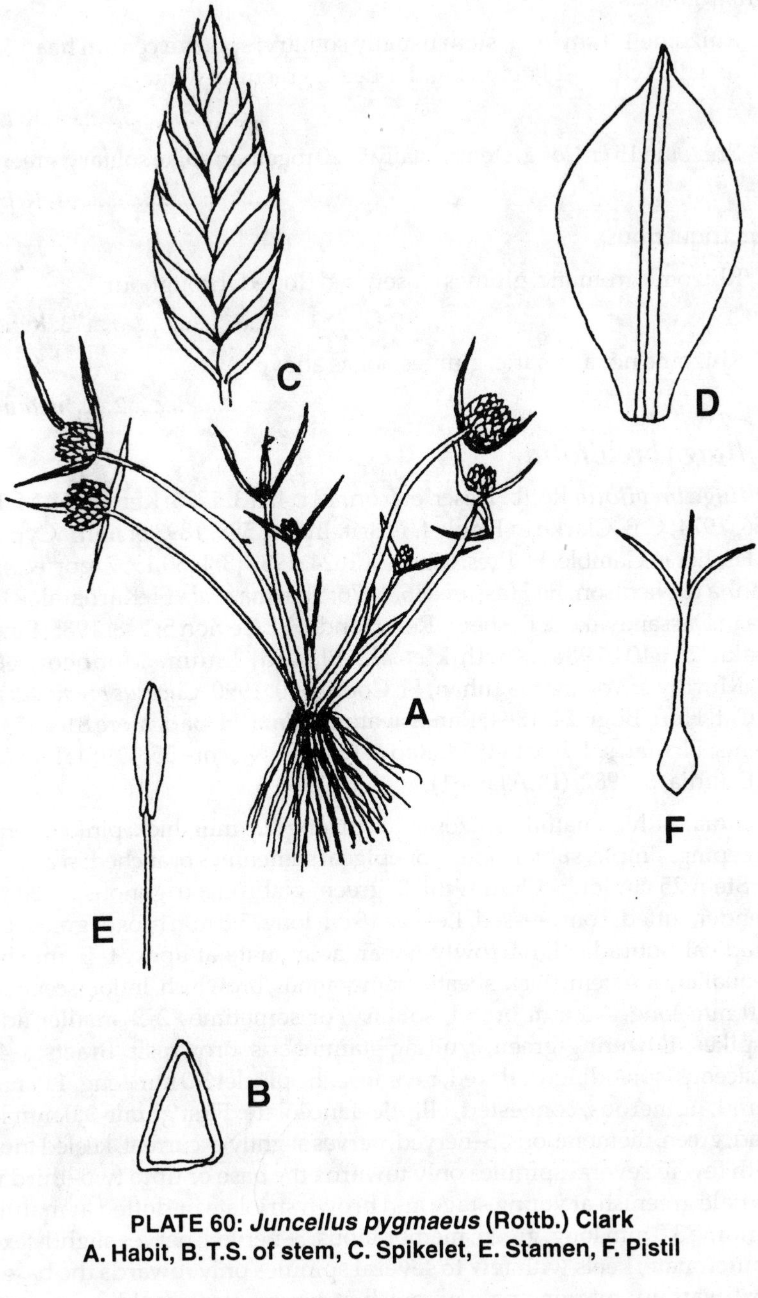

PLATE 60: *Juncellus pygmaeus* (Rottb.) Clark
A. Habit, B. T.S. of stem, C. Spikelet, E. Stamen, F. Pistil

8. *Kyllinga* Rottb.

KEY TO THE SPECIES

1a. Stem trigonous:

2a. Rhizome 1.0 cm long; stem usually solitary; spikes a central head & usually with 1–2 lateral smaller head, greenish-white .... **4. *K. tenuifolia***

2b. Rhizome 15 cm long; stem usually 5–20 together; spike solitary, green .... **1. *K. brevifolia***

1b. Stem triquetrous:

3a. Rhizome aromatic, glumes closely red dotted throughout .... **3. *K. nemoralis***

3b. Rhizome not aromatic, glumes not as above .... **2. *K. melanosperma***

1. *Kyllinga brevifolia*

Kyllinga brevifolia Rottb. Descr. et Icon. 13 t. 4. f. 3. 1773; Kern in Fl. Males. ser 1. 7 (3): 656. 1974; C.B. Clarke in Hook. f. F. Brit. Ind. 6: 588. 1893 et in III. Cyp. t. 1. f. 1–4. 1909; Fischer in Gamble, Fl. Pres. Madras. 1624. 1931 (3: 1130.1957 repr. ed.); Hooper in Saldanha & Nicolson, Fl. Hassan 686. 1976; Sharma *et al.*, Fl. Karnataka 312. 1984; Koyama in Dassanayake & Fosberg. Rev. Handb. Fl. Ceylon 5: 248. 1985; Singh, Fl. E. Karnataka 2: 640. 1988; Karthik *et al.*, Fl. Ind. Enum. Monocot. 60. 1989; KeshavaMurthy & Yoganarasimhan, Fl. Coorg 510. 1990. *Cyperus brevifolius* (Rottb.) Hassk., Cat. Hort. Bogr. 24. 1884; Ramaswamy & Razi, Fl. Bangalore 81. 1973; Kern in van Steenis, Fl. Males, 1.7: 656. 1974; Rao & Razi, Fl. Mysore 558. 1981; Rao & Verma, Cyp. N.E. India. 6. 1982. (PLATE–61).

Perennials. Rhizomatous; rhizome 15 cm long, 2.0 mm thick, pinkish or purplish black, creeping, simple, subterranean or epigeal sometimes branched; scales ultimate, fibrous. Stem 25 cm long, 1.5 mm thick, green, glabrous, trigonous, 5–20 together, erect, slender, tufted, compressed. Leaves 18 cm long, 3.0 mm broad, green, glabrous, fleshy, radical–subradical, narrowly linear, acuminate at apex, 1–5, much shorter than to equalling the stem, thick, sheath membranous, brownish. Inflorescence capitate, spike 8.0 mm long, 4.0 mm broad, solitary, or sometimes 2–3 smaller additional lateral spikes, flowering green, fruiting stamineous–brownish. Bracts 3–4, 10 cm long, foliaceous, spreading, deflexed; rays absent. Spikelet 3.0 mm long, 1.0 mm broad, 1–flowered, numerous, congested, elliptic–lanceolate. First glume 3.0 mm long, 1.0 mm broad, green, membranous, 3–nerved, nerves slightly excurrent, keeled mucronate; keels with few to several spinules only towards the base or upto two–third way up; margins pale greenish at young stage and brown striolate or dotted at mature stage. Second glume 3.0 mm long, green, membranous, 3–nerved, nerves slightly excurrent, keeled mucronate; keels with few to several spinules only towards the base or upto two–third way up; margins pale greenish at young stage and brown striolate or dotted at mature stage. Stamens 3; anther 1.0 mm long, brown, lanceolate–ovate,

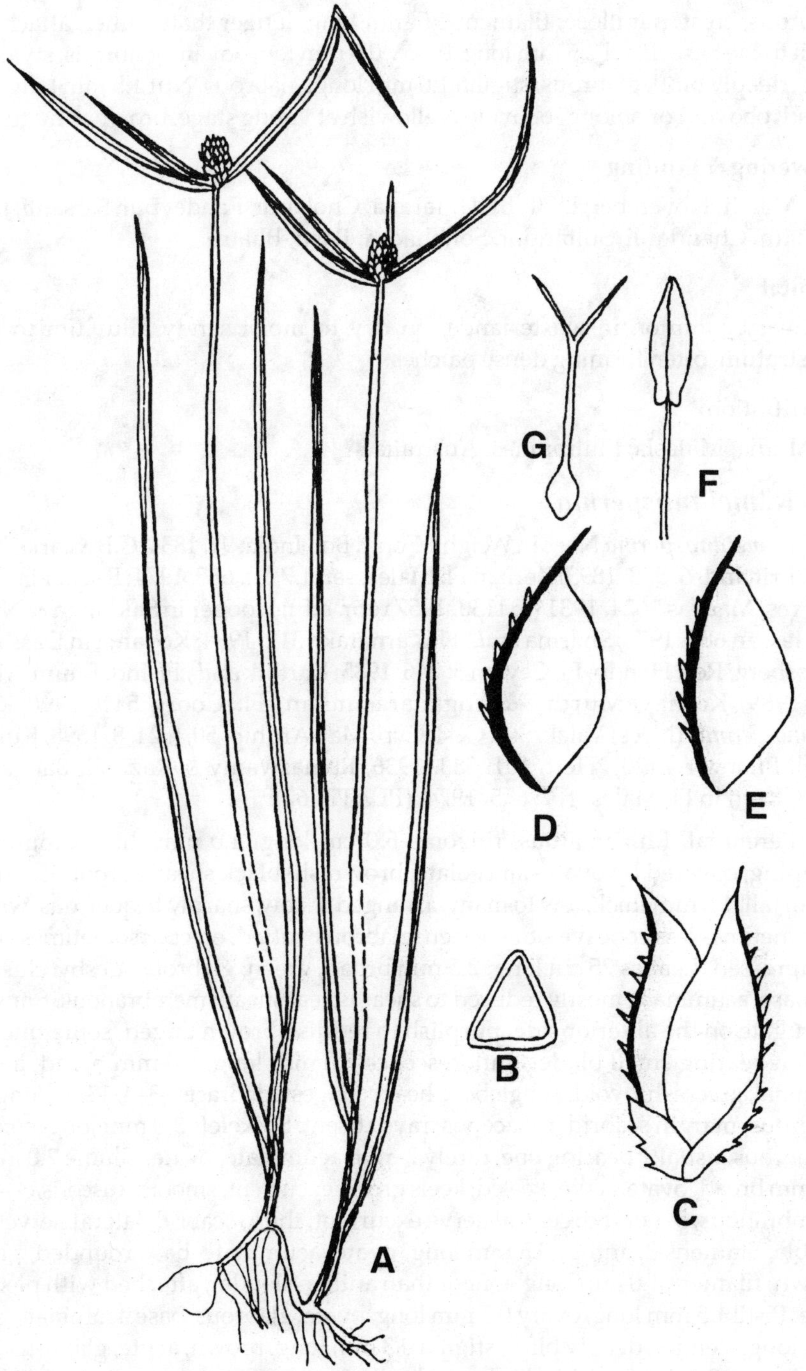

PLATE 61: ***Kyllinga brevifolia*** **Rottb.**
A. Habit, B. T.S. of stem, C. Spikelet, D. Ist Glume, E. IInd glume, F. Stamen, G. Pistil

glabrous, crest, papillose; filament 1.5 mm long, longer than anther, attached with pistil to the base. Pistil 2.5 mm long; ovary 0.5 mm long, ovate, glabrous; style 1.0 mm long, deeply bifid, glabrous; stigma 1.0 mm long, glabrous. Nut 1.5 mm long, 1.0 mm broad, obovoid or oblong–obovoid, yellowish at young stage, brown at mature stage.

Flowering & Fruiting

March–November; UPC, BHU, Tarana, Cholapur, Pandeypur, Sarsauli, Jaunpur, Sarnath, Chandauli, Sultanpur, Sonbhadra, Basti, Bhatni.

Habitat

Very common in waste lands, on dry to moist sandy alluvium to gravely substratum, often forming dense patches.

Distribution

India, Malesia, Philippines, Australia.

2. *K. melanosperma*

K. melanosperma Nees in Weight. Contr. Bot. India. 91. 1834; C.B. Clarke in Hook. f. Fl. Brit. Ind. 6: 588. 1893; Kern in Fl. Males. ser. 1.7 (3): 655. 1974; Fischer in Gamble, Fl. Pres. Madras 1624. 1931 (3: 1130. 1957 repr. ed.); Hooper in Saldanha & Nicolson, Fl. Hassan 687. 1976; Sharma *et al.*, Fl. Karnataka 312. 1984; Koyama in Dassanayake & Fosberg, Rev. Handb. Fl. Ceylon 5: 246. 1985; Karthik *et al.*, Fl. Ind. Enum. Monocot. 60. 1989; KeshavaMurthy & Yoganarasimhan, Fl. Coorg 511. 1990. *Cyperus melanospermus* (Nees) Valck. Sur. Gesl. Cyp. Mal. Ardhip. 50, f. 2 f. 8. 1898; Kukenth in Engl. Pflanzer 4 (20), Heft, 101. 583. 1936; Ramaswamy & Razi, Fl. Bangalore 90. 1973; Kern in Fl. Males. 1, 7: 655. 1974. (PLATE–62).

Perennial. Rhizomatous; rhizome 5.0 cm long, 4.0 mm thick, stout, woody, creeping, covered by ovate–lanceolate, brownish–black sheaths; root fibrous. Stem 40 cm tall, 1.5 mm thick, few to many, arranged in row, sharply triquetrous, with more or sometimes less concave sides, green, glabrous, tufted, erect or sometimes suberect, compressed. Leaves 25 cm long, 2.5 mm broad, green, glabrous, fleshy, clustered at the base, acuminate, mostly reduced to sheaths; leaf sheath membranous, transversely corrugate on the anterior side, purplish to reddish brown tinged, sometimes upper sheath bearing small blades. Inflorescence 7.0 mm long, 6.0 mm broad, a solitary, terminal, greenish, ovoid, subglobose head, congested. Bracts –3–4, 13 cm long, green, glabrous, margin scabrid, foliaceous; rays absent. Spikelets 3.0 mm long, congested, numerous, usually bearing one, rarely 2–nut, acuminate, ovate. Glume 2.0 mm long, 1.0 mm broad, ovate, acute, keeled; keels green, excurrent, smooth, fuscous, 3–4 striate, membranous, 5–nerved; central nerve excurrent, thick, scabrid, lateral nerves faintly visible. Stamens 3; anther 0.5 mm long, ovate, acuminate, base rounded, glabrous, brown; filament 1.0 mm long, longer than anther, slender, attached with pistil to the base. Pistil 1.5 mm long, ovary 0.5 mm long, ovate, glabrous, base acuminate; style 0.7 mm long, slender, deeply bifid; stigma 0.3 mm long, brown, acute, glabrous. Nut 1.5 mm long, 1.0 mm broad, black, obovoid, elliptic, oblong, trigonous, apiculate, laterally compressed.

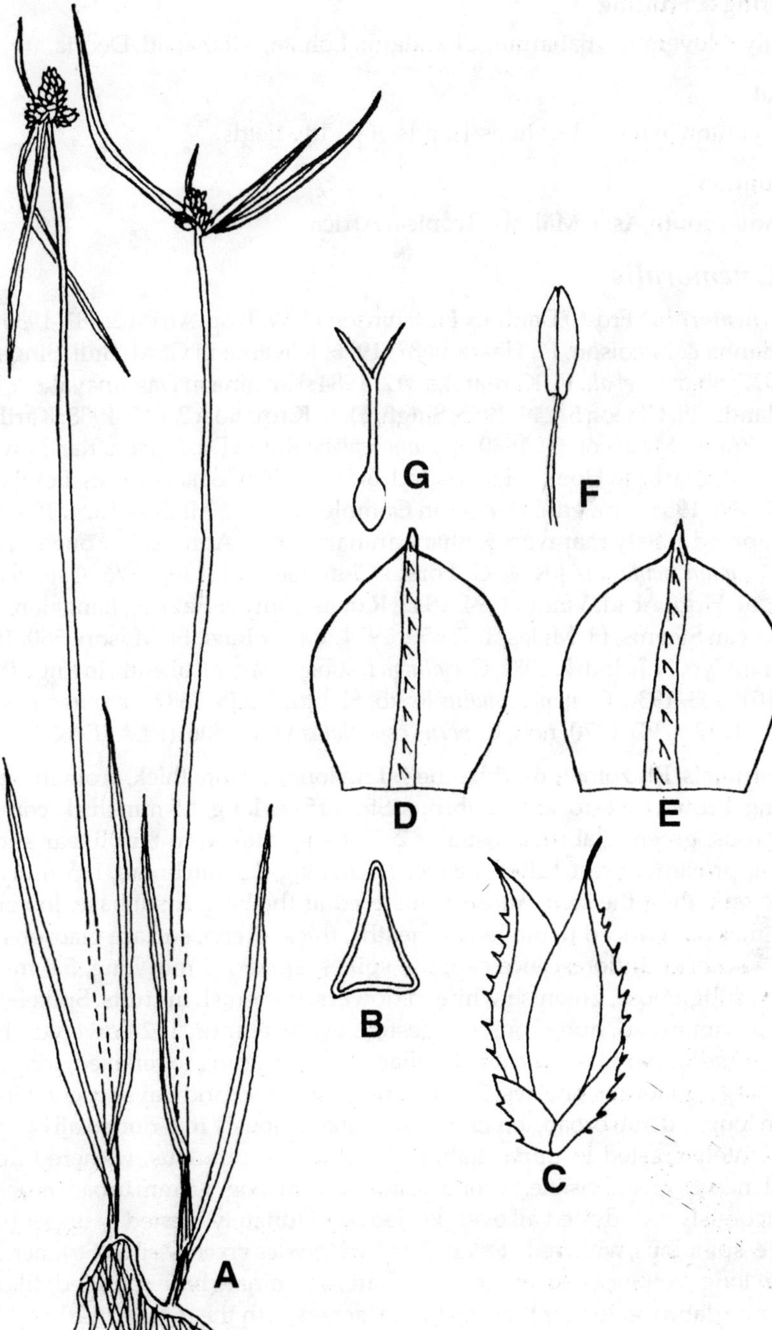

PLATE 62: *Kyllinga melanosperma* Nees
A. Habit, B. T.S. of stem, C. Spikelet, D. & E. Glume, F. Stamen, G. Pistil

Flowering & Fruiting

July–November; Babatpur, Chandauli, Lohata, Allahabad, Deoria.

Habitat

Common in moist localities, bunds of paddy fields.

Distribution

India, South Asia, Malesia, Tropical Africa.

3. *K. nemoralis*

K. nemoralis (Frost.) Danly ex Hutchinson Fl. W. Trop. Africa 2: 487. 1936; Hooper in Saldanha & Nicolson, Fl. Hassan 687. 1976; Koyama in Gard. Bull. Singapore 30: 163. 1977; Sharma *et al.*, Fl. Karnataka 312. 1984; Koyama in Dassanayake & Fosberg, Rev. Handb. Fl. Ceylon 5, 249. 1985; Singh, Fl. E. Karnataka 2: 640. 1988; Karthik. *et al.*, Fl. Ind. Enum. Monocot. 60. 1989. *K. monocephala* Rottb. Descr. Icon. Rar. Nov. Pl. 13. t. 4, f. 4. 1773; Clarke in Hook. f. Fl. Brit. Ind. 6: 588. 1893; Cooke, Fl. Pres. Bombay 2: 876. 1908(3: 390. 1958 repr. ed.); Fischer in Gamble, Fl. Pres. Madras 1624. 1931 (3: 1130. 1957 repr. ed.); Satyanarayan & Shankaranarayan in Ann. Arid Zone 2: 146. 1963. *Thryocephalon nemoralis* J.R. & G. Forst., Char. Gen. Pl. 130. 1776. *Cyperus kyllinga* Endl. Cat. Hort. Acad. Vindb. 1: 94. 1842; Ramaswamy & Razi, Fl. Bangalore 89. 1983; Kern in van Steenis, Fl. Males. 1. 7: 659. 1974; Rao & Razi, Fl. Mysore 560. 1981; Rao & Verma, Cyp. N.E. Ind. 7. 1982. *C. kyllinga* f. subsp. *triceps* Kukenth. in Engl. Pflanzenr. Heft. 101: 608. 1936. *C. monocephalus* Roxb. Fl. Ind. 2: 248. 1832. *Scirpus cephalotes* Jaq. Vindob. 1: 42. t. 97. 1770, non, *Cyperus cephalotes* Vahl 1806. (PLATE–63).

Pernnials. Rhizomatous; rhizome 4.0 cm long, 4.0 mm thick, aromatic, creeping, blacking–brown root stock; root fibrous. Stem 15 cm long, 1.5 mm thick, compressed, triquetrous, green, glabrous, usually 2–3, sometimes upto 8 in linear succession, rather approximet, erect, tufted. Leaves 17 cm long, 2.5 mm broad, 1–5, much shorter than to equalling the stem, mostly clustered at the base & subbase, lowermost or sometimes all reduced to bladeless sheaths, thick, green, surface glabrous, margin slightly scabrid. Inflorescence capitate spikes; spike 7.5 mm long, 5.0 mm broad, solitary, subglobose, greenish white in flowers, brownish in fruits. Spikelet 3.0 mm long, 1.0 mm broad, numerous, congested, ovate, elliptic, 1–2 flowered, the upper flower sometimes male. Bracts 3–4, foliaceous, spreading to reflexed, longest 20 cm long, margin glabrous, shortest 5.0 cm long, margin scabrid; rays absent. First glume 2.0 mm long, 1.0 mm broad, mucronulate, conspicuously red–dotted all over; keeled; keel lunately crested in upper half; crest ciliolate–spinulous, withered dotted; 5–nerved; nerves green, visible. Second glume 3.0 mm long, 1.5 mm broad, mucronulate, conspicuously red–dotted all over; keeled; keel lunately crested in upper half; crest ciliolate–spinulous, withered, dotted; 7 nerved; nerves green, visible. Stramen 3; anther 1.0 mm long, yellow, crest, ovate, lanceolate, acuminate base rounded; filament 1.5 mm long, glabrous, longer than anther, attached with the pistil to the base. Pistil 2.5 mm long; ovary 0.5 mm long, glabrous, ovate; style 1.0 mm long, deeply bifid, glabrous; stigma 1.0 mm long, glabrous. Nut 2.0 mm long, 1.0 mm broad, obovoid–oblong, lenticular, blackish–brown; laterally compressed, biconvex.

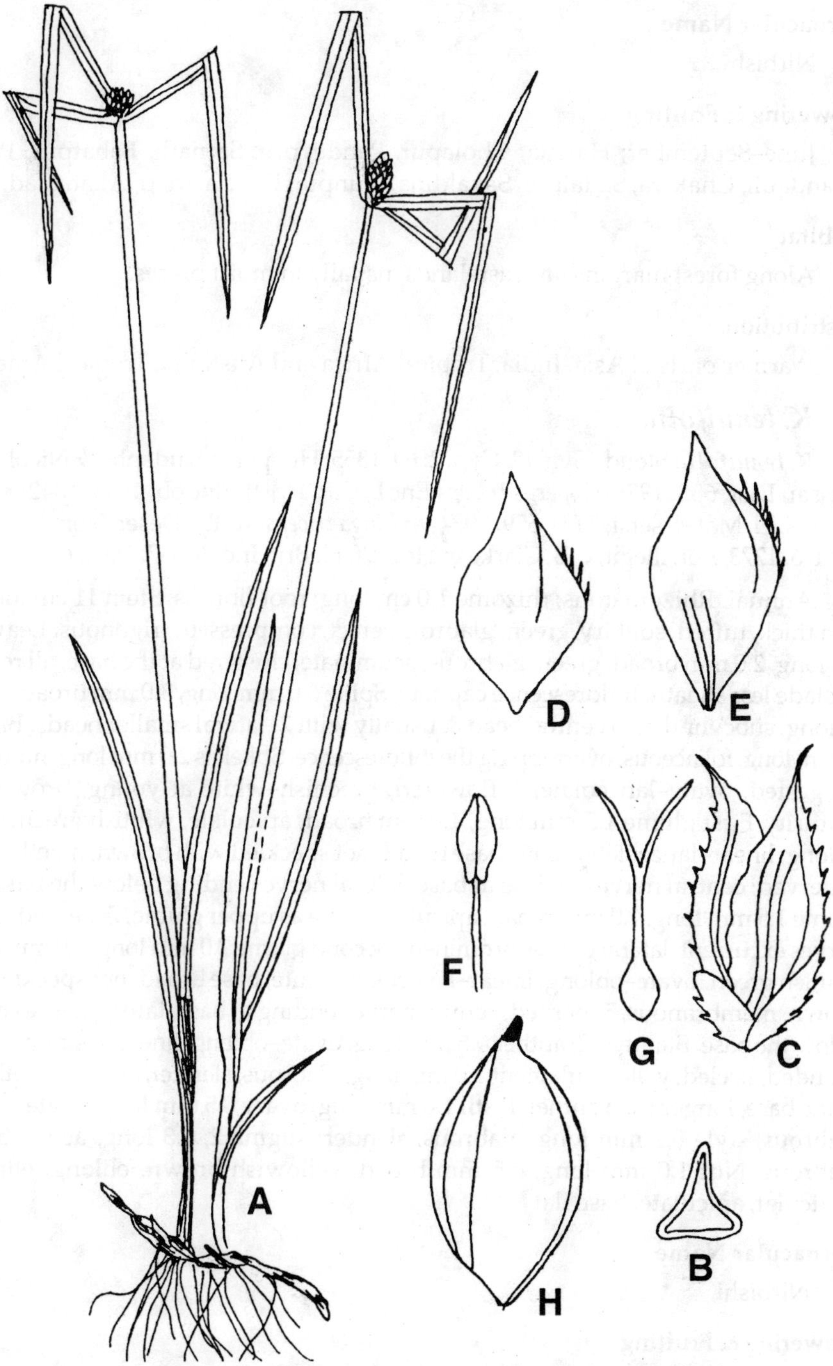

PLATE 63: *Kyllinga nemoralis* (Forster) Dandy ex Hutcinson
A. Habit, B. T.S. of stem, C. Spikelet, D. Ist glume, E. IInd glume, F. Stamen, G. Pistil, H. Nut

Vernacular Name

Nirbishi.

Flowering & Fruiting

June–September; Harrua, Cholapur, Pandeypur, Sarnath, Babatpur, Pindra, Chandauli, Chakiya, Sadalpur, Sakaldiha, Jaunpur, Deoria, Mau, Allahabad.

Habitat

Along forest margins, in wastelands, usually in moist places.

Distribution

Warmer parts of Asia, India, Tropical Africa and Australia, Tropical America.

4. *K. tenuifolia*

K. tenuifolia Steud., Syn. Pl. Cyp. 2: 69. 1855; Hooper in Saldanha & Nicolson, Fl. Hassan Dist. 688. 1976. *Cyperus triceps* Endl., Cat. Horti Vindob. 1: 94. 1842; Kern in Steeins, Fl. Males. ser. I. 7 (3): 659. 1974. *Kyllinga triceps* Rottb., Deser. Icon. Rar. Pl. 14. t. 4. f. 6. 1773, non illegit; C.B. Clarke in Hook. f. Fl. Brit. Ind. 6: 587. 1893. (PLATE–64).

Annual. Rhizomatous; rhizome 1.0 cm long; root fibrous. Stem 11 cm long, 1.0 mm thick, tufted, solitary, green, glabrous, erect, compressed, trigonous. Leaves 7.0 cm long, 2.0 mm broad, green, glabrous, acuminate, clustered at the base, all reduced to blade less sheath. Inflorescence capitate. Spike 6.0 mm long, 3.0 mm broad, ovoid–oblong, subcylindric, a central head & usually with 2–lateral smaller heads. Bracts 4, 8.0 cm long, foliaceous, overtopping the inflorescence. Spkelets 2.0 mm long, numerous, congested, ovate–lanceolate, 1–flowered, greenish–white at young, brownish at maturity. First glume 1.5 mm long, 1.0 mm broad, apiculate, whitish green, ovate–oblong, linear–lanceolate, acute, base broad, not speckled with brown, membranous, 5–nerved; central nerves ending at base, lateral nerves ending below the base. First glume 1.5 mm long, 1.0 mm broad, apiculate, same as upper glume, 7–nerved; central nerves excurrent, lateral nerves prominent. Second glume 2.0 mm long, 1.0 mm broad, whitish green, ovate–oblong, linear–lanceolate, acute, base broad, not speckled with brown, membranous, 5–nerved; central nerves ending at base, lateral nerves ending below the base. Stamens 2; anther 0.5 mm long, ovate–oblong, linear, acuminate, base rounded, keeled, yellow; filament 1.0 mm long, glabrous, slender, attached with pistil to the base, longer than anther. Pistil 1.0 mm long; ovary 0.5 mm long, ovate, oblong, glabrous; style 0.2 mm long, glabrous, slender; stigma 2; 0.3 long, acute, brown, glabrous. Nut 1.0 mm long, 0.5 mm broad, yellowish brown, oblong, ellipsoid, lenticular, apiculate, base flat.

Vernacular Name

Nirbishi.

Flowering & Fruiting

August–December; Chaukhandi, Jaunpur, Chandauli, Sonbhadra, Basti.

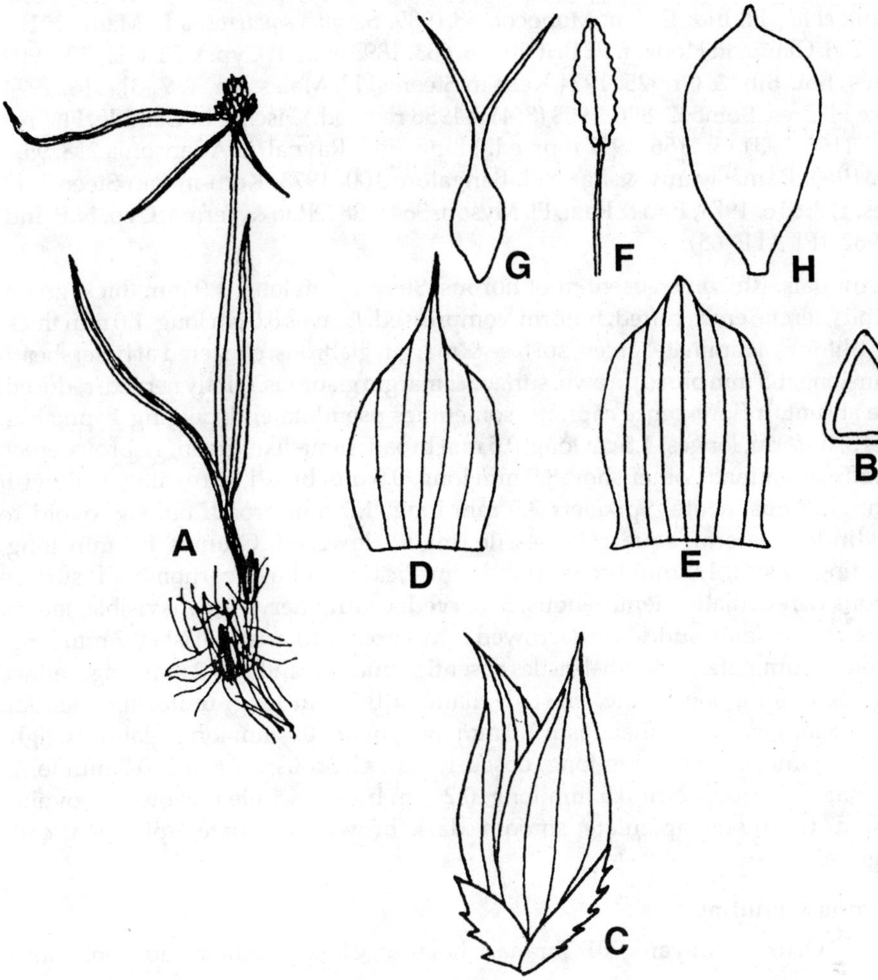

PLATE 64: ***Kyllinga tenuifolia*** **Steud.**
A. Habit, B. T.S. of stem, C. Spikelet, D. Ist glume, E. IInd glume, F. Stamen, G. Pistil, H. Nut

Habitat

Common in moist places, open fields, bunds of agricultural fields.

Distribution

India, Africa, Europe, Asia.

9. *Lipocarpha* R. Br.

Lipocarpha squarrosa (L.) Goetgh. in Blumea 4: 167. 1957. *Rikliella squarrosa* L.J. Raynal in Adansonia ser. 2. 13: 154. 1973; Hooper in Saldanha & Nicolson, Fl. Hassan dist. 696. 1976; Sharma *et al.*, Fl. Karnataka 215. 1984; Koyama in Dassanayake & Fosberg, Rev. Handb. Fl. Ceylon 5: 254. 1985; Singh, Fl. E. Karnataka 2: 645. 1988;

Karthik *et al.*, Fl. Ind. Enum. Monocot. 68.1989. *Scirpus squarrosus* L. Mant. 2: 181. 1771; C.B. Clarke in Hook. f. Fl. Brit. Ind. 6: 663. 1893 et in III Cyp. t. 52. f. 11–13. 1909; Haines, Bot. Bih. & Or 925. 1924 Kern in Steenis, Fl. Males. ser. 1. 7 (3): 516. 1974; Cooke Fl. Pres. Bomb. 2: 896. 1908 (3: 410. 1958 repr. ed.) Fischer in Gamble, Fl. Pres. Madr. 1166. 1931 (3: 1156. 1957 repr. ed.); Linn., in J. Raynal. in Adansonia 2, 8: 95. f. 9–12. 1968; Ramaswamy & Razi, Fl. Bangalore 100. 1973; Kern in van Steenis, Fl. Males. 1, 7: 516. 1974; Rao & Razi, Fl. Mysore 564. 1981; Rao & Verma, Cyp. N.E. Ind. 45. 1982. (PLATE–65).

Annuals. Rhizome absent, root fibrous. Stem 10 cm long, 1.0 mm thick, green, glabrous, terete, erect, tufted, filiform, compressed. Leaves 8.0 cm long, 1.0 mm thick, erect, filiform, acuminate, green, surface & margin glabrous, clustered at base; sheath 8.0 mm long, 1.5 mm broad, brown, surface & margin glabrous, faintly nerved, reduced; ligule absent. Inflorescence capitate, sometimes pseudolateral, bearing 4 spikelets. Bracts 3; filiform, longest 1.5 cm long, 0.5 mm broad, stem–like, green, glabrous, erect, flexuous, acuminate, other short 8.0 mm long, 0.5 mm broad, spreading, yellowish green, glabrous, acute. Spikelets 3.0 mm long, 1.0 mm broad, oblong, ovoid to subcylindric, lateral, squarrose sessile, many flowered. Glumes 1.5 mm long, (including arista), 1.0 mm broad, tightly imbricating, obovate–rhomboid, surface glabrous & reticulate, membranous, 5–nerved; central nerve green, visible, lateral nerves brown, faint, suddenly narrowed into a excurved arista; arista 0.7 mm long, glabrous, acuminate. Perianth–bristles absent. Stamens 1; anther 0.2 mm long, yellow, glabrous, oblong, lanceolate, crested; filament 0.5 mm long, uniformly slender, glabrous, longer than anther. Pistil 1.0 mm long; ovary 0.5 mm long, glabrous, light green, obovate; style 0.3 mm long, obscurly flat, glabrous; stigma 3, 0.2 mm long, acuminate, glabrous. Nut 0.5 mm long, 0.2 mm broad, sessile narrowly obovoid–ellipsoid, trigonous, apiculate, smooth, dark–brown, reticulate; epidermal cells hexagonal.

Flowering & Fruiting

July–October; Mayank 901; Tarana, Cholapur, Chakiya, Allahabad, Sonbhadra.

Habitat

Very common in moist places, rice field, along road side.

Distribution

India, Africa, Malesia, Asia.

10. *Mariscus* Vahl

KEY TO THE SPECIES

1a. Anthela evolute, capitate; glumes with long mucro

.... 2. *M. squarrosus*

1b. Anthela simple, compound or decompound; glumes with short mucro:

2a. Anthela simple; spike cylindrical; spikelet 2–flowered, strongly compressed

.... 3. *M. sumatrensis*

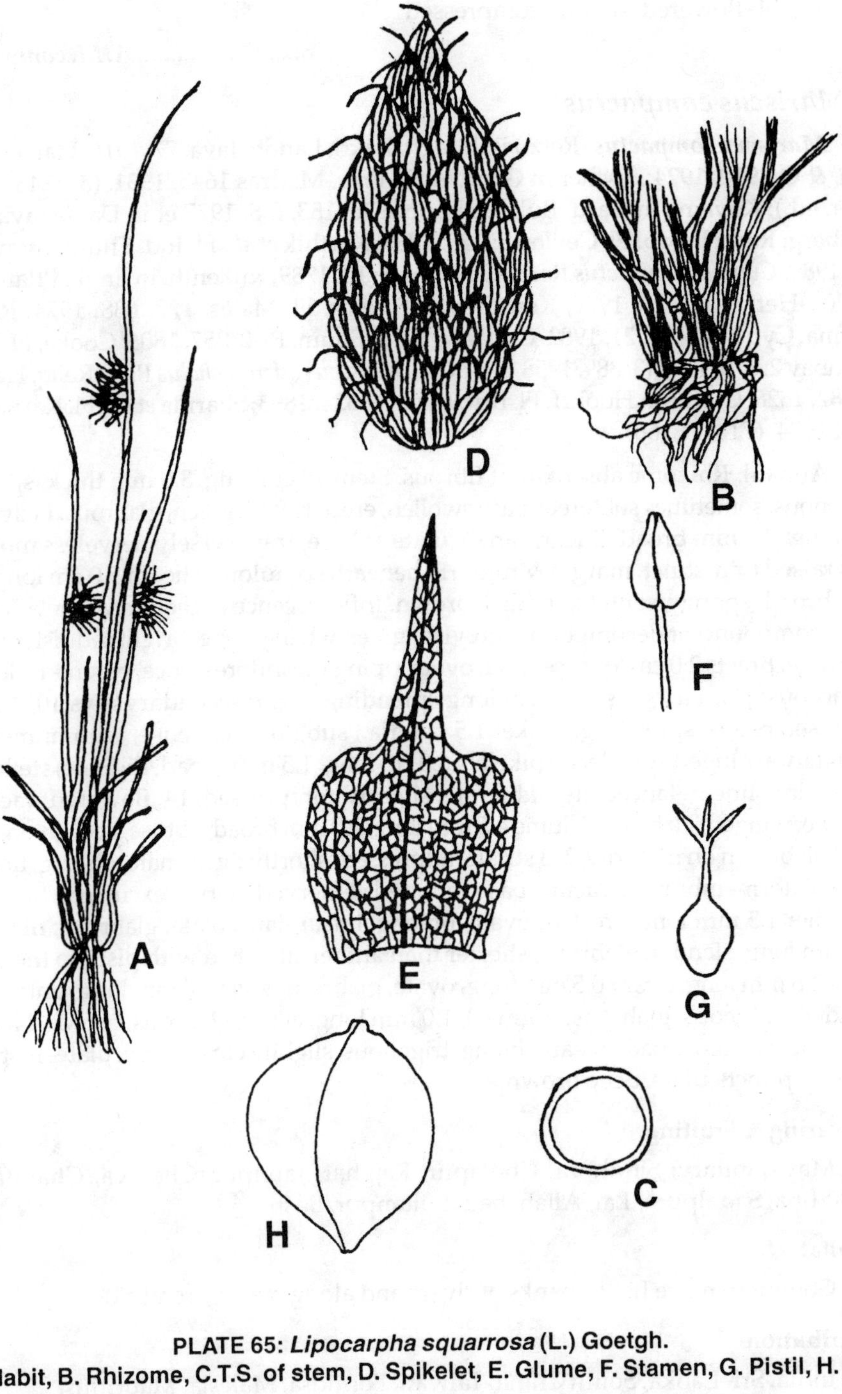

PLATE 65: *Lipocarpha squarrosa* (L.) Goetgh.
A. Habit, B. Rhizome, C.T.S. of stem, D. Spikelet, E. Glume, F. Stamen, G. Pistil, H. Nut

2b. Anthela compound or decompound; spikes sub-globose; spikelets 14–flowered, slightly compressed .... 1. *M. compactus*

1. *Mariscus compactus*

Mariscus compactus (Retz.) Boldingh, Zakfl. Landb. Java. 77. 1916; Haines, Bot. Bih. & Or. 910. 1924; Fischer in Gamble, Fl. Pres. Madras 1645. 1931. (3: 1143. 1957, repr. ed.); Koyama in Gard. Bull. Singapore 30: 153, f. 8. 1977 et in Dassanayake & Fosberg, Rev. Handb. Fl. Ceylon 5: 228. 1985; Karthik *et al.*, Fl. Ind. Enum. Monocot. 62. 1989. *Cyperus compactus* Retz., Obs. Bot. 5: 10. 1789; Kukenth. in Engl., Pflanzenr. 4 (20), Heft. 101. 423. 1936; Kern in van Steenis, Fl. Males. 1, 7: 638. 1974; Rao & Verma. Cyp N.E. Ind. 21. 1982. *C. dilutus* Vahl, Enum. Pl. 2: 357. 1806; Cooke, Fl. Pres. Bombay 2: 875. 1908 (3: 389. 1958, repr. ed.). *Mariscus microcephalus* Pres. Reliq. Hanek. 1: 182. 1828; Clarke in Hook. f. Fl. Brit. India. 6: 624. 1893; Sharma *et al.*, Fl. Karnataka 313. 1984. (PLATE–66).

Annaul. Rhizome absent; root fibrous. Stem 18 cm long, 3.0 mm thick, spongy, trigonous, sometimes subterete, base swollen, erect, tufted, green, glabrous. Leaves 15 cm long, 3.0 mm broad, linear, canaliculate at base, transversely nerveless more or less raised or distinct, margin & mid–rib beneath spinulous; sheath 5.0 cm long, 4.0 mm broad, spongy, septate, reddish brown. Inflorescence anthelate; anthela 6.0 cm long, compound or decompound, greyish–green when young, bright reddish brown with age. Bracts 2.0 cm long, several, overtopping the inflorescence, patent–reflexed, foliaceous; primary rays 8, 10 cm long spreading, rigid; secondary rays 10, 4.0 cm long, setaceous, spreading. Spike 1.5 cm dia., subglobose, dense, with numerous stellately arranged spikelets. Spikelets 6.0 cm long, 1.5 mm broad, digitate, stellately spreading, linear–lanceolate, subterete, slightly compressed, 14. flowered; rachilla hyaline, winged, glabrous. Glume 4.5 mm long, 2.0 mm broad, obtuse, keel green, side reddish brown on maturity, 2–3 striate, adpressed, shortly mucronate, oblong, linear–lanceolate, membranous, acute, base rounded, 3–nerved; nerves excurrent. Stamens 3; anther 1.5 mm long, crested, ovate, oblong, linear–lanceolate, glabrous; filament 0.5 mm long, slender, glabrous, shorter than anther, attached with pistil to the base. Pistil 2.5 mm long; ovary 0.5 mm long, ovate, glabrous; style 1.0 mm long, uniformly slender, caducous, glabrous; stigma 3, 1.0 mm long, acute, glabrous, brown. Nut 1.5 mm long, 1.0 mm broad, linear–oblong, trigonous, slightly curved, apiculate–rostrate, densely puncticulate, deep brown.

Flowering & Fruiting

May–January; Sindhora, Cholapur, Rajghat, Jaunpur, Chakiya, Chandauli, Sakaldiha, Sadalpura, Lar, Allahabad, Sultanpur, Basti.

Habitat

Common in rice fields, banks of rivers and along marshy road sides.

Distribution

India, Sri–Lanka, South China, Taiwan, Formosa, Malesia, Mauritius.

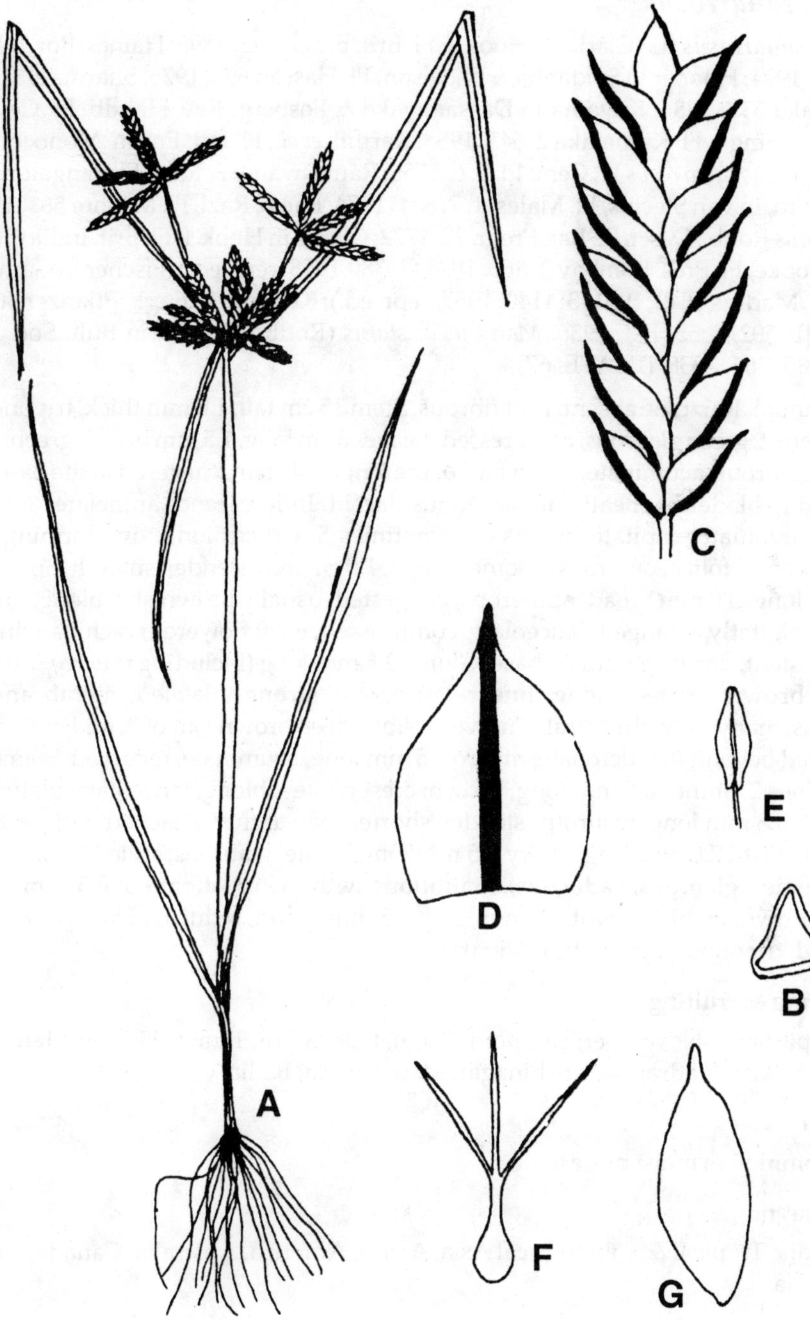

PLATE 66: ***Mariscus compactus*** **(Retz.) Boldingh**
A. Habit, B. T.S. of stem, C. Spikelet, D. glume, E. Stamen, F. Pistil, G. Nut

2. *M. squarrosus*

M. squarrosus (L.) Clarke in Hook. f. Fl. Brit. India 6: 23. 1893; Haines, Bot. Bih. & Or. 909. 1924; Hooper in Saldanha & Nicolson, Fl. Hassan 690. 1976; Sharma *et al.*, Fl. Karnataka 313. 1984; Koyama in Dassanayake & Fosberg, Rev. Handb. Fl. Ceylon 229. 1985; Singh, Fl. Karnataka 2: 643. 1988; Karthik *et al.*, Fl. Ind. Enum. Monocot. 64. 1989. *Cyperus squarrosus* L., Cent. Pl. 2: 6. 1756; Ramaswamy & Razi, Fl. Bangalore 94. 1973; Kern, in van Steenis, Fl. Males. 1. 7: 631. 1974; Rao & Razi, Fl. Mysore 561. 1981. *C. aristatus* Rottb., Descr. Pl. Rar. Progr. 22. 1772; Clarke in Hook. f. Fl. Brit. India 6: 606. 1893; Cooke, Fl. Pres. Bombay 2: 866. 1908 (3: 380.1958, repr. ed.); Fischer in Gamble, Fl. Pres. Madras 1640. 1931 (3: 1140. 1957, repr. ed.); Kukenth in Engl., Pflanzer 4(20), Heft 101: 502, f. 55, F–J. 1936. *Mariscus aristatus* (Rottb.) Cherm. in Bull. Soc. Bot. France 85: 366. 1938. (PLATE–67).

Annual. Rhizome absent; root fibrous. Stem 15 cm tall, 1.0 mm thick, trigonous, tufted, erect, green glabrous, compressed. Leaves 8 cm long, 1.5 mm broad, green, flat, linear, glabrous, acuminate, about half of the length of stem, clustered at the base, all reduced to bladeless sheath, membranous flaccid. Inflorescence anthelate: anthela simple, evoluate, capitate. Bracts 4, sometimes 5, 6.0 cm long, overtopping the inflorescence, foliaceous, rays 3, sometimes 5, 1.0 cm long slender, smooth. Spikelets 6.0 mm long, 3.0 mm broad, numerous, congested, usually brownish, oblong, linear, sessile, digitatly arranged, lanceolate, compressed, 6–20 flowered; rachilla winged by persistent, decurrent glume base. Glume 3.5 mm long (including mucro), 1.0 mm broad, brown, ovate–oblong, linear–lanceolate, strongly keeled, membranous, glabrous, many nerved; central 3–nerves visible, deep brown, out of 3, mid one nerve produced beyond it; mucronate; mucro 1.5 mm long, acuminate, recurved. Stamen 1 sometimes 2; anther 0.5 mm long, brown, crest ovate–oblong, lanceolate, glabrous; filament 0.3 mm long, glabrous, slender, shorter than anther, attached with pistil to the base. Pistil 2.0 mm long; ovary 0.5 mm long, ovate, glabrous; style 1.0 mm long, flat, slender, glabrous, caducous, continuous with ovary; stigma 3; 0.5 mm long, acute, brown, glabrous. Nut 1.0 mm long, 0.5 mm broad, reddish brown, narrowly obovoid, trigonous, apiculate, lenticular.

Flowering & Fruiting

September–November; Sindhora, Rajghat, Babatpur, Tarana, Harraua, Jaunpur, Chakiya, Basti, Padrauna, Kushinagar, Mau, Deoria, Ballia.

Habitat

Common in moist places.

Distribution

India, Tropical & sub–tropical Asia, Africa, America, Australia, Canada, Chile, Argentina.

3. *M. sumatrensis*

M. sumatrensis (Retz.) J. Raynal. in Adansonia 15: 110. 1995; Hooper in Saldanha & Nicolson, Fl. Hassan 690. 1976; Koyama in Gard. Bull. Singapore 30: 154. 1977;

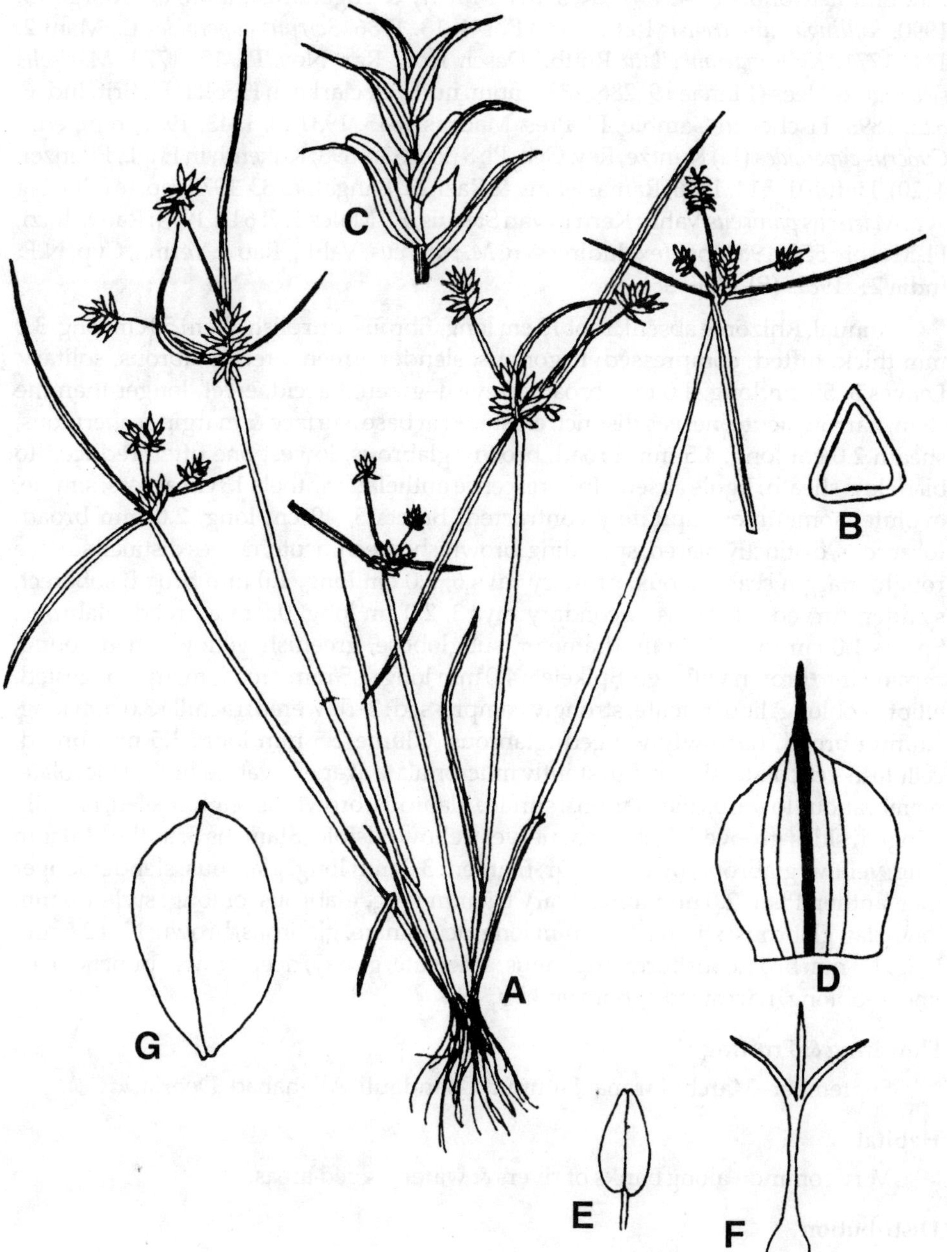

PLATE 67: ***Mariscus squarrosus*** **(L.) Clarke**
A. Habit, B. T.S. of stem, C. Spikelet, D. Glume, E. Stamen, F. Pistil, G. Nut

Sharma *et al.*, Fl. Karnataka 313. 1984; Koyama in Dassanayake & Fosberg, Rev. Handb. Fl. Ceylon 5.232. 1985; Singh, Fl. E. Karnataka 2: 643. 1988; Karthik *et al.*, Fl. Ind. Enum. Monocot. 64. 1989; Keshava Murthy & Yoganarasimhan, Fl. Coorg 513. 1990. *Kyllinga sumatrensis* Retz., Obs. Bot. 4: 13. 1786. *Scirpus cyperoides* L., Mant 2: 181. 1771. *Kyllinga umbellata* Rottb., Descr. Icon. Rar. Nov. Pl. 15. 1773. *Mariscus sieberianus* Nees (Linnaea 9: 286. 1835, nom. nud.) ex clarke in Hook f. Fl. Brit. Ind. 6: 622. 1893; Fischer in Gamble, Fl. Pres. Madras 1645. 1931 (3: 1143. 1957, repr. ed.). *Cyperus cyperoides* (L.) Kuntze. Rev. Gen. Pl. 3 (2). 333. 1898; Kukenth in Engl., Pflanzer. 4 (20), Heft 101: 514. 1936; Ramaswamy & Razi, Fl. Bangalore 83. 1973, pp. (excluding syn. *Mariscus paniceus* Vahl); Kern in van Steenis, Fl. Males 1, 7: 642. 1974; Rao & Razi, Fl. Mysore 558. 1981, pp. (excluding syn. *M. paniceus*, Vahl.); Rao & Verma, Cyp. N.E. India 21. 1982. (PLATE–68).

Annual. Rhizome absent; root 15 cm long, fibrous, purplish. Stem 50 cm long, 3.0 mm thick, tufted, compressed, trigonous, slender, green, erect, glabrous, solitary. Leaves 3, 55 cm long, 4.0 mm broad, greyish–green, flaccid, erect, longer than the stem, narrow, acute, nerves distinct, clustered at base, surface & margin scaberulous; sheath 2.0 cm long, 4.5 mm broad, brown, glabrous, lowest one often reduced to bladeless sheath; ligule absent. Inflorescence anthelate; anthela 13 cm across, simple, evolute, sometimes capitately contracted. Bracts 5, 40 cm long, 2.0 mm broad, foliaceous, distinctly placed, spreading, brownish–green, acute, nerves distinct, surface rough, margin scaberulous, primary rays 6, 7.0 cm long, 1.0 mm broad, suberect, slender, curved, glabrous; secondary rays 3, 2.0 cm long, 0.5 mm broad, glabrous. Spikes 1.0 cm long, 8.0 mm diameter, subglobose, greenish–yellow when young, turning into brown with age. Spikelets 4.0 mm long, 1.5 mm broad, many, congested, elliptic–oblong, linear, acute, strongly compressed, 2–flowered; rachilla 3.0 mm long, 1.0 mm broad, narrowly winged, glabrous. Glume 2.5 mm long, 1.5 mm broad, cellulose–reticulate, shortly but stoutly mucronulate at apex ovate, elliptic, lanceolate, membranous towards the margins, surface glabrous, brown, strongly keeled, broadly winged, sides grooved, 3–nerved; nerves yellow, visible. Stamens 3; anther 1.0 mm long, yellow, glabrous, ovate–linear; filament 3.0 mm long, glabrous, slender longer than anther. Pistil 2.0 mm long; ovary 0.5 mm long, glabrous, oblong; style 1.0 mm long, flat, glabrous; stigma 3, 0.5 mm long, acuminate, glabrous, brown. Nut 2.5 mm long, 1.5 mm broad, fusiform, trigonous, apiculate, glossy, deep brown, puncticulate, linear–oblong, narrowed at both ends.

Flowering & Fruiting

September–March; Tarana, Jaunpur, Chandauli, Allahabad, Deoria.

Habitat

Very common along banks of rivers & water logged areas.

Distribution

India, Tropical Africa, Madagascar.

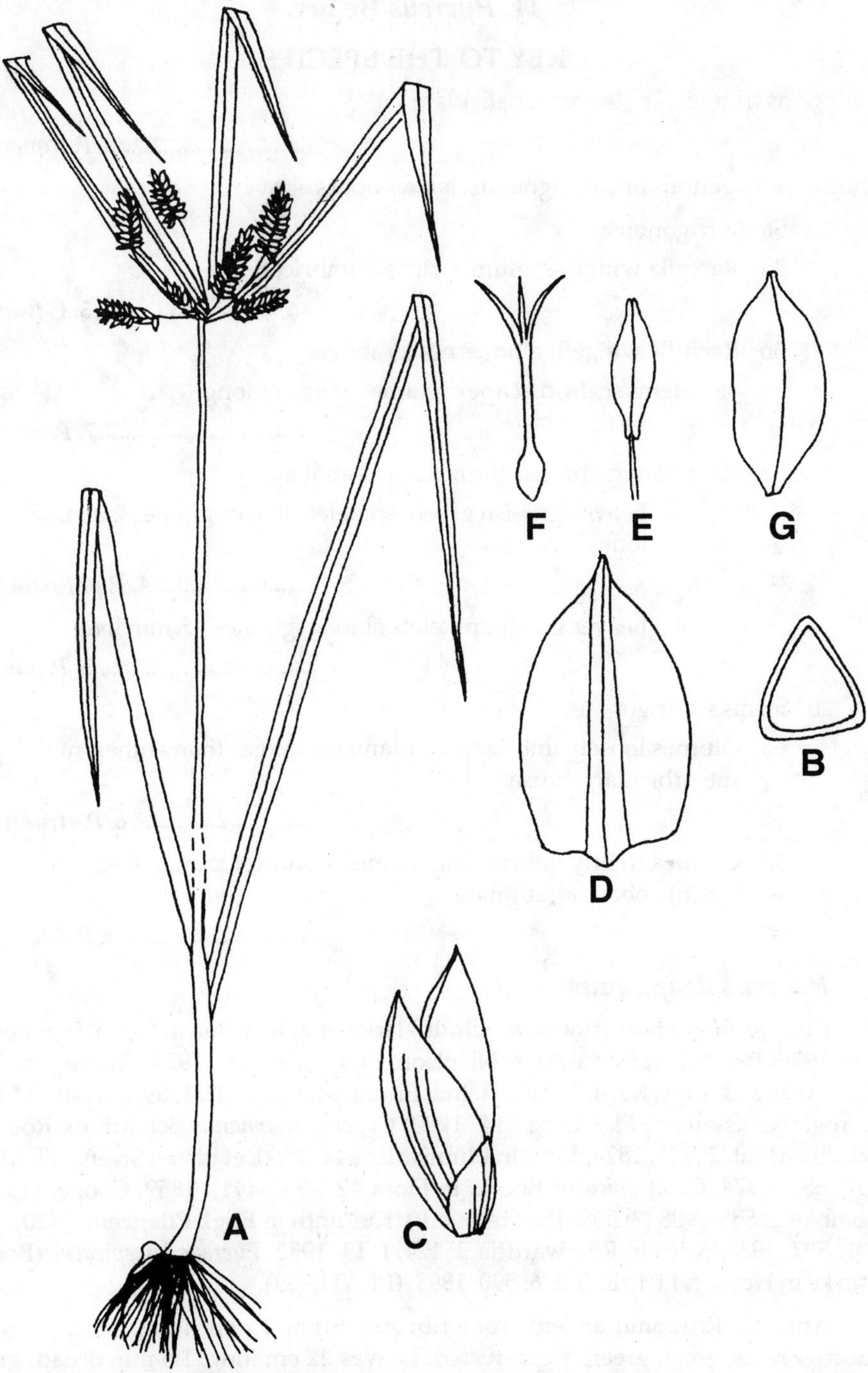

PLATE 68: ***Mariscus sumatrensis*** **(Retz.) J. Raynal**
A. Habit, B. T.S. of stem, C. Spikelet, D. Glume, E. Stamen, F. Pistil, G. Nut

11. *Pycreus* Beauv.

KEY TO THE SPECIES

1a. Stems triquetrous; leaves canaliculate .... **2. *P. flavescens***

1b. Stems trigonous or subtrigonous; leaves not as above:

- 2a. Stems trigonous:
  - 3a. Rachilla wingless; glumes closely imbricating .... **3. *P. flavidus***
  - 3b. Rachilla winged; glumes not as above:
    - 4a. Stems scabrid at apex; glumes acute, oblong .... **7. *P. sulcinux***
    - 4b. Stems glabrous; glumes not as above:
      - 5a. Leaves greyish green; spikelets linear; glumes 2.0 mm long .... **4. *P. polystachyos***
      - 5b. Leaves green; spikelets oblong; glumes 1.5 mm long .... **5. *P. pumilus***
- 2b. Stems subtrigonous:
  - 6a. Glumes loosely imbricating; filaments longer than anther; nut sub orbicular, obtuse .... **6. *P. stramineus***
  - 6b. Glumes tightly imbricating; fiaments shorter than anther; nut broadly obovoid, stipitate .... **1. *P. diaphanus***

1. *Pycreus diaphanus*

Pycreus diaphanus (Roem. & Schult.) Hooper & Koyama in Journ. Jap. Bot. 51: 316. 1976; Hooper in Saldanha & Nicolson. Fl. Hassan 691. 1976; Sharma *et al.*, Fl. Karnataka 313. 1984; Karthik *et al.*, Fl. Ind. Enum. Monocot. 65. 1989; KeshavaMurthy & Yoganarrasimhan, Fl. Coorg 514. 1990. *Cyperus diaphanus* Schard. ex Roem. & Schult., Mant. 2: 477. 1824; Kern in Blumea 10: 644. 1960. et in van Steenis Fl. Males. 1, 7: 651. 1974. *C. latespicatus* Boeck. in Flora 42. 433. (441). 1859; Cooke, Fl. Pres. Bombay 2: 855.1908 (3: 368. 1958 repr. ed.); Kukunth in Engl. Pflanzenr 4 (20), Heft 101: 392. 1936; Kern in Reinwardtia 2: 124, f. 13. 1952. *Pycreus latespicatus* (Boeck.) Clarke in Hook. f. Fl. Brit. Ind. 6: 590. 1893. (PLATE–69).

Annual. Rhizome absent, root fibrous. Stem 25 cm long, 1.5 mm thick, subtrigonous, erect, green, rigid, tufted. Leaves 22 cm long, 1.5 mm broad, green, shorter than the stem, stiff, acute, linear, glabrous, clustered at base; sheath reduced; ligule absent. Inflorescence anthelate; anthela simple, contracted or evolute, sometimes

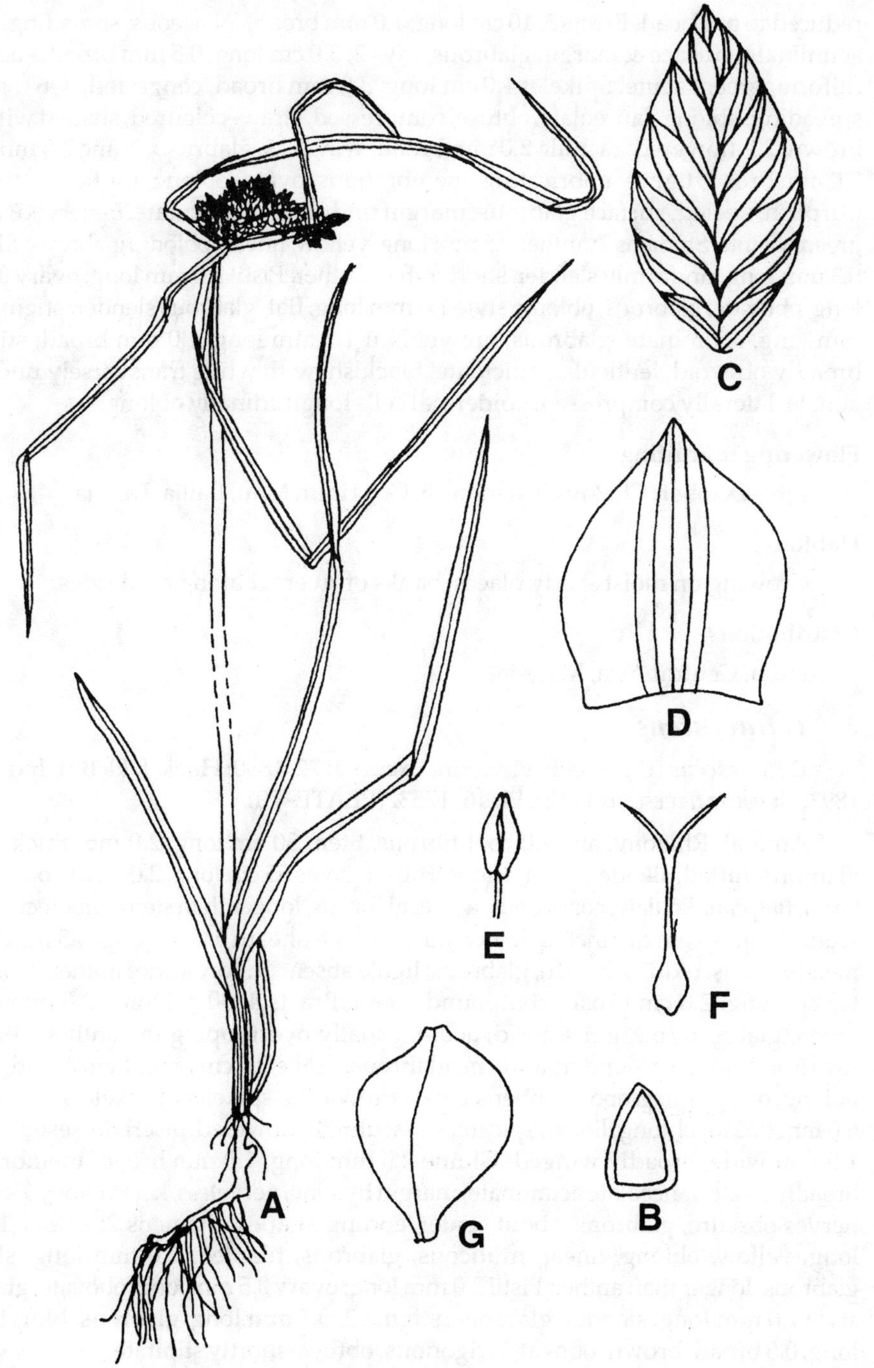

PLATE 69: ***Pycreus diaphanus*** **(Roem. & Schutt.) Hooper & Koyama**
A Habit, B. T.S. of stem, C. Spikelet, D. Glume, E. Stamen, F. Pistil, G. Nut

reduced to one head. Bracts 3, 10 crn long, 1.0 mm broad, foliaceous, spreading, green, acuminate, surface & margin glabrous; rays 2, 3.0 cm long, 0.5 mm broad, sub erect, filiform, green, acute. Spikelet 1.0 cm long, 3.0 mm broad, congested, 3–6 together, spreading, oblong–lanceolate, obtuse, compressed, straw–coloured, shaded with red–brown, 20–flowered; rachilla 2.0 mm broad, wingless, glabrus. Glume 2.5 mm long, 2.0 mm broad, tightly imbricating, membranous, ovate–oblong, muticous, fulvous, purplish, keeled, surface glabrous, margin undulating, apiculate, 5–nerved; nerves green, visible. Stamens 2; anther 0.5 mm long, yellow, linear–oblong, glabrous; filament 0.3 mm long, uniformly slender, shorter than anther. Pistil 2.0 mm long; ovary 0.5 mm long, obovate, glabrous, oblong; style 1.0 mm long, flat, glabrous slender; stigma 2, 0.5 mm long, acuminate, glabrous, brown. Nut 1.5 mm long, 1.0 mm broad, stipitate, broadly obovoid, lenticular, apiculate, blackish, with white transversely undulatly zonate, laterally compressed, epidermal cells longitudinally oblong.

Flowering & Fruiting

July–October; Chakiya, Chandauli, Ghazipur, Mau, Ballia, Deoria.

Habitat

Growing on moist shady places, banks of rivers & along road sides.

Distribution

India, Central Asia, Malesia.

2. *P. flavescens*

P. flavescens (L.) Reichb. Fl. Germ. Excurs 1: 72. 1830; Hook. f. Fl. Brit. Ind. 6: 589. 1893. *Cyperus flavescens* L. Sp. Pl. 46. 1753. (PLATE–70).

Annual. Rhizome absent; root fibrous. Stem 50 cm long, 2.0 mm thick, green, glabrous, tufted, slender, erect, triquetrous. Leaves 55 cm long, 2.0 mm broad, green, basal, flat, canaliculate, coriaceous, acute, glabrous, longer than stem, suberect, margin scabrid, mid–vein distinct on lower surface; leaf–sheath 5.0 cm long, 3.5 mm broad, membranous, reddish brown, glabrous; ligule absent. Inflorescence anthelate; anthela 6.0 cm long, 2.0 cm broad, compound, open. Bracts 4, 30 cm long, 2.0 mm broad, green, glabrous, margin scabrid, acute, usually overtopping the anthela; primary rays 5, 4.0 cm long, slender, filiform, glabrous. Spike 2.0 cm long, 1 cm broad, linear, oblong, ovate, sub globose, suberect, pale brown, 20–spikelets. Spikelets 1.0 cm long, 6.0 mm broad, oblong, linear, spicate, subacute, 20–flowered, biseriate, sesile; rachilla 1.0 mm wide, broadly winged. Glume 2.0 mm long, 1.0 mm broad, membranous, broadly ovate, lanceolate acuminate, margin hyaline, nerveless, keel yellow, 3–nerved; nerves obscure, glabrous, about center, ending at apex. Stamens 2; anther 1.0 mm long, yellow, oblong, linear, muticous, glabrous; filament 2.0 mm long, slender, glabrous, longer than anther. Pistil 2.0 mm long; ovary 0.5 mm long, obovate, glabrous; style 1.0 mm long, slender, glabrous; stigma 2, 0.5 mm long, glabrous. Nut. 1.0 mm long, 0.5 broad, brown, obovate, trigonous, obtuse, shortly stipitate.

Flowering & Fruiting

August–October; Ghazipur, Sindhora, Deoria, Basti, Sultanpur, Allahabad.

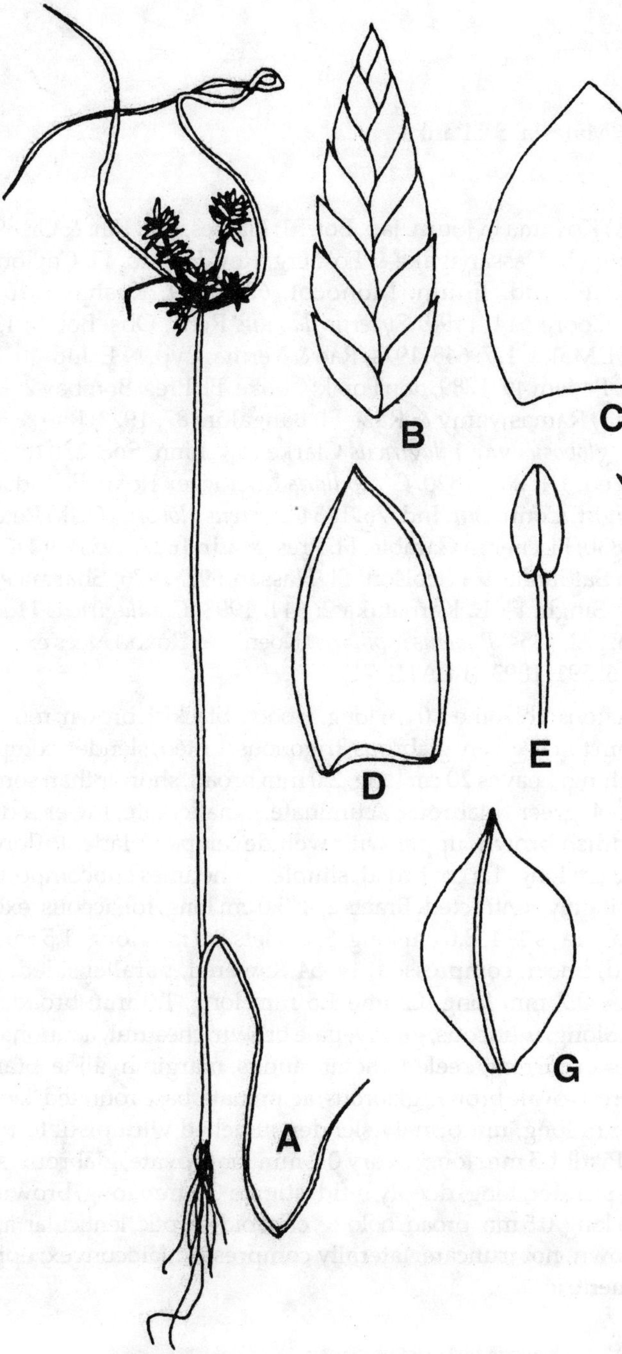

PLATE 70: *Pycreus flavescens* (L.) Reichb.
A. Habit, B. Spikelet, C & D. Glume, E. Stamen, F. Pistil, G. Nut

Habitat

Along the rice fields.

Distribution

India, Australia, Malesia, Sri Lanka.

3. *P. flavidus*

P. flavidus (Retz.) Koyama in Journ. Jap. Bot. 51: Haines, Bot. Bih. & Or. 893. 1924 (10): 313. 1976; Koyama in Dassanayake & Fosberg, Rev. Handb. Fl. Ceylon 5: 222. 1985; Karthik *et al.*, Fl. Ind. Enum. Monocot. 65. 1989; KeshavaMurthy & Yoganarasimhan, Fl. Coorg 514. 1990. *Cyperus flavidus* Retz., Obs. Bot. 5: 13. 1788; Kern in van Steenis, Fl. Males. 1, 7: 648. 1974; Rao & Verma, Cyp. N.E. Ind. 10. 1982. *C. globosus* All., Auct. Fl. Pedem 49. 1789, non Forsk; Cooke. Fl. Pres. Bombay 2: 857 1908 (3: 370. 1958, repr. ed.); Ramaswamy & Razi, Fl. Bangalore 87. 1973; Rao & Razi, Fl. Mysore 559. 1981. *C. globosus* var. *nilagiricus* Clarke in J. Linn. Soc. 21: 49. 1884. *C. strictus* Roxb. Fl. Ind. ed. 1.1: 203. 1820. *C. capillaris* Koenig. ex Roxb. Fl. Ind. ed. 1, 1: 198. 1820; Nees in Wight, Contr. Bot. Ind. 76. 1834. *Pycreus globosus* (All.) Reichb., Fl. Germ. Excurs. 140. 1830; Fischer in Gamble, Fl. Pres. Madr. 1627. 1931 (3: 1132. 1957, repr. ed.); Hooper in Saldanha & Nicolson, Fl. Hassan 692. 1976; Sharma *et al.*, Fl. Karnataka 314. 1984; Singh, Fl. E. Karnataka 2: 644. 1998. *C. nilagiricus* Hochst. ex Steud., Syn. Pl. Glum. 2: 2. 1854. *Pycreus capillaris* (Koenig ex Roxb.) Nees ex Clarke in Hook. f. Fl. Brit. Ind. 6: 591. 1893. (PLATE–71).

Annual. Rhizomatous; rhizome 1.0 cm long, woody, blackish brown; root fibrous. Stem 30 cm tall, 2.0 mm thick, green, glabrous, trigonous, tufted, slender, compressed, base brownish, sheathing. Leaves 20 cm long, 2.0 mm broad, shorter than sometimes equalling the stem, 1–4, green, glabrous, acuminate, canaliculate, lower reduced to bladeless sheath, reddish brown, upper with well developed blade. Inflorescence anthelate; anthela 2.0 cm long, 1.5 cm broad, simple, sometimes subcompound, lax, dense, evolute or capitately contracted. Bracts 2–4, 3.0 cm long, foliaceous, exceeding the anthela, spreading; rays 1–4, 8.0 cm long. Spikelets 8.0 mm long, 1.5 mm broad numerous, congested, linear, compressed, 14–54 flowered, parallel sided; rachilla wingeless, internodes 0.7 mm long. Glume 1.5 mm long, 1.0 mm broad, closely imbricating, ovate–oblong, muticous, glossy, pale brown, chestnut, acuminate, base flat, 3–nerved; nerves excurrent; keeled, membranous, margin hyaline. Stamens 2; anther 1.0 mm long, crest–ovate brown, glabrous, acuminate, base rounded, lanceolate, linear; filament 0.2 mm long, uniformaly slender, attached with pistil to the base, shorter than anther. Pistil 1.5 mm long; ovary 0.2 mm long, ovate, glabrous; style 0.8 mm long, uniformly slender, long, deeply bifid, stigma 0.5 mm long, brown, acute, glabrous. Nut 1.0 mm long, 0.5 mm broad, oblong, obovoid, elliptic, lenticular, apiculate at apex, blackish–brown, not truncate, laterally compressed, bioconvex; epidermal cells minute, isodiamentric.

Flowering & Fruiting

September–November; Chanduli, Chakiya, Jaunpur, Sakaldiha, Kushinagar.

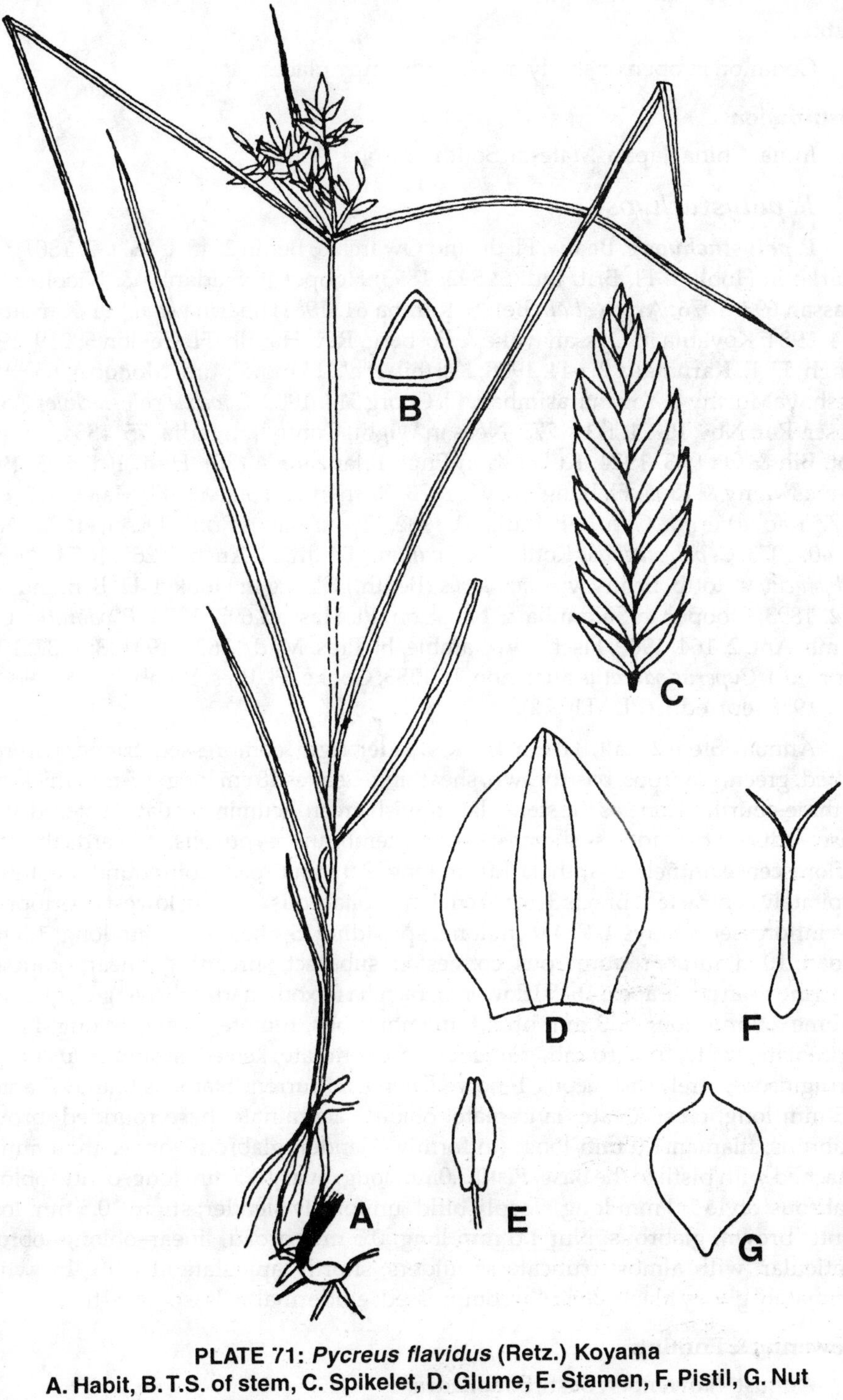

PLATE 71: ***Pycreus flavidus*** **(Retz.) Koyama**
A. Habit, B. T.S. of stem, C. Spikelet, D. Glume, E. Stamen, F. Pistil, G. Nut

Habitat

Common in open or shady moist to swampy places.

Distribution

India, China, Japan, Malesia, South–Europe, Africa.

4. *P. polystachyos*

P. polystachyos P. Beauv. Fl. de and Owarch & Benin 2: 48. t. 86. f. 2. 1807; C.B. Clarke in Hook. f. Fl. Brit. Ind. 6: 592. 1893; Hooper in Saldanha & Nicolson, Fl. Hassan 693. 1976; Arora *et al.*, Bot. S. Karana 61. 1981; Sharma *et al.*, Fl. Karnataka 314. 1984; Koyama in Dassanayake & Fosberg, Rev. Handb. Fl. Ceylon 5: 219. 1985; Singh, Fl. E. Karnataka 2: 644. 1988; Karthik *et al.*, Fl Ind. Enum. Monocot. 65. 1989; KeshavaMurthy & Yoganarasimhan, Fl. Coorg 514. 1900. *Cyperus polystachyos* Rottb. Descr. Rar. Nov. Pl. 21. f. 1. 1773. Nees in Wight, Contr. Bot. India. 75. 1834; Haines, Bot. Bih & Or. 903. 1924. Kukenth in Engl. Pflanzenr. 4 (20). Heft: 101, 365. 1936; Ramaswamy & Razi, Fl. Banglore 91. 1973; Kern in van Steenis, Fl. Males. 1, 7: 649. 1977; Rao & Verma, Cyp. N.E. India 10. 1982. *C. paniculatus* Rottb. Descr. et. Rar. Nov. Pl. 40. 1073. *C. polystachyos* Rottb. var. *laxiflorus* Benth. Fl. Austr 7: 261. 1878. *Pycerus polystachyos* Rottb. Beauv. var. *laxiflorus* (Benth.) Clarke in Hook. f. Fl. Brit. India. 6: 592. 1893; Hooper in Saldanha & Nicolson, Fl. Hassan 693. 1776. *P. odoratus* Urb. Symb. Ant. 2: 164. 1900; Fischer in Gamble, Fl. Pres. Madr. 1627. 1931 (3: 1132. 1957, repr. ed.). *Cyperus odoratus* auct. non. L. 1753; Cooke, Fl. Pres. Bomb. 2: 859. 1908 (3: 372. 1958 repr. ed.). (PLATE–72).

Annual. Stem 25 tall, 3.0 mm thick, slender, rigid, compressed–trigonous, erect, tufted, green, glabrous, base brown, sheathing. Leaves 18 cm long, 1.5 mm thick, half to three–fourth as long as the stem, flat, greyish green, acuminate, flat, clustered at the base, reduced to bladeless sheaths; sheath membranous, purplish towards the base. Inflorescence anthelate; anthela 1.0 cm long, 9.0 mm broad, compound, evolute or capitately contracted. Bracts 3–6, 5.0 cm long, foliaceous, usually lowest overtopping the inflorescence; rays 1–7, 4.0 cm long, spreading. Spikelets 8.0 mm long, 1.5 mm broad, 10 in number, numerous, congested, suberect, spreading, linear, gradually tapering to an acute apex, 16–50 flowered; rachilla flexous, narrowly winged, glabrous. Glume 2.0 mm long, 1.0 mm broad, membranous, remote, ovate–oblong, linear–lanceolate, acute, base round, muticous, mucronulate, keeled, stramineous or pale ferrugineous, rarely castaneous, 1–nerved; nerves excurrent. Stamens 1 rarely 2; anther 0.5 mm long, crest, ovate, lanceolate, oblong, acuminate, base rounded, brown, glabrous; filament1.0 mm long, uniformly slender, glabrous longer than anther, attached with pistil to the base. Pistil 2.0mm long; ovary 0.5 mm long, ovate–oblong, glabrous; style 1.0 mm long, deeply bifid, uniformly slender; stigma 0.5 mm long, acute, brown, glabrous. Nut 1.0 mm long, 0.5 mm broad, linear–oblong, obtuse, lenticular, with almost truncate shoulders, shortly apiculate at apex, brown or ultimately glossy black, laterally compressed, epidermal cells isodiametric.

Flowering & Fruiting

August–November; Basti, Sonbhadra.

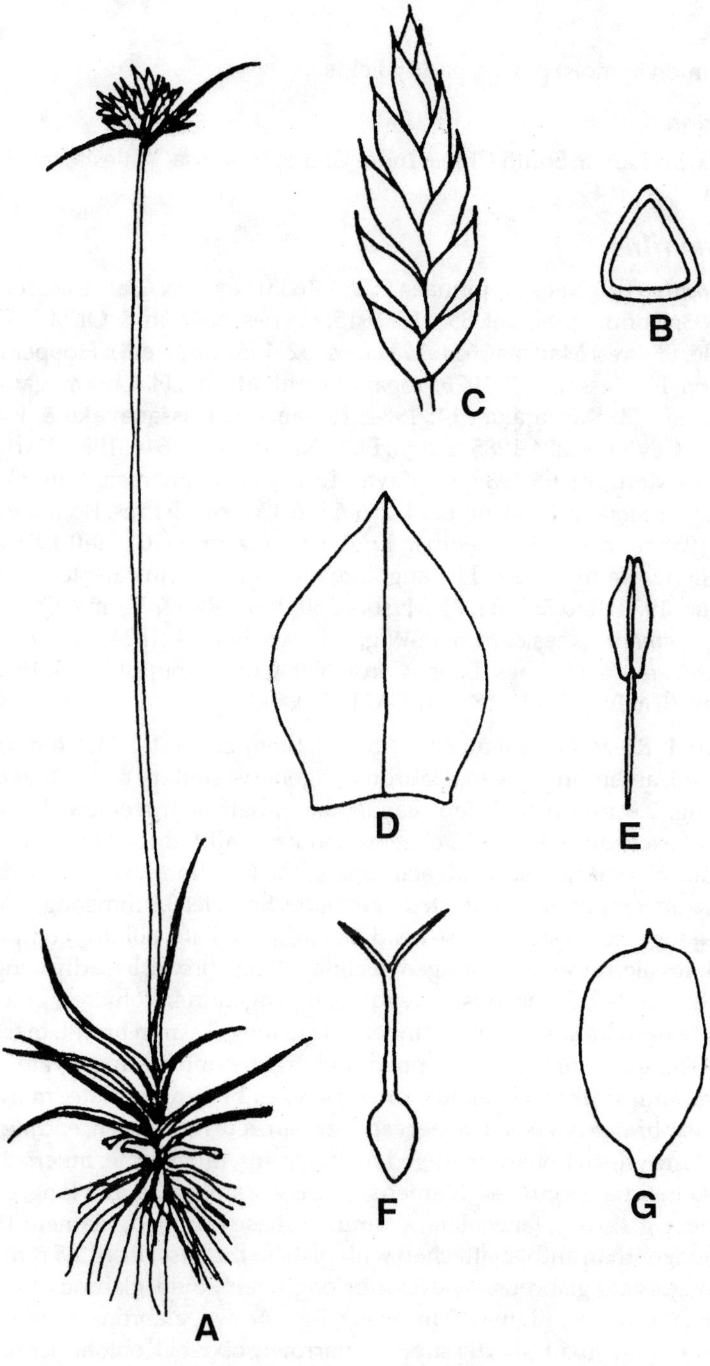

PLATE 72: *Pycreus polystachyos* (Rottb.) Clarke
A. Habit, B. T.S. of stem, C. Spikelet, D. Glume,
E. Stamen, F. Pistil, G. Nut

Habitat

Common in moist places, paddy fields.

Distribution

India, Sri Lanka, South China, Indo–China, Formosa, Malesia, tropical Africa & Australia.

5. *P. pumilus*

P. pumilus (L.) Nees in (Linnaea 9: 283. 1835) Nees, ex Clarke in Hook. f. Fl. Ind. 6: 591. 1893; Domin. Bibl. Bot. 85: 417. 1915; Haines, Bot. Bih. & Or. 905. 1924; Fischer in Gamble, Fl. Pres. Madras 1624. 1931 (3: 1132. 1957 Repr. ed.); Hooper in Saldanha & Nicolson, Fl. Hassan 693. 1976; Yoganarasimhan *et al.*, Fl. Chikmagalur 360. 1981; Sharma *et al.*, Fl. Karnataka 314. 1984; Koyama in Dassanayake & Fosberg, Rev. Handb. Fl. Ceylon 6: 224.1985; Singh, Fl. E. Karnataka 2: 644. 1988; Karthik *et al.*, Fl. Ind. Enum. Monocot. 66.1989; Keshava Murthy & Yoganarasimhan, Fl. Coorg 515. 1990. *Cyperus pumilus* L., Cent. L. Pl. 2: 6. 1756; Cooke, Fl. Pres. Bombay 2: 857. 1908 (30: 370. 1958 repr. ed.); Kunkenth in Engler. Pflanzenr. 4 (20), Heft 101: 375, t. 44. A–E. 1936; Ramaswamy & Razi, Fl. Bangalore 92. 1973; Kern in van Steenis, Fl. Males. 1: 7: 650, f. 66. 1974; Rao & Razi, Fl. Mysore 560. 1981; Rao & Verma, Cyp. N.E. Ind. 9: 1982. *C. pulvinatus* Nees & May in Wight, Contr. Bot. 74. 1834. *Pycreus nitens* Nees, Nova Actorum. Acad. Caes. Leop–Carol. Nat. Cur. 19. Suppl. 1: 53. 1843; Clarke in Hook. f. Fl. Brit. Ind. 6: 591. 1893. (PLATE–73).

Annual. Rhizome absent; root fibrous. Stem 25 cm tall; 1.0 mm thick tufted, compressed, ascending, green, glabrous, trigonous, solitary or 1–5 together. Leaves 18 cm long, 2.0 mm broad, few, canaliculate, half or three–fourth as long as or sometimes exceeding the stem, clustered at base, all reduced to blade less sheath, green, glabrous margin scabrid, acuminate, thick. Inflorescence anthelate; anthela simple or subcompound, contracted or evolute. Spikelet 3.0 mm long, 1.5 mm broad, congested together in subglobose head, linear, lanceolate, oblong, compressed, 8–20 flowered, sessile, digitately arranged; rachilla 1.0 mm broad, broadly winged, margin hyaline. Bracts 3–5, foliaceous, spreading, arranged in involucres, green; outer 7.0 cm long, 1.0 mm broad, glabrous; inner 3.0 cm long, 1.0 mm broad, margin scabrid, hyaline. Glume 1.5 mm long, 1.0 mm broad, tightly imbricating, ovate, truncate on shoulders, uniformly stramineus, or sometimes brown striolate, margin hyaline, surface membranous, keeled, 5–nerved; nerves excurrent, green, ending beyond the apex of glume, distichously arranged on the rachis, mucronate; mucro 0.3 mm long, erect or sometime subcurved. Stamens 1, rarely 2; anther 0.5 mm long, crest, brown, ovate, linear, glabrous, lanceolate, acuminate, base rounded; filament 1.0 mm long, slender, longer than anther, attached with pistil to the base. Pistil 1.5 mm long; ovary 0.5 mm long, ovate, glabrous; style 0.5 mm long, deeply bifid, glabrous, flat, continuous with ovary, caducous; stigma 0.5 mm long, light brown, glabrous, acuminate. Nut 1.0 mm long, 0.5 mm broad, shortly stipitate, narrowly obovoid, oblong, lenticular, obtuse and apiculate, blackish brown, laterally compressed, biconvex, epidermal cells iscodiametric.

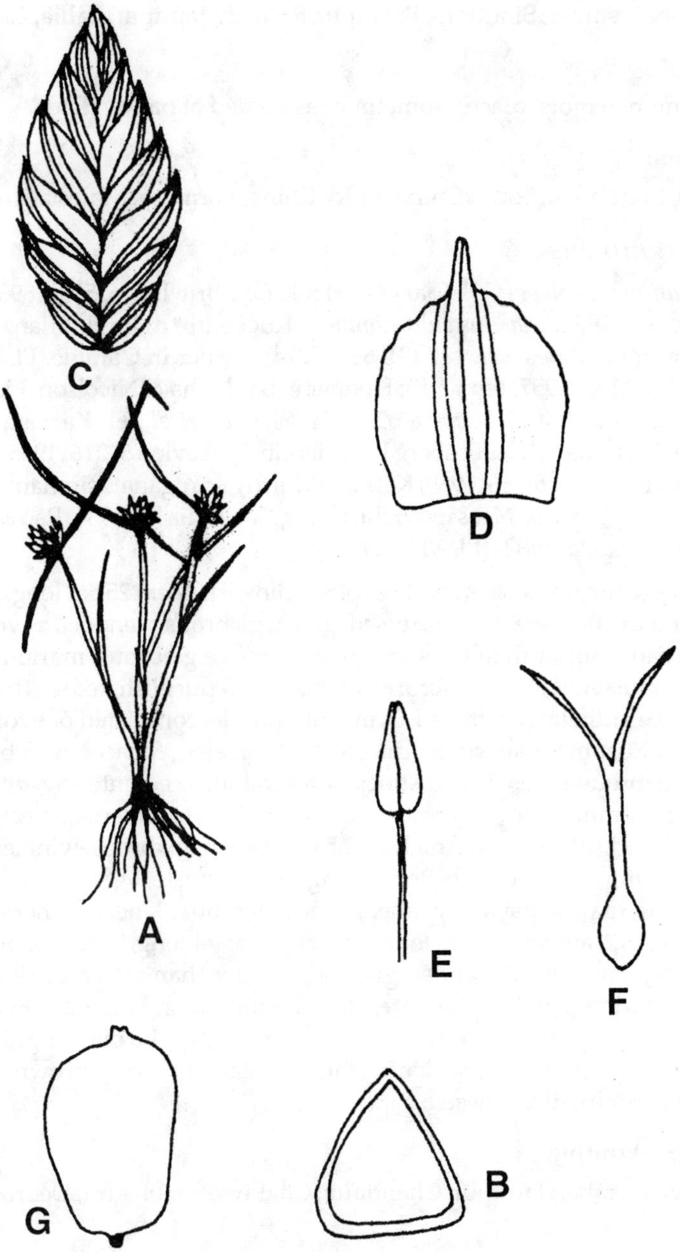

PLATE 73: *Pycreus pumilus* (L.) Nees
A. Habit, B. T.S. of stem, C. Spikelet, D. Glume, E. Stamen, F. Pistil, G. Nut

Flowering & Fruiting

June–November; Sindhora, Babatpur, Sarnath, Jaunpur, Ballia, Gorakhpur.

Habitat

Common in moist places, sometimes as a weed of paddy fields.

Distribution

India, Sri–Lanka, South China, Indo–China, Formosa, Malesia, Tropical Africa.

6. *P. stramineus*

P. stramineus (Nees) C.B. Clarke in Hook. f. Fl. Brit. Ind. 6: 589. 1893; Haines, Bot. Bih. & Or. 904. 1924. *Cyperus substramineus* Kuekenth. in Engl. Pflanzenr. 101: 398. 1936; Kern in Fl. Males. ser. 1. 7 (3): 653. 1974; Fischer in Gamble, Fl. Pres. Madras 1627. 1931 (3: 1132. 1957, repr. ed.); Hooper in Saldanha & Nicolson, Fl. Hassan. 694. 1976; Arora *et al.*, Bot. S. Kanara 62. 1981; Sharma *et al.*, Fl. Karnataka 314. 1984; Koyama in Dassanayake & Fosberg Rev. Handb. Fl. Ceylon 5: 216. 1985; Karthik *et al.*, Fl. Ind. Enum. Monocot. 66. 1989; KeshavaMurthy & Yoganarasimhan, Fl. Coorg 515. 1990. *Cyperus stramineus* Nees in Wight, Contr. Bot. India 74. 1834; Rao & Verma, Cyp. N.E. India 7. f. 2–2a. 1982. (PLATE–74).

Annual. Rhizome absent; root fibrous, yellowish. Stem 25 cm long, 1.5 mm thick, subtrigonous, tufted, erect, compressed, green, glabrous, slender. Leaves 20 cm long, 1.5 mm broad, shorter than the stem, green, surface glabrous, margin scaberulous, acuminate, clustered at base, filiform, erect, sheath purplish, reduced; ligule absent. Inflorescence anthelate; anthela 1.0 cm long, simple; contracted or evolute. Bracts 3, 8.0 cm long, 1.0 mm broad, subtending to next spikelets, foliaceous, suberect, surface & margin glabrous, green, vein distinct on ventral surface. acute; rays absent. Spikelet 1.0 cm long, 3.0 mm broad, suberect, linear–oblong, 2–6 together, suberect, compressed, straw–coloured, 20–flowered; rachilla 2.5 mm broad, glabrous, wingless. Glume 2.0 mm long, 1.5 mm broad, closely imbricating, keeled, ovate, light brown, mucronulate, membranous, margin partly hyaline, surface glabrous, 3–nerved; nerves excurrent, green. Stamens 2; anther 0.5 mm long, brown, ovate–oblong, linear, glabrous; filament 1.0 mm long, uniformaly slender, glabrous, longer than anher. Pistil 1.5 mm long; ovary 0.5 mm long, glabrous, ovate; style 0.5 mm long, flat, glarbous, deely bifid; stigma 2, 0.5 mm long, brown, acuminate, glabrous. Nut 1.0 mm long, 0.5 mm broad, obovoid, subordicular, obtuse, lenticular, apiculate, blackish–brown, with whitish transversely undulating lines.

Flowering & Fruiting

July–November; Jaunpur, Chandauli; Chakiya, Sonbhadra, Deoria.

Habitat

Common in waste places & along road side.

Distribution

India, Sri–Lanka, Myanmar, Indo–China, Malesia.

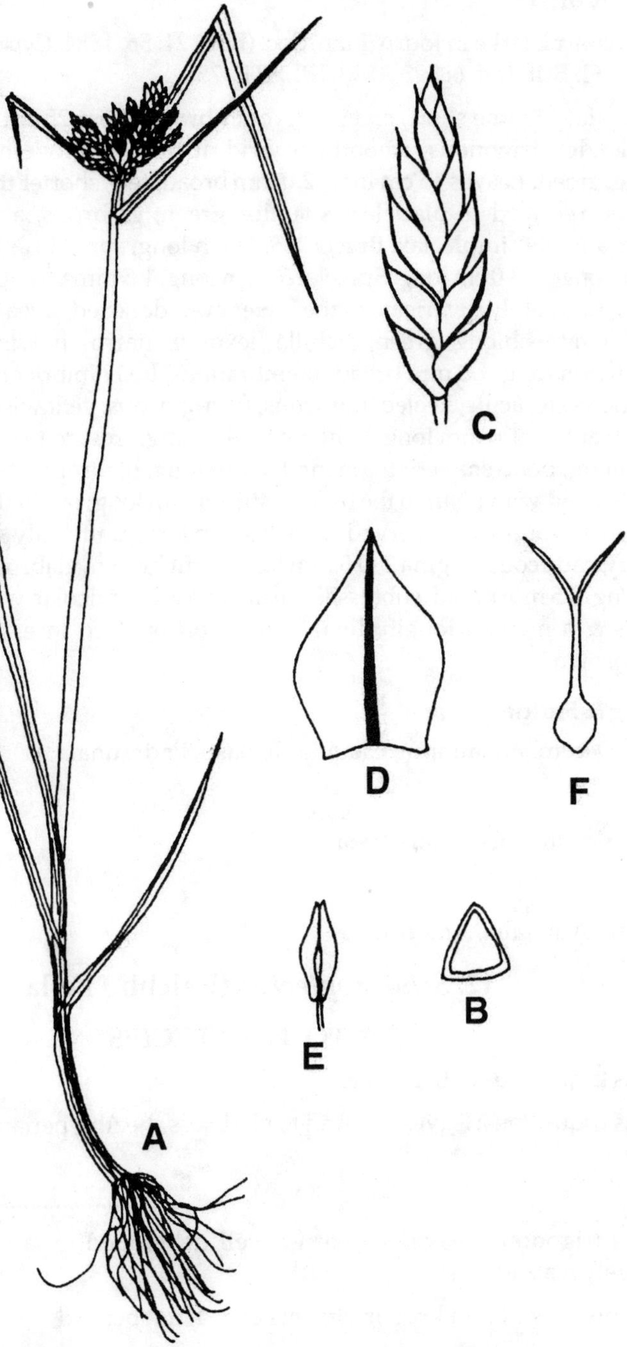

PLATE 74: *Pycreus stramineus* (Nees) Clarke
A. Habit, B. T.S. of stem, C. Spikelet, D. Glume, E. Stamen, F. Pistil

7. *P. sulcinux*

P. sulcinux Clarke in Journ. Linn. Soc. (Bot.) 21: 56. 1884. *Cyperus sulcinux* Clarke in Hook. f. Fl. Brit. Ind. 6: 593. 1893. (PLATE–75).

Annual. Rhizome short, creeping; root fibrous. Stem 25 cm tall, 3.0 mm thick, tufted, slender, trigonous, smooth, scabrid at tip, 3–10 together, rarely solitary, ascending, green. Leaves 15 cm long, 2.0 mm broad, few, shorter than stem, clustered at base, all reduced to bladeless sheaths, green, glabrous, acute. Inflorescence anthelate; anthela simple, lax. Bracts 3–5, 11 cm long, spreading, foliaceous; rays 3–8, slender, longest 8.0 cm long. Spikelet 3.0 cm long, 1.5 mm broad, shortly spicate, 3–8 together, ultimately divaricate or the lower even deflexed, linear–lanceolate, 16–50 flowered, ovate–oblong, green; rachilla flexuous, narrowly winged when young. Glume 2.0 mm long, 1.0 mm broad, membranous, laxly imbricating, ovate, oblong, linear–lanceolate, acute, keeled, muticous, ferrugineous, yellowish brown, glabrous. Stamens 1; anther 1.0 mm long, light green at young, crest red at maturity, glabrous, linear–oblong, boat–shaped; filament 0.5 mm long, glabrous, slender, shorter than anther, attached with pistil to the base. Pistil 1.5 mm long; ovary 0.5 mm long, ovate, glabrous, base rounded, 2–nerved; style 0.5 mm long, uniformly slender continuous with ovary, caducous; stigma 2, 0.5 mm long, light brown, glabrous, acuminate. Nut 1.5 mm long, 0.5 mm broad, subsessile, linear–oblong, lenticular, with almost truncate, shoulders & a median longitudinal depression on each face, apiculate, smooth, blackish brown.

Flowering & Fruiting

July–December; Jaunpur, Chandauli, Basti, Padrauna.

Habitat

Common in moist & sandy soil.

Distribution

Indian, Australia, America.

12. *Schoenoplectus* (Reichb.) Palla

KEY TO THE SPECIES

1a. Stems triquetrous or trigonous:

2a. Stems triquetrous; leaves reduced to bladeless sheaths; perianth bristles scaly

.... **4.** ***S. litoralis***

2b. Stems trigonous; leaves absent or well developed; prianth bristles retrosely scabrid:

3a. Spikelets 12 mm long, in clusters of 2–4; nut beaked

.... **2.** ***S. lacustris***

3b. Spikelets 15 mm long; in clusters of 3–8; nut not beaked

.... **5.** ***S. maritimus***

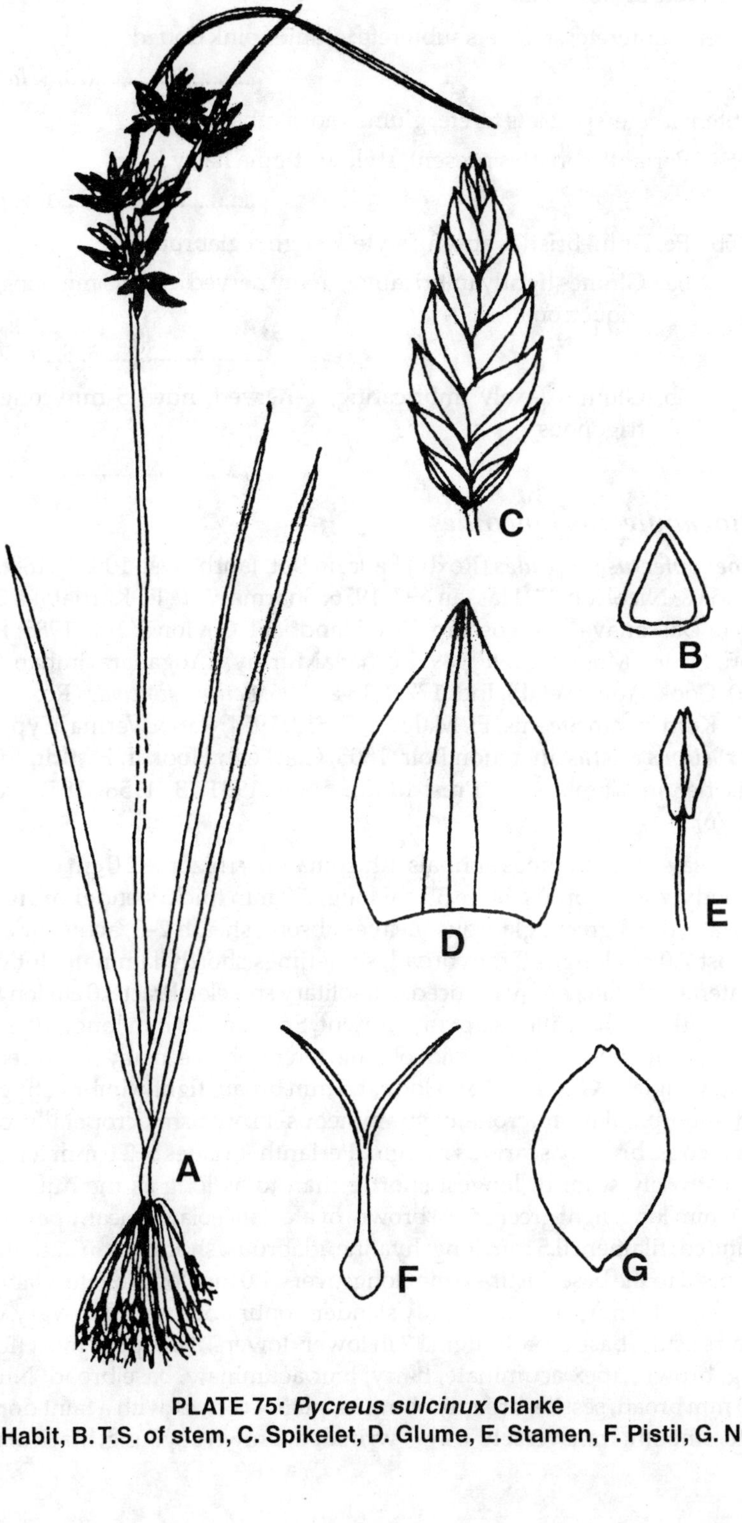

PLATE 75: ***Pycreus sulcinux*** **Clarke**

A. Habit, B. T.S. of stem, C. Spikelet, D. Glume, E. Stamen, F. Pistil, G. Nut

1b. Stems terete or subterete:

4a. Stem subterete; spikelets subterete; glumes pink dotted .... **3. *S. lateriflorus***

4b. Stem terete; spikelets terete; glumes not as above:

5a. Perianth–bristles present; style & stigma hairy .... **1. *S. juncoides***

5b. Perianth–bristles absent; style & stigma glabrous:

6a. Glumes tightly imbricating, many nerved, nut 2.0 mm long, triquetrous .... **6. *S. prolongatous***

6b. Glumes laxely imbricating, 1–nerved; nut 1.5 mm long, trigonous .... **7. *S. roylei***

1. *Schoenoplectus juncoides*

Schoenoplectus juncoides (Roxb.) Palla. in Bot. Jahrb. Syst. 10: 299.1888; Hooper in Saldanha & Nicolson, Fl. Hassan 697. 1976; Sharma *et al.*, Fl. Karnataka 315. 1984; Koyama in Dassanayake & Fosberg, Rev. Handb. Fl. Ceylon 5: 160. 1985; Karthik *et al.*, Fl. Ind. Enum. Monocot. 69. 1989; KeshavaMurthy & Yoganarasimhan, Fl. Coorg 516. 1990; Cook, Aqu. Wet. Pl. Ind. 177. f. 184. 1996. *Scirpus juncoides* Roxb. Fl. Ind. 1: 218. 1820; Kern in van Steenis, Fl. Males. 1, 7: 512. 1974; Rao & Verma, Cyp. N.E. Ind. 42. 1982. *Scirpus erectus* auct. non Poir. 1805; Clarke in Hook. f. Fl. Brit. Ind. 6: 656. 1893; Fischer in Gamble, Fl. Pres. Madr. 1666. 1931 (3: 1156. 1957. repr. ed.). (PLATE–76).

Perennials, or sometimes annuals. Rhizomatous; rhizome 2.0 cm long, creeping, short, woody; roots fibrous. Stem 35 cm long, 3.0 mm thick, slender or rigid, terete, finely striate, deed green, glabrous. Leaves absent; sheath 2–3, acute, or cuspidate, upper most 7.0 cm long, 4.0 mm broad, sometimes shortly laminate. Inflorescence pseudolateral, capitate, rarely reduced to a solitary spikelet. Bract 8.0 cm long, solitary, erect, stem–like with a callous tip; rays absent. Spikelet 5.0 mm long, 3.0 mm broad, usually 2–4, sometimes upto 8, ovoid–oblong, terete, obtuse, many–flowered; rachilla persistent, wingless. Glume 3.0 mm long, 1.5 mm broad, tightly imbricating, broadly ovate to suborbicular, mucronate, stramineous–brownish, acropetally caducous, membranceous, broadly scarious margin. Perianth–bristles 5, 2.0 mm long unequal brown, restrosely scabrid, longest shorter than to as long as the nut. Stamens 3; anther 1.0 mm long, light green, crest brown, ovate–lanceolate, linear, apex acuminate, base rounded; filament 0.5 mm long, hyaline, glabrous, shorter than anther, attached with the pistil to the base. Pistil 4.0 mm long; overy 1.0 mm long, ovate, glabrous, base rounded; style 1.0 mm long, uniformly slender, continuous with the ovary, caducous hairy; hairs acute, base broad; stigma 2 in lower flowers, 3 in upper most flowers, 2.0 mm long, brown, apex acuminate, hairy; hair acuminate, base broad. Nut 2.0 mm long, 2.0 mm broad, sessile, broadly obovoid, plano–convex with a faint dorsal ridge, apiculate, smooth, transversely wavy–wrinkled, brown at young, black at maturity.

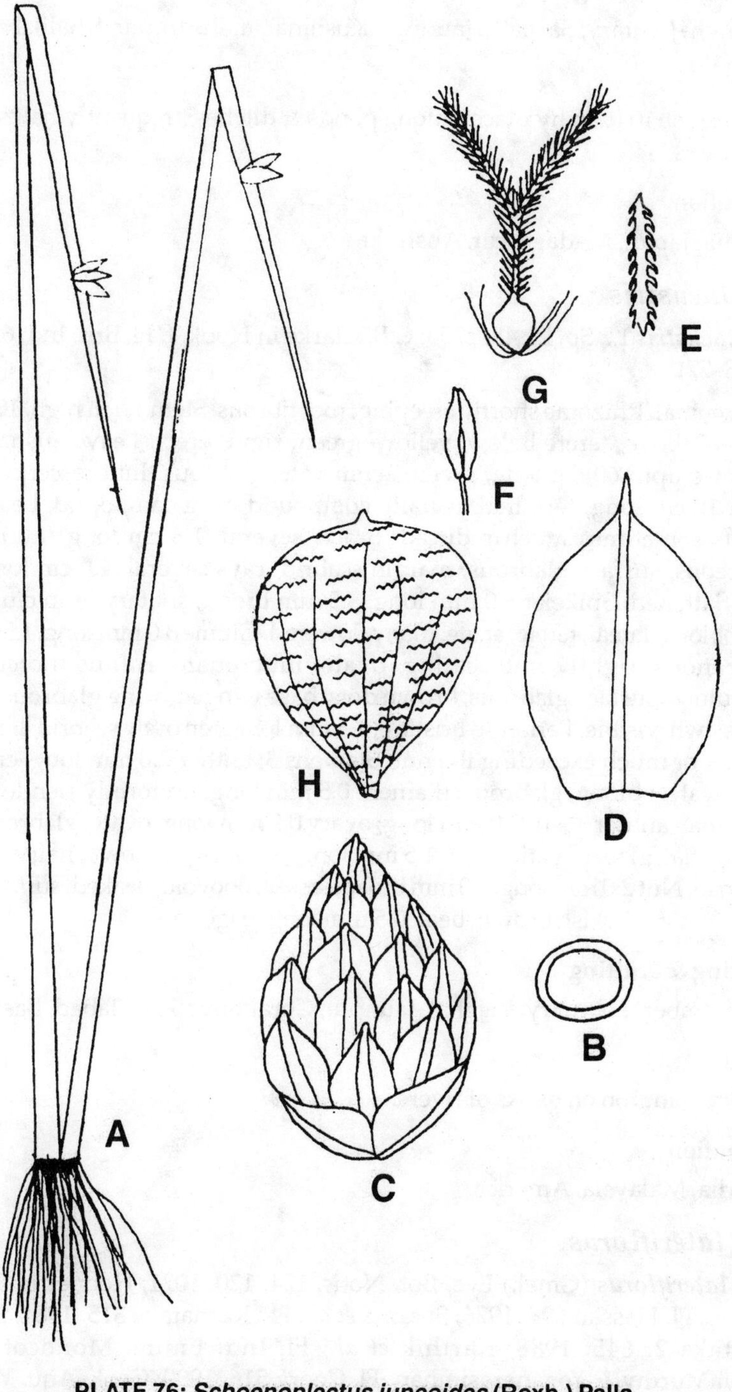

PLATE 76: ***Schoenoplectus juncoides*** **(Roxb.) Palla**
A. Habit, B. T.S. of stem, C. Spikelet. D. Glume, E. Bristle, F. Stamen, G. Pistil, H. Nut

Flowering & Fruiting

March–January; Sarnath, Jaunpur, Kushinagar, Rudrapur, Khalilabad.

Habitat

Common in marshy places, along ponds & ditches, frequently half submerged during rains.

Distribution

India, Japan, Madagascar, Australia.

2. *S. lacustris*

S. lacustris L., Sp. Pl. 48. 1753; C.B. Clarke in Hook. f. Fl. Brit. Ind. 6: 658. 1893. (PLATE–77).

Perennial. Rhizome shortly creeping; root fibrous. Stem 1.5 m high, 10 mm thick, trigonous above, terete below, yellow–green, rigid, erect. Leaves usually absent, sometimes upto 60 cm long, green, acuminate, glabrous. Inflorescence anthelate; anthela 10 cm long, terminal, usually compound or decompound, bearing many spikelets sometimes much reduced. Bracts several, 1.8 cm long, 1.0 mm broad, glumaceous, surface glabrous, margin scabrid; rays several, 4.0 cm long, 1.2 mm broad, flattened. Spikelet 8.0 mm long, 3.5 mm broad, solitary or in cluster of 2–4, ovoid, oblong, linear, terete, acute, many flowered. Glume 3.0 mm long, 1.5 mm broad, membranous, tightly imbricating, ovate, mucronate, rufous, notched at tip, puberculous, surface glabrous, brown dots, base winged; wing glabrous, 1–nerved; nerve brown, visible. Perianth–bristle 6, 1.5 mm long, retrosely scabrid, longest much equalling or much exceeding the nut. Stamens 3; anther 2.0 mm long, crest brown, setose, ovate, oblong, glabrous; filament 0.5 mm long, uniformly slender, glabrous, shorter than anhter. Pistil 3.0 mm long; ovary 0.5 mm long, ovate, glabrous; style 1.0 mm long, flat, glabrous stigma; 3, 1.5 mm long, acuminate, brown, hairy; hairs acute, base broad. Nut 2.0 mm long, 1.0 mm broad, sessile, obovoid, beaked, surface glabrous, margin wavy, blackish brown; beak 0.5 mm long, trigonous.

Flowering & Fruiting

December–February; Rajghat, Jaunpur, Gorakhpur, Khalilabad, Basti, Lar.

Habitat

Very common on banks of rivers.

Distribution

India, Malaysia, America.

3. *S. lateriflorus*

S. lateriflorus (Gmel.) Lye, Bot. Notis. 124. 120. 1071; Hooper in Saldanha & Nicolson, Fl. Hassan 698. 1976; Sharma *et al.*, Fl., Karnataka 315. 1984; Singh, Fl. E. Karnataka 2: 645. 1988; Karthik *et al.*, Fl. Ind. Enum. Monocot. 69. 1989; KeshavaMurthy & Yoganarasimhan, Fl. Coorg 516. 1990; Cook, Aqu. Wet. Pl. Ind. 178. f. 185 a–e. 1996. *Scirpus lateriflorus* Gmel. Syst. Veg. 1: 127. 1791; Kern in van

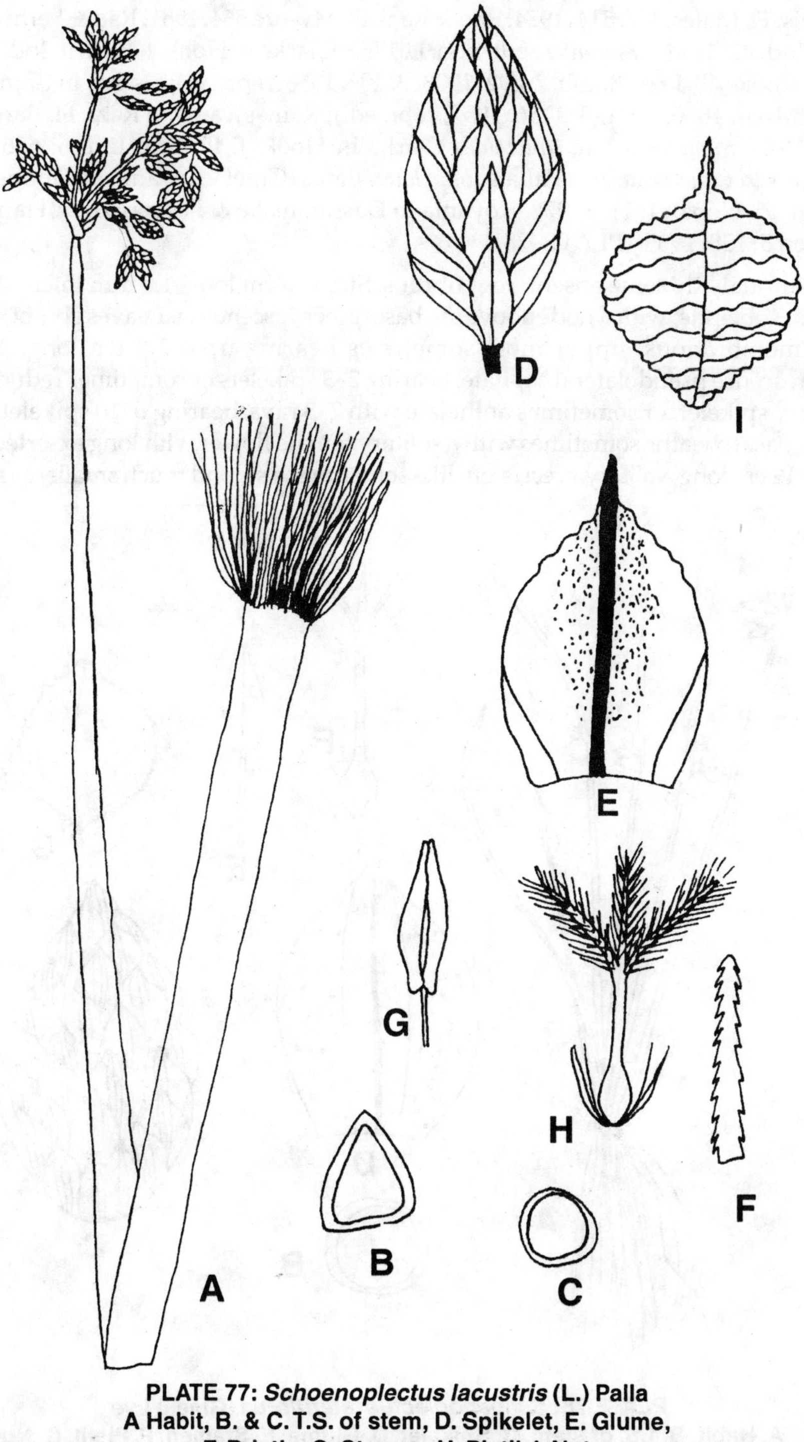

PLATE 77: *Schoenoplectus lacustris* (L.) Palla
A Habit, B. & C. T.S. of stem, D. Spikelet, E. Glume, F. Bristle, G. Stamen, H. Pistil, I. Nut

Steenis, Fl. Males. 1, 7: 514. 1974; Rao & Razi, Fl. Mysore 564. 1981; Rao & Verma, Cyp. N.E. Ind. 42. 1982. *S. supinus* acut. non. L. 1753; Clarke in Hook. f. Fl. Brit. Ind. 6: 655. 1893; Cooke, Fl. Pres. Bomb. 2: 892. 1908 (3: 406. 1958, repr. ed.); Fischer in Gamble, Fl. Pres. Madr. 1666. 1931 (3: 1156. 1957, repr. ed.); Ramaswamy & Razi, Fl. Bangalore 100. 1973. *S. supinus* var. *unioloides* Clarke in Hook. f. Fl. Brit. Ind. 6: 656. 1893. *Schoenoplectus supinus* (L.) Palla subsp. *lateri florus* (Gmel.) Koyama in Hooper et al; Enum. Pl. Nepal. 1: 119. 1978. Koyama in Dassanayake & Fosberg. Rev. Handb. Fl. Ceylon 5: 158. 1985. (PLATE–78).

Annual. Rhizome absent; root fibrous. Stem 15 cm long, 1.0 mm thick, slender, tufted, subterete, with a node above the base, green, ascending. Leaves absent; sheath 2–3, membranous, upper most sometimes bearing upto 2.0 cm long lamina. Inflorescence pseudolateral capitate, bearing 2–3 spikelets or sometimes reduced to a solitary spikelet, or sometimes anthelate with 2–3 rays, bearing 6–10 spikelets; axils of the basal sheaths sometimes with a solitary female flower with long exserted style. Bract 12 cm long, solitary, erect, stem like sometimes a second much smaller bract also

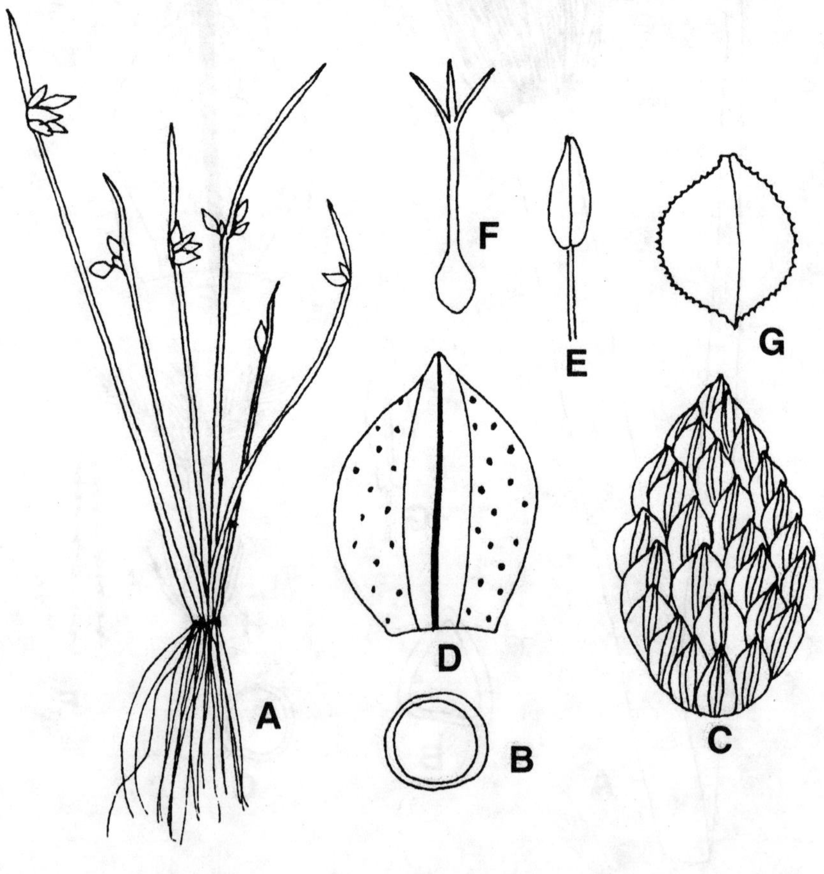

PLATE 78: *Schoenoplectus lateriflorus* (Gmel.) Lye
A. Habit, B. T.S. of stem, C. Spikelet, D. Glume, E. Stamen, F. Pistil, G. Nut

present. Spikelet 5.0 mm long, 3.0 mm broad, ovoid–oblong, subterete, acute, clustered, many flowered; raçhilla persistent, wingless. Glume 2.0 mm long, 1.0 mm broad, laxly imbriating, ovate, mucronate, stramineous with purple dots along the midrib, acropetably caducous, lower 1–3 empty, 3–nerved; central nerve visible, green, leteral nerves faintly visible, side almost nerveless. Perianth absent. Stamens 3; anther 0.5 mm long, apiculate, base rounded, glabrous, linear–lanceolate, oblong, yellow; filament 1.0 mm long, slender, glabrous, longer than anther, attached with pistil to the base. Pistil 2.5 mm long; ovary 0.5 mm long, ovate, glabrous; style 1.0 mm long, uniformly slender, continuous with ovary, entire, decidous; stigma 3, 1.0 mm long, acuminate, light brown. Nut 1.0 mm long, 1.0 mm broad, sessile, broadly ovoid, trigonous, shortly apiculate, stramineus–brownish at young, blackish at maturity, transversely wavy, wrinkled, rugose, apex remnant with style base,

Flowering & Fruiting

August–October; Babatpur, Harraua, Chandauli, Mau, Ballia.

Habitat

Common in moist localities.

Distribution

India, China, Malesia, Formosa, Australia.

4. *S. litoralis*

S. litoralis (Schrad.) Palla Koyama in Dassanayake & Fosberg, Rev. Handb. Fl. Ceylon 5: 157. 1985; Karthik *et al.*, Fl. Ind. Enum. Monocot. 69. 1989. *Scirpus subulatus* Vahl, Enum. Pl. 2: 268. 1806. *S. litoralis* Schrad. Fl. Germ. 1: 142. t. 5. f. 7. 1806; C.B. Cl. in Hook. f. Fl. Brit. Ind. 6: 659. 1893 et in I 11. Cyp. t. 50. f. 1909; Haines, Bot. Bih. & Or. 926. 1924; Kern in Fl. Males. ser. 1.7 (3): 5410. 1974: Clarke in Hook. f. Fl. Brit. Ind. 6: 659. 1893; Cooke, Fl. Pres. Bomb. 2: 894. 1908 (3: 408. 1958 repr. ed.); Fischer in Gamble, Fl. Pres. Madras 1667. 1931 (3: 1122. 1957 repr. ed.); Singh, Fl. E. Karnataka 646. 1988. *Schoenoplectus litoralis* subsp. *thermalis* (Trabut) Murbeck, Contrib. Fl. Maroc. 1: 18. 1922; Sharma *et al.*, Fl. Karnataka 315. 1984. *S. litoralis* var. *subulatus* (Vahl) Chiov. Pl. Nov. 16. 1928; Rao & Razi, Fl. Mysore 564. 1981. (PLATE–79).

Perennial, aquatic. Rhizomatous; rhizome 2.0 cm long, root stock, creeping, with stolons; stolons slender, woody; root fibrous. Stem 2.0 m tall, 1.0 cm thick, stout, erect, green, triquetrous above, terete below, gradually narrow to apex. Leaves 10 cm long, 5.0 mm broad, clustered at base, all reduced to blade less sheath. Inflorescence anthelate; anthela 5.0 mm long, lateral, compound, rarely decompound. Bracts 6.0 cm long, erect, as though a continuation of stem; rays 4.0 cm long slender. Spikelet 1.5 cm long, 5.0 mm broad, 22 in number, solitary, oblong, cylindrical, ovoid; rachilla 1.0 cm long, 1.5 mm broad, peresistent, glabrous, wingless. Glumes 4.0 mm long, 1.5 mm broad, spiral, whitish brown at young, reddish brown on maturity, elliptic, oblong, notched, cuspidate, with scarious margin, membranous, 1–nerved, nerve green, visible, aristate; arista 1.0 mm long, glabrous. Perianth 2.5 mm long, 1.0 mm broad, 7 in number, scale like, plumose, segments longer than nut, rarely bristly. Stamens 3; anther 2.0 mm long, yellow, glabrous, linear, oblong, crested; filament 0.5 mm long,

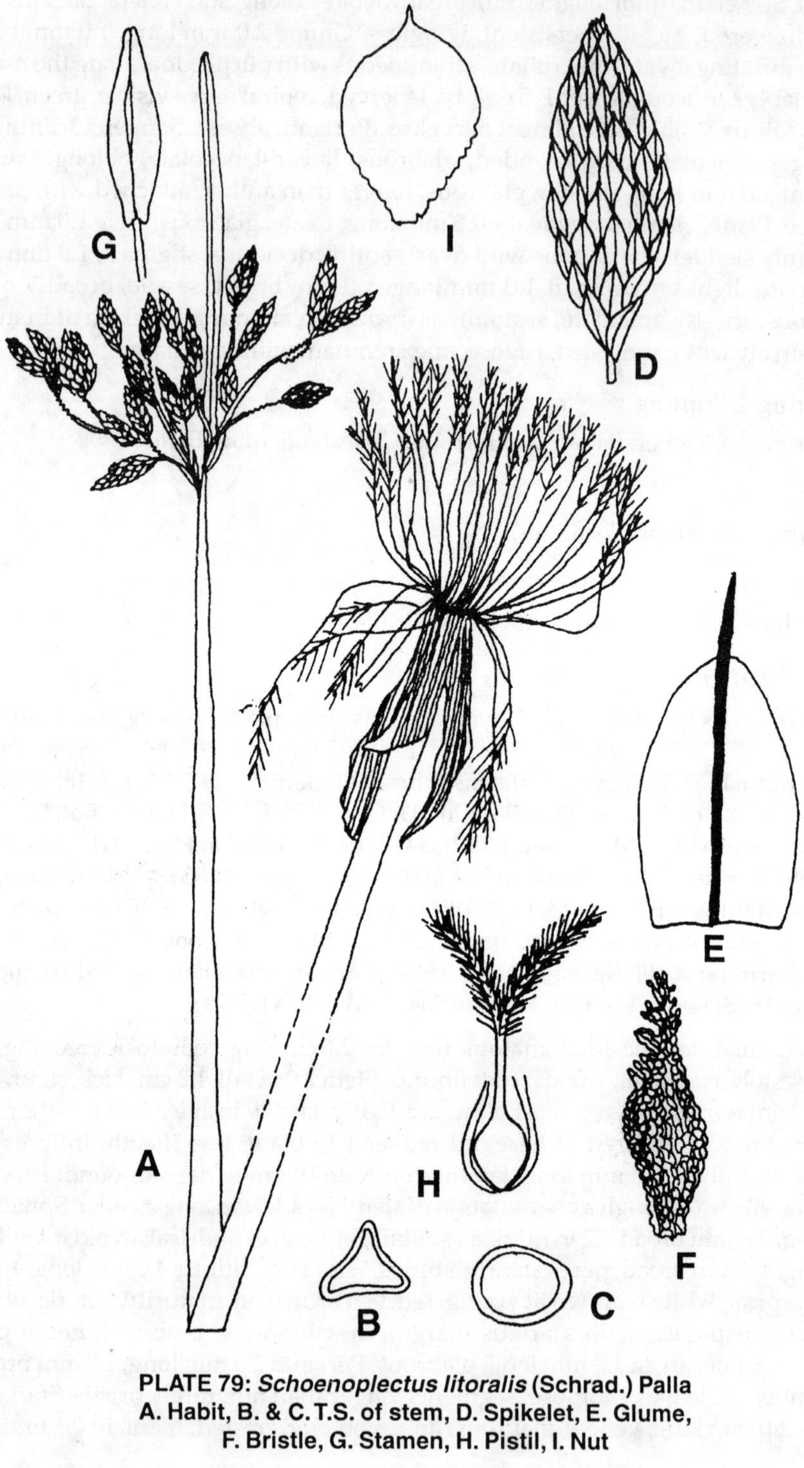

PLATE 79: ***Schoenoplectus litoralis*** **(Schrad.) Palla**
A. Habit, B. & C. T.S. of stem, D. Spikelet, E. Glume, F. Bristle, G. Stamen, H. Pistil, I. Nut

slender, shorther than anther. Pistil 3.5 mm long; ovary, 1.0 mm long, oblong, glabrous; style 1.0 mm long, uniformly slender, entire; stigma 2, 1.5 mm long, hairy; hairs acuminate, base broad. Nut 2.0 mm long, 1.5 mm broad, deepbrown, plano–convex apiculate, smooth.

Flowering & Fruiting

December–April; Cholapur, Sindhora, Allahabad, Lar, Sonbhadra.

Habitat

Common along, marshy areas, standing water.

Distribution

India, Sri–Lanka, Pakistan, Malesia, Micronesia, Tropical Africa, Madagascar.

5. *S. maritimus*

S. maritimus L. Sp. Pl. 51. 1753; Haines, Bot. Bih. & Or. 926. 1924; Kern in Steenis, Fl. Males. ser. 1. 7: 499. 1974; C.B. Clarke in Hook. f. Fl. Brit. Ind. 6: 658. 1893; Beetle in Amer. J. Bot. 29: 84. 1942 Lakshminarasimhan & Sharma, Fl. Nasik 495. 1991. *S. maritimus* L. var. *affinis* (Roth) Clarke in Hook. f. Fl. Brit. Ind. 6: 659. 1893; Cooke Fl. Pres. Bomb. 2: 893. 1908 (3: 408. 1958, rep. ed.); Fischer in Gamble, Fl. Pres. Madr. 1666. 1931. (3: 1156. 1957, rep. ed.) *S. strobilinus* (Roxb.) Hort. Beng. 6. 1814. nom, nud.), Fl. Ind. eds. Carey & Wall. 1: 222. 1820. *Bulboschoenus affinis* (Roth) Drobov. Tridy Bot. Muz. Imp. Akad. Nauk. 16. 139. 1916. *B. maritimus* (L.) Palla in Koch. Syn. Deutch. Schwiz, Fl. ed. 3: 2532. 1904; subsp. *affinis* (Roth) Koyama in Brittonia 31: 284. 1979. *Scirpus affinis* Roth in Roem. & Shult., Syst. Veg. 2: 140. 1817; Kern in Reinwardtia 6: 33. 1961; Karthik *et al.*, Fl. Ind. Enum. Monocot. 70. 1989. (PLATE–80).

Perennials. Rhizomatous; rhizome creeping, forming ovoid, oblong ellipsoid tubers at the nodes; root fibrous. Stem 45 cm tall, 2.0 mm thick below the inflorescence, 6.0 mm near the base, stout, slender, trigonous, green, suberect, glabrous. Leaves 40 cm long, 6.0 mm broad, subcoriaceous, long sheathing, as long as the stem, caudate, acuminate, green, glabrous, mid–vein prominent at lower surface. Inflorescence anthelate; anthela terminal, corymb–like, very variable in development, usually simple or compound, bearing clusters 2–6, sessile, spikelets, often reduced to a single spikelets. Bracts 3, 7.0 cm long, one–several, foliaceous, exceeding the anthela. Spikelet 2.0 cm long, 7.0 mm broad, solitary or in clusters of 3–8, ovoid, oblong to subcylindric, acutish; rachilla persistent, wingless. Glume 8.0 mm long, 4.0 mm broad, laxly imbricating, ovate, oblong, surface glabrous, margin scaberulous, greenish, brown, notched at apex, lacerate, awned; awn 1.5 mm long, slightly curved, antrorsely scabrid, 1–nerved; nerves visible, green. Perianth–bristle 3, 1.5 mm long, unequal, brown, retrosely scabrid, longest shorter than to as long as the nut. Stamens 3; anther 2.0 mm long, yellow, oblong–lanceolate, acuminate, base rounded, crest, setose, glabrous; filament 0.5 mm long, shorter than anhter, attached with pistil to the base. Pistil 5.0 mm long; ovary 0.5 mm long, ovate, glabrous, base acuminate; style 2.0 mm long, uniformaly slender, caducous; stigma 2, 2.5 mm long, acute, brown. Nut 3.0 mm long, 2.0 mm broad, sessile, narrowly obovoid, trigonous, beaked, smooth, brown, surface

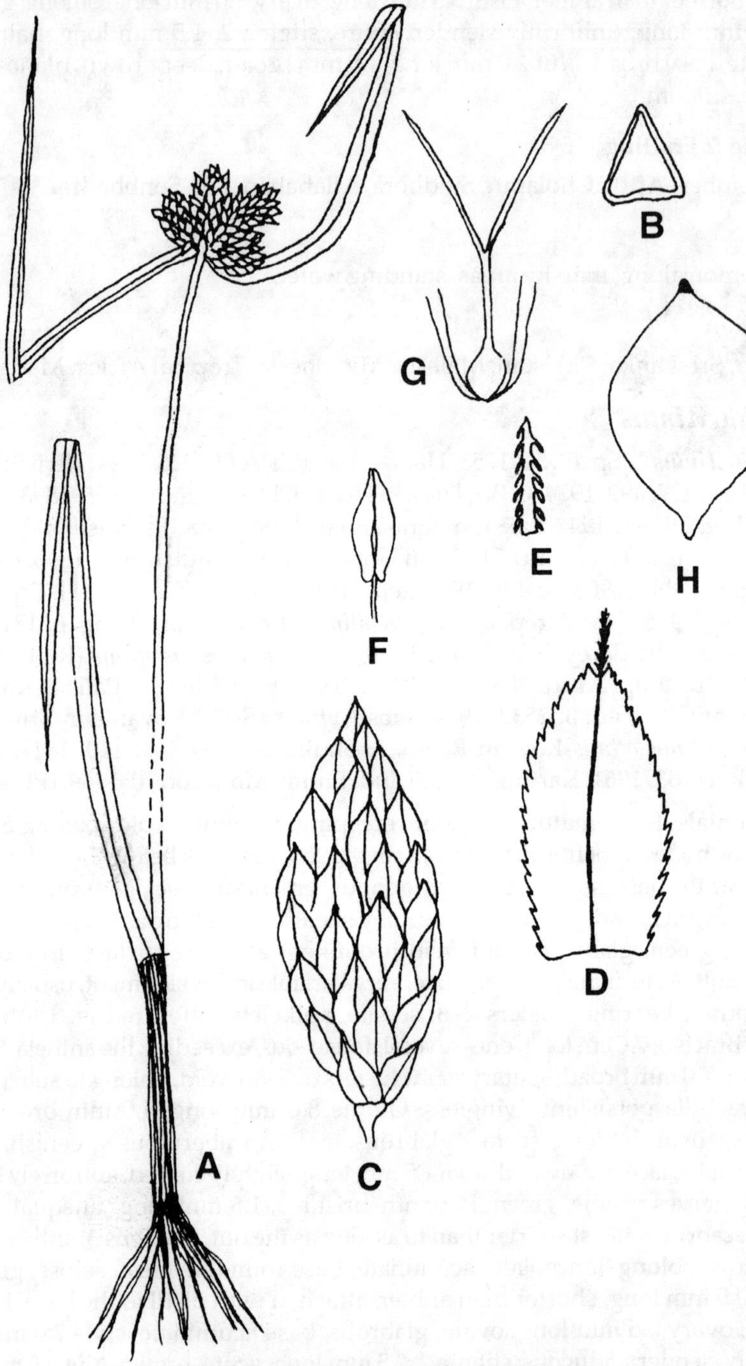

PLATE 80: ***Schoenoplectus maritimus*** **L.**
A. Habit, B. T.S. of stem, C. Spikelet, D. Glume, E. Bristle, F. Stamen, G. Pistil, H. Nut

cells isodimetric, prominent, giving a honeycomb–like apearance to the face, strongly compressed, very rarely with one face slightly convex, prominently apiculate at apex.

Flowering & Fruiting

February–August; Cholapur, Jaunpur, Allahabad, Ballia, Deoria.

Habitat

Very common in moist places, banks of rivers.

Distribution

India, Central, South and South–East Asia.

6. *S. prolongatus*

S. prolongatus Poir. Encyl. 6: 764. 1804; Kern in van Steenis, Fl. Males. ser. 1.7. (3): 514. 1974; Poir, J. Royal in Adansonia 2, 16: 148. 1976. *S. articulatus* (L.) Palla, Bot. Jahrab. Syst. 10: 299. 1888. *Scirpus articulatus* L. Sp. Pl. 47. 1753; C.B. Clarke in Hook. f. Fl. Brit. Ind. 6: 656. 1893; Kern, in van Steenis, Fl. Males. ser. 1, 7 (3): 513. 1974. (PLATE–81).

Annual. Rhizome absent; root fibrous. Stem 35 cm long, 8.0 mm thick, green, erect, stout, terete, spongy, transversely septate, densely tufted, glabrous. Leaves absent; sheath 3, 10 cm long, 9.0 mm broad, enveloping the stem, stramineus–brown, soon disintegrating, rarely bearing 2.0 cm long lamina. Inflorescence pseudolateral capitate, 3.0 cm across, bearing 15–60 spikelets. Bracts 20 cm long, solitary, erect, stem like, transversely septate; rays absent. Spikelets 8.0 mm long, 4.0 mm broad, ovoid–oblong, terete, acutish, many flowered; rachilla persistent, wingless. Glume 3.0 mm long, 3.0 mm broad, membranaceous, tightly imbricating, broadly ovate to suborbicular, acute, base rounded, mucronulate, brown, many nerved, acropetably caducous. Perianth–bristles absent. Stamens 3; anther 0.5 mm long, oblong, lanceolate, glabrous, whitish–yellow; filament 1.0 mm long, glabrous, slender, longer than anther, attached pistil to the base. Pistil 3.0 mm long; ovary 0.5 mm long, ovate, glabrous; style 2.0 mm long, uniformly slender, caducous, glabrous; stigma 3, 1.0 mm long, brown, acute, glabrous. Nut 2.0 mm long, 1.0 mm broad, sessile, obovoid, triquetrous, apiculate, transversely wavy–wrinkled, brown at young, black at maturity.

Flowering & Fruiting

November–January; Sarnath, Allahabad, Sonbhadra.

Habitat

Rare along wast places & road sides.

Distribution

India, South Asia, Australia, Africa.

7. *S. roylei*

S. roylei (Nees). Ovczi et. Czukavina, i. Syst. Veg. 2: 140. 1817. *Scirpus roylei* (Nees) Parker in Duthie, Fl. Upp. Gang. Pl. 3: 361. 1929; Beetle in Amer. J. Bot. 29: 655.

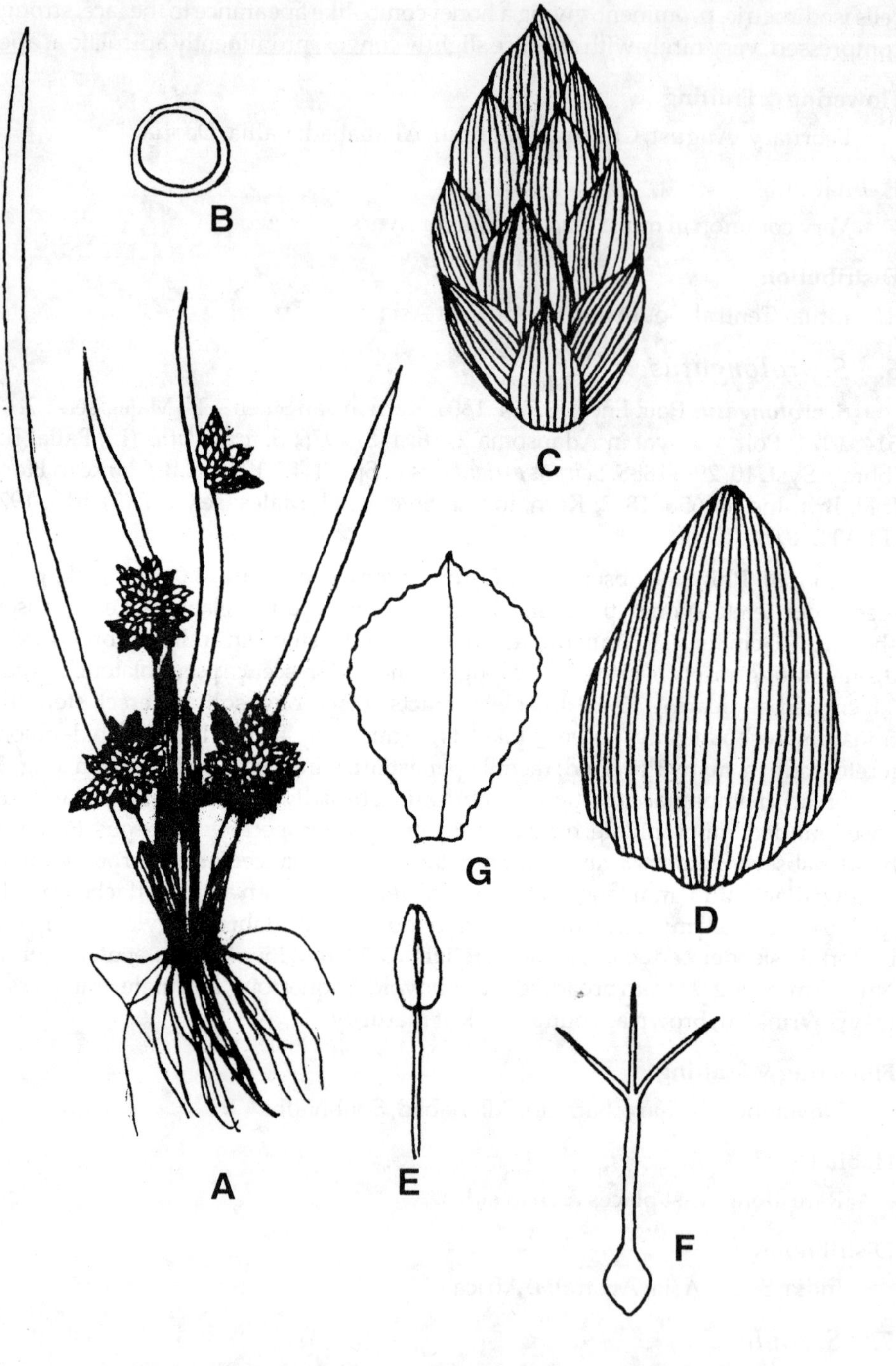

PLATE 81: *Schoenoplectus prolongatus* (Poir) J. Roynal
A. Habit, B. T.S. of stem, C. Spikelet, D. Glume,
E. Stamen, F. Pistil, G. Nut

1942. Maheswari in Sci. & Cult. 31: 195. 1965. *Isolepis roylei* Nees in Wt. Contr. Bot. Ind. 107. 1834. *Scirpus quinquefarius* Buch. Ham. ex Boeck. in Linnaea 36: 701. 1870; C.B. Clarke in Hook. f. Fl. Brit. Ind. 6: 663. 1893 et in Ill. cyp. t. 48 f. 8–10. 1909. (PLATE–82).

Annual. Rhizome absent; root fibrous. Stem 15 cm long, 1.5 mm thick, slender, terete, smooth, often septate, green, tufted, erect, patched, filiform to stout. Leaves absent; sheath 2–3, 4.0 cm long, 1.5 mm broad, loosely enveloping the stem, green at young, brownish at maturity. Inflorescence 8.0 mm long, pseudolateral, capitate. Bracts 13 cm long, stem–like, erect, foliaceous, green. Spikelet 5.0 mm long, 2.0 mm broad, terete, oblong–lanceolate, acutish, clustered, polygonal, many flowered; rachilla persistent, wingless. Glume 2.0 mm long, 1.5 mm broad, particularly in fruits, very laxly imbricating, oblong, lanceolate, acute, membranaceous, margin green, brown near nerve, several, striate, spiral, acropetally caducous, 1–nerved; nerve green, excurrent; aristate; arista 1.0 mm long, antrosely scabrid. Periant–bristles absent. Stamens 3; anther 1.0 mm long, linear–oblong, muticous, yellow, glabrous, lanceolate; filament 2.0 mm long, slender, glabrous, longer than anther, attached with pistil to the base. Pistil 2.0 mm long; ovary 0.5 mm long, ovate, glabrous, wavy, base rounded; style 1 mm long, uniformaly slender, continuous with ovary, usually entire, deciduous; stigma 3, 0.5 mm long, glabrous, light brown, acuminate. Nut 1.5 mm long, 1.0 mm broad, sessile, narrowly obovoid, trigonous, transversely wavy–wrinkled, yellow brown at young, blackish at maturity, oblong, apiculate a apex, acute at base.

Flowering & Fruiting

July–December; Varanasi, Deoria, Ballia, Mau.

Habitat

Rare, along marshy places.

Distribution

India, Africa, Asia.

13. *Scirpus* L.

KEY TO THE SPECIES

1a. Stems slender, inflorescence a solitary terminal head

.... **1. *S. kyllingoides***

1b. Stem trigonous or triquetrous; inflorescence anthelate:

2a. Stems trigonous; perianth bristles present; nut blackish, trigonous

.... **3. *S. ternatanus***

2b. Stem triquetrous; perianth bristles absent; nut brown, triquetrous

.... **2. *S. michelianus***

1. *Scirpus kyllingoides*

Scirpus kyllingoides (A. Rich.) Boek. in Linnaea 36: 733. 1870; Clarke in Hook. f. Fl. Brit. Ind. 6: 662. 1893; Cooke, Fl. Bombay 2: 895. 1908 (3: 409. 1958 repr. ed.);

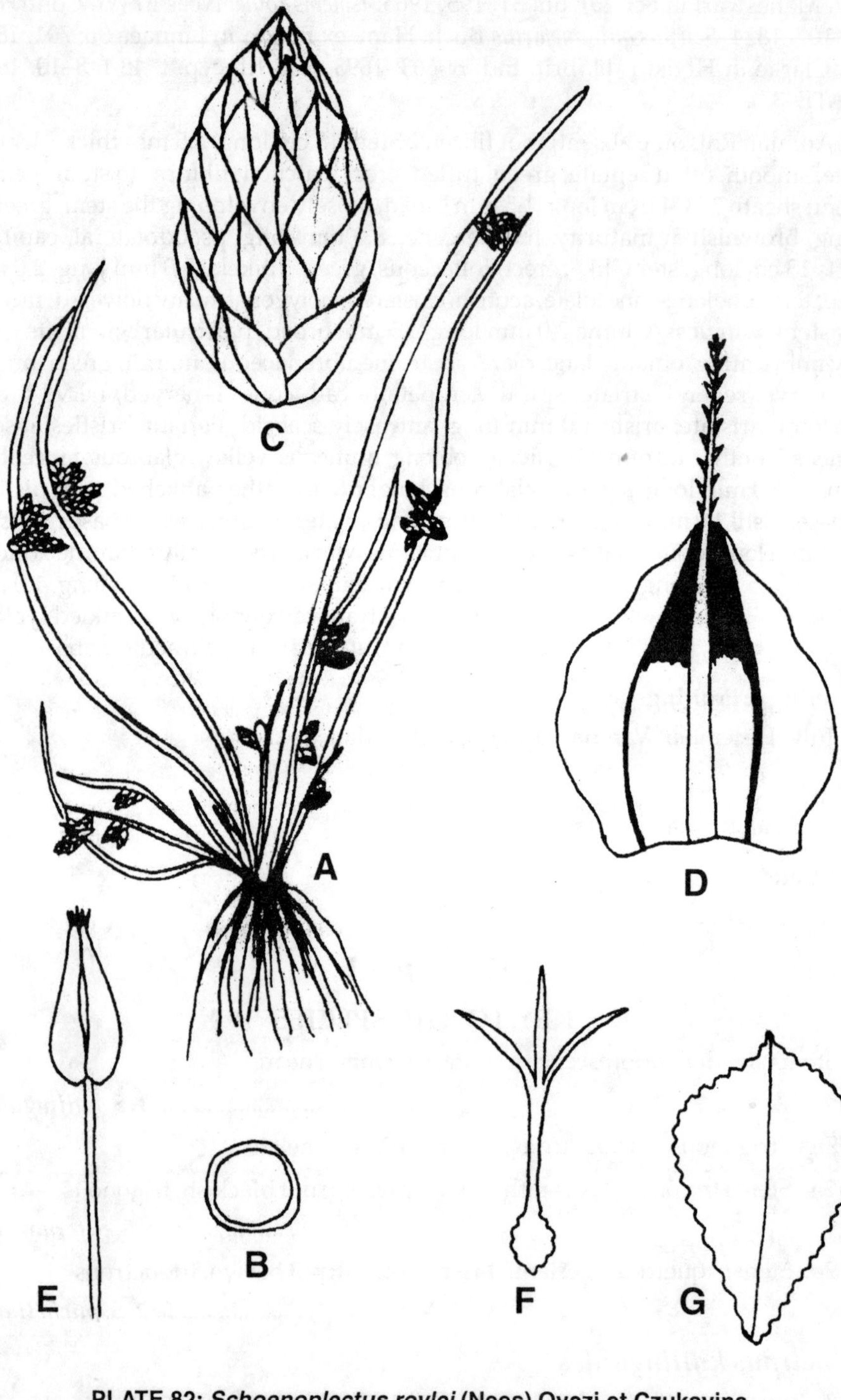

PLATE 82: ***Schoenoplectus roylei*** **(Nees) Ovczi et Czukavina**
A. Habit, B. T.S. of stem, C. Spikelet, D. Glume, E. Stamen, F. Pistil, G. Nut

Wadood Khan in Rheedea 23 (2): 119. 2000. *Kyllinga microcephala* (Steud.) R. Haines & Lye in Bot. Notis. 131: 176. f. 1 1978; Sharma *et al.*, Fl. Karnataka 323. 1984; Karthik *et al.*, Fl. Ind. Enum. Monocot. 61. 1989. *Kyllinga microcephala* Steud in Flora 25: 597. 1842. *Scirpus microcephalus* (Steud.) Dandy in F.W. Andrews, Fl. Pl. Sudan. 3: 366. 1956. *Isolepis kyllingoides* A. Rich in Tent. Fl. Abyss. 2: 502. 1851. (PLATE–83).

Perennials. Rhizomatous, bearing long, slender stolons; root fibrous. Stem 25 cm long, 2.0 mm thick, slender, thickened at base, green, glabrous, erect, tufted, compressed. Leaves 5.0 cm long, 1.0 mm broad, green, surface glabrous, margin scabrid, acute reduced to sheath. Inflorescence a solitary terminal, head bearing 5–15 spikelets. Bracts 9.0 cm long, foliaceous.Spikelets 7.0 mm long, 2.0 mm broad, ovate–lanceolate, many- flowered, terete; rachilla persistent, glabrous, wingless. Glume 3.0 mm long, 1.5mm broad, tightly imbricating, brown, membranaceous, stramineous, sometimes with purple dots, oblong, linear,lanceolate, keeled, acute, base broad, many nerved; mid-nerve deeply visible, thick, lateral nerves slightly visible. Perianth absent. Stamens 3; anther 0.5 mm long, oblong, lanceolate, glabrous, yellow; filament 1.0 mm long, slender, glabrous, longer than filament, attached with pistil to the base. Pistil 2.0 mm long; ovary 0.5 mm long, ovate, glabrous, acuminate, style 1.0 mm long, uniformly slender, caducous; stigma 3, 0.5 mm long brown, glabrous, acuminate. Nut 1.0 mm long, 0.5 mm broad, blackish, obovoid–oblong, trigonous, subcompressed, minutely apiculate, puncticulate, transversely wavy ridges.

Flowering & Fruiting

October–December; Navalpur, Natinia Dai, Jaunpur, Chandauli, Babatpur, Chakiya, Allahabad, Mardah, Ballia, Lar, Deoria, Sardar Nagar.

Habitat

Very common in moist places.

Distribution

India, Malayasia, Africa, Asia.

2. *S. michelianus*

S. michelianus L. Sp. Pl. 52. 1773; Hook. f. Fl. Brit. Ind. 76: 662. 1994; Kern in Fl. Males. ser 1.7: 635. 1974. Misra and Verma, Fl. Allad. 405. 1992. (PLATE–84).

Perennial. Rhizomatous; rhizome 3.0 cm long, creeping; root fibrous. Stem 35 cm long, 3.0 mm thick, green, glabrous, erect, tufted, trigoncus, stout, compressed. Leaves absent; sheath 8.0 cm long, 4.0 mm broad, reddish–brown, faintly nerved, acute, surface & margin glabrous; ligule absent. Inflorescence, anthelate; anthela 1.0 cm long, terminal, decumpound– supradecumpound, bearing numerous spikelets. Bracts 5.0 cm long, leafy, foliaceous, dilated at base, exceeding the anthela; rays 1.0 cm long, divericate, many, rigid, slender. Spikelets 6.0 mm long, 1.5 mm broad, oblong, suberect, glabrous, acutish, solitary, arranged in cluster, green at young brownish tinged with age. Glume 2.0 mm long, 1.5 mm broad, membranous, broadly ovate, lanceolate, acute, glabrous, margin hyaline, tightly imbricating, 3–nerved; nerves green, obscure, glabrous, ending at apex. Penianth–bristles absent. Stamens 3; anther 1.0 mm long,

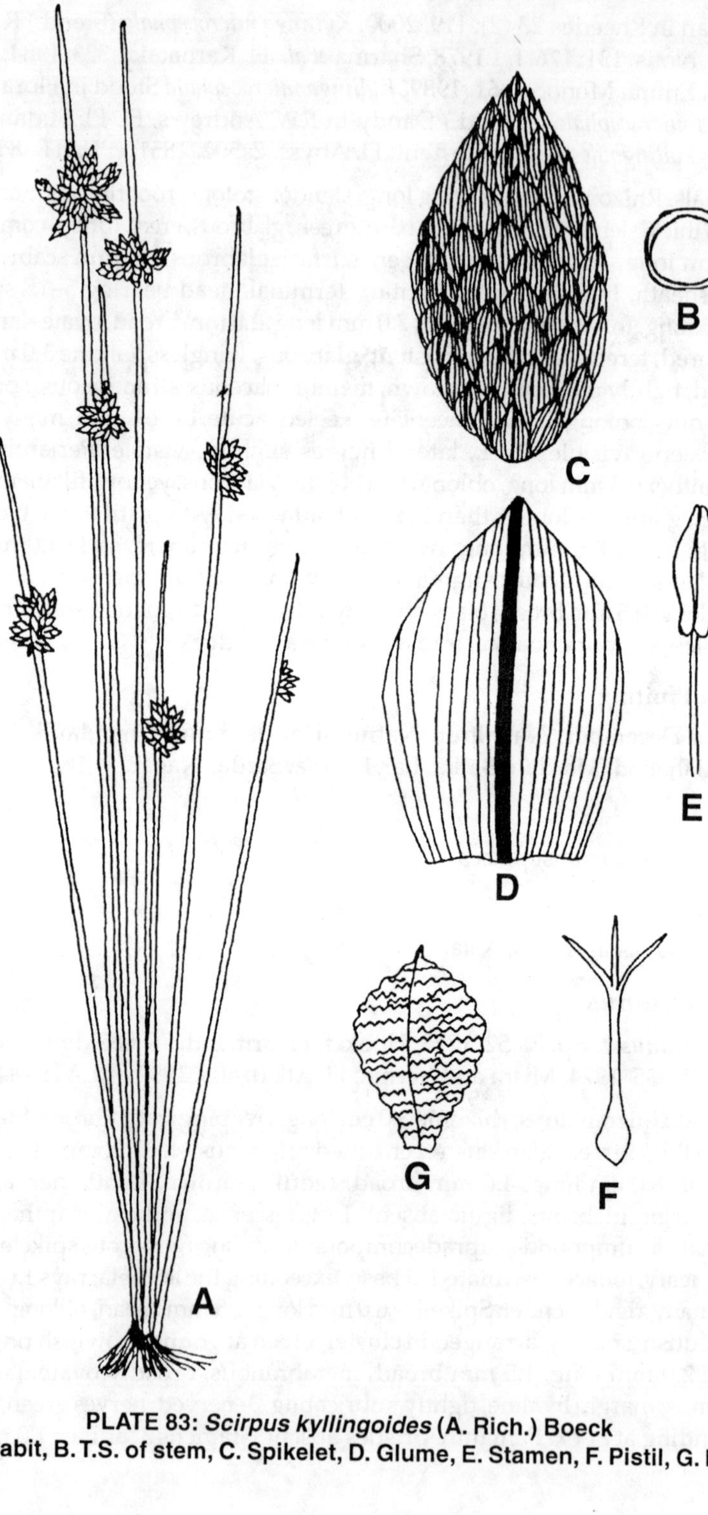

PLATE 83: *Scirpus kyllingoides* (A. Rich.) Boeck
A. Habit, B. T.S. of stem, C. Spikelet, D. Glume, E. Stamen, F. Pistil, G. Nut

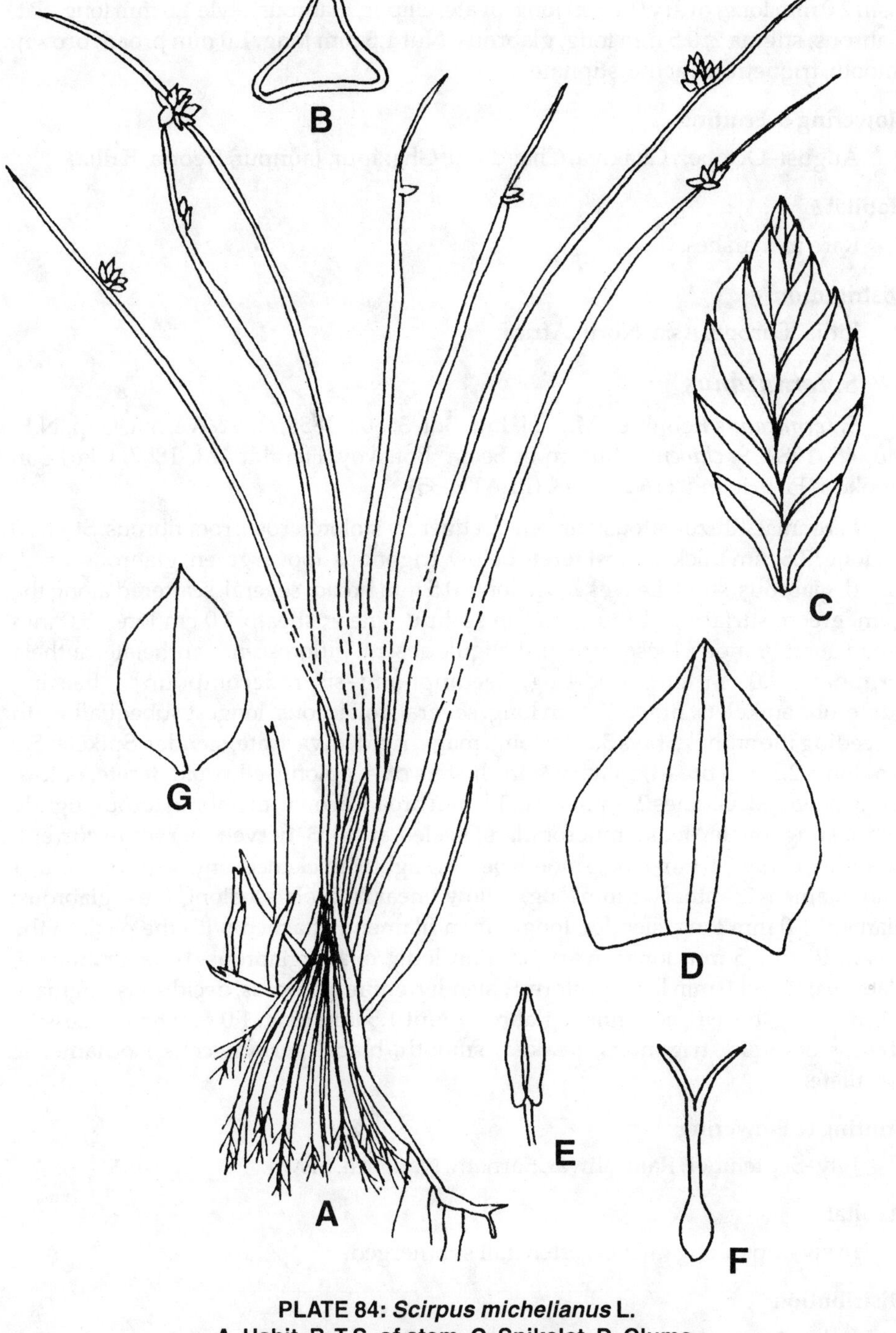

PLATE 84: *Scirpus michelianus* L.
A. Habit, B. T.S. of stem, C. Spikelet, D. Glume, E. Stamen, F. Pistil, G. Nut

yellow, linear, glabrous; filament 0.5 mm long, slender, glabrous, shorter than anther. Pistil 2.0 mm long; ovary 0.5 mm long, ovate, elliptic, glabrous; style 1.0 mm long, flat, glabrous; stigma 2, 0.5 mm long, glabrous. Nut 1.5 mm long, 1.0 mm broad, brown, smooth, triquetrous, acute, stipitate.

Flowering & Fruiting

August–October; Chakiya, Chandauli, Ghazipur, Jaunpur, Deoria, Ballia.

Habitat

Rare in marshes.

Distribution

India, Europe, Asia, North Africa.

3. *S. ternatanus*

S. ternatanus Reinw. ex Mig. Fl. Ind. Bot. 3: 307. 1856; Rao & Verma, Cyp. N.E. Ind. 45. 1982. *S. chinensis* Munro in Seem, Bot. Voy. Herald. 423. 1857; Clarke in Hooke. f. Fl. Brit. Ind. 6: 662. 1893. (PLATE–85).

Perennial. Rhizomatous; rhizome frequently stoloniferous; root fibrous. Stem 20 cm long, 3.5 mm thick, almost terete below, trigonous above, green, glabrous, erect, tufted, glabrous, stout. Leaves 20 cm long, 0.5 mm broad, several, scattered along the stem, green, surface glabrous, margin scabrid, linear; sheath 3.0 cm long, 5.0 mm broad, membranous, loose, brownish; ligule absent. Inflorescence anthelate; anthela terminate, 9.0 cm long, nodding, decumpound–supradecompounds, bearing numerous spikelets. Bracts 7.0 cm long, several, foliaceous, longest subequalling to exceeding the anthela; rays 3.0 cm long, many, rigid, divaricate, slender. Spikelet 5.0 mm long, 2.5 mm broad, solitary & in clusters of 4, oblong–ellipsoid, terete, obtuse many flowered. Glumes 2.0 mm long, 1.5 mm broad, brown, membranaceous, tightly imbricating, ovate, acute, mucronulate, keeled at tip, 3–nerved; nerves excurrent. Prianth–bristles 2.0 mm long, smooth below, slightly thickened, antrorsely scabrid at apex. Stamens 2; anther 0.5 mm long, yellow, linear lanceolate, oblong, crest, glabrous; filament 1.0 mm long, slender, longer than filament, attached with the pistil to the based. Pistil 2.5 mm long; ovary 0.5 mm long, ovate, glabrous, base acuminate, glabrous; style 1.0 mm long, uniformly slender, entire, glabrous, deciduous; stigma 3, 1.1 mm long, brown, acuminate glabrous. Nut 1.5 mm long, 1.0 mm broad, sessile, oblong–obovoid, trigonous, beaked, smooth, black, surface–cells isodiametric apiculate.

Fruiting & Flowering

July–September; Rameshwar, Sarnath, Naugarh, Siswa.

Habitat

In swamps along ditches, often half submerged.

Distribution

India, Africa, Australia.

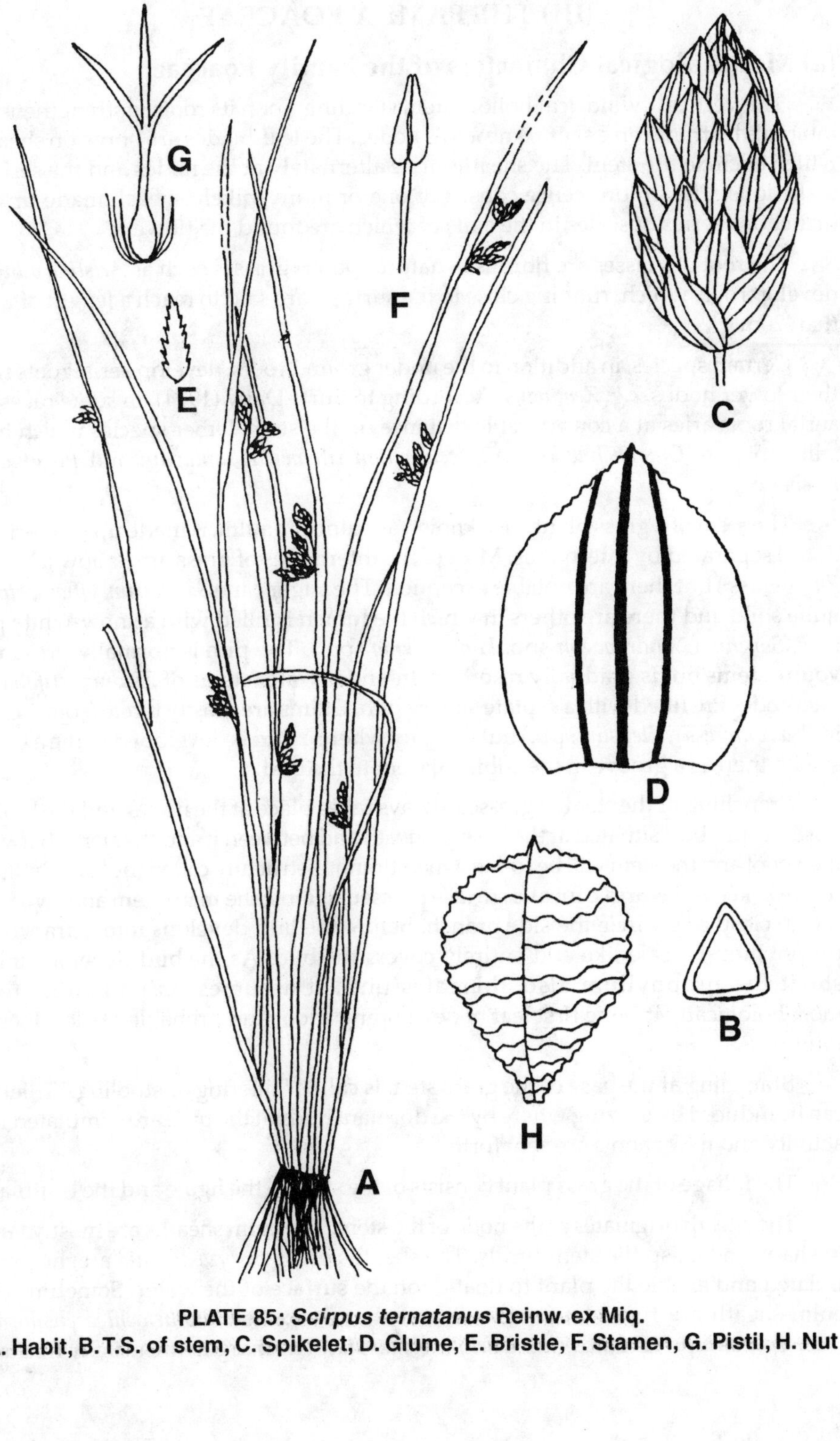

PLATE 85: *Scirpus ternatanus* Reinw. ex Miq.
A. Habit, B. T.S. of stem, C. Spikelet, D. Glume, E. Bristle, F. Stamen, G. Pistil, H. Nut

(III) THE FAMILY POACEAE

(*a*) Morphological Characters of the Family Poaceae

The vertical cylindrical hollow stem standing upon its roots is strengthened at intervals by transverse septa, known as nodes. The leaf-blades are borne on sheaths, which encircle the stem. The sheaths arise alternately at the nodes and the series is distichous. The inflorescence consist of one or more spikelets, each made up of a distichous series of scales in the axils of which are found the flowers.

The root of grasses are fibrous in nature. Desert grasses such as *Aristida pungens* develop roots which, running close to the surface, are said to reach a length of more than 20m.

Certain species, in addition to the under ground roots, develop aerial roots from their lower nodes. *e.g. Zea mays*. According to Burtt-Davy (1914), in some races the aerial roots aries at a considerable distance up the stem. Other species which have stilt roots are *Chionachne koenigii, Ischaemum ritchiei, I. santapaui* and *Pogonachen racemosa*.

The stem of a grass plant, also known as culm or haulm, is made up of a series of nodes separated by internodes. Most of the internodes of grass are hallow (*Arundo, Phragmites*) but there are notable exception. The sugar cane *Saccharum officinarum* is quite soild and there are others in which the lumen is filled with a snow-white pith (*Zea, Sorghum, Andropogon* spp. *Hyparrhenia* spp.). The pith is normally present in young stems but is gradually resorbed. In under water stems of *Zizania latifolia* the internodes are filled with a septate sugary pith. Culms are mostly terete (compressed in *Poa compressa, Eleusine* spp.) but in many where a bud is developed in the axil of a sheath there is a groove in the culm adjacent to the bud.

Branching of the stem in grasses always takes place at the nodes and the branch arises from a bud situated at the base of sheath and between it and the stem. Between the shoot and the stem will be found a membranous structure called the "prophyllum" It is two-keeded with a concave surface pressed against the main stem and two flaps which clasp or encircle the side branch. Before the bud develops into a branch the prophyllum is scale-like and entirely covers the bud. As the bud develops into a shoot the prophyllum also elongates until it assumes its palea-like form morphologically it is the first leaf of developing shoot and probable blades does no form.

Branching at the base of the grass stem is called "tillering or stooling." Tillering can be induced by grazing where by the dormant buds at the base are stimulated into activity and new shoots are sent forth.

The foliage of the grass plant consists of the sheath, the ligule and the leaf-blade.

The sheath originates at the node of the stem. The culm sheaths are mostly terete in shape and clasp the stem firmly. The sheaths of *Hygroryza aristata* are, however, inflated and enable the plant to float upon the surface of the water. Sometimes the culm-sheath are firmly compressed and sharply keeled (*Bothriochloa ensiformis, Chrysopogon asper, C. lancearius, Dactylis glomerata*). The margins are usually free from one another and overlap, the outer margins being as a rule ciliate. In some species of following genera, *Festuca, Melica, Bromus, Glyceria*, the margin of the sheath are joined together and form a closed tube sometimes up to mouth of the sheath. The surface of sheath is usually ridged and very often hairy and or scabrid. The basal sheaths are crowded owing to basal nodes being very close together. The sheats growing in this position are under extreme pressure and are frequently forced, upon and fall away the stem.

At the top of the sheath, where it meets the blade, there is usually a line of demarcation which in some species of *Garnotia* and *Aristida cyanantha*, marks an abscission line where the leaf-blade falls away. The collar may be horizontol, as in most species, so that the leaf-blade stands away from the stem.

The upper leaf sheaths have reduced blades and some times the blade is not developed at all. The sheath then acts as a spathe, enclosing and protecting the inflorescence. (*Lygeum spartum, Hyparrhenia* spp, *Cymbopogon* spp.) An even more extraordinary metamorphosis is the so-called involucre of Job's Tears (*Coix lacryma-jobi*). The ivory, osseous, or brittle container in which the female flower of this grass is situated, is morphologically a metamorphosed sheath in the axil of which the inflorescence is developed.

At the junction of sheath and blade, on the inside, closely an appressed to the culm is a structure called the ligule. The ligule of grasses exhibit great variation in shape and texture.

The ligules may be very delicate, hyaline membranes (*Poa* spp.) or very firmly chartaceous (*Deyeuxia* spp.) or coriaceous (*Sclerostachya fusca*) structures or they may consist merely of a fringe of hairs (*Isachne* spp.). They may be completely absent (*Echinochloa crus-galli, E. colona, Hubbardia heptaneuron*) or a ligule composed of a fringe of hairs may be present in the lower sheaths and completely absent above (*Echinochloa stagnina*). It may vary from none to several centimeters in lèngth (*Deyeuxia simlensis*). It may be glabrous or hairy on the dorsal surface or ciliate on the upper margin.

At the top of the leaf-sheath is leaf-blade. The bases of some leaves are provided with auricles, that is, curved tongue-like out growths from the basal tissue which encircle the stem.

Some leaf-blades are jointed to the sheath, known as petiole.

The tips of leaves are usually acute or acuminate and very scabrid. The tip of the leaves of most species of *Poa* are, however, boat shaped or hooded. *Stenotaphrum dimidiatum* and *Schizachyrium bervifolium* have oblong leaves, which are character istically rounded at the tips. The upper surface of the leaf is often rough, more hairy and dark green than the lower surface. The roughness is due either to the nerves being scabrid or to the surfaces being covered with tubercle-based hairs. The margins of some leaves are furnished with very small teeth so that they are razor-sharp. The margins of other leaves are comparatively quite smooth. Other again are decorted with spaced bulbous-based hairs, a very common phenomenon in *Arthraxon*. The leaves of forest grasses are soft and in periods of drought hang limp and flaccid. In the species of the desert and dry habitats the leaves are more rigid and fold up or roll up to prevent desiccation. Such leaves are grooved between the vascular bundles and at the bottom of the groove certain large epidermal cells are developed. The vascular system of the leaf-blade is visible as a rule on the surface as the veins. The midrib is always quite definite but the lateral strands are often quite faint, *e.g. Cyrtococcum*. The midrib is occasionally so broad that it occupies more than two- thirds of the breadth of the leaf-blades *e.g. Saccharum*. The lateral veins can be of two kinds (a). those that diverge at the base and come together again at the tip. (*Oplismenus*) and (b). those that diverge at the base and without bending towards the apex proceed in a straight line to join a strong marginal nerve *e.g. Triplopogon ramosissimus*. The cross nerves called tesselate nervation are not usually visible but in certain species are quite definite. Grasses which exhibit tesselates nervation are *Centotheca lappacea, Lophatherum gracil, Streptogyna crinita* and two species of *Leptaspis*. The section Ptychophyllum of the genus *Setaria* exhibits a curious form of nervation that has been termed palmate. The leaves in the bud are compressed so that the soft green tissue between the arcuate nerves is folded longitudinally. When the leaf expands this tissue becomes flattened but the leaf as a whole nerves loses its resemblance to a palm-leaf.

The inflorescence of the grass plant is produced on shoots, which may be terminal, and axillary or very often only terminal. Some species, *Eragrostis barrelieri* and *Cleistogenes gateacrei* produce inflorescences from the very lowest nodes and these inflorescences are half concealed within the sheaths. Inflorescences may be divided in to (1) Spikes (2) Racemes (3) Spike-like racemes (4) Panicles (True and false).

Spikes

The spikelets are sessile or subsessile on a jointed or continuous axis. *e.g. Lolium, Tripogon, Thuarea, Eragrostiella, Microchloa, Agrophyron, Elymus* etc. The spikes may be solitany and straight as in *Tripogon, Lolium,* and *Eragrostella*, curved as in *Microchloa*, paired as in spp. of *Paspalum, Dimeria* and *Digitaria*, digitate as in *Cynodon, Dactyloctenium henrardianum* and spp. of *Leptochloa*. The spikelets can be unilateral as in *Microchloa* and *Thuarea*, or arranged alternately on each side of the rachis, as in *Lolium, Elymus* and *Agropyron*.

Racemes

The spikelets are pedicelled. Species of *Brachypodium*, many species of *Digitaria, Bromus himalacius, Nardurus* spp. *Sclerochloa dura, Streptogyna crinita* and *Pogonachne racemosa* have racemose inflorescence. The last is a reduced form in which the sessile spikelets of the usual pairs in Andropogoneae have been suppressed. In *Digitaria*, the arrangement is a series of short pedicelled spikelets, acompanied by one or two or the three longer pedicelled spikelets, attached to the under surface of a continuous rhachis.

Spike-like Racemes

This type of arrangement is exceedingly common in the Andropogoneae. It consists of a jointed axis at each articulation of which is a sessile spikelet accompanied by a pedicelled spikelet. When the inflorescence breaks up the sessile spikelet falls accompanied by the joint of the axil and the pedicel of pedicelled spikelet. This arrangment of the spikelets is very typical of *Andropogon, Saccharum, Bothriochloa, Narenga* and a most of other genera. The pedicelled spikelet may be reduced to the pedicel as in *Apocopis,* represented by an arista on the pedicel as in *Eremochloa* or reduced to a mere rudiment of the pedicel as in some species of *Arthraxon.*

Panicles

This type of inflorescence can be further subdivided into (i) true panicles (ii) false or compound panicles.

True Panicle

This is a type of inflorescence arise verticillately, irregularly in groups, or solitary and alternate from a central axis. There are no axillant sheaths. The branches, branch and re-branch and finally end in the pedicelled spikelets, a common arrangement being a short-pedicelled spikelet accompanied by a long pedicelled spikelet. Verticillate branching is shown by *Chrysopogon gryllus* and *Sporobolus collettii.* In many genera, *e.g.* in *Poa, Agrostis,* etc. the branches may be in pairs in 3,4 or more, at one point. The panicle may be effuse (*Eragrostis aspera,* many spp. of *Panicum*) or contracted. In the latter the spikelets are crowded together, the branchlets short and often spiculate to the base *e.g. Agrostis semiverticillata* and *Eragristis diarrhena.* Further contraction and reduction lead to spike-like panicles in which the branching is much reduced. *Eragrostis coarctata, E. ciliaris, Phleam* spp., *Alopecurus* sp., *Phalaris* spp. and *Arundinella spicata* have spike-like panicles.

False Panicle

The unit of this panicle consists essentially of a group of spikelets seated on a peduncle issuing from a spathe-like sheath, which terminates the branchlets of (a) an involucre of four male or neuter spikelets supporting a sessile or pedicelled hermaphrodite spikelet, accompanied by two pedicelled spikelets (b) a pair of typical andropogonaceous spike-like racemes, one pedicelled, the other sessile, which may be epinastic or not, or (c) a triplet of spikelets consisting of one hermaphrodite sessile, one pedicelled male and one pedicel, and (d) in which the female spikelet at the base of a raceme is protected by a metamorphosed sheath and the terminal male spikelets are racemose. Exa. Of (a) are *Themeda* spp. and *Iseilema* spp. of (b) *Cymbopogon* spp, some spp. of *Andropogon* and *Hyparrhenia* spp, of (c). *Apluda mutica* and of (d). *Coix lacryma-jobi.* The branching of some species of *Cymbopogon* and *Hyparrhenia* may be exceedingly complex. There are monoecious and dioecious grasses.

The Spikelet

The spikelet is the unit of the inflorescence. It consists esssentially of a pair of distichous alternate bracts at the base, called glumes, followed by a floret or florets, the latter distichously arranged and alternate on a jointed or tough axis, the rachilla, each floret consisting of two bracts enclosing a flower. The outer bract is called the lemma. The inner the palea. The spikelet may have from none to forty or more florets. The sexes may be together in the same floret (hermaphrodite) or in different florets in the same spikelets or the sexes may be in different spikelets on the same plant (monoecious) or they may be on different plant (dioecious).

The Rhachilla

This is the axis of the spikelet upon which the glumes and florets (in several-flowered spikelets) are distichously arranged. It is sometimes tough, that is, mature florets fall away from it, *e.g.* in a section of the genus *Eragrostis*, but generally speaking it breaks up either above or below the glumes and between the florets, and the florets fall away with the joint of the floret next above attached. In the species *Streptogyna crinita* this joint of the rachilla is part of the dispersal mechanism.

In many or several-flowered spikelets the terminal joint of the rhachilla is usually crowned by the rudiment of a floret. *e.g. Poa, Chloris, Melica, Tripogon, Eragrostiella,* and many others. In some genera, *e.g. Phalaris, Hierochloe* and *Anthoxanthum*, the reduction takes place in the other *direction* and the rudiments of the floret are below the uppermost which is the hermaphrodite floret.

In the one-flowered spikelet of some genera the rachilla is produced as a naked or ciliate bristle which very rarely bears a rudimentary floret *e.g. Agrostis*, all species of *Deyeuxia* and one species of *Calamagrostis*. The terminal joint of the rhachilla of the species *Danthonidium gammiei* is worthy of note. The rhachilla here is flattened and spathulate with the minute process at the tip which presumbly represents the rudiments of another floret. In one species of *Panicum*, a genus with spikelets containing two florets, *P. fischeri* has a definite prolongation of the rhachilla above the second floret, an unusal feature in the Paniceae.

The glumes: There are two bracts at the base, distichousy arranged, one inserted on the rhachilla (spikelet-axis) just above the other and designated as the lower and the upper glume. The spikelet of *Lygeum spartum* has no glumes, that of *Lolium, Zoysia* and some species of *Digitaria* and *Paspalum*, only one. The glumes of *Parapholis incurva* are placed side to side instead of being distichous and alternate.

In texture, there is also great variation from the hyaline glumes of the Aristideae and Stipeae to the coriaceous glumes of the Andropogoneae and Zoisieae. If the protection of the grain is taken over by lemma and palea the glumes tend to be membranous or hyaline. If the glumes themsclves are the protecting scales they tend to be firmer.

The lower glumes of some genera of Andropogoneae are crustaceous and are sculptured, grooved, wrinkled, pitted or provided with appendages for which it is hard to find a purpose (*Ratzeburgia pulcherrima, Hackelochloa granularis, Manisuris divergens, M. forficulata, Ischaemum rugosum* and many other).

In the Andropogoneae the glumes are often more or less the same size. In the Pooideae the lower glumes tends to be shorter with fewer nerves. Even this is reversed in some species of *Stipa* and *Aristida*.

In the Andropogoneae the spikelets are generally dorsally compressed, hence the lower glume is often two keeled and the upper glume naricular. In the Pooideae, on the other hand, the spikelets are often laterally compressed and the glumes and lemmas are then keeled. Even there are exception for the spikelets of *Cyrtococcum*, one of the Paniceae, show extreme lateral compression, while those of *Isachne*, a member of the Pooideae, are almost globular or even dorsally compressed.

The glumes are sometimes oblique and the lower glume of *Tripogon* has lobe or tooth on one side, while the glume of *Lopholepis ornithocephala* and *Latipes senegalensis* are so *fantastic* as to defy any rotational explanation. In the genus *Eriochloa* the lower glume is minute and scale-like and is wrapped round the swollen basal rhachilla joint. In the genus *Oryza* the upper and lower glumes are represented by two semilunar ledges at the top of the pedicels.

The glumes are usually free from one another but in the genus *Alopecurus* the glumes are connate, atleast at the base. Some glumes are awned *e.g. Oplismenus* where both glumes are awned; in *Lophopogon tridentatus* the lower glume is unawned, the upper awned. In the genus *Aegilops* the glumes may be unawned in the lower part of the inflorescence while they are one to several-awned towards the tip.

In the lower glumes of some species of *Bothriochloa*, *Dicanthium* and *Eremopogon foveolatus* are to be found curious depression, or pits which stand out as knobs on the inner surface of the glume.

The lemmas: These are the flower bearing bracts. In the Andropogoneae, the plan is that the glumes take over the protection of the grain and so are firm in texture. In most of which the lemmas are hyaline in texture. The lower of two lemmas may be absent altogether or very much reduced. The upper lemma is usually awned but the awn is absent in *Imperata, Sclerostachya, Eriochrysis, Saccharum,* etc. Awn-bearing lemmas are often two-lobed or bifid at the tip. As long as the spikelet or very small (*Microstegium* spp.) or so much reduced in width that they are merely the hyaline base of the awn.

In the Paniceae, the scheme is also a lower and upper lemma but here the protection of the grain is undertaken by the upper lemma and the palea, hence the upper is much firmer than the other scales and may be coriaceous or crustaceous. The lower lemma in the Paniceae is rarely awned but it is so in *Oplismenus.* The upper lemma in *Brachiaria* and *Urochloa* is often shortly awned. The surface of the lemma is mostly smooth and polished but in some spp. of *Brachiaria, Panicum maximum* and most species of *Setaria* the upper lemma is transversely rugose. The lower lemma of Paniceae is often like the upper glume with perhaps different nervation. In *Ottochloa,* the glumes are very much smaller than the lower lemma. In *Holcolemma* the lower lemma has a marked furrow on the dorsal wing-like appendages at the base of lemma, suggesting stipules (Arber 1934).

In the tribes of Pooideae the lemma shows greet variation. In the those tribes with one floret the lemma is usually hyaline but can be much firmer and scabrid, e. g. in *Deyeuxia scabrescens* and *Aulacolepis treutleri.* The lemmas may be glabrous or hairy, shorter or longer than the glumes, awned from the tip, middle of the back or from the base. They may be acute or truncate. *Agrostis wardii* is unique in having a dorsal awn as well as the other four nerves produced as awns from the truncate tip of the lemma. In the Aveneae most species have a strongly geniculate dorsal awn & in some species the lobes of lemma are also awned. The lemmas are mostly coriaceous or crustaceous. In *Anthoxanthum* and *Hierochloe* the two lower florets are different in sex from the upper. Their lemmas, too, are much more delicate in texture. The upper floret is coriaceous and polished and much smaller than other two. In *Chloris* and *Lophatherum* the upper lemmas are very much reduced & appear as a small bunch of awns. In *Jansenella griffithiana* of the Arundinelleae, the upper lemma is deeply bifid with a very strong geniculate awn issuing from between the lobes. The lobes themselves are branched at the base. In *Aristida* and *Stipa*, the lemma becomes coriaceous & clasps the grain very firmly. The callus of the lemma is often sharp & barbed so that it can pierce the flesh of an animal or be driven in to soil. In *Trikeraria*, of the Stipeae, the median awn is flanked on each side by two firm scabrid shorter awns. In *Enneapogon* the lemmas are nine-nerved & each nerve is produced beyond the upper margin as a scabrid ciliate awn. In Phalarideae the lower two lemmas are usually represented by two empty scales, but none-or both may be absent. In the genus *Oryza* the lower two lemmas are reduced to scales which may be small or nearly as long as spikelet. The spikelet of *Lygeum spartum* is two-flowered. In this species the margins of the lemmas of the two floret are fused to one another in the lower half or third. The two paleas are fused together, back to back. The lemma in the dioecious genus *Leptaspis*, is shaped like an utricle & is closed except for an apical or lateral pore, through which the styles are thrust out.

The Palea

The palea is the scale borne on the floral axis directly facing the lemma, and between them is to be found the flower. It is the first bract borne on the flower axis and is homologus, with the prophyllum which is the first leaf of the axillary shoot arising from a bud. Like the prophyllum, it has, except in a very few species, only two nerves. The palea is almost away two-keeled in the Pooideae, the adaxial surface is concave and the two flaps embrace the flower. The arrangement and structure suggest that the cramped condition in the floral bud determine the shape and nervation of the palea.

In the Andropogoneae, the palea is hyaline in texture and two nerved. In many species it is not conspicuously two- keeled and is not a few completely absent. In the Paniceae the palea is often firmer as it, with the lemma, assumes the protection of the grain. If confirms closely to the shape of upper lemma and if the latter is transversely rugose, the palea is also transversely rugose. In the Pooideae, where there are often many florets in spikelet, the palea has greater scope for variation. It is usually two-keeled, but it may be completely absent in Agrostideae or a nerveless hyaline scale. The keels may be toothed or ciliate or ciliate below and dentate above. The two nerves are produced as two firm awns in the genus *Massia* of the Aveneae.

The adaxial surface may be glabrous or shortly hairy, or as in *Jansenella griffithiana* covered with the finger like protrusions which increase in length from below upwards. The grain is often closely inverted by the lemma and palea, while in the Bromeae the latter is firmaly adherent to the caryopsis.

The flower consists of gynaeceum, androeceum and lodicules. The gynaeceum consists of the pistil & its contents. The pistil is made up of the ovary, ovule, style & stigmas. The ovary is considered to be from three carpels fused together in to a hollow ovary in which is found a solitary ovule attached to the adaxial inner surface. The vascular system of pistil is exactly that of a tricarpellary ovary (Arber, 1934).

The ovule is attached to the ovary wall with or without a minute funiculus & is anatropus, hemitropus or slightly campylotropus. It has two integuments each of two cells in thickness. The inner integument eventually becomes hardened as the testa of the seed. The point of attachment of the ovule to the ovary wall is distinctly visible on the surface of the mature grain as the hilum.

At the top of the ovary there is to be found, in several tribes, a fleshly appendage which is usually hairy. This is common in the Brachypodieae and Bromeae, where the styles issue from the top of the ovary laterally below the appendage.

The number of styles is usually two but the genus *Pseudodenthonia* has three, & the genera *Duthiea*, *Zea*, *Leptaspis* and *Nardus* have only one. In some of the genera *e.g.* in *Urochodra*, the bases of the styles are connate in to a beak which is persistent and attached to the pericarp.

The style of *Zea mays* may be 15-30 cm long and may grow even longer if no pollen falls upon them (Burtt-Davy,1914). The style in this species is divided into two short arms at the top, usually called the stigmas, but the style is propably receptive to pollen for some distance below the bifurcation, where it is covered with hairs to which pollen grains adhere.

The styles end in stigmata, which may be papillose on the surface or plumose. The papillose surface is viscid so that the pollen grains adhere; in the latter the grain are cought among the hairs.

The styles and stigma may appear at the sides of the floret between lemma and palea, this being the more usual mode, or may be thrust out at the tip of the spikelet as in Maydeae and Lygeeae.

The androeceum consists of the stamens which may number one (*Vulpia*), two (some *Sporoboli*), three (most grasses)or six (*Leersia*, *Oryza*, *Leptaspis*). Each stamens consists of a long slender filament carrying a two- celled anther. The anther- cells are some times minutely bearded at the apex (*Pennisetum*, *Duthiea*). The anther cells may open by longitudinal slits (most grasses) or by pores (Bambuseae).

The anthers are usually thrust out between the lemma and palea and discharge the pollen to be carried away by the wind. In some species the florets are cleistogamous in which state the stamens will be found firmly pressed against the stigmas at the apex of the ovary (*Castellia*, *Cleistogenes*). To ensure pollination some plants are partially chasmogamous and partially cleistogamous. Plants which continue to flower late in the season are inclined to be cleistogamous.

The colour of the anthers is bright yellow or deep purple or even mottled. The yellow colour is sometimes diagnostic as in *Alopecurus*.

There are small to fairly large (in comparison with the lemma) scales, oblong, obcuneate-truncate or lanceolate in shape, or lobed in various ways, glabrous, hairy, two or three number or completely absent. When three in number two are found at the margin of the palea at the base and the third on the abaxial surface of the palea. The two lodicules in *Oryza sativa* and *O. rufipogon* are adnate to the base of palea. The lodicules are capable of becoming very turgid at anthesis, particularly at their bases, and by this means the lemma and palea are thrust apart so that the stigmas and stamens may emerge. There are three lodicules in the genus *Stipa* a genus in which the lemma tightly enfolds the palea. To separate lemma and palea must necessitate considerable force, hence the need for an extra lodicule. The third lodicule is, however not always particularly well developed, at least in *Stipa pennata*, where the third lodicule is rudimentary and non-vascular.

The dry, one-seeded, indehiscent fruit of the Poaceae consists of the mature fertilized ovary surrounded by the pericarp or remains of the ovary wall. On the abaxial surface at the base of the caryopsis will be found a depressed elliptic or ovate area where the embryo is situated. On the other side, diametrically opposite, is the hilum which represents the point or area of attachment of the ovule. The hilum is darker in colour than the rest of the grain and may, in shape, be a mere point, elliptic, oblong or a thin line which is sometimes as long as the grain and occasionally is found at the bottom of a groove. The hilum is sometimes diagnotic of certain genera such as *Festuca*, *Streptogyna*, *Helictotrichon*, and *Agropyron*, where the hilum is along this line. In the Brachypodieae & Bromeae, the appendage to the ovary is still evident on the caryopsis. Some caryopsis such as those of the genus *Eragrostis* fall naked, others are enclosed in the lemma and palea, which sometimes become firmer in texture (*Phalaris*). Those of the Bromeae are adherent to the palea and only with difficulty separable from it. In the Andropogoneae the caryopsis are enclosed in the complete spikelet which falls entire.

In some genera, *Urochondra*, *Sporobolus* and *Crypsis*, the pericarp is transformed in to mucilage if the moistened. In *Urochondra* and some species of *Sporobolus* the gelatinous pericarp splits down one side and the naked seed is ejected but remains attached to the envelope.

In most of the Grasses one or few outermost layers of the endosperm become highly specialized morphologically and constitute the aleurone tissue. In Barley this tissue is 3 or 4–layered (Varner, 1972). Several morphologically distinct types of aleurone cells occur in Grasses. In *Setaria* the aleurone cells show wall ingrowths near placental vascular supply. The aleuron cells are characterized by the presence of thick walls and non-vacuolated cytoplasm. In *Zea mays*, patches of two different colours have sometimes been observed, forming a sort of irregular mosaic pattern, or part of the endosperm is starchy and part is sugary.

The embryo of grasses is very much different from that of other monocots in its structure. The embryogeny of grasses is so different from that of other groups that Batygina (1969) remarked: "The unique mode of embryogenesis in Gramineae may allow the separation of a new type of embryogeny-Graminad Type."

The most conspicuous part of th grass embryo is the scutellum. Coulter (1915) interpreted the scutellum as the functional cotyledon arising from the peripheral cotyledonary ring. Arber (1934) considers that the scutellum and coleoptile jointly constitute the cotyledon, the coleoptile representing the cotyledonary sheath. Avery (1930) interprets the scutellum as the cotyledon; the coleoptile as the second leaf; and the elongated structure between the cotyledone and the coleoptile, called the "mesocotyl" (PLATE–2B).

(*b*) Systematic Treatments of the Family Poaceae

KEY TO THE SUBFAMILIES

1a. Culms woody; leaves dimorphic, of the main culm with short deciduous blades, rarely bladeless, of the branches with a well developed blade & pseudopetiole between sheath & blade

.... **2. Bambusoideae**

1b. Culms not woody; leaves not dimorphic, with developed blade & without pseudopetiole between sheath & blade:

2a. Tall reeds; spikelets with usually two or more fertile awnless, florets or if with one fertile floret then sterile floret above it; rachilla joints or lemma with long silky hairs

.... **1. Arundinoideae**

2b. Small herb; spikelets with one fertile awned floret; rachilla joints or lemma not hairy:

3a. Spikelets arising in open or contracted or spike like panicles less often in racemes or spike; ovary hairy or glabrous; caryopsis longitutinally groved with the linear hilum

.... **5. Pooideae**

3b. Spikelets not as above; ovary glabrous; caryopsis not as above:

4a. Leves narrow; inflorescence effuse or spiciform, digitate or racemosely arranged in spikes; spikelets 1 to many flowered, breaking up at maturity above the more or less peristent glumes; glumes awned or awnless; caryopsis usually with an adhering or free pericarp

.... **3. Chloridoideae**

4b. Leaves broad; inflorescence panicle or diffuse, sub digitate racemes; spikelets one or two flowered, breaking up at maturity from the pedicels; glumes awnless; caryopsis with free pericarp

.... **4. Panicoideae**

1. SUB FAMILY – ARUNDINOIDEAE

KEY TO THE TRIBES

1a. Spikelets 1-flowered, arranged in short effuse or contracted panicle; lemmas 3-awned

.... **1. Aristideae**

1b. Spikelets 2 to 6 flowered, arranged in the long, diffuse or lax, plumose panicle; lemma 1–awned:

2a. Spikelets 2 to 6 flowered; some what laterally compressed, unisexual or hermaphrodite; glumes as long as lemma; lodicules 2–3 nerved

.... **2. Arundineae**

2b. Spikelets 2– rarely 3 flowered, hermaphrodite; glumes about half of the length of the lemma, lodicules without nerve
.... **3. Thysanolaeneae**

1. Tribe–Aristideae

1. *Aristida* L.

KEY TO THE SPECIES

1a. Panicle effuse:

2a. Culms up to 1.5 m long; ligule membranous; spikelets deep purple
.... 2. *A. cyanantha*

2b. Culms 80 cm long; ligule a row of hairs or very narrow fimbriate membranes; spikelets yellowish or straw coloured
.... 4. *A. hystrix*

1b. Panicle contracted:

3a. Leaf–blades hairy; palea scabrid at the upper part of the margin
.... 3. *A. funiculata*

3b. Leaf–blades glabrous; palea not as above:

4a. Glumes awned; caryopsis longer than the lemma
.... 5. *A. setacea*

4b. Glumes awneless; caryopsis equalling the lemma
.........................1. *A. adscensionis*

1. *Aristida adscensionis*

Aristida adscensionis Linn. Sp. Pl. ed. 1: 82. 1753; Kunth. Enum. Pl. 1: 190. 1838; Steud. Syn. Pl. Glum. 139. 1855; Hook. f. Fl. Brit. Ind. 7: 224. 1896; Cooke, Fl. Bomb. 2: 1008. 1908; Rang & Tad., Hand. S. Ind. Grass. 223. t. 171–172. 1921; Haines, Bot. Bih. & Or. 977. 1924; Bor, Grass. Bur. Cey. Ind. Pak. 407. t. 43. 1960; U.C. Bhattacharyya, Bull. Bot. Sur. Ind. 6: 209. 1964; Raizada & Jain, Ind. For. 92: 638. 1966; Panigrahi & Rajagopal, J. Ind. Bot. Soc. 47: 223. 1968; Singh, Bull. Bot. Sur. Ind. 11: 20. 1969; Cope in Nasir & Ali, Fl. W. Pak 143: 41. 1982; Raizada *et al.*, Grass. Upp. Gang. Pl. pt. 3: 10. 1983; Roy, Grass. M.P. 28. 1984; Blatter & McCann, Bomb. grass. 209. t. 138. 1984; Sreekumar & Nair, Fl. Kerala Grass. 326. 1991; Misra & Verma, Fl. Allaha. 414. 1992. Moulik Grass. Bamb. Ind. 2: 423. 1997. *A. adscensionis* Linn. var. *aethiopica* Hook. f. Fl. Brit. Ind. 7: 225. 1896. *A. adscensionis* Linn. var. *festucoides* Henrard in Meded. Rijks. Herb. 1: 177. 1926. *A. adscensionis* Linn. var. *ehrenbergii* Henrard in Meded. Rijks. Herb, 1: 158. 1926. *A. adscensionis* Linn. var. *angustifolia* Pilger in Henrard in Meded. Rijks. Herb, 1: 9. 1926. *A. adscensionis* Linn. var. *typica* Stapf. in Hook. f. Fl. Brit. Ind. 7: 224. 1896. *A. adscensionis* var. *bromoides* Henrard in Meded. Rijks. Herb. 1: 62. 1926. *A. adscensionis* Linn. subsp. *guineensis* Henrard in Meded. Rijks. Herb. 1: 216. 1926. *A. adscensionis* Linn var. *humilis* Henrard. in Meded. Rijk. Herb. 1: 247. 1927. *A. adscensionis* Linn. var. *condensata* Henrard in Meded. Rijks. Herb. 2: 318. 1927. *A. depressa* Retz.

Obs. Bot. 4: 22. 1786; Fischer in Gamble, Fl. Pres. Madr. 1809. 1934 (Repr. ed. 3: 1251. 1957); Bor, Grass. Bur. Cey. Ind. Pak. 409. 1960. *A. abyssinica* Trin. & Rupr. Sp. Gram. Stip. in Act. Acad. Petrop. 6: 134. 1842. *A. canariensis* Willd. Enum. 99. 1809. *A. modatica* Steud. Syn. Pl. Glum. 139. 1855. *A. curvata* Nees. var. *abyssinica* Rich. Tent. Fl. Abyss. 2: 392. 1851. *A. divaricata* Jacg. Ecol. Gram. 7, t. 6. 1813. *A. hermanni* Regel. in Act. Hort. Petrop. 7 (2): 649. 1881. *A. aethiopica* Trin. & Rupr. Sp. Gram. Stip. in Act. Acad. Petrop 6: 136. 1842. *A. ehrenbrgii* Trin. & Rupr. Sp. Gram. Stip. in Act. Acad. Petrop. 6: 136. 1842. *A. festucoides* Poir. Encyl. 1: 453. 1810. *A. coarctata* H.B.K. Nov. Gen. et. Sp. 1: 122. 1815. *A. debilis* Mez. in Fedde. Rep. 17: 151. 1921. *A. fasciculata* Torrey in Ann. Lyc. Nat. Hist. New. York 1: 154. t. 1. 1824. *A. grisebachiana* Fournier Nex. Pl. Gram. pt. 2. 78. 1881. *A. hermanni* Mez. in Fedde. Rep. 17: 153. 1921. *A. macrochloa* Hochst in Flora. 38. 200. 1855. *A. maritima* Steud. Syn. Pl. Glum. 137. 1855. *A. mauritiana* Hochst. ex. A. Rich. Tent. Fl. Abyss. 2: 392. 1851. *A. mongholica* Trin. et Rupr. Sp. Gram. Stip. in Act. Acad. Peterop 6: 133. 1842. A. nana Steud. Syn. Pl. Glum. 137. 1855. *A. nigrescens* Presl. Reliq. Haenk. 1: 223. 1830. *A. coerulescens* Desf. Fl. Alt. 1: 109. t. 21. fig. 2. 1798. *A. chaetophylla* Steud. Syn. Pl. Glum. 420. 1855. *A. jacquiniana* Tausch. in Flora. 2: 508. 1836. *A. paniculata* Forskal, Fl. Aegypt–Arab. 25. 1775. *Chaetaria canariensis* P. Beauv. Agrost. 30. 1812. *C. coerulescens* P. Beauv. Roem. & Schult. Syst. Veg. 2: 294. 1817. (PLATE–86).

Annual. Rhizome absent; root fibrous. Culm 45 cm tall, 1.0 mm thick, green, glabrous, erect, densely tufted, very slender, geniculate at base, shortly decumbent, ascending, leafy, rooting at base, branching at lower nodes; branches erect, nodes glabrous; internodes 8.0 cm long. Leaf–blade 20 cm long, 1.5 mm broad, green, sub–coriaceous, convolute, rarely flat, filiform, narrow, linear, lanceolate, acute, base tapering, surface glabrous below, scabrid above also with hyaline margin, mid–vein distinct on lower surface, 2–lateral veins on each side of mid–vein; lateral veins distinct on upper surface; leaf–sheath 6.0 cm long, 2.0 mm broad, green, membranous, glabrous, margin hyaline, slighty keeled, loose, erect, compressed, thinly striate, very much shorter than internodes, veins distinct on both surface; ligule 1.0 mm long, membranous, a fring of hairs at apex; hairs 0.5 mm long, white, acute, base broad; auricle rounded, glabrous. Inflorescence consist of panicle; panicle 10 cm long, 6.0 cm broad, contracted, erect, subsecuned, flabelliform, lax, open; rachis filiform, smooth. Branches 5.0 cm long, filiform, erect, nodding, flexuous, scabrous, branching from the base to the middle. Spikelets 7.0 mm long, 1.5 mm broad, linear, lanceolate, erect, pale green at young, turning into purplish with age, 1–flowered, awned, pedicelled; pedicel 1.0 mm long, capillary. Lower glume 5.0 mm long, 1.0 mm broad, membranous, oblong, linear, lanceolate, acute, surface glabrous except the nerve, also with margin, 1–nerved; nerves green, visible, scabrid, ending at apex. Upper glume 6.0 mm long, 1.0 mm broad, membranous, oblong, linear, lanceolate, bifid & apiculate at apex, surface glabrous except the nerves, also with margin, 1–nerved; nerve green, visible, scabrid, ending at apex. Lemma 7.0 mm long, 1.5 mm broad, subcoriaceous, oblong, linear, laterally compressed, surface glabrous except the nerve, margin scabrid, 1–nerved; nerves green, slightly visible, scabrid, ending at apex, 3–awned; awn slender, antrosely scabrid, central one 11 mm long, lateral 2– awn, 9.0 mm long; callus 0.5 mm long, pointed, hairy at base; hairs 1.0 mm long, white, acute, base broad. Palea 5.0 mm

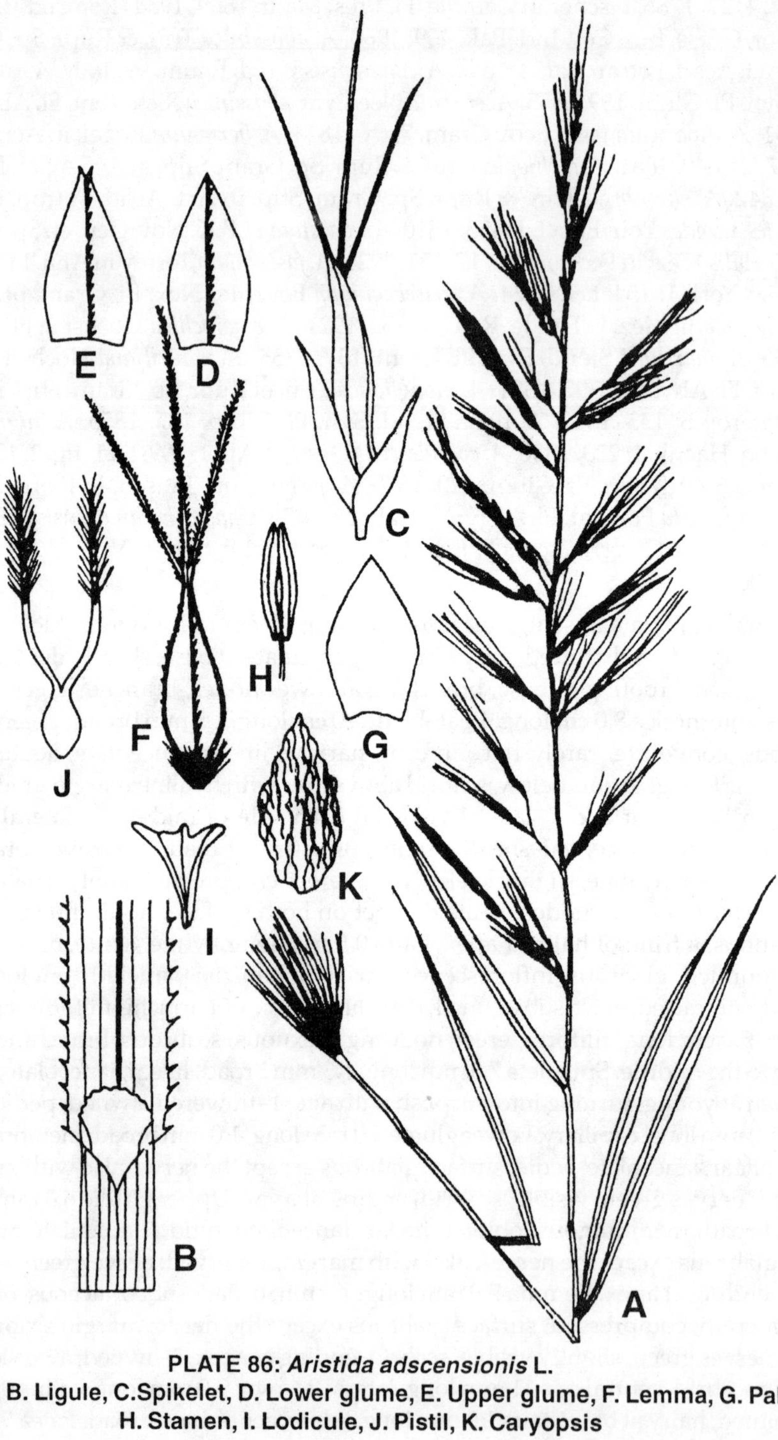

PLATE 86: ***Aristida adscensionis*** **L.**

A. Habit, B. Ligule, C.Spikelet, D. Lower glume, E. Upper glume, F. Lemma, G. Palea, H. Stamen, I. Lodicule, J. Pistil, K. Caryopsis

long, 1.0 mm broad, hyaline, oblong, retuse, acute, glabrous, nerveless. Stamens 3; anther 1.5 mm long, yellow, linear, glabrous; filament 0.5 mm long, slender, glabrous, shorter than anther. Lodicules 2, 0.4 mm long, 0.2 mm broad, hyaline, glabrous, oblanceolate, cuneate at base, wavy at apex, 3–nerved; nerves colourless, glabrous, faint, ending at apex. Pistil 3.5 mm long; ovary 0.5 mm long, ovate, oblong, glabrous; style 1.0 mm, slender, glabrous, bifid from base; stigma 2.0 mm long, purple, hairy; hairs acute, base broad, in 2–rows. Caryopsis 6.5 mm long, 1.0 mm broad, purplish, cylindric, linear, margin wavy, surface smooth.

Vernacular Names

Motti–burri, Longi–kussal, Lani.

Flowering & Fruiting

August–December; Bhojubir, Chandmari, Chaukhandi, Lohata, Chandauli, Chakiya, Sakaldiha, Allahabad, Basti, Deoria, Mau.

Habitat

Very common along the walls of old buildings and dry sandy habitat.

Distribution

India, Pakistan, Europe, Asia, America.

2. *A. cyanantha*

A. cyanantha Nees ex Steud. Syn. Pl. Glum. 1: 141. 1854. Bor, Grass. Bur. Cey. Ind. Pak. 409. 1960; Roy, Grass. M.P. 29. 1984; Moulik, Grass. Bomb. Ind. 2: 424.1997. (PLATE–87).

Perennial. Rhizomatous; rhizome 1.0 cm long, creeping; root fibrous. Culm 1.5 cm tall, 3.0 mm thick, green, glabrous, erect, densely tufted, stout, ascending, decumbent, leafy, rooting at base, branching at lower nodes; branches suberect; nodes glabrous; internodes 10 cm long. Leaf–blade 20 cm long, 5.0 mm broad, green, glacous, convolute, narrow, linear–lanceolate, acuminate, base rounded, coriaceous, glabrous margin sub–membranous, scabrid, mid–vein distinct on lower surface, 2–lateal veins on each side of mid–vein; lateral veins distinct on upper surface; leaf–sheath 11 cm long, 5.0 mm broad, green, sub–coriaceous, glabrous, margin sub–membranous, slightly keeled, erect, loose, slightly comp. ssed, longer than internodes, veins distinct on outer surface; ligule 1.0 mm long, membranous, lacerate, truncate; auricle glabrous. Inflorescence consists of panicle; panicle 20 cm long, 2.0 cm broad, erect, open, effuse, dense, linear; rachis scabrid, slender. Branches 6.0 cm long, suberect, densely crowded, flexuous, nodding. Spikelets 8.0 mm long, (exculiding awn), 1.0 mm broad, purplish, awned, linear, lanceolate, narrow, erect, subsessile, 1–flowered. Lower glume 5.0 mm long, 1.0 mm broad, membranous, linear, lanceolate, surface glabrous except the nerves, margin scabrid, 1–nerved; nerves green, slightly visible, scabrid, ending at apex, awned; awn 1.0 mm long, slender, antrorsely scabrid, from the apex of glume. Upper glume 8.0 mm long, 1.0 mm broad, membranous, linear, lanceolate, slightly oblong, 2–lobed at apex, surface glabrous, margin scabrid, 1–nerved; nerve green, visible, glabrous, ending at apex, awned; awn 1.5 mm long, slender, antrorsely scabrid,

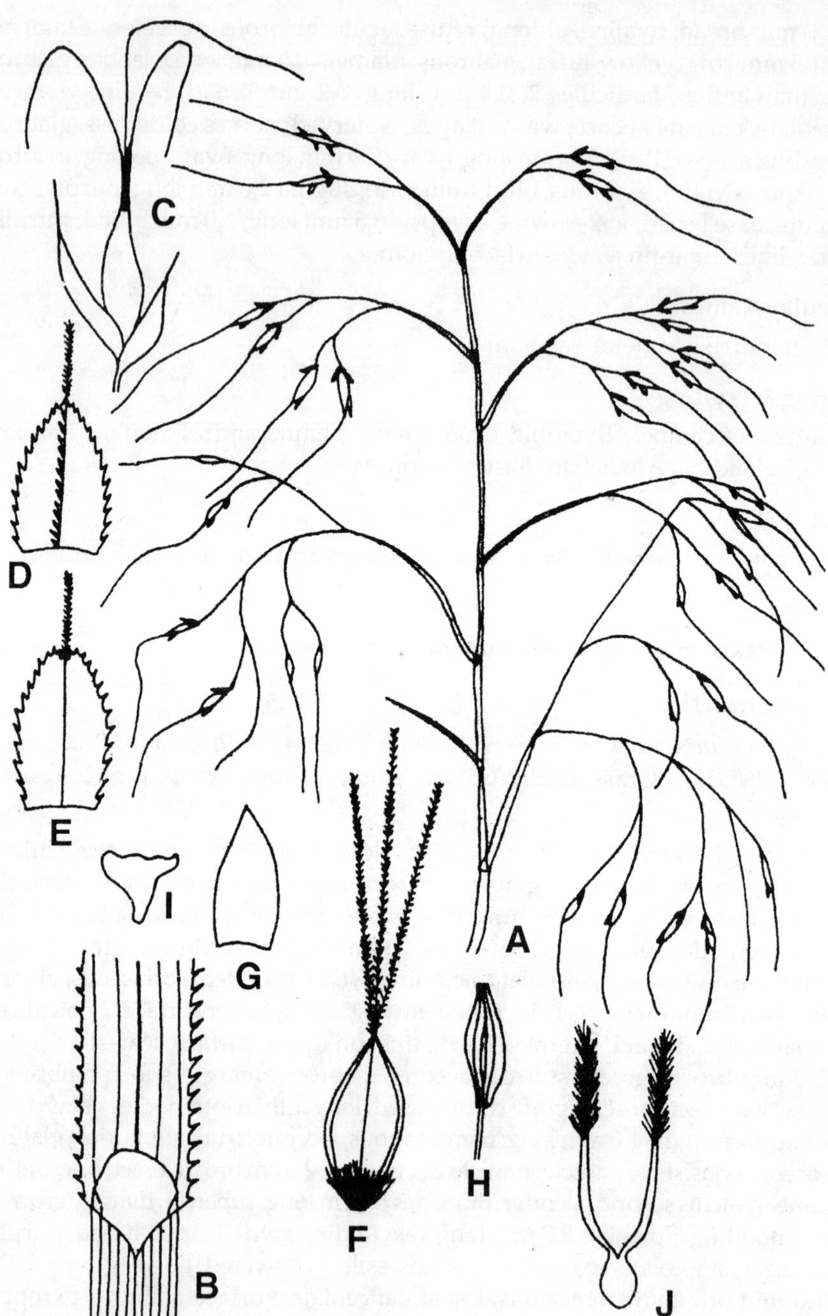

PLATE 87: ***Aristida cyanantha*** **Nees ex Steud.**
A. Inflorescence, B. Ligule, C. Spikelet, D. Lower glume, E. Upper glume, F. Lemma, G. Palea, H. Stamen, I. Lodicule, J. Pistil

from the base of lobe. Lemma 8.0 mm long, 1.5 mm broad, subcoriaceus, oblong, linear, lanceolate, laterally compressed, glabrous, 3–nerved; nerves green, slightly visible, glabrous, ending at apex, 3–awned; awn antrorsely scabrid, slender, from the apex, central one 20 mm long, lateral awn 18 mm long; callus 1.0 mm long, densely hairy; hairs 1.5 mm long, white, acute, base tubercled, no articulation between lemma & awn. Palea 4.0 mm long, 1.0 mm broad, hyaline, oblong, acuminate, glabrous, linear, nerveless. Stamens 3; anther 5.0 mm long, yellow, oblong, linear, glabrous; filament 1.0 mm long, slender, glabrous, shorter than anther. Lodicule 2, 0.3 mm long, o.1 mm broad, hyaline, glabrous, oblanceolate, nerveless, cuneate at base, wavy at apex. Pistil 6.5 mm long; ovary 1.5 mm long, ovate, elliptic, glabrous; style 3.0 mm long, slender, glabrous, bifid from base; stigma 2.0 mm long, purple, hairy; hairs acute, base broad, in 2–rows. Caryopsis 8.0 mm long, 1.0 mm broad, terete.

Flowering & Fruiting

September–February; Jaunpur, Ghazipur, Deoria, Basti, Sonbhadra.

Habitat

Rare, along the roadsides.

Distribution

India, Europe, Asia, Africa, Philippine Islands.

3. *A. funiculata*

A. funiculata Trin. & Rupr. Sp. Gram. Stip. 159. 1842; Trin., Mem. Acad. Petersb. ser. 6, 7: 159. 1849; Aitchis, cat. Pun. Pl. 164. 1869; Boiss. Fl. Or. 5: 492. 1884; Duthie, Grass. N. Ind. 47. 1883; Hook. f. Fl. Brit. Ind. 7: 226. 1896; Cooke, Fl. Bomb. 2: 1010. 1908; Haines, Bot. Bih. & Or. 978. 1924. Bor, Grass. Bur. Ceyl. Ind. Pak. 410. 1960; Bhattacharya, Bull. Bot. Surv. Ind. 6: 209. 1964; Roy, Grass. M.P. 30. 1984; Blatter & McCann, Bomb. Grass. 214. t. 142. 1984. Moulik, Grass. Bamb. Ind. 2: 424. 1997. *A. funiculata* Trin. & Rupr. var. *mallica* Henrard in Medd. Rijks Herb. 2: 328. 1927. *A. funiculata* Trin. & Rupr. var. *paradoxa* Henrard in Medd. Rijks Herb. 2: 425. 1927. *A. macratheca* Rich. Tent. Fl. Abyss. 2: 393. 1851; Boiss, Fl. Or. 5: 493. 1884. *A. mallica* Edgew. in J. Proc. Linn. Soc. 6: 206. 1862. *A. paradoxa* Steud et Schimidt. Fl. Cap. Verd. 140. 1852. *A. funicularis* Trin. ex Steud. Nom. ed. 2: 131. 1841. *A. kotschyi* Hochst ex Steud. Syn. Pl. Glum. 1: 142. 1854. *A. stipacea* Ehrenb. et. Hempr. Trin et Rupr. Sp. Gram. Stip. 159. 1842. (PLATE–88).

Perennial. Rhizomatous; rhizome 1.0 cm long, creeping; root fibrous. Culm 60 cm tall, 1.0 mm thick, green, glabrous, erect, densely tufted, very slender, leafy, geniculately ascending, rooting at base, simple; nodes glabrous; internodes 6.0 cm long. Leaf–blade 15 cm long, 2.0 mm broad, green, convolute, narrow, linear, lanceolate, acute, base rounded, sparsely hairy at upper surface; hairs acute, base tubercled; glabrous at lower surface, margin, hyaline, scabrid, mid–vein distinct on lower surface, 2–lateral veins on each side of mid–vein; lateral veins distinct on upper surface; leaf sheath 5.0 cm long; 2.0 mm broad, green, surface membranous, glabrous, margin hyaline, keeled, loose, erect, compressed, shorter than internodes, veins distinct on outer surface; ligule 0.5 mm long, a ring of white hairs; hairs acute,

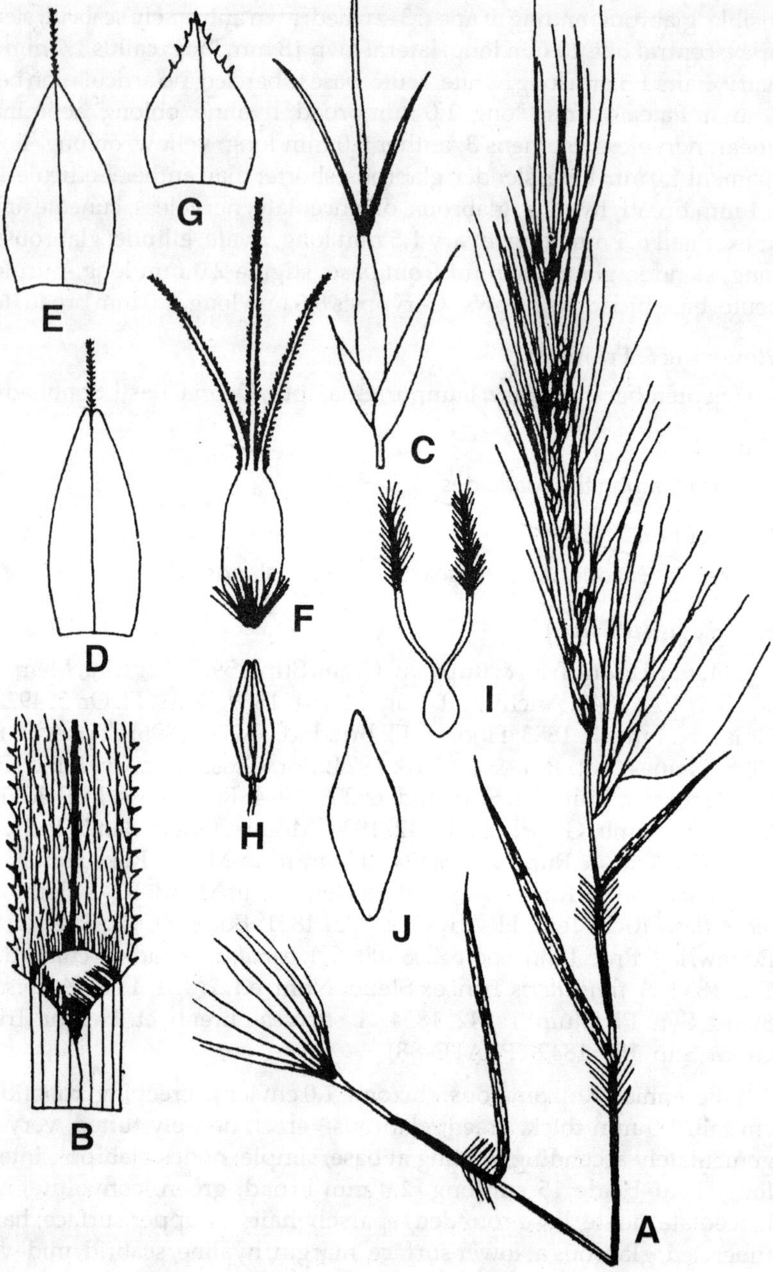

PLATE 88: ***Aristida funiculata*** **Trin. & Rupr.**
A. Habit, B.Ligule, C.Spikelet, D. Lower glume, E. Upper glume, F. Lemma, G. Palea, H. Stamen, I. Pistil, J. Caryopis

base broad; auricle absent. Inflorescence consists of panicle; panicle 15 cm long, 5.0 cm broad, narrow, lax, erect, open; rachis angular, scabrid. Branches 6.0 cm long, capillary, erect, scabrid. Spikelet 10 mm long, (excluding awn) 1.0 mm broad, linear, lanceolate, narrow, straw coloured, awned, 1–flowered. Lower glume 5.0 mm long, 1.3 mm broad, membranous, linear, lanceolate, glabrous, bilobed at apex; lobe acute; 1–nerved; nerve green, visible, glabrous, ending at apex; awnd; awn 3.0 mm long, antrorsely scabrid, from the sinus of lobes. Upper glume 7.0 mm long, 1.3 mm broad, membranous, linear, lanceolate, acute, surface glabrous, margin scabrid at apex, glabrous below, 1–nerved; nerves green, visible, glabrous, ending at apex, awned; awn 3.0 mm long, antrosely scabrid. Lemma 5.0 mm long, 0.5 mm broad, membranous, narrow, linear, glabrous, articulation distinct, nerveless, 3–awned; awn slender, antrosely scabrid, central one 4.5 cm long, lateral 4.0 cm long; callus 1.5 mm long, hairy; hairs 1.5 mm long, white, acute, base tubercled. Palea 4.5 mm long, 1.0 mm broad, hyline, surface glabrous, margin scabrid at apex, glabrous below, linear, lanceolate, acute, nerveless. Stamens 3; anther 1.0 mm long, yellow, linear, glabrous; filament 0.5 mm long, slender, glabrous, shorter than anther. Lodicules absent. Pistil 1.5 mm long; ovary 0.2 mm long, ovate, glabrous; style 0.5 mm long, slender, glabrous, bifid from base; stigma 0.8 mm long, brown, hairy; hairs acute, base broad, in 2–rows. Caryopsis 3.0 mm long, 1.0 mm broad, cylindric, black.

Vernacular Names

Pandhrikusal, Bushikussal.

Flowering & Fruiting

August–December; Ghazipur, Chandauli, Chakiya, Sultanpur, Khalilabad.

Habitat

Rare, along the walls.

Distribution

India, Baluchistan, Arabia, Tropical Africa.

4. *A. hystrix*

A. hystrix Linn. f. Suppl. 113. 1781; Roxb. Fl. Ind. 1: 350. 1820; Grah. Cat. Bomb. Pl. 335. 1839; Dalz. & Gibs., Bomb. Fl. 295. 1861; Duthie, Grass. N. Ind. 47. t. 31. 1888; Baker, Fl. Maurit. 451. 1877. Hook. f. Fl. Brit. Ind. 7: 225. 1896; Rang. & Tad. Handb. S. Ind. Grass. 226. t. 174. 1921; Cooke, Fl. Bomb. 2: 1009. 1908; Fischer in Gamble, Fl. Pres. Madr. 1809. 1934 (Repr. ed. 3: 1252. 1957); Bor, Grass. Bur. Cey. Ind. Pak. 410. 1960; Raizada & Jain, Ind. For. 92: 638. 1966; Raizada *et al.*, Grass. Upp. Gang. Pl. pt. 3: 14. 1983. Blatter & McCann, Bomb. Grass. 211. t. 140. 1984; Sreekumar & Nair, Fl. Kerl. Grass. 327. 1991; Misra & Verma, Fl. Allaha. 414. 1992; Moulik, Grass. Bamb. Ind. 2: 424. 1997. *Chaetaria hystrix* P. Beauv. Agrost. 30. 1812. *Aristida rigescens* Roem. ex Schult. Syst. Veg. 2: 100. 1817. *A. rigida* Roth. Nov. Pl. Sp. 42. 1821. (PLATE–89).

Perennial. Rhizomatous; rhizome creeping; root fibrous. Culms 50 cm tall, 1.0 mm thick, green, glabrous, erect, densely tufted, very slender, ascending, very leafy, rooting at base, branching at lower nodes; branches suberect, dense; nodes glabrous;

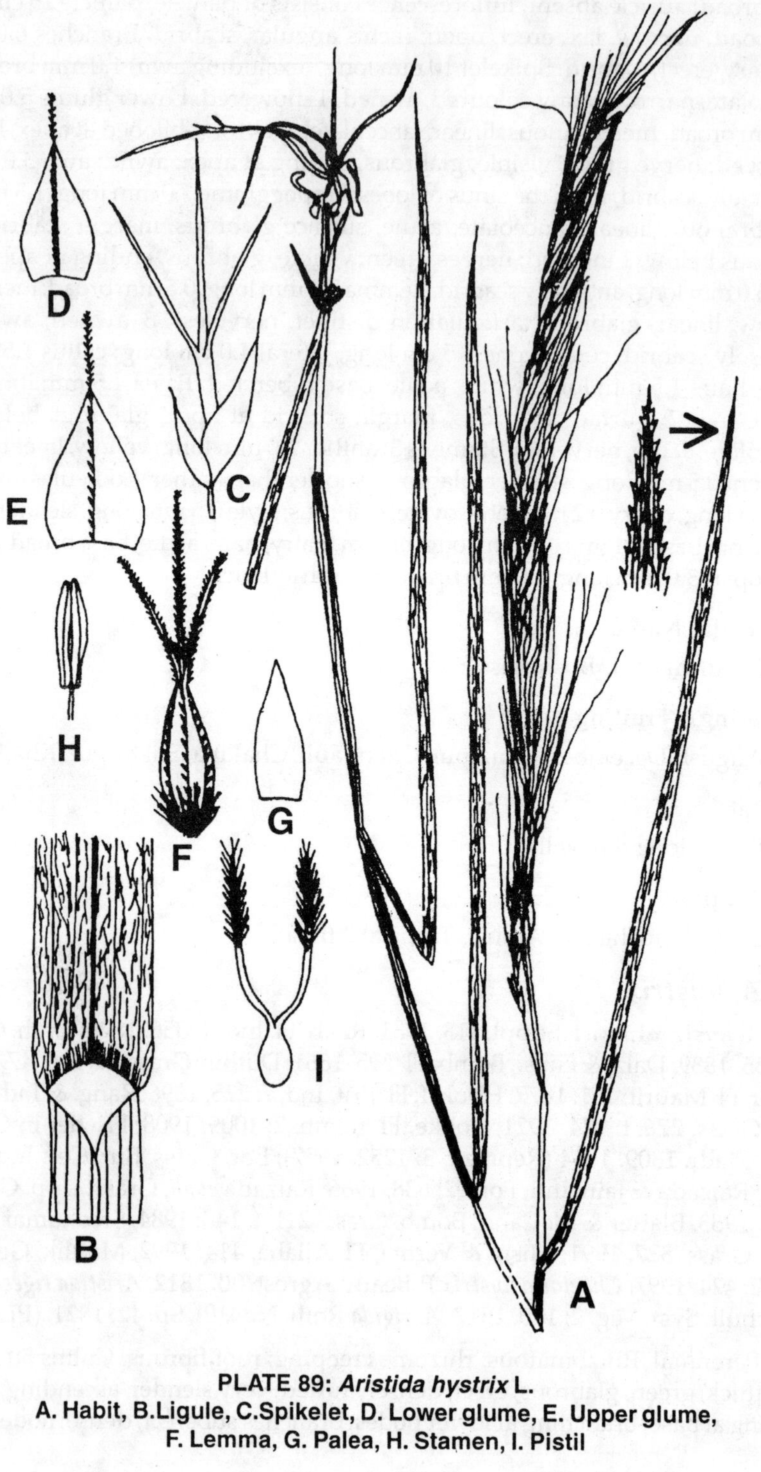

PLATE 89: *Aristida hystrix* L.
A. Habit, B.Ligule, C.Spikelet, D. Lower glume, E. Upper glume, F. Lemma, G. Palea, H. Stamen, I. Pistil

internodes 6.0 cm long. Leaf blade 10 cm long, 2.0 mm broad, green, convolute, narrow, linear–lanceolate, acuminate, base rounded, membranous, upper surface sprasely hairy; hair 2.0 mm long; white, acute, base tubercled, lower surface glabrous, margin hyaline, glabrous below, scabrid at apex, mid–vein distinct on lower surface, 2–lateral veins on each side of mid–vein; lateral veins distinct on upper surface; leaf–sheath 5.0 cm long, 2.0 mm broad, green, surface membranous, glabrous, margin hyaline, slightly keeled, erect, loose, compressed, shorter than internodes, veins distinct on both surface; ligule 1.0 mm long, a ring of white hairs; hairs acute, base broad; auricle absent. Inflorescence consist of panicle; panicle 10 cm long, 1.0 cm broad, erect, open, effuse, as broad as long, rigid; rachis stout, flexuous and angular. Branches 5.0 cm long, filiform, erect, nodding, smooth. Spikelets 1.0 cm long (excluding awn), 1.5 mm broad, linear, lanceolate, narrow, 1–flowered, awned, purplish–straw coloured. Lower glume 9.0 mm long, 1.0 mm broad, membranous, linear, lanceolate, slightly ovate, acute, surface glabrous except the nerves, also with margin, 1–nerved; nerve green, visible, scabrid, ending at apex, awned; awn 3.0 mm long, slender, antrorsely scabrid. Upper glume 1.0 cm ong, 1.5 mm broad, membranous, linear, lanceolate, acute, 1–nerved; nerve green, visible, scabrid, ending at apex, awned; awn 4.0 mm long, slender, antrorsely scabrid. Lemma 3.0 mm long, 1.0 mm broad, membranous, lanceolate, surface glabrous between the nerves, also with margin, 3–nerved; nerves green, visible, scabrid, ending at apex; callus 1.5 mm long, pointed, hairy; hairs 0.5 mm long, white, acute, base broad, 3–awned; awn slender, antrorsely scabrid, central one 5.5 cm long, lateral 4.0 cm long. Palea 2.5 mm long, 1.0 mm broad, sub–membranous, linear, oblong, glabrous, subacute, nerveless. Stamens 3; anther 1.5 mm log, yellow, linear, glabrous; filament 0.5 mm long, slender, glabrous, shorter than anther. Lodicules 2, 0.3 mm long, 0.1 mm broad, hyaline, oblanceolate, nerveless, glabrous. Pistil 2.5 mm long; ovary 0.5 mm long, obovate, oblong, glabrous; style 1.0 mm long, slender, glabrous, bifid from base; stigma 1.0 mm long, purplish, hairy; hairs acute, base broad, in 2–rows.

Vernacular Names

Kusal, Pawn–burri, Matari–Kussal, Kalkussal.

Flowering & Fruiting

July–November; Rajghat, Chakiya, Chandauli, Sonbhadra, Kushinagar.

Habitat

Common along railway tracks & walls.

Distribution

India, Sri Lanka, Australia.

5. *A. setacea*

A. setacea Retz. Obs. Bot. 4: 22. 1786; Roxb. Fl. Ind. 1: 349. 1820; Grah. Cat. Bomb. Pl. 218. 1839; Dalz. & Gibs. Bomb. Fl. 295. 1861; Duthie, Grass. N.W. Ind. 27. 1883; Lisboa, Bomb. Grass. 91. 1896; Hook. f. Fl. Brit. Ind. 7: 225. 1896. Cook, Fl. Bomb. 2: 1008. 1908; Rang. & Tad., Handb. S. Ind. Grass 225. f. 173. 1921; Haines, Bot. Bih. & Or. 979. 1924; Fischer in Gamble, Fl. Pres. Madr. 1809. 1934 (Repr. ed. 3: 1252. 1957);

Bor, Grass. Bur. Cey. Ind. Pak. 412. 1960; Raizada & Jain, Ind. For. 92: 638. 1966; Gilliland, Fl. Malaya 3: 112. 1971; Raizada *et al.*, Grass. Upp. Gang. Pl. pt. 3: 17. 1983; Roy, Grass. M.P. 30. 1984; Blatter & McCann. Bomb. Grass. 211. t. 139. 1984; Sreekumar & Nair, Fl. Kera. Grass. 328. 1991; Misra & Verma, Fl. Allaha. 415. 1992; Moulik, Grass. Bamb. Ind. 2: 426. 1997. *A. quinqueseta* Steud. Syn. Pl. Glum. 420. 1855. (PLATE–90).

Perennial. Rhizomatous; rhizome 1.0 cm long, vertical; root fibrous. Culm 75 cm tall, 2.0 mm thick, green, glabrous, robust, tufted, slender, geniculately ascending, leafy, rooting at base, branching at lower nodes; branches sub erect; nodes glabrous; internodes 8.0 cm long. Leaf–blade 20 cm long, 2.0 mm broad, green, inrolled, filiform, acuminate, base rounded, surface coriaceous, margin scabrid, sub–membranous, mid–vein distinct on lower surface, 2–lateral veins on each side of mid–vein; lateral veins distinct on upper surface; leaf–sheath 6.0 cm long, 3.5 mm broad, green, coriaceous, glabrous, margin keeled, erect, loose, compressed, striate, shorter than internodes, veins distinct on outer surface; ligule 0.5 mm long, a ring of white hairs; hairs acute, base broad; auricle absent. Inflorescence consists of panicle; panicle 15 cm long, 6.0 cm broad, inclined, open, contracted; rachis slender, glabrous. Branches 5.0 cm long, mostly fascicled, erect, ascending, scabrid. Spikelets 16 mm long (excluding awn), 1.0 mm broad, erect, straw coloured at young, purplish with age, linear. Lower glume 13 mm long, 0.7 mm broad, membranous, linear, lanceolate, surface glabrous except the nerve, margin scabrid at apex, glabrous below, 1–nerved; nerves green, visible, scabrid, ending at apex, awned; awn 3.0 mm long, slender, antrorsely scabrid. Upper glume 1.6 mm long, 1.0 mm broad, membranous, linear, lanceolate, slightly ovate, surface glabrous except the nerves, margin scabrid at apex, glabrous below, bifid at apex, 1–nerved; nerves green, visible, scabrid, ending at apex, awned; awn 3.5 mm long, slender, antrorsely scabrid, from the base of lobe. Lemma 12 mm long, 1.0 mm broad, membranous, linear, lanceolate, convolute, glabrous, 3– nerved; nerves, green, visible, glabrous, ending at apex, column indistinct, 3–awned; awn slender, antrosely scabrid, central one 3.0 cm long, lateral 2.5 cm long; callus 1.5 mm long, sharp, hairy; hairs 1.5 mm long, white, acute, base tubercled. Palea 11 mm long, 1.0 mm broad, hyaline, linear, lanceolate, acute, glabrous, nerveless. Stamens 3; anther 2.0 mm long, yellow, oblong, linear, glabrous; filament 0.5 mm long, slender, glabrous, shorter than anther. Lodicules absent. Pistil 2.5 mm long; ovary 0.5 mm long, ovate, glabrous; style 1.0 mm long, slender, glabrous, bifid from base; stigma 1.0 mm long, hairy; hairs acute, base broad, in 2–rows. Caryopsis 15 mm broad, cylindric, brown.

Vernacular Name

Mothi–kusal.

Flowering & Fruiting

November–February; Chakiya, Chandauli, Padrauna, Sonbhadra.

Habitat

Very common along road sides & hill slopes.

Distribution

India, Mascarene Islands.

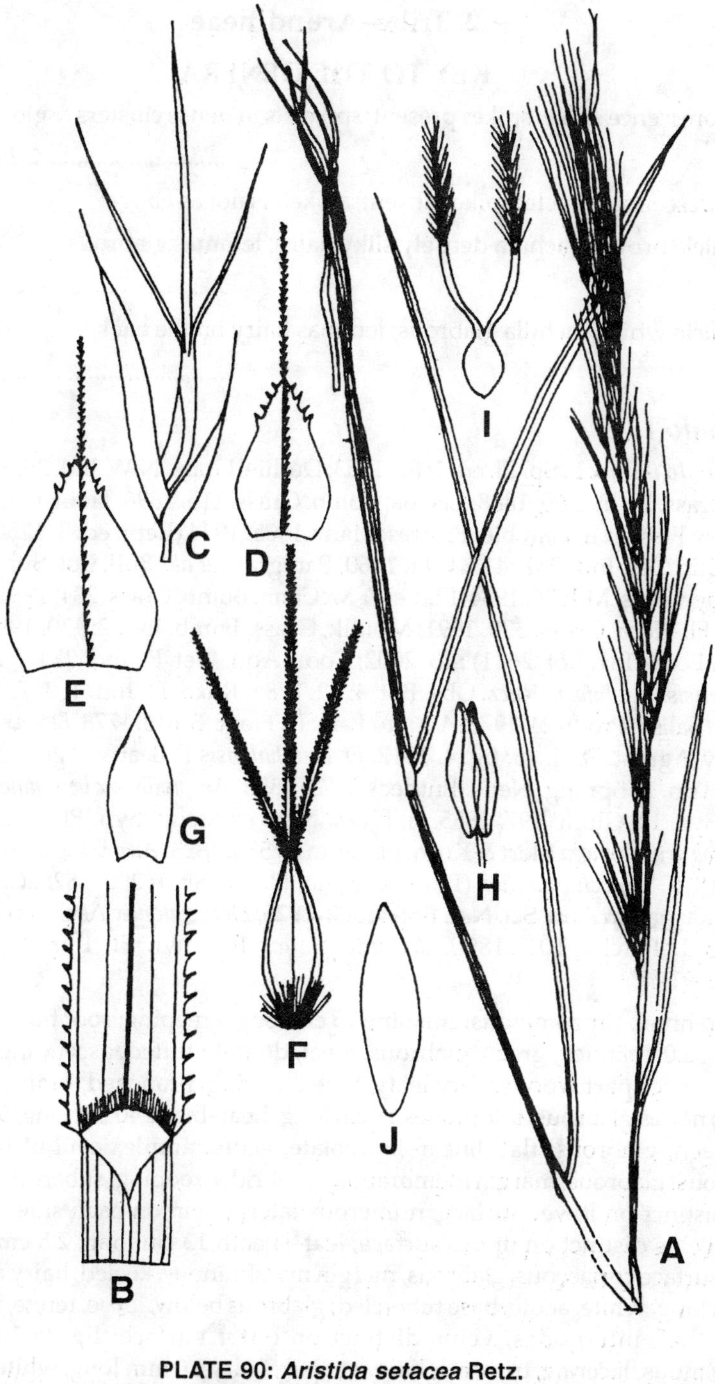

PLATE 90: *Aristida setacea* Retz.
A. Habit, B.Ligule, C. Spikelet, D. Lower glume, E. Upper glume, F. Lemma, G. Palea, H. Stamen, I. Pistil, J. Caryopsis

2. Tribe–Arundineae

KEY TO THE GENERA

1a. Inflorescence spike; spikes present; spikelets in dense clusters & globose
.... 2. ***Elytrophorus***

1b. Inflorescence panicle; spikes absent; spikelets not as above:

2a. Panicle brown; rachilla densely silky hairy; lemmas glabrous
.... 3. ***Phragmites***

2b. Panicle white; rachilla glabrous; lemmas hairy on the back
.... 1. ***Arundo***

1. *Arundo* L.

Arundo donax L. Sp. Pl. ed. 1: 81. 1753; Duthie, Grass. N.W. Ind. 35. 1883; Duthie, Fodd. Grass. N. Ind. 60. 1888; Lisboa, Bomb. Grass. 115. 1896; Hook f. Fl. Brit. Ind. 7: 302. 1896; Fischer in Gamble, Fl. Pres. Madr. 1805. 1934 (Repr. ed. 3: 1250. 1957); Bor, Grass. Bur. Cey. Ind. Pak. 413. t. 44. 1960; Panigrahi *et al.*, Bull. Bot. Sur. Ind. 11: 111. 1969; Roy, Grass. M.P. 36. 1984; Blatter & McCann, Bomb. Grass. 204. 1984; Sreekumar & Nair, Fl. Kera. Grass. 330. 1991; Moulik, Grass. Bamb. Ind. 2: 430. 1997; Athar Ali Khan, J. Econ. Tax. Bot. 26 (1) : 43. 2002; Cook, Aqu. Wet. Pl. Ind. 284. f. 292. 1996. *A. benghalensis* var. *bifaria* Retz. Obs. Bot. 4: 22. 1786; Roxb. Fl. Ind. 1: 347, 348. 1832. *A. longifolia* Salisb. Prodr. 24. 1976. *A. sativa* Lam. Fl. Franc. 3: 616. 1778. *Donax arundinaceus* P. Beauv. Agrost. 78. t. 16. fig. 4. 1812. *D. benghalensis* P. Beauv. Agrost. 78. 1812. *D. bifarius* Trin. in Spreng. Neue. Entdeck 2: 73. 1850. *Amphidonax benghalensis* Nees ex Steud. Syn. Pl. Glum. 197. 1855. *A. bifaria* Nees ex Steud. Syn. Pl. Glum 410. 1855. *Scolochloa arundinacea* Mert & Koch. Fl. Germ. 1: 530. 1823. *Aira bengalensis* Gmel. Syst. 1: 174. 1810. *Scolochloa donax* (Linn.) Gandin, Fl. Helvet. 1: 202. 1828. *Cynodon donax* (Linn.) Raspail in Ann. Sci. Nat. Bot. 5: 302. 1825. *Donax donax* Aschers et Graebn, Fl. Nardostd. Flachl. 101. 1890. *Arundo glauca* Bubawi, Fl. Pyr. 4: 1303. 1901. (PLATE–91).

Perennial. Rhizomatous; rhizome 15 cm long, creeping; root fibrous. Culm 3.0 mm tall, 2.0 cm thick, green, glabrous, erect, densely tufted, ascending, upper part hollow, lower part woody, very leafy, base creeping, branched; branches suberect, hollow; nodes glabrous; internodes 10 cm long. Leaf–blade 40 cm long, 5.0 cm broad, dull green, glabrous, flat, linear–lanceolate, acute, amplexicaul at base, surface coriaceous, glabrous, margin membranous, scabrid, drooping, suberect, striate, mid–vein distinct on lower surface, numerous lateral vein on each side of mid–vein; lateral veins distinct on upper surface; leaf–sheath 13 cm long, 2.5 cm broad, dull green, surface coriaceous, glabrous, margin membranous, keeled, hairy at apex; hairs 2.0 mm long, white, acute, base tubercled; glabrous below, loose, terete, erect, striate, longer than internodes, veins distinct on outer surface; ligule 5.0 mm long, membranous, lacerate, truncate, hairy at apex; hairs 1.0 mm long, white, acute, base broad; auricle glabrous. Inflorescence composed of panicle; panicle 50 cm long, 15 cm broad, compact, erect, open, whitish at young, turnning into purplish with age,

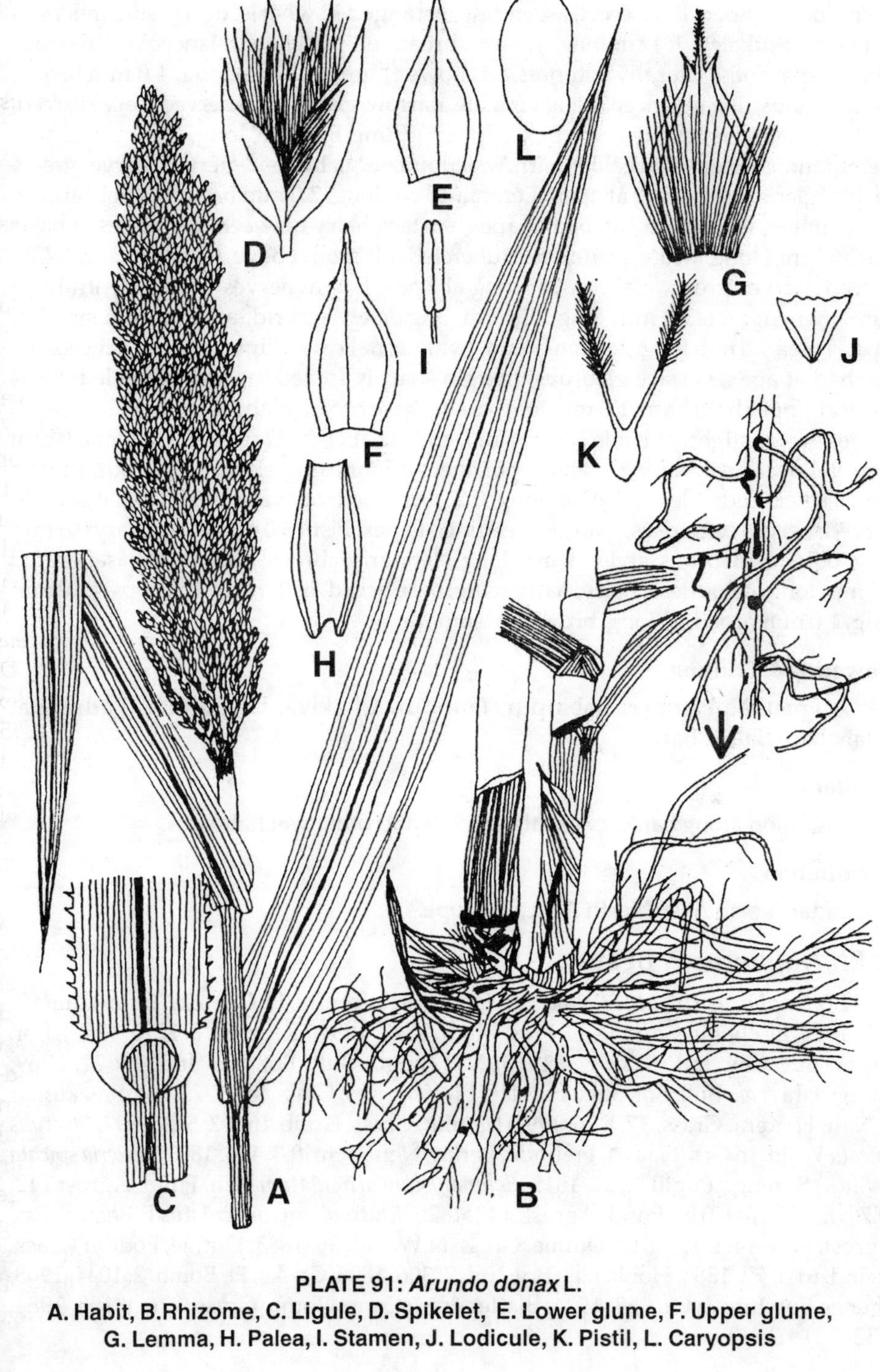

PLATE 91: ***Arundo donax*** **L.**

A. Habit, B. Rhizome, C. Ligule, D. Spikelet, E. Lower glume, F. Upper glume, G. Lemma, H. Palea, I. Stamen, J. Lodicule, K. Pistil, L. Caryopsis

cylindrical, pyramid; rachis terete, smooth, nodding, internodes 6.0 cm long. Branches 20 cm long, scaberulous, erect, ascending, arranged in whorls, decreasing in length upwards. Spikelets 1.0 cm long, 1.0 mm broad, elliptic, linear–lanceolate, narrow, awned, glabrous, laterally compressed. Lower glume 7.0 mm long, 1.0 mm broad, membranous, linear–lanceolate, acuminate, narrow, glabrous, 3–nerved; nerves green, visible, glabrous, ending at apex. Upper glume 6.0 mm long, 1.0 mm broad, membranous, linear–lanceolate, narrow, acuminate, glabrous, 3–nerved; nerves green, visible, glabrous, ending at apex. Lemma 10 cm long, 2.0 mm broad, membranous, ovate, linear, lanceolate, bilobed at apex, surface hairy between the nerves at base; hairs 6.0 mm long, white, acute, base tubercled; glabrous above, margin glabrous, 3–nerved; nerves green, glabrous, ending at apex, lateral nerves visible, central one faint, awned; awn 3.0 mm long, filiform, antrorsely scabrid, arising from sinus of lobes. Palea 5.0 mm long, 2.0 mm broad, hyaline, delicate, elliptic, oblong, lanceolate, notched at apex, surface glabrous, margin sharply folded inwards, 2 keeled; keels hairy at upper half; hairs 1.0 mm long, acute, base broad, glabrous below, 2–nerved; nerves green, slightly visible & marginal, ending at apex. Stamens 3; anther 2.0 mm long, yellow, linear, oblong, glabrous; filament 0.5 mm long, slender, glabrous, shorter than anther. Lodicules 2, 0.5 mm long, 0.2 mm broad, obovate, hyaline, glabrous 3–nerved; nerves colourless, glabrous, ending at apex. Pistil 3.0 mm long; ovary 0.5 mm long, oblong, glabrous; style 1.0 mm long, slender, glabrous, bifid from base; stigma 1.5 mm long, colourless, hairy; hairs acute; base broad, in 2–rows. Caryopsis 2.0 mm long, 1.0 mm broad, oblong, brownish, smooth.

Flowering & Fruiting

August–November; Babatpur, Jaunpur, Chakiya, Chandauli, Padrauna, Allahabad, Ballia, Basti.

Habitat

Common along banks of canals, ditches and other wet habitates.

Distribution

India, North Asia. North Africa, Europe.

2. *Elytrophorus* P. Beauv.

Elytrophorus spicatus (Willd.) A. Camus in Lecomte, Fl. Gen. del. Indo–China 7: 547. 1923; Fischer in Gamble, Fl. Pres. Madr. 1843. 1934. (Repr. ed. 3: 1275. 1957); Bor. Grass. Bur. Cey. Ind. Pak. 493. 1960; N.P. Singh, Bull. Bot. Sur. Ind. 11: 21. 1969; Panigrhahi *et al.*, Bull. Bot. Surv. Ind. 11: 112. 1969; Roy, Grass. M.P. 77. 1984; Sreekumar & Nair, Fl. Kera. Grass. 37. t. 76. 1991; Moulik, Grass. Bamb. Ind. 2: 592. 1997. *Dactylis spicata* Willd. in Ges. Naturf. Freunde. Berlin, Neue Schrift 3: 416. 1801. *Sesleria spicata* (Willd.) Spreng., Pugill. 2: 21. 1815. *Echinalysium articulatum* Trin. Fund. Agrost 142. 1820. *E. strictum* Trin. Fund. Agrost. 142. 1820. *Elytrophorus articulatus* P. Beauv., Ess. Agrost. 67. t. 14. f. 12. 1812; Duthie, Grass. N.W. Ind. 36. 1883; Duthie, Fodder Grass. N. Ind. 61. t. 72. 1888; Hook. f. Fl. Brit. Ind. 7: 306. 1896; Cooke, Fl. Bomb. 2: 1044. 1908; Haines Bot. Bih. & Or. 963. 1924; Blatter & McCann, Bomb. Grass. 276. t. 188. 1984. (PLATE–92).

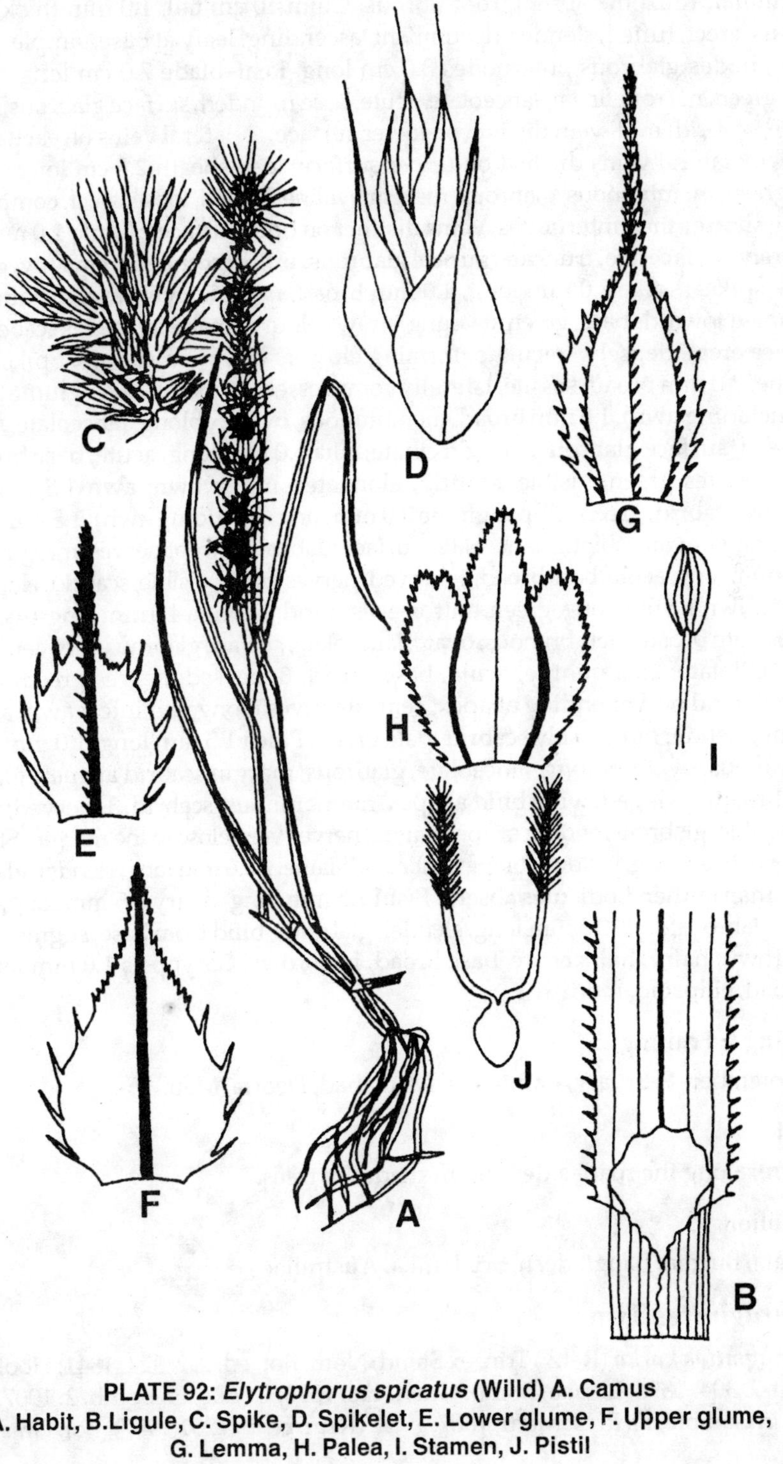

PLATE 92: ***Elytrophorus spicatus*** **(Willd) A. Camus**
A. Habit, B. Ligule, C. Spike, D. Spikelet, E. Lower glume, F. Upper glume, G. Lemma, H. Palea, I. Stamen, J. Pistil

Annual. Rhizome absent; root fibrous. Culm 10 cm tall, 1.0 mm thick, green, glabrous, erect, tufted, slender, decumbent, ascending, leafy at base, simple, rooting at base; nodes glabrous; internodes 3.0 cm long. Leaf–blade 7.0 cm long, 3.0 mm broad, green, narrow, linear, lanceolate, acute, base rounded, surface glabrous, margin hyaline, scabrid, mid–vein distinct on lower surface, 2–lateral veins on each sides of mid–vein; lateral veins distinct on upper surface; leaf–sheath 2.5 cm long, 1.5 mm thick, green, membranous, glabrous, margin hyaline, keeled, loose, erect, compressed, slightly shorter than internodes, veins distinct on both surfaces; ligule 1.0 mm long, membranous, lacerate, truncate; auricle glabrous. Inflorescence 6.0 cm long, consists of false spikes; spike 7.0 mm long, 4.0 mm broad, narrow, cylindric, distant, erect, interrupted towards base, green at young, straw colour at maturity; rachis scaberulous. Branches erect, densely spiculate, forming clusters of varied shapes. Spikelets 4.0 mm long, 2.0 mm broad, sessile, laterally compressed, awned. Lower glume 3.0 mm long (including awn), 1.0 mm broad, membranous, ovate, oblong, lanceolate, slightly appressed, surface glabrous, margin ciliate; cilia 1.0 mm long, acute, base broad, 1–nerved; nerves green, visible, scabrid, elongated into a awn; awn 1.5 mm long, antrorsely scabrid, green. Upper glume 3.0 mm long (including awn), 1.5 mm broad, membranous, ovate, oblong, lanceolate, surface glabrous except nerves, margin ciliate; cilia 1.0 mm long, acute, base broad, 1–nerved; nerves green, visible, scabrid, elongating into awn; awn 1.5 mm long, green, antrorsely scabrid. Lemma 4.0 mm long (including awn), 1.0 mm broad, membranous, ovate, lanceolate, surface glabrous between nerves, margin ciliolate; cilia minute, acute, base broad, 3–nerved; nerves green, visible, scabrid, lateral nerves ending at apex, central nerves elongated into a awn; awn 1.5 mm long, slender, antrorsely scabrid, pale green. Palea 1.5 mm long, 1.0 mm broad, membranous, ovate, oblong, lanceolate, glabrous, margin scabrid at apex, glabrous below, broadly winged; wing bifid at apex, membranous, scabrid; 3–nerved; nerves green, visible, glabrous, ending at apex, lateral nerves very close to the margin. Stamens 3; anther 0.3 mm long, yellow, oblong, glabrous; filament 0.6 mm long, slender, glabrous, shorter than anther. Lodicules absent. Pistil 1.5 mm long; ovary 0.5 mm long, ovate, elliptic, glabrous; style 0.5 mm long, slender, glabrous, bifid from base; stigma 0.5 mm long, brown, hairy; hairs acute, base broad, in 2–rows. Caryopsis 1.0 mm long, 0.5 mm broad, ellipsoid, reddish.

Flowering & Fruiting

September–February; Ramnagar, Allahabad, Deoria, Mau.

Habitat

Rare, along the road–sides and in shades of trees.

Distribution

India, Burma, Bangladesh, Sri–Lanka, Australia.

3. *Phragmites* Trin.

Phragmites karka (Retz.) Trin. ex Steud. Nom. Bot. ed. 2, 2: 324. 1841; Hook. f. Fl. Brit. Ind. 7: 304. 1896; Lisboa, Bomb. Grass. 116. 1896; Cooke, Fl. Bomb. 2: 1007. 1908; Fischer in Gamble, Fl. Pres. Madr. 1807. 1934. (Repr. ed. 3: 1251.1957); Bor, Grass. Bur.

Cey. Ind. Pak. 416. 1960; Panigrahi *et al.*, in Bull. Bot. Sur. Ind. 11: 113. 1969; Cope in Nasir & Ali, Fl. W. Pak. 143. 26. 1982; Manilal & Sivaraj, Fl. Calic. 345. 1982; Roy, Grass. M.P. 131. 1984; Sreekumar & Nair, Fl. Ker. Grass. 331. 1991; Moulik, Grass. Bamb. Ind. 2: 433. 1997. *P. communis* Trin, Fund. Agrost. 134. 1820; Boiss, Fl. Or. 563. 1884; Duthie, Grass. N.W. Ind. 35. 1883; Duthie, Fodd. Grass. N. Ind. 601. 1888; Lisboa, Bomb. Grass. 116. 1896; Muschler Man. Fl. Egypt. 1: 115. 1912. *P. bifaria* Wight. Herb. no. 3310. 1805; Hook. f. Fl. Brit. Ind. 7: 304. 1896. *P. chilensis* Steud. nom. ed. 23: 234. 1841. *P. hispanica* Nees in Nov. Act. Nat. Cur. 19, Suppl. 1: 174. 1843. *P. humilis* Not. in Cat. Hort. Genuen. 27. 1846. *P. muranica* Kunth. Rev. Gram. 1: 80, 277. t. 50. 1829; Kunth. Enum. Pl. 1: 251. 1838. *P. nepalensis* Nees ex Steud. Syn. Pl. Glum. 196. 1854. *P. pumila* Griseb. in Goett. Nachr. 76. 1863. *Arundo karka* Retz. Obs. Bot. 4: 21. 1786. *Arundo maxima* Forsk; Fl. Aegypt–Arab. 24. 1775. (PLATE–93).

Perennial. Rhizomatous; rhizome 10 cm long, creeping; root fibrous. Culm 2 m tall, 1.0 cm thick, green, glabrous, erect, densely tufted, fistular, reed–like, ascending, decumbent, stout, simple, very leafy, base creeping; nodes glabrous; inter node 15 cm long. Leaf–blade 60 cm long, 2.0 cm broad, green, coriaceous, lower surface scabrid, hairy on both surface; hairs acute, base tubercled; margin hyaline, scabrid, mid–vein distinct on lower surface & scabrid, 9–lateral veins on each side of mid–vein; lateral vein distinct on upper surface; leaf–sheath 18 cm long, 1.5 cm broad, green, coriaceous, hard, glabrous, margin membranous, glabrous below, hairy and keeled at apex; hairs 5.0 mm long, white, acute, base tubercled, loose, erect, very much longer than internodes, veins distinct on outer surface; ligule 1.0 mm long; membranous, lacearate, hairy at apex; hairs 6.0 mm long, white, acute, base broad; auricle hairy; hairs 3.0 mm long, acute, base tubercled. Inflorescence made up of panicle; panicle 35 cm long, 20 cm broad, open, pyramidal, lax, linear, brownish, oblong, wooly; lower nodes hairy; hairs 5.0 mm long, white, acute, base broad; rachis fragile, angular, smooth, grooved. Spikelets 1.2 cm long, 1.0 mm broad, linear–lanceolate, narrow, 6–flowered; pedicles 6.0 mm long, densely hairy; hairs 1.0 cm long, brown, acute, base broad, silky. Lower glume 3.0 mm long, 1.0 mm broad, membranous, brown, ovate, oblong, acuminate, linear, lanceolate, surface glabrous between the nerves, also with margin, 3–nerved; nerves brown, visible, scabrid, central one ending at apex, lateral nerves ending half of the length of glume. Upper glume 4.0 mm long, 1.5 mm broad, membranous, brown, elliptic, oblong, lanceolate, linear, surface glabrous between the nerves, also with margin, 3–nerved; nerves brown, visible, scabrid, central one ending at apex, lateral nerves ending below the apex. Lower lemma 7.0 mm long, 1.0 mm broad, membranous, brown, narrowly elliptic, linear, lanceolate, acute, empty, paleate, glabrous, 3–nerved; nerves brown, slightly visible, glabrous, central one ending at apex, lateral nerves ending very below at apex. Lower palea 2.5 mm long, 1.5 mm broad, hyaline, delicate, oblong, ovate, lanceolate, glabrous, margin sharply folded inwards, 2–lobed at apex; lobes excurrent, acute, transversely scabrid at apex, glabrous below, 2–nerved; nerves green, slightly visible, glabrous & marginal, ending at apex.Upper lemma 7.0 mm long, 1.0 mm broad, membranous, brown, elliptic, linear, lanceolate, acute, surface glabrous, margin slightly scabrid at apex, glabrous below, bisexual, paleate, 3–nerved; nerves brown, slightly visible, glabrous, ending at apex. Upper palea 2.0 mm long, 0.5 mm broad, hyaline, delicate, oblong, elliptic, lanceolate,

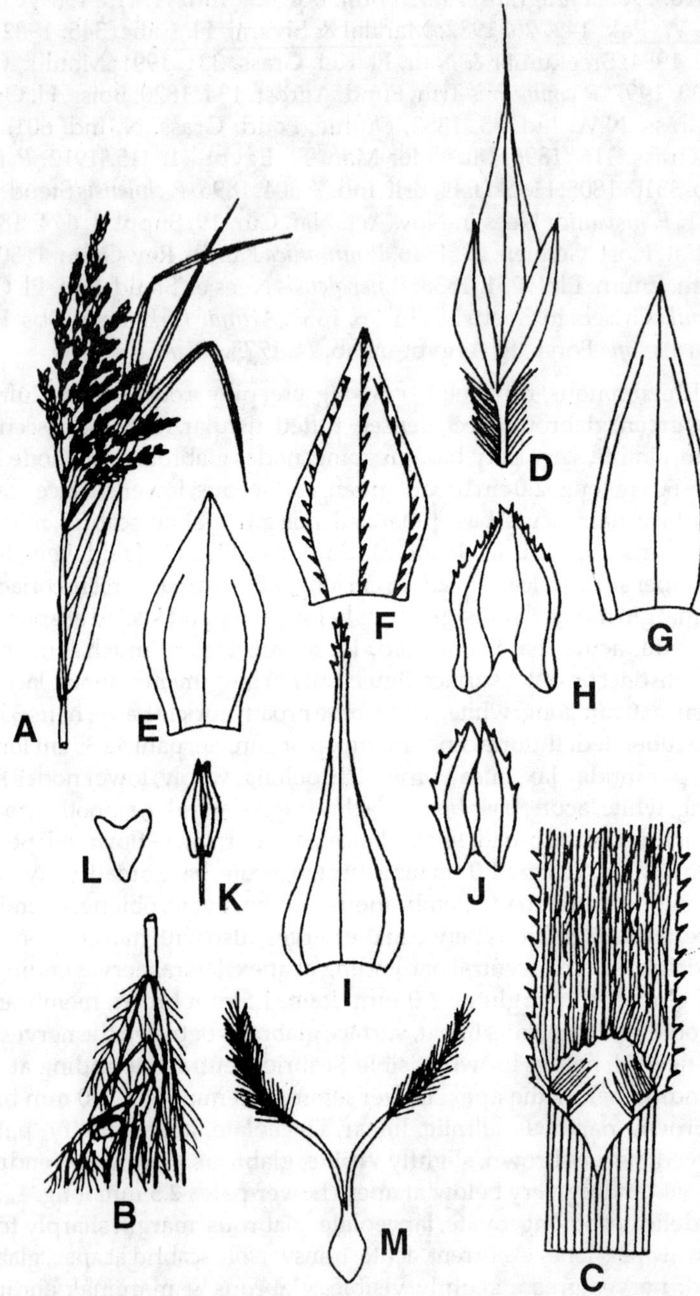

PLATE 93: ***Phragmites karka*** **(Retz.) Trin. ex Steud.**
A. Habit, B.Rhizome, C. Ligule, D. Spikelet, E. Lower glume, F. Upper glume, G. Lower lemma, H. Lower palea, I. Upper lemma, J. Upper palea, K. Stamen, L. Lodicule, M. Pistil

2–lobed at apex; lobes striaght, acute, glabrous, margin sharply folded inwards, transversely slightly scabrid from middle to apex, glabrous below, 2–keeled at apex, 2–nerved; nerves green, slightly visible, glabrous & marginal, ending at apex. Stamens 3; anther 1.0 mm long, yellow, oblong, linear, glabrous; filament 0.5 mm long, slender, glabrous, shorter than anther. Lodicule 2, 0.3 mm long, 0.1 mm broad, hyaline, glabrous, nerveless, oblanceolate, wavy at apex. Pistil 2.5 mm long; ovary 0.5 mm long, ovate, elliptic, glabrous; style 1.0 mm long, slender, glabrous slightly connate at base, bifid at apex; stigma 1.0 mm long, brown, hairy; hairs apiculate, base broad, in 2–rows. Caryopsis 1.5 mm long, 0.7 mm broad, purplish, narrowly elliptic, smooth.

Vernacular Names

Nainarakula, Nar, Narkul, Nulanara, Narkat.

Flowering & Fruiting

September–November; Jaunpur, Lar, Allahabad, Khalilabad, Ballia.

Habitat

Rare along the banks of rivers and waterlogged sandy soils.

Distribution

India, South–East Asia, Burma, Sri Lanka, Australia.

3. TRIBE–THYSANOLAENEAE

Thysanolaena Nees.

Thysanolaena maxima (Roxb.) O. Ktze Rev. Gen. Pl. 2: 794. 1891; Fischer in Gamble, Fl. Pres. Madr. 1854. 1934 (Repr. ed. 3: 1244. 1957); Bor, Grass. Bur. Cey. Ind. Pak. 650. 1960; Panigrahi *et al.*, Bull. Bot. Sur. Ind. 11: 113. 1969. Roy, Grass. M.P. 161. 1984; Sreekumar & Nair, Fl. Kera. Grass. 444. 1991; Moulik, Grass. Bamb. Ind. 2: 558. 1997. *T. agrostis* Nees in Edinb. N. Phil. Journ. 218: 180. 1835; Hook. f. Fl. Brit. Ind. 7: 61.1896; Cooke, Fl. Bomb. 2: 1006. 1908; Haines, Bot. Bih. & Or. 982. 1924. *T. procera* Mez. in Janowski Bot. Archiv. 1: 27. 1922; Blatter & McCann, Bomb. Grass. 201. t. 132. 1984. *T. acarifera* Nees. & Arn. in Nov. Act. Leopold 19, Suppl. 1: 181. 1843; Lisboa, J. Bomb. Nat. Hist. Soc. 5: 347. 1890; Duthie, Grass. N.W. Ind. 13. 1883; Duthie, Fodder Grass. N. Ind. 21. 1888. *Agrostis maxima* Roxb. Fl. Ind. 1: 317. 1820. *A. procera* Retz. Obs. Bot. 4: 19. 1786. *A. latifolia* Heyne ex Hook. f. Fl. Brit. Ind. 7: 61. 1896. *Melica latifolia* Roxb. Hort. Beng. 8. 1814. *Panicum acariferum* Trin. Sp. Gram 1: 87. 1828. *Myriachaeta arundinacea* Zoll. & Mor. Syst. Verz. Zoll, 101. 1845–46. *M. glauca* Mor. ex. Steud. Syn. Pl. Glum. 1: 404. 1855. (PLATE–94).

Perennial. Rhizomatous; rhizome 10 cm long, horizontal, root fibrous. Culm 3.0 m tall, 2.0 cm thick, green, glabrous, erect, densely tufted, stout, ascending, polished, leafy thoughout, branching at lower nodes; branches suberect, leafy up to inflorescence, rooting at base; nodes glabrous; internodes 8 cm long. Leaf–blade 45 cm long, 6.0 cm broad, green, flat, linear, lanceolate, acuminate, base cordate, glaucous, surface glabrous, coriaceous, margin membranous, scabrid, mid–vein distinct on lower surface, lateral veins numerous, distinct on upper surface; leaf–sheath 12 cm

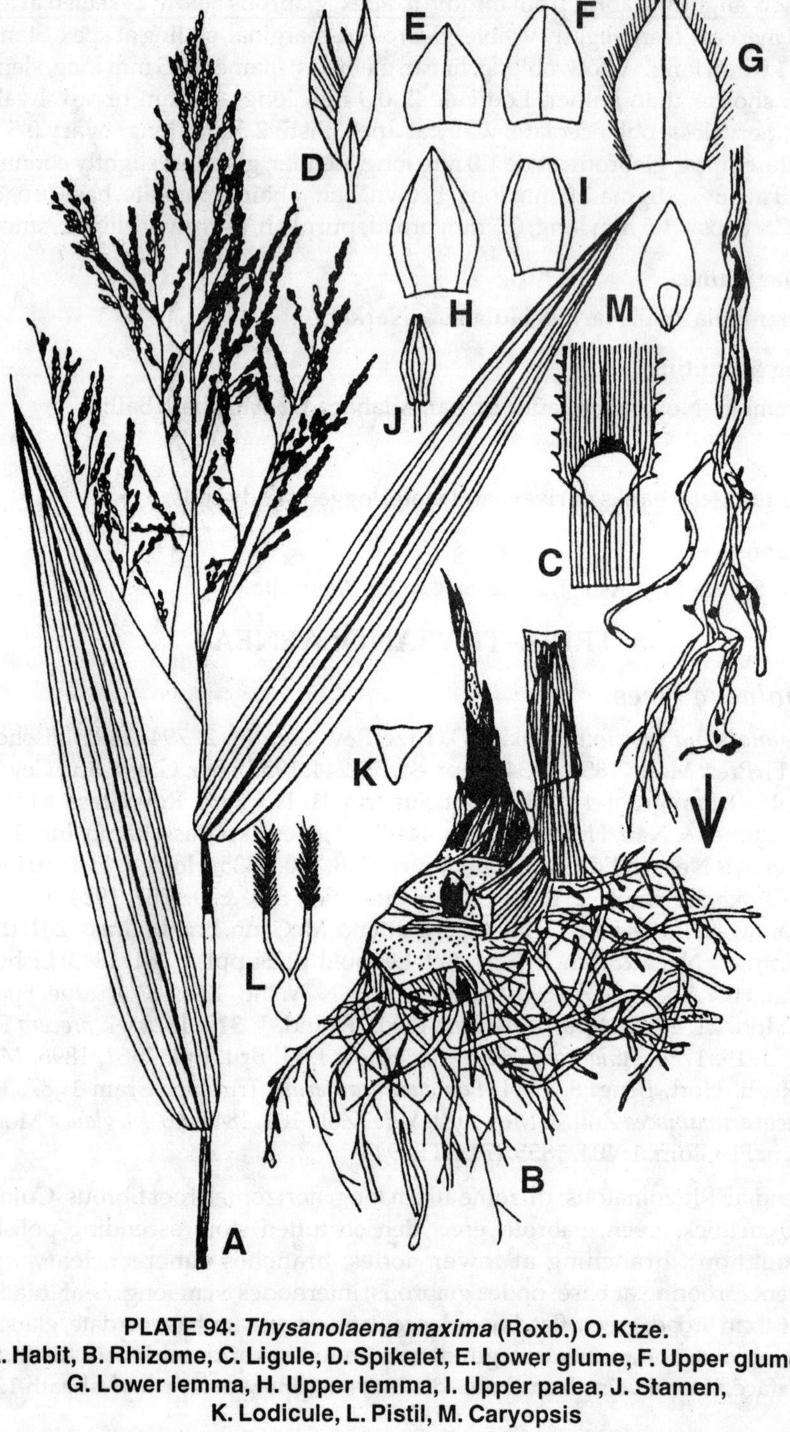

PLATE 94: ***Thysanolaena maxima*** **(Roxb.) O. Ktze.**
A. Habit, B. Rhizome, C. Ligule, D. Spikelet, E. Lower glume, F. Upper glume, G. Lower lemma, H. Upper lemma, I. Upper palea, J. Stamen, K. Lodicule, L. Pistil, M. Caryopsis

long, 3.0 cm broad, green, coriaceous, glabrous, margin membranous & keeled, loose, erect, striate, very much longer than internodes, basal portion enclosed by lower sheath; veins distinct on outer surface; ligule 1.5 mm long, membranous, truncate, hairy at apex; hairs 0.5 mm long, white, acute, base broad; auricle glabrous. Inflorescence made up of panicle; panicle 50 cm long, 25 cm broad, soft, glabrous, erect, open, dropping, pyramidal. Branches 15 cm long, numerous, suberect, densely crowded, filiform, glabrous with many short branchlets carring small spikelets. Spikelets 1.5 mm long, 1.0 mm broad, green–yellow at young, turning into pale brown with age, neumerous, ovate–lanceolate, binate, 2–flowered, pedicellate; pedicel 0.5 mm long, smooth. Lower glume 0.5 mm long, 0.3 mm broad, hyaline, ovate, lanceolate, acute, glabrous, 1–nerved; nerves colourless, glabrous, ending at apex. Upper glume 1.0 mm long, 0.5 mm broad, hyaline, ovate, lanceolate, oblong, cuneate, glabrous, 1–nerved; nerve colourless, glabrous, ending at apex. Lower lemma 2.0 mm long, 1.0 mm broad, membranous, lanceolate, acute, surface glabrous, margin hairy; hairs 1.0 mm long, acute, base tubercled, empty, epaleate, 1–nerved; nerves green, visible, glabrous, ending at apex. Upper lemma. 1.5 mm long, 1.0 mm broad, submembranous, ovate, lanceolate, acute, glabrous, bisexual, paleate, 3–nerved; nerves green, slightly visible, glabrous, ending at apex. Upper palea 1.0 mm long, 0.5 mm broad, hyaline, oblong, lanceolate, notched at apex, glabrous, 2–nerved; nerves green, slightly visible, glabrous & marginal, ending at apex. Stamens 2; anther 0.8 mm long, yellowish–green, oblong, glabrous; filament 0.2 mm long, slender, glabrous, shorter than anther. Lodicules 2, 0.2 mm long, 0.1 mm broad, hyaline, glabrous, obovate, cuneate at base, wavy at apex, nerveless. Pistil 1.5 mm long; ovary 0.3 mm long, ovate, glabrous; style 0.6 mm long, slender, glabrous, bifid from base; stigma 0.6 mm long, brown, hairy; hairs acute, base broad, in 2–rows. Caryopsis 0.5 mm long, 0.2 mm broad, ovoid, reddish brown, smooth.

Vernacular Name

Phoolbaharan, Baruch.

Flowering & Fruiting

November–May; Sarnath, Allahabad, Padrauna, Sonbhadra.

Habitat

Rare, along the dry habitat.

Distribution

India, Penang, New Guinea.

2. Sub-family–Bambusoideae

KEY TO THE TRIBES

1a. Inflorescence branched panicle with clusters of spikes along the branches; spikelets 1 to many flowered; stamens 3; stigma 1–3 or more; lodicules usually 3

............................**1. Bambuseae**

1b. Inflorescence simple panicle; spikes absent; spikelets 1–flowered; stamens 6; stigma 2; lodicules 2 .... **2. Oryzeae**

1. Tribe–Bambuseae

KEY TO THE SPECIES

1a. Spikelets many flowered; palea scabrid; ovary hairy at apex; pericarp adnate to the caryopsis .... **1. *Bambusa***

1b. Spikelets 2–6 flowered; palea hairy & scabrid; ovary entirely hairy; pericarp separable from the caryopsis .... **2. *Dendrocalamus***

1. *Bambusa* Schreber.

Moulik, Grass. Bamb. Ind. 1: 21. 1997. (PLATE–95).

Perennials. Rhizome well developed, stout. Culm 35 m tall, 15 cm thick, green, glabrous, densely tufted, erect, dropping at the top, arborescent, liginified, segmented; nodes prominent, rooting at lower node, giving out horizontal shoots armed at nodes with 2–3 recurved spines & with few leaves. Leaves dimorphic; culm sheath upto 30 cm long, 20 cm broad, coriaceous, triangular, without blade, petiole usually absent, auricles & oral setae often present, orange–yellow when young, often striped with green or red, striate, thickly ciliate with golden hairs when young; hairs 2.5 mm long, acute, base broad; foliage leaf–blade 30 cm long 5.0 cm broad, oblong, linear, lanceolate, acuminate, base rounded, surface hairy; hairs minute, acute, base tubercled; margin scabrid, mid–vein narrow, 6–lateral veins on each side of mid–vein with 7–9 intermediate veins; leaf–sheath 10 cm long, 1.0 cm broad, ending in a thick callus with a short bristing auricle; ligule 1.5 mm long, membranous. Flowering often gregarious & at long intervals, usually dies down after flowering. Spikelets 1–many–flowered, arranged in panicles, racemes dense clusters or terminal heads; flowers bisexual. Fruit a caryopsis, achene or berry–like.

Vernacular Name

Bans.

Flowering & Fruiting

Not seen; Jaunpur, Chandauli, Chakiya, Ghazipur, Mirzapur, Sonbhadra, Basti.

Habitat

Planted in most part of Eastern U.P.

Distribution

Tropical & sub tropical parts of the world.

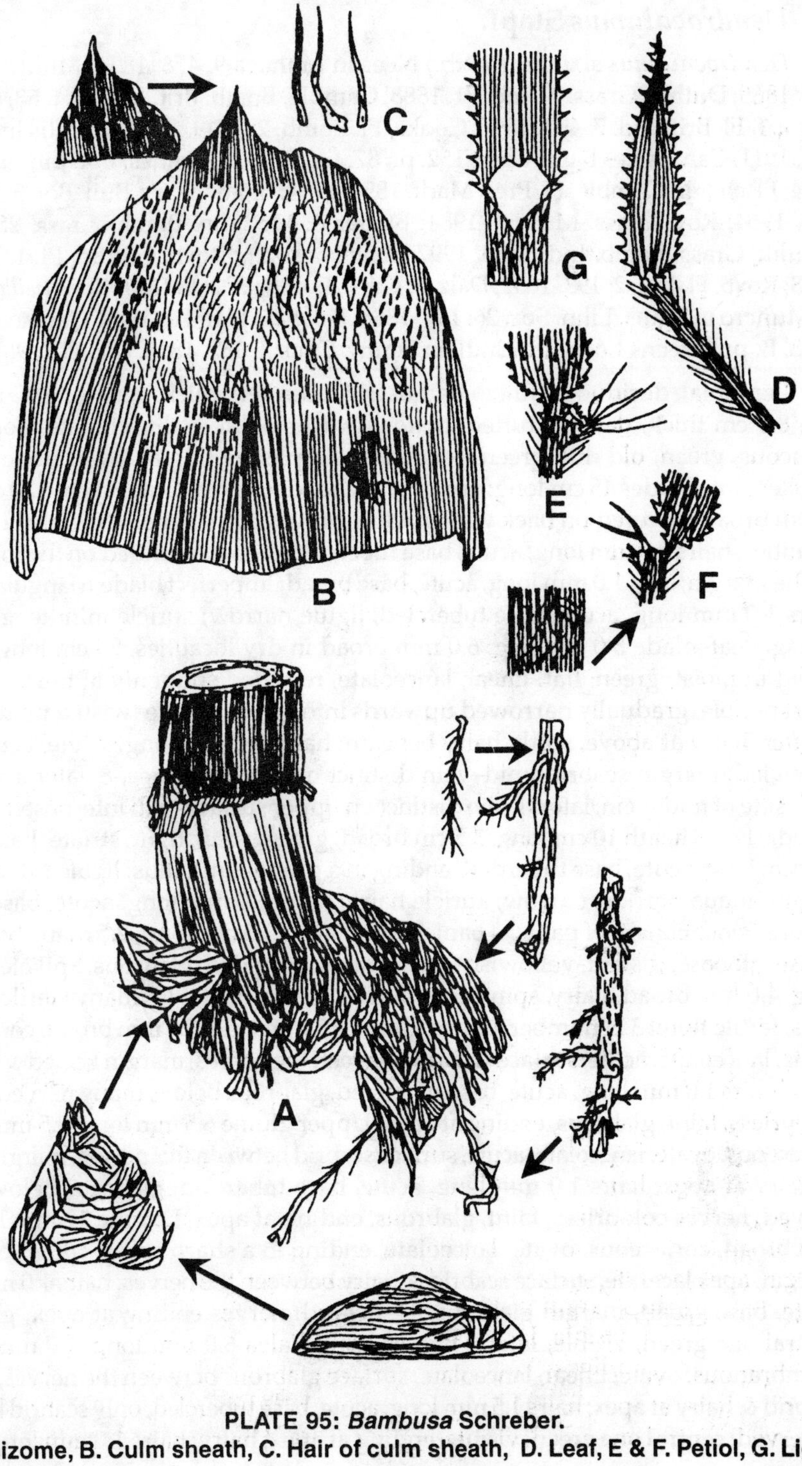

PLATE 95: *Bambusa* Schreber.

A. Rhizome, B. Culm sheath, C. Hair of culm sheath, D. Leaf, E & F. Petiol, G. Ligule

2. *Dendrocalamus* Stapf.

Dendrocalamus strictus (Roxb.) Nees in Linnaea 9: 476. 1834; Munro. Mongr. 147. 1865; Duthei, Grass. N. Ind. 71. 1888; Gamble, Bomb. Brit. Ind. 78. t. 68, 69. 1896; Hook. f. Fl. Brit. Ind. 7: 404. 1896; Cooke, Fl. Bomb. 2: 1049. 1908; Brandis Ind. Trees. 675. 1911; Camus Les Bambusees 152. pt. 87. fig. B. 1913; Haines, Bot. Bih. & Or. 947. 1924; Fischer in Gamble, Fl. Pres. Madr. 1858. 1934; Bhattachary, Bull. Bot. Sur. Ind. 6: 209. 1964; Roy, Grass. M.P. 63; 1984; Blatter & McCann, Bomb. Grass. 285. 1984; Moulik, Grass. Bamb. Ind. 1: 30. 1997. *Bambusa stricta* Roxb. Corom. Pl. 1: 58. t. 80. 1798; Roxb. Fl. Ind. 2: 193. 1832; Dalz. & Gibs., Bomb. Fl. 299. 1861. *B. verticillata* Rottl. ex Muncro in Trans. Linn. Soc. 26: 147. 1868. *B. tanaea* Ham. ex Wall. Cat. no. 5038. A. 1828. B. pubescens Lodd. ex Lindl. in Penny. Cycl. 3: 357. 1835. (PLATE–96).

Perennial, deciduous. Rhizomatous; rhizome very thick, very long. Culm 15 m tall, 8.0 cm thick, densely tufted, hollow in wet, solid in dry climate, young one glaucous, green, old dull green–yellowish, simple; rooting at lower nodes; nodes swollen; internodes 45 cm long; upper branches decurved. Culm sheath 30 cm long, 15 cm broad, covered on back with golden–brown stiffy hairs, but glabrous in dry localities; hairs 2.0 mm long, acute, base tubercled; striate, rounded on the top, ciliate on the margin; cilia 1.0 mm long, acute, base broad; imperfect blade triangular, hairy; hairs 1.0 mm long, acute, base tubercled; ligule narrow; auricle minute, glabrous. Foliage leaf–blade 5.0 cm long, 6.0 mm broad in dry localities, 25 cm long, 2.0 cm broad in moist, green, flat, linear, lanceolate, rounded suddenly at the base into a short petiole, gradually narrowed upwards into an acuminate twisted point, rough & often hairy at above, softly hairy beneath; hairs 2.0 mm long, white, acute, base tubercled; margin scabrid, mid–vein distinct on lower surface, 8–lateral veins on each side of mid–vein; lateral vein distinct on upper surface with interposed pellucid glands; leaf–sheath 10 cm long, 3.5 cm broad, green, erect, tight, striate, hairy; hairs 2.0 mm long, acute, base tuberded, ending in a prominant callus; ligule 1.0 mm long, membranous, serrate, narrow; auricle hairy; hairs 1.0 mm long, acute, base broad. Inflorescence consist of panicle; panicle very large, 3.5 cm apart, 2.5 diam; branched, dense, globose, distant, yellowish–green; rachis rounded, glabrous. Spikelet 12 mm long, 4.0 mm broad, hairy, spinescent, fertile, intermixed with many sterile smaller ones, fertile floret 3 in number. Lower glume 6.5 mm long, 3.0 mm broad, coriaceous, ovate, lanceolate, acute, surface scabrid between the nerves, margin keeled & hairy at apex; hairs 1.0 mm long, acute, base tubercled, glabrous below, many nerved; nerves colourless, faint, glabrous, ending at apex. Upper glume 6.5 mm long, 2.5 mm broad, coriaceous, ovate, lanceolate, acute, surface scabrid between the nerves, margin keeled & hairy at apex; hairs 1.0 mm long, acute, base tubercled, glabrous below, many nerved; nerves colourless, faint, glabrous, ending at apex. Lemma 12 mm long, 8.0 mm broad, coriaceous, ovate, lanceolate, ending in a sharp spine; spine scabrid at margin, apex lacerate, surface scabrid & hairy between the nerves; hairs 1.0 mm long, acute, base broad, margin glabrous, 11–nerved; nerves ending at apex, glabrous, central one green, visible, lateral nerves faint. Palea 6.0 mm long, 1.0 mm broad, membranous, ovate, linear, lanceolate, surface glabrous between the nerves, margin scabrid & hairy at apex; hairs 1.5 mm long, acute, base tubercled, only scabrid beneath, 6–nerved; central one green, visible, ending at apex, hairy; hairs 1.0 mm long, acute,

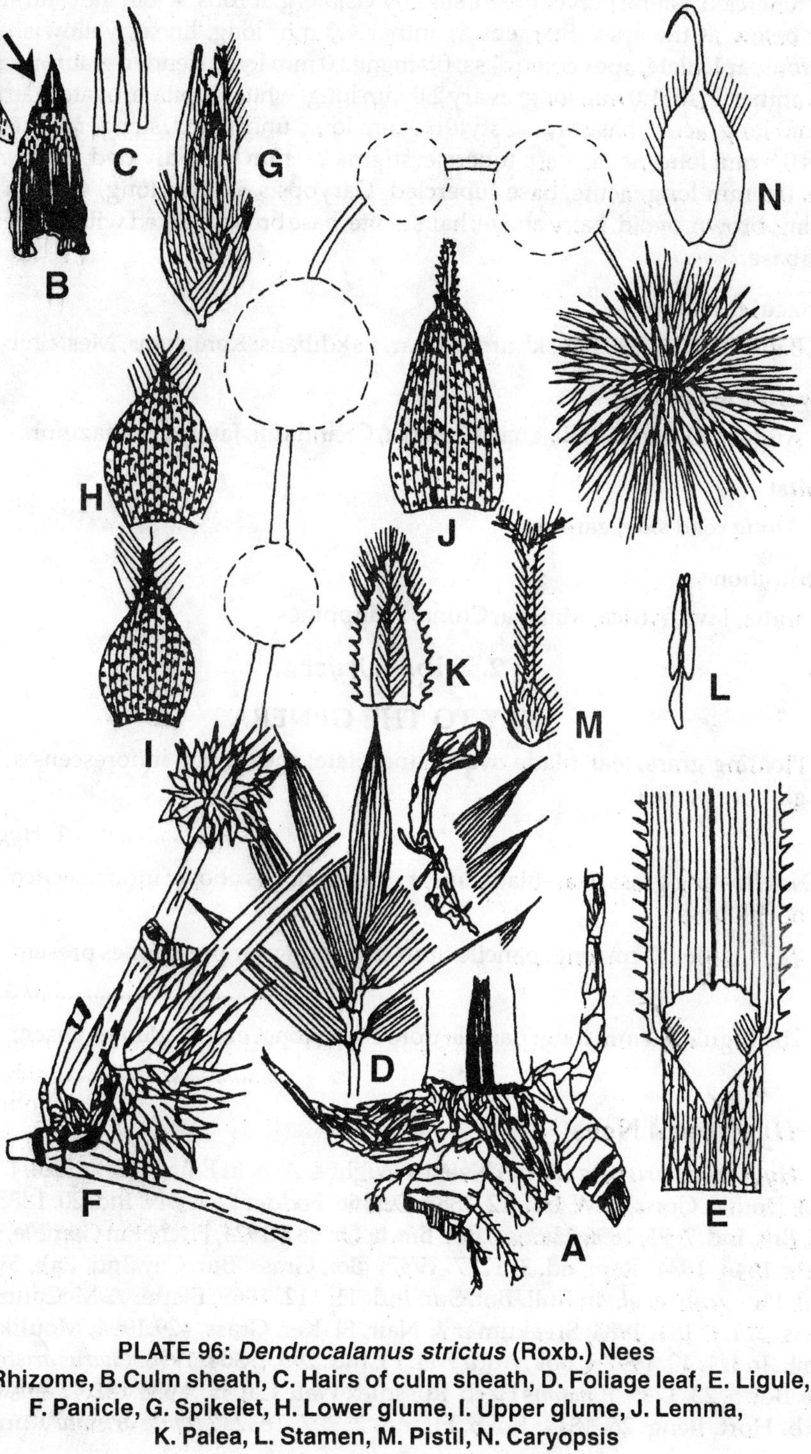

PLATE 96: ***Dendrocalamus strictus*** **(Roxb.) Nees**
A. Rhizome, B.Culm sheath, C. Hairs of culm sheath, D. Foliage leaf, E. Ligule, F. Panicle, G. Spikelet, H. Lower glume, I. Upper glume, J. Lemma, K. Palea, L. Stamen, M. Pistil, N. Caryopsis

base tubercled; lateral nerves green, slightly visible, glabrous, & join the central nerves very below at the apex. Stamens 6; anther 4.0 mm long, linear, yellowish–green, glabrous, apiculate, apex colourless; filament 2.0 mm long, slender, glabrous, shorter than anther. Pistil 1.0 mm long; ovary 2.0 mm long, white, stipitate, ovate, hairy; hairs 0.5 mm long, acute, base broad; style 6.0 mm long, undivided, slender, white, hairy; hairs 0.5 mm long, acute, base tubercle; stigma 2.0 mm long, divided, purple, hairy; hairs 0.5 mm long, acute, base tubercled. Caryopsis 8.0 mm long, 4.0 mm broad, shining, brown, ovoid, hairy above; hairs acute, base broad, beaked with the persistant style base.

Vernacular Names

Bans, Banskaban, Banskhurd, Kopar, Lakdibans; Kaniabans, Mes, Shib, Basa.

Flowering & Fruiting

August–December; Rajghat, Chakiya, Chandauli, Jaunpur, Ghazipur.

Habitat

Along road side, gardens.

Distribution

India, Java, Africa, Malaya, China, Philippines.

2. Tribe–*Oryzeae*

KEY TO THE GENERA

1a. Floating grass; leaf–blade ovate, lanceolate; sheath flat; inflorescens a globose panicle .... 1. ***Hygroryza***

1b. Not floating grass; leaf–blade linear; sheath not as above; inflorescence not globose:

2a. Ligule 1.7 cm long; pancile upto 25 cm long; empty glumes present .... 3. ***Oryza***

2b. Ligule 1.0 mm long; panicle upto 10 cm long, empty glumes absent .... 2. ***Leersia***

1. *Hygroryza* Nees

Hygroryza aristata (Retz.) Nees ex Wight & Arn. in Edinb. New. Phil. J. 15. 380. 1833; Duthie, Grass. N.W. Ind. 12. 1883; Duthie, Fodder Grass. N. Ind. 20. 1888; Hook. f. Fl. Brit. Ind. 7: 95. 1896; Haines, Bot. Bih. & Or. 981. 1924; Fischer in Gamble, Fl. Pres. Madr. 1846. 1934 (Repr. ed. 3: 1277. 1957); Bor, Grass. Bur. Cey. Ind. Pak. 597. t. 67. 1960. Panigrahi *et al.*, in Bull. Bot. Sur. Ind. 11: 112. 1969; Blatter & McCann, Bomb. Grass. 271. t. 184. 1984; Sreekumar & Nair, Fl. Ker. Grass. 429.1994; Moulik, Grass. Bamb. Ind. 1: 43. 1997; Cook, Aqu. Wet. Pl. Ind. 295. f. 304. 1996. *Pharus aristate* Retz. Obs. Bot. 5: 23. 1789. *P. natans* Herb. Russdlex Wall. Cat. no. 8638.1828. *Leersia aristata* Roxb. Hort. Beng. 26. 1814; Roxb. Fl. Ind. 2: 207. 1832. *Zizania aristata* Kunth. Rev.

Gram 1: 8. 1829; Kunth. Enum. Pl. 1: 10. 1838. *Z. retzii* Spreng. Syst. 2: 136. 1825. *Potamochloa retzii* Griff. Journ. As. Soc. Beng. 5: 571. t. 24, fig. 2. 1836. (PLATE–97).

Annual. Rhizome absent; root whorled, fibrous. Culms 20 cm long, 2.0 mm thick, green, glabrous, spongy, tufted, spreading, ascending, rarely geniculate, slender, leafy, base creaping, branching at lower nodes; branches suberect; nodes glabrous; internode 2.0 cm long, rooted. Leaf–blade 2.5 cm long, 1.0 cm broad, green, ovate, oblong, lanceolate, flat, rounded at apex, trnucate at base, glaucous, subcoriaceous, upper surface scabrid, lower surface glabrous also with margin, mid–vein prominent, lateral vein numerous, very slender, prominent; leaf–sheath 3.0 cm long, 8 mm broad, green, spongy, compressed, suberect, inflated, loose, glabrous, margin keeled, veins absent, longer than blade & internode, forming floats; ligule 1.0 mm long, membranous, truncate, lacerate; auricle absent. Inflorescence consists of panicle; panicle 5.0 cm long, 5.0 cm broad, globose, erect, spreading, loose, ovate, triangular, glabose; rachis of panicle slender, stiff, smooth. Racemes 1.5 cm long, suberect, distant, 6 in number, decreasing in length upwards, with 2–3 spikelets; rachis of raceme, angular, subterete, slender, stiff, glabrous. Spikelets 17 mm long, 1.5 mm broad, one flowered, narrow, lanceolate, compressed, linear, awned, pedicelled; pedicel 0.5 mm long, glabrous, with a red band belcw the midde. Glumes absent. Lemma 7.0 mm long, 3.0 mm broad, chartaceous, linear–lanceolate, laterally compressed, surface hairy between the nerves; hairs minute, acute, base broad; margin scabird, 5–nerved; nerves visible, scabrid upwards, deep green, towards at apex, yellowish green base, ending at apex, but central nerve prolongated in to awned; awn 10 mm long, slender, stiff, antrorsely scabrid. Palea 7.0 mm long, 3.0 mm broad, chartaceous, slightly ovate, elliptic, linear, lanceolate, acute, laterally compressed, surface glabrous between the nerves, also with introlled margin, 3–nerved; nerves green, visible, central one dentate upwards, lateral nerves glabrous, ending at apex. Stamens 6; anther 4.0 mm long, yellow, linear, glabrous; filament 1.0 mm long, slender, glabrous, shorther than anther. Lodicules 2, 0.4 mm long, 0.2 mm broad, hyaline, glabrous, globose, slightly wavy at base, 3–nerved; nerves lightly green, slightly visible, ending below at the apex. Pistil 4.5 mm long; ovary 1.0 mm long, ovate, oblong, 2–nerved; nerves green, visible, glabrous & marginal; style 1.5 mm long, slender, glabrous, bifid from base; stigma 2.0 mm long, purplish, hairy; hairs acute, base broad, in 2–rows. Caryopsis 3.5 mm long, 1.0 mm broad, oblong, linear, lanceolate, acuminate, shortly stipitate, smooth, deep brown.

Vernacular Names

Jangalidal, Tili, Deobhat, Urodhan.

Flowering & Fruiting

October–December; Barahikala (Varanasi), Basti, Padrauna.

Habitat

Free floating grass found in ponds.

Distribution

India, Tonkin. Burma, Sri Lanka, Malaya.

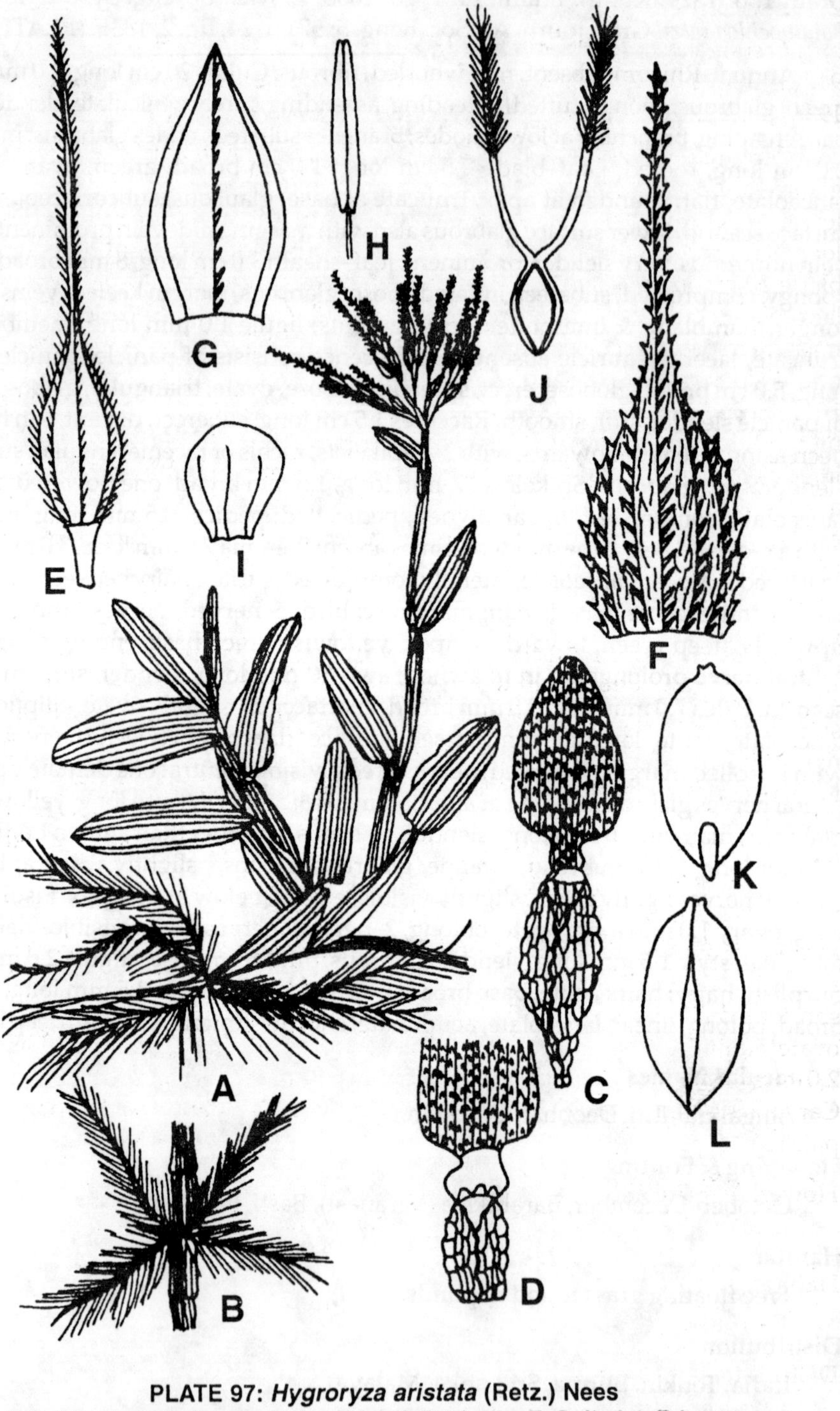

PLATE 97: *Hygroryza aristata* (Retz.) Nees
A. Habit, B. Node & internode, C. Leaf, D. Ligule, E. Spikelet, F. Lemma, G. Palea, H. Stamen, I. Lodicule, J. Pistil, K & L. Caryopsis

2. *Leersia* Soland.

Leersia hexandra Swartz. Prodr. Veg. Ind. Occ. 21. 1788; Hook. f. Fl. Brit. Ind. 7: 94. 1896. Moulik, Grass. Bamb. Ind. 2: 45. 1997; Cook, Aqu. Wet. Pl. Ind. 302. f. 310. 1996. (PLATE–98).

Perennial. Rhizome absent; root fibrous. Culm 1.0 m tall, 4.0 mm thick, green, glabrous, erect, densely tufted, slender, ascending, leafy, creeping at base, rooting at lower nodes, branched; branches suberect, leafy upto inflorescence; nodes 5.0 mm long, hairy; hairs 1.0 mm long, white, acute, base broad; internodes 20 cm long. Leaf-blade 20 cm long, 1.5 cm broad, green, flat, coriaceous, narrow, glaucous, linear-lanceolate, acute, base rounded, scabrid upwards on both surface, also with hyaline margin, mid–vein scabird, distinct on lower surface, 4–lateral veins on each side of mid–vein; lateral veins slender, very distant, distinct on upper surface; leaf–sheath 10 cm long, 5.0 mm broad, green, sub–coriaceous, slightly scabird, margin membranous, glabrous, tight, compressed, terete, erect, very much shorter than internodes, veins distinct on outer surface; ligule 1.0 mm long, membranous, lanceolate; auricle glabrous. Inflorescence consists of panicle; panicle 8.0 cm long, 4.0 cm broad, erect, pyramidal, ovate, contracted, open, terminal; rachis of panicle grooved, scabrid, triquetrous. Racemes 5.0 cm long, suberect, spreading, distant, discreasing in length upwards, alternate, pedicelled; pedicle 0.5 mm long; rachis of racemes scabrid, grooved, angular. Spikelet 4.0 mm long, 1.0 mm broad, oblong, linear, compressed, closely, imbricate, glabrous, sessile, bisexual. Glumes absent. Lemma 4.0 mm long, 2.0 mm broad, ovate, lanceolate, acute, whitish, cartilaginous, laterally compressed, surface glabrous between the nerves, margin hyaline & scabrid upwards, 5–nerved; nerves green, visible, ending at apex, central one dentate upwards; denta 1.0 mm long, pectinate, whitish–green, lateral 2–nerves near central one glabrous, marginal 2–nerves ciliate; cilia 1.5 mm long, acute, base broad. Palea 3.8 mm long, 1.5 mm broad, membranous, linear, oblong, lanceolate, dimidiate, surface glabrous, between the nerves, laterally compressed, 3–nerved; nerves green, visible, ending at apex, lateral nerves thin, glabrous, central one thick, dentate upwards, elongated into short awned; denta 1.0 mm long, pectinate, white; awn 0.5 mm long, smooth. Stamen 6; anther 3.0 mm long, yellow, glabrous, linear; filament 0.5 mm long, slender, glabrous, shorter than anther. Lodicules absent. Pistil 3.5 mm long; ovary 0.5 mm long, green, ovate, elliptic, glabrous; style 1.0 mm long, slender, glabrous, bifid from base; stigma 2.0 mm long, hairy, whitish–gray, shining; hairs acute, base broad, in 2–rows. Caryopsis 4.0 mm long, 1.0 mm broad, glabrous, linear, hardly enclosed by lemma & palea.

Flowering & Fruiting

September–November; Mirzapur, Mau, Padrauna, Khalilabad.

Habitat

Rare, along ditches & ponds.

Distribution

India, Europe, Burma, Sri Lanka, Malaya.

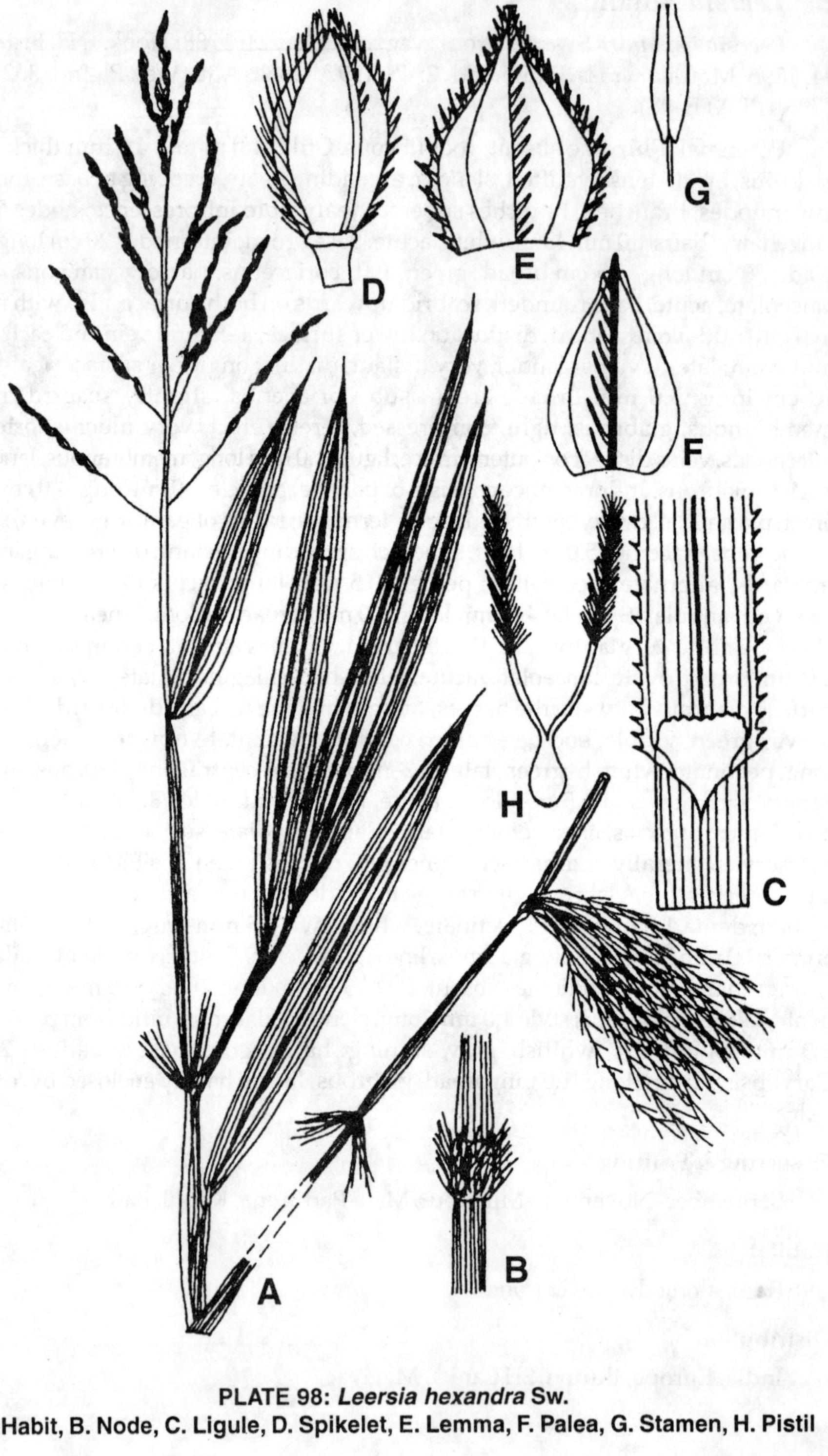

PLATE 98: ***Leersia hexandra*** **Sw.**
A. Habit, B. Node, C. Ligule, D. Spikelet, E. Lemma, F. Palea, G. Stamen, H. Pistil

3. *Oryza* L.

KEY TO THE SPECIES

1a. Spikelets with long awned; anther 6, 5.0 mm long .... 1. *O. rufipogon*

1b. Spikelets awnless, rarely short awned; anther 3, 2.0 mm long .... 2. *O. sativa*

1. *Oryza rufipogon*

Oryza rufipogon Griff. Notul. Pl. Asiat. 3: 5. 1851; Bor. Grass. Bur. Cey. Ind. Pak. 605. 1960; Bhattacharya in Bull. Bot. Surv. Ind. 6: 209. 1964; Panigrahi *et al.*, in Bull. Bot. Surv. Ind. 11: 113. 1969; Gilliland in Fl. Malaya 3: 100. 1971; Raizada *et al.*, Grasses Upp. Gang. Pl. pt. 3: 101. 1983; Moulik, Grass. Bamb. Ind. 1: 47. 1997; SreeKumar & Nair, Fl. Kerla. Grass. 432. 1991; Roy in Grass. M.P. 117. 1984; Cook, Aqu. Wet. Pl. Ind. 304. f. 313. c–d. 1996. *Oryza sativa* sensu. Hook. f. Fl. Brit. Ind. 7: 92. 1896. (PLATE–99).

Annual. Rhizomatous; rhizome 5.0 cm long, creeping, stout, smooth; root fibrous. Culm 45 cm tall, 5.0 mm thick, green, glabrous, erect, densely tufted, compressed, slightly geniculate, simple, leafy, spongy at lower part, rooting at lower nodes; nodes glabrous; internodes 8.0 cm long. Leaf–blade 20 cm long, 1.5 cm broad, green, flat, coriaceous, linear–lanceolate, acuminate & bidentat at apex; base rounded, surface scabrid also with hyaline margin, mid–vein scabrid & distinct on lower surface, 7–lateral veins on each side of mid–vein; lateral veins distinct on upper surface & scabrid; leaf–sheath 9.0 cm long, 1.0 cm broad, green, coriaceous, glabrous, margin membranous, keeled, terete, loose, compressed, striate, erect, longer than internodes, veins distinct on outer surface; ligule 1.7 cm long, membranous, ovate, truncate, erose, spliting at apex; auricle glabrous. Inflorescence consists of panicle; panicle 20 cm long, oblong, lax, exserted, compound; branches sub erect, spreading, decreasing in length at upwards, angular, slender, scabrid, dilated at apex. Spikelets 7.0 mm long, 2.0 mm broad, oblong, compressed, awned, 1–flowered, falling with age, articulation above, minute, semi–circular glumes. Lower glume 3.0 mm long, 1.5 mm broad, chartaceous, ovate, oblong, acute, surface glabrous, margin scabrid, 1–nerved; nerves colourless, glabrous, slightly visible, ending at apex. Upper glume 3.0 mm long, 1.5 mm broad, chartaceous, ovate, oblong, surface glabrous, margin scabrid, truncate, 1–nerved; nerves colourless, slightly visible, glabrous, ending at apex; Lemma 7.0 mm long, 3.5 mm broad, chartaceous, ovate, oblong, laterally compressed, surface hairy between the nerves; hairs 1.0 mm long, acute, base broad, margin scabird, 3–nerved; nerves green, slightly visible, ending at apex, scabrid, awned; awn 10 cm long, purplish, antrorsely scabrid, slender. Palea 6.0 mm long, 3.0 mm broad, coriaceous, oblong, laterally compressed, margin slightly scabrid at apex, glabrous below, surface scabrid, 1–nerved; nerves green, slightly visible, glabrous, ending at apex, awned; awn 6 mm long, antrorsely scabrid. Stamens 6; anther 5.0 mm long, yellow, linear, oblong; glabrous; filament 1.5 mm long, slender, glabrous, shorter than anther. Lodicule 2, 1.0 mm long, 0.5 mm broad, oblong, glabrous, 3–nerved; nerves

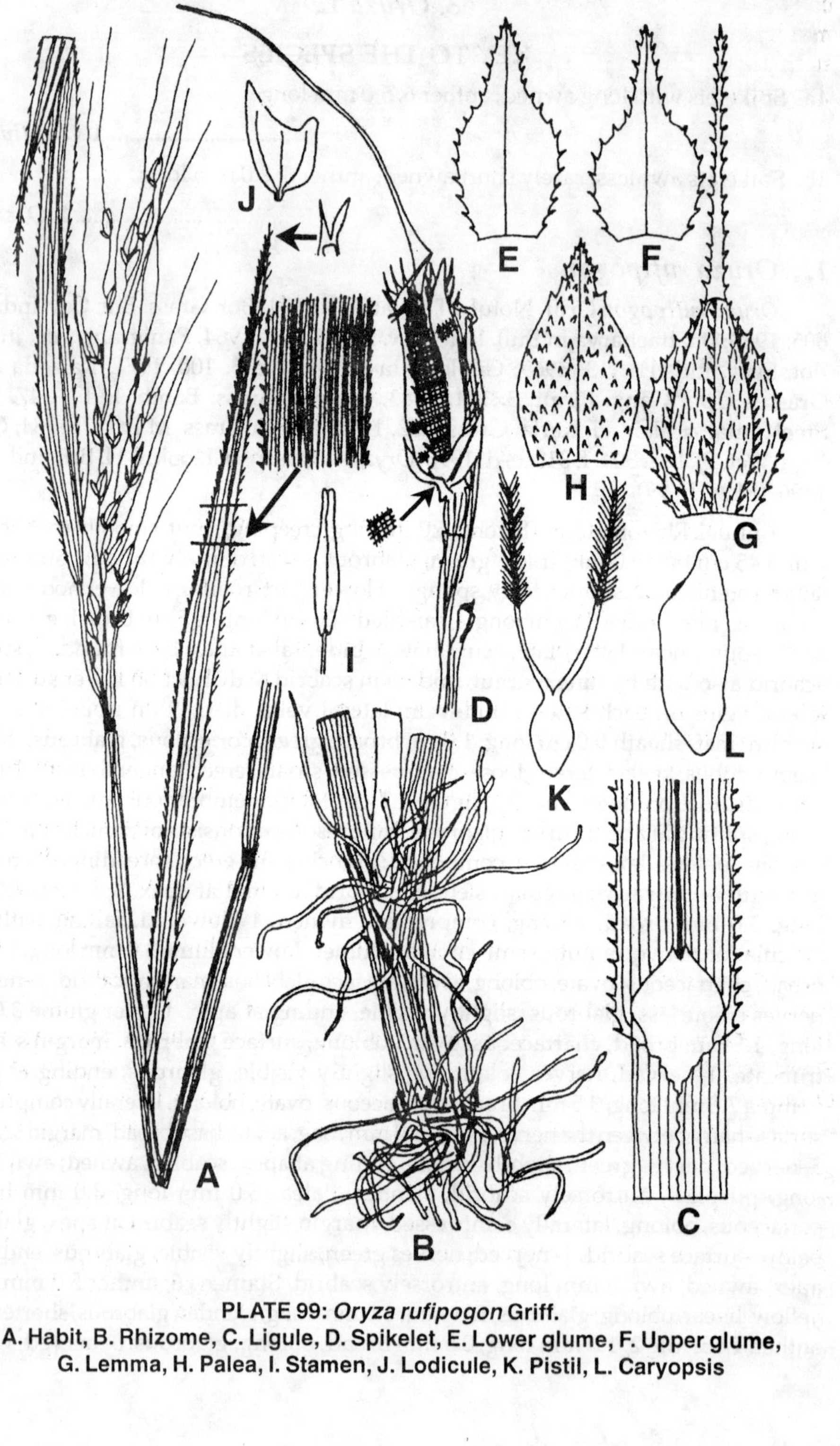

PLATE 99: ***Oryza rufipogon*** **Griff.**

A. Habit, B. Rhizome, C. Ligule, D. Spikelet, E. Lower glume, F. Upper glume, G. Lemma, H. Palea, I. Stamen, J. Lodicule, K. Pistil, L. Caryopsis

colourless, glabrous, ending slightly below at the apex. Pistil 4.0 mm long; ovary 0.5 mm long, oblong, glabrous; style 1.5 mm long, slender, glabrous, bifid from base; stigma 2.0 mm long, cream coloured, hairy; hairs acute, base broad, in 2–rows. Caryopsis 5.0 mm long, 1.5 mm broad, elliptic, oblong, pale brown.

Vernacular Name

Pasai, Lamera, Tinni.

Flowering & Fruiting

September–November; Tarana, Sarnath, Harraua, Navalpur, Barahikala, Sindhora, Chandmari, Jaunpur, Chakiya, Chandauli, Allahabad, Gorakhpur, Basti.

Habitat

Very common along edges of ponds, ditches.

Distribution

India. Asia.

2. *O. sativa*

O. sativa Linn. Sp. Pl. 333. 1753; Roxb. Fl. Ind. 2: 200. 1832; Duthie, Grass. N.W. Ind. 2: 1883; Duthie, Field & Gard. 15. t. 4. 1885; Duthie, Fodder Grass. N. Ind. 20. 1888; Hook. f. Fl. Brit. Ind. 7: 72. 1896; Cooke. Fl. Bomb. 2: 1043. 1908; Haines, Bot. Bih. & Or. 980. 1924; Fischer in Gamble, Fl. Pres. Madr. 1844. 1935 (Repr. ed. 3: 1276. 1957); Bor, Grass. Bur. Cey. Ind. Pak. 605. 1960; Panigrahi *et al.*, in Bull. Bot. Surv. Ind 11: 113. 1969; Moulik Grass. Bamb. Ind. 2: 48. 1997; Cook, Aqu. Wet. Pl. Ind. 304. f. 313. 1996. *O. communissima* Lour. Fl. Cohin. 267. 1790. *O. glutinosa* Lour. Fl. Cohin. 267. 1790. *O. montana* Lour. Fl. Cohin. 267. 1790; Ham. in Wall. Cat. no. 8633. 1828. *O. praecox* Lour. Fl. Cohin. 267. 1790. *O. perennis* Much. Meth. 197. 1794. *O. palustris* Hamilt. Prodr. 25. 1796. *O. latifolia* P. Beauv. Agrost. 27. 1812. *O. parviflora* P. Beauv. Agrost. 27. 1812. *O. denudata* Desv. ex. Steud. Nomencl. ed. 1: 577. 1821. *O. elongata* Desv. ex. Steud. Nomencl. ed. 1: 577. 1821. *O. marginata* Desv. ex Steud. Nomencl. ed. 1: 577. 1821. *O. pubescens* Desv. ex Steud. Nomencl. ed. 1: 1821. *O. rubribaribis* Desv. ex Steud. Nomencl. ed. 1: 577. 1821. *O. emarginata* Steud. Nomencl. ed. 2: 234. 1841. *O. pumila* Hoschst. ex. Steud. Nomencl. ed. 2: 234. 1841. *O. rufipogon* Griff. Not. 3: 5. 1851. *O. glumacepatula* Hoschst. ex Steud. Syn. Pl. Glum. 1: 3. 1854. *O. nepalensis* Don. ex Steud. Syn. Pl. Glum. 1: 3. 1854. *O. repens* Herb. Ham. ex Steud. Syn. Pl. Glum. 1: 3. 1854. *O. segetalis* Russ. ex Steud. Syn. Pl. Gulm 1: 3. 1854. *O. sorghoides* Desv. ex Steud. Syn. Pl. Glum. 1: 3. 1854. *O. caudata* Trin. ex Doell. in Mart. Fl. Bras. 2: 8. 1871. (PLATE–100).

Annual. Rhizome absent; root fibrous. Culm 1.0 m tall, 5.0 mm, thick, green, glabrous, erect, densely tufted, compressed, ascending, decumbent, leafy, simple, rooting at node; nodes glabrous; internodes 10 cm long. Leaf–blade 40 cm long, 1.5 cm broad, green, flat, narrow, linear–lanceolate, acuminate, base rounded, surface rough, margin, hyaline, scabrid upwards, mid–vein distinct on lower surface, 5–lateral veins on each side of mid vein; lateral veins slender, distant, distinct on upper surface; leaf–sheath 11.0 cm long, 7.0 mm broad, green, coriaceous, glabrous, margin membranous, striate, compressed, loose, longer than internodes, veins distinct on

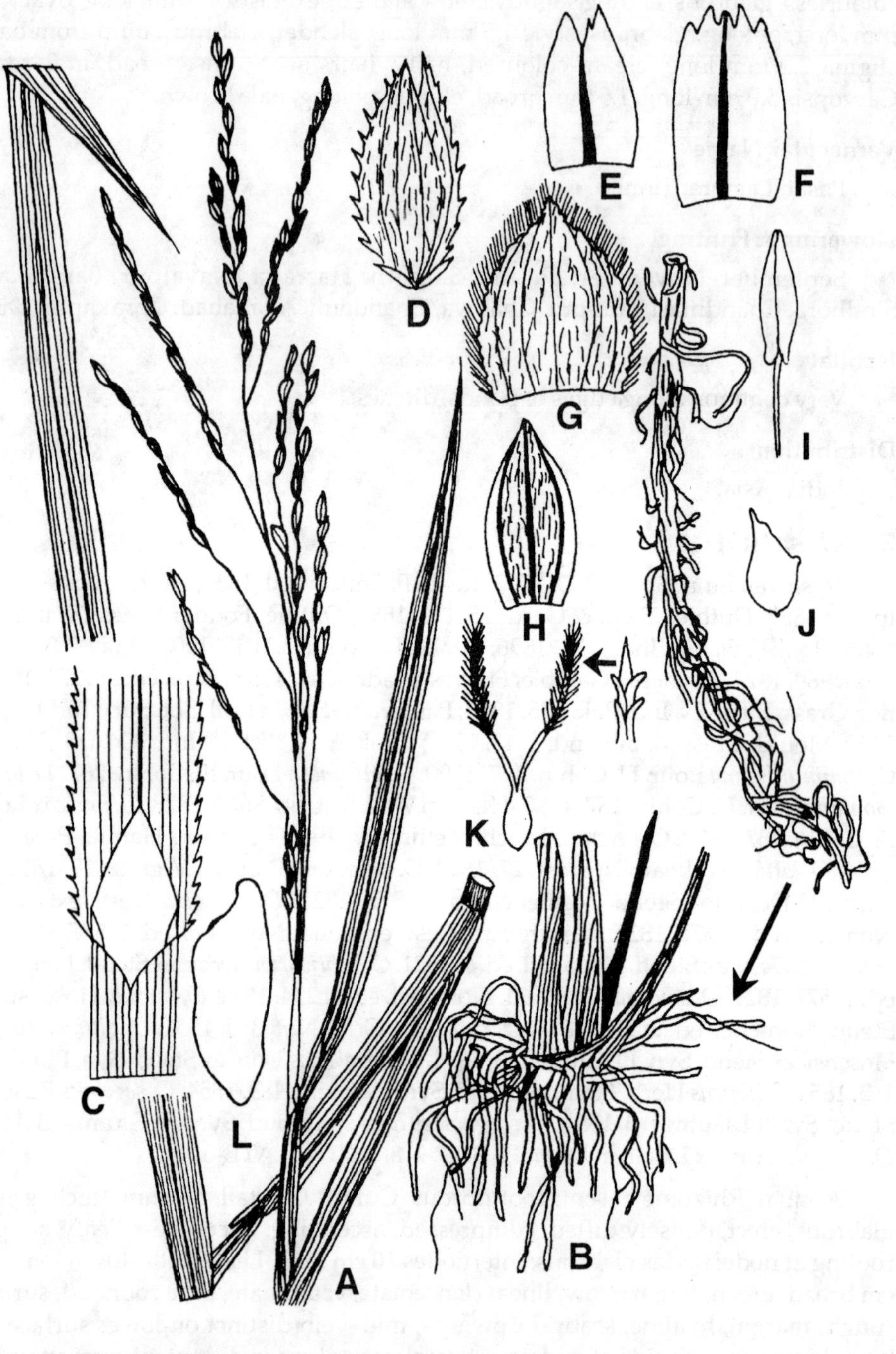

PLATE 100: *Oryza sativa* L.

A. Habit, B.Rhizome, C. Ligule, D. Spikelet, E. Lower glume, F. Upper glume, G. Lemma, H. Palea, I. Stamen, J. Lodicules, K. Pistil, L. Caryopsis

outer surface; ligule 1.5 cm long, membranous, ovate, acute; auricle glabrous. Inflorescence consists of panicle; panicle 20 cm long, 5.0 cm broad, shining, closely imbricating, spreading, compound, open, lax; rachis of panicle grooved, angular, slender. Branches 16 cm long, suberect, distant decreasing in upward, angular, scabrid. Spikelet 8.0 mm long, 2.0 mm broad, ovate, oblong, laterally compressed. subacute, awnless, yellowish. Lower glume 1.5 mm long, 1.0 mm broad, coriaceous, ovate, glabrous, bifid at apex, 1–nerved; nerves green, visible, glabrous, thicker at base, thin towards the apex, ending at apex. Upper glume 2.0 mm long, 1.0 mm broad, coriaceous, lip–like, glabrous, 5–lobed at apex, 1–nerved; nerves green, visible, glabrous, ending at apex, thicker towards the base, thiner towards at apex. Lemma 7.0 mm long, 4.0 mm broad, coriaceous, oblong, ovate, lanceolate, acuminate, laterally compressed, surface hairy; hairs 1.0 mm long, acute, base broad, margin slightly in rolled, hairy, scabrid at apex, keeled below; hairs 2.0 mm long, acute, base broad; 5–nerved; nerves green, slightly visible, glabrous, ending at apex. Palea 6.5 mm long, 2.5 mm broad, coriaceous, ovate, oblong, linear, acute, surface hairy between the nerves; hairs 1.0 mm long, acute, base broad, margin glabrous, 3–nerved; nerves green, visible, glabrous, ending at apex. Stamens 6; anther 2.5 mm long, yellow, oblong, linear, glabrous; filament 2.0 mm long, slender, glabrous, shorter than anther. Lodicules 2, 1.0 mm long, 0.5 mm broad, hyaline, glabrous, nerveless. Pistil 3.0 mm long; ovary 0.5 mm long, ovate, oblong, glabrous, green; style 1.0 mm long, slender, glabrous, bifid from the base; stigma 1.5 mm long, brown, hairy; hairs acute, base broad, in 2–rows. Caryopsis 6.0 mm long, 2.0 mm broad, white, linear, glabrous, apicular.

Vernacular Name

Chaur, Dhan.

Flowering & Fruiting

August–November; throughout study area.

Habitat

Cultivated crop.

Distribution

The whole world.

3. Sub-family–Chloridoideae

KEY TO THE TRIBES

1a. Spikelets 1–many flowered, slightly compressed; glume herbaceous; lateral nerves of lemmas close to the margin & hairy; palea narrowly winged on keels

.... **1. Cynodonteae**

1b. Spikelets 2–many flowered; strongly compressed, glumes membranous or coriaceous; lateral nerves of lemma distant to margin & glabrous; palea occasionally winged on keels

.... **2. Eragrostideae**

1. Tribe–Cynodonteae

KEY TO THE GENERA

1a. Lower leaf blade curved; spike recurved

.... 4. *Microchloa*

1b. Lower leaf–blade not curved; spike erect:

2a. Spikelets awned:

3a. Spike digitate:

4a. Spikes 1–3; lower lemma broadly cuneiform

.... 6. *Tetrapogon*

4b. Spikes 2–10; lower lemma broadly elliptic

.... 1. *Chloris*

3b. Spike not digitate:

5a. Rachis filiform, scaberulous; spikelets 2–flowered; empty glumes four, densely hairy; lemma & palea hairy

.... 3 *Melanocenchris*

5b. Rachis sub terete, glabrous; spikelets 1–flowered; empty glume two, glabrous; lemma & palea glabrous

.... 5. *Perotis*

2b. Spikelets awnless:

6a. Upper glume with rows of stout hooked spines along the nerves

.... 7. *Tragus*

6b. Upper glume not as above:

7a. Glumes membranous; lemma hairy; palea scabrid on keels, lodicules obovate, cuneate

.... 2. *Cynodon*

7b. Glumes cartilagenous; lemma glabrous; palea smooth on keels; lodicules absent

.... 8. *Zoysia*

1. *Chloris* Sw.

KEY TO THE SPECIES

1a. Spikes 3–6, spathiform; rachis pilose

.... 4. *C. montana*

1b. Spikes 4–17, open; rachis scabird:

2a. Empty lemmas 2–4:

3a. Culms 75 cm long; lower glume 1.0 mm long, awned; empty lemmas awned

.... 1. *C. barbata*

3b. Culm 1.5 m long; lower glume 1.5 mm long, awnless; empty lemmas awnless

.... 3. *C. gayana*

2b. Empty lemma solitary:

4a. Spikes spreading; spikelets 7.0 mm long; caryopsis 4.0 mm long, subtrigonous

.... 2. *C. dolichostachya*

4b. Spikes crowded; spikelets 4.0 mm long; caryopsis 1.5 mm long, trigonous

.... 5. *C. virgata*

1. *Chloris barbata*

Chloris barbata Sw. Fl. Ind. Occ. 1: 200. 1797; Kunth. Enum. Pl. 1: 264. 1838; Nees Agrost. Bras. 421. 1829; Steud. Syn. Gram. 204. 1854; Roxb. Fl. Ind. 1: 329. 1832; Grah. Cat. 234. 1839; Aitchis, Cat. Punj. Pl. 167. 1869; Duthie, Grass. N.W. Ind. 33. 1883; Duthie, Fodd. Grass. N. Ind. 53. t. 34. 1888; Lisboa, J. Bomb. Nat. Hist. Soc. 7: 367. 1893; Griseb. Fl. Brit. W. Ind. 539. 1864; Beth. Fl. Hongk. 429. 1861; Benth. Fl. Austr. 7: 613. 1878; Doell in Mart. Fl. Bras. 2, 3: 67. 1890; Hook. f. Fl. Brit. Ind. 7: 292. 1896; Prain, Beng. Pl. 1227. 1903; Watt. Dict. Econ. Prodr. 2: 269. 1903; Cooke, Fl. Bomb. 2: 1035. 1908; Achariyar, S. Ind. Grass. 264. 1924; Haines, Bot. Bih. & Or. 969. 1924; Fischer in Gamble, Fl. Pres. Madr. 1838. 1934 (Repr. ed. 3: 1272. 1957); Bor, Grass. Bur. Cey. Ind. Pak. 465. 1960; Bhattacharya, Bull. Bot. Sur. Ind. 6: 209. 1964; ChienChang. Hsu, Taiwan Grass. 365. t.78. 1975; Manilal &Sivaraj, Fl. Calic. 338. 1982; Roy, Grass. M.P. 50. 1984; Blatter & McCann, Bomb. Grass. 256. t. 171.1984; Sreekumar& Nair, Fl. Kera. Grass. 357. 1991; Moulik, Grass. Bamb. Ind. 2: 563. 1997. *Andropogon barbatus* Linn. Pl. Jam. Pugil. 30. 1792. (PLATE–101).

Perennial. Rhizomatous; rhizome 4.0 cm long, creeping; root fibrous. Culm 60 cm tall, 2.0 mm thick, green, glabrous, erect, tufted, compressed, slender, stout, geniculately ascending, leafy, rooting at lower nodes; branched; branches erect, quite smooth; nodes glabrous; internodes 7.0 cm long, shining. Leaf–blade 15 cm long, 6.0 mm broad, green, flat, folded in xeric condition, narrow, linear, lanceolate, acuminate, base rounded, surface glabrous, margin hyaline, scabrid, mid vein distinct on lower surface, 4–lateral veins on each side of mid–vein; lateral veins distinct on upper surface; leaf–sheath 4.5 cm long, 4.0 mm broad, green, surface membranous, glabrous, margin hyaline, keeled, loose, striate, compressed, very much shorter than internode, hairy at mouth; hairs 4.0 mm long, white, acute, base tubercled; veins distinct on both surface; ligule 0.5 mm long, membranous, narrow, truncate; auricle glabrous. Inflorescence composed of spike–like racemes; raceme 6.0 cm long, 1.5 mm broad, 6 in number, slender, erect, densely spiculate, digitately arranged in a truncate fascicle on a slender peduncle, purplish green; rachis slender, angular, scaberulous. Spikelets 3.0 mm long, 1.5 mm broad, subsessile, obovate, 2–flowered, biseriate, imbricate, 3–awned, lower floret perfect, upper imperfect, rachilla disarticulate above the glumes with a tuft of hairs at base; hairs 1.0 mm long, acute, base broad. Lower glume 1.5 mm long, 0.5 mm broad, hyaline, linear–lanceolate, acute, glabrous, 1–nerved; nerves green, slightly visible, scabrid, ending at apex, awned; awn 1.0 mm long, slender,

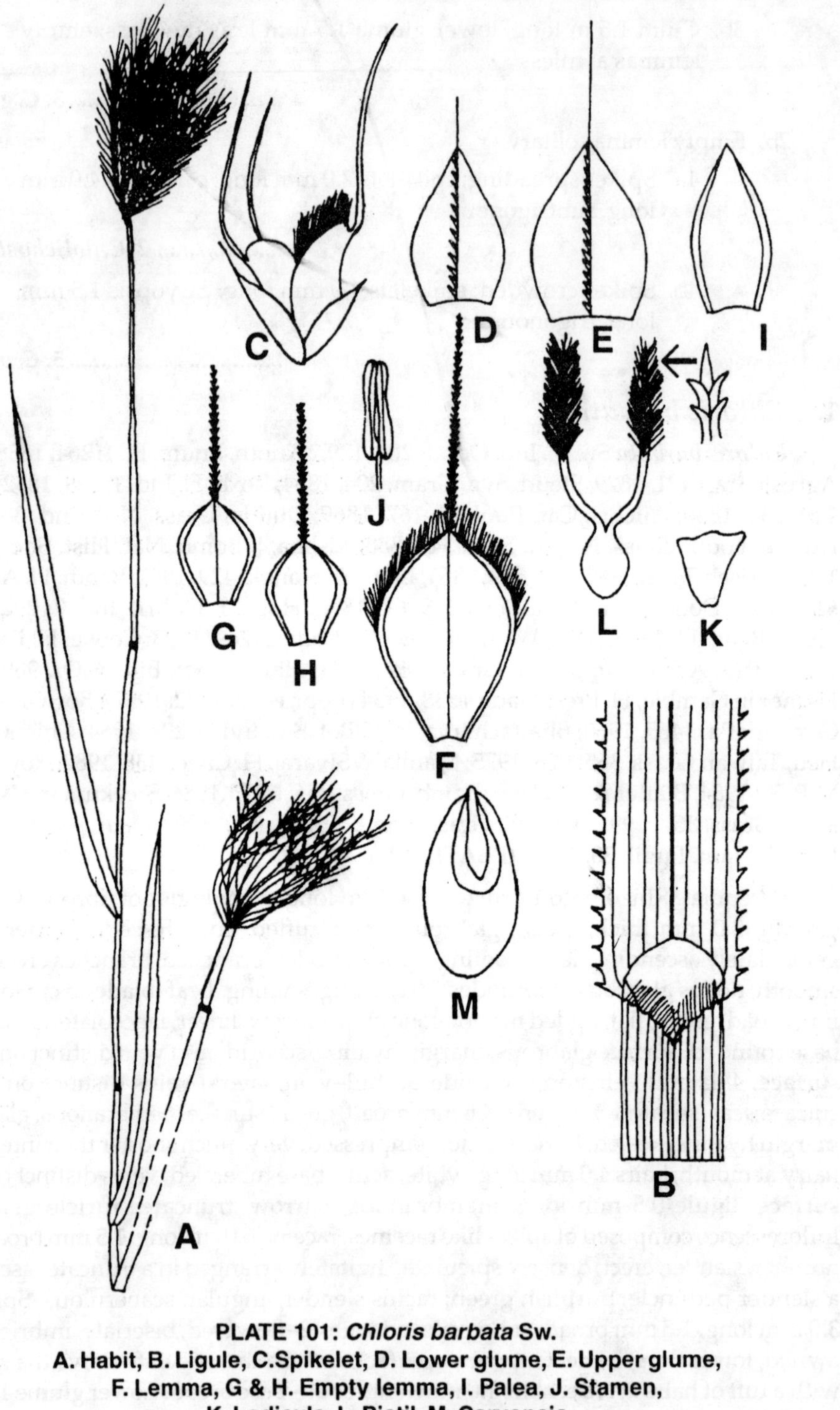

PLATE 101: ***Chloris barbata*** **Sw.**
A. Habit, B. Ligule, C.Spikelet, D. Lower glume, E. Upper glume, F. Lemma, G & H. Empty lemma, I. Palea, J. Stamen, K. Lodicule, L. Pistil, M. Caryopsis

glabrous. Upper glume 2.5 mm long, 1.0 mm broad, hyaline, linear–lanceolate, acute, glabrous, 1–nerved; nerves green, slightly visible, scabrid, ending at apex, awned; awn 4.0 mm long, slender glabrous. Lemma 3.0 mm long, 1.5 mm broad, membranous, broadly ovate, elliptic, shortly apiculate, surface smooth, margin hairy at apex, glabrous below; hairs 2.0 mm long, white, acute, base tubercled; purplish, bisexual, paleate, 3–nerved; nerves green, visible, glabrous, ending at apex, lateral nerves close to the margin, awned; awn 4.5 mm long, slender, antrorsely scabrid. Empty lemma 2, both 0.7 mm long, 0.3 mm broad, membranous, obovate, truncate, glabrous, emarginate on a stalk; stalk 0.5 mm long, epaleate, 3–nerved; nerves green, visible, glabrous, ending at apex, awned; awn 3.0 mm long, slender, glabrous. Palea 2.0 mm long, 1.0 mm broad, hyaline, elliptic, oblanceolate, glabrous, 2–keeled, 2–nerved; nerves green, slightly visible, glabrous & marginal, ending at apex. Stamens 3; anther 1.5 mm long, yellow, linear, glabrous; filament 0.5 mm long, slender, glabrous, shorter than anther. Lodicules 2, 0.3 mm long, 0.2 mm broad, hyaline, obovate, nerveless, glabrous, cuneate at base, wavy at apex. Pistil 2.0 mm long; ovary 0.5 mm long, elliptic, glabrous; style 0.5 mm long, slender, glabrous, bifid from base; stigma 1.0 mm long, hairy, purple; hairs acute, base broad, in 2–rows. Caryopsis 2.0 mm long, 0.5 mm broad, brownish, fusiform, trigonous, apiculate, base rugose.

Flowering & Fruiting

August–January; BHU, Navalpur, Rajghat, Sindhora, Babatpur, Harraua, Rameshwar, Chandauli, Chakiya, Khadda, Sonbhadra, Allahabad.

Habitat

Very common along the roadsides, disturbed places, bunds of cultivated fields, banks of wetlands.

Distribution

India, Tropical America, Asia.

2. *C. dolichostachya*

C. dolichostachya Lagasca, Gen. Sp. Pl. 5. 1816; Bor, Grass. Bur. Cey. Ind. Pak. 466. 1960; Bhattacharya, Bull. Bot. Sur. Ind. 6: 209. 1964; Panigrahi *et al.*, Bull. Bot. Sur. Ind. 11. 111. 1969; Cope in Nasir & Ali, Fl. W. Pak. 143: 120. 1982; Roy, Grass. M.P. 51. 1984; Sreekumar & Nair, Fl. Kera. Grass. 358. 1991. *C. incompleta* Roth. Nov. Pl. Sp. 60. 1821; Steud, Syn. Pl. Glum 207. 1855; Hook. f. Fl. Brit. Ind. 7: 290. 1896; Cooke, Fl. Bomb. 2: 1034. 1908; Rang. & Tad., Handb. S. Ind. Grass. 258. t. 196. 1921; Haines, Bot. Bih. & Or. 968. 1924; Fischer in Gamble, Fl. Pres. Madr. 1938. 1934. (Repr. ed. 3: 1272. 1957); Blatter & McCann, Bomb. Grass. 253. f. 168. 1984. *C. digitata* Steud. Syn. Pl. Glum. 210. 1855; Duthie, Grasses N.W. Ind. 23. 1883; Lisboa in J. Bomb. Nat. Hist. Soc. 7: 369. 1893. *C. radiata* Heyne ex Roth. Nov. Pl. Sp. 61. 1821. *C. roxburghii* Edgew. in J. As. Soc. Beng. 21: 160, 183. 1853; Duthie, Fodder Grass. N. Ind. 54. t. 65. 1885; Lisboa, J. Bomb. Nat. Hist. Soc. 7: 368. 1893. *C. tetrameris* Trin. Gram. Unif. 235. 1824; Steud. Syn. Pl. Glum. 206. 1855. *Digitaria elongata* Spreng. Syst. 1: 271. 1825. *Gymnopogon digitatus* Nees in Wight. Cat. no. 1753.1833; Steud. Nom. ed. 2, 1: 713. 1840. *Melica digitata* Roxb. Fl. Ind. 1: 326. 1820; Kunth. Enum. Pl. 1: 37. 1838. *Tenium digitatum* Spreng. Syst. 1: 274. 1825. *Cynodon elongatus* Trin. in Spreng. N. Entdeck. 2: 64. 1830. (PLATE–102).

Perennial. Rhizome absent; root fibrous. Culm 1.5 m tall, 2.0 mm thick, green, glabrous, erect, procumbent, willowy, tufted, slender, ascending, decumbent, leafy, rooting at lower nodes, branched; branches suberect, nodes glabrous, internodes 15 cm long. Leaf–blade 20 cm long, 7.0 mm broad, green, flat, glaucous, narrow, linear–lanceolate, acuminte, base rounded, upper surface sparsely hairy; hairs 3.0 mm long, white, acute, base tubercled, lower surface glabrous, margin hyaline, scabrid, mid–vein prominent on both surface, 4–lateral veins on each side of mid–vein; lateral veins distinct on upper surface; leaf–sheath 17 cm long, 5.0 mm broad, green, subcoriaceous, glabrous, margin submembranous, keeled, loose, compressed, striate, longer than internodes, veins distinct on outer surface, hairy at mouth; hairs 6.0 mm long, white, acute, base tubercled; ligule 1.5 mm long, a ring of white, dense, hairs; hairs acute, base broad; auricle absent. Inflorescence composed of spike like racemes; raceme 20 cm long, 1.5 mm broad, erect, digitate, densely spiculate, in terminal whorls; rachis filiform, scabrid. Spikelets 5.0 mm long, 1.5 mm broad, oblong, lanceolate, linear, narrowly fusiform, imbricating, erect, closely appressed, 2–flowered, biseriate, subsessile; callus 1.5 mm long, hairy; hairs minute, acute, base broad. Lower glume 1.5 mm long, 0.5 mm broad, membranous, linear–lanceolate, acute, glabrous, keeled, 1–nerved; nerves green, visible, scabrid, ending at apex. Upper glume 5.0 mm long, 1.0 mm broad, membranous, linear–lanceolate, acuminate, glabrous, keeled, 1–nerved; nerves green, visible, glabrous, ending at apex, awned; awn 1.0 mm long, slender, antrorsely scabrid. Lemma 4.0 mm long, 1.0 mm broad, membranous, oblong, linear–lanceolate, bifid at apex, glabrous between the nerves, bisexual, paleate, 3–nerved; nerves green, visible, lateral nerves scabrid at apex, glabrous below, ending at apex; central one glabrous, produced into a awn; awn 1.0 cm long, antrorsely scabrid, slender. Empty lemma 6.0 mm long, 1.5 mm broad, membranous, oblong, linear, lanceolate, bifid at apex, epaleate, 3–nerved; nerves green, visible, ending at apex, central one glabrous, lateral 2–nerves scabrid, awned; awn 6.0 mm long, slender, arising from the sinus of lobes, antrorsely scabrid. Palea 4.0 mm long, 1.0 mm broad, hyaline, delicate, oblong, linear, lanceolate, acuminate, glabrous, margin folded inwards, 2–keeled, 2–nerved; nerves green, visible, glabrous & marginal, ending at apex. Stamens 3; anther 1.0 mm long, yellow, oblong, linear, glabrous; filament 0.2 mm long, slender, glabrous, shorter than anther. Lodicule 2, 0.3 mm long, 0.1 mm broad, hyaline, glabrous, oblanceolate, cuneate at base, wavy at apex, nerveless. Pistil 2.0 mm long; ovary 0.5 mm long, elliptic, oblong, glabrous; style 0.5 mm long, slender, glabrous, bifid from base; stigma 1.0 mm long, colourless, hairs acute, base broad, in 2–rows. Caryopsis 4.0 mm long, 1.0 mm broad, brown, smooth, oblong, lanceolate, linear, subtrigonous.

Flowering & Fruiting

September–December; Chaukaghat, Rajghat, Chakiya, Rajdari, Sonbhadra, Khadda.

Habitat

Common along the shade of trees and hilly soils.

Distribution

India, Tropical Africa, America.

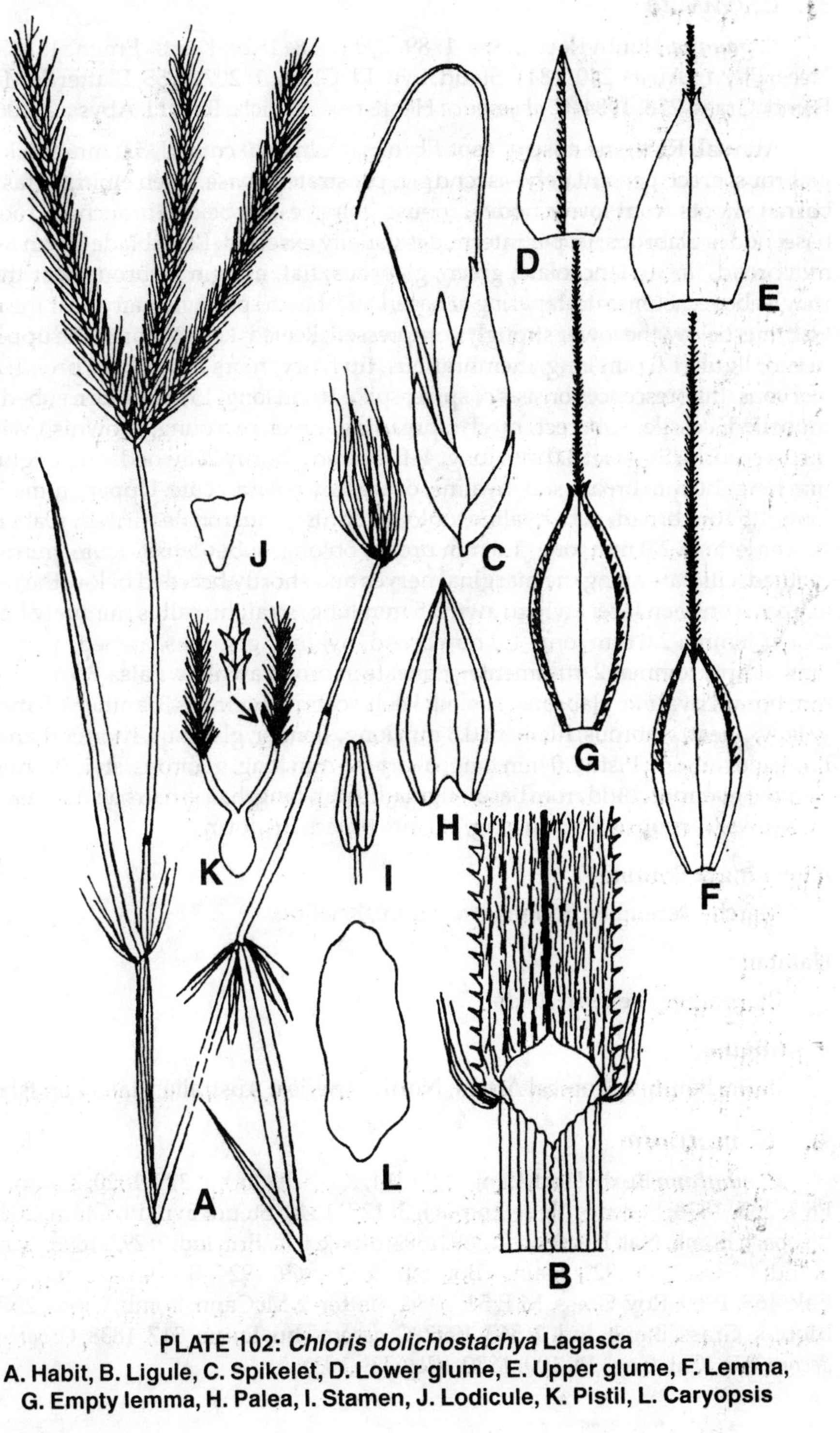

PLATE 102: *Chloris dolichostachya* Lagasca
A. Habit, B. Ligule, C. Spikelet, D. Lower glume, E. Upper glume, F. Lemma, G. Empty lemma, H. Palea, I. Stamen, J. Lodicule, K. Pistil, L. Caryopsis

3. *C. gayana*

C. gayana Kunth Rev. Gram. 1: 89, 293. t. 58. 1829; Kunth Enum. 1: 267. 1838; Nees, Fl. Afr. Austr. 240. 1841; Steud. Syn. Pl. Glum. 1: 207. 1855; Blatter & McCann, Bomb. Grass. 258. 1984. *C. abyssinica* Hochst ex. A. Rich. Tent. Fl. Abyss. 2: 406. 1851.

Annual. Rhizome absent; root fibrous. Culm 120 cm tall, 5.0 mm thick, green glabrous, erect, geniculately–ascending, prostrate at base, often emitting fasicles of barren shoots from lower nodes, robust, compressed below, branched, rooting at base; nodes glabrous; upper internodes usually exserted. Leaf–blade 30 cm long, 8.0 mm broad, linear, lanceolate, green, glabrous, flat, margin glabrous near the base, rough above, acuminate, tapering at base; leaf–sheath sparingly hairy near the mouth, glabrous below, the lower strongly compressed, keeled; keels scabrid, the upper most tumid; ligule 1.0 mm long, membranous, tip hairy; hairs acute, base broad; auricle glabrous. Inflorescence consists of spike; spike 10 cm long, 15 in number, sub–digitate, umbelled, sessile, suberect, rarely spreading, green at young, brownish with age; rachis scabrid. Spikelet 3.0 mm long, 4–flowered, shortly 2–awned. Lower glume 1.5 mm long, 1.0 mm broad, sub–hyaline, ovate, lanceolate, acute. Upper glume 3.0 mm long, 1.5 mm broad, sub–hyaline, oblong, obtuse, mucronate, firmer, scaberulous. Lower lemma 2.0 mm long, 1.0 mm broad, oblong, sub–obtuse, acute, minutely 2–toothed, ciliolate along the marginal nerves and shortly bearded below the tip, with a groove on each face, awned; awn 3.5 mm long, straight; callus minutely bearded. Upper lemma 2.0 mm long, 1.0 mm broad, hyaline, glabrous, awned; awn 2.0 mm long. Empty lemmas 2, rudimentry, cuneate in profit, awnless. Palea 1.8 mm long, 1.0 mm borad, hyaline, glabrous, keeled; keels scabrid. Stemens 3; anther 1.5 mm long, yellow, linear, glabrous; filament 0.3 mm long, slender, glabrous, shorter than anther. Lodicules absent. Pistil 2.0 mm long; ovary 0.5 mm long, glabrous; style 0.5 mm long, slender, glabrous, bifid from base; stigma 1.0 mm long, hairy; hairs acute, base broad, in 2–rows.Caryopsis 1.5 mm long, 1.0 mm broad, fusiform.

Flowering & Fruiting

March–December; Sonbhadra, Dudhi, Khadda.

Habitat

Rare, along the open fields.

Distribution

India, South & Tropical Africa, North America, Australia, Hawaiian Islands.

4. *C. montana*

C. montana Roxb. Hort. Beng. 82. 1814; Roxb. Fl. Ind. 1: 329. 1820; Kunth. Enum. Pl. 1: 235. 1838; Spreng, Neue Entdeck 3: 127. 1840; Steud. Syn. Pl. Glum. 204. 1855; Lisboa, J. Bomb. Nat. Hist. Soc. 7: 369. 1893; Hook. f. Fl. Brit. Ind. 7: 292. 1896; Achariyar, S. Ind. Grass. 270. 1921; Haines Bot. Bih. & Or. 969. 1924; Bor, Grass. Bur. Cey. Ind. Pak. 466. 1960; Roy, Grass. M.P. 51. 1984; Blatter & McCann, Bomb. Grass. 257. 1984; Moulik, Grass. Bamb. Ind. 2: 563. 1997. *C. decora* Thu. Enum. 317. 1838. *C. barbata* var. *decora* Trin. Cat. Ceyl. Pl. 109. 1880. (PLATE–103).

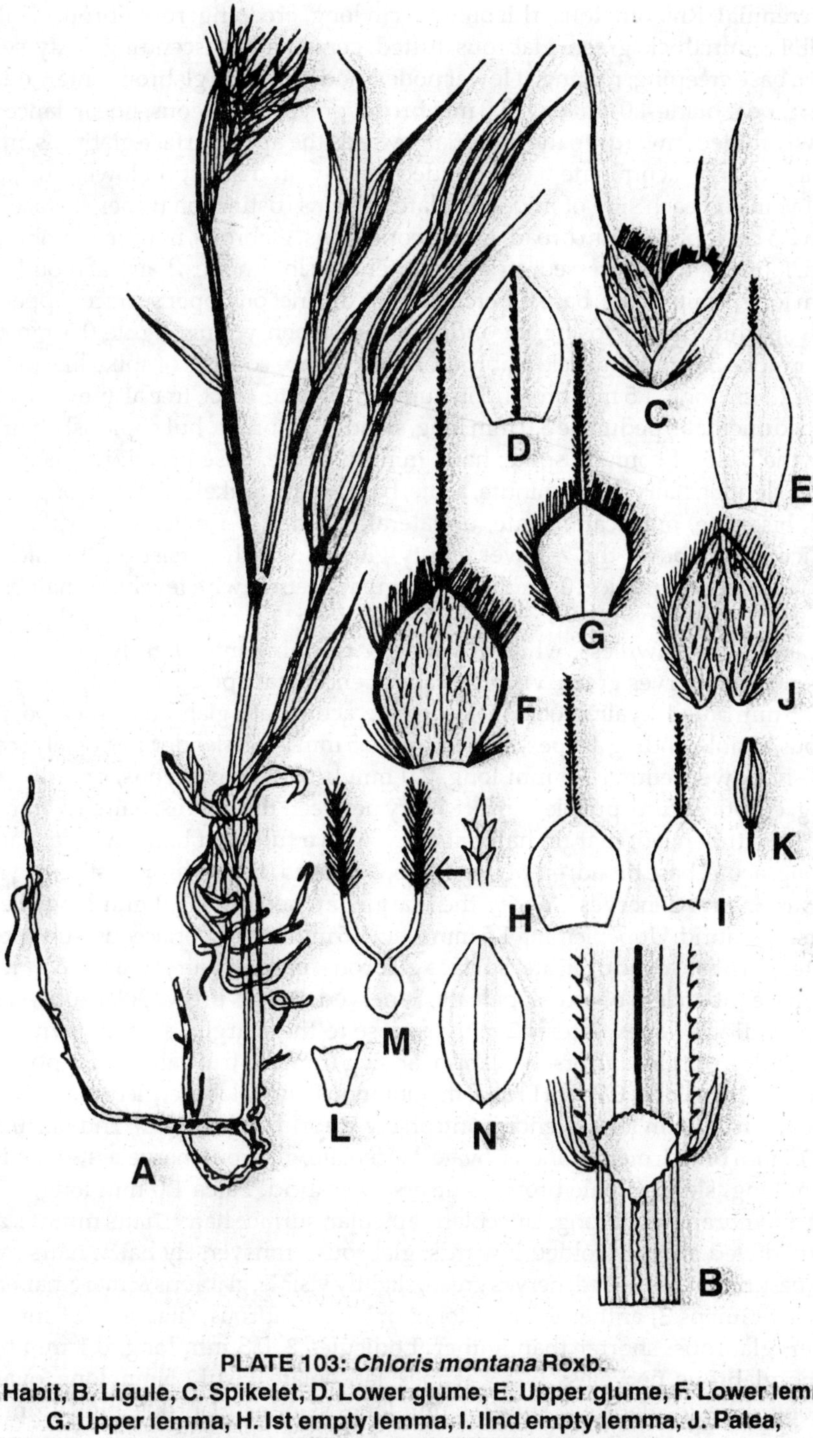

PLATE 103: ***Chloris montana*** **Roxb.**

A. Habit, B. Ligule, C. Spikelet, D. Lower glume, E. Upper glume, F. Lower lemma, G. Upper lemma, H. Ist empty lemma, I. IInd empty lemma, J. Palea, K. Stamen, L. Lodicule, M. Pistil, N. Caryopsis

Perennial. Rhizomatous; rhizome 1.0 cm long, creeping; root fibrous. Culm 20 cm tall, 1.5 mm thick, green, glabrous, tufted, geniculately ascending, leafy, simple, slender, base creeping, rooting at lower nodes; nodes brown, glabrous; internodes 3.5 cm long. Leaf blade 4.0 cm long, 3.0 mm broad, green, coriaceous, linear, lanceolate, narrowly folded inwards to the base, flat towards the apex, surface glabrous, margin hyaline, scabrid, acuminate base rounded, mid–vein distinct on lower surface; 2–lateral veins on each side of mid–vein; lateral veins distinct on upper surface; leaf–sheath 2.5 cm long, 3.0 mm broad, green, coriaceous, glabrous, margin membranous, keeled, loose, flat, compressed, striate, shorter than internodes, hairy at mouth; hairs 3.0 mm long, white acute, base tubercled, veins distinct on upper surface, upper most sheath spathiform, enclosing the inflorescence when young; ligule 0.5 mm long, membranous, lacerate; auricle glabrous. Inflorescence consists of spike like racemes; raceme 2.5 cm long, 1.5 mm broad, 3 in number, digitate, erect, free at apex, connate at base, pedunculate; peduncle 3.0 mm long, slender, glabrous, but copiously hairy just below the base of connate spike; hairs minute, acute, base broad; rachis angular, scabrid, slender, hairy; hairs minute, acute, base broad. Spikelet 3.0 mm long, 1.5 mm broad, biseriate, imbricate, ovate, unilateral, slender, 1–flowered, purple tinged, disarticulating above, the 2–lower empty glumes which persist on the rachis, 4–awned, pedicelled; pedicel 0.5 mm long, angular, scabrid with few hiars; hair 0.5 mm long, acute, base broad. Lower glume 2.0 mm long, 1.0 mm broad, hyaline, ovate, lanceolate, acute, awnless, white at young & turnning into slightly purplish with age, 1–nerved; nerves green, visible, scabrid, ending at apex. Upper glume 2.5 mm long, 1.5 mm broad, hyaline, oblong–lanceolate, acuminate, glabrous, 1–nerved; green, glabrous, visible, ending at apex, awned; awn 0.5 mm long, slender, antrorsely scabrid, purplish. Lower lemma 2.5 mm long, 2.0 mm broad, chartaceous, broadly ovate, oblong, elliptic, acute, purple, surface hairy between the nerves; hairs 1.5 mm long, acute, base tubercled; margin hairy at apex, with a tufted of hairs at base, hairs 2.0 mm long, acute, base broad; empty, epaleate, 3–nerved; nerves brown, slightly visible, glabrous, lateral 2–nerves close to the margin, awned; awn 7.0 mm long, slender, antrorsely scabrid. Upper lemma 1.5 mm long, 1.5 mm broad, coriaceous, sub–globose, cuniate, purplish, conduplicate, surface glabrous, margin hairy; hairs 1.0 mm long, acute, base tubercled, bisexual, paleate, 3–nerved; nerves brown, glabrous, slightly visible, ending at apex, lateral 2–nerves close to the margin, awned; awn 3.0 mm long, slender, purple, antrorsely scabrid, arising from the truncate tip. Empty lemma 2, 1st one 1.0 mm long, 0.7 mm broad, membranous, sub–globose, glabrous, nerveless, awned; awn 2.0 mm long, slender, antrorsely scabrid, pale brown; 2nd one 0.8 mm long, 0.4 mm broad, membranous, ovate, lanceolate, glabrous, nerveless, awned; awn 1.0 mm long, slender, pale brown, antrorsely scabrid. Palea 1.0 mm long, 0.5 mm broad, membranous, oblong, lanceolate, apicular, surface hairy; hairs minute, acute, base tubercled, margin folded inwards, glabrous, transversely hairs; hairs minute acute, base broad, 2–nerved; nerves green, slightly visible, glabrous & marginal, ending at apex. Stamens 3; anther 0.5 mm long, oblong, glabrous; filament 0.1 mm long, slender, glabrous, shorter than anther. Lodicules 2, 0.3 mm long, 0.1 mm broad, hyaline, glabrous, nerveless, wavy at apex, lanceolate. Pistil 1.5 mm long; ovary 0.2 mm long, elliptic, glabrous; style 0.6 mm long, slender, glabrous, bifid from base; stigma 0.7 mm long, brown, hairy; hairs acute, base broad, in 2–rows. Caryopsis 1.5 mm long, 0.5 mm broad, brown, trigonous, oblong, acuminate, shortly stipitate.

Flowering & Fruiting

August–January; Chaukhandi, Chandauli, Padrauna, Basti.

Habitat

Rare, along the open fields.

Distribution

India, Asia, Tropical America.

5. *C. virgata*

C. virgata Sw. Ind. Occ. 1: 203. 1797; Doell in Mart, Fl. Bras. 2: 3.1890; Hook. f. Fl. Brit. Ind. 7: 291. 1896; Achariyar, S. Ind. Grass. 260. 1921; Haines, Bot. Bih.&Or. 968. 1924; Bor, Grass. Bur. Cey. Ind. Pak. 468. 1960; Raizada& Jain, Indian For. 92: 639. 1966; Panigrahi&Rajagopal, J. Indian Bot. Soc. 47: 230. 1967; N.P. Singh, Bull. Bot. Sur. Ind. 11: 20. 1969; Chien–Chang Hsu, Taiwan Grass. 371. t. 81. 1975; Cope in Nasir& Ali, Fl. W. Pak. 143. 121. 1982; Raizada *et al.*, Grass. Upp. Gang. Pl. pt. 3: 24. 1983; Roy, Grass. M.P. 52. 1984; Blatter & McCann, Bomb. Grass. 255. t. 170. 1984. *C. compressa* DC Cat. Hort. Monsp. 94. 1813; Nees Agrost. Bras. 421. 1829; Nees, Fl. Afr. Austr. 240. 1841; Steud. Syn. Pl. Glum. 204. 1855. *C. caudata* Trin. ex Bunge Enum. Pl. 1: 70. 1838. *C. cryptostachya* Steud in Schmidt. Fl. Cap. 148. 1898. *C. decora* Nees ex Steud. Syn. Pl. Glum. 205. 1855. *C. elegans* Kunth. Enum. Pl. 1: 264. 1838. *C. meccana* Hochst & Steud. ex Schult. Ind. Sem. Hort. Hal. 7. 1843; Steud. Syn. Pl. Glum. 1: 265. 1838; Boiss, Fl. Or. 5: 554. 1881; Duthie, Grass. N.W. Ind. 33. 1883. *C. montana* Griseb in Goett. Nachr. 84. 1868; Duthie, Grass. N.W. Ind. 34: 1883. *C. pallida* Link. Hort. Berol. 1: 56. 1827; Haines, Bot. Bih. & Or. 967. 1924. *C. penicillata* Hort. ex Nees, Agrost. Bras. 423. 1829. *C. polydactyla* Durand. Diss. Chlorid. 14: 22. 1808. *Heterolepis elegans* Ehrt. ex. Boiss. Fl. Or. 5: 560. 1881. *Rabdochloa virgata* D. Beavu. Agrost. 84. 1 8 1 2 (PLATE–104).

Annual. Rhizome absent; root fibrous. Culm 70 cm tall, 2.0 mm thick, green, glabrous, erect, tufted, usually flattened, ascending, decumbent, leafy at base, creeping at base, rooting at lower nodes, profusely branched; branches sub erect, leafy upto inflorescence; nodes glabrous; internodes 12 cm long. Leaf–blade 15 cm long, 4.0 mm broad, green, flat, narrow, linear, lanceolate, acuminate, base rounded, young blade hairy on upper surface, hairs 3.0 mm long, white, acute, base tubercled, glabrous at lower surface, old ones glabrous on both surface, margin hyaline, scabrid at apex, keeled & hairy near throat; hairs 5.0 mm long, white, acute, base tubercled, midvein distinct on lower surface, 3–lateral veins on each side of mid–vein; lateral veins distinct on upper surface; leaf–sheath 7.0 cm long, 3.5 mm broad, green, glabrous, subcoriaceous, margin submembranous, slightly keeled, loose, striate, compressed, upper one somewhat inflated, very much shorter than internodes, veins distinct on outer surface; ligule 1.0 mm long, membranous, narrow, lacerate; auricle glabrous. Inflorescence made up of spike–like racemes; raceme 6.0 cm long, 1.5 mm broad, 6– in number, erect, digitate, dull green, crowded at the end of peduncle; rachis capillary, scabrid, angular, scaberulous on edges. Spikelets 3.0 mm long, 1.5 mm broad, subsessile, in 2–rows, 3–awned, 1–flowered. Lower glume 1.5 mm long, 1.0 mm

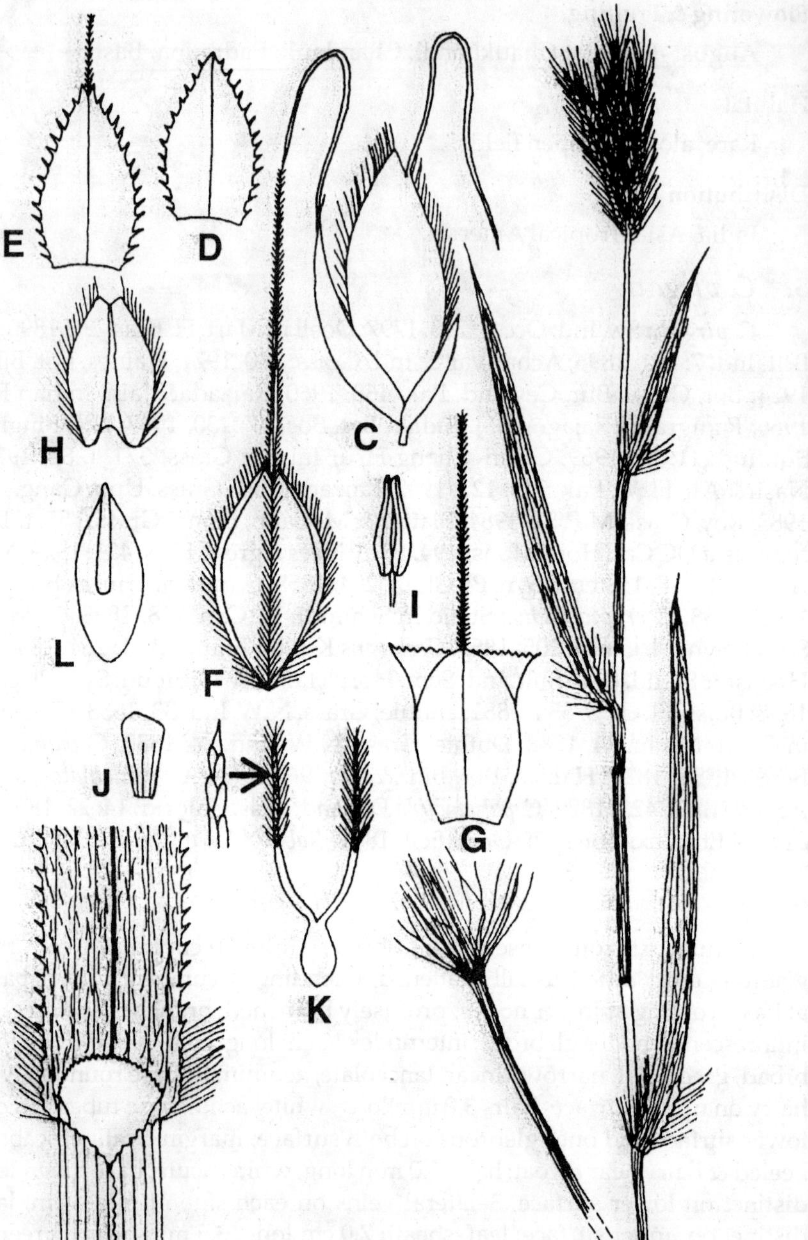

PLATE 104: ***Chloris virgata*** **Sw.**

A. Habit, B. Ligule, C. Spikelet, D. Lower glume, E. Upper glume, F. Lemma, G. Empty lemma, H. Palea, I. Stamen, J. Lodicule, K. Pistil, L. Caryopsis

broad, membranous, ovate, linear, lanceolate, acute, surface glabrous, margin scabrid, 1–nerved; nerves green, slightly visible, glabrous ending at apex. Upper glume 2.5 mm long, 1.5 mm broad, membranous, ovate, oblong, lanceolate, bifid at apex, surface glabrous, margin scabrid, 1–nerved; nerves green, slightly visible, glabrous, elongated into awn; awn 1.0 mm long, slender, antrorsely scabrid. Lemma 3.0 mm long, 1.5 mm broad, subcoriaceous, oblong, ovate, cymbiform, glabrous, bifid at apex, margin slightly minutely hairy upto about the middle & then closely long hairs almost to the tip; hairs 2.0 mm long, acute, base tubercled, 3–nerved; nerves green, ending at apex, visible, lateral nerves close to margin, glabrous, central nerve hairy; hairs 1.0 mm long, acute, base broad, bisexual, paleate, awned; awn 7.0 mm long, slender, arising from sinus of lobes, antrorsely scabrid. Empty lemma 1.8 mm long, 1.0 mm broad, membranous, oblanceolate, truncate, winged at apex; wing hyaline, scabrid; epaleate, glabrous, 1–nerved; nerves green, glabrous, slightly visible, ending at apex, awned; awn 5.0 mm long, slender, antrorsely scabrid. Palea 2.0 mm long, 1.0 mm broad, membranous, obovate, narrow, linear, notched at apex, nerveless, surface glabrous, margin folded inwards, transversely hairy; hairs 1.0 mm long, acute, base tubercled. Stamens 3; anther 0.5 mm long, yellow, oblong, glabrous; filament 0.2 mm long, slender, glabrous, shorter than anther. Lodicules 2; 0.3 mm long, 0.1 mm broad, hyaline, glabrous, 3–nerved; nerves colourless, glabrous, ending below the apex, oblanceolate. Pistil 1.5 mm long; ovary 0.3 mm long, oblong, glabrous; style 0.6 mm long, slender, bifid from base, glabrous; stigma 0.6 mm long, brown, hairy; hairs acute, base broad, in 2–rows. Caryopsis 1.0 mm long, 0.5 mm broad, pale–yellow, shining, fusiform, obovate, trigonous, slightly curved.

Vernacular Name

Kagarfagar, Sawum, Gauraka.

Flowering & Fruiting

August–December; Bhojubeer, Navalpur, Babatpur, Barahikala, Sindhora, Sarsauli, Sarnath, Chakiya, Chandauli, Jaunpur, Duddhi, Chunar, Gorakhpur.

Habitat

Common along the roadsides, old walls, gardens.

Distribution

India, Algeria, Taiwan, Mongolia, America, Tropical & South Africa.

2. *Cynodon* Rich

KEY TO THE SPECIES

1a. Leaf–blade 5–veined; leaf–sheath slightly compressed; palea smooth .... **3.*C. dactylon***

1b. Leaf–blade usually 7–veined; leaf–sheath slightly or strongly compressed; palea hairy or scabrid:

2a. Spikes erect; palea with clavate hairs .... **2. *C. barberi***

2b. Spikes ascending; palea scaberulous

1. *Cynodon arcuatus*

Cynodon arcuatus J.S. Presl. ex. C.B. Presl. Haenk. 1: 290. 1830; Bor, Grass. Bur. Cey. Ind. Pak. 469. 1960; Chase & Niles, Index. Grass. Suppl. 1: 516. 1962; Jain, Indian For. 92: 201. 1966; Jain, Kheti 19: 39. 1966; Jain, Bull. Bot. Surv. Ind. 8: 204. 1966; Jain, Bull. Bot. Surv. Ind. 9: 145. 1967; Roy, Grass M.P. 60. 1984; Sreekumar & Nair, Fl. Ker. Grass. 359. 1991. Moulik, Grass. Bamb. Ind. 2: 565. 1997. *C. dactylon* (L.) Pers. var. *intermedius* (Rang. et Tad.) Fischer in Gamble, Fl. Pres. Madr. 1835. 1934. (Repr. ed. 3: 1270. 1957); Rhind, Grass. Burma 36, 1945; Senaratna, Grass. Ceyl. 90. 1956; Chase & Niles, Index Grass Spp. 1: 517. 1962. *C. dactylon* (L.) Pers. var. *suberectus* Haines, Bot. Bih. & Or. 967. 1924; Chase & Niles, Index Grass Spp. 1: 517. 1962. *C. intermedius* Rang. et. Tad. in J. Bomb. Nat. Hist. Soc. 26: 304. 1918; Achariyar & Mudaliyar, S. Indian Grass. 252. 1921; Chase & Niles, Index Grass. Suppl. 1: 518. 1962. (PLATE–105).

Perennial. Stoloferous; stolons creeping, root fibrous. Culms 40 cm tall, 1.0 mm thick, green, glabrous, erect, tufted, compressed, ascending, geniculate, leafy throughout, branched; branches suberect, rooting at lower nodes; nodes brown, swollen; internodes 4.0 cm long. Leaf–blade 6.0 cm long, 4.0 mm broad, green, glabrous, flat, narrow, linear–lanceolate, acuminate, rounded at base, glaucous, alternate, margin hyaline & scabrid, mid–vein scabrid below, glabrous towards the apex, distinct on lower surface, 3–lateral veins on each side of mid–vein; lateral veins closed, distinct on upper surface; leaf–sheath 4.5 mm long, 2.0 mm broad, green, glabrous, membranous, margin hyaline, slightly keeled, striate, slightly compressed below, loose towards the apex, hairy at mouth; hairs 1.5 mm long, white, acute, base tubercled, longer than internodes, veins distinct on both surface; ligule 1.0 mm long, membranous, truncate, lacerate; auricle, glabrous. Inflorescence consists of spike like racemes; raceme 7.0 cm long, 1.0 mm broad, 4 in number, digitate, flexuous, slender, slightly drooping, erect, spreading with smooth peduncle; rachis slender, somewhat compressed, angular, scabrid along the margin, tumid and hairy at base; hairs 0.5 mm long, white, acute, base broad. Spikelets 2.0 mm long, 1.0 mm broad, elliptic, lanceolate, 1–flowered, narrow, awnless, sessile, arranged in two rows on the rachis, and appressed to it; rachilla produced to about half of the length of spikelet as a slender naked bristle behind the palea. Lower glume 1.0 mm long, 0.5 mm broad, membranous, lanceolate, acute, curved, glabrous, 1–nerved; nerves green, visible, glabrous, ending at apex. Upper glume 1.5 mm long, 0.5 mm broad, membranous, linear–lanceolate, acute, glabrous, 1–nerved; nerves green, visible, glabrous, ending at apex. Lemma 2.0 mm long, 1.5 mm broad, chartaceous, broadly ovate, obliquely oblong, obtuse when spread, boat shaped, acute at compressed, surface glabrous, margin keeled & hairy; hairs 2.0 mm long, clavate; 3–nerved; nerves green, visible, ending at apex, central nerves hairy; hairs 1.0 mm long, acute, base tubercled, lateral nerves glabrous. Palea 1.8 mm long, 1.0 mm broad, membranous, oblong, linear, lanceolate, 2–keeled; keels scaberulous, 2–nerved; nerves green, slightly visible, glabrous & marginal, ending at apex. Stamens 3; anther 1.0 mm long, yellow, linear, oblong, glabrous; filament 0.1 mm long, slender, glabrous, shorter than anther.

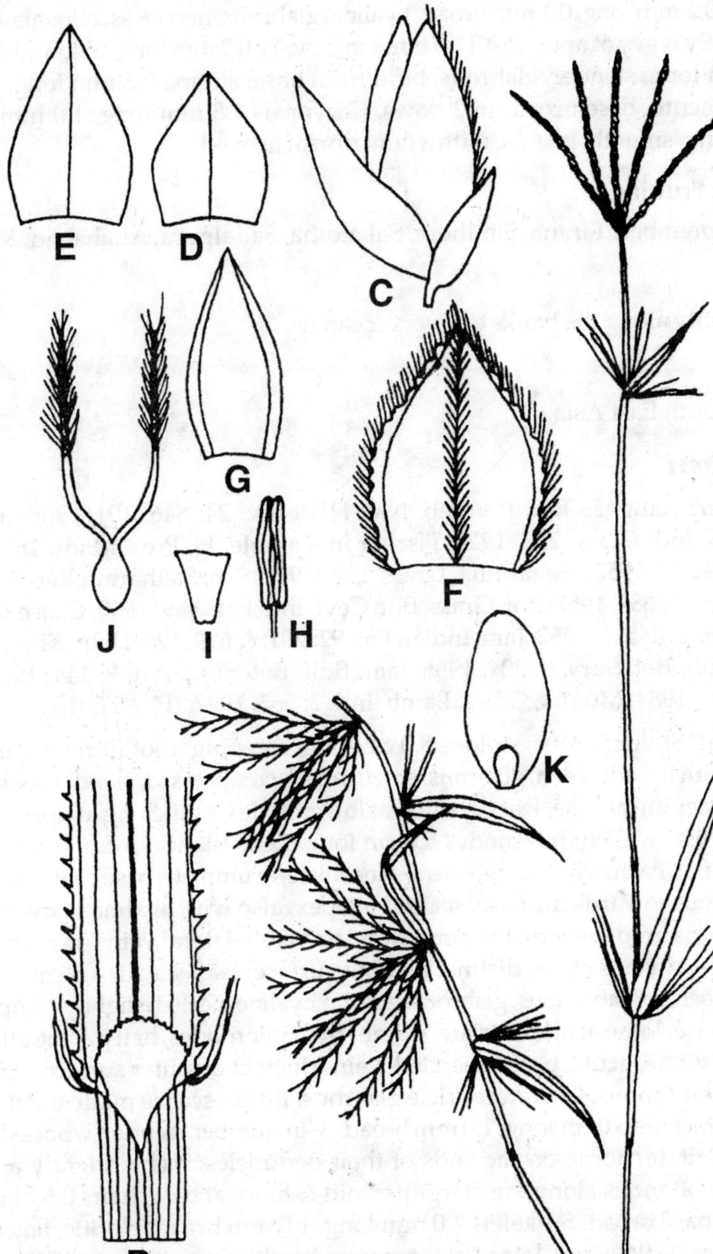

PLATE 105: *Cynodon arcuatus* **J.S. Presl**
A. Habit, B. Ligule, C. Spikelet, D. Lower glume, E. Upper glume, F. Lemma, G. Palea, H. Stamen, I. Lodicule, J. Pistil, K. Caryopsis

Lodicules 2, 0.2 mm long, 0.1 mm broad, hyaline, glabrous, nerveless, obovate, cuneate at base, slightly wavy at apex. Pistil 1.5 mm long; ovary 0.2 mm long, oblong, glabrous; style 0.5 mm long, slender, glabrous, bifid from base; stigma 0.8 mm long, purple, hairy; hairs acute, base broad, in 2–rows. Caryopsis 1.5 mm long, 1.0 mm broad, elliptic, oblong, smooth, brown, with a faint dorsal groved.

Flowering & Fruiting

July–September; Tarana, Sindhora, Sakaldiha, Sadalpura, Allahabad, Khadda.

Habitat

Commonly along the banks of rivers & canals.

Distribution

India, South East Asia.

2. *C. barberi*

C. barberi Rang. & Tad. J. Bomb. Nat. Hist. Soc. 24: 846. 1916; Achariyar & Mudaliyar, S. Ind. Grass. 255. 1921; Fischer in Gamble, Fl. Pres. Madr. 1835. 1934 (Repr. ed. 3: 1270. 1957); Senaratna, Grass. Ceyl. 91. 1956; Sakharam Rao, J. Bomb. Nat. Hist. Soc. 54: 687. 1957; Bor, Grass. Bur. Ceyl. Ind. Pak. 469. 1960; Chase & Niles, Index Grass Spp. 1: 516. 1962; Jain, Indian For. 92: 201, & 699. 1966; Jain, Kheti. 19: 39. 1966; Jain, Bull. Bot. Surv. 8: 204. 1966; Jain, Bull. Bot. Surv. Ind. 9: 144. 1967; Roy, Grass. M.P. 60. 1984. Moulik, Grass. Bamb. Ind. 2: 565. 1997. (PLATE–106).

Perennial. Stoloniferous; stolons 8.0 cm long, creeping; root fibrous. Culms 30 cm long, 1.0 mm broad, green, glabrous, erect, tufted, compressed, slender, ascending, geniculate, creeping at base, leafy, rooting at lower nodes, branching at upper nodes; nodes glabrous, swollen; internodes 5.0 cm long. Leaf–blade 3.5 cm long, 4.0 mm broad, green, flat, narrowly oblong, linear–lanceolate, acuminate, base rounded, upper surface glabrous towards the base, scabrid at apex, also with hyaline margin, lower surface glabrous, mid–vein distinct on lower surface, 3–lateral veins on each side of mid–vein; lateral veins close, distinct on both surface; leaf–sheath 4.5 cm long, 2.0 mm broad, green, membranous, glabrous, margin hyaline, keeled, slightly compressed towards the base, loose at apex, striate, shorter than internodes, hairy at mouth; hairs 2.0 mm long, white, acute, base tubercled; veins distinct on outer surface; ligule 1.0 mm long, membranous, lacerate; auricle glabrous. Inflorescence made up of spike–like racemes; raceme 6.0 cm long, 1.0 mm broad, 4–in number, slender, whorled, erect, spreading, digitate, borne on the ends of their peduncles; rachis slightly angular, slender, scaberulouous along the margin, tumid & hairy at base; hairs 0.5 mm long, white, acute, base broad. Spikelets 2.0 mm long, 1.0 mm broad, elliptic, lanceolate, sessile, awnless; 1–flowered, laterally compressed, imbricate, alternately biseriate on ventral side of rachis & appressed to it; rhachilla about half of length of the spikelet disarticulating above the glumes & produced beyond into a bristle behind the palea. Lower glume 1.2 mm long, 0.6 mm broad, membranous, narrowly ovate, lanceolate, acute, glabrous, keeled, with a strong nerve; nerve green, visible, scabrid, ending at apex. Upper glume 2.0 mm long, 1.0 mm broad, membranous, ovate, lanceolate, acute, surface glabrous, margin scabrid at apex, glabrous below, 1–nerved; nerves green,

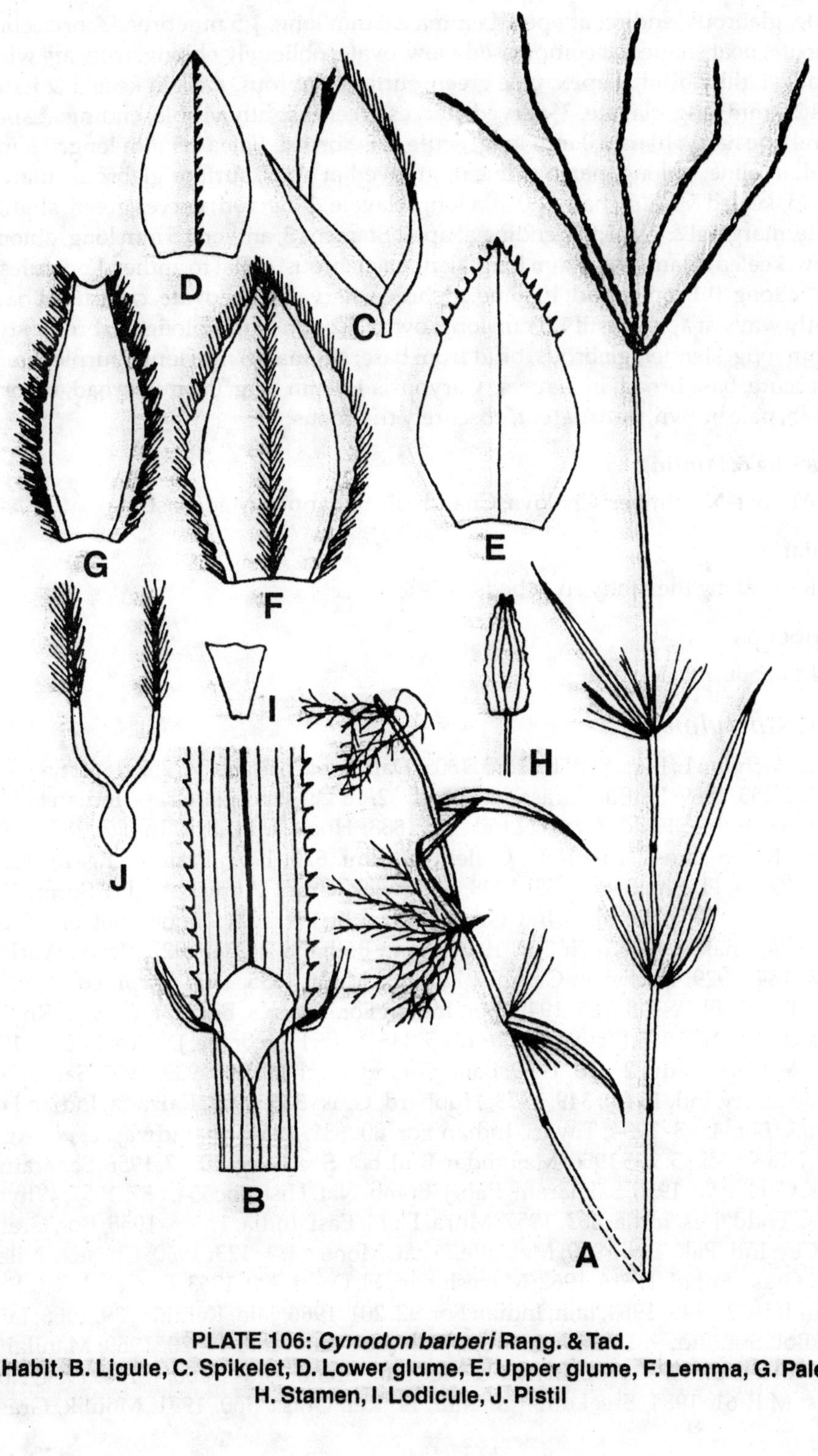

PLATE 106: ***Cynodon barberi*** **Rang. & Tad.**
A. Habit, B. Ligule, C. Spikelet, D. Lower glume, E. Upper glume, F. Lemma, G. Palea, H. Stamen, I. Lodicule, J. Pistil

visible, glabrous, ending at apex. Lemma 2.0 mm long, 1.5 mm broad, coriaceous, subacute, boat shaped at compressed view, ovate, obliquely oblong, truncate when spread, slightly bifid at apex, pale green, surface glabrous, margin keeled & hairy; hairs 2.0 mm long, clavate, 3–nerved; nerves green, slightly visible, ending at apex, central one hairy; hairs 1.0 mm long, acute, base broad. Palea 1.5 mm long, 0.5 mm broad, hyaline, oblong, narrow, linear, grooved at apex, surface glabrous, margin scabrid, keeled & hairy; hairs 1.0 mm long, clavate, 2–nerved; nerves green, slightly visible, marginal & glabrous, ending at apex. Stamens 3; anther 0.5 mm long, oblong, yellow, keeled; filament 0.5 mm long, slender, glabrous, equal to anther. Lodicules 2, 0.2 mm long, 0.1 mm broad, hyaline, glabrous, nerveless, obovate, cuneate at base, slightly wavy at apex. Pistil 1.5 mm long, ovary 0.2 mm long, oblong, glabrous; style 0.5 mm long, slender, glabrous, bifid from base; stigma 0.8 mm long, purple, hairy; hairs acute, base broad, in 2–rows. Caryopsis 1.3 mm long, 0.5 mm broad, oblong, smooth, pale brown, transparent, obscurely trigonous.

Flowering & Fruiting

August–November; Chakiya, Chandauli, Sultanpur, Mau, Basti.

Habitat

Rare, along the sandy river beds.

Distribution

India, Burma, Malaya.

3. *C. dactylon*

C. dactylon L. Pers. Syn. Pl. 1: 85. 1805; Dalz & Gibs, Bomb. Fl. 297. 1861; Stewart, Punj. Pl. 253. 1869; Duthie, Grass. N.W. Ind. 32. 1883; Symonds, Grass. Indian Penin. 12. 1884; Duthie, Fodd. Gras. N.Ind. 52. 1888; Hook. f. Fl. Brit. Ind. 7: 280. 1896; Lisboa, Bomb. Grass. 104. 1896; Collett, Fl. Siml. 620. 1902; Prain, Beng. Pl. 1227. 1903; Cooke, Fl. Bomb. 2: 1032. 1908; Sedgwick, Grass. Ahmedabad & Surat. 116. 1914; Rang. & Tad., Handb. S. Ind. Grass. 250. t. 190–191. 1921; Haines, Bot. Bih. & Or. 966. 1924; Thakar, Pl. Kutch 293. 1926; Stent, Bothalia 2: 274. 1927; Bews, Worlds, Grass. 184, 1929; Fischer in Gamble, Fl. Pres. Madr. 1835. 1934 (Repr. ed. 3: 1270. 1957); Bor. in Fl. Ass. 5: 125. 1940; Bor, Indian For. Rec. n.s. Bot. 2: 110. 1941; Rhind, Grass. Burma 36. 1945; Stewart, Brittonia 5: 446. 1945; Hurcombe, J.S. Afr. Bot. 13: 107. 1947; Wealth of India 2: 420. 1950; Bailey, Stand. Cycl. Hort. 2: 939. 1953; Santapau, Rec. Bot. Surv. Ind. 16 (1): 349. 1953; Hubbard, Grass 335. 1954; Raizada, Indian For. Rec. n.s. Bot. 4: 93. 1954; Tiwari, Indian For. 80: 681. 1954; Bharadwaja *et al.*, Agra Univ. J. Res. (Sci.) 5: 295 1956; Majumdar, Bull. bot. Soc. Beng. 10: 37. 1956; Senaratna, Grass. Ceylon 90. 1956; Sakharam, Rao, J. Bomb. Nat. Hist. Soc. 54: 687. 1957; Whyte, Grassl. Fodd. Res. India. 363. 1957; Mitra. Fl. Pl. East. India. 1: 166. 1958; Bor, Grass. Bur. Cey. Ind. Pak. 469. 1960; Metcalfe, Anat. Monocot. 1: 123. 1960; Chase & Niles, Index Grass. Suppl. 1: 516. 1962; Maheshwari, Fl. Delhi, 390. 1963; Puri et al; Rec. Bot. Sur. Ind. 19 (1): 149. 1964; Jain, Indian For. 92: 201. 1966; Jain, Kehti 19: 39. 1966; Jain, Bull. Bot. Sur. Ind. 9: 139. 1967; Singh, Bull. Bot. Surv. Ind. 11: 20. 1969; Manilal & Sivaraj, Fl. Calic. 339. 1982; Blatter & McCann, Bomb. Grass. 250. t. 166. 1984; Roy, Grass, M.P. 61. 1984; Sreekumar & Nair, Fl. Ker. Grass. 360. 1991; Moulik, Grass.

Bamb. Ind. 2: 566. 1997. *C. erectum* Presl. Reliq. Haenk. 1: 290. 1830. *C. filiformis* Voigt Hort. Calc. 712. 1829. *C. linearis* Willd. Enum. Hort. Berol. 1: 90. 1827. *C. maritimus* H.B. & K. Gen. et. Sp. 1: 170. 1816. *C. radiatus* Roxb. Nov. Pl. Sp. 38. 1821. *C. tenuis* Trin. Spreg. Neue. Entd. 2: 63. 1821. *C. oxidentalis* Willd. ex Steud. Nom. Bot. ed. 2, 1: 463. 1840. *C. portoricensis* Willd. ex Steud. Nom. Bot. ed. 2, 1: 463. 1840. *Chloris cynodon* Trin. Gram. Uniff. 229. 1824. *C. maritima* Trin. Gram. Uniff. 226. 1824. *Fibichia umbellata* Koel. Gram. Gall et. Gram. 308. 1802. *Digitaria dactylon* Scop. Fl. Carn. Ed. 2. 1: 52. 1772. *D. litoralis* Salisb. Prodr. 19. 1796. *D. stolonifera* Schard. Fl. Germ. 1: 165. 1806. *D. maritima* (H.B.K.) Spreng. Syst. Veg. 1: 272. 1825. *Panicum dactylon* Linn. Sp. Pl. 58. 1753; Grah. Cat. Bomb. Pl. 236. 1829; Roxb. Fl. Ind. 2: 289. 1832. *P. lineare* Burm. Fl. Ind. 25. t. 10. fig. 2. 1768. *Paspalum dactylon* DC. Fl. Franc. 3: 16. 1740; Lamk. Tab. Encyl. Meth. Bot. 1: 179. 1791. *Agrostis burnmudiana* Tussac ex Kunth. Enum. Pl. 1: 259. 1833. *A. filiformis* Koen ex Kunth. Enum. Pl. 1: 261. 1833. *A. linearis* Retz. Obs. Bot. 4: 19. 1786. *Milium dactylon* L. Moench., Meth. Pl. Suppl. 67.1802. *Capriola dactylon*. (L.) O. Kuntze. Rev. Gen. Pl. 2: 764. 1891. (PLATE–107).

Perennail. Rhizomatous; rhizome 20 cm long, creeping, scaly, stoloniferous, forming tufts, matted; root fibrous. Culm 30 cm long, 1.0 mm thick, green, glabrous, tufted, slender, decumbent, ascending, geniculate, simple, leafy, rooting at lower nodes; nodes glabrous; internodes 3.0 cm long. Leaf–blade 5.0 cm long, 3.0 mm broad, green, flat, narrow, convolute, glaucous, acuminate, base rounded, linear–lanceolate, lower surface glabrous, upper surface scabrid also with hyaline margin, often conspicuously distichous on lower parts of culm & barren shoots, mid–vein distinct on lower surface, 2–lateral nerves on each side of mid–vein; lateral veins distinct on both surface; leaf–sheath 3.5 cm long, 1.5 mm broad, green, membranous, glabrous, margin hyaline & slightly keeled, striate, tight, compressed, longer than internodes, hairy on mouth; hairs 2.0 mm long, white, acute, base tubercled; veins distinct on both surface; ligule 0.5 mm long, a ring of white hairs; hairs acute base broad; auricle absent. Inflorescence consists of spikes like recemes; receme 6.0 cm long, 1.0 mm wide, 6–in number, digitate, green at young, turning into purplish with age, erect, spreading, smooth, radiating from the top of a slender, smooth penducle; rachis slender, compressed, angular, flat, scabrid along the margin, tumid and hairy at base; hairs 0.5 mm long, white, acute, base broad. Spikelets 3.0 mm long, 1.5 mm broad, oblong, elliptic, lanceolate, laterally compressed, 1–flowered, awnless, sessile in two rows on one sides of the rachis and appressed to it; rachilla 1.5 mm long, very slender, disarticulating above the glumes and produced beyond the palea as a slender naked bristle. Lower glume 1.5 mm long, 1.0 mm broad, membranous, narrowly oblong, lanceolate, acute, surface glabrous, margin scabrid & keeled; keel glabrous 1–nerved; nerves green, visible, glabrous, ending at apex. Upper glume 2.0 mm long, 1.0 mm broad, membranous, narrowly oblong, lanceolate, acute, surface glabrous, margin scabrid & keeled; keel glabrous, 1–nerved; nerves green, visible, glabrous, ending at apex. Lemma 3.0 mm long, 2.0 mm broad, membranous, boat–shaped, ovate, obliquely oblong, strongly compressed, acute, & scabrid at apx, surface glabrous, margin keeled & hairy; hairs 2.0 mm long, white, acute, base broad, 3–nerved; nerves glabrous, ending at apex, green, central one obscure, but lateral two nerves slightly visible & closed to the margin. Palea 2.0 mm long, 1.0 mm broad, membranous, boat–shaped, oblong, linear, obtuse, 2–keeled; keels scaberulous, 2–nerved; nerves green, slightly visible, glabrous & marginal,

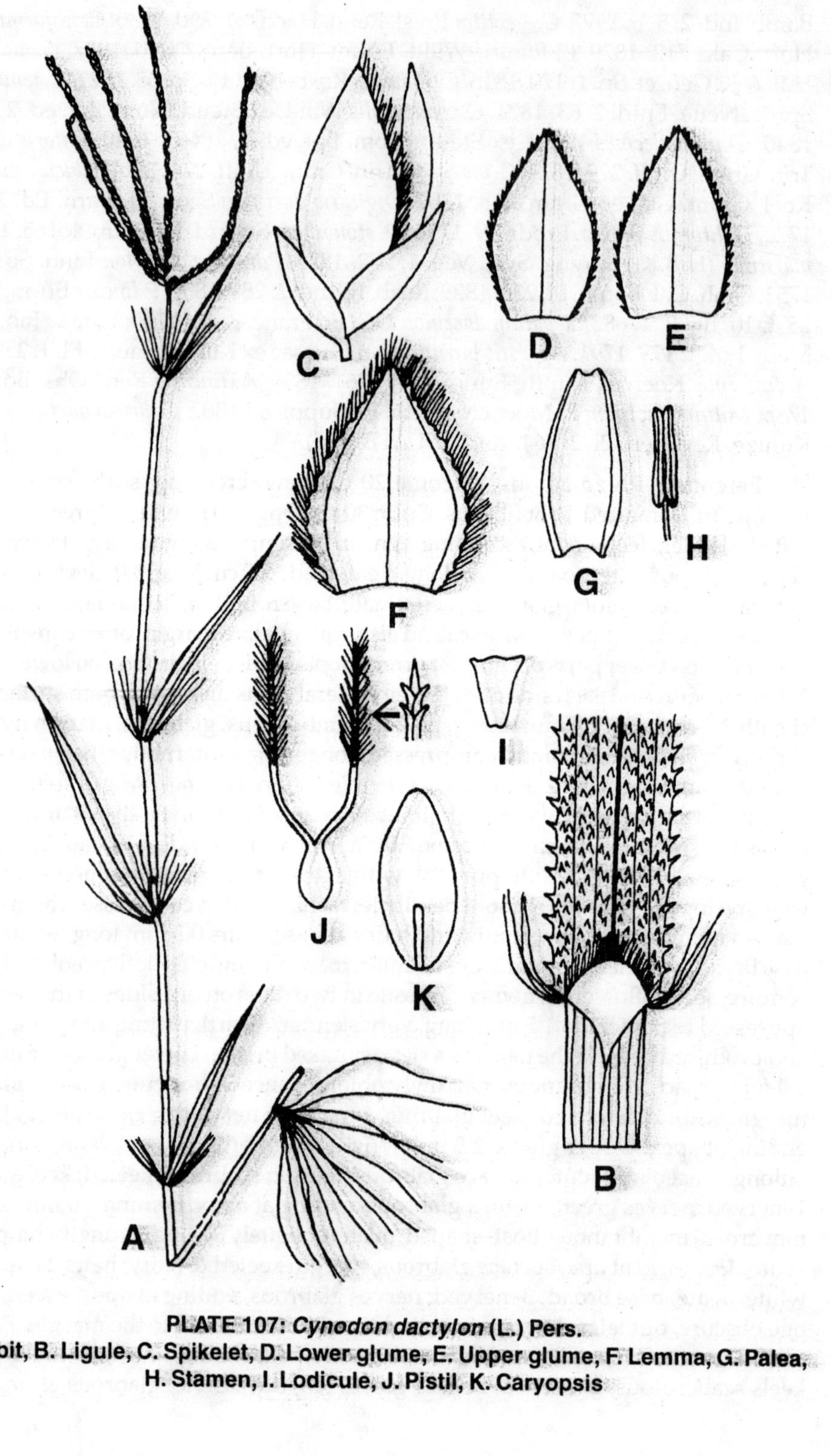

PLATE 107: ***Cynodon dactylon* (L.) Pers.**
A. Habit, B. Ligule, C. Spikelet, D. Lower glume, E. Upper glume, F. Lemma, G. Palea, H. Stamen, I. Lodicule, J. Pistil, K. Caryopsis

ending at apex. Stamens 3; anther 1.0 mm long, yellow, oblong, linear, glabrous; filament 0.1 mm long, slender, glabrous, shorter than anther. Lodicules 2, 0.2 mm long, 0.1 mm broad, hyaline, glabrous, nerveless, obovate, cuniate at abse, slightly wavy at apex. Pistil 1.7 mm long; ovary 0.2 mm long, oblong, glabrous; style 0.5 mm long, slender, glabrous, bifid from base; stigma 1.0 mm long, purple hairy; hairs acute, base broad, in 2–rows. Caryopsis 1.0 mm long, 0.5 mm broad, oblong, smooth, pale brown.

Vernacular Name

Dub, Durba.

Flowering & Fruiting

Winter season; Throughout the study area.

Habitat

Very common along the roadside, gardens, cultivated fields, banks of rivers & other moist localities.

Distribution

C. dactylon is a cosmopolitan species mostly in tropical & temperate region.

3. Melanocenchris Nees.

Melanocenchris jacquemontii Jaub. & Spach, 3: Pl. Or. 4: 36. t. 325. 1851; Nair & Nayar, Bull. Bot. Sur. Ind. 16: 142. 1974; Bor, Grass. Bur. Cey. Ind. Pak. 473. 1960; Roy, Grass. M.P. 110. 1984. *M. royleana* Nees in Proc. Linn. Soc. 1: 95. 1841; Aitchis. Cat. Punjab. Pl. 168. 1869; Duthie, Grass. N.W. Ind. 33. 1883; Duthie, Fodder Grass N. Ind. 54. t. 67. 1888; Lisboa, J. Bom. Nat. Hist. Soc. 7: 370. 1893; Moulik, Grass. Bamb. Ind. 2: 574. 1997. *Gracilea royleana* Hook. f. Fl. Brit. Ind. 7: 284. 1896; Prain, Beng. Pl. 1226. 1903; Cooke, Fl. Bomb. 2: 1031. 1908; Haines, Bot. Bih. & Or. 965. 1924; Blatter & McCann, Bomb. Grass. 248. t. 165. 1984. *Pommereulla royleana* Steud. Nom. ed. 2, 2: 379. 1840. *Pennisetum plumosum* Hochst ex Steud. Syn. Gram. 201. 1854. *Ptilonema plumosa* Steud. Syn. Gram. 202. 1854. *Eutriana abyssinica* R.Br. ex Fresen. in Mus. Senkenb. 2: 142. 1837. *Melanocenchris plumosa* Jaub & Spach. 3, Pl. Or. 4: 37. 1851; Hochst in Flora. 273. 1855. (PLATE–108).

Annual. Rhizome absent; root fibrous. Culm 7.0 cm long, 1.0 mm broad, green, glabrous, erect, densely tufted, very slender, decumbend, leafy at base, rooting at base, branching at lower nodes; branches suberect, leafy up to inflorescence, nodes glabrous; internodes 1.5 mm long. Leaf–blade 4.0 cm long, 2.0 mm broad, green, flat, narrow, linear, lanceolate, sharply acute, base rounded, upper surface sparsely hairy; hairs 2.0 mm long, white, acute, base tubercled, lower surface glabrous, margin hyaline, scabrid, keeled & hairy; hairs 2.0 mm long, acute, base tubercled; mid–vein distinct on lower surface, 2–lateral veins on each side of mid–vein; lateral veins close, distinct on upper surface; leaf- sheath1.0 cm long, 2.0 mm broad, green, membranous, glabrous, margin hyaline, keeled, loose, striate, compressed, shorter than internodes, veins, distinct on both surface; ligule 0.2 mm long, a ring of white, fringed hairs; hairs acute, base broad; auricle absent. Inflorescence made up of spike; spike 1.5 cm long, spicate, filiform, flexuous, often curved, rarely erect; rachis trigonous, angular, smooth.

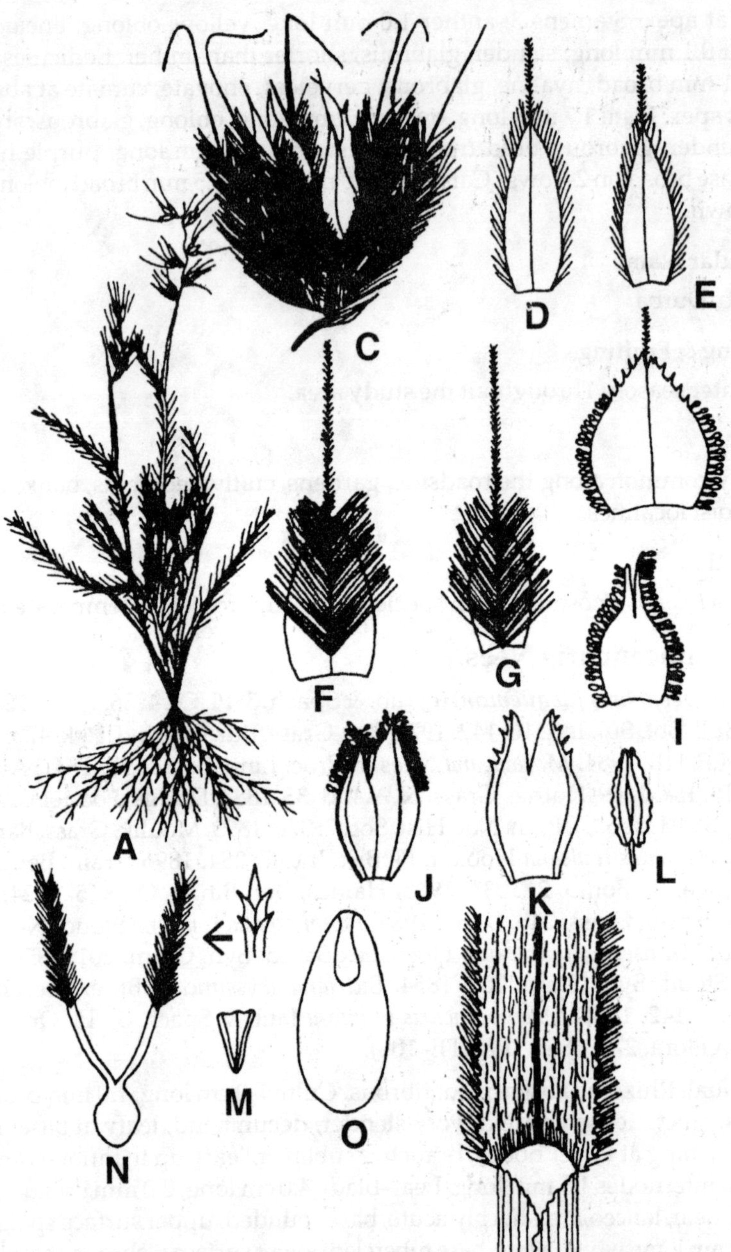

PLATE 108: ***Melanocenchris jacquemontii*** **Jaub. & Spach.**
A. Habit, B. Ligule, C. Spikelet, D. Lower glume, E. II Lower glume, F. I Upper glume, G. II Upper glume, H. Lower lemma, I. Lower palea, J. Upper lemma, K. Upper palea, L. Stamen, M. Lodicule, N. Pistil, O. Caryopsis

Spikelets 6.0 mm long, 3.0 mm broad, cluster of 3, green at young, turning into purplish with age, sessile, awned, densely hairy. Lower glume 2, 1st one 4.0 mm long, 0.5 mm broad, coriaceous, linear, narrow, acute, surface glabrous, margin hairy; hairs 5.0 mm long, white, acute, base tubercled; 1–nerved; nerve green, slightly visible, glabrous, ending at apex, awned; awn 2.0 mm long, slender, antrorsely scabrid. 2nd one 3.5 mm long, 0.3 mm broad, coriaceous, linear, narrow, acute, surface glabrous, margin hairy; hairs 3.0 mm long, white, acute, base tubercled, 1–nerved; nerve green, slightly visible, glabrous, ending at apex, awned; awn 1.0 mm long, slender, antrorsely scabrid. Upper glume 2, 1st one 3.0 mm long, 1.0 mm broad, coriaceous, linear, narrow, glabrous, 1–nerved; nerve green, slightly visible, ending at apex, glabrous towards the base, densely hairy above; hairs 5.0 mm long, white, acute, base tubercled, awned; awn 4.0 mm long, slender, antrorsely scabrid. 2nd one 2.0 mm long, 0.5 mm broad, coriaceous, linear, narrow, acuminate, glabrous, 1–nerved; nerve green, slightly visible, ending at apex, densely hairy; hairs 5.0 mm long, white, acute, base tubercled; awned; awn 3.5 mm long, slender, antrorsely scabrid. Lower lemma 3.0 mm long, 1.5 mm broad, subcoriaceous, broadly ovate, lanceolate, surface glabrous, margin scabrid at apex, hairy below; hairs 1.0 mm long, clavate, bifid at apex, empty, paleate, 3–nerved; nerves green, glabrous, ending at apex, central one visible, lateral nerves slightly visible, awned; awn 1.0 mm long, slender, antrorsely scabrid, arising from sinus of lobes. Lower palea 2.5 mm long, 1.5 mm broad, membranous, oblong, nerveless, surface glabrous, margin hairy; hairs 0.5 mm long; clavate, bifid at apex; one lobes acute, long, second one apiculate, short. Upper lemma 2.0 mm long, 1.0 mm broad, membranous, ovate, elliptic, oblong, surface glabrous, margin hairy at apex; hairs 1.0 mm long; clavate, glabrous below, 3–lobed at apex, bisexual, paleate, awnless 3–nerved; nerves green, visible, glabrous, ending at tip of lobes. Upper palea 2.0 mm long, 1.0 mm broad, membranous, obovate, oblanceolate, surface glabrous, margin scabrid at apex, glabrous below, bilobed at apex, lobes acute, equal, 2–nerved; nerves green, slightly visible, ending at apex,glabrous & marginal. Stamens 3; anther 1.0 mm long, yellow, linear, oblong, keeled; filament 0.5 mm long, slender, glabrous, shorter than anther. Lodicule 2, 0.2 mm long, 0.1 mm broad, hyaline, glabrous, oblanceolate, cuneate at base, wavy at apex, 3–nerved; nerves colourless, glabrous, ending at apex. Pistil 1.5 mm long; ovary 0.2 mm long, obovate, glabrous; style 0.5 mm long, slender, glabrous, bifid at base; stigma 0.8 mm long, purple, hairy; hairs acute, base broad, in 2–rows. Caryopsis 1.5 mm long, 0.5 mm broad, linear, oblong, polished, grey–brown.

Vernacular Names

Guli, Beradi, Dongri.

Flowering & Fruiting

September–November; Chakiya, Sonbhadra, Dudhi, Sonaghati.

Habitat

Rare, along the hills.

Distribution

India, Socotra, Nubia, Arabia, Abyssinia.

4. *Microchloa* R. Br.

Microchloa indica (L.F.) P. Beauv. Ess. Agrost. 20. 1812; O. Kuntze Rev. 3: 356. 1893; Hackel in Fedde. Report. Nov. Sp. 7: 373. 1909; Merril in Philipp. J. Soc. Bot. 7: 74. 1912; Druce in Rep. Bot. Exch. Cl. Brit. Isles 635. 1916; Fischer in Gamble, Fl. Pres. Madr. 1830. 1934; Blatter & McCann, Bomb. Grass. 248. t. 164. 1984; Bor, Grass. Bur. Cey. Ind. Pak. 473. 1960; Saldana & Nicolson, Fl. Hassan 741. 1976; Moulik, Grass. Bamb. Ind. 2: 95. 1997. *M. setacea* R. Br. Prodr. 208. 1810; H.B. & K. Nov. Gen. et Sp. 1: 84. t. 22. 1816; P. Beauv. Agrost. 115. t. 20. fig. 8. 1812; Nees Agrost. Bras. 441. 1829; Nees, Fl. Afr. Austr. 247. 1841; Kunth Enum. Pl. 1: 258. 1838; Doell in Mart. Fl. Brass. 2, 3: 76. t. 21. 1850; Steud. Syn. Pl. Glum. 202. 1855; Benth. Fl. Hongk. 428. 1861; Benth. Fl. Austr. 7: 608. 1878; Hook. f. Fl. Brit. Ind. 7: 283. 1896; Cooke, Fl. Bomb. 2: 1031. 1908; Prain, Beng. Pl. 1226. 1903; Haines, Bot. Bih. & Or. 964.1924. *Nardus indica* Linn. f. Suppl. 105. 1781. *Rottboellia setacea* Roxb. Fl. Ind. 1: 357. 1820. (PLATE–109).

Perennial. Rhizome absent; root fibrous. Culm 15 cm tall, 1.0 mm thick, green, glabrous, densely tufted, filiform, erect, compressed below, densely leafy at base; rooting at base, branching at lower node; branches suberect; nodes glabrous; internodes 2.5 cm long. Leaf–blade 3.0 cm long, 1.0 mm broad, green, surface, glabrous & membranous, margin hyaline & scabrid, narrow, linear, plicate, acute, base rounded, setaceous, lower blade often curved, mid vein distinct on lower surface, 1–lateral veins on each side of mid vein; lateral veins distinct on both surface; leaf–sheath 3.0 cm long, 1.5 mm broad, green, surface membranous & glabrous also with hyaline margin, compressed, loose, erect, longer than internodes; ligule 0.5 mm long, thinly membranous, ridge; auricle absent. Inflorescence consistsof spike; spike 7.5 cm long, solitary, very slender, usually curved, often purplish; rachis jointed at base & pubescent among the spikelets. Spikelets 2.5 mm long, 1–flowered, secund, erect, glabrous, unilateral, crowded on the flattened rachis, dorsally compressed. Lower glume 2.0 mm long, 1.0 mm broad, membranous, lanceolate, oblong, acute, glabrous, flattened on the back, 1–nerved; nerves strongly visible, glabrous, ending at apex. Upper glume 2.5 mm long, 1.5 mm broad, membranous, lanceolate, oblong, acute, glabrous, flattened on the back, 1–nerved; nerve strongly visible, glabrous, ending at apex. Lemma 1.5 mm long, 1.0 mm broad, hyaline, delicate, abruptly acuminate, oblong, ovate, glabrous, 3–nerved; nerves slightly visible, ending at apex, hairy; hairs minute, white, acute, base tubercled. Palea 2.0 mm long, 1.0 mm broad, hyaline, delicate, bifid, oblong, glabrous, 2–keeled, 2–nerved; nerves slightly visible, ending at apex, marginal & hairy; hairs minute, white, acute, base tubercled. Stamens 3; anther 1.0 mm long, yellow, linear, glabrous; filament 0.3 mm long, slender, glabrous, shorter than anther. Lodicules 2; 0.3 mm long, 0.1 mm broad, hyaline, glabrous, truncate, cuneate, nerveless. Pistil 2.0 mm long; ovary 0.5 mm long; oblong, glabrous; style 0.5 mm long, slender, glabrous, bifid from base; stigma 1.0 mm long, hairy; hairs acute, base broad. Caryopsis 1.2 mm long, 0.7 mm broad, oblong, ovate, subfusiform, smooth, free with in the glumes.

Flowering & Fruiting

August–September; Naugarh, Mirzapu, Allahabad, Bhatpar.

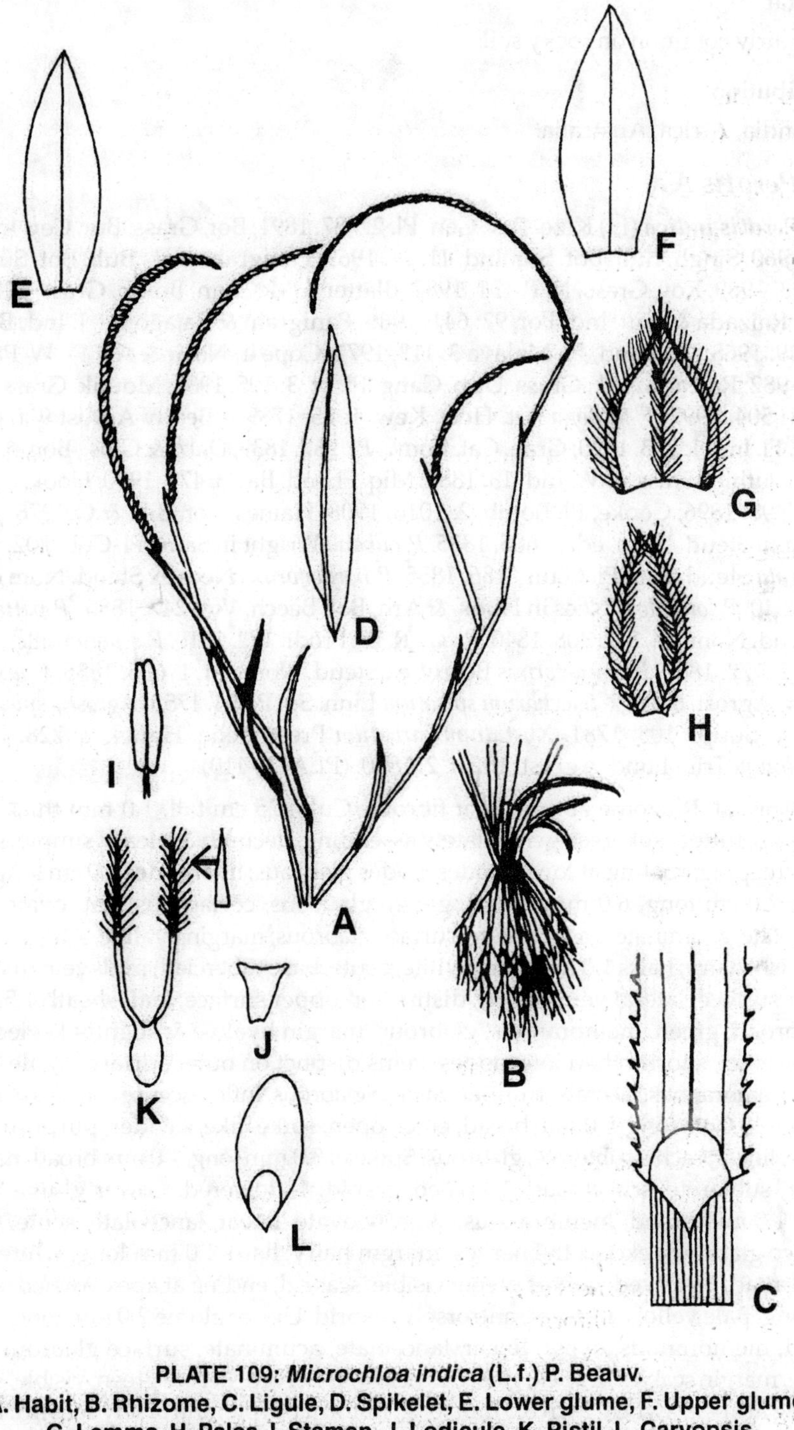

PLATE 109: *Microchloa indica* (L.f.) P. Beauv.
A. Habit, B. Rhizome, C. Ligule, D. Spikelet, E. Lower glume, F. Upper glume, G. Lemma, H. Palea, I. Stamen, J. Lodicule, K. Pistil, L. Caryopsis

Habitat

Fairly common on rocky soil.

Distribution

India, Africa, Australia.

5. *Perotis* Ait.

Perotis indica (L.) Ktze. Rev. Gen. Pl. 2: 787. 1891; Bor, Grass. Bur. Cey. Ind. Pak. 611. 1960; Singh, Bull. Bot. Sur. Ind. 11: 21. 1969; Panigrahi *et al.*, Bull. Bot. Surv. Ind. 11: 113. 1969; Roy, Grass. M.P. 131. 1984; Blatter & McCann, Bomb. Grass. 220 t. 147. 1984; Raizada & Jain, Ind. For. 92: 641. 1966; Panigrahi & Rajagopal, J. Ind. Bot. Soc. 47: 239. 1968; Gilliland, Fl. Malaya 3: 113. 1971; Cope in Nasir & Ali, Fl. W. Pak. 143: 148. 1982; Raizada *et al.*, Grass. Upp. Gang. Pl. pt. 3: 105. 1983; Moulik, Grass. Bamb. Ind. 1: 504. 1997. *P. latifolia* Ait. Hort. Kew. 1: 85. 1789; P.Beauv Agrost 9. t. 4. 1812; Roxb. Fl. Ind. 1: 233. 1820; Grah. Cat. Bomb. Pl. 237. 1839; Dalz. & Gibs., Bomb. Fl. 296. 1861; Duthie, Grass. N.W. Ind. 13. 1883; Miq. Fl. Ind. Bat. 3: 479. 1900; Hook. f. Fl. Brit. Ind. 7: 98. 1896; Cooke, Fl. Bomb. 2: 1016. 1908; Haines, Bot. Bih. & Or. 978. 1924. *P. burmanii* Steud. Nom. ed. 1: 605. 1835. *P. cubana* Wright in Sauv. Fl. Cub. 202. 1873. *P. glabarata* Steud. Syn. Pl. Glum. 186. 1855. *P. hordeiformis* Nees ex Steud. Nom ed. 2, 2: 306. 1840. *P. longiflora* Nees in Hook. & Arn. Bot. Beech. Voy. 247. 1880. *P. patula* Nees ex Steud. Nom. ed. 2, 2: 308. 1840. *P. rara* R. Br. Prodr. 172. 1810. *P. scabra* Willd. ex Trin. Diss. 2: 172. 1830. *P. spicaeformis* Beauv. ex Steud. Nom. ed. 1: 605. 1835. *P. laxifolia* P. Beauv. Agrost. 6: 1812. *Saccharum spicatum* Linn. Sp. Pl. 54. 1753. *Agrostis spicaeformis* Linn. F. Suppl. 108. 1781. *Xystidium barbatum* Prest. Reliq. Hanek. 1: 228. 1830. *X. maritimum* Trin. Fund. Agrost. 102. t. 2. 1820. (PLATE–110).

Annual. Rhizome absent; root fibrous. Culm 25 cm tall, 1.0 mm thick, green, glabrous, tufted, suberect, geniculately ascending, decumbent, leafy, simple, slender, base creeping, rooting at lower nodes; nodes glabrous; internodes 2.0 cm long. Leaf-blade 2.0 cm long, 6.0 mm broad, green, glaucous, coriaceous, flat, ovate, linear, lanceolate, acuminate, base cordate, surface glabrous, margin hyaline, scabrid, keeled, purplish, hairy; hairs 1.0 mm long, white, acute, base tubercled, mid–vein distinct on lower surface, lateral veins many, distinct on upper surface; leaf–sheath 1.5 cm, 1.2 mm broad, green, membranous, glabrous, margin hyaline & slightly keeled, tight, terete, erect, shorter than internodes, veins distinct on outer surface; ligule 0.2 mm long, membranous, lacerate, truncate; auricle glabrous. Inflorescence consists of raceme; raceme 15.0 cm long, 5.0 mm broad, erect, open, spike–like, slender, purplish–white, dense, linear; rachis subterete, glabrous. Spikelet 3.0 mm long, 1.0 mm broad, narrowly linear, subterete, scaberlous, 2–awned, hispid, 1–flowered. Lower glume 3.0 mm long, 1.0 mm broad, membranous, slightly ovate, linear, lanceolate, acute, purple, surface glabrous, except the nerves, margin hairy; hairs 1.0 mm long, white, acute, base broad, 1–nerved; nerves green, visible, scabrid, ending at apex, awned; awn 1.0 cm long, pale yellow, filiform, antrorsely scabrid. Upper glume 2.0 mm long, 1.0 mm broad, membranous, ovate, linear, lanceolate, acuminate, surface glabrous except nerve, margin scabird at apex, glabrous below, 1–nerved; nerves green, visible, scabrid, ending at apex, awned; awn 8.0 mm long, filiform purple, antrorsely scabrid. Lemma

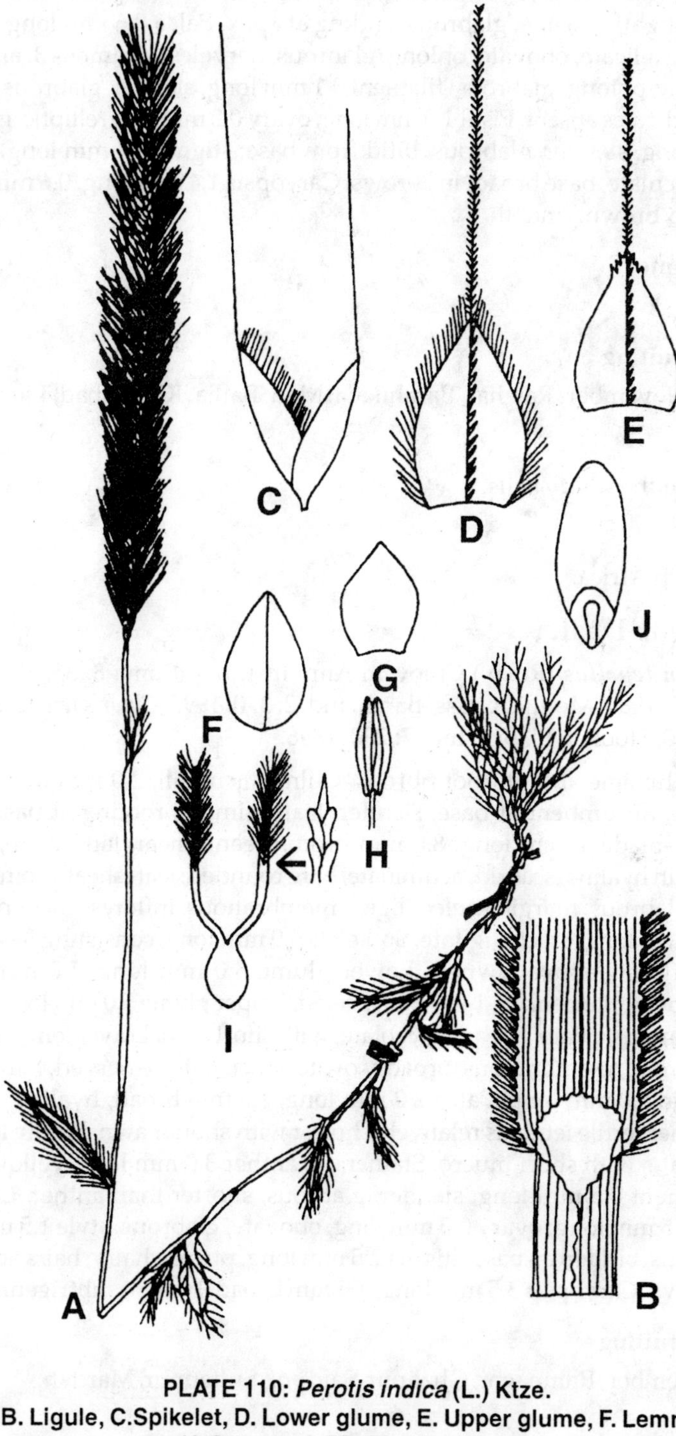

PLATE 110: *Perotis indica* (L.) Ktze.
A. Habit, B. Ligule, C.Spikelet, D. Lower glume, E. Upper glume, F. Lemma, G. Palea, H. Stamen, I. Pistil, J. Caryopsis

1.0 mm long, 0.5 mm broad, hyaline, delicate, ovate, lanceolate, glabrous 1–nerved; nerves green, slightly visible, glabrous, ending at apex. Palea 0.6 mm long, 0.3 mm broad, hyaline, delicate, obovate, oblong, glabrous, nerveless. Stamens 3; anther 0.4 mm long, yellow, oblong, glabrous; filament 0.1 mm long, slender, glabrous, shorter than anther. Lodicules absent. Pistil 1.0 mm long; ovary 0.2 mm long, elliptic, glabrous; style 0.4 mm long, slender, glabrous, bifid from base; stigma 0.4 mm long, purple, hairy; hairs apiculate, base broad, in 2–rows. Caryopsis 1.4 mm long, 0.6 mm broad, terete, cylindric, brown, smooth.

Vernacular Name

Kuras.

Flowering & Fruiting

October–November; Rajghat, Barahikala, Mau, Ballia, Khalilabad, Lar.

Habitat

Rare, along the sandy soils.

Distribution

India, South Africa.

6. *Tetrapogon* Desf.

Tetrapogon tenellus (Roxb.) Chiov. in Ann. Inst. Bot. Roma 8: 352. 1908; Roy, Grass. M. P. 158. 1984. Moulik, Grass. Bamb. Ind. 2: 370. 1997. *Chloris tenella* Roxb. Fl. Ind. 1: 330. 1820; Hook. f. Fl. Brit. Ind. 7: 290. 1896.

Annual. Rhizome absent; root fibrous. Culm 50 cm tall, 2.0 mm thick, green, glabrous, erect, decumbent at base, slender, leafy, simple, rooting at base; nodes glabrous. Leaf–blade 15 cm long, 8.0 mm brod, green, linear, lanceolate, surface glabrous, margin hyaline, scabrid, acuminate, base rounded; leaf–sheath compressed, erect, surface glabrous, margin keeled; ligule membranous. Inflorescence consists of spike; spike 10 cm long, erect, digitate. Spikelets 6.0 mm long, consisting 3–4 perfect, & 1–3 sterile florets above, awned. Lower glume 3.0 mm long, 1.5 mm broad, membranous, ovate, lanceolate, 1–nerved, keeled. Upper glume 5.0 mm long, 2.0 mm broad, membranous, ovate, linear, lanceolate, with short awn. Lower lemma 5.5 mm long, 3.0 mm broad, membranous, broadly ovate, lanceolate, 3–nerved, hairy; callus crustaceus; awn 8.0 mm long. Palea 2.0 mm long, 1.0 mm broad, hyaline, delicate, keels hairy. Other fertile lemmas relatively shorter with shorter awn. Empty lemma 2, obovate, cuneate, with short mucro. Stamens 3; anther 3.0 mm long, yellow, linear, glabrous; filament 0.5 mm long, slender, glabrous, shorter than anther. Lodicules absent. Pistil 4.5 mm long; ovary 0.5 mm long, obovate, glabrous; style 1.5 mm long, slender, glabrous, bifid from base; stigma 2.5 mm long, purple, hairy; hairs acute base broad, in 2–rows. Caryopsis 3.5 mm long, 1.5 mm broad, brown, subtrigenous.

Flowering & Fruiting

July–November; Ramnagar, Ghajipur, Saidpur, Sultanpur, Mardah.

Habitat

Common along the road sides & moist habitats.

Distribution

India, Arabica, Africa.

7. *Tragus* Scop.

Tragus roxburhii Panigrahi in Kew Bull. 29 (3): 496. 1794. Bor, Grass. Bur. Cey. Ind. Pak. 1960; Raizada & Jain, Indian Forester 92: 641. 1966; Panigrahi & Rajagopal, J. Ind. Bot. Soc. 47: 244. 1968; Cope in Nasir & Ali, Fl. W. Pak. 143: 150. 1982; Raizada *et al.*, Grass. Upp. Gang. Pl. 3: 122. 1983; Roy, Grass. M. P. 162. 1984; Moulik, Grass. Bamb. Ind. 2: 506. 1997. *T. racemosus* Scop. Introd. Hist. Nat. 73. 1777; Desf. Fl. Atlant. 2: 386. 1805; Duthie, Grass. N. W. Ind. 13. 1888; Duthie, Indig. Fodd. Grass. t. 14. 1886; Duthie, Fodd. Grass. N. Ind. 22. 1888; Hook. f. Fl. Brit. Ind. 7: 97. 1896; Cooke, Fl. Bomb. 2: 1014. 1908; Haines, Bot. Bih & Or. 979. 1924. *T. biflorus* Schult. in Schult. & Schult. f. Mantissa 2: 205. 1834. *T. brevicaulis* Boiss. Diag. Pl. Or. ser. 1. 13: 44. 1850. *Nazia racemosa* Kuntze, Rev. Gen Pl. 3: 357. 1898; Hitchcock, Genera of Grass. Unit. St. in Unit. St. Dept. Agric. Bull. 772: 165. 1920; Blatter & McCann, Bomb Grass. 217. t. 144. 1984. *Lappago racemosa* Honck. Syn. Pl. Germ. 1: 440. 1792; Host. Gram. Austr. 1: 36. 1816; Sibth. Fl. Grace 2: t. 101. 1840; Aitchis. Cat. Punj. Pl. 163: 1869. *L. biflora* Roxb. Fl. Ind. 1: 281. 1820; Grah. Cat. Bomb. Pl. 234. 1839. *L. aliena* Dalz. & Gibs., Bomb. Fl. 295. 1861. *L. occidentalis* Nees in Schimp. Pl. Arab. Fed. ed. 2: 793. 1870. *Cenchrus racemosus* Linn. Sp. Pl. 1049. 1753. *C. linearis* Lam. Fl. Franc 3: 631. 1791. *Phalaris muricata* Forsk. Fl. Aeg–Arab. 302. 1775. (PLATE–111).

Annual. Rhizome absent; root fibrous. Culm 10 cm long, 1.0 mm thick, green, glabrous, erect, tufted, decumbent, geniculately–ascending, leafy below, rooting at base, branching at lower nodes; branches sub erect, often fascicled, densely leafy; nodes glabrous; internodes 2.0 cm long. Leaf–blade 2.0 cm long, 3.0 mm broad, green, flat, narrow, linear, lanceolate, acuminate, base rounded, surface glabrous, very glaucous, suppungent, margin hyaline, scabrid & hairy; hairs 1.5 mm long, white, acute, base tubercled; closely striate, mid–vein distinct on lower surface; lateral veins many, distinct on upper surface; leaf–sheath 1.5 cm long, 2.0 mm broad, green, membranous, glabrous, margin hyaline, keeled, loose, striate, erect, compressed, shorter than internodes, hairy at mouth; hairs 5.0 mm long, white, acute, base tubercled, veins distinct on both surface; upper most sheath spathiform, tumid, usually embracing the base of the panicle; ligule 0.5 mm long, membranous, truncate, tip hairy; hairs minute white, acute, base broad. Inflorescence consists of spike like racemes; raceme 2.0 cm long, 5.0 mm broad, terminal, erect, simple, cylindrical, linear, binate, green at young, turning into brownish with age; rachis slender, straight, slightly undulated, ending with spikelets, pubescent. Spikelets 4.0 mm long, 1.5 mm broad, sessile, acute, binate, spikelets of pairs facing each other giving appearance of a single spikelets with 2–equal echinate glumes. Lower glume 0.5 mm long, 0.2 mm broad, hyaline, scale–like, glabrous, nerveless. Upper glume 4.0 mm long, 1.5 mm broad, membranous, ovate, lanceolate, slightly curved, involute, envelopping the floral glume, surface hairy between the nerves; hairs 1.0 mm long, acute, base tubercled,

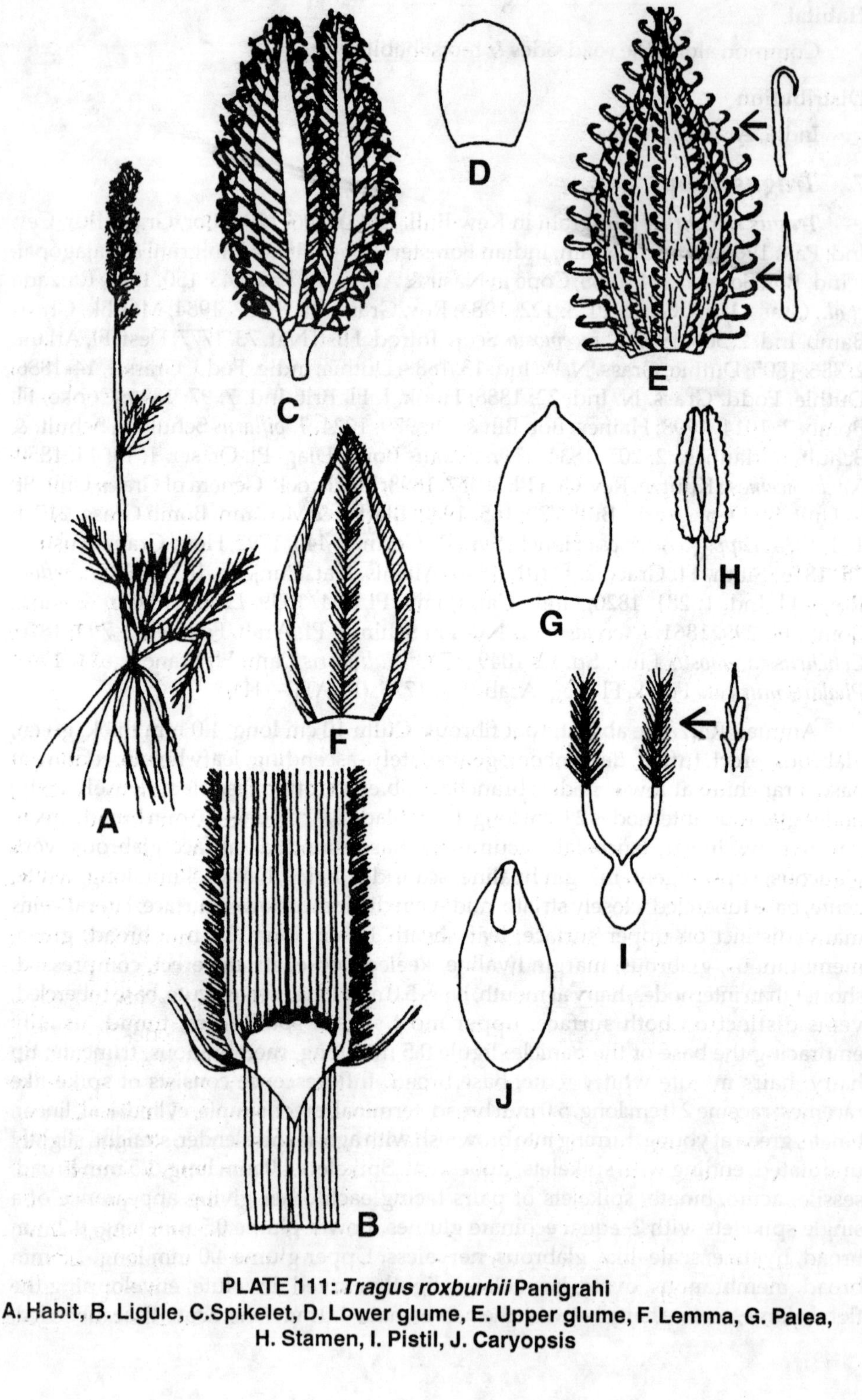

PLATE 111: *Tragus roxburhii* Panigrahi
A. Habit, B. Ligule, C. Spikelet, D. Lower glume, E. Upper glume, F. Lemma, G. Palea, H. Stamen, I. Pistil, J. Caryopsis

margin keeled & spiny; spines 1.0 mm long, hook shaped, echinate, stout, 5–nerved; nerves green, visible, ending at apex, spiny; spines 1.0 mm long, hook shaped, echinate, stout. Lemma 3.0 mm long, 1.5 mm broad, hyaline, oblong, lanceolate, acute, surface glabrous between the nerves, also with margin, 3–nerved; nerves green, slightly visible, ending at apex, hairy; hairs 1.0 mm long, acute, base broad; lateral nerves close to the margin. Palea 2.0 mm long, 1.0 mm broad, hyaline, delicate, oblong, lanceolate, truncate, apiculate, glabrous, nerveless. Stamens 3; anther 0.7 mm long, yellow, oblong, keeled; filament 0.2 mm long, slender, glabrous, shorter than anther. Lodicules absent. Pistil 1.0 mm long; ovary 0.3 mm long, obovate, elliptic, glabrous; style 0.3 mm long, slender, glabrous, bifid from base; stigma 0.4 mm long, colourless, hairy; hairs acuminate, base broad, in 2–rows. Caryopsis 1.5 mm long, 1.0 mm broad, obovate, dull white, smooth.

Vernacular Name

Barchente.

Flowering & Fruiting

July–September; Chakiya, Rajdari, Naugarh, Sonbhadra, Mirzapur.

Habitat

Rare, along the hill slopes.

Distribution

India, Asia, Europe, Africa, North & South America.

8. *Zoysia* Willd.

Zoysia matrella (Linn.) Merr. in Philipp. J. Sci. Bot. 7: 230. 1912; Fischer in Gamble, Fl. Pres. Madr. 1815. 1934 (Repr. ed. 3: 1257. 1957); Bor, Grass. Bur. Cey. Ind. Pak. 684. 1960; Moulik, Grass. Bamb. Ind. 2: 508. 1997. *Agrostis matrella* Linn. Mant. Pl. 2: 185. 1771. *Zoysia pungens* Willd. in Gest. Naturf. Freunde Berlin Neue Schr. 3: 441. 1801; Hook. f. Fl. Birt. India. 7: 99. 1896; Haines, Bot. Bih. & Or. 979. 1924. *Osterdamia matrella* Linn. O. Ktze. Rev. Gen. Pl. 2: 781. 1891. (PLATE–112).

Perennial. Rhizomatous; rhizome stoloniferous, vertical; root fibrous, forming black mat. Culm 15 cm long, 1.5 mm thick, green, rigid, densely tufted, erect, ascending, geniculate & creeping at base, very slender, densely leafy at base, branching and rooting at lower nodes; branches suberect; nodes glabrous; internodes 3.0 cm long. Leaf–blade 4.0 cm long, 3.0 mm broad, green, flat, narrow, convolute, acicular, linear–lanceolate, acuminate, base rounded, surface glabrous, margin hyaline, antrorsely scabrid, mid–vein distinct on lower surface, 2–lateral veins of each side of mid–vein; lateral veins closed, distinct on upper surface; leaf–sheath 2.5 cm long, 2.0 mm broad, green, membranous, glabrous, margin hyaline, slightly keeled, compressed, tight, slightly shorter than internodes, veins distinct on outer surface; ligule 1.0 mm long, membranous, rounded, hairy at apex; hairs 0.3 mm long, white, acute, base tubercled; auricle glabrous. Inflorescence consists of panicle; panicle 2.5 cm long, 4.0 mm broad, spiciform, open, erect, cylindrical, whitish–purple rachis of panicle glabrous, groved. Raceme 1.0 cm long, sub–erect, closely arranged, dense flowered, sessile; rachis of

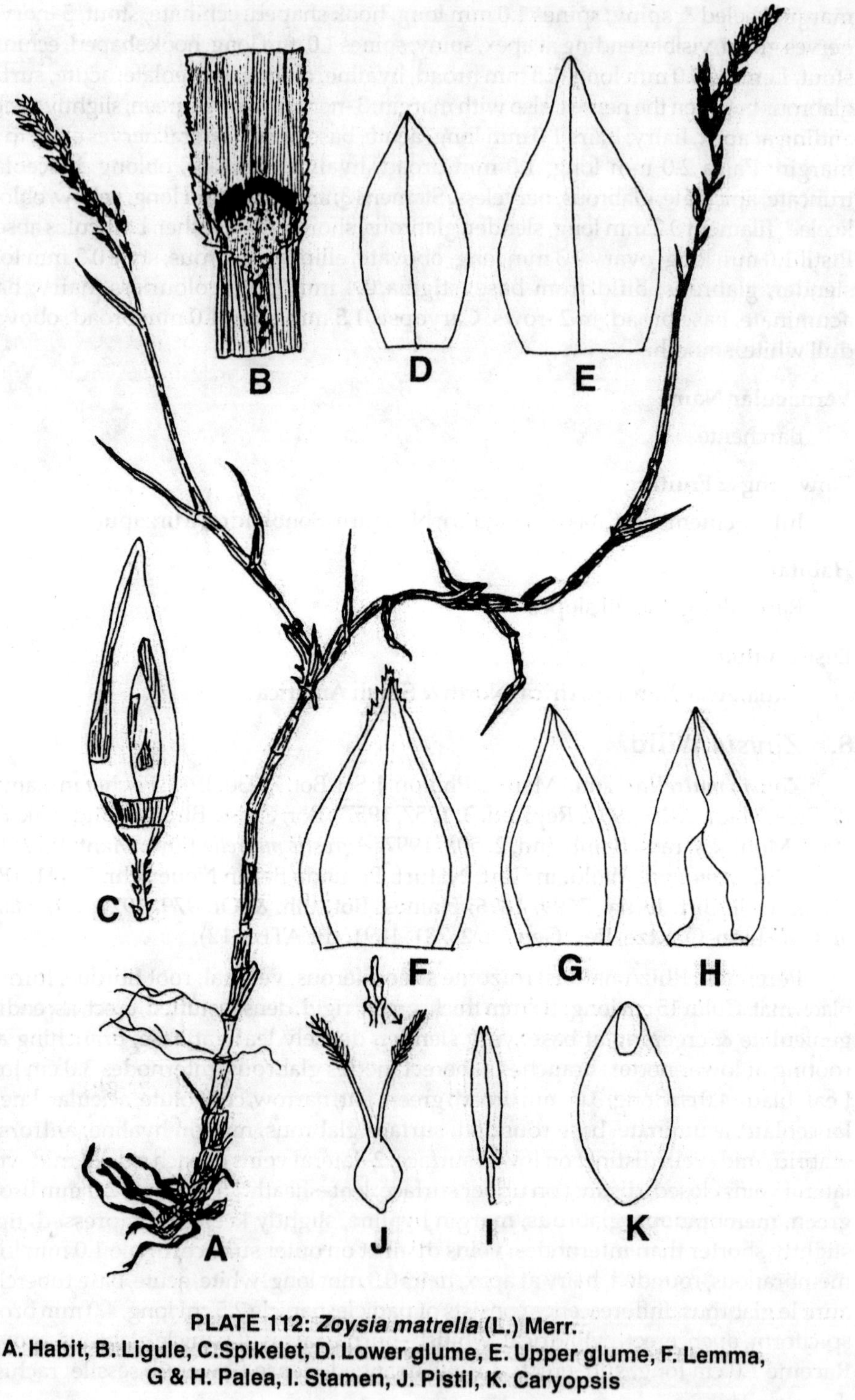

PLATE 112: ***Zoysia matrella*** **(L.) Merr.**
A. Habit, B. Ligule, C. Spikelet, D. Lower glume, E. Upper glume, F. Lemma, G & H. Palea, I. Stamen, J. Pistil, K. Caryopsis

racemes scabrid. Spikelets 2.0 mm long, 1.0 mm broad, oblong, lanceolate, acute, 1–flowered, glabrous, shortly pedicelled; pedicel 0.5 mm long, scabrid, dilated at apex, narrow to wards the base. Lower glume 1.0 mm long, 0.5 mm broad, membranous, oblong, lanceolate, acute, glabrous, faintly 1–nerved; nerve colourless, glabrous, ending at apex. Upper glume 1.5 mm long, 1.0 mm broad, membranous, ovate, oblong, lanceolate, acuminate, glabrous, margin slightly inrolled, apiculate when spread, faintly 1–nerved, nerve colourless, glabrous, ending at apex. Lemma 1.5 mm long, 1.0 mm broad, membranous, broadly ovate, lanceolate, acute, surface glabrous, margin slightly inrolled, scabrid at apex, glabrous below, faintly 3–nerved; nerves colourless, glabrous, ending at apex. Palea 1.5 mm long, 0.7 mm broad, thinly membranous, oblong, lanceolate, acute, surface glabrous, margin folded inward, transversely smooth, 2–keeled at centre, 2–nerved; nerves colourless, glabrous & marginal, ending at apex. Stamens 3; anther 1.0 mm long, green–yellow, linear, glabrous; filament 0.1 mm long, slender, minute, shorter than anther. Lodicules absent. Pistil 1.0 mm long; ovary 0.2 mm long, oblong, glabrous; style 0.5 mm long, slender, glabrous, bifid from base; stigma 0.3 mm long, white, hairy; hairs acuminate, base broad, in 2–rows. Caryopsis 1.0 mm long, 0.5 mm broad, oblong, ovoid, whitish, smooth.

Flowering & Fruiting

July–September; Lamhi, Basti, Khadda.

Habitat

Rare, along the drying ponds.

Distribution

India, Asia, America.

2. Tribe – Eragrostideae

KEY TO THE GENERA

1a. Inflorescence a solitary or 2–to many digitate or subdigitate spike:

2a. Lemma aristate:

3a. Rachis ending in a sharp point .... 3. *Dactyloctenium*

3b. Rachi not as above:

4a. Spikelets 2–3 flowered; closely imbricate; lower glume aristate .... 5. *Dinebra*

4b. Spikelets 30 flowered; not imbricate; lower glume notched .... 13. *Tripogon*

2b. Lemma not as above:

5a. Spike 2 to many:

6a. Spikes scattered; spikelets 8–12 flowered; lateral nerves of lemma ending in small teeth .... 1. *Acrachne*

6b. Spikes in whorld; spikelets 3–6 flowered; lateral nerves of lemma not as above .... 7. ***Eleusine***

5b. Spikes solitary:

7a. Culm upto 60 cm long; Spikelets many flowered; rachis angular .... 8. ***Eragrostiella***

7b. Culm upto 10 cm long; spikelets 1-flowered; rachis tetragonous .... 11. *Oropetium*

1b. Inflorescence panicle:

8a. Rhizomatous grasses:

9a. Panicle consists of many spikes; spikelets 3–6 flowered, purple-brown .... 4. ***Desmostachya***

9b. Panicle not as above; spikelet 6–10 flowered, grey or olive–green .... 6. ***Diplachne***

8b. Grasses not rhizomatous:

10a. Culm 15 cm long, prostrate; major portion of panicle concealed in the sheath .... 2. ***Crypsis***

10b. Culm more than 15 cm long, erect; panicle open:

11a. Branches of panicle short; spikelets 1–flowered; anther 2 .... 12. ***Sporobolus***

11b. Branches of panicle long; spikelets many flowered; anther 2 or 3:

12a. Leaf–blade hairy; ligule membranous; caryopsis trigonous .... 10. ***Leptochloa***

12b. Leaf–blade glabrous; ligule hairy or membranous; caryopsis globose or subglobose .... 9. ***Eragrostis***

1. *Acrachne* Wt. & Arn.

Acrachne racemosa (Heyne) Ohwi in Bull. Tokyo Sci Mus. n. 18. 1. 1947; Bor, Grass.Bur. Cey. Ind. Pak. 487. 1960; Raizada & Jain, Ind. For. 92: 638. 1966; Panigrahi & Rajagopal, J. Ind. Bot. Soc. 47: 232. 1968. Cope in Fl. W. Pak. 143: 104. 1982; Raizada *et al.*, Grasses Upp. Gang. Pl. pt. 3: 65. 1983; Roy, Grass. M. P. 25. 1984; Misra & Verma, Fl.Allah. 412. 1992; Moulik, Grass. Bamb. Ind. 2: 584. 1997. *A. verticillata* (Roxb.) Lindley ex Chiov. in Ann. Istit. Bot. Rom. 8: 361. 1908. *Leptochloa racemosa* (Heyne) Kunth. Rev. Gram, 1: 91. 1829. *L. verticillata* (Roxb.) Kunth. Rev. Gram. 1: 91. 1829.

Eleusine racemosa Heyne ex Roem. et Schult, Syst. Veg. 2: 583. 1817; Roth. Nov. Pl. Sp. 80. 1821. *E. verticillata* Roxb. Fl. Ind. 1: 346. 1820; Aitchis. Cat. Punjab. Pl. 168. 1869; Duthie, Grass. N.W. Ind. 34. 1883; Duthie, Grass. N. Ind. 58. t. 70. 1888; Lisboa, J. Bomb. Nat. Hist. Soc. 7: 377. 1893; Hook, f. Fl. Brit. Ind. 7: 295. 1896; Saxton & Sedgwick, Pl. N. Guj. in Rec. Bull. Bot. Surv. Ind. 6: 322. 1918; Blatter & McCann, Bomb. Grass. 260. t. 174. 1984. *Acrachne eleusinoides* Wight & Arn. in Wight Cat. no. 1760. 1885. (PLATE–113).

Annual. Rhizome absent; root fibrous. Culm 80 cm tall, 2.0 mm thick, green, glabrous, erect, tufted, stout towards the base, slender at apex, soft, ascending, decumbent, branched; branches sub erect, slender, rooting at lower nodes; nodes brown, glabrous; internodes 15 cm long. Leaf–blade 15 cm long, 5.0 mm broad, green, flat, glabrous, narrow, acute, base rounded, linear–lanceolate, margin hyaline, scabrid, mid–vein distinct on lower surface; 4–lateral veins on each side of mid–vein, lateral veins closed, distinct on both surface; leaf–sheath 16 cm long, 4.0 mm broad, green, glabrous, margin hyaline, striate, loose, compressed, longer than internodes, veins distinct on both surface; ligule 1.5 mm long, a ring of white hairs; hairs acute, base broad; auricle absent. Inflorescence consists of spikes; spike 12 cm long, 4.0 mm broad, 15 in number, sub–erect, linear, scattered upwards, whorled below, whitish–green, shining, sub–digitate, spikelets densely arranged, opposite; rachis of spike flexuous, trigonous, flattened on back. Spikelets 5.0 mm long, 2.5 mm broad, sessile, 8–flowered, erect, crowded, imbricate, sub–sessile. Lower glume 1.0 mm long, 0.7 mm broad, membranous, broadly ovate, lanceolate, surface glabrous, margin scabrid from middle to apex, glabrous below, 1–nerved; nerve green, glabrous, visible, ending at apex; awned; awn 1.0 mm long, antrorsely scabrid. Upper glume 2.0 mm long, 1.0 mm broad, membranous, broadly ovate, lanceolate, surface glabrous, margin scabrid from middle to apex, glabrous below, 1–nerved; nerves green, visible, glabrous, ending at apex; awned; awn 1.5 mm long, antrorsely scabrid. Lemma 3.0 mm long, 1.5 mm broad, membranous, broadly ovate, lanceolate, bifid at apex; bifid part scabrid, 3–nerved; nerves green, visible, lateral nerves thin, glabrous, close to margin & ending at apex, central one thick, scabrid, prolongated into awn; awn 1.0 mm long, antrorsely scabrid. Palea 2.0 mm long, 1.0 mm broad, hyaline, ovate, oblong lanceolate, slightly bifid at apex, surface glabrous, margin antrorsely scabrid at apex, retrorsely scabrid towards the base, 2–nerved; nerves green, slightly visible, marginal & glabrous, ending at apex. Stamens 3; anther 0.6 mm long, yellow, oblong, glabrous; filament 0.1 mm long, slender, glabrous, shorter than anther. Lodicules absent. Pistil 1.0 mm long; ovary 0.2 mm long, oblong, elliptic, glabrous; style 0.4 mm long, slender, glabrous, bifid from base; stigma 0.5 mm long, brown, hairy; hairs acute, base broad. Caryopsis 1.0 mm long, 0.5 mm broad, linear, rod–shaped, rugose, brown, pericarp caducous.

Vernacular Name

Chinki.

Flowering & Fruiting

August–November; Chakiya, Jaunpur, Basti, Ballia, Padrauna.

Habitat

Rare, along the road sides.

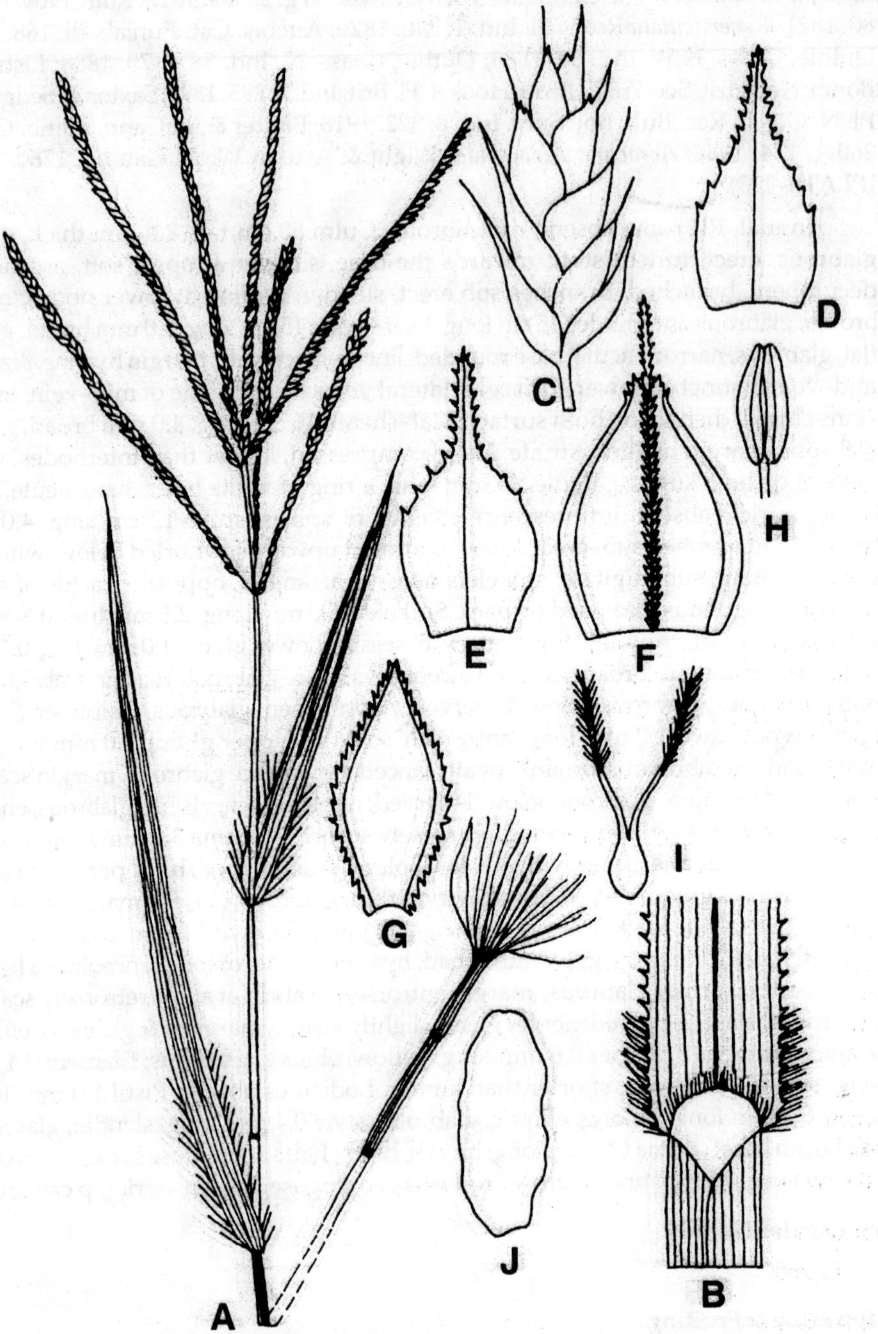

PLATE 113: *Acrachne racemosa* (Heyne) Ohwi
A. Habit, B. Ligule, C. Spikelet, D. Lower glume, E. Upper glume, F. Lemma, G. Palea, H. Stamen, I. Pistil, J. Caryopsis

Distribution

India, Tropical Africa, South–East Asia, Australia.

2. *Crypsis* Ait.

Crypsis schoenoides (L.) Lam. Tab. Encyl. Meth. Bot. 1: 166. 1791; Bor, Grass. Bur. Cey. Ind. Pak. 622. 1960; Raizada & Jain, Indian Forester 92: 639. 1966; Roy, Grass. M.P. 57. 1984; Misra & Verma, Fl. Allah. 426. 1992; Cope in Nasir & Ali, Fl. W. Pak. 143. 132. 1982; Raizada *et al.*, Grass. Upp. Gang. Pl. pt. 3: 108. 1983; Moulik, Grass. Bamb. Ind. 2: 475. 1997. *Phleum schoenoides* L. Sp. Pl. 60. 1753. *Helochloa schoenoides* Host, Gram. Austr. 1: 23. 1801; Hook. f. Fl. Brit. Ind. 7: 235. 1896. (PLATE–114).

Annual. Rhizome absent; root fibrous. Culm 15 cm tall, 2.0 mm thick, green, glabrous, prostrate, striate, patch forming, multicauline, rigid. Leaf blade 3.0 cm long, 3.0 mm broad, green, glabrous, margin scabrid, narrowly lanceolate, acuminate, base rounded; leaf–sheath 2.0 cm long, 3.0 mm broad, green, inflated, deeply furrowed, margin ciliate; ligule 0.6 mm long, a ring of white hairs; hairs acute, base broad; auricle absent. Inflorescence consists of panicle; panicle spike like, ovoid, oblong, major portion concealed in the sheath. Spikelet 3.2 mm long, 1.6 mm broad. Lower glume 2.5 mm long, 1.2 mm broad, membranous, linear, lanceolate, folded, keel ciliate, scarious, 1–nerved. Upper glume 2.0 mm long, 1.0 mm broad, membranous, ovate, lanceolate, acute, keel 1–nerved, ciliate. Lemma 3.0 mm long, 2.0 mm broad, membranous, broadly ovate, acute, keel glabrous, 1–nerved. Palea 3.2 mm long, 1.6 mm broad, hyaline, oblong, obtuse, glabrous. Stamens 3; anther 0.5 mm long, yellow, glabrous; filament 1.5 mm long, slender, glabrous, longer than anther. Lodicules absent. Pistil 2.5 mm long; ovary 0.5 mm long, oblong, glabrous; style 1.0 mm long, slender, glabrous, bifid from base; stigma 1.0 mm long, hairy; hairs acute, base broad, in 2–rows. Caryopsis 1.2 mm long, 0.6 mm broad, oblong, ellipsoid.

Flowering & Fruiting

November–April; Barahikala, Dudhi, Allahabad, Jhusi, Salempur.

Habitat

Rare, along the dried ditches.

Distribution

India, Afghanistan, East Africa.

3. *Dactyloctenium* Willd.

KEY TO THE SPECIES

1a. Culms prostrate; ligule hairy; spikes forming a cross .... **3. *D. sindicum***

1b. Culms erect; ligule membranous; spikes not forming a cross:

2a. Spikes solitary; empty lemma present; caryopsis globose .... **2. *D. aristatum***

2b. Spikes 3–6; empty lemma absent; caryopsis reniform .... **1. *D. aegyptium***

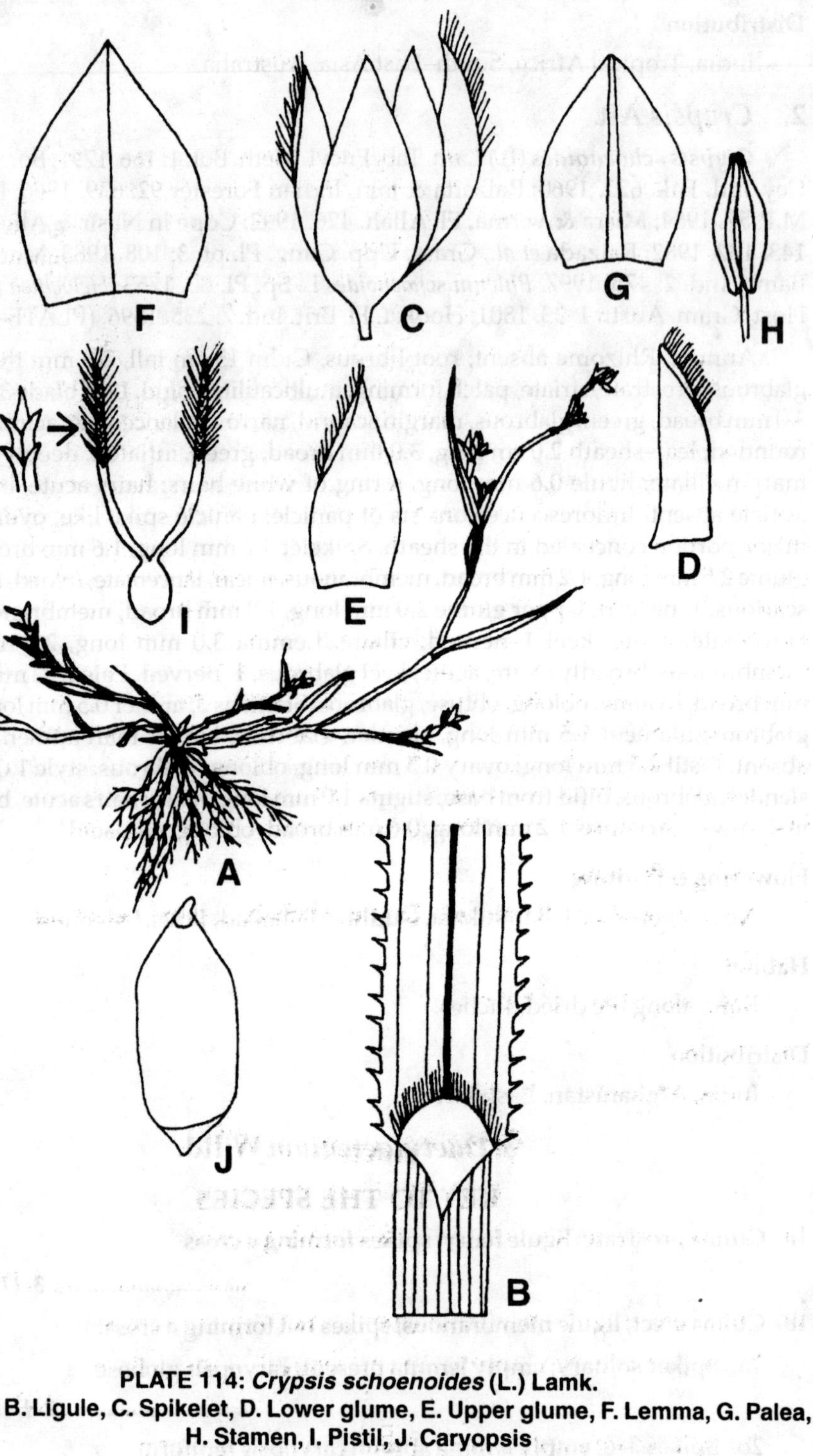

PLATE 114: ***Crypsis schoenoides*** **(L.) Lamk.**

A. Habit, B. Ligule, C. Spikelet, D. Lower glume, E. Upper glume, F. Lemma, G. Palea, H. Stamen, I. Pistil, J. Caryopsis

1. *Dactyloctenium aegyptium*

Dactyloctenium aegyptium (L.) P. Beauv. Ess. Agrost. Expl. Pl. 15. 1812; Richt. Pl. Europ. 1: 68. 1889; Muschler, Fl. Egypt. 1: 108. 1992; Willd. Enum. Pl. Hort. Berol. 1029. 1809; Fischer in Gamble, Fl. Pres. Madr. 1840. 1934 (Repr. ed. 3: 1273. 1957); Bor, Grass. Bur. Cey. Ind. Pak. 489. t. 54. 1960; Bhattacharya, Bull. Bot. Surv. Ind. 6: 209. 1964; Raizada & Jain, Ind. For. 92: 639. 1966; Panigrahi & Rajagopal, J. Ind. Bot. Soc. 47: 233. 1968; Singh, Bull. Bot. Surv. Ind. 11: 20. 1969; Panigrahi *et al.*, Bull. Bot. Surv. Ind. 11: 112. 1969; Gililand, Fl. Malaya 3: 81. 1971; Manilal & Sivaraj, Fl. Calic. 337. 1982; Cope in Nasir & Ali, Fl. W. Pak. 143: 106. 1982; Raizada *et al.*, Grass. Upp. Gang. Pl. pt. 3: 73. 1983; Roy, Grass. M.P. 62. 1984; Blatter & McCann, Bomb. Grass. 262. t. 176. 1984; Sreekumar & Nair, Fl. Kera. Grass. 367. 1991; Misra & Verma, Fl. Allah. 428. 1992; Moulik, Grass. Bamb. Ind. 2: 585. 1997. *D. aegyptiacum* Willd. Enum. Pl. 1029. 1809; P. Beauv. Agrost. 72. t. 15. fig. 2. 1812; Kunth. Enum. Pl. 1: 261. 1838; Grah. Cat. Bomb. Pl. 235. 1839; Dalz. & Gibs, Bomb. Fl. 297. 1861; Aitchis. Cat. Punjab Pl. 167. 1869; Boiss, Fl. Or. 5: 556. 1884. *D. distachyum* Bojer Hort. Maurit. 370. 1837. *D. figarei* Notar. in Ann. Sc. Nat. ser. 3, 9: 325. 1848. *D. meridionale* Ham. Prodr. Pl. Ind. Occ. 6. 1850. *D. muronatum* Willd. Enum. Pl. Hort. Berol. 1029. 1809. *D. prostratum* Willd. Enum. Pl. Hort. Berol. 1029. 1809. *D. radulans* Beauv. Agrost. 72. 1812; Kunth. Enum. Pl. 1: 262. 1838. *Eleusine aegyptiaca* Desf. Fl. Atlant. 1: 85. 1798; Roxb. Fl. Ind. 1. 344. 1820; Duthie, Grass. N.W. Ind. 34. 1883; Duthie, Fodder Grass. N. Ind. 56. t. 36.. 1888; Hook. f. Fl. Brit. Ind. 7: 295. 1896; Lisboa, J. Bomb. Nat. Hist. Soc. 7: 374. 1893; Prain, Bengl. Pl. 1229. 1903; Cooke, Fl. Bomb. 2: 1038. 1908; Achariyar, S. Ind. Grass. 276. 1921; Haines, Bot. Bih. & Or. 970. 1924. *E. ciliata* Rafin, Desv. Journ. Bot. 4: 273. 1814. *E. cruciata* Lamk. Encycl. Meth. Bot. 31, 1: 203. t. 48. fig. 2. 1816. *E. mucronata* Stokes Bot. Mat. Med. 1: 150. 1880; Lisboa, J. Bomb. Nat. Hist. Soc. 7: 376. 1893. *E. pectinata* Moench, Meth. Pl. Suppl. 68. 1802. *E. prostrata* Spreng. Syst, 1: 350. 1825. *E. radulans* R. Br. Prodr. 186. 1810. *Cynosurus aegyptius* Linn. Sp. Pl. 72. 1753. *C. distachyus* Rottl. ex Steud. Nom. ed. 2, 1: 465. 1810. *Chloris mucronata* Mich, Fl. Bor. Amer. 1: 59. 1803. *Cenchrus aegyptius* Beauv. Agrost. 157. 1812. *Rhabdochloa mucronata* Beauv. Agrost. 157. 1812. *Aegilops saccharinus* Walt. Fl. Carol. 1: 249. 1725. (PLATE–115).

Annual. Rhizome absent; root fibrous, Culm 50 cm tall, 2.0 mm thick, green, glabrous, erect, densely tufted, decumbent, geniculately ascending, compressed, leafy, branched; branches suberect, terete, rooting at lower nodes, creeping at base; nodes glabrous; internodes 8.0 cm long. Leaf–blade 12 cm long, 6.0 mm broad, green, glaucous, oblong, flat, linear, lanceolate, acuminate, base rounded, coriaceous, hairy on upper surface; hairs 2.0 mm long, white, acute, base tubercled; glabrous on lower surface, margin hyaline & scabrid, mid–vein distinct on lower surface; 4–lateral veins on each side of mid–vein; lateral veins distinct on upper surface; leaf–sheath 7.0 cm long, 5.0 mm broad, green, membranous, sparsely hairy; hairs 1.0 mm long, white acute, base tubercled; membranous, margin hyaline & slightly keeled, compressed, loose, erect, shorter than internodes, hairy at mouth; hairs 5.0 mm long, white, acute, base tubercled; veins distinct on both surface; ligule 1.0 mm long, membranous, ovate, truncate, hairy at apex; hairs minute, acute, base broad; auricle glabrous. Inflorescence consists of spikes; spike 4.0 cm long, 5.0 mm broad, 6 in number, digitately radiating; rachis stout, trigonous, dorsally flattened, rigid, projected

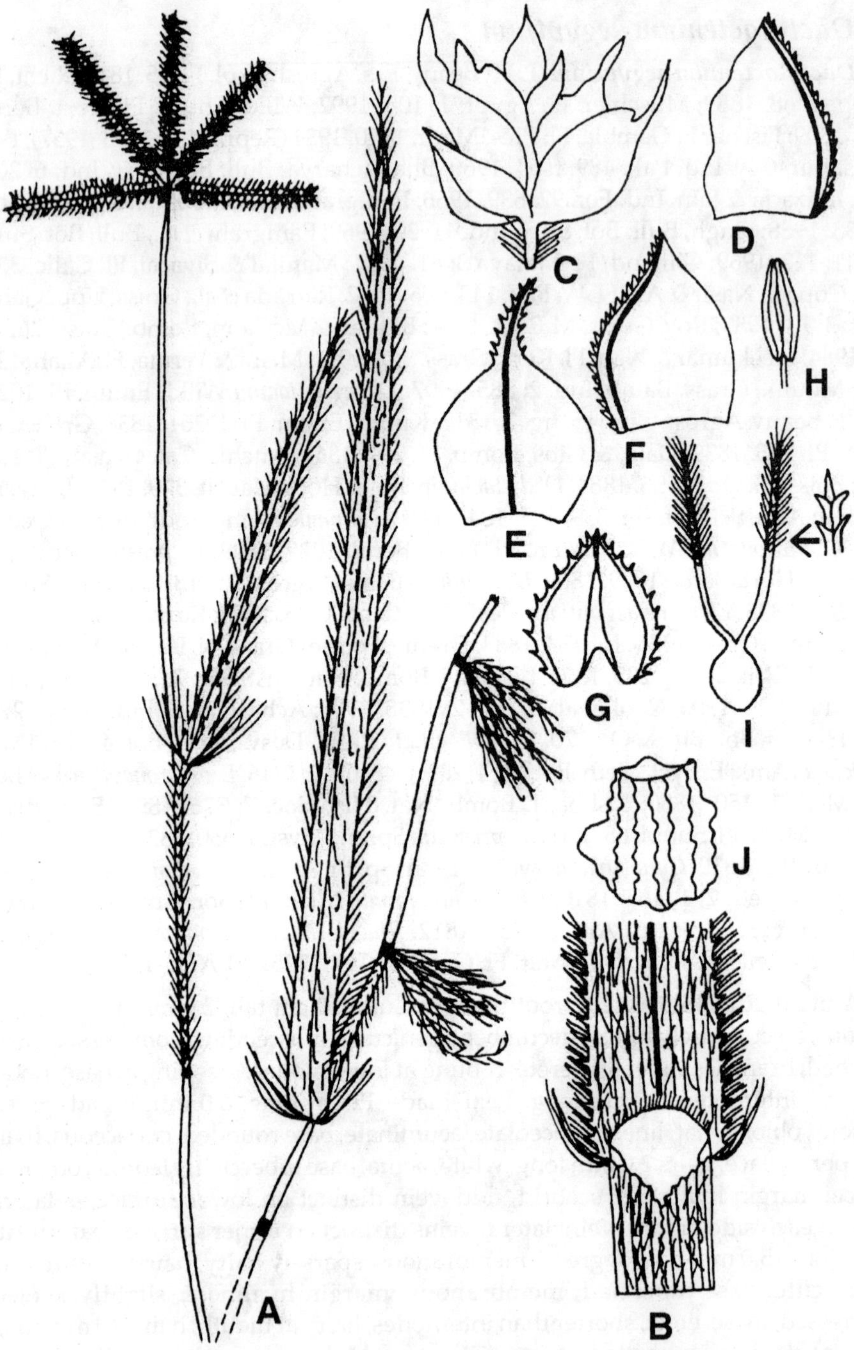

PLATE 115: *Dactyloctenium aegyptium* (L.) P. Beauv.
A. Habit, B. Ligule, C. Spikelet, D. Lower glume, E. Upper glume, F. Lemma, G. Palea, H. Stamen, I. Pistil, J. Caryopsis

as sharp pointed, pungent, mucro beyond the spikelets, mucro 2.5 mm long, rigid. Spikelets 4.0 mm long, 2.5 mm broad, 4–flowered, ovate, oblong, laterally compressed, densely crowded, sessile, spreading at right angle to the rachis. Lower glume 3.0 mm long, 1.5 mm broad, coriaceous, boat shaped at compressed view, ovate, acute, when spread, keeled; keel narrowly winged, scabrid, 1–nerved; nerve green, visible, glabrous, ending at apex. Upper glume 2.0 mm long, 1.5 mm broad, coriaceous, ovate, elliptic, sub–orbicular, cuspidate at apex, keeled; keel scabrid, 1–nerved; nerve green, visible, glabrous, produced into a usually curved awn; awn 1.5 mm long, antrorsely scabrid. Lemma 3.0 mm long, 1.5 mm broad, chartaceous, gibbously ovate, lanceolate, keeled; keel minutely winged & scabrid, 1–nerved; nerve green, visible, glabrous, ending below the apex, awned; awn 1.0 mm long, antrorsely scabrid. Palea 1.5 mm long, 1.0 mm broad, membranous, ovate, oblong, obtuse, lanceolate, glabrous, margin folded inwards, transversely scabrid, 2–nerved; nerves green, slightly visible, glabrous, marginal, ending at apex. Stamens 3; anther 0.5 mm long, yellow, oblong, glabrous; filament, 0.2 mm long, slender, glabrous, shorter than anther. Lodicules 2, 0.3 mm long, 0.2 mm broad, hyaline, obovate, glabrous, nerveless, wavy at apex. Pistil 1.5 mm long; ovary 0.2 mm long, ovate, oblong, glabrous; style 0.5 mm long, slender, glabrous, bifid from base; stigma 0.8 mm long, brown, hairy; hairs acute, base broad, in 2–rows. Caryopsis 1.0 mm long, 0.5 mm broad, obovate, globose, transversely rugose, reniform, whitish–brown.

Vernacular Name

Makra, Gandhi, Anchi, Manchi.

Flowering & Fruiting

August–November; Chakiya, Chandauli, Ghazipur, Jaunpur, Basti.

Habitat

Very common along the lawns, road sides, railway tracks.

Distribution

India, America, Western Peninsula, Baluchistan, Afghanistan, Arbia, Nubia.

2. *D. aristatum*

D. aristatum Link. Hort. Berol. 1: 59. 1827; Bor, Grass. Bur. Cey. Ind. Pak. 489. 1960; Roy, Grass. M. P. 62. 1984; Moulik, Grass. Bamb. Ind. 2: 587. 1997. *Eleusine aristata* sensu. Hook. f. Fl. Brit. Ind. 7: 296. 1896. (PLATE–116).

Annual. Rhizome absent; root fibrous. Culms 10 cm long, 1.0 mm thick, erect, tufted, green, glabrous, very leafy, ascending, simple, compressed, geniculate, rooting at base; nodes brown, glabrous; internodes 3.0 cm long. Leaf blade 5.0 cm long, 4.0 mm broad, green, membranous, flat, narrow, linear–lanceolate, acuminate, base rounded, hairy on both surface; hairs 1.0 mm long, white, acute, base tubercled; margin hyaline, keeled, hairy, scabrid; hairs 1.0 mm long, white, acute, base broad; mid–vein distinct on lower surface, 3–lateral veins on each side of mid–vein; lateral vein distinct on both surface; leaf sheath 1.5 cm long, 3.0 mm broad, green, surface membranous, hairy; hairs, 1.0 mm long, white, acute, base tubercled; margin hyaline,

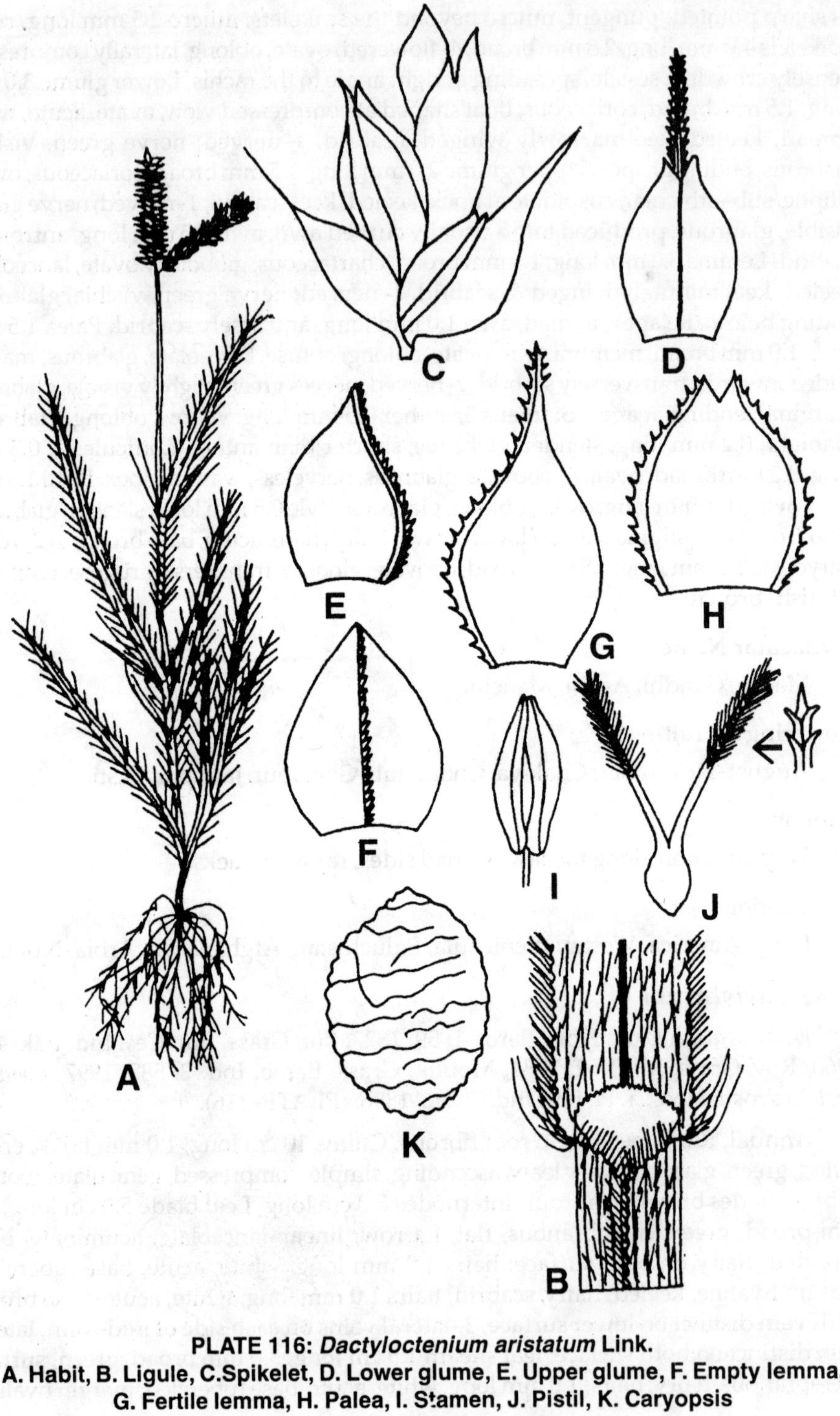

PLATE 116: ***Dactyloctenium aristatum*** **Link.**
A. Habit, B. Ligule, C.Spikelet, D. Lower glume, E. Upper glume, F. Empty lemma, G. Fertile lemma, H. Palea, I. Stamen, J. Pistil, K. Caryopsis

keeled, hairy; hairs 1.0 mm long, acute, base tubercled; loose, compressed, very much shorter than internodes, hairy on mouth; hair 5.0 mm long, white, acute, base tubercled; veins distinct on both surface; ligule 0.1 mm long, membranous, lacerate, slightly truncate; auricle glabrous. Inflorescence consists of solitary spikes; spikes 2.5 cm long, 6.0 mm wide, erect, digitate, linear, spikelets densely arranged, white at young, dull yellowish with age, rachis slender, hairy at base; hairs 3.0 mm long, white, acute, base tubercled, apical projection 3.0 mm long, beyond the spikelets. Spikelets 3.0 mm long, 1.5 mm broad, suberect, densely crowded, laterally compressed, sessile, distichous, 3–flowered. Lower glume 2.5 mm long, 1.5 mm broad, coriaceous, oblong, lanceolate, glabrous, bifid at apex, keeled; keel winged, laterally compressed, 1–nerved; nerves green, visible, glabrous, prolonged into a awn; awn 2.0 mm long, antrorsely scabrid, incurved. Upper glume 2.0 mm long, 1.0 mm broad, coriaceous, oblong, ovate, acute, laterally compressed, glabrous, winged; wing hyaline, scabrid, 1–nerved; nerve green, visible, glabrous, ending at apex. Empty lemma 3.0 mm long, 1.5 mm broad, hyaline, solitary, ovate, lanceolate, acute, glabrous, 1–nerved; nerves green, slightly visible, scabrid, ending at apex. Fertile lemma 4.0 mm long, 1.5 mm broad, membranous, linear, oblong, elliptic, acute, laterally compressed, keeled, winged; wing hyaline, scabrid; 1–nerved; nerves green, visible, glabrous, ending at apex. Palea 3.0 mm long, 1.5 mm broad, hyaline, ovate, oblong, bifid at apex, glabrous, margin scabrid, 2–nerved; nerves green, slightly visible, glabrous & marginal ending at apex. Stamens 3; anther 0.5 mm long, yellow, oblong, glabrous; filament 0.1 mm long, slender, glabrous, shorter than anther. Lodicules absent. Pistil 1.2 mm long; ovary 0.2 mm long, oblong, elliptic, glabrous; style 0.5 mm long, slender, glabrous, bifid from base; stigma 0.5 mm long, colourless, hairy; hairs acute, base broad, in 2–rows. Caryopsis 1.5 mm long, 1.0 mm broad, whitish, globose, rugose.

Flowering & Fruiting

July–October; Tarana, Babatpur, Jaunpur, Ghazipur, Allahabad, Padrauna

Habitat

Common along the moist localities, bunds of agricultural fields & rocky soil.

Distribution

India, North–East Africa, Europe, America.

3. *D. sindicum*

D. sindicum Boiss. Diagh. ser. 2, fasc. 4: 131. 1859; Boiss, Fl. Or. 5: 557. 1884; Bor, Grass. Bur. Cey. Ind. Pak. 489. 1960; Ray, Grass. M. P. 63. 1984; Blatter & McCann, Bomb. Grass. 264. 1984; Moulik, Grass. Bamb. Ind. 2: 587. 1997. *D. glaucophyllum* Courb. in Ann. Sc. Nat. ser. 4; 18: 133. 1862. *Eleusine sindica* Duthie, Fodder Grass. N. Ind. 58. 1888; Lisboa in J. Bomb. Nat. Hist. Soc. 7: 377. 1893. *E. glaucophylla* Munro ex Benth. in J. Linn. Soc. 19: 107. 1881. *E. aristata* Ehrenb. ex Boiss. Fl. Or. 5: 557. 1884; Hook. f. Fl. Brit. Ind. 7: 296. 1896; Cooke, Fl. Bomb. 2: 1039. 1908.

Perennial. Rhizome absent; root fibrous. Culm 40 cm long, 2.0 mm broad, green, glabrous, prostrate, stiff, widely spreading, slender, leafy, profusely branched;

branches at lower nodes, rooting at the distant, thickened, woody nodes; nodes glabrous; internodes 5.0 cm long. Leaf blade 10 cm long, 2.5 mm broad, green, glabrous, acuminate, base rounded, linear, lanceolate, surface & margin sparsely hairy; hairs acute, base tubercled; sheath glabrous; ligule obsolete. Inflorescence consists of spike, spike 1.5 cm long, 3 in number, spreading, decurved, forming a cross, breaded at the base; rachis excurrent into a pungent point. Spikelets 4.0 mm long, sessile, crowded. Lower glume 2.5 mm long, 1.0 mm broad, ovate, acute, 1–nerved, with ciliate keel. Upper glume 3.0 mm long, 2.0 mm broad, sub–orbicular, awned. Lemma 4.0 mm long, 2.0 mm broad, broadly ovate, acute, cuspidate, glabrous, keeled, membranous, with a strong mid–nerve and weak lateral nerves. Palea 3.5 mm long, 2.0 mm broad, keels, scaberulous. Stamens 3; anther 0.8 mm long, yellow, glabrous; filament 0.5 mm long, slender, glabrous, shorter than anther. Lodicules absent. Pistil 1.5 mm long; ovary 0.3 mm long, glabrous; style 0.5 mm long, slender, glabrous, bifid from base; stigma 0.7 mm long, plumose. Caryopsis 0.5 mm long, 0.3 mm broad, ovoid, rugose.

Flowering & Fruiting

September–January; Varanasi, Vindhayachal, Banshi.

Habitat

Common along the moist localities.

Distribution

India, Europe, Pakistan, Nepal, Malaya.

4. *Desmostachya* Stapf.

Desmostachya bipinnata (L.) Stapf, Fl. Cap. 7: 632. 1900; Bor, Grass. Bur. Cey. Ind. Pak. 491. 1960; Bhattachrya, Bull. Bot. Surv. Ind. 6: 209. 1964; Raizada & Jain, Ind. For. 92: 639. 1966; Panigrahi & Rajagopal, J. Ind. Bot. Soc. 47: 234. 1968; Singh, Bull. Bot. Surv. Ind. 11: 21. 1969; Cope in Nasir & Ali, Fl. W. Pak. 143. 109. 1982; Raizada *et al.*, Grass. Upp. Gang. Pl. pt. 3: 68. 1983; Roy, Grass. M. P. 64. 1984; Blatter & McCann, Bomb. Grass. 244. t. 162. 1984; Moulik, Grass. Bamb. Ind. 2: 588. 1997. *Briza bipinnata* Linn. Syst. Nat. ed. 10: 875. 1759. *Uniola bipinnata* Linn. Sp. Pl. ed. 2: 104. 1762. *Leptochloa bipinnata* (Linn.) Hochst in Flora. 38: 422. 1855. *Cynosurus durus* Forssk, Fl. Aegypt.–Arab. 21. 1775. *Stapfiola bipinnata* (Linn.) O. Ktze. in Post et. O. Ktze, Lexic, Gen. Phan. 532. 1903. *Pogonarthria bipinnata* (Linn.) Chiov. in Ann. Ist. Bot. Roma. 8: 362. 1908. *Poa cynosuroides* Retz. Obs. Bot. 4: 20. 1786; Roxb. Fl. Ind. 1: 333. 1820; Del. Fl. d' Egypte. 159. t. 10. 1812; Grah. Cat. Bomb. Pl. 236. 1839; Kunth. Enum. 1: 227. 1838. *Desmostachya cynosuroides* Stapf in Haines, Bot. Bih. & Or. 962. 1924; Blatter, McCann, & Sabnis, J. Ind. Bot. 7: 76. 1927. *Eragrostis cynosuroides* P. Beauv. Agrost. 162. 1812; Steud. Syn. Pl. Glum. 264. 1855; Wight, Cat. no. 1774. 1881; Trin, in Mem. Acad. Petersb. ser. 6, 1: 415. 1831; Dalz. & Gibs, Bomb. Fl. 298. 1861; Aitchis, Cat. Punjab Pl. 169. 1869; Duthie, Grass, N.W. Ind. 37. 1883; Duthie, Fodd. Grass. N. Ind. 62. t. 40. 1888; Boiss, Fl. Or. 5: 583. 1884; Lisboa, J. Bomb. Nat. Hist. Soc. 7: 387. 1893; Hook. f. Fl. Brit. Ind. 7: 324. 1896; Cooke, Fl. Bomb. 2: 1028. 1908; Prain, Beng, Pl. 1221. 1903. (PLATE–117).

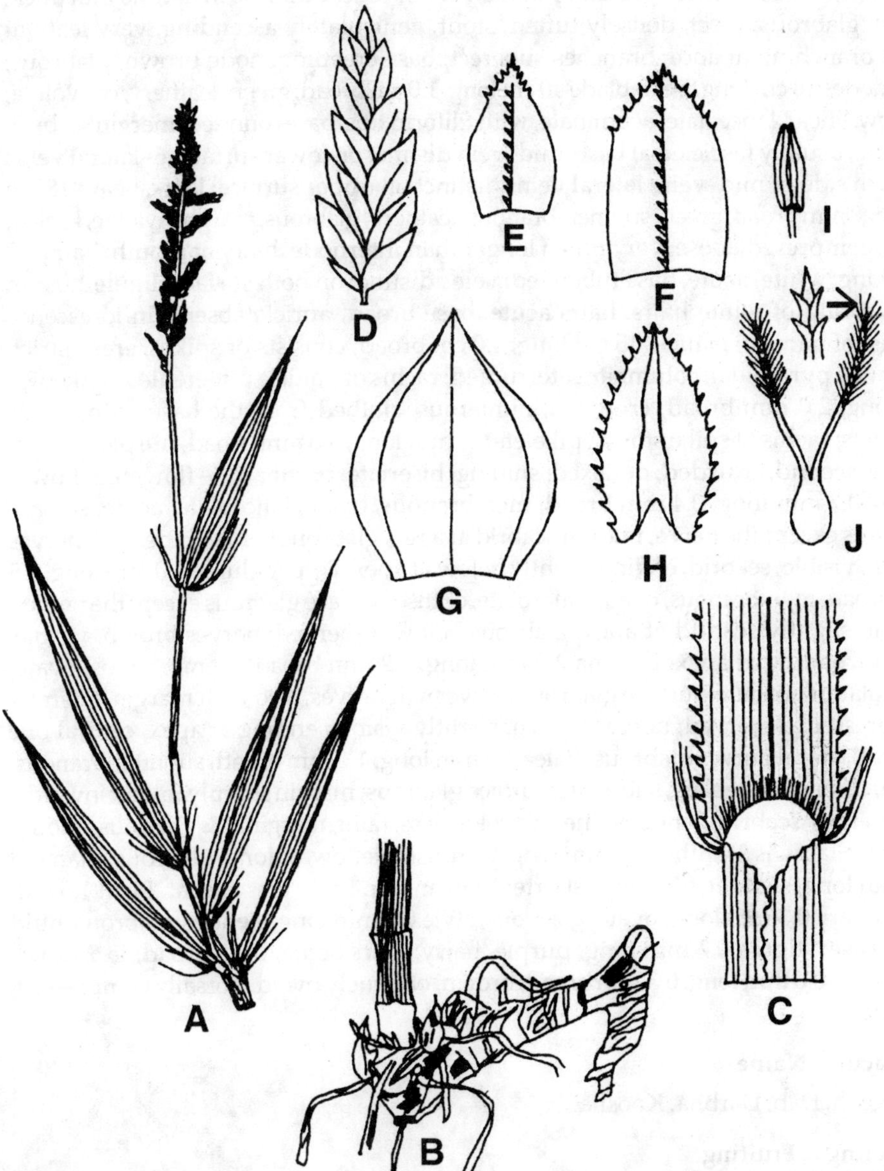

PLATE 117: ***Desmostachya bipinnata* (L.) Stapf.**
A. Habit, B. Rhizome, C. Ligule, D. Spikelet, E. Lower glume, F. Upper glume, G. Lemma, H. Palea, I. Stamen, J. Pistil

Strongly perennial. Rhizomatous; rhizome 6.0 cm long, horizontal, stout, covered by scales; scale 1.6 cm long, membranous; root fibrous. Culm 1.0 m tall, 3.0 mm thick, green, glabrous, erect, densely tufted, stout, geniculately ascending, very leafy at base, branching at abse; branches suberect; base creeping; node brown, glabrous; internodes 10 cm long. Leaf–blade 40 cm long, 1.0 cm broad, green, leathery, convolute, narrow, linear, lanceolate, acuminate, with filiform tips, base rounded, margin scabrid, hyaline, usually fascicled at base, mid–vein distinct on lower surface, 6–lateral veins on each side of mid–vein; lateral veins distinct on upper surface; leaf–sheath 18 cm long, 5.0 mm broad, green, submembranous, leathery, glabrous, margin hyaline, keeled, erect, compresed, loose, very much longer than internode, hairy at mouth; hairs 5.0 mm long, white, acute, base tubercled; veins distinct on both surface; ligule 1.0 mm long, a ring of white hairs; hairs acute, base broad, auricle absent. Inflorescence consist of panicle; panicle 45 cm long, 4.0 cm broad, consists of spikes, erect, strict, narrowly pyramidal, columnar, interrupted; rachis of panicle pubercules. Spikes 2.0 cm long, 2.0 mm broad, crowded, numerous, clothed from the base with sessile spikelets; rachis of spike stout. Spikelets 6.0 mm long, 3.0 mm broad, purple–brown, sessile, secund, crowded, deflexed, shining, biseriate, pectinate, 6–flowered. Lower glume 0.7 mm long, 0.4 mm broad, membranous, ovate, lanceolate, acute, surface glabrous except the nerve, margin scabrid at apex, glabrous below, 1–nerved; nerves brown, visible, scabrid, ending slightly below at apex. Upper glume 1.0 mm long, 0.5 mm broad, membranous, ovate, lanceolate, obtuse, surface glabrous except the nerves, margin slightly scabrid at apex, glabrous below, 1–nerved; nerves brown, visible, scabrid, ending at apex. Lemma 2.0 mm long, 1.2 mm broad, membranous, ovate, lanceolate, subacute, surface glabrous between th nerves, also with margin, laterally compressed, 3–nerved; nerves brown, slightly visible, ending at apex, central one scabrid, lateral nerves glabrous. Palea 1.5 mm long, 1.0 mm broad, submembranous, ovate, oblong, lanceolate, sub acute, surface glabrous, margin sharply folded inwards, transversely scabrid, 2–nerved; nerves colourless, faint, marginal & glabrous, ending at apex. Stamens 3; anther 1.0 mm long, purplish–yellow, oblong, glabrous; filament 0.2 mm long, slender, glabrous, shorter than anther. Lodicules absent. Pistil 1.5 mm long; ovary 0.3 mm long, ovate, glabrous; style 0.5 mm long, slender, glabrous, bifid from base; stigma 0.7 mm long, purple, hairy; hairs acute, base broad, in 2–rows. Caryopsis 0.6 mm long, 0.3 mm broad, brown, obliquely ovoid, dorsally compressed, smooth.

Vernacular Name

Kush, Dab, Darbha, Koosha.

Flowering & Fruiting

August–October & March–April; Bavanbigha, Navalpur, Cholapur, Sarnath, Harraua, Tarana, Chakiya, Chandauli, Jaunpur, Mirzapur, Deoria, Nichlaul.

Habitat

Very common along bunds of cultivated fields & road sides.

Distribution

India, Syria, North, Africa, Egypt. Nubia, Peris, Arabia.

5. *Dinebra* R. Br.

Dinebra retroflexa (Vahl.) Panzer in Denkscher. Acad. Munch 270. t. 12. 1814; Boiss, Fl. Or. 5: 557. 1824; Muschler, Fl. Egypt. 1: 106. 1912. Blatter & McCann, Bomb. Grass. 264. t. 177. 1984; Roy, Grass. M. P. 72. 1984; Bor, Grass. Bur. Cey. Ind. Pak. 491. 1960; Moulik, Grass. Bamb. Ind. 2: 589. 1997. *D. arabica* Jacq. Frangn. 77. 1809; P. Beauv. Agrost. 98. t. 16. fig. 2. 98. 1812; Duthie, Grass. N. W. Ind. 34. 1883; Duthie, Fodder Grass. N. Ind. 55. 1888; Lisboa in J. Bomb. Nat. Hist. Soc. 7: 372. 1893; Hook. f. Fl. Brit. 7: 297. 1896; Prain, Bengl. Pl. 1230. 1903; Watt. Dict. Econ. Prod. 3: 115. 1890. Cooke, Fl. Bomb. 2: 1039. 1908; Achariyar, S. Ind. Grass. 279. 1921; Haines, Bot. Bih. & Or. 971. 1924. *D. aegyptica* Del. Fl. d' Eg. 25. t. 11. fig. 3. 1812. *Cynosurus retroflexus* Vahl. Symb. Bot. 2: 20. 1791. *Leptochloa arabica* Kunth. Rev. Gram. 1: 91. 1829; Kunth. Enum. Pl. 1: 271. 1838. Wight. Cat. no. 1756. 1880. Aitchis. Cat. Punjab. Pl. 167. 1869. *Leptochloa calycina* Kunth Rev. Gram. 1: 91. 1829; Kunth. Enum. Pl. 1: 272. 1838; Dalz. & Gibs. Bomb. Fl. 297. 1861. *Eleusine calycina* Roxb. Fl. Ind. 1: 346. 1820. *Dactylis paspaloides* Willd. Enum. Hort. Berol. 111. 1827. (PLATE–118).

Annual. Rhizome absent; root fibrous. Culms 60 cm tall, 3.0 mm thick, green, glabrous, erect, densely tufted, stout, geniculately–ascending, leafy throughout, simple, rooting at base; nodes brown, glabrous; internodes 10 cm long. Leaf blade 10 cm long, 5.0 mm broad, green, flat, linear–lanceolate, flaccid, acuminate, base rounded, surface glabrous, margin hyalin, scabrid, mid–vein distinct on lower surface, 3–lateral veins on each side of mid–vein; lateral veins distinct on upper surface; leaf–sheath 11 cm long, 5.0 mm broad, green, thinly membranous, glabrous, margin hyaline, slightly keeled, loose, compressed, striate, longer than internodes, veins distinct on both surface, hairy at mouth; hairs 3.0 mm long, white, acute, base tubercled; ligule 1.0 mm long, membranous, narrow, lacerate; auricle glabrous. Inflorescence consists of raceme; raceme 12 cm long, 2.5 cm broad, erect, spike like, yellowish–green when young, purplish, deflexed with age, oblong, linear; rachis of raceme slender, smooth. Spikes 1.5 cm long, 3.0 mm broad, single – fascicles of 3 together, alternate, opposite above, whorled below; rachis of spikes 0.5 mm wide, angular, dorsally flattered, ventrally trigonous, margin hyaline, scabrid. Spikelets 5.0 mm long, 2.5 mm broad, 3–flowered, closely imbricate, alternate, sessile, awned. Lower glume 2.0 mm long, 1.0 mm broad, hyaline, ovate, lanceolate, acute, surface glabrous, margin scabrid, slightly recurved, 1–nerved; nerves green, visible, glabrous, ending at apex, prolongated into awn; awn 1.5 mm long, antrorsely scabrid. Upper glume 2.5 mm long, 1.5 mm broad, hyaline, broadly ovate, lanceolate, acute, surface glabrous between nerve, margin scabrid at apex, glabrous below, 1–nerved; nerves green, visible, scabrid, ending at apex, prolongated into awn; awn 2.0 mm long, antrorsely scabrid. Lemma 2.0 mm long, 1.0 mm broad, membranous, ovate, oblong, subacute, white, surface glabrous, margin glabrous at upper half, hairy at lower half; hairs 1.0 mm long, acute, base tubercled, 3–nerved; nerves green, visible, ending at apex, central one, glabrous at upper half, hairy at lower half; hairs 1.0 mm long, acute, base tubercled; lateral nerves glabrous, shortly awned; awn 0.5 mm long, antrorsely scabrid. Palea 1.5 mm long, 1.0 mm broad, hyaline, delicate, oblong, bifid at apex, surface glabrous, margin scabrid at upper half, hairy at lower half; hairs 1.0 mm long, acute, base broad; 2–nerved;

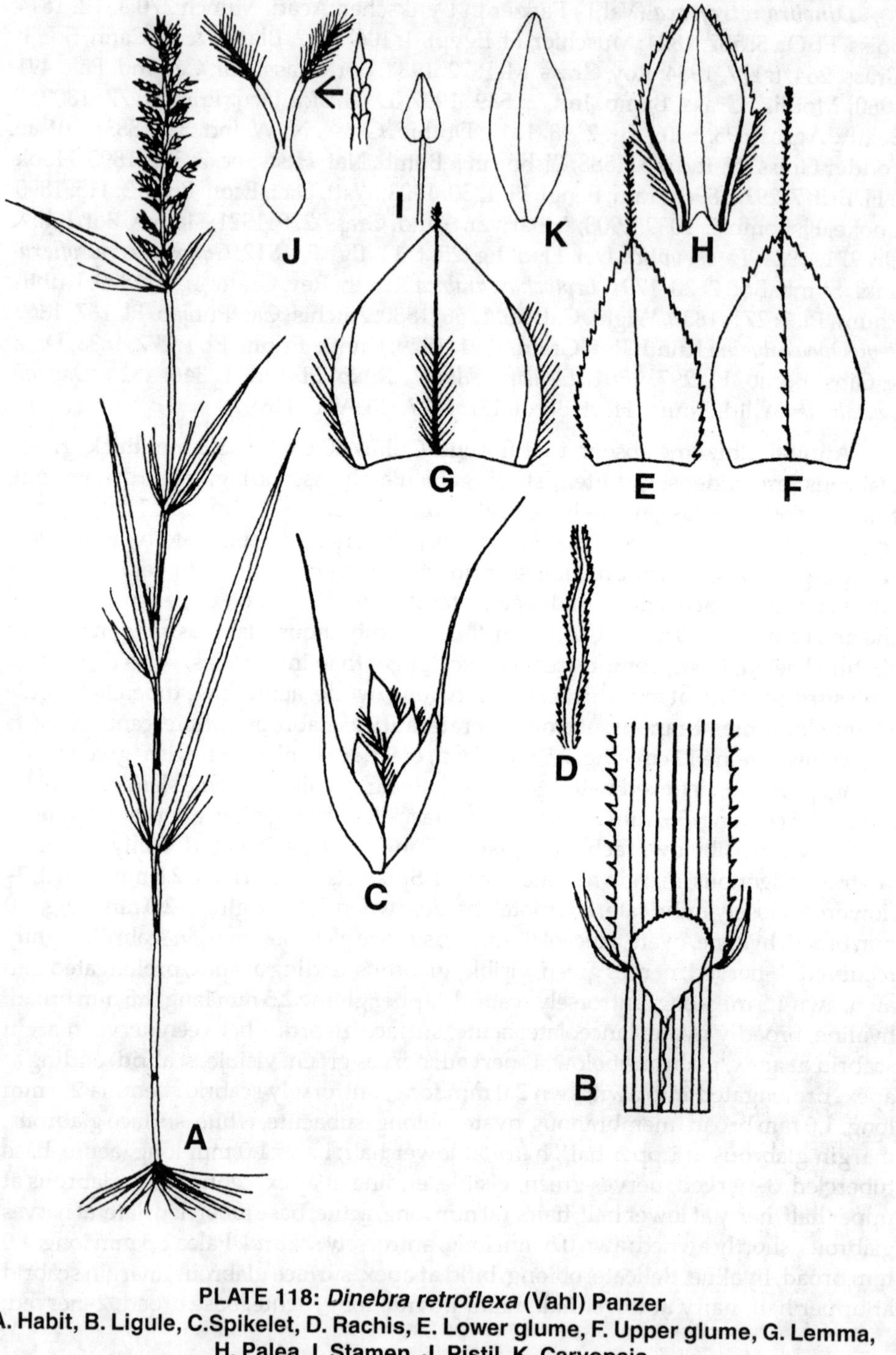

PLATE 118: *Dinebra retroflexa* (Vahl) Panzer
A. Habit, B. Ligule, C.Spikelet, D. Rachis, E. Lower glume, F. Upper glume, G. Lemma, H. Palea, I. Stamen, J. Pistil, K. Caryopsis

nerves green, slightly visible, glabrous & marginal, ending at apex. Stamens 3; anther 0.5 mm long, yellow, glabrous; filament 0.7 mm long, slender, glabrous, longer than anther. Lodicules absent. Pistil 1.5 mm long; ovary 0.2 mm long, oblanceolate, glabrous; style 0.5 mm long, slender, bifid from base, glabrous; stigma 0.8 mm long, brown, hairy; hairs acute, base broad, in 2–rows. Caryopsis 1.5 mm long, 0.7 mm broad, pale brown, ellipsoid, oblong, trigonous, smooth.

Vernacular Name

Kalikauli, Kharia, Lona.

Flowering & Fruiting

August–November; Ghazipur, Mirzapur, Dudhi, Jhusi, Lar, Mau.

Habitat

Rare, along the moist road sides.

Distribution

India, Afghanistan, Egypt, Senegal, East Africa.

6. *Diplachne* P. Beauv.

Diplachne fusca (L.) P. Beauv. ex. Roem. & Schult. Syst. Veg. 2: 615. 1817; P. Beauv. Ess. Agrost. 80: 165. 1812; Hook. f. Fl. Brit. Ind. 7: 329. 1896; Cooke, Fl. Bomb. 2: 1030. 1908; Haines, Bot. Bih. & Or. 962. 1924. Fischer in Gamber, Fl. Pres. Madr. 1829. 1934 (Repr. ed. 3: 1266. 1957); Bor, Grass. Bur. Cey. Ind. Pak. 492. 1960; Cope in Nasir & Ali, Fl. W. Pak. 143: 72. 1982; Batter & McCann, 246. t. 163. 1984; Roy, Grass. M. P. 73. 1984; Sreekumar & Nair, Fl. Kera. Grass. 368–370. t. 75. 1991. *Festuca fusca* Linn. Sp. Pl. ed. 2: 109. 1753; Del. Fl. d' Egypt. 24. f. 11. fig. 11. 1812. *F. indica* Retz. Obs. Bot. 4: 21. 1786; Kunth. Enum. 412. 1838. *Uralepis fusca* Steud. Syn. Pl. Glum. 247. 1855. *U. drummondii* Steud. Syn. Pl. Glum. 245. 1855. *Tridens indicus* Nees in Wight. Cat. no. 1794. 1828. *Eragrostis procera* Steud. Syn. Pl. Glum. 266. 1855. *Poa procera* Roxb. Fl. Ind. 1: 332. 1820. *P. contracta* Retz. Obs. Bot. 3: 11. 1783. *Triodia ambigya* R. Br. Prodr. 183. 1810. *Bromus polystachios* Forsk. Fl. Aeg.–Arab. 23. 1775. *Leptochloa fusca* Kunth. Rev. Gram. 1: 91. 1829. *Diplachne indica* Spreng. Syst. 1: 351. 1825. (PLATE–119).

Perennial. Rhizomatous; rhizome 3.0 cm long, creeping; root fibrous. Culms 70 cm tall, 3.0 mm thick, green, terete, glabrous, erect, tufted, compressed, geniculately–ascending, leafy, polished, branched; branches suberect, rooting at lower nodes; nodes glabrous; internodes 15 cm long. Leaf–blade 15 cm long, 4.0 mm broad, green, glaucous, flat, involute, strict, narrow, linear–lanceolate, acute, base rounded, glabrous upwards, hairy below; hairs 2.0 mm long, acute, base tubercled; margin hyaline & scabrid, mid–vein distinct on lower surface, 3–lateral veins on each side of mid–vein; lateral veins distinct on lower surface; leaf–sheath 9.0 cm long, 5.0 mm broad, green, glabrous, membranous, margin hyaline, keeled, loose, erect, shorter than internodes, hairy at mouth; hairs 3.0 mm long, white, acute, base tubercled; veins distinct on outer surface; ligule 0.5 mm long, oblong, membranous, lacerate; auricle glabrous. Inflorescence 15 cm long, 4.0 cm broad, erect, narrowly oblong, densely spiculate;

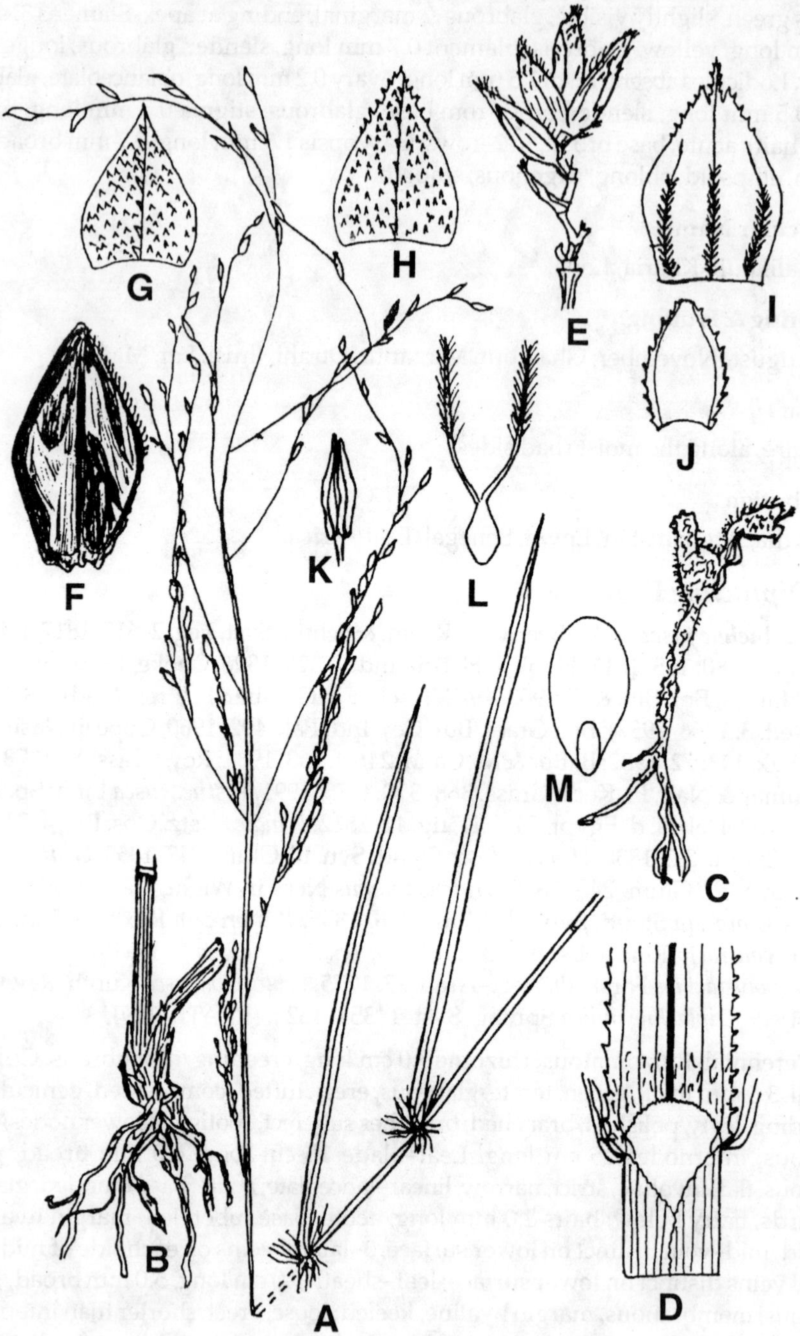

PLATE 119: *Diplachne fusca* (L.) Beauv.Roem. ex Schult.
A. Habit, B. Rhizome, C. Root hair, D. Ligule, E. Spikelet, F. Floret, G. Lower glume, H. Upper glume, I. Lemma, J. Palea, K. Stamen, L. Pistil, M. Caryopsis

rachis of panicle smooth. Branches 10 cm long, spike–like, sub erect, lower whorled, upper alternate, distant. Spikelets 6.0 mm long, 2.5 mm broad, erect, rather distant, linear, lanceolate, oblong, elliptic, greyish–green at young, olive with age, 6–flowered, shortly pedicelled; pedicel 0.5 mm long, slender, glabrous; rachis of spikelets slender, fragile. Lower glume 2.0 mm long, 1.0 mm broad, membranous, oblong, lanceolate, acute, surface scabrid, margin glabrous, 1–nerved; nerves green, slightly visible, scabrid, ending at apex. Upper glume 2.5 mm long, 1.5 mm broad, membranous, ovate, oblong, elliptic, acute, surface scabrid also with margin at apex, glabrous below, 1–nerved; nerves green, slightly visible, scabrid, ending at apex. Lemma 3.0 mm long, 1.5 mm broad, membranous, oblong, elliptic, linear, 3–lobed at apex, surface glabrous between the nerved, margin scabrid at apex, glabrous below, 3–nerved; nerves green, visible, central on ending at apex, lateral nerves ending below at the apex, all nerves glabrous in upper half, hairy in lower half; hairs 1.0 mm long, acute, base broad; shortly awned; awn 1.0 mm long, antrorsely scabrid. Palea 1.5 mm long, 1.0 mm broad, hyaline, delicate, oblong, elliptic, glabrous, 2–keeled; keel scabrid at apex, 2–nerved; nerves green, visible, glabrous & marginal, ending at apex. Stamens 3; anther 1.0 mm long, yellow, oblong glabrous; filament 0.2 mm long, slender, shorter than anther. Lodicule absent. Pistil 2.0 mm long; ovary 0.5 mm long, oblanceolate, glabrous; style 0.5 mm long, slender, glabrous bifid from base; stigma 1.0 mm long, brown, hairy; hairs acute, base broad, in 2–rows. Caryopsis 1.0 mm long, 0.5 mm broad, brownish, ovoid–oblong, subtrigonous.

Vernacular Name

Shotigandar.

Flowering and Fruiting

January–April; Jaunpur, Ghazipur, Ballia, Khalilabad, Khadda.

Habitat

Rare, along the muddy soil.

Distribution

India, Western Peninsula, Egypt, Tropical Asia, Africa, Australia.

7. *Eleusine* Gaertn

KEY TO THE SPECIES

1a. Spikes slender, straight, glabrous at base; caryopsis oblong, trigonous
.... 2. *E. indica*

1b. Spikes stout, incurved, hairy at base; caryopsis globose
.... 1. *E. coracana*

1. *Eleusine coracana* (L.) Gaertn. Fruct. 1: 8. t. 1. 1788; Lamk. Tab. Encycl. Meth. Bot. Suppl. 3.1.28. 1791; Trin. Sp. Gram. 70. 1826; Panz. in Muench Denkschr. 4: 8. 1814; Roxb. Fl. Ind. 1: 342. 1820; Grah. Cat. Bomb. Pl. 235. 1839; Dalz. & Gibs. Bomb. Fl. 97. 1861; Duthie, Grass. N. W. Ind. 34. 1883; Duthie, Fodd. Grass. N. Ind. 57. t. 69. 1888; Duthie, Field & Gard. Crops 15. t. 28.1890; Lisboa, J. Bomb. Nat. Hist. Soc. 7: 373.

1893; Hook. f. Fl. Brit. Ind. 7: 294. 1896; Prain, Beng. Pl. 1229. 1903; Cooke, Fl. Bomb. 2: 1039. 1908; Haines, Bot. Bih. & Or. 970. 1924; Fischer in Gamble, Fl. Pres. Madr. 1839. 1934. (Repr. ed. 3: 1273. 1957); Bor, Grass. Bur. Cey. Ind. Pak. 492. 1960; Phillips, Kew Bull. 27: 254. 1972; Manilal & Sivaraj, Fl. Calic. 339. 1982; Roy, Grass. M. P. 76. 1984; Blatter & McCann, Bomb. Grass. 260. t. 173. 1984; Sreekumar & Nair, Fl. Kera. Grass. 371. 1991; Moulik, Grass. Bamb. Ind. 2: 590. 1997. *Eleusine coracana* var. *stricta* (Roxb.) Nees, Fl. Afr. Austr. 1: 251. 1814. *E. coracana* (L.) Gae. var. *tocussa* (Fresen.) Bull. Soc. Hist. Nat. Autun. 8: 377. 1895. *E. coracana* var. *alba* Koern; Handb. Get. 1: 329.1885. *E. luco* Welw; Apont. Phyto. Geog. 591. 1858. *E. indica* var. *stricta* (Roxb.) Chiov. in Nuov. Giorn. Bot. Ital. N. S. 26: 83. 1919. *E. stricta* var. *rufoabbreviata* Cif.in Alti Inst. Bot. Univ. Pavia, ser, 5, 2: 172. 1944. *E. pilosa* Gilli. in Ann. Natur. Mus. Wien. 69: 50. 1965. *E. cerealis* Salisb. Prodr. 19. 1796. *E. sphaerosperma* Stokes. Bot. Mat. Meth. 1: 419. 1812. *E. stricta* Roxb. Fl. Ind. 1: 343. 1820. *E. tocussa* Fresen. in Mus. Senkenb. 2: 141. 1837. *Cynosurus coracanus* Linn. Syst. Nat. ed. 10, 2: 875. 1759. (PLATE–120).

Annual. Rhizome absent; root fibrous. Culm 60 cm tall, 3.0 mm thick, green, glabrous, tufted, stout, leafy, geniculately–ascending, decumbent, rooting at base, branching at lower nodes; branches suberect, node glabrous; internodes 7.0 cm long. Leaf–blade 25 cm long, 5.0 mm broad, green, loosely folded at base, flat above, linear, lanceolate, glabrous, margin hyaline, scabrid, usually overtopping the culm, acuminate, base rounded, mid–vein distinct at lower surface, 3–lateral veins on each side of mid–vein; lateral veins distinct on upper surface; leaf–sheath 10 cm long, 5.0 mm broad, green, sub coriaceous, glabrous, margin membranous & strongly keeled, compressed, loose, erect, very much longer than internodes, veins distinct on outer surface; ligule 1.0 mm long, a ring of white hairs; hairs acute, base broad; auricle absent. Inflorescence consists of spikes; spike 5.0 cm long, 7.0 mm broad, 3 in number, sub–digitate, compact, suberect, densely spiculate, linear, oblong, clustered at top of culm, incurved at the tips and forming a tight cluster, green at young, turning into pale–dark brown with age; rachis stout, sub trigonous, flattened on back, scabrid, hairy at base; hairs 1.0 mm long, white, acute, base broad. Spikelets 5.0 mm long, 2.5 mm broad, ovate, much congested, 6–flowered, sessile, laterally compressed, biserate, very closely over lapping on the rachis, not shattering at maturity, only caryopsis falling from the persistent florets. Lower glume 1.5 mm long, 1.0 mm broad, membranous, broadly ovate, oblong, lanceolate, acute, surface glabrous except the nerves, margin scabrid, 1–nerved; nerves green, visible, scabrid, ending at apex. Upper glume 2.0 mm long, 1.0 mm broad, membranous, broadly ovate, narrowly oblong, lanceolate, acute, surface glabrous except the nerves, margin scabrid, 1–nerved; nerves green, visible, scabrid, ending at apex. Lemma 3.0 mm long, 1.5 mm broad, membranous, ovate, lanceolate, subacute, laterally compressed, surface glabrous between the nerves, also with margin, 3–nerved; nerves, green, visible, central one thick, scabrid, ending at apex, lateral nerves glabrous, thin, ending very below at the apex, close to the central nerves. Palea 2.0 mm long, 1.5 mm broad, hyaline, ovate, oblong, sub–globose, bifid at apex, glabrous, margin folded inwards, transversely scabrid, plicate, nerveless. Stamens 3; anther 1.0 mm long, yellow, oblong, glabrous; filament 0.3 mm long, slender, glabrous, shorter than anther. Lodicules 2, 0.2 mm long, 0.1 mm broad, hyaline, glabrous, nerveless, cuneate, oblanceolate. Pistil 2.0 mm long; ovary 0.5 mm long, ovate, elliptic, glabrous; style 1.0 mm long, slender glabrous,

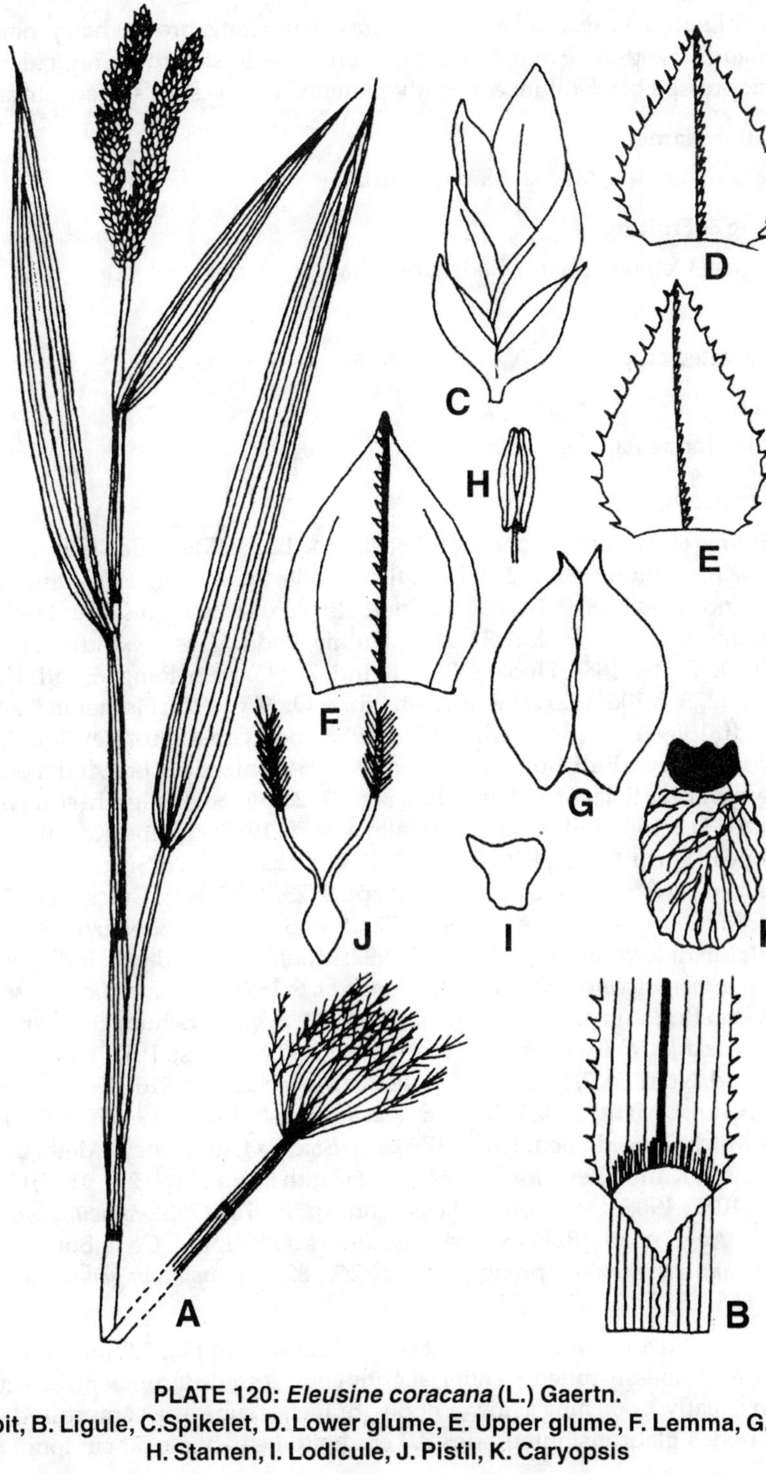

PLATE 120: *Eleusine coracana* (L.) Gaertn.
A. Habit, B. Ligule, C.Spikelet, D. Lower glume, E. Upper glume, F. Lemma, G. Palea, H. Stamen, I. Lodicule, J. Pistil, K. Caryopsis

slightly connate at base, bifid above; stigma 0.5 mm long, brown, hairy; hairs acute, base broad. Caryopsis 1.0 mm long, 0.5 mm broad, reddish–brown, orbicular, rugose, with a depressed black hilum & slightly flattened on one side, surface striate.

Vernacular Names

Makra, Mandua, Marua, Mand, Murua.

Flowering & Fruiting

August–October; Bhatni, Padrauna, Bhatpar, Nagra.

Habitat

Cultivated crop.

Distribution

India, Tanzania.

2. *E. indica*

E. indica (L.) Gaertn. Fruct. 1: 8. 1788; Lamk. Encyl. Meth. Bot. Supp. 3, 1: 203. t. 48. 1791; Kunth Enum. Pl. 1: 273. t. 16. fig. 4. 1838; Steud. Syn. Pl. Glum. 211. 1855; Roxb. Fl. Ind. 1: 345. 1820; Grah. Cat. Bomb. Pl. 235. 1839; Aichis. Cat. Punj. Pl. 168. 1869; Duthie, Grass. N. W. Ind. 34. 1883; Duthie, Fodd. Grass. N. Ind. 57. t. 69. 1888; Boiss, Fl. Or. 5: 555. 1881; Hook. f. Fl. Brit. Ind. 7: 293. 1896; Rang. & Tad. Handb. S. Ind. Grass. 273. t. 206. 1921; Haines, Bot. Bih. & Or. 970. 1924; Fischer in Gamble, Fl. Pres. Madr. 1839. 1934 (Repr. ed. 3: 1273. 1957); Bor, Grass. Bur. Cey. Ind. Pak. 493. 1960; Bhattacharya, Bull. Bot. Sur. Ind. 6: 209. 1964; Raizada & Jain, Ind. For. 92: 639. 1966; Panigrahi & Rajagopal, J. Ind. Bot. Soc. 47. 234. 1968; Panigrahi *et al.*, Bull. Bot. Sur. Ind. 11: 112. 1969; Gilliland in Fl. Malaya. 3: 78. 1997; Phillips, Kew Bull. 27: 256. 1972; Cope in Nasir & Ali, Fl. W. Pak. 143: 103. 1982; Manilal & Sivaraj, Fl. Calic. 340. 1982; Raizada *et al.*, Grass. Upp. Gang. Pl. pt. 3: 73. 1983; Roy, Grass. M. P. 76. 1984; Blatter & McCann, Bomb. Grass. 259. t. 172. 1984; Sreekumar & Nair, Fl. Kera. Grass. 371. 1991; Misra & Verma, Fl. Allah. 437. 1992; Moulik, Grass. Bamb. Ind. 2: 590. 1997. *E. indica* var. *monostachya* F.M. Bailey, Queensl. Fl. 6: 1898. 1902. *E. indica* var. *sandaensis* Vanderyst in Bull. Agr. Congo. Belg. 11: 122. 1920. *E. glabra* Schum., Besker. Guin. Pl. 53. 1827. *E. textiles* Welw. in Murray, Journ. Trav. and Nat. Hist. 1: 31. 1868. *E. distachya* Trin ex Steud. Nom. ed. 2, 1: 549. 1840. *E. distans* Moench. Meth. 210. 1794. *E. domingensis* Sieb. ex Schult.. Mant. 2: 323. 1824. *E. gracilis* Salisb. Prodr. 19. 1796. *E. marginata* Lindl. in Mitch. Three Exped. 1: 319. 1790. *E. tristachya* Lamk. Encyl. Meth. Bot. Supp. 3. 1: 203. 1791; Kunth. Rev. Gram, 1: 92. 1829; Kunth Enum. Pl. 1: 273. 1838; Cooke, Fl. Bomb. 2: 1037. 1908. *Cynosurus indicus* Linn. sp. Pl. 72. 1753. *Panicum compressum* Forssk. Fl. Aeg.–Arab. 18. 1775. *Paspalum dissectum* Kniphof. Cent. Bot. in Orig. 11. 1780. *Triticum geminatum* Spreng. Syst. 1: 326. 1825. *Agropynum geminatum* Schult. Mant. 2: 655. 1817. (PLATE–121).

Annual. Rhizome absent; root fibrous. Culm 70 cm tall, 2.0 mm thick, green, glabrous, erect, loosely tufted, slender, slightly compressed, decumbent, geniculately–ascending, leafly, branching at lower nodes; branches sub erect, compressed, rooting at base; nodes glabrous; internodes 7.0 cm long. Leaf–blade 30 cm long, 8.0 mm

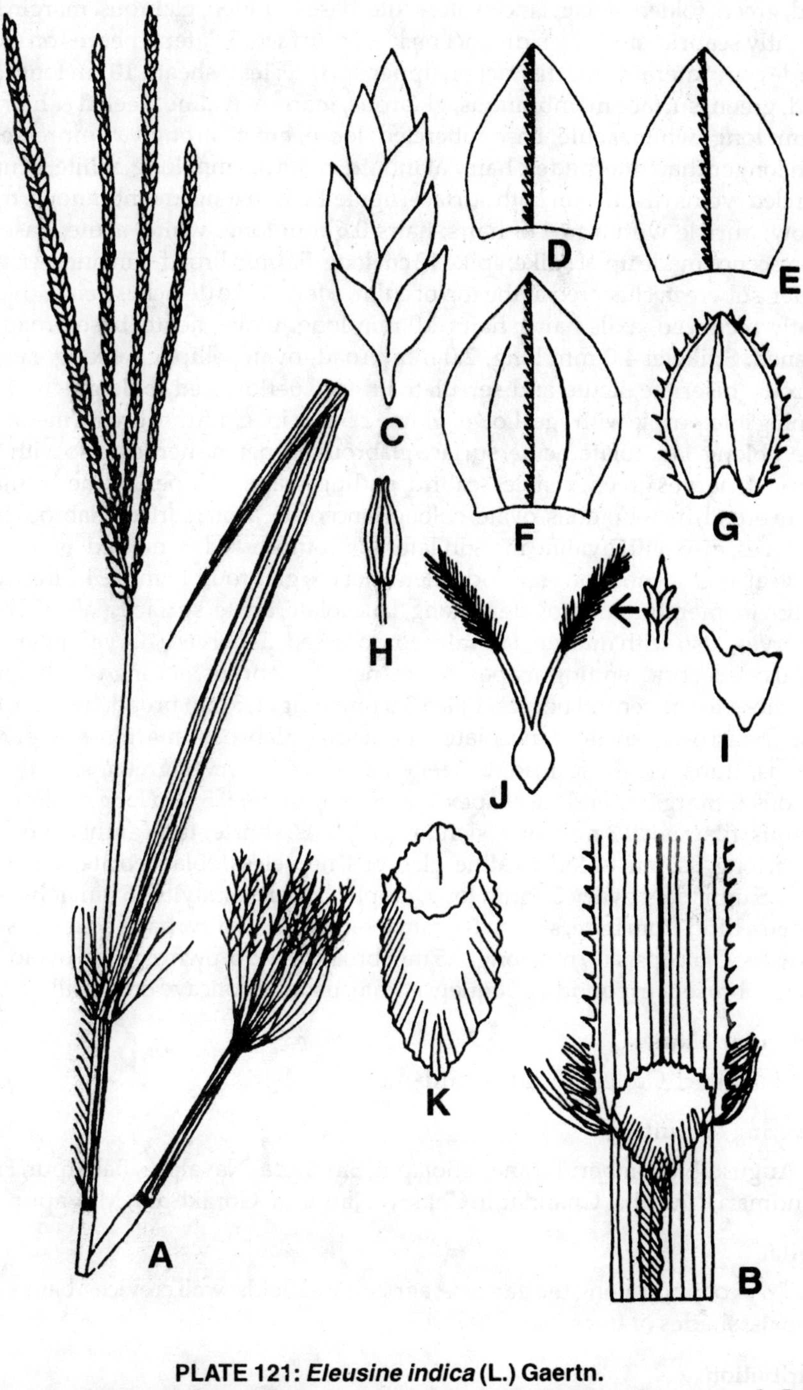

PLATE 121: ***Eleusine indica*** **(L.) Gaertn.**
A. Habit, B. Ligule, C. Spikelet, D. Lower glume, E. Upper glume, F. Lemma, G. Palea, H. Stamen, I. Lodicule (0.2), J. Pistil, K. Caryopsis

broad, green, folded, linear, lanceolate, acute, base rounded, glabrous, margin hyaline & slightly scabrid, mid–vein distinct on lower surface, 5–lateral nerves on each side of mid–vein; lateral veins distinct on upper surface; leaf–sheath 10 cm long, 5.0 mm broad, green, surface membranous, glabrous, margin hyaline, keeled & hairy; hairs 1.0 mm long, white, acute, base tubercled; loose, erect, strongly compressed, very much longer than internodes, hairy at mouth; hairs 5.0 mm long, white, acute, base tubercled; veins distinct on both surface; ligule 1.0 mm long, membranous, truncate, narrow; auricle with a tuft of hairs; hairs 0.6 mm long, white, acute, base broad. Inflorescence made up of spike; spike 12 cm long, 5.0 mm broad, 8 in number, digitate, slender, suberect, clustered at the top of culm, a few spike detaches below the umbel, slightly recurved, axils hairy; hairs 1.0 mm long, white, acute, base broad; rachis flattened. Spikelet 4.0 mm long, 2.0 mm broad, ovate, elliptic, oblong, lanceolate, glabrous, biseriate, acute and serrulate on top, 6–flowered, pale green at young, turnning into purple with age. Lower glume 2.5 mm long, 1.5 mm broad, membranous, ovate, oblong, lanceolate, acute, surface glabrous excpet the nerves, also with margin, 1–nerved; nerves green, visible, scabrid, ending at apex. Upper glume 4.0 mm long, 2.0 mm broad, membranous, ovate, oblong, lanceolate, acute, surface glabrous between the nerves, also with hyaline margin,laterally compressed, 3–nerved, green, visible, ending at apex, central one scabrid, lateral nerves glabrous. Lemma 4.0 mm long, 2.0 mm broad, membranous, ovate, oblong, lanceolate, acute, surface glabrous between the nerves, also with margin, laterally compressed, 3–nerved; nerves green, visible, central one scabrid, ending at apex, lateral nerves glabrous, ending very below at the apex, close to the central nerves. Palea 3.0 mm long, 1.5 mm broad, hyaline, oblong, elliptic, narrowly ovate, lanceolate, sub acute, glabrous, margin sharply folded inwards, transversely scabrid, 2–keeled, 2–nerved; nerves green, slightly visible, glabrous & marginal, ending at apex. Stamens 3; anther 0.5 mm long, yellow, oblong, glabrous; filament 0.3 mm long, slender, glabrous, shorter than anther. Lodicules 2; 0.3 mm long, 0.2 mm broad, hyaline, glabrous, nerveless, oblanceolate, wavy at apex. Pistil 1.5 mm long; ovary 0.3 mm long, elliptic, glabrous; style 0.6 mm long, slender, glabrous, bifid from base; stigma 0.6 mm long, purple, hairy; hairs acute, base broad, in 2–rows. Caryopsis 1.0 mm long, 0.5 mm broad, deep brown, rugose, ovoid, elliptic, oblong, obtusely trigonous, obliquely striate, usually concave on the hilum side.

Vernacular Names

Malankuri, Guder, Khurd, Khurd–mendi.

Flowering & Fruiting

August–November; Tarana, Cholapur, Sindhora, Navalpur, Babatpur, Harraua, Chandmari, Murdha, Chandauli, Chakiya, Jaunpur, Gorakhpur, Mirzapur.

Habitat

Very common along the gardens, agricultural fields, wall crevices, banks of rivers & canals, shades of trees.

Distribution

India, Europe, Asia, Sri–Lanka, Malaysia, Australia, Newzealand.

8. *Eragrostiella* Bor

KEY TO THE SPECIES

1a. Basal leaves very long, rolled or folded reaching to the base of the inflorescence:

2a. Keels of palea broadly winged; wings hyaline & smooth .... **3. *E. leioptera***

2b. Keels of palea narrowly winged; wings shortly ciliate .... **4. *E. nardoides***

1b. Basal leaves short, narrow, not reaching to the base of the inflorescence:

3a. Ligule membranous; caryopsis ± 0.5 mm long, broadly ovate .... **1. *E. bifaria***

3b. Ligule hairy; caryopsis ± 1.0 mm long, ellipsoid .... **2. *E. brachyphylla***

1. *Eragrostiella bifaria*

Eragrostiella bifaria (Vahl.) Bor. in Ind. For. 66: 270. 1940; Bor, Grass. Bur. Cey. Ind. Pak. 494. 1960; Raizada & Jain, Ind. For 92: 640. 1966; Raizada *et al.*, Grasses Upp. Gang. Pl. pt. 3: 76. 1983; Roy, Grass. M. P. 79. 1984; Sreekumar & Nair, Fl. Kera. Grass. 374. 1991; Moulik, Grass. Bamb. Ind. 2: 594. 1997. *Poa bifaria* Vahl., Sym. Bot. 2: 19. 1791. *P. coromandeliana* Koen. ex. Rottl. in Gen. Natur. Freunde. Berlin, Neue Schrift. 4: 191. 1803. *Eragrostis coromandeliana* (Koen ex Rottl.) Trin. in Mem. Acad. Sci. Petersb. ser. 6, 1: 415. 1830. *E. bifaria* (Vahl.) Wight ex Steud., Nom. Bot. ed. 2, 1: 562. 1840; Steud. Syn. Pl. Glum. 1: 264. 1854; Hook. f. Fl. Brit. Ind. 7: 325. 1896. Rang. & Tad. Handb. S. Ind. Grass. 307. t. 227.. 1921; Fischer in Gamble, Fl. Pres. Madr. 1828. 1934 (Repr. ed. 3: 1265. 1957). *Catapodium bifarium* Link., Hort. Berol. 2: 194. 1833. *C. coromandelianum* Link., Hort. Berol. 2: 194. 1833. (PLATE–122).

Perennial. Rhizome absent; root fibrous. Culms 50 cm tall, 1.0 mm thick, green, glabrous, very slender, erect, tufted, ascending, decumbent, young leaves at apex, old remain at base, simple, rooting at base; nodes brown, glabrous; internodes 10 cm long. Leaf–blade 15 cm long, 3.0 mm broad, green, membranous, glabrous, flat, convolute, narrow, linear–lanceolate, acute, base rounded, surface glabrous, margin hyaline & scabrid, mid–vein distinct on lower surface, 2–lateral veins on each side of mid–vein; lateral veins distinct on both surface; leaf–sheath 8 cm long, 2.0 mm thick, green, coriaceous, glabrous, margin membranous, keeled, striate, erect, loose, compressed, shorter than internodes, hairy at mouth; hairs 5.0 mm long, white, acute, base tubercled; veins distinct on outer surface; ligule 0.1 mm long, membranous, truncate, ovate; auricle glabrous. Inflorescence consists of raceme; raceme 20 cm long, 4.0 mm broad, erect, linear; rachis of raceme terete, smooth. Spikelets 10 mm long, 4.0 mm broad, suberect, linear, elliptic, 25 flowered, biseriate, 15 in number, secund, overlapping by half of their length, distant; rachis of spikelets angular, smooth. Lower glume 2.0 mm long 1.0 mm broad, membranous, lanceolate, glabrous, acute, 1–nerved; nerves green, visible, scabrid, ending at apex. Upper glume 2.0 long, 1.0 mm broad,

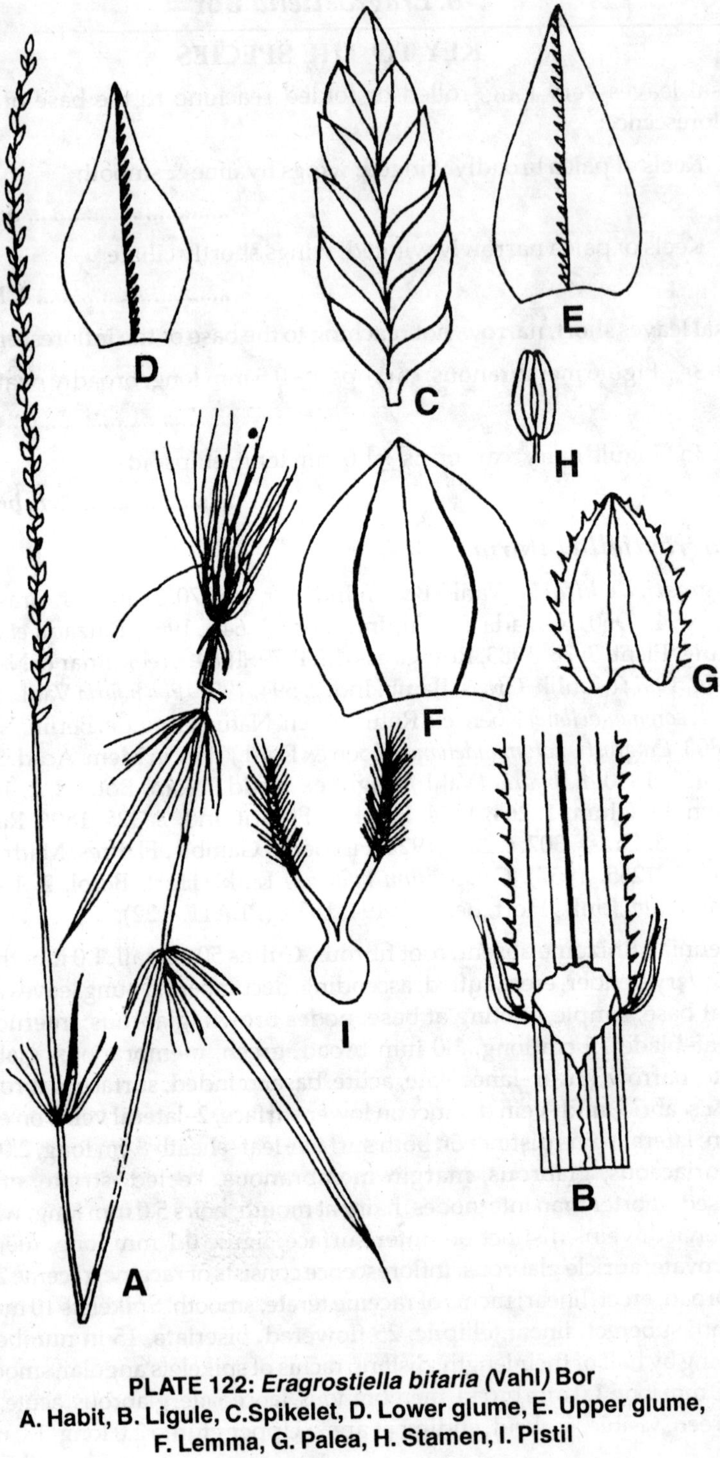

PLATE 122: ***Eragrostiella bifaria*** **(Vahl) Bor**
A. Habit, B. Ligule, C.Spikelet, D. Lower glume, E. Upper glume, F. Lemma, G. Palea, H. Stamen, I. Pistil

membranous, narrowly ovate, lanceolate, acuminate, glabrous, 1–nerved; nerves green, slightly visible, scabrid, ending at apex. Lemma 2.5 mm long, 1.5 mm broad, membranous, broadly ovate, sub–orbicular, acute, laterally compressed, 3–nerved; nerves green, glabrous, ending at apex, central one slightly visible, lateral nerves visible from base to below the apex, faint above, very distant with each other & also with margin. Palea 1.0 mm long, 0.5 mm broad, winged along the keel, margin scabrid, 2–nerved; nerves green, slightly visible, glabrous & marginal, ending at apex. Stamens 3; anther 0.3 mm long, yellow, oblong, glabrous; filament 0.1 mm long, slender, glabrous, shorter than anther. Lodicules absent. Pistil 1.0 mm long; ovary 0.3 mm long, ovate, glabrous; style 0.3 mm long, slender, bifid from base, glabrous; stigma 0.4 mm long, brown, hairy; hairs acute, base broad. Caryopsis 0.5 mm long, 0.2 mm broad, broadly ovate, pale brownish.

Flowering & Fruiting

August–October; Rajdhari, Sonbhadra, Sonaghati, Vindhayachal.

Habitat

Rare, along hill slopes.

Distribution

India, Burma, Sri Lanka.

2. *E. brachyphylla*

E. brachyphylla (Stapf.) Bor in Ind. For. 66: 270. 1940; Bor, Grass. Bur. Cey. Ind. Pak. 494. 1960; Roy, Grass. M. P. 80. 1984; Moulik, Grass. Bamb. Ind. 2: 594. 1997. *Eragrostis brachyphylla* Stapf apud. Hook. f. Fl. Brit. Ind. 7: 327. 1896. (PLATE–123).

Perennial. Rhizome absent; root fibrous. Culm 45 cm tall, 1.0 mm thick, green, glabrous, erect, tufted, very slender, ascending, leafy at base, rooting at base, simple; nodes glabrous, brown; internodes 8.0 cm long. Leaf–blade 10 cm long, 2.0 mm broad, green, flat, narrow, linear–lanceolate, acuminate, base rounded, surface glabrous, margin hyaline, scabrid, mid–vein distinct on lower surface; 3–lateral veins on each side of mid–vein; lateral veins distinct on both surface; leaf–sheath 7.0 cm long, 1.5 mm broad, green, membranous, glabrous, margin hyaline & keeled, slightly tight, compressed, terete, erect, shorter than internodes, veins distinct on both surface, hairy at mouth; hairs 5.0 mm long, white, acute, base tubercled; ligule 1.5 mm long, a ring of white hairs; hairs acute, base broad; auricle absent. Inflorescence consists of raceme; raceme 20 cm long, 4.0 mm broad, erect, open, compressed, linear; rachis of raceme triquetrous, smooth. Spikelet 8.0 mm long, 1.5 mm broad, linear, oblong, secund, biseriate, suberect, 12–flowered, 18 in number; rachis of spikelets smooth, compressed, flat. Lower glume 1.5 mm long, 1.0 mm broad, membranous, ovate, lanceolate, acute, surface glabrous between the nerves also with margin, 1–nerved; nerves green, visible, ending at apex, scabrid. Upper glume 2.0 mm long, 1.0 mm broad, membranous, ovate, lanceolate, acute, glabrous between the nerves also with margin; 1–nerved; nerves green, visible, scabrid, ending at apex. Lemma 2.0 mm long; 1.5 mm broad, membranous, broadly ovate, oblong, lanceolate, apex 4–dentate, glabrous, laterally compressed, 3–nerved; nerves green, visible, glabrous, ending at

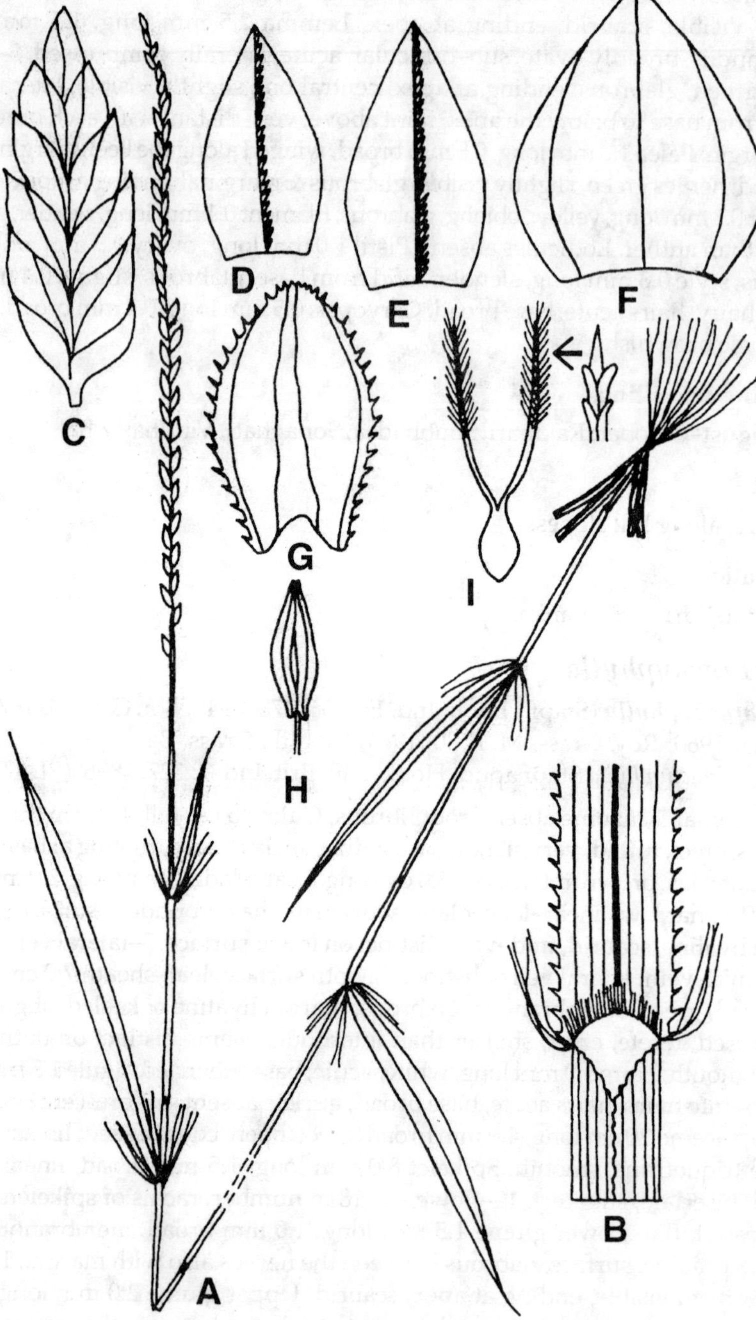

PLATE 123: ***Eragrostiella brachyphylla*** **(Stapf) Bor**
A. Habit, B. Ligule, C. Spikelet, D. Lower glume, E. Upper glume, F. Lemma, G. Palea, H. Stamen, I. Pistil

apex, very distant with each other & also with margin. Palea 2.0 mm long, 1.0 mm broad, hyaline, delicate, concave, obovate, surface glabrous, margin scabrid, also with apex, 2–nerved; nerves green, visible, glabrous & marginal, ending at apex. Stamens 3; anther 1.0 mm long, yellow, linear, oblong, glabrous; filament 0.2 mm long, slender, glabrous, shorter than anther. Lodicules absent. Pistil 1.5 mm long; ovary 0.3 mm long, ovate, elliptic, glabrous; style 0.6 mm long, slender, slightly connate at base, bifid at apex; stigma 0.6 mm long, hairy, brown; hairs apiculate, base broad, in 2–rows. Caryopsis 1.0 mm long, 0.5 mm broad, ellipsoid, sub–trigonous, pale brown.

Flowering & Fruiting

August–December; Chakiya, Duddhi, Meza Road, Mirzapur.

Habitate

Rare, along the rocky slopes.

Distribution

India, Africa, Australia.

3. *E. leioptera*

E. leioptera (Stapf) Bor in Ind. For. 66: 270. 1940: Bor, Grass. Bur. Cey. Ind. Pak. 495. 1960; Roy, Grass. M.P. 80. 1984; Moulik, Grass. Bamb. Ind. 2: 594. 1997. *Eragrostis leioptera* Stapf in Hook. f. Fl. Brit. Ind. 7: 325. 1896. (PLATE–124).

Perennial. Rhizome absent; root fibrous. Culms 60 cm tall, 1.0 mm thick, green, glabrous, erect, very slender, densely tufted, very leafy at base, simple, decumbent, rooting at base; nodes glabrous; internodes 10 cm long. Leaf blade 20 cm long, 2.0 mm broad, green, convolute, narrow, linear, lanceolate, acute, base rounded, glabrous, margin hyaline, scabrid, mid–vein distinct on outer surface, 2–lateral veins on each side of mid–vein; lateral veins very close, distinct on upper surface; leaf sheath 5.0 cm long, 2.0 mm broad, green, membranous, glabrous, margin hyaline, terete, slightly compressed, loose, striate, very much shorter than internodes, hairy at mouth; hairs 3.0 mm long, white, acute, base tubercled, veins distinct on outer surface; ligule 0.5 mm long, a narrow ring of white, tufted, hairs; hairs acute, base broad; auricle absent. Inflorescence consists of racemes; racemes 25 cm long, 6.0 mm broad, erect, open, falling upwards to downward at maturity, yellowish at young, purplish with age; rachis of raceme angular, glabrous, slender. Spikelets 10 mm long, 3.0 mm broad, linear, suberect, alternate, 40 in number, secund, distichous, distant, 15–flowered, sub–sessile. Lower glume 1.5 mm long, 1.0 mm broad, membranous, ovate, lanceolate, acuminate, glabrous, 1–nerved; nerves greenish–brown, visible, scabrid upwards, ending below at the apex. Upper glume 2.0 mm long, 1.0 mm broad, membranous, ovate–lanceolate, acute, dorsally rounded, glabrous 1–nerved; nerves green, visible, antrorsely scabrid, ending at apex. Lemma 2.5 mm long, 1.5 mm broad, membranous, broadly ovate, lanceolate, obtuse, subacute, laterally compressed, glabrous, 3–nerved; nerves green, visible, glabrous, central one ending at apex, lateral 2–nerves ending very below at the apex. Palea 2.5 mm long, 1.5 mm broad, hyaline, delicate, ovate, elliptic, oblong, acuminate, folded & enclosed by lemma, surface glabrous, margin scabrid, 2–nerved; nerves green, slightly visible, glabrous marginal, ending at apex.

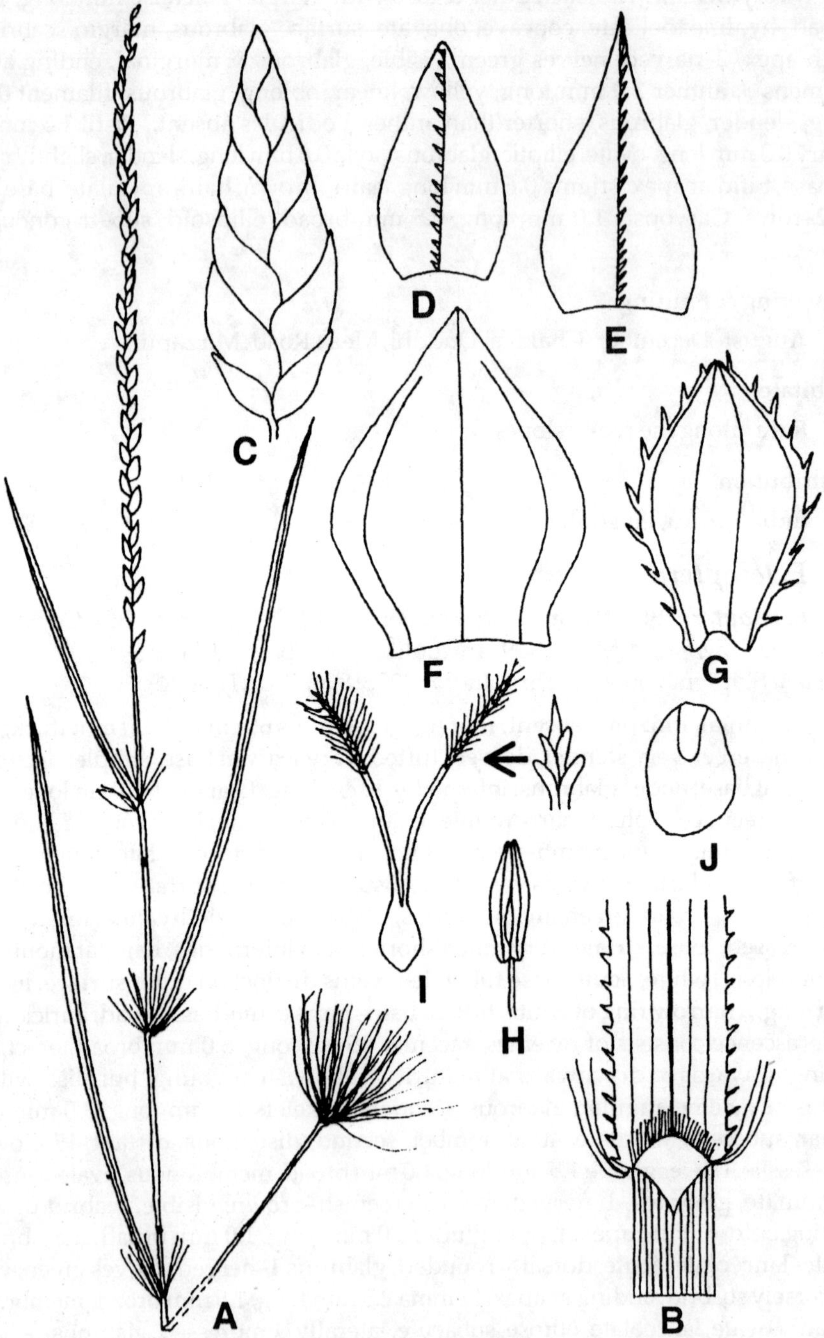

PLATE 124: ***Eragrostiella leioptera*** **(Stapf) Bor**
A. Habit, B. Ligule, C. Spikelet, D. Lower glume, E. Upper glume, F. Lemma, G. Palea, H. Stamen, I. Pistil, J. Caryopsis

Stamens 3; anther 1.0 mm long, yellow, oblong, linear, glabrous; filament 0.2 mm long, slender, glabrous, shorter than anther. Lodicules absent. Pistil 2.0 mm long; ovary 0.5 mm long, ovate, oblong, glabrous; style 1.0 mm long, slender, glabrous, slightly connate at the base, bifid at apex; stigma 0.5 mm long, brown, hairy; hairs acute, base broad, in 2–rows. Caryopsis 1.0 mm long, 0.5 mm broad, narrowly elliptic, brown, smooth.

Flowering & Fruiting

September–November; Rajdhari, Naugarh, Bherafarm, Chandauli.

Habitat

Rare, along rocky slopes.

Distribution

India, Africa, Australia.

4. *E. nardoides*

E. nardoides (Trin.) Bor Ind. For. 66: 270. 1940; Bor, Grass. Bur. Cey. Ind. Pak. 495. 1960; Roy, Grass. M. P. 81. fig. 8. 1984; Moulik, Grass. Bamb. Ind. 2: 594. 1997. *Eragrostis nardoides* Trin. in Mem. Acad. Sci. Petersb ser. 6, 1: 415. 1830; Hook. f. Fl. Brit. Ind. 7: 326.1896. *Catapodium nepalense* Link. Hort. Bot. Berol. 2: 194. 1833. (PLATE–125).

Perennial. Rhizome absent; root fibrous. Culm 50 cm tall, 1.0 mm thick, green, glabrous, erect, densely tufted, ascending, very slender, simple, leafy at base, rooting at base; nodes glabrous; internodes 7.0 cm long. Leaf–blade 45 cm long, 3.0 mm broad, green, convolute, linear, lanceolate, narrow, acuminate, base rounded, upper surface glabrous above, sparsely hairy at base; hairs 3.0 mm long, white, acute, base tubercled; lower surface entirely glabrous, margin hyaline, scabrid, mid–vein distinct on lower surface; 2–lateral veins on each side of mid–vein; lateral veins distinct on upper surface; leaf–sheath 6.0 cm long, 1.5 mm broad, green, membranous, glabrous, margin, hyaline, keeled, slightly compressed, loose, erect, shorter than internodes, veins distinct on both surface; ligule 0.3 mm long, membranous, truncate, lacerate; auricle glabrous. Inflorescence consists of raceme; raceme 20 cm long, 5.0 mm broad, erect, linear, slender, sub–flexuous; rachis of racemes terete, smooth. Spikelets 8.0 mm long, 1.5 mm broad, suberect, linear, oblong, closed, 20–flowered, 10–in number, relatively closed; rachis of spikelet glabrous. Lower glume 1.5 mm long, 1.0 mm broad, membranous, ovate, lanceolate, acute, glabrous, between the nerves also with margin; 1–nerved; nerves green, slightly visible, ending at apex, scabrid. Upper glume 2.0 mm long, 1.0 mm broad, membranous, ovate, lanceolate, acute, glabrous between the nerves also with margin, 1–nerved; nerves green, slightly visible, scabrid, ending at apex. Lemma 2.0 mm long, 1.5 mm broad, membranous, broadly ovate, lanceolate, subacute, glabrous, laterally compressed, 3–nerved; nerves colourless, glabrous, ending at apex, distant with each other also with margin. Palea 2.0 mm long, 1.0 mm broad, hyaline, delicate, ovate, oblong, apiculate, surface glabrous, margin scabrid, 2–nerved; nerves colourless, glabrous & marginal, ending at apex. Stamens 3; anther 0.8 mm long, yellow, linear, oblong, glabrous; filament 0.2 mm long, slender, glabrous, shorter than anther. Lodicules absent. Pistil 1.0 mm long; ovary 0.2 mm long, oblong, glabrous; style 0.4 mm long, slender, slightly connate at base, bifid at apex; stigma 0.4

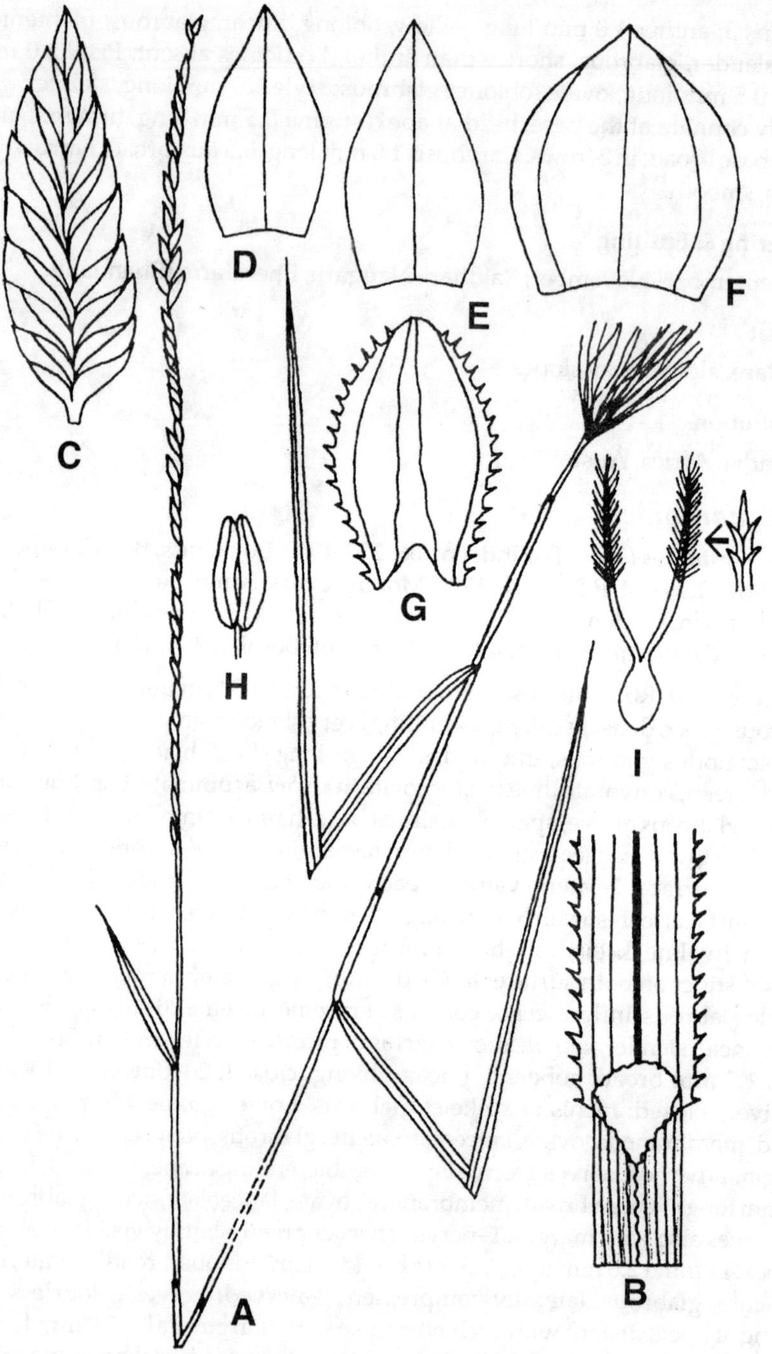

PLATE 125: *Eragrostiella nardoides* **(Trin.) Bor**
A. Habit, B. Ligule, C.Spikelet, D. Lower glume, E. Upper glume, F. Lemma, G. Palea, H. Stamen, I. Pistil

mm long, brown, hairy; hairs acute, base broad, in 2–rows. Caryopsis 0.5 mm long, 0.2 mm broad, brown, subovoid–subglobose, embryo end pointed.

Flowering & Fruiting

August–October; Chakiya, Chandauli, Vindhyachal, Sonaghati.

Habitat

Rare, along rocky slopes.

Distribution

India, Sri Lanka.

9. *Eragrostis* Wolf

KEY TO THE SPECIES

1a. Ligule membranous:

2a. Anther 3:

3a. Panicle 30 cm long, lax; caryopsis polished .... **2. *E. atrovirens***

3b. Panicle 15 cm long, contracted; caryopsis not as above:

4a. Lodicules oblanceolate; caryopsis spindle shaped .... **17. *E. tef***

4b. Lodicules absent; caryopsis elliptic, sub–globose .... **7. *E. curvula***

2b. Anther 2:

5a. Ligule 1.0 mm long; panicle linear, lanceolate; spikelets 1.5 mm long .... **10. *E. japonica***

5b. Ligule 0.1–0.5 mm long; panicle oblong; spikelets more than 1.5 mm long:

6a. Perennial; anther yellow; stigmatic hairs simple .... **8. *E. diarrhena***

6b. Annual; anther pink; stigmatic hairs in 2–rows:

7a. Culm 90 cm long; panicle pyramidal; spikelets 10 flowered .... **13. *E. namaquensis***

7b. Culms 60–75 cm long, panicle ovate, oblong; spikelets with more than 10 flowers:

8a. Auricle present; spikelets 4.0 mm long, gray .... **9. *E. gangetica***

8b. Auricle absent; spikelets 12 mm long, purple .... **21. *E. unioloides***

1b. Ligule hairy:

9a. Lemma margin hairy:

10a. Culm 15 cm long; panicle cylindric, purplish; caryopsis ovoid
.... 5. *E. ciliata*

10b. Culm 35 cm long; panicle subspiciform; light green; caryopsis obovoid
.... 6. *E. coarctata*

9b. Lemma margin not hairy; scabrid or glabrous:

11a. Palea transversely hairy:

12a. Panicle compact:

13a. Panicle spike like, wooly, cylindric; caryopsis 0.5 mm long
.... 4. *E. ciliaris*

13b. Panicle contracted, not wooly, sub cylindric; caryopsis 1.0 mm long
.... 16. *E. riparia*

12b. Panicle effuse:

14a. Branches and branchlets of panicle hairy on the axils; stigma purple
.... 22. *E. viscosa*

14b. Branches and branchlets of panicle not hairy on the axils; stigma cream coloured
.... 18. *E. tenella*

11b. Palea transversely scabrid or glabrous:

15a. Branche of panicle hairy on the axils:

16a. Lower branches of panicle whorled; lower glume hyaline; stigmatic hairs simple
.... 15. *E. pilosa*

16b. Lower branches of panicle not as above; lower glume membranous; stigmatic hairs not as above:

17a. Plant glandular; spikelets 4.0 mm long; anther violet
.... 11. *E. maderaspatana*

17b. Plant eglandular; spikelets 8.0 mm long; anther yellow
.... 12. *E. minor*

15b. Branches of panicle not hairy on the axils:

18a. Panicle thyriform; pedicels of spikelets 2.0 mm long
.... 1. *E. aspera*

18b. Panicle oblong; pedicels of spikelets 1.0 mm long:

19a. Spikelets 40 flowered; lemmas wavy & hyaline at the margin .... **20. *E. tremula***

19b. Spikelets less then 40 flowered; lemmas not as above:

20a. Upper glume 3–nerved; anthers 0.5 mm long .... **3. *E. cilianensis***

20b. Upper glume 1–nerved; anther more than 0.5 mm long:

21a. Caryopsis compressed, dorsally grooved .... **19. *E. tenuifolia***

21b. Caryopsis not as above:

22a. Spikelets blue–grey; lemma membranous; stamens keeled, apex diverticate .... **14. *E. nutans***

22b. Spikelets deep brown; lemma sub-membranous; stamens not keeled, apex not diverticate .... **23. *E. zeylanica***

1. *Eragrostis aspera*

Eragrostis aspera (Jacq.) Nees, Fl. Afr. Austr. 408. 1841; Bor, Grass. Bur. Cey. Ind. Pak. 502. f. 55. 1960; Moulik, Grass. Bamb. Ind. 2: 5597. 1997. *Poa aspera* Jacq. Hort. Vindob. 3: 32. 1776. *P. paniculata* Roxb. Fl. Ind. 1: 341. 1820. *Eragrostis paniculata* (Roxb.) Steud. Syn. Pl. Glum. 1: 226. 1854. *E. laxiflora* Schard. in Linnaea 12: 451. 1838. (PLATE–126).

Annual. Rhizome absent; root fibrous. Culm 35 cm tall, 1.5 mm thick, green, glabrous, erect, densely tufted, very slender, ascending, simple, rooting at base, leafy, nodes glabrous; internodes 6.0 cm long. Leaf blade 10 cm long, 2.0 mm broad, green, glabrous, convolute, acute, base rounded, linear–lanceolate, margin hyaline, scabrid, mid–vein distinct on lower surface, 4–lateral veins on each side of mid–vein; lateral vein distinct on upper surface; leaf–sheath 4.0 cm long, 2.5 mm broad, green, membranous, glabrous, margin hyaline & slightly keeled, loose, slightly compressed, striate, veins distinct on outer surface, shorter than internodes, hairy at mouth; hairs 5.0 mm long, white, acute with tubercled base; ligule 0.2 mm long, a ring of white dense hairs; hairs acute, base broad; auricle absent. Inflorescence consists of panicle; panicle 10 cm long, 1.5 cm broad, loose, erect, pyramidal, purplish– slaty, thyrisiform, ovate, elliptic, falling upwards from downwards at maturity; rachis of panicle scabrid, fragile, grooved. Branches 1.0 cm long, suberect, whorled, scabrid, decreasing in length upwards, densely arranged. Spikelets 4.0 mm long, 1.5 mm broad, distichous, linear, ovate, pedicelled, 10–flowered; pedicels 2.0 mm, scabrid, slender; rachis of spikelets scabrid, filiform, angular. Lower glume 0.7 mm long, 0.4 mm broad, membranous, ovate, lanceolate, acuminate, surface glabrous, margin scabrid, 1–nerved; nerves green, visible, glabrous, ending at apex. Upper glume 1.0 mm long, 0.5

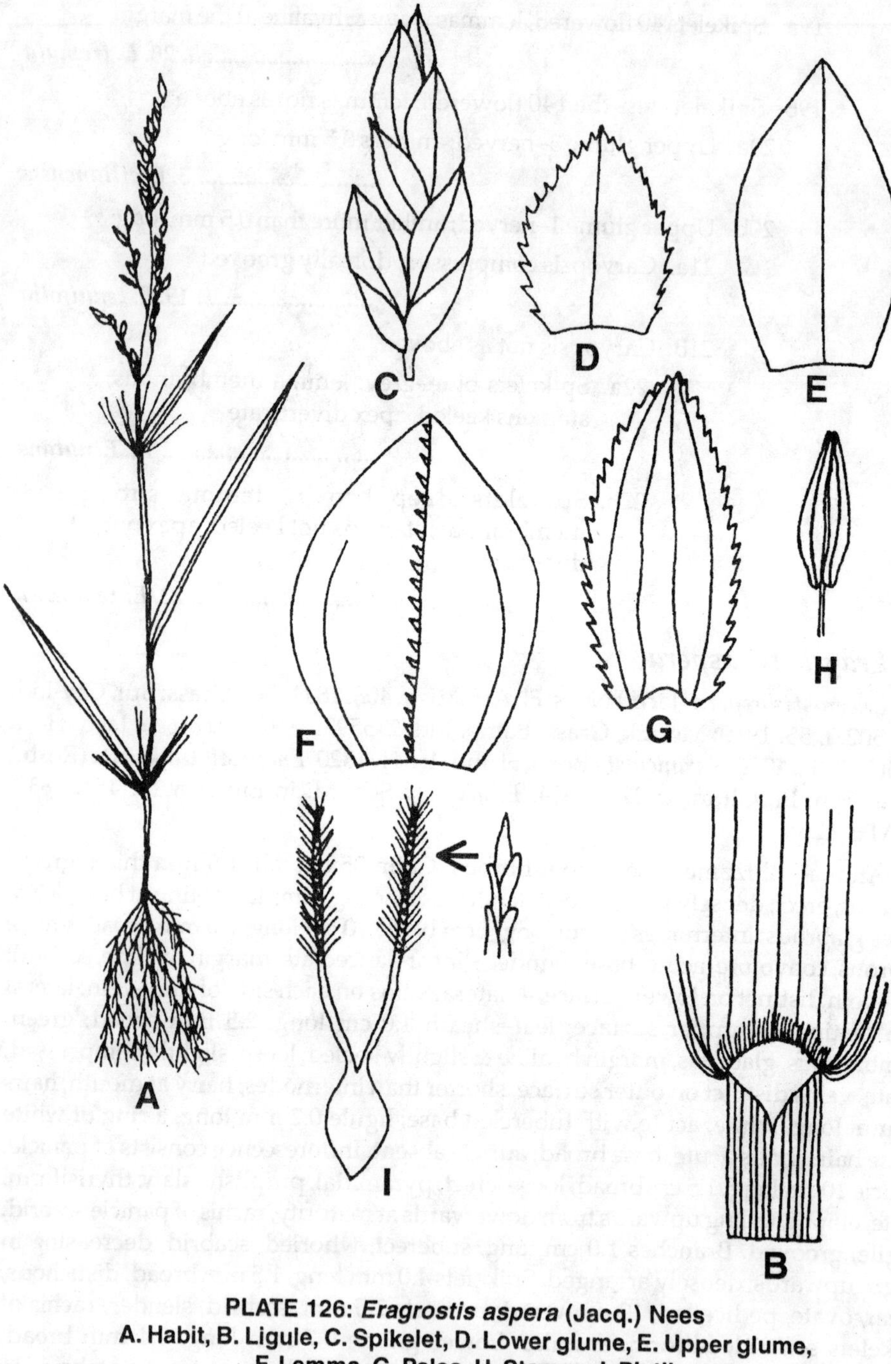

PLATE 126: ***Eragrostis aspera*** **(Jacq.) Nees**
A. Habit, B. Ligule, C. Spikelet, D. Lower glume, E. Upper glume, F. Lemma, G. Palea, H. Stamen, I. Pistil

mm broad, membranous, ovate, lanceolate, acute, glabrous, 1–nerved; nerves green, visible, glabrous, ending at apex. Lemma 1.5 mm long, 1.0 mm broad, membranous, broadly ovate, obtuse, lanceolate, acuminate, laterally comprssed, glabrous, 3–nerved; nerves green, visible, central one antrorsely scabrid, ending at apex, lateral 2–nerves glabrous, ending very below at apex. Palea 1.5 mm long, 1.0 mm broad, hyaline, delicate, oblong, ovate, elliptic, folded & enclosed by lemma, surface glabrous, margin scabrid, 2–nerved; nerves green, slightly visible, glabrous & marginal, ending at apex. Stamens 3; anther 0.3 mm long, yellowish–brown, glabrous; filament 0.1 mm long, slender, glabrous, shorter than anther. Lodicules absent. Pistil 1.0 mm long; ovary 0.2 mm long, obovate, glabrous; style 0.3 mm long, slender, glabrous, bifid from base; stigma 0.5 mm long, brown, hairy; hairs acute, base broad, in 2–rows. Caryopsis 0.5 mm long, 0.3 mm broad, sub–globose, pale brown, smooth.

Flowering & Fruiting

August–October; Chakiya, Chandauli, Bhatpur, Dohrighat, Patnaghat.

Habitat

Along banks of rivers & on rocky soils.

Distribution

India, Africa, Mascarene Islands.

2. *E. atrovirens*

E. atrovirens (Desf.) Trin. ex. Steud. Nom. Bot. ed. 2 (1): 562. 1840; Raizada & Jain, Indian Forester 92: 640. 1966; Gilliland in Fl. Malaya 3: 68. 1971; Cope in Fl. W. Pak. 143. 92. 1982; Raizada *et al.*, Grasses Upper. Gang. Pl. 3: 81. 1983; Roy, Grass. M. P. 85. 1984; Misra & Verma, Fl. Allah. 439. 1992; Moulik, Grass. Bamb. Ind. 2: 597. 1997. Bor, Grass. Bur. Cey. Ind. Pak. 503. 1960. *Poa atrovirens* Desf. Fl. Atlant 1: 73. 1798. (PLATE–127).

Strongly perennial. Rhizome absent; root fibrous. Culms 45 cm tall, 2.0 mm thick, erect, rigid, slender, rooting at base, green, glabrous; node brown, glabrous; internodes 8.0 cm long. Leaf–blade 7.0 cm long, 1.5 mm broad, green, narrow, flat, acuminate base rounded, surface glabrous, margin scabrid, linear–convolute, mid–vein distinct on lower surface; 2–lateral veins on each side of mid vein; lateral vein distinct on upper surface; leaf–sheath 7.0 cm long, 2.0 mm broad, glabrous, compressed, green, shorter than internodes, hairy at mouth; hair 5.0 mm long, white, acute, base tubercled; ligule 0.3 mm long, membranous, truncate; auricle glabrous. Inflorescence a panicle, panicle 30 cm long, lax, ovate, oblong. Branches 3.0 cm long, suberect, spreading, scabrid, decreasing in length upwards. Spikelet 10 mm long, 2.0 mm broad, 6–12–flowered, grey–straw, ovate, oblong, glaucous, pedicellate. Lower glume 0.8 mm long, 0.4 mm broad, membranous, ovate–lanceolate, acute, 1–nerved; nerve green, visible, glabrous, ending at apex. Upper glume 1.0 mm long, 0.5 mm broad, membranous, glabrous, boat shaped, lanceolate, acute, 1–nerved; nerve green, visible, glabrous, ending at apex. Lemma 2.0 mm long, 1.0 mm broad, broadly ovate, sub–acute, 3–nerved; nerves green, visible, lateral nerve glabrous; mid–nerve scabrid. Palea 1.5 mm long, 1.0 mm broad, narrowly ovate, 2–nerved; scabrid along the nerves. Stamen

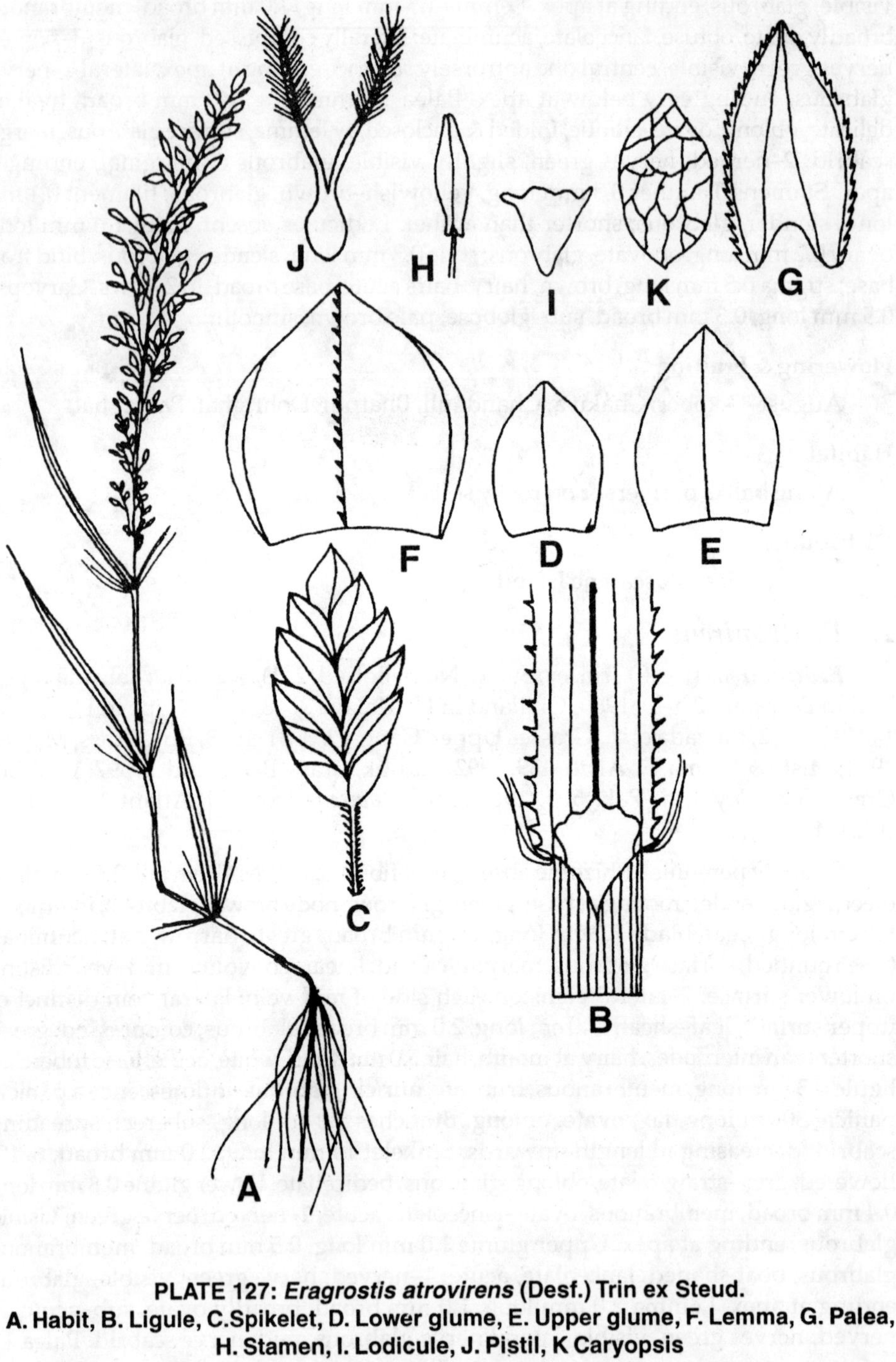

PLATE 127: ***Eragrostis atrovirens*** **(Desf.) Trin ex Steud.**
A. Habit, B. Ligule, C.Spikelet, D. Lower glume, E. Upper glume, F. Lemma, G. Palea, H. Stamen, I. Lodicule, J. Pistil, K Caryopsis

3; anther 1.0 mm long, brownish yellow, glabrous; filament 0.1 mm long, slender, glabrous, shorter than anther. Lodicule 2, 0.5 mm long, 0.3 mm broad, glabrous, hyaline, broadly truncate. Pistil 1.5 mm long; ovary 0.4 mm long, ovate, glabrous; style 0.5 mm long, bifid from base, slender, glabrous; stigma, 0.6 mm long, brown, hairy; hairs, acute, base broad. Caryopsis 1.0 mm long, 0.5 mm broad, red brown, polished, cylindrical.

Flowering & Fruiting

September–March; Chakiya, Babatpur, Sindhora, Ghazipur, Naini, Mau, Basti.

Habitat

Found on river-beds, margins of agricultural fields & other damp places.

Distribution

India, Africa, Mascarene Islands, Burma, Malaysia.

3. *E. cilianensis*

E. cilianensis (All.) Vignolo–Lutani in Malpighia 18: 386. 1904; Bor, Grass. Bur. Cey. Ind. Pak. 503. 1960; Fischer in Gamble, Fl. Pres. Madr. 1827. 1934. (Repr. ed. 3: 1264. 1957); Cope in Nasir & Ali, Fl. W. Pak. 143: 93. 1982; Blatter & McCann, Bomb. Grass. 237. t. 157. 1984; Roy, Grass. M. P. 86. 1984; Sreekumar & Nair, Fl. Kera. Grass. 379. 1991. Moulik, Grass. Bamb. Ind. 2: 598. 1997. *E. major* (Linn.) Host. Gram. Austr. 4: 14.t. 24. 1809. *E. polymorpha* (Koenig ex Rottler) Roem. ex. Schult., Syst. Veg. 2: 575. 1817. *E. megastachya* (Koel.) Link, Hort. Berol. 1: 187. 1827. *E. vulgaris* var. *megastachya* (Koel.) Coss. et Germ., Fl. Env. Pairs. 641. 1845. *E. flexuosa* (Roxb.) Steud. Syn. Pl. Glum. 1: 266. 1854. *E. poaeoides* var. *megastachya* (Koel.) A. Gray, Man. Bot. North. Unit. States. ed. 2: 563. 1856. *E. minor* var. *major* (Host.) Beck, Fl. Nied. Ost. 1: 88. 1890. *E. eragrostis* (Linn.) Mac Mill, Metasperm. Minnesota Vall. 75. 1892. *E. megastachya* var. *cilianensis* (All.) Aschers. et Graen., Syn. Mittleleurop. Fl. 2, Abt. 1, 371. 1900. *E. minor* var. *megastachya* (Koel). Burtt–Davy. in Jepson, Fl. West. Midd. Calif. 60. 1901. *E. costata* F. Lurner in Proc. Linn. Soc. N.C. Wales. 30: 91. 1905. *E. vulgaris* subsp. *major* Host. Rouy, Fl. France 14: 262. 1913. *E. articulata* DeWild. in Bull. Jard. Bot. Brux. 6: 58. t. 2. 1919. *E. starosselskyi* Grossheim in Not. Syst. Herb. Hort. Bot. Petrop 4: 18. 1923. *E. schweinfurthiana* Jedwabnick in Mez. Bot. Archis 5: 194. 1924. *Briza eragrostis* Linn., Sp. Pl. ed. 1: 70. 1753. *B. oblonga* Monech. Meth. Pl. 185. 1794. *B. megastachya* (Koel.) Hort. ex Roem. et Schult., Syst. Veg. 2: 585. 1817. *Poa cilianensis* All., Fl. Pedem. 2: 246. t. 91. f. 2. 1785. *P. megastachya* Koel. Descr. Gram. 181. 1802. *P. polymorpha* Koenig ex Rott. in Ges. Natur. Freunde, Berlin. 4: 194. 1803. *P. eragrostis* (Linn.) Brot. Fl. Lust. 1: 103. 1804. *P. oblonga* Moench. Baumg. Enum. Strip. Transs. 3: 238. 1816. *P. flexuosa* Roxb. Fl. Ind. 1: 340. 1820. *P. roxburghiana* Schult., Syst. Veg. 2: Mant. 314. 1824. *P. tortuosa* Spreng. Syst. Veg. 1: 345. 1825. *P. chachetica* Schumacher, Besker. Guin. Pl. 66. 1827. *Erosion ciliare* (All.) Lunell in Amer. Midl. Nat. 4: 221. 1915. *Megastachya oblonga* (Moench) P. Beauv., Ess. Agrost. 74. 1812. *M. eragrostis* (Linn.) Roem. et Schult. Syst. Veg. 2: 575, 584. 1817. (PLATE–128).

Annual. Rhizome absent; root fibrous. Culms 50 cm tall, 1.0 mm thick, green, glabrous, erect, densely tufted, very slender, ascending, decumbent, geniculate at

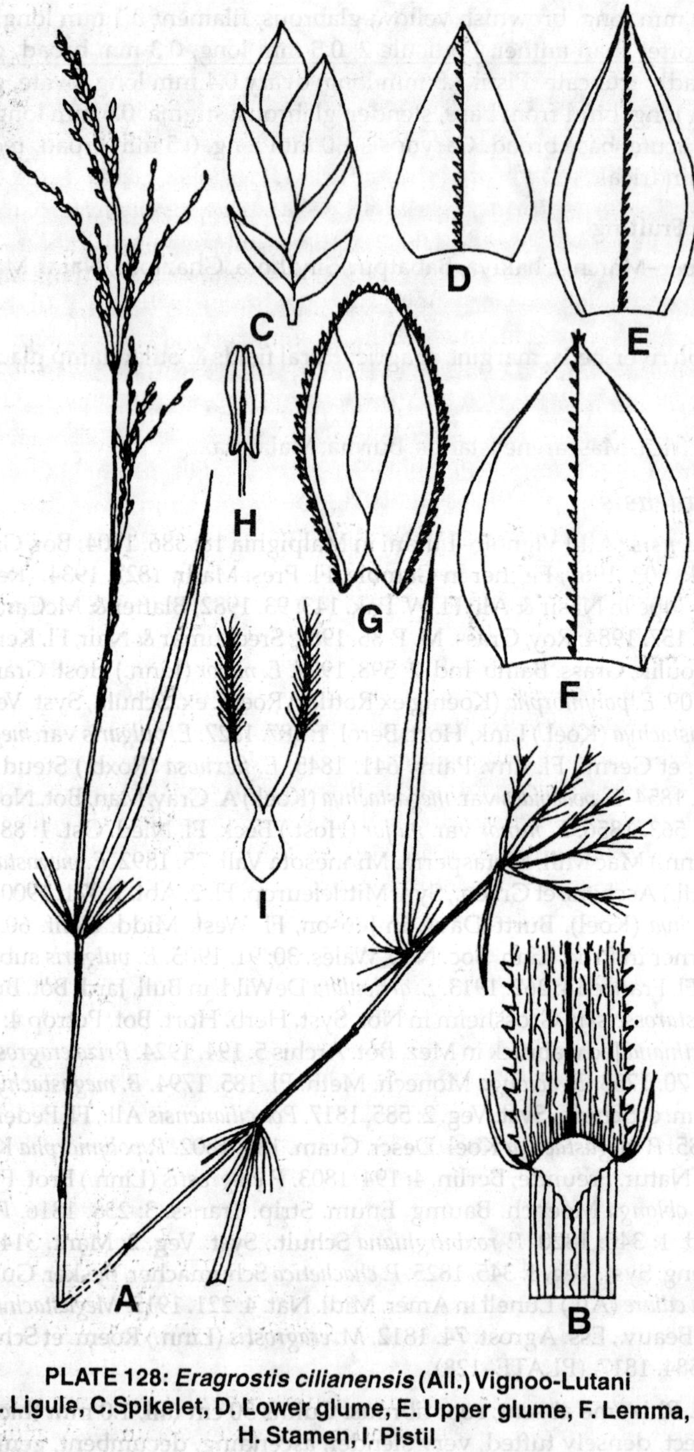

PLATE 128: *Eragrostis cilianensis* (All.) Vignoo-Lutani
A. Habit, B. Ligule, C.Spikelet, D. Lower glume, E. Upper glume, F. Lemma, G. Palea, H. Stamen, I. Pistil

base, simple, leafy, rooting at base; nodes brown, glabrous; internodes 7.0 cm long. Leaf–blade 10 cm long, 4.0 mm broad, green, narrow, flat, linear, lanceolate, acuminate, base rounded, submembranous, sparsely hairy on upper surface; hairs 2.0 mm long, white, acute, base broad; glabrous on lower surface, margin hyaline, scabrid, mid–vein distinct on lower surface, 2–lateral veins on each side of mid–vein; lateral veins distinct on upper surface; leaf–sheath 6.0 cm long, 2.0 mm broad, green, membranous, glabrous, margin hyaline, slightly keeled, loose, erect, shorter than internodes, hairy at mouth; hairs 5.0 mm long, white, acute, base tubercled; veins distinct on both surface; ligule 0.1 mm long, a ring of hairs; hairs white, acute, base broad; auricle absent. Inflorescence consists of pancile; panicle 8.0 cm long, 3.0 cm broad, open, erect, condensed, oblong, pyramidal; rachis of panicle scabrid. Branches 6.0 cm long, scabrid, lax, erect, slender. Spikelets 3.5 mm long, 1.5 mm broad, ovate, oblong, lanceolate, 6–flowereded, distichous, pedicellate; pedicel 1.0 mm long, slender, glabrous; rachis of spikelets angular, glabrous. Lower glume 1.0 mm long, 0.5 mm broad, membranous, ovate, lanceolate, acute, surface glabrous except the nerves, also with margin, 1–nerved; nerves colourless, scabrid, ending at apex. Upper glume 2.0 mm long, 1.0 mm broad, membranous, ovate, lanceolate, acute, surface glabrous except the nerves, also with margin, 3–nerved; nerves green, visible, ending at apex, central one scabrid, lateral nerves glabrous. Lemma 2.0 mm long, 1.5 mm broad, membranous, broadly ovate, lanceolate, acute and tridentate at apex, laterally compressed, surface glabrous except the nerves, also with margin, 3–nerved; nerves green, visible, ending at apex, central one scabrid, lateral nerves glabrous, slightly close to the margin. Palea 1.5 mm long, 1.0 mm broad, hyaline, delicate, oblong, ovate, lanceolate, obtuse, glabrous, margin, sharply folded inwards, transversely scabrid, 2–nerved; nerves green, visible, glabrous & marginal, ending below at apex. Stamens 3; anther 0.5 mm long, yellow–brown, oblong, glabrous; filament 0.3 mm long, slender, glabrous, shorter than anther. Lodicules absent. Pistil 1.0 mm long; ovary 0.2 mm long, ovate, glabrous; style 0.3 mm long, slender, glabrous, bifid from base; stigma 0.5 mm long, brown, hairy; hairs acute, base broad. Caryopsis 0.5 mm long, 0.5 mm broad, globose, reddish–brown, rugulose.

Flowering & Fruiting

September–November; Chakiya, Basti, Ballia, Varanasi, Mirzapur.

Habitat

Rare, along banks of canals.

Distribution

India, Tropical & Sub tropical Asia.

4. *E. ciliaris*

E. ciliaris (L.) R. Br. in Tuckey, Narr. Exp. Cong. App. 478. 1812; Link. Hort. Berol. 1: 192. 1827; Boiss, Fl. Or. 5: 582. 1884; Baker, Fl. Maurit 456. 1877; Duthie, Grass. N. W. Ind. 37. 1883; Duthie, Fodd. Grass. N. Ind. 62. 1888; Aitchis, Cat. Punj. Pl. 169. 1869; Lisboa, J. Bomb. Nat. Hist. Soc. 7: 380. 1893; Hook. f. Fl. Brit. Ind. 7: 314. 1896; Cooke, Fl. Bomb. 2: 1023.1908; Bor, Grass. Bur. Cey. Ind. Pak. 506. 1960; Raizada &

Jain, Ind. For. 92: 640.1966; Panigrahi & Rajagopal, J. Ind. Bot. Soc. 47: 236. 1968; Chien–Chan Hsu, Taiwan Grass. 395. t. 93. 1975; Cope in Nasir & Ali, Fl. W. Pak. 143: 86. 1982; Raizada *et al.*, Grass. Upp. Gang. Pl. pt. 3: 82.1982; Roy, Grass. M.P. 86. 1984; Blatter & McCann, Bomb. Grass. 231. t. 154. 1984; Misra & Verma, Fl. Allah. 440. 1992; Moulik, Grass. Bamb. Ind. 2: 598. 1997. *E. ciliaris* var. *ciliaris* (Linn.) Stapf in Hook. f. Fl. Brit. Ind. 7: 314. 1896. *E. boryana* (Willd.) Steud. Nom. Bot. ed. 2: 562. 1840. *E. lobata* Trin. in Mem. Acad. Petersb. ser. 6, 1: 396. 1831. *E. lepida* Hochst. ex A. Rich. Tent. Fl. Abyss. 2: 424. 1851. *E. plumosa* Boiss, Fl. Or. 5: 582. 1884. *E. pulchella* Parl. in Hook. Niger. Fl. 186. 1841. *E. arabica* Jaub. & Spach. Pl. Or. 4: 31. t. 322. 1810. *Poa ciliaris* Linn. Syst. Nat. ed. 10, 2: 875. 1759; Linn. Sp. Pl. 102. 1753. *P. amboinica* Linn. sec. Retz. Obs. Bot. 4: 20. 1786. *P. boryana* Willd., Enum. Pl. Hort. Berol. 109. 1809. *Megastachya ciliaris* (Linn.) P. Beauv., Ess. Agrost. 74. 1812. *M. boryana* (Willd.) Roem. et Schult., Syst. Veg. 2: 592. 1817. *Cynodon ciliaris* (Linn.) Rasp. in Ann. Sci. Nat. 5: 302. 1825. (PLATE–129).

Annual. Rhizome absent; root fibrous. Culm 35 cm tall, 1.0 mm thick, green, glabrous, tufted, ascending from a geniculate base, slender, leafy, simple, rooting at base; nodes glabrous; internodes, 6.0 cm long. Leaf–blade 6.0 cm long, 2.0 mm broad, green, narrow, convolute, linear, lanceolate, acute, base rounded, surface sub–coriaceous, glabrous, margin hyaline, scabrid, mid–vein distinct on lower surface, 2–lateral vein on each side of mid–vein; lateral vein distinct on upper surface; leaf–sheath 4.0 cm long, 3.0 mm broad, green, surface sub–coriaceous, glabrous, margin sub membranous, slightly keeled, loose, erect, striate, compressed, shorter than internodes, hairy at mouth; hairs 5.0 mm long, white, acute, base tubercled; veins distinct on outer surface; ligule 0.2 mm long, a ring of white hairs; hairs acute, base broad; auricle absent. Inflorescence consist of panicle; panicle 5.0 cm long, 5.0 mm broad, spike–like, linear, narrow, erect, open, cylindrical, contracted, spiciform, lobed, interrupted, compact, appears wooly due to long hairs of the palea keels; rachis of panicle glabrous. Branches very short, divided from base. Spikelets 4.5 mm long, 2.0 mm broad, strongly compressed, clustered, 6–flowered, pale, biseriate, pedicelled; pedicels 0.5 mm long, glabrous; rachis of spikelet fragile. Lower glume 1.0 mm long, 0.5 mm broad, membranous, ovate, lanceolate, acute, glabrous, 1–nerved; nerves green, slightly visible, glabrous, ending at apex. Upper glume 1.2 mm long, 0. 6 mm broad, membranous, ovate, lanceolate, acute, glabrous, 1–nerved; nerves green, visible, glabrous, ending at apex. Lemma 1.0 mm long, 0.5 mm broad, membranous, oblong, subtruncate, mucronulate, spreading, laterally compressed, glabrous, 3–nerved; nerves green, visible, glabrous, ending at apex, lateral nerves close to the margin. Palea 1.0 mm long, 0.5 mm broad, hyaline, delicate, oblong, lanceolate, glabrous, nervesless, margin folded inwards, transversely keeled; keel hairy; hairs 2.0 mm long, colourless, acute, base tubercled. Stamens 3; anther 0.3 mm long, light pink, oblong, glabrous; filament 0.2 mm long, slender, glabrous, shorter than anther. Lodicules absent. Pistil 1.0 mm long; ovary 0.2 mm long, obovate, glabrous; style 0.3 mm long, slender, glabrous, bifid from base; stigma 0.5 mm long, purplish, hairy; hairs acute, base broad, in 2–rows. Caryopsis 0.5 mm long, 0.3 mm broad, ovate, elliptical, blackish–brown, smooth, embryo shiny & yellow.

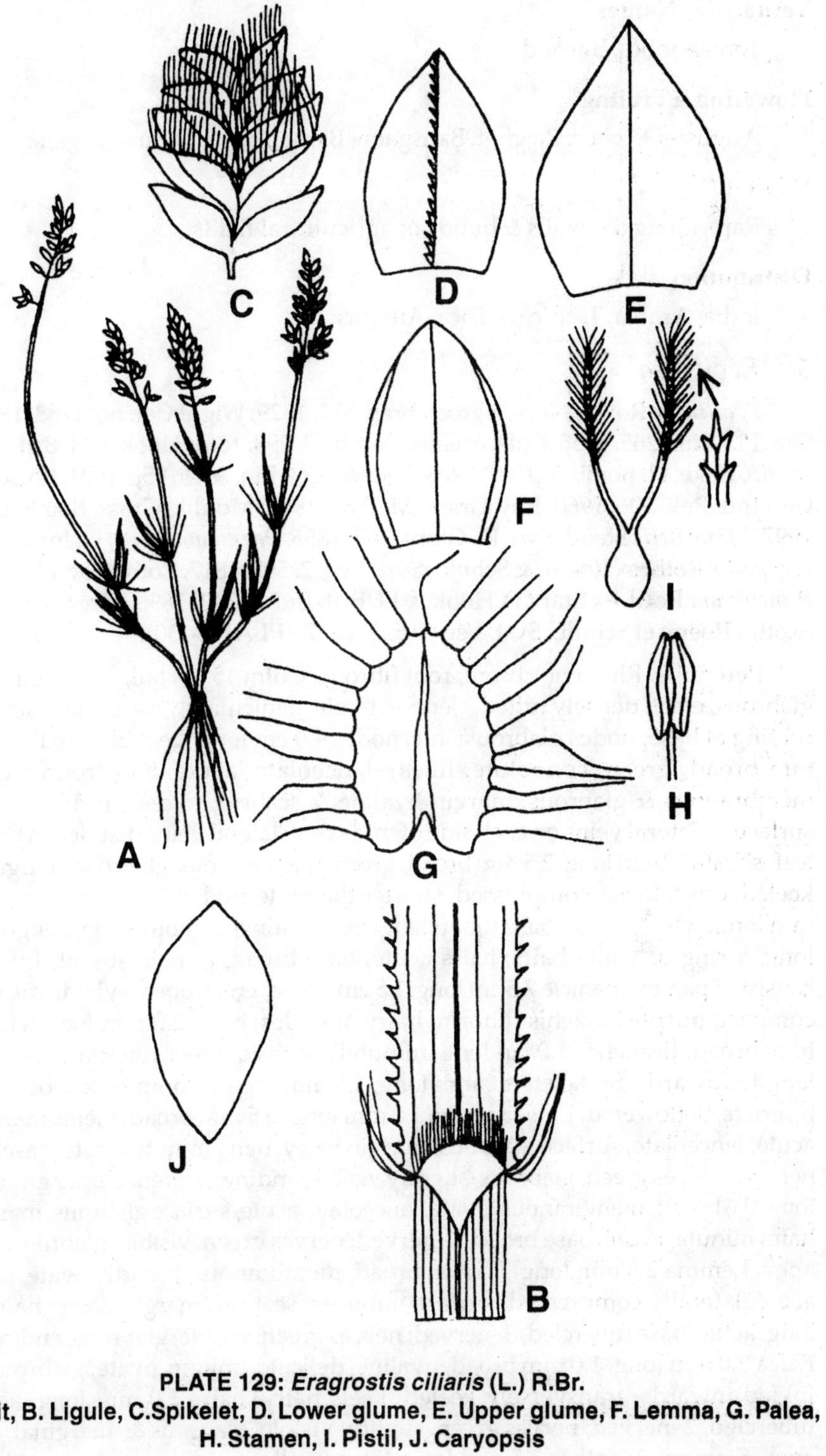

PLATE 129: *Eragrostis ciliaris* (L.) R.Br.

A. Habit, B. Ligule, C.Spikelet, D. Lower glume, E. Upper glume, F. Lemma, G. Palea, H. Stamen, I. Pistil, J. Caryopsis

Vernacular Names

Torchandbol, Burbud.

Flowering & Fruitng

August–October; Rajghat, Bansgaon, Barhaj, Basti, Dumariyaganj.

Habitat

Rare, along the walls & bunds of agricultural fields.

Distribution

India, Arabia, Tropical Africa, America.

5. *E. ciliata*

E. ciliata (Roxb.) Nees, Agrost. Bras. 512. 1829; Wight Cat. no. 1788. 1840; Steud. Syn. Pl. Glum 265. 1855; Dalz. & Gibs. Bomb. Fl. 298. 1861; Hook. f. Fl. Brit. Ind. 7: 313. 1896; Cooke, Fl. Bomb. 2: 1022. 1908; Haines, Bot. Bih. & Or. 956. 1924; Bor, Grass. Bur. Cey. Ind. Pak. 506. 1960; Roy, Grass. M. P. 86. 1984; Moulik, Grass. Bamb. Ind. 2: 601. 1997. *E. rupestris* Steud. Syn. Pl. Glum. 265. 1855. *Poa ciliata* Roxb. Fl. Ind. 1: 336. 1820. *P. rupestis* Roth. ex Roem. & Schult., Syst. Veg. 2: 592. 1817; Roth. Nov. Pl. Sp. 71. 1821. *P. incurvata* Rottl. ex Stapf in Hook. f. Fl. Brit. Ind. 7: 313. 1896. *Megastachya rupestris* (Roth.) Roem. et Schult., Syst. Veg. 2: 592. 1817. (PLATE–130).

Perennial. Rhizome absent; root fibrous. Culm 15 cm tall, 1.0 mm thick, green, glabrous, erect, densely tufted, slender, terete, geniculately ascending, simple, leafy, rooting at base; nodes glabrous; internodes 4.0 cm long. Leaf blade 6.0 cm long, 2.5 mm broad, green, convolute, linear–lanceolate, acute, base rounded, surface membranous & glabrous, margin hyaline & scabrid, mid–vein distinct on lower surface, 2–lateral veins on each side of mid–vein; lateral veins distinct on both surfae; leaf–sheath 3.0 cm long, 2.5 mm broad, green, membranous, glabrous, margin, hyaline keeled, erect, loose, compressed, shorter than internodes, hairy at mouth; hairs 5.0 mm long, white, acute, base tubercled; veins distinct on both surface; ligule 0.2 mm long, a ring of white hairs; hairs acute, base broad; auricle absent. Inflorescence consist of panicle; panicle 7.5 cm long, 1.5 cm broad, erect, open, cylindrical, spiciform, compact, purplish, rachis filiform, hairy at nodes; hairs 2.0 mm long white, acute, base broad. Branches 1.0 cm long, minutely scabrid, erect, alternate, decreasing in length upwards. Spikelets 4.0 mm long, 2.5 mm broad, compressed, oblong, ovate, biseriate, 8–flowered. Lower glume 1.0 mm long, 0.5 mm broad, membranous, ovate, acute, lanceolate, surface glabrous, margin hairy; hairs minute, acute, base broad, 1–nerved; nerves green, glabrous, slightly visible, ending at open. Upper glume 1.2 mm long, 0.6 broad, membranous, ovate, lanceolate, acute, surface glabrous, margin hairy; hairs minute, acute, base broad, 1–nerved; nerves green, visible, glabrous, ending at apex. Lemma 2.5 mm long, 1.5 mm broad, membranous, broadly ovate, lanceolate, acute, laterally compressed, surface minutely scabrid, margin hairy; hairs 1.0 mm long, acute, base tubercled, 3–nerved; nerves green, visible, glabrous, ending at apex. Palea 2.0 mm long, 1.0 mm broad, hyaline, delicate, oblong, ovate, glabrous, margin folded inwards, transversely keeled; keels hairy; hairs 1.0 mm long, acute, base tubercled, 2–nerved; nerves green, slightly visible, glabrous & marginal, ending at apex. Stamens 3; anther 1.5 mm long, linear, yellow, oblong, glabrous; filament 0.3

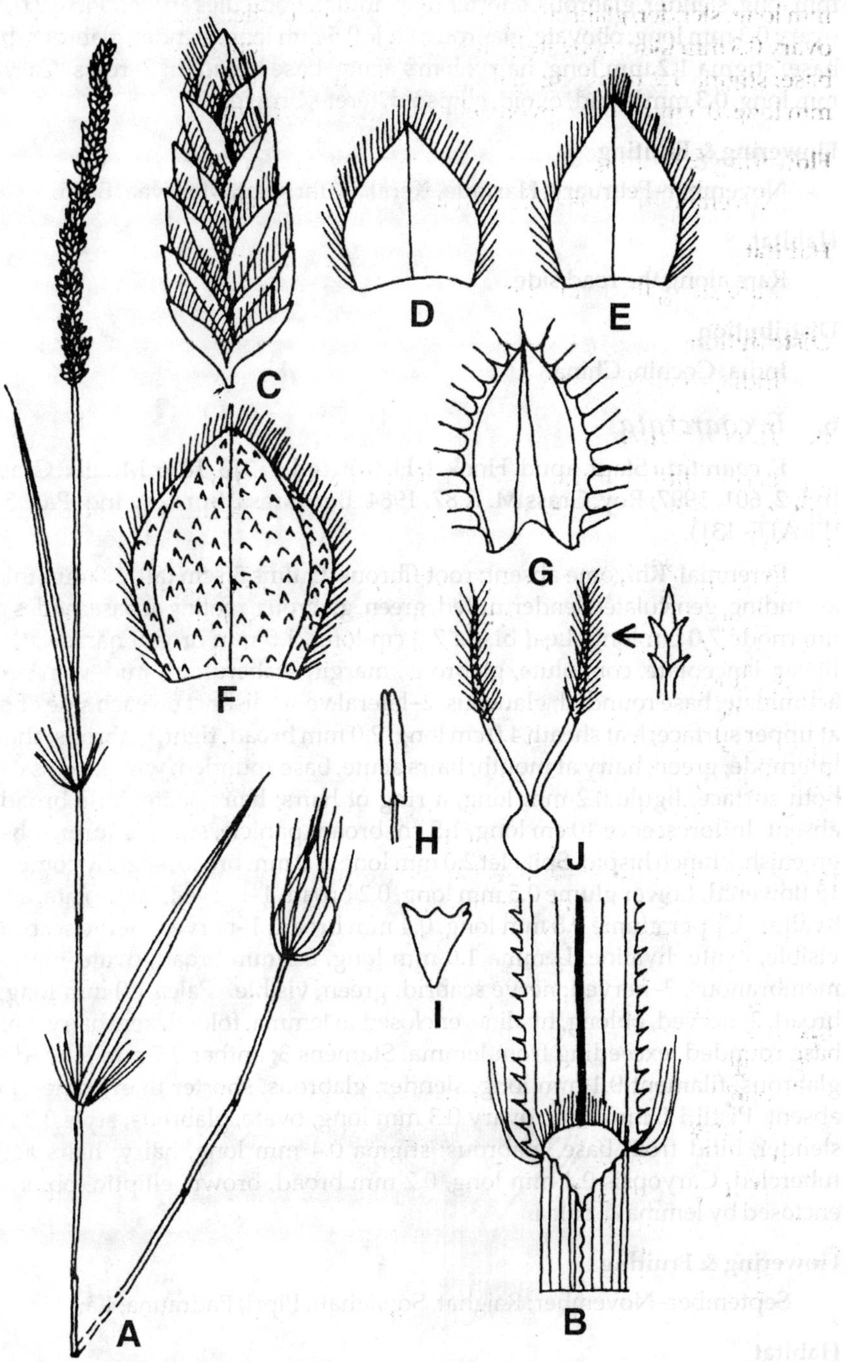

PLATE 130: ***Eragrostis ciliata*** **(Roxb.) Nees**
A. Habit, B. Ligule, C. Spikelet, D. Lower glume, E. Upper glume, F. Lemma, G. Palea, H. Stamen, I. Lodicule, J. Pistil

mm long, slender, glabrous, shorter than anther. Lodicules absent. Pistil 2.0 mm long; ovary 0.3 mm long; obovate, glabrous; style 0.5 mm long, slender, glabrous, bifid from base; stigma 1.2 mm long, hairy; hairs acute, base broad, in 2–rows. Caryopsis 0.6 mm long, 0.3 mm broad, ovoid, ellipsoid, terete, brown.

Flowering & Fruiting

November–February; Harraua, Kerakat, Tarighat, Mairwaa, Basti.

Habitat

Rare along the road side.

Distribution

India, Cochin, China.

6. *E. coarctata*

E. coarctata Stapf. apud. Hook. f. Fl. Brit. Ind. 7: 313. 1896; Moulik, Grass. Bomb. Ind. 2. 601. 1997; Roy, Grass M. P. 87. 1984; Bor, Grass. Bur. Cey Ind. Pak. 507. 1960. (PLATE–131).

Perennial. Rhizome absent; root fibrous. Culms 35 cm tall, 1.0 mm thick, erect, ascending, geniculate, slender, tufted, green, glabrous, rooting at base; nodes glabrous; internode 7.0 cm long. Leaf blade 7.0 cm long, 1.5 mm broad, narrow, flat, green, linear, lanceolate, convolute, glabrous, margin scaberulous, mid–vein prominent, acuminate, base rounded, glaucous, 2–lateral veins distinct on each side of mid–vein at upper surface; leaf sheath 4.0 cm long, 2.0 mm broad, tight, glabrous, shorter than internode, green, hairy at mouth; hairs acute, base rounded, white; veins distinct on both surface; ligule 0.2 mm long, a ring of hairs; hairs acute, base broad; auricle absent. Inflorescence 10 cm long, 1.5 cm broad, panicle, subspiciform, lobed or lax, greenish, branch hispid. Spikelet 2.0 mm long, 1.0 mm broad, slightly compressed. 5–12 flowered. Lower glume 0.5 mm long, 0.2 broad, 1–nerved, lanceolate, acuminate, hyaline. Upper glume 0.8 mm long, 0.4 mm broad, 1–nerved; nerve scabrid, green, visible, ovate, hyaline. Lemma 1.0 mm long, 0.5 mm broad, ovate–obtuse, acute, membranous, 3–nerved; nerve scabrid, green, visible. Palea 1.0 mm long, 0.5 mm broad, 2–nerved, oblong, hyaline, enclosed in lemma. folded, keel hairy; hairs acute, base rounded, exceeding from lemma. Stamens 3; anther 0.3 mm long, deep pink, glabrous; filament 0.1 mm long, slender, glabrous, shorter than anther. Lodicules absent. Pistil 1.0 mm long; ovary 0.3 mm long, ovate, glabrous; style 0.3 mm long, slender, bifid from base, glabrous; stigma 0.4 mm long, hairy; hairs acute, base tubercled. Caryopsis 0.4 mm long, 0.2 mm broad, brown, elliptid, oblong, tightly enclosed by lemma & palea.

Flowering & Fruiting

September–November; Rajghat, Sonaghati, Pipri, Padrauna, Lar.

Habitat

Common on sandy soil.

Distribution

India, Africa, Burma, Bangladesh.

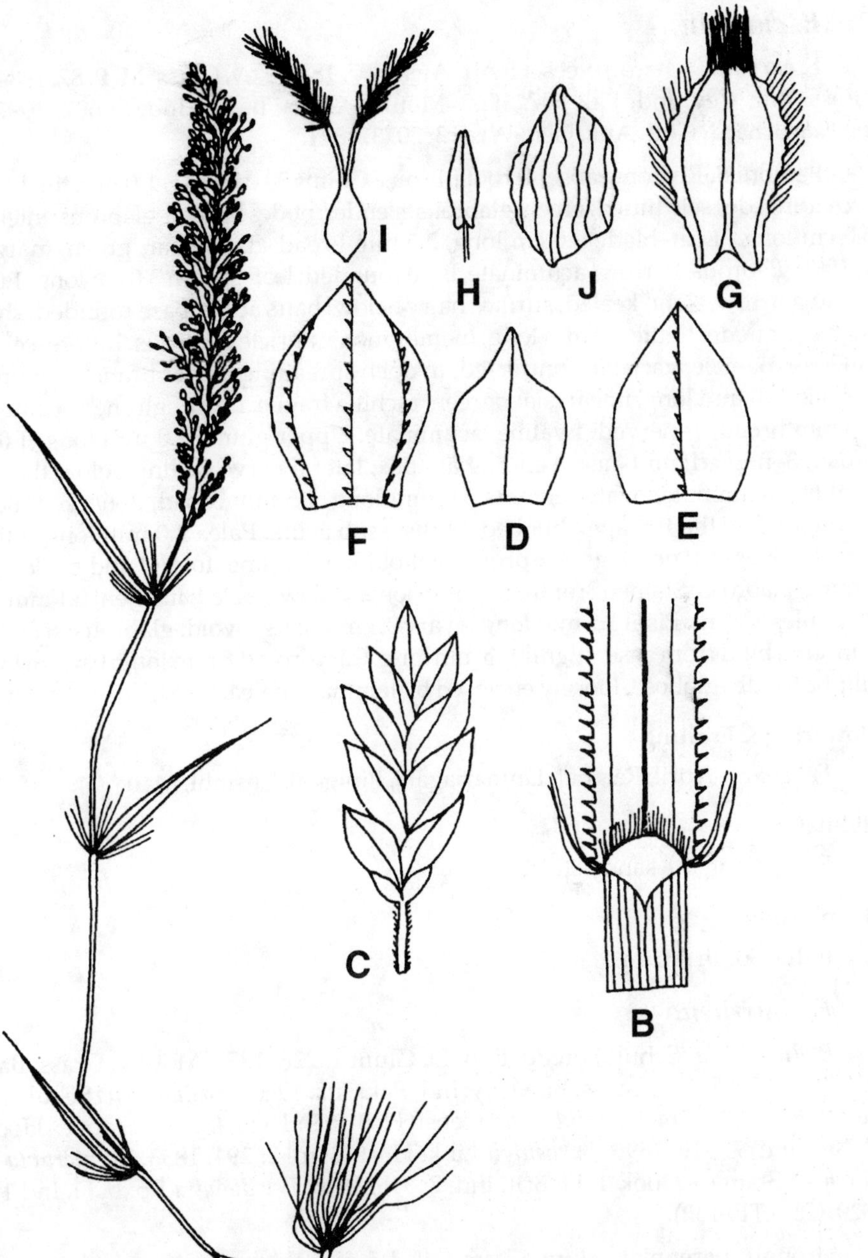

PLATE 131: ***Eragrostis coarctata*** **Stapf**

A. Habit, B. Ligule, C. Spikelet, D. Lower glume, E. Upper glume, F. Lemma, G. Palea, H. Stamen, I. Pistil, J. Caryopsis

7. *E. curvula*

E. curvula (Schard.) Nees, Fl. Afr. Aust. 397. 1841; Roy, Grass. M. P. 87. 1984; Bor, Grass. Bur. Cey. Ind. Pak 507. 1960; Moulik, Grass. Bamb. Ind. 2: 602. 1997. *Poa curvula* Schard. Gott. Anz. Gen. Wiss. 3; 2073. 1821.

Perennial. Rhizome absent; root fibrous. Culms 30 cm long, 1.0 mm thick, erect, ascending, densely tufted, green, glabrous, slender; nodes brown, glabrous; internode 5.0 cm long. Leaf–blade 9.0 cm long, 2.5 mm broad, flat, linear, green, margin & surface glabrous, narrow, acuminate, base rounded; leaf–sheath 3.0 cm long, 1.0 mm broad, narrow, tight, keeled, surface hairy below; hairs acute, base rounded; shorter than inter node; ligule 1.0 mm long, membranous; auricle glabrous. Inflorescence 15 cm long, panicle, variable contracted, to open spreading, lower branches whorled. Spikelet 8.0 mm long, linear, slaty green; rachilla fragile. Lower glume 1.5 mm long, 1.0 mm broad, 1–nerved, hyaline, acuminate. Upper glume 2.0 mm long, 1.0 mm broad, 3–nerved; mid–nerve ending at apex, lateral nerve ending below the apex, hyaline, acuminate, ovate. Lemma 3.0 mm long, 1.5 mm broad, 3–nerved; nerves ending below the tip, apex bilobed, obtuse–sub acute. Palea 2.0 mm long, 1.0 mm broad, 2–nerved; nerve green, prominent oblong, hyaline, folded and enclosed by lemma, glabrous. Stamen 3; anther 1.0 mm long, yellow, keeled; filament 0.1 mm long. Lodicules absent. Pistil 1.5 mm long; ovary 0.3 mm long, ovoid, glabrous; style 2, 0.5 mm long, bifid from base; stigma 0.5 mm long. Caryopsis 0.8 mm long, 0.6 mm broad, elliptic to sub–globose, loosely enclosed by lemma & palea.

Flowering & Fruiting

February–April; Rajghat, Dumariaganj, Pipiganj, Kusmhi, Mau.

Habitat

Rare, along the sandy soils.

Distribution

India, South Africa.

8. *E. diarrhena*

E. diarrhena (Schult.) Steud. Syn. Pl. Glum 1: 226. 1854; Moulik, Grass. Bamb. Ind. 2: 602. 1997; Bor, Grass. Bur. Cey. Ind. Pak. 507. 1960. *Poa diarrhena* Schult. Syst. Veg. 2: 616. 1927. *Eragrostis interrupta* sensu Fl. Brit. Ind. var. *koenigii* Stapf ex Hook. f. Fl. Brit. Ind. 7: 316. 1896. *E. koenigii* Link. Hort. Berol 2: 294. 1833. *E. interupta* var. *diarrhena* Stapf ex Hook. f. Fl. Brit. Ind. 7: 316. 1816. *Poa diandra* Roxb. Fl. Ind 1: 33. 1820. (PLATE–132).

Strongly perennial. Clum 65 cm tall, 1.5 mm thick, erect, ascending, green, glabrous, nodes brown, glabrous; internodes 10 cm long. Leaf–blade 9.0 cm long, 1.5 mm broad, narrow, convolute, green, glabrous, margin scabrid, acute, base rounded, mid–vein prominent, 2–lateral veins on each side of mid vein; lateral veins distinct on upper surface, leaf–sheath 8.0 cm long, 2.0 mm broad, loose compressed, glabrous, green, shorter than internodes, hairy at mouth; hair white, acute, base broad; ligule 0.1 mm long, membranous, lacerate; auricle glabrous. Inflorescence a panicle; panicle

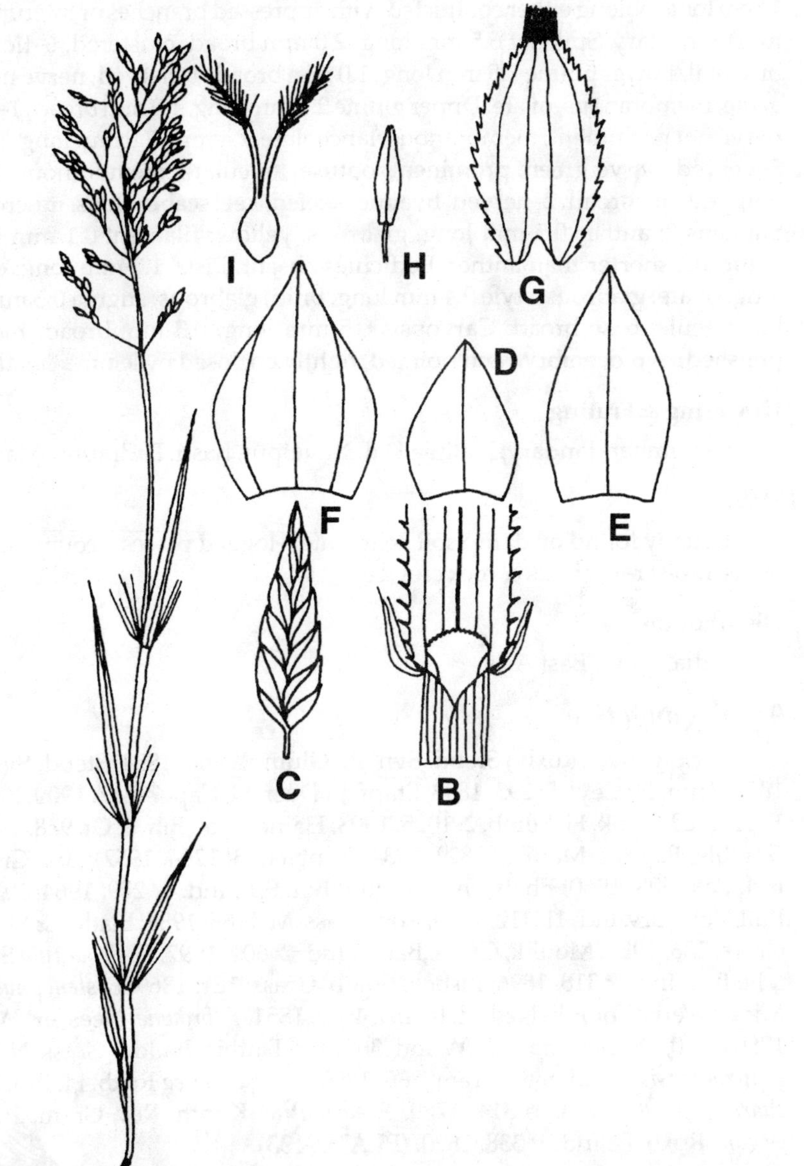

PLATE 132: ***Eragrostis diarrhena*** **(Schult.) Steud.**
A. Habit, B. Ligule, C. Spikelet, D. Lower glume, E. Upper glume, F. Lemma, G. Palea, H. Stamen, I. Pistil

10 cm long, oblong either contracted with appressed branches or interupted, branches mostly solitary. Spikelet 6.5 mm long, 2.0 mm broad, clustered, 6–flowered, ovate, bisexual. Lower glume 1.5 mm long, 1.0 mm broad, 1–nerved, nerve not prominent, acute, membranous, ovate. Upper glume 2.0 mm long, 1.0 mm broad, 1–nerved; acute nerve not prominent, membranous, lanceolate. Lemma 2.0 mm long, 1.0 mm broad, 3–nerved, nerve green, prominent, obtuse, apiculate, membranous. Palea 1.8 mm long, 1.0 mm broad, 2–nerved, hyaline, keeled, keel scaberulous, mucronate, oblong. Stamens 2; anther 0.5 mm long, glabrous, yellow; filament 0.1 mm long, slender, glabrous, shorter than anther. Lodicules absent. Pistil 1.2 mm long; ovary 0.3 mm long, ovate, glabrous; style 0.4 mm long, bifid, glabrous; stigma 0.5 mm long, hairy; hairs acute, base broad. Caryopsis 0.5 mm long, 0.3 mm broad, reddish brown, polished, ovoid, embryo end pointed, tightly enclosed by lemma & palea.

Flowering & Fruting

September–January; Natinia Dai, Navalpur, Basti, Padrauna, Mau.

Habitat

Usually found on damp soil near water–logged places, around large perennial tanks, moist rever beds & rock crevices.

Distribution

India, South East Asia.

9. *E. gangetica*

E. gangetica (Roxb.) Steud. Syn. Pl. Glum. 1: 266. 1854; Steud. Syn. Gram. 266. 1855; Trim. Fl. Ceyl. 5: 293. 1898; Stapf. in Dyer, Fl. Cap. 7: 617. 1900; Prain, Beng. Pl. 1221. 1903; Cook, Fl. Bomb. 2: 1025. 1908; Haines, Bot. Bih. & Or. 958. 1924. Fischer in Gamble, Fl. Pres. Madr. 3: 1826. 1934 (Repr. ed. 3: 1264. 1957); Bor, Grass. Bur. Cey. Ind. Pak. 508. 1960; Bhattacharya, Bull. Bot. Sur. Ind. 6: 209. 1964; Panigrahi *et al.*, Bull. Bot. Surv. Ind. 11: 112. 1969; Roy, Grass. M. P. 88. 1984; Blatter & McCann, Bomb. Grass. 236. 1984; Moulik, Grass. Bamb. Ind. 2: 602. 1997. *E. elegantula* Stapf in Hook. f. Fl. Brit. Ind. 7: 318. 1896; Lisboa, Bomb. Grass. 121. 1869. *E. stenophylla* Hochst. ex Miq. in Verh. Konink–Nederl. Inst. 3, 4: 39. 1851. *E. brownei* Nees. ex. Wight. Cat. no. 1780. 1840; Duthie, Grass. N.W. Ind. 36. 1883; Duthie, Fodder Grass. N. Ind. 62: 1888. *E. luzoniensis* Steud. Syn. Gram. 266. 1855. *Poa gangetica* Roxb. Fl. Ind. 1: 41. 1820. *P. chariis* Schult. Mant. 1: 314. 1789. *P. elegantula* Kunth. Rev. Gram. 1: 340. 1829. *P. elegans* Roxb. Fl. Ind. 1: 338. 1820. (PLATE–133).

Annual. Rhizome absent; root fibrous. Culm 75 cm tall, 2.0 mm thick, green, glabrous, stout, tufted, shining, ascending, leafy, rooting at base, branched; branches sub erect, slender; nodes glabrous; internodes 9.0 cm long. Leaf–blade 15 cm long, 3.0 mm broad, green, glaucous, convolute, narrow, linear, lanceolate, acuminate, base rounded, surface glabrous, margin hyaline, scabrid, mid vein distinct on lower surface, 2–lateral veins on each side of mid–vein; lateral veins distinct on upper surface; leaf–sheath 7.0 cm long, 4.0 mm broad, green, glabrous, shorter than internodes; ligule 0.5 mm long, membranous, truncate; auricle glabrous. Inflorescence panicle; panicle 15 cm long, ovate, oblong, lanceolate, nodding, lower branch not whorled, usually solitary,

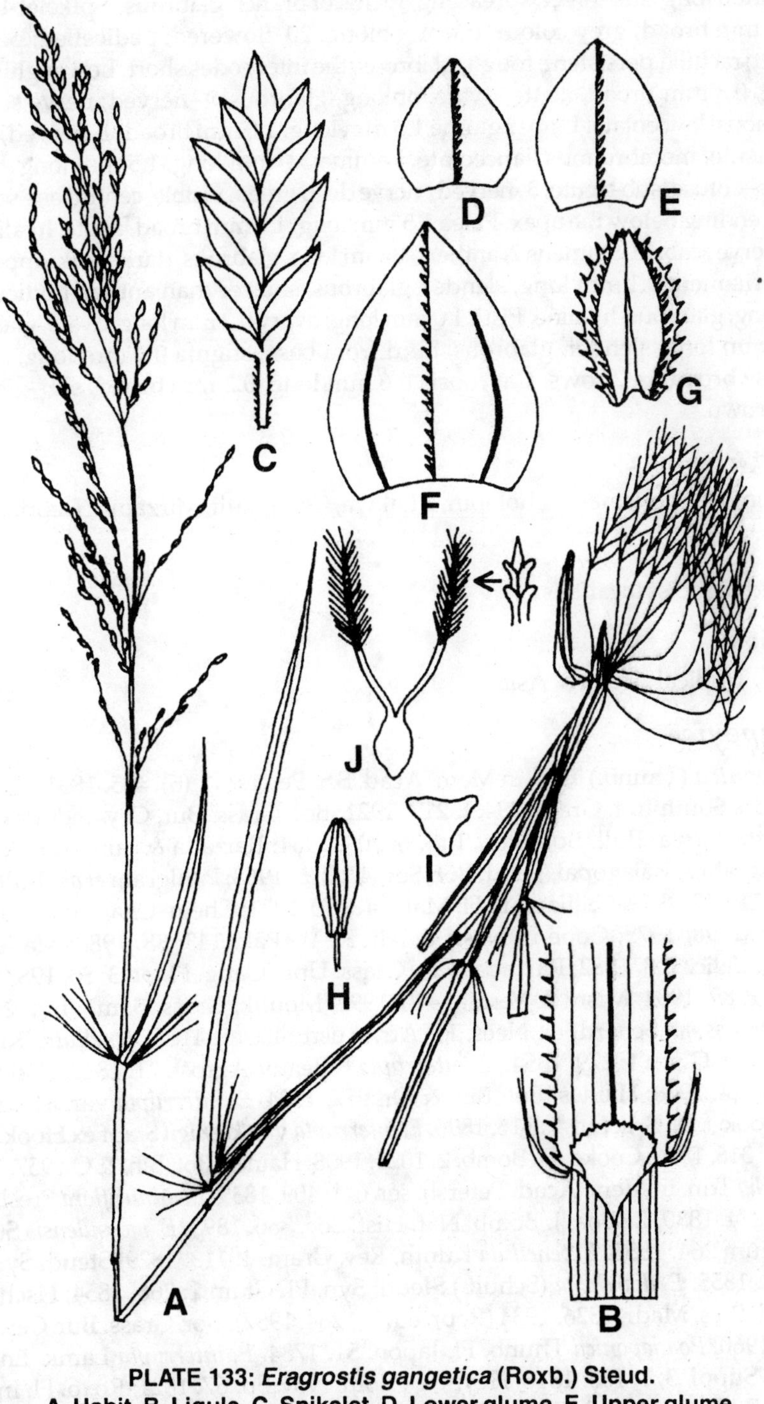

PLATE 133: ***Eragrostis gangetica*** **(Roxb.) Steud.**
A. Habit, B. Ligule, C. Spikelet, D. Lower glume, E. Upper glume, F. Lemma, G. Palea, H. Stamen, I. Lodicule, J. Pistil

distant, often long, sub–erect, spreading; rachis of branch glabrous. Spikelet 4.0 mm long, 2.0 mm broad, grey colour, linear, oblong, 20–flowered, pedicelled; pedicels very short; rachilla persistent, tough, glabrous, the internodes short. Lower glume 1.2 mm long, 0.8 mm broad, acute, ovate, oblong, glabrous, 1–nerved; nerve scabrid, membranous, lanceolate. Upper glume 1.5 mm long, 1.0 mm broad, 1–nerved, nerve scabrid, acute, membranous, lanceolate. Lemma 3.0 mm long, 1.5 mm long, broad, ovate, sub–obtuse, sub–acute, 3–nerved, nerve deep green, visible, central one scabrid, marginal ending below the apex. Palea 1.5 mm long, 1.0 mm broad, acute, hyaline, 2–nerved, nerve scabrid. Stamens 2; anther 0.3 mm long, glabrous, dark–pink, appearing blackish; filament 0.1 mm long, slender, glabrous, shorter than anther. Lodicules 2, 0.4 mm long, glabrous, hyaline. Pistil 1.0 mm long; ovary 0.2 mm long, ovate, glabrous; style 0.4 mm long, slender, glabrous, bifid from base; stigma 0.4 mm long, brown acute, base broad, in 2 rows. Caryopsis 0.6 mm long, 0.2 mm broad, sub–globase, oblong, brown.

Flowering & Fruiting

September–November; Cholapur, Chakiya, Chandauli, Mirzapur, Deoria.

Habitat

Found in moist localities.

Distribution

India, Tropical Africa & Asia.

10. *E. japonica*

E. japonica (Thunb.) Trin. in Mem. Acad. Sci. Petersb. 1 (6): 405. 1831; Rang. & Tad. Handb. South Ind. Grass. 294. t. 217. 1921; Bor, Grass. Bur. Cey. Ind. Pak. 509. 1960; Bhattacharya, Bull. Bot. Surv. Ind. 6: 209. 1964; Raizada & Jain, Ind. For. 92: 1966; Panigrahi & Rajagopal, J. Ind. Bot. Soc. 47: 236. 1968; Panigrahi *et al.*, Bull. Bot. Surv. Ind. 11: 112. 1969; Gilliland in Fl. Malaya 3: 70. 1971; Chein–Chan, Hsu, Taiwan Grass. 405. t. 98. 1975; Cope in Nasir & Ali, Fl. W. Pak. 143: 88. 1982; Manilal & Sivaraj, Fl. Calic. 348. 1982; Raizada *et al.*, Grass. Upp. Gang. Pl. pt. 3: 85. 1983; Roy, Grass. M. P. 89. 1984; Misra & Verma, 440. 1992; Moulik, Grass. Bamb. Ind. 2: 603. 1997. *E. tenuissima* Schard. ex Nees, Fl. Afr. Austr. 409. 1841. *E. tenelluta* (Kunth) Steud. Syn. Pl. Glum 1: 279. 1854. *E. interrupta* P. Beauv. Agrost. 71. 1812; Cooke, Fl. Bomb. 2: 1024. 1908; Haines, Bot. Bih. & Or. 957. 1924. *E. interrupta* var. *tenuissima* Stapf ex Hook. f. Fl. Brit. Ind. 7: 316. 1896. *E. interrupta* var. *koenigii* Stapf ex Hook. f. Fl. Brit. Ind. 7: 316. 1896; Cooke, Fl. Bomb. 2: 1024. 1908; Haines, Bot. Bih. & Or. 957. 1924. *E. hapalantha* Trin. in Mem. Acad. Petersb. ser. 6, 1: 409. 1839. *E. minutiflora* Presl. Rel. Haenk. 1: 274. 1830; Lisboa, J. Bomb. Nat. Hist. Soc. 386. 1893. *E. mossulensis* Steud. Syn. Pl. Glum 264. 1855. *E. tenellula* Kunth, Rev. Gram. 1: 113. 1829; Steud. Syn. Pl. Glum. 279. 1855. *E. diarrhena* (Schult.) Steud. Syn. Pl. Glum 1: 266. 1854; Fischer in Gamble, Fl. Pres. Madr. 1826. 1934 (Repr. ed. 3: 1264. 1957); Bor, Grass. Bur. Cey. Ind. Pak. 507. 1960. *Poa japonica* Thunb. Fl. Japon. 51. 1784. *P. interrupta* Lamk. Encycl. Meth. Bot. Suppl. 3, 1: 185. 1816; Heyne ex Roth. Nov. Sp. 67. 1821; Roxb. Fl. Ind. 1: 335. 1820. *P. koenigii* Kunth Enum. Pl. 1: 346. 1838. *P. tenella* R. Br. Prodr. 181. 1810;

Roxb. Fl. Ind. 1: 337. 1832. *P. tenelluta* Kunth. Enum. Pl. 1: 338. 1838. *P. diarrhena* Schult. Syst. veg. 2. Mant. 616. 1824. (PLATE–134).

Annual. Rhizome absent; root fibrous. Culm 70 cm tall, 1.0 mm thick, green, glabrous, erect, densely tufted, geniculately–ascending, decumbent, slender, very leafy, simple, rooting at base; nodes brown, glabrous; internode 7.0 cm long. Leaf–blade 15 cm long, 4.0 mm broad, green, flat, narrow, linear–lanceolate, acuminate, base rounded, surface membranous, glabrous, margin hyaline, scabrid, mid–vein distinct on lower surface, 3–lateral veins on each side of mid vein; lateral veins distinct on both surface; leaf–sheath 6.0 cm long, 1.5 mm broad, green, surface membranous, glabrous, margin, hyaline, keeled, tight, compressed, erect, shorter than internodes, hairy at mouth; hairs 5.0 mm long, white, acute, base tubercled; veins distinct on both surface; ligule 1.0 mm long, membranous, truncate; acuricle glabrous. Inflorescenc made up of panicle; panicle 30 cm long, 5.0 mm broad, linear, lanceolate, spiciform, contracted with appressed branches, interrupted with many stout, spreading, sub–whorled, branches, suberect, open; rachis of panicle slender, scabrid. Branches 3.0 cm long, spreading, scabrid, veriticillate, distant, densely crowded. Spikelet 1.5 mm long, 1.0 mm broad, ovate, oblong, lanceolate, 6–flowered, scattered, loose, biseriate, green at young, turnning into purplish with age. Lower glume 0.5 mm long, 0.2 mm broad, submembranous, ovate, oblong, lanceolate, obtuse, glabrous, 1–nerved; nerves green, slightly visible, glabrous, ending at apex. Upper glume 1.0 mm long, 0.5 mm broad, submembranous, ovate, oblong, obtuse, lanceolate, glabrous, 1–nerved; nerve green, visible, glabrous, ending at apex. Lemma 1.0 mm long, 0.5 mm broad, membranous, broadly ovate, obtuse, lanceolate, laterally compressed, bifid at apex, glabrous; 3–nerved; nerves green, visible, glabrous, ending at apex. Palea 1.0 mm long, 0.5 mm broad, hyaline, delicate, oblong, ovate, lanceolate, glabrous, margin folded inwards, transversely scabrid at middle, 2–nerved; nerves green, slightly visible, marginal & glabrous, ending at apex. Stamens 2; anther 0.3 mm long, light yellow, oblong, glabrous; filament 0.1 mm long, slender, glabrous, shorter than anther. Lodicules absent. Pistil 1.0 mm long; ovary 0.2 mm long, obovate, glabrous; style 0.3 mm long, slender, glabrous, bifid from base; stigma 0.5 mm long, purple, hairy; hairs acute, base tubercled. Caryopsis 0.7 mm long, 0.3 mm broad, reddish brown, elliptic, obovoid, polished.

Flowering & Fruiting

Septemebr–January; Tarana, Harraua, Bajrangnagar, Tarighat, Barhalganj.

Habitat

Rare, along the moist places.

Distribution

India, Tropical Asia.

11. *E. maderaspatana*

E. maderaspatana Bor, Grass. Bur. Cey. Ind. Pak. 509. 1960; Moulik, Grass. Bamb. Ind. 2: 604. 1997. *E. willdenowiana* Arn., Bot. Beechey Voy 252. 1838; Arn., Nov. Act. Nat. Cur. 19. Suppl. 1: 205. 1843. (PLATE–135).

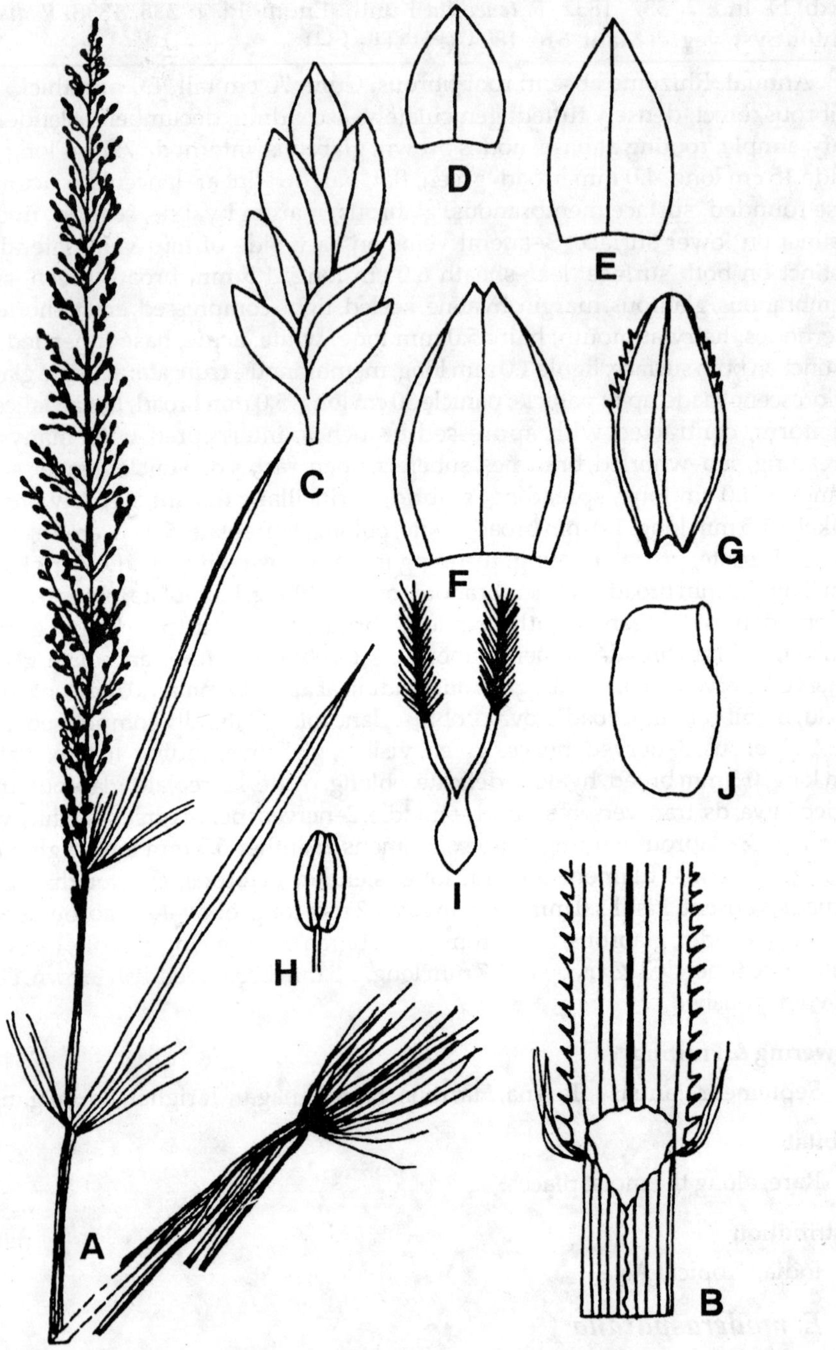

PLATE 134: ***Eragrostis japonica*** **(Thunb.) Trin.**
A. Habit, B. Ligule, C. Spikelet, D. Lower glume, E. Upper glume, F. Lemma, G. Palea, H. Stamen, I. Pistil, J. Caryopsis

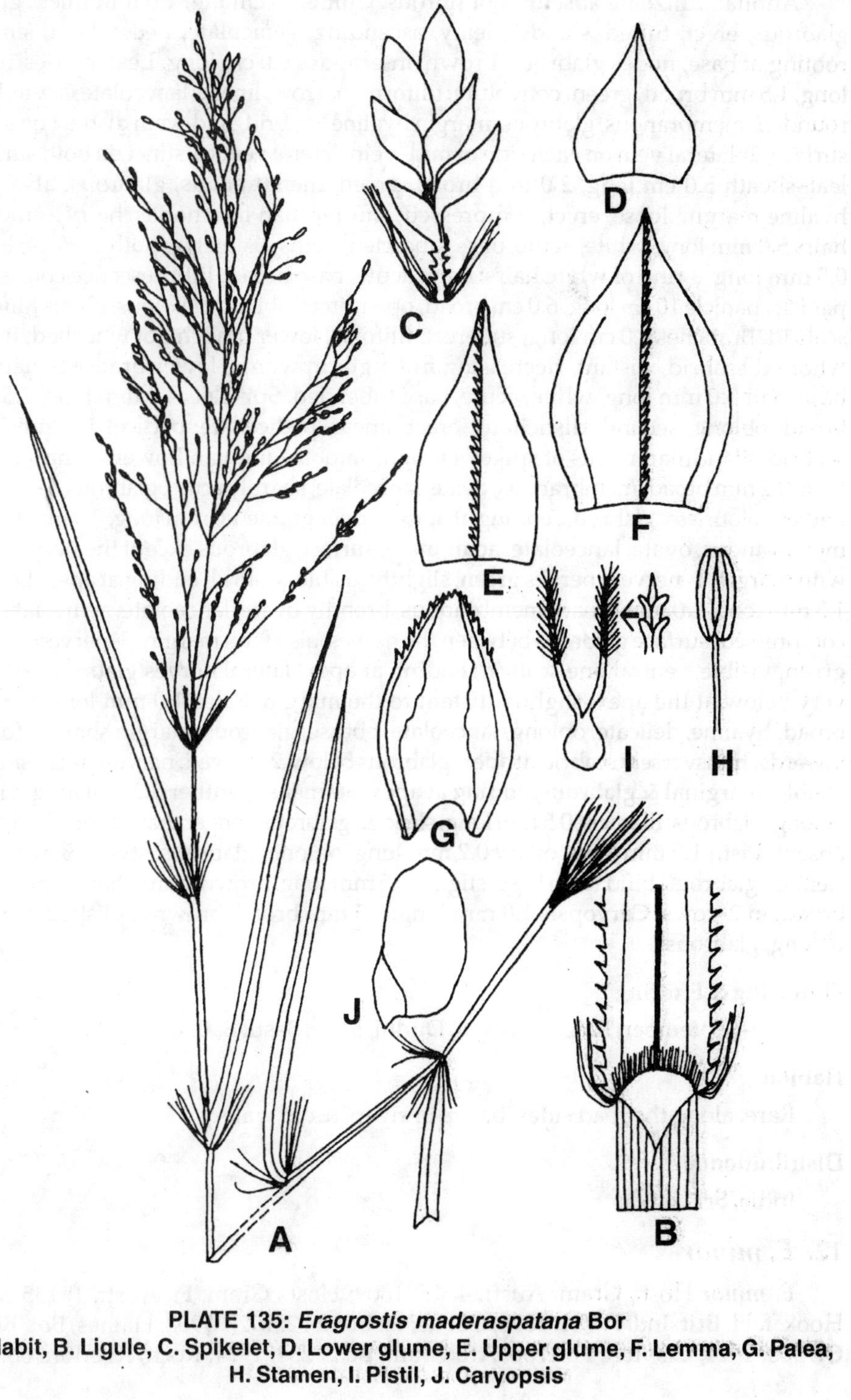

PLATE 135: ***Eragrostis maderaspatana*** **Bor**
A. Habit, B. Ligule, C. Spikelet, D. Lower glume, E. Upper glume, F. Lemma, G. Palea, H. Stamen, I. Pistil, J. Caryopsis

Annual. Rhizome absent; root fibrous. Culms 45 cm tall, 1.0 mm thick, green, glabrous, erect, tufted, slender, leafy, ascending, geniculate, decumbent, simple, rooting at base; nodes glabrous, brown; internodes 6.0 cm long. Leaf–blades 10 cm long, 1.5 mm broad, green, convolute, filiform, narrow, linear, lanceolate, acute, base rounded, membranous, glabrous, margin hyaline, scabrid, mid–vein distinct on lower surface, 1–lateral vein on each side of mid–vein; lateral veins distinct on both surface; leaf–sheath 5.0 cm long, 2.0 mm broad, green, membranous, glabrous, also with hyaline margin, loose, erect, compressed, shorter than internodes, hairy at mouth; hairs 5.0 mm long, white, acute, base tubercled; veins distinct on both surface; ligule 0.5 mm long, a ring of white hairs; hairs acute, base broad. Inflorescence consists of panicle; panicle 10 cm long, 6.0 cm broad, open, erect, oblong, flexuous; rachis filiform, scabrid. Branches 8.0 cm long, suberect, filiform, lower branch not whorlled, upper whorled scabrid, distant, decreasing in length upwards; lower branches hairy at base; hair 3.0 mm long, white, acute, base tubercled. Spikelets 4.0 mm long, 1.5 mm broad, oblong, secund, distichous, erect, linear, pedicellate; pedicel 1.0 mm long, scabrid, glandular; rachis of spikelet tough, smooth, angular. Lower glume 0.5 mm long, 0.2 mm broad, membranous, ovate, lanceolate, sharply acute, glabrous, 1–nerved; nerves colourless, glabrous, ending at apex. Upper glume 1.0 mm long, 0.5 mm broad, membranous, ovate, lanceolate, acuminate, surface glabrous except the nerves, also with margin, 1–nerved; nerves green, slightly visible, scabrid, ending at apex. Lemma 1.5 mm long, 1.0 mm broad, membranous, broadly ovate, lanceolate, acute, laterally compressed, surface glabrous between the nerves, also with margin, 3–nerved; nerves green, visible, central one scabrid, ending at apex, lateral nerves glabrous, ending very below at the apex, slightly distant to the margin. Palea 1.0 mm long, 0.5 mm broad, hyaline, delicate, oblong, lanceolate, obtuse, glabrous, margin sharply folded inwards, transversely scabrid at apex, glabrous below, 2–nerved; nerves green, slightly visible, marginal & glabrous, ending at apex. Stamens 3; anther 0.2 mm long, violet, oblong, glabrous; filament 0.5 mm long, slender, glabrous, longer than anther. Lodicules absent. Pistil 1.0 mm long; ovary 0.2 mm long, oblong, glabrous; style 0.3 mm long, slender, glabrous, bifid from base; stigma 0.5 mm long, brown, hairy; hairs acute, base broad, in 2–rows. Caryopsis 1.0 mm long, 0.5 mm broad, brown, polished, elliptic, oblong, glabrous.

Flowering & Fruiting

July–September; Rajghat, Naini, Dudhi, Mau, Basti, Ballia.

Habitat

Rare, along the road sides, banks of rivers & old walls.

Distribution

India, Sri Lanka.

12. *E. minor*

E. minor Host., Gram. Austr. 4: 15. 1809; Host., Gram. Fl. Austr. 1: 135. 1827; Hook. f. Fl. Brit. Ind. 7: 321. 1896; Cooke, Fl. Bomb. 2: 1027. 1908; Haines, Bot. Bih. & Or. 960. 1924; Clayton, Fl. Trop. Afr. Gram. pt. 2: 234. 1974; Koch, Rhodora 80: 397.

1978; Cope in Nasir & Ali, Fl. W. Pak. 143: 95. 1982; Blatter & McCann, Bomb. Grass. 238. t. 158. 1984; Sreekumar & Nair, Fl. Kera. Grass. 383. 1991; Misra & Verma, Fl. Allah. 441. 1992. Moulik, Grass. Bamb. Ind. 2: 604. 1997. *E. poaeoides* Beauv. Agrost. 76. 1812; Duthie, Grass. N. W. Ind. 38. 1883; Duthie, Fodder Grass. N. Ind. 65. 1888; Aitchis. Cat. Punj. Pl. 170. 1869; Lisboa, J. Bomb. Nat. Hist. Soc. 7: 387. 1893; Boiss. Fl. Or. 5: 580. 1884; Fischer in Gamble Fl. Pres. Madr. 1827. 1934. (Repr. ed. 3: 1264. 1957); Bor, Grass. Bur. Cey. Ind. Pak. 512. 1960. Raizada & Jain, Ind. For. 92: 640. 1966; Panigrahi & Rajagopal, J. Ind. Bot. Soc. 47: 237. 1968; Singh, Bull. Bot. Sur. Ind. 11: 21. 1969; Chein–chang Hsu, Taiwan Grass. 419. t. 105. 1975; Raizada *et al.*, Grass. Upp. Gang. Pl. pt. 3: 88. 1983; Roy, Grass. M.P. 91. 1984. *E. eragrostis* (Linn.) P. Beauv., Ess. Agrost. 71: t. 14. fig. 11. 1812. *E. vulgaris* var. *microstachya* Coss. et. Germ., Fl. Env. Paris, 641. 1845. *E. megastachya* subsp. *poaceides* (P. Beauv.) Hushot, Gram. France, Belg 55. 1898. *E. vulgaris* var. *minor* (Host.) Rouy, Fl. France 14: 263. 1913. *E. minor* var. *typica* Beck, Fl. Nied.–Ost. 1: 88. 1890. *E. poaeformis* Link. Hort. Berol. 1: 188. 1827. *E. poaeoides* Trin. in Memb. Acad. Petersb. ser. 6, 1: 404. 1831. *Poa eragrostis* Linn. Sp. Pl. 68. 1753; Kunth. Enum. Pl. 1: 332. 1883. *Briza eragrostis* Vill. Fl. Delph. 2: 50. 1840. (PLATE–136).

Annual. Rhizome absent; root fibrous. Culm 40 cm tall, 1.0 mm thick, green, glabrous, erect, tufted, compressed, slender, geniculately ascending, slender, leafy at base, simple, rooting at base; nodes brown, glabrous; internodes 5.0 cm long. Leaf–blade 6.0 cm long, 4.0 mm broad, green, flat, narrow, linear–lanceolate, acuminate, base rounded, coriaceous, surface glabrous, margin hyaline, scabrid, mid–vein distinct on lower surface, 2–lateral veins on each side of mid–vein; lateral veins distinct on upper surface; leaf–sheath 4.0 cm long, 2.0 mm broad, green, coriaceous, glabrous, margin submembranous, keeled, loose, erect, striate, compressed, shorter than internodes, hairy at mouth; hairs 5.0 mm long, white, acute, tubercled, veins distinct on outer surface; ligule 1.0 mm long, membranous, erose, truncate, hairy at apex; hairs minute, acute, base broad; auricle glabrous. Inflorescence consist of panicle; panicle 8.0 cm long, 5.0 cm broad, stiff, ovate, oblong, crateriform, erect; rachis glabrous. Branches 4.0 cm long, sub–erect, spreading, slightly angular, scabrid, decreasing in length upwards, hairy on the axils; hairs 5.0 mm long, white, acute, base tubercled. Spikelets 8.0 mm long, 1.5 mm broad, linear, ovate, greyish–green, secund, distichous, pedicelled; pedicel 6.0 mm long, slender, scabrid; rachis of spikelets angular, smooth. Lower glume 1.2 mm long, 0.6 mm broad, membranous, ovate, lanceolate, acute, glabrous, 1–nerved; nerve green, visible, glabrous, ending at apex. Upper glume 1.5 mm long, 1.0 mm broad, membranous, ovate, lanceolate, acute, glabrous, 1–nerved; nerve green, visible, glabrous, ending at apex. Lemma 2.5 mm long, 1.5 mm broad, membranous, broadly ovate, acute, laterally compressed, surface glabrous between the nerves, also with margin, 3–nerved; nerves green, visible, ending at apex, central one scabrid, lateral glabrous & slightly close to the margin. Palea 1.5 mm long, 1.0 mm broad, hyaline, delicate, ovate, oblong, glabrous, margin folded inwards, transversely scabrid, enclosed by lemma, 2–nerved; nerves green, visible, glabrous & marginal, ending at apex. Stamens 3; 0.3 mm long, yellow, oblong, glabrous; filament 0.5 mm long, slender, glabrous, shorter than anther. Lodicules absent. Pistil 1.2 mm long; ovary 0.2 mm long, ovate, glabrous; style 0.5 mm long, slender, glabrous,

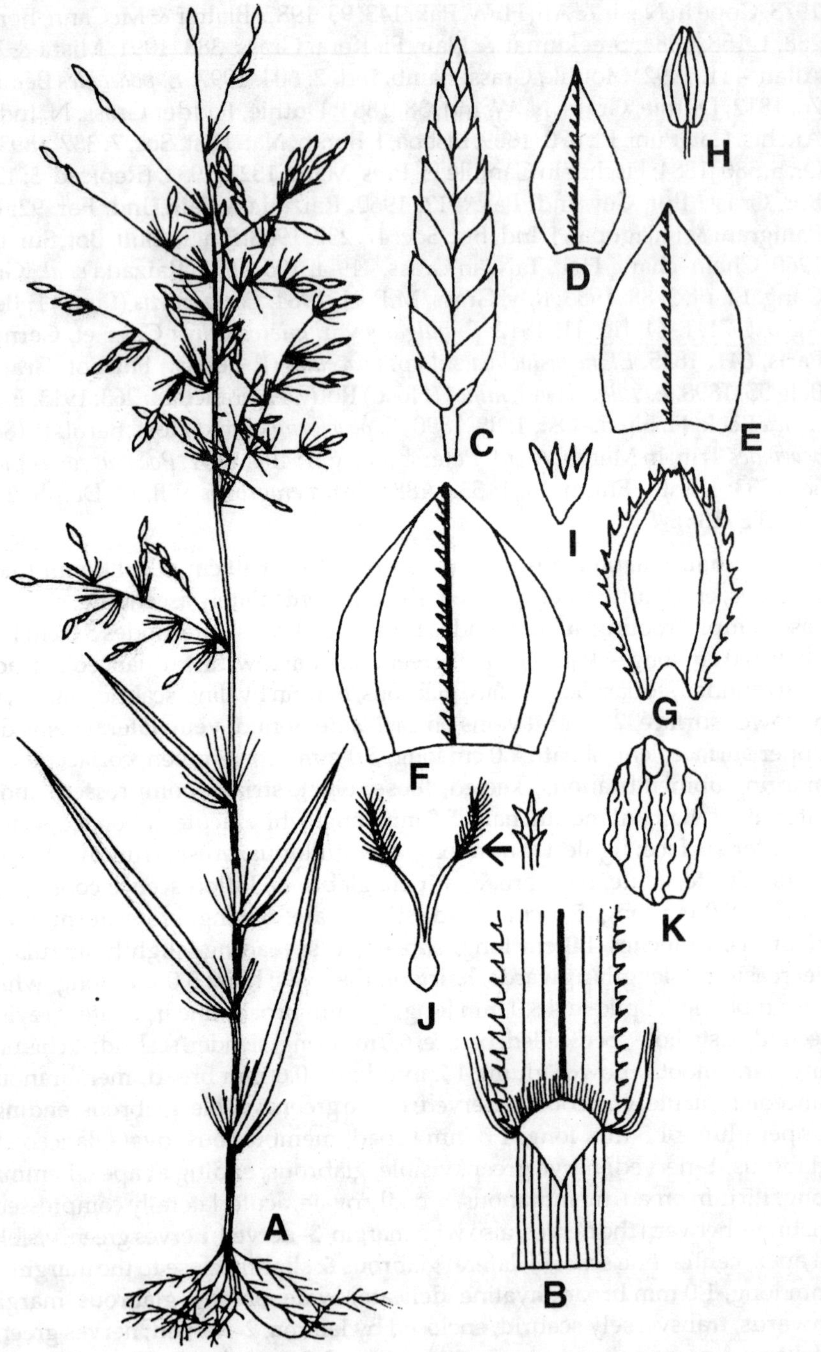

PLATE 136: ***Eragrostis minor*** **Host.**

A. Habit, B. Ligule, C. Spikelet, D. Lower glume, E. Upper glume, F. Lemma, G. Palea, H. Stamen, I. Lodicule, J. Pistil, K. Caryopsis

bifid from base; stigma 0.5 mm long, brown, hairy; hairs acute, base broad, in 2–rows. Caryopsis 0.7 mm long, 0.4 mm broad, oblong, red–brown, slightly flat on dorsal surface.

Flowering & Fruiting

September–February; Rajghat, Chakiya, Chandauli, Allahabad, Ghazipur, Basti.

Habitat

Common along banks of rivers & rocky soils.

Distribution

India, Pakistan, Afghanistan, Tropical Africa.

13. *E. namaquensis*

E. namaquensis (Schard) W.D. Clayton in Kew Bull. 25: 251. 1971. *E. diplachnoides* Steud. Syn. Pl. Glum. 1: 268. 1854; Bor, Grass. Bur. Cey. Ind. Pak. 508. 1960; Moulik, Grass. Bamb. Ind. 2: 605.1997. *E. interrupta* var. *diplachnoides* (Steud.) Stapf in Hook. Fl. Brit. Ind. 7: 316. 1896. *E. leprieurii* Steud. Syn. Pl. Glum. 1: 269. 1854. (PLATE–137).

Annual. Rhizome absent; root fibrous. Culms 90 cm tall, 1.0 mm thick, green, glabrous, erect, tufted, very slender, ascending, leafy, simple, rooting at base, branching at lower nodes; branches suberect, leafy upto panicle; nodes brown, glabrous; internodes 8.0 cm long. Leaf–blade 10 cm long, 3.0 mm broad, green, narrow, convolute, linear–lanceolate, acute, base rounded, glabrous, margin hyaline, scabrid, mid–vein distinct on lower surface, 2–lateral veins on each side of mid–vein; lateral veins distinct on both surface; leaf–sheath 8.0 cm long, 2.0 mm broad, green, glabrous, membranous, margin hyaline, slightly compressed, loose, striate, equal to internodes, hairy at mouth; hairs 3.0 mm long, white, acute, base tubercled; veins distinct on outer surface; ligule 0.3 mm long, membranous, truncate, lacerate; auricle glabrous. Inflorescence made up of panicle; panicle 10 cm long, 8.0 cm broad, linear, pyramidal, open, erect; rachis of panicle scabrid. Branches 6.0 cm long, erect, lower branch in whorl, upper branch alternate, distant, decreasing in length upwards at base, hairy at base; hiars 2.0 mm long, white, acute, base tubercled. Spikelets 3.0 mm long, 1.5 mm broad, ovate, oblong, glabrous, 10–flowered, distichous, compressed, secund, pedicelled; pedicels 1.0 mm long, slender, glabrous; rachis of spikelets angular, glabrous, fragile. Lower glume 0.5 mm long, 0.2 mm broad, membranous, ovate–lanceolate, surface glabrous, margin scabrid, acute, 1–nerved; green, visible, scabrid upwards, ending at apex. Upper glume 1.0 mm long, 0.5 mm broad, membranous, ovate, lanceolate, acute, surface glabrous, margin scabrid, 1–nerved; green, visible, scabrid, upwards, ending at apex. Lemma 1.0 mm long, 0.7 mm broad, membranous, broadly ovate, lanceolate, acuminate, obtuse, glabrous, ending at apex, lateral nerves light green, faint, central one deep green, visible. Palea 1.0 mm long, 0.5 mm broad, hyaline, delicate, oblong, ovate, surface glabrous, margin folded inwards, enclosed by lemma, 2–keeled; keels scabrid upwards, 2–nerved; nerves green, glabrous slightly visible, marginal, ending at apex. Stamens 2; anther 0.3 mm long, pink, oblong, glabrous; filament 0.1 mm long, slender, glabrous, shorter than anther. Lodicules absent. Pistil 1.0 mm long; ovary 0.2 mm long, elliptic, obovate, glabrous; style 0.4 mm

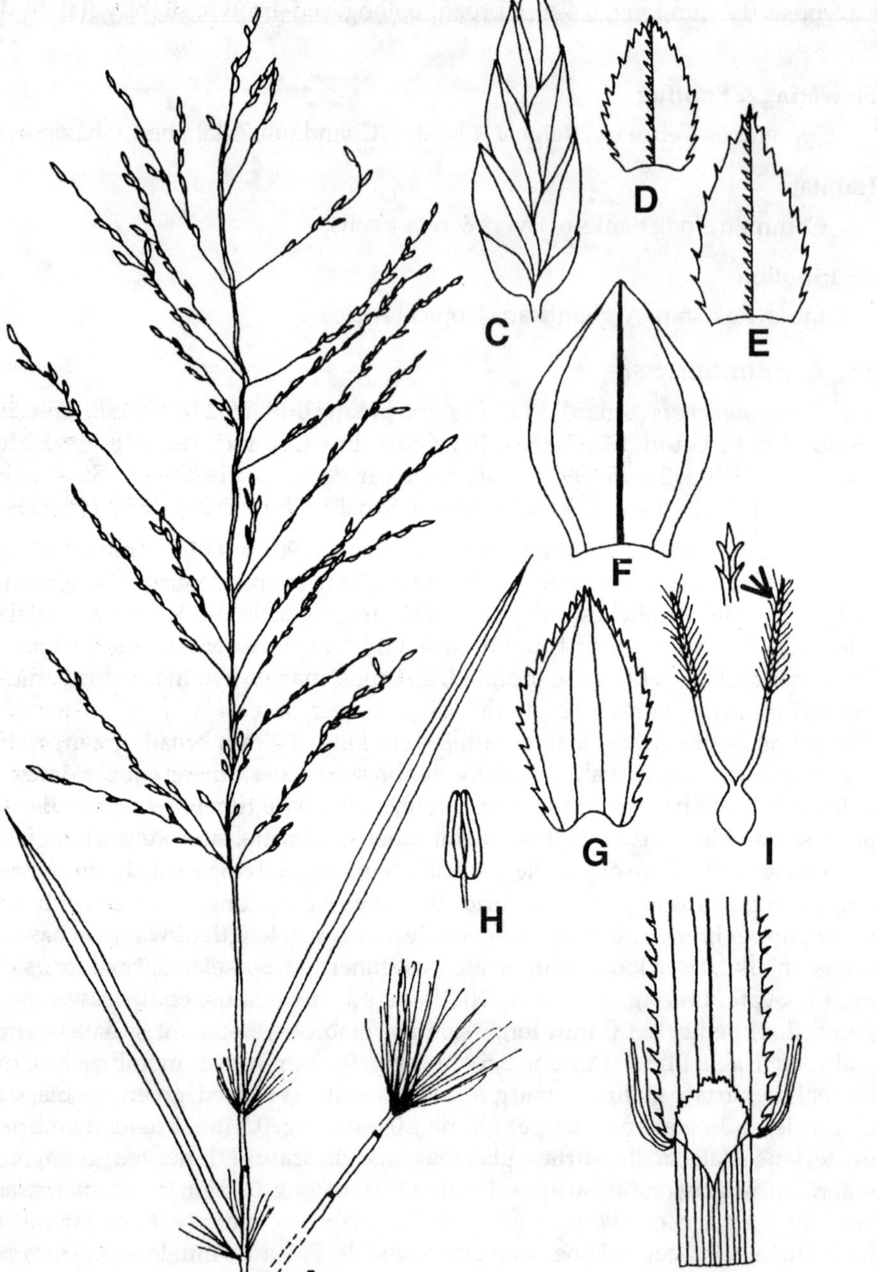

PLATE 137: ***Eragrostis namaquensis*** **Schrad.**
A. Habit, B. Ligule, C. Spikelet, D. Lower glume, E. Upper glume, F. Lemma, G. Palea, H. Stamen, I. Pistil

long, slender, bifid from base, glabrous; stigma 0.4 mm long, brown, hairy, hairs acute, base broad, in 2–rows. Caryopsis 0.5 mm long, 0.3 mm broad, ellipsoid, glabrous, brown.

Flowering & Fruiting

August–January; Sarsauli, Taktakpur, Dudhi, Jhusi, Dumariyagunj.

Habitat

Common along the walls & road sides.

Distribution

India, Tropical and sub–tropical Asia & Africa.

14. *E. nutans*

E. nutans (Retz.) Nees ex Steud. Nom. Bot. ed. 2. 563. 1840; Fischer in Gamble, Fl. Pres. Madr. 1826. 1934 (Repr. ed. 3: 1264. 1957); Bor, Grass. Bur. Cey. Ind. Pak. 511. 1960; Bhattacharya, Bull. Bot. Surv. Ind. 209. 1964; Roy, Grass. M. P. 90. 1984; Blatter & McCann, Bomb. Grass. 237. 1984; Sreekumar & Nair, Fl. Kera. Grass. 385. 1991; Moulik, Grass. Bamb. Ind. 2: 605. 1997. *E. elegantula* Nees ex Steud., Syn. Pl. Glum. 1: 266. 1854; Duthie, Grass. N. W. Ind. 37. 1883; Duthie, Fodder Grass. N. Ind. 63. t. 74. 1888. *E. stenophylla* Hocshst ex Miq. Anal. Bot. Ind. 2: 27. 1851; Lisboa, J. Bomb. Nat. Hist. Soc. 7: 385. 1893; Hook. f. Fl. Brit. Ind. 7: 318. 1896; Prain, Beng. Pl. 1221. 1903; Cooke, Fl. Bomb. 2: 1026. 1908; Haines, Bot. Bih. & Or. 959. 1924. *E. chariis* (Schult.) Hitchc. in Lingnan. Sci. 7: 1930. 1931. *Poa nutans* Retz. Obs. Bot. 4: 19. 1786. *P. elegans* Roxb., Hort. Beng. 82. 1814; Roxb. Fl. Ind. 1: 339. 1820. *P. chariis* Schult. Syst. Veg. 2, Mant. 314. 1824. *P. elegantula* Kunth., Rev. Gram. 1: 114. 1829. *Eragrostis brownei* Nees in Wight. Cat. no. 1780. 1833. (PLATE–138).

Annual. Rhizome absent; root fibrous, culm 50 cm tall, 1.0 mm thick, green, glabrous, erect, densely tufted, compressed, ascending, decumbent, leafy at base, slender, simple, rooting at base; nodes brown, glabrous; internodes 6.0 cm long. Leaf–blade 10 cm long, 4.0 mm broad, green, flat, narrow, linear, lanceolate, submembranous, glabrous, margin hyaline scabrid, acuminate, base rounded, mid–vein distinct on lower surface, 2–lateral veins on each side of mid–vein; lateral veins distinct on upper surface; leaf–sheath 5.0 cm long, 2.0 mm broad, green, membranous, glabrous, margin hyaline, keeled, loose, erect, compressed, shorter than internodes, hairy at mouth; hairs 5.0 mm long, white, acute, base tubercled, veins distinct on both surface; ligule 0.3 mm long, a ring of hairs; hairs white, acute, base broad; auricle absent. Inflorescence consists of panicle; panicle 10 cm long, 5.0 cm broad, open, erect, oblong; rachis of panicle, slender, smooth. Branches 6.0 cm long, suberect, alternate, distant, filiform, scabrid, decreasing in length upwards. Spikelets 7.0 mm long, 1.5 mm broad, erect, linear, oblong, 10–flowered, secund, distichous, sub–sessile, blue–grey; rachis of spikelets tough, filiform, angular, smooth. Lower glume 0.7 mm long, 0.4 mm broad, membranous, ovate, lanceolate, acute, glabrous, 1–nerved; nerves blue, slightly visible, glabrous, ending at apex. Upper glume 1.0 mm long, 0.5 mm broad, membranous, ovate, lanceolate, acuminate, glabrous, 1–nerved; nerves blue, slightly visible, glabrous, ending at apex. Lemma 2.0 mm long, 1.5 mm broad, sub–membranous,

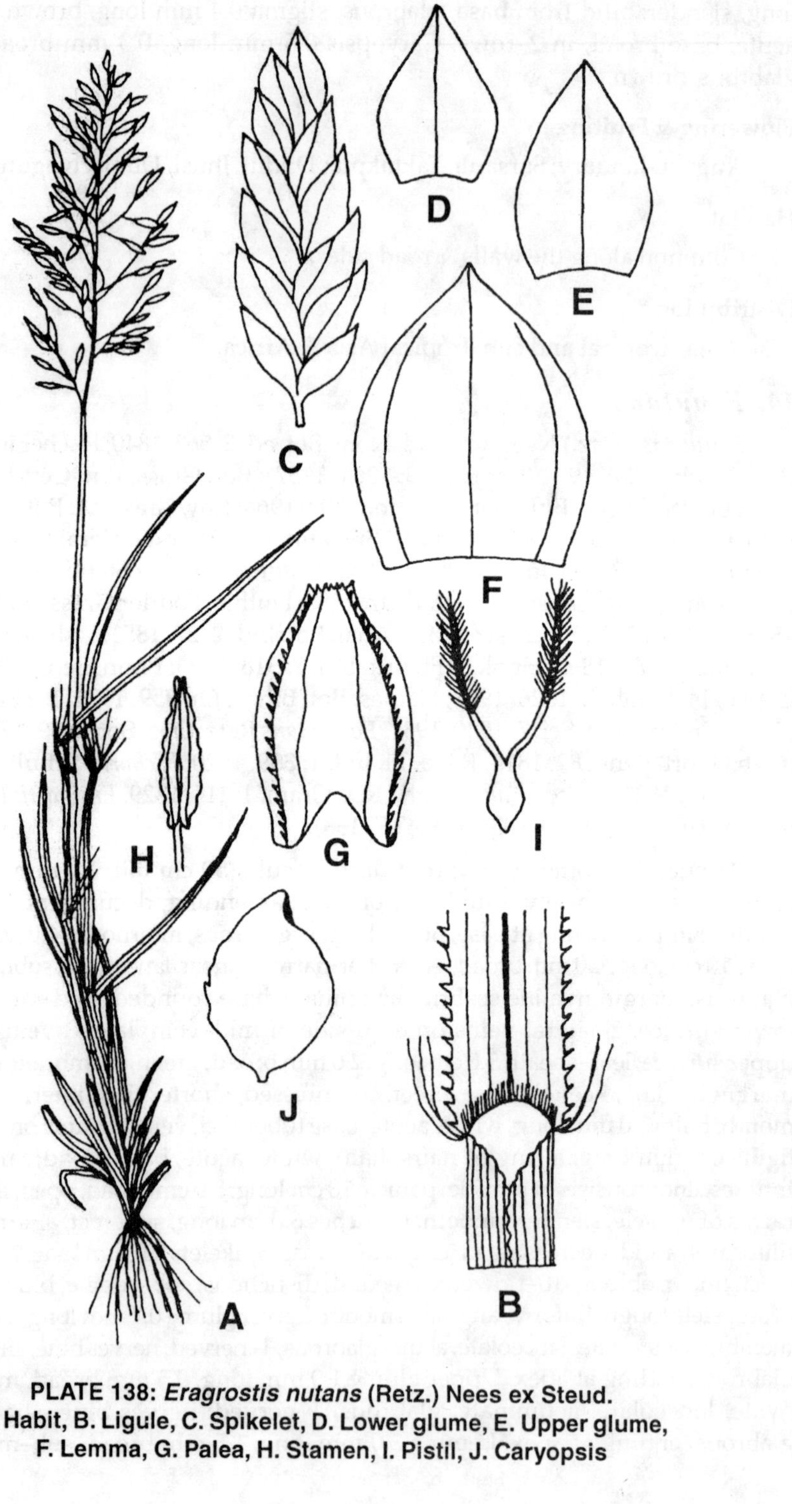

PLATE 138: ***Eragrostis nutans*** **(Retz.) Nees ex Steud.**
A. Habit, B. Ligule, C. Spikelet, D. Lower glume, E. Upper glume, F. Lemma, G. Palea, H. Stamen, I. Pistil, J. Caryopsis

broadly ovate, lanceolate, acuminate, glabrous, laterally compressed, 3–nerved; nerves glabrous, central one blue, visible, ending at apex, lateral nerves grey, slightly visible, ending slightly below at apex, distant to margin. Palea 1.5 mm long, 1.0 mm broad, hyaline, delicate, oblong, surface glabrous between the nerves, margin folded inwards, apex lacerate, 2–nerved; nerves blue, slightly visible, scabrid, marginal, ending at apex. Stamens 3; anther 0.7 mm long, yellow, keeled, oblong, apex divericate; filament 0.2 mm long, slender, glabrous, shorter than anther. Lodicule absent. Pistil1.0 mm long; ovary 0.2 mm long, obovate, glabrous; style 0.3 mm long, slender, glabrous, bifid from base; stigma 0.5 mm long, brown, hairy; hairs acute, base broad. Caryopsis 1.0 mm long, 0.5 mm broad, brown, elliptic, sub–globose, rugulose.

Flowering & Fruiting

September–December; Naugarh, Sonaghati, Bhatpar, Jhusi.

Habitat

Rare, along the moist places.

Distribution

India, China, Sri Lanka.

15. *E. pilosa*

E. pilosa (L.) P. Beauv. Ess. Agrost. 71: 162–175. 1812; Duthie, Grass. N.W. Ind. 38. 1883; Duthie, Fodd. Grass. N. Ind. 64. 1884; Aitchis. Cat. Punj. Pl. 170. 1869; Boiss. Fl. Or. 5: 581. 1884; Lisboa, J. Bomb. Nat. Hist. Soc. 7: 381. 1893; Stapf in Hook. f. Fl. Brit. Ind. 7: 323. 1896; Cooke, Fl. Bomb. 2: 1028. 1908; Rang. & Tad., Handb. S. Ind. Grass. 305. t. 255. 1921; Haines, Bot. Bih. & Or. 960. 1924; Fischer in Gamble, Fl. Pres. Madr. 1827. 1934 (Repr. ed. 3: 1265. 1957); Bor, Grass. Bur. Cey. Ind. Pak. 512. 1960; Raizada & Jain, Ind. For. 92: 640. 1966; Panigrahi & Rajagopal, J. Ind. Bot. Soc. 47: 236. 1968; Gilliland in Fl. Malaya 3: 69. 1971; Chien–Chang, Hus, Taiwan Grass. 413. t. 102. 1975; Cope in Nasir & Ali, Fl. W. Pak. 143: 91. 1982; Manilal & Sivaraj, Fl. Calic. 348. 1982; Raizada *et al.*, Grass. Upp. Gang. Pl. pt. 3: 87. 1983; Roy, Grass. M.P. 90. 1984; Blatter & McCann, Bomb. Grass. 241. t. 160. 1984; Sreekumar & Nair, Fl. Kera. Grass. 385. 1991; Misra & Verma, Fl. Allaha. 441. 1992; Moulik, Grass. Bamb. Ind. 2: 606. 1997. *E. filiformis* Link, Hort. Berol. 1: 191. 1827. *E. linkii* Steud. Syn. Pl. Glum. 1: 273. 1854. *E. tenuiflora* Rupr. ex Steud. Syn. Pl. Glum. 1: 268. 1854. *E. senegalensis* Chev. in Rev. Bot. Appl. 14: 120. 1934. *E. indica* Steud. Syn. Pl. Glum. 264. 1855. *E. parviflora* Trin. in Mem. Acad. Petersb. ser. 6, 1: 411. 1831. *E. pellucida* Steud. Syn. Pl. Glum. 279. 1855. *E. punctata* Link. ex Steud. Nom. ed. 2, 1: 561. 1840; Syn. Pl. Glum. 264. 1855. *E. verticillata* Roem. & Schult. Syst. Veg. 2: 575. 1817; Aitchis. Cat. Punjab Pl. 170. 1869. *Poa eragrostis* Walt., Fl. Carol. 80. 1788. *P. linkii* Kunth., Rev. Gram 1: 113. 1829. *P. senegalensis* Desv., Opusc. 100. 1831. *P. pilosa* Linn. Sp. Pl. 68. 1753; Host. Gram. Austr. 2: 168. t. 68. 1869. *P. indica* Koen et Rottl. in Ges. Naturf. Fl. Berl. Neue Schrift. 4: 194. 1803. *P. parviflora* var. *pellucida* R. Br. Prodr. 180. 1810. *P. punctata* Linn. f. Suppl. 109. 1781; Kunth. Enum. Pl. 1: 330. 1838; Roxb. Fl. Ind. 1: 338. 1820. *P. verticillata* Kunth. Enum. Pl. 1: 329. 1838. (PLATE–139).

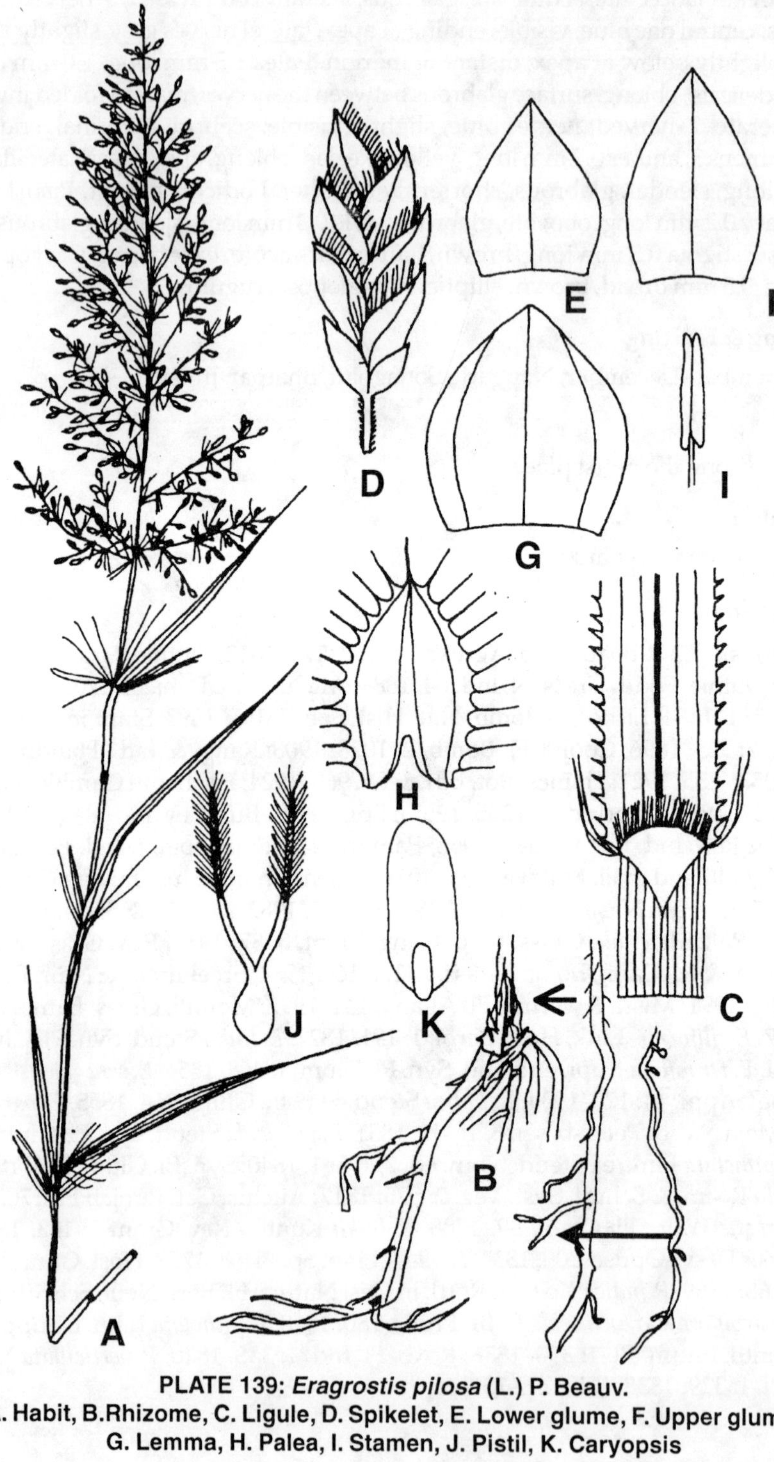

PLATE 139: ***Eragrostis pilosa*** **(L.) P. Beauv.**
A. Habit, B.Rhizome, C. Ligule, D. Spikelet, E. Lower glume, F. Upper glume, G. Lemma, H. Palea, I. Stamen, J. Pistil, K. Caryopsis

Annual. Rhizome absent; root fibrous. Culm 50 cm tall, 1.0 mm thick, green, glabrous, erect, densely tufted, very slender, geniculately ascending, leafy, rooting at base; nodes brown, glabrous; internodes 5.0 cm long. Leaf–blade 10 cm long, 3.0 mm broad, green, convolute, narrow, linear, lanceolate, acuminate, rounded at base, surface membranous, glabrous, margin hyaline, scabrid, mid–vein distinct on lower surface, 2–lateral veins on each side of mid–vein; lateral veins distinct on both surface; leaf–sheath 3.0 cm long, 2.0 mm broad, green, surface membranous, glabrous, margin hyaline, keeled, loose, erect, compressed, striate, shorter than internodes, hairy at mouth; hairs 5.0 mm long, white, acute, base tubercled, veins distinct on both surface; ligule 0.3 mm long, a ring of white hairs; hairs acute, base broad; auricle absent. Inflorescence consists of panicle; panicle 10 cm long, 3.0 cm broad, pyramidal, elliptic, ovate, oblong, lanceolate, erect, open, contracted, greyish–green at young, turning into purplish with age; rachis of panicle scabrid, slender. Branches 2.0 cm long, suberect, capillary, much divided, scabrid, upper alternate, lower fascicled, subwhorled with glands, hairy in the axils of branches; hairs 3.0 mm long, white, acute, base tubercled. Spikelets 3.0 mm long, 1.5 mm broad, scattered, narrow, oblong, linear, lanceolate, straight, biseriate; rachis of spikelet tough, smooth, angular. Lower glume 0.7 mm long, 0.3 mm broad, hyaline, narrowly ovate, lanceolate, acute, 1–nerved; nerve green, slightly visible, glabrous, ending at apex. Upper glume 1.0 mm long, 0.5 mm broad, sub–membranous, ovate, oblong, lanceolate, acute, glabrous, 1–nerved; nerve green, visible, glabrous, ending at apex. Lemma 1.5 mm long, 1.0 mm broad, membranous, broadly ovate, lanceolate, truncate, laterally compressed, glabrous, 3–nerved; nerves green, visible, glabrous, ending at apex. Palea 1.0 mm long, 0.5 mm broad, hyaline, delicate, oblanceolate, smooth, margin folded inwards, 2–nerved; nerves green, slightly visible, glabrous & marginal, ending at apex. Stamens 3; anther 0.3 mm long, deep pink, oblong, glabrous; filament 0.1 mm long, slender, glabrous, shorter than anther. Lodicules absent. Pistil 1.0 mm long; ovary 0.2 mm long, ovate glabrous; style 0.3 mm long, slender, glabrous, bifid from base; stigma 0.5 mm long, purple, hairy; hairs acute, base broad. Caryopsis 1.0 mm long, 0.5 mm broad, reddish brown, ellipsoid, obovate, cuneate.

Flowering & Fruiting

September–December; Basti, BHU, Pindra, Ramnagar, Deoria, Jaunpur, Chakiya.

Habitat

Very common on moist open grounds, waste places, along road side & old walls.

Distribution

India, Pakistan, Africa, Australia.

16. *E. riparia*

E. riparia (Willd.) Nees, Agrost. Bras. 512. 1829; Fischer in Gamble, Fl. Pres. Madr. 1825. 1934. (Repr. ed. 3: 1263. 1957.); Bor, Grass. Bur. Cey. Ind. Pak. 513. 1960; Manilal & Sivaraj, Fl. Calic. 347. 1982; Sreekumar & Nair, Fl. Ker. Grass. 386. 1991. *Poa riparia* Willd. in Gen. Natur. Freunde. Berlin Neue Schrift 4: 185. 1803. *Eragrostis tenella* (L.) P. Beauv. var. *riparia* Stapf in Hook. f. Fl. Brit. India 7: 315. 1896. *Poa ambonica*

Retz. Obs. Bot. 4: 20.1786 (non Linn. 1771). *Megastachya riparia* (Willd.) Nees Agrost. Bras. 512. 1829. (PLATE–140).

Perennial. Rhizomatous; rhizome creeping; root fibrous. Culms 40 cm tall, 1.0 mm thick, densely tufted, erect, green, glabrous, simple, leafy, rooting at base; nodes brown, glabrous; internodes 7.0 cm long. Leaf blade 7.0 cm long, 2.0 mm wide, linear–lanceolate, acute, base rounded, green, glabrous, margin scabrid, mid–vein distinct on lower surface; 1–lateral vein on each side of mid–vein; lateral veins distinct on both surface; leaf–sheath 4.5 cm long, 2.0 mm broad, surface glabrous, margin slightly keeled, compressed, green, shorter than internodes, hairy at throat; hairs white, acute, base rounded; ligule 0.3 mm long, a short ring of hairs; hairs acute, base rounded; auricle absent. Inflorescence panicle; panicle 8.0 cm long, 2.0 cm broad, delicate, contracted, narrow, sub–cylindric. Branches scabrid. Spikelet 6.0 mm long, crowded, oblong, 7–flowered; rachilla disarticulating above the glume. Lower glume 1.0 mm long 0.5 mm broad, ovate, lanceolate, acute, margin scaberulous, 1–nerved; nerves green, visible, glabrous, ending at apex. Upper glume 1.5 mm long 1.0 mm broad, margin scabrid, acute, membranous, lanceolate, 1–nerved; nerves green, visible, glabrous, ending at apex. Lemma 2.5 mm long, 1.0 mm broad, rounded at apex, membranous, margin scaberulous, globose, oblong, ovate, 3–nerved; nerves green, visible, glabrous, ending at apex. Palea 1.5 mm long, 1.0 mm broad, hyaline, mucronate, keeled, keel hairy; hairs acute, base broad. Stamens 3; anther 0.3 mm long, glabrous, pink; filament 0.1 mm long, slender, glabrous, shorter than anther. Lodicules absent. Pistil 1.5 mm long; ovary 0.3 mm long, ovate, glabrous; style 0.5 mm long, bifid from base, slender, glabrous; stigma 0.5 mm broad, brown, hairy; hairs acute, base broad in 2–rows. Caryopsis 1.0 mm long, 0.5 mm broad, brown, ovoid, loosely enclosed by lemma & palea.

Flowering & Fruiting

August–November; Sindhora, Mau, Ballia, Barhalganj, Deoria.

Habitat

Rare, along the railway tracks.

Distribution

India, Burma, Sri Lanka.

17. *E. tef*

E. tef (Zucc.) Trotter in Bull. Soc. Bot. Ital. 62. 1918; Bor, Grass. Bur. Cey. Ind. Pak. 513. 1960; Roy, Grass. M.P. 92. 1984; Sreekumar & Nair, Fl. Kera. Grass. 388. t. 80. 1991; Moulik, Grass. Bamb. Ind. 2: 608. 1997. *Eragrostis abyssinica* (Jacq.) Link. Hort. Berol. 1: 192. 1827. *P. tef* Zucc., Diss. Ist. Planta, Panizz. Abiss. 1774. *Poa abyssinica* Jacq., Misc. Austr. 2: 364. 1781. (PLATE–141).

Annual. Rhizome absent; root fibrous. Culm 45 cm long, 1.0 mm thick, green, glabrous, erect, tufted, compressed, slender, ascending, decumbent, leafy at base, branching at lower nodes; branches sub erect, spreading, leafy upto panicle, rooting at base; nodes glabrous, brown; internodes 4.0 cm long. Leaf–blade 10 cm long, 6.0 mm broad, green, flat, narrow, linear–lanceolate, acuminate, base rounded, coriaceous,

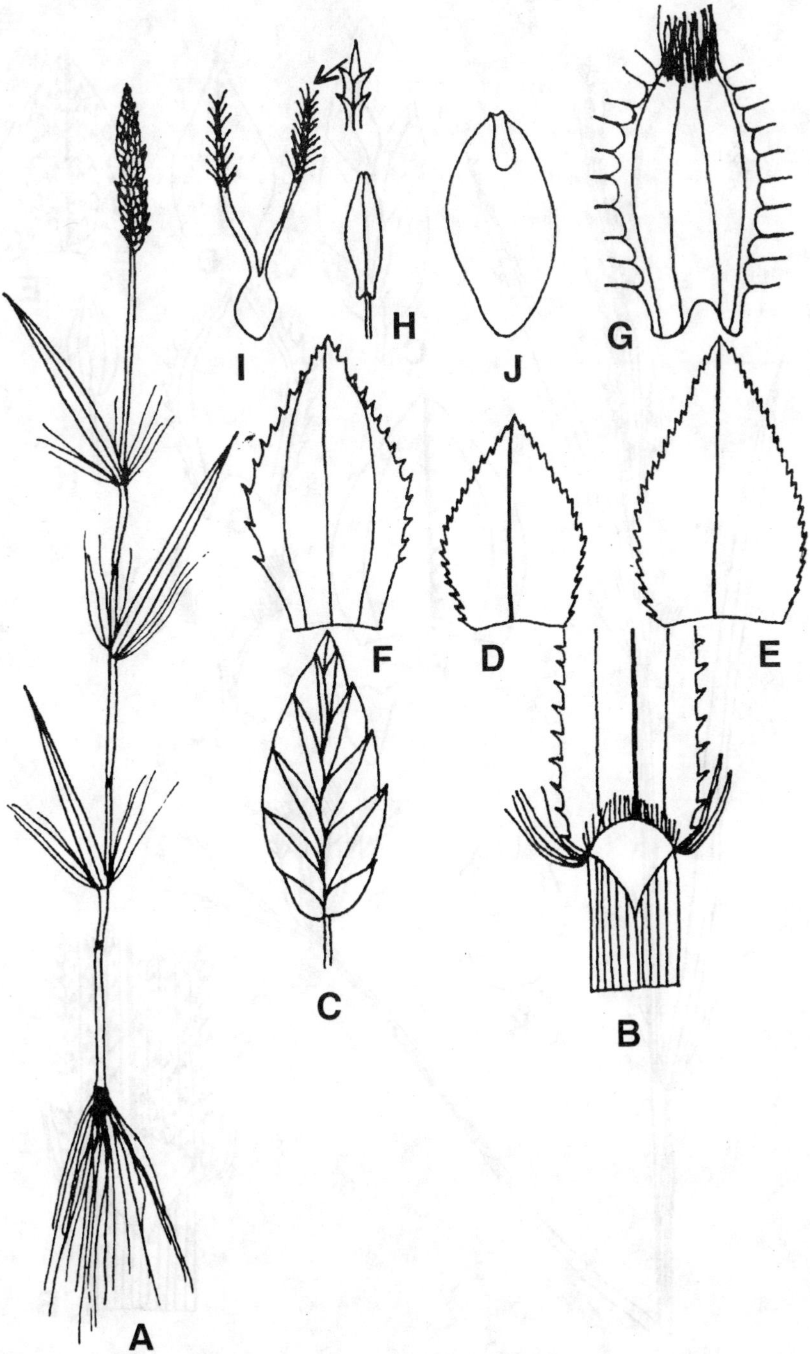

PLATE 140: *Eragrostis riparia* (Willd.) Nees
A. Habit, B. Ligule, C. Spikelet, D. Lower glume, E. Upper glume, F. Lemma, G. Palea, H. Stamen, I. Pistil, J. Caryopsis

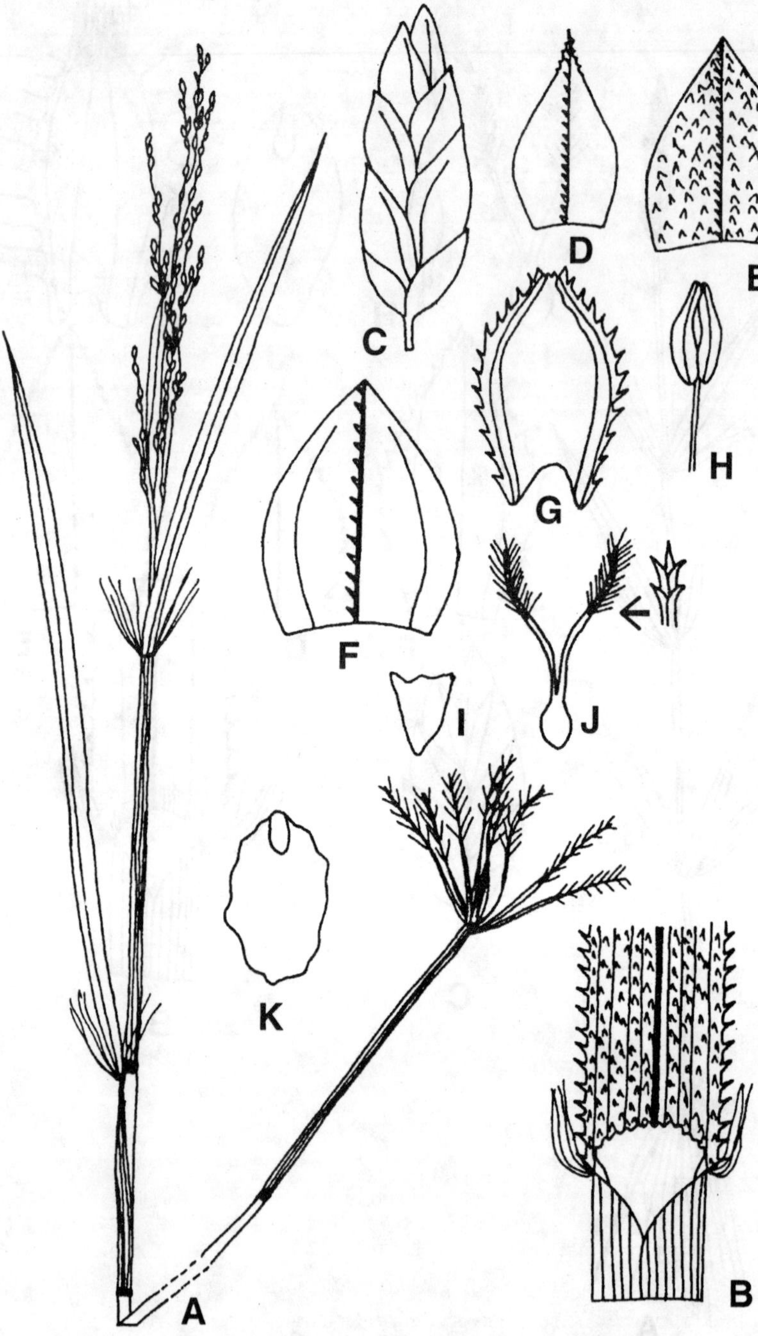

PLATE 141: *Eragrostis tef* (Zucc.) Trotter

A. Habit, B. Ligule, C. Spikelet, D. Lower glume, E. Upper glume, F. Lemma, G. Palea, H. Stamen, I. Lodicules, J. Pistil, K. Caryopsis

upper surface scabrid, lower surface glabrous, margin hyaline scabrid, mid–vein distinct on lower surface, 4–lateral veins distinct on upper surface; leaf–sheath 5.0 cm long, 2.0 mm broad, green, coriaceous, glabrous, margin sub membranous, keeled, loose, compressed, erect, striate, longer than internodes, hairy at mouth; hairs 5.0 mm long, white, acute, base, tubercled, veins distinct on outer surface; ligule 0.5 mm long, membranous, lacerate; auricle glabrous. Inflorescence made up of panicle; panicle 10 cm long, 7.0 cm broad, erect, oblong, ovate, pyramidal, open; rachis terete, glabrous. Branches 5.0 cm long, suberect, flexuous, spreading, angular, scabrid. Spikelets 10 mm long, 1.5 mm broad green at young turnning into purple with age, linear, ovate, secund, biseriate, erect, 15–flowered, florets persistent; rachis of spikelets angular, glabrous. Lower glume 1.5 mm long, 1.0 mm broad, membranous, ovate, lanceolate, sharply acute, surface glabrous except the nerves, margin slightly lacerate at apex, glabrous below, 1–nerved; nerves green, visible, scabrid, ending at apex. Upper glume 2.0 mm long, 1.0 mm broad, membranous, ovate, lanceolate, acute, surface scabrid, margin glabrous, 1–nerved; nerves green, visible, scabrid, ending at apex. Lemma 2.5 mm long, 1.5 mm broad, membranous, broadly ovate, lanceolate, acute, surface glabrous between the nerves, also with margin, laterally compressed, 3–nerved; nerves green, central nerve visible, scabrid, ending at apex, lateral nerves slightly visible, ending slightly below at the apex, very distant to margin. Palea 2.0 mm long, 1.0 mm broad, hyaline, delicate ovate, oblong, lacerate at apex, folded & enclosed by lemma, glabrous, margin sharply folded inwards, transversely scabrid, 2–nerved; nerves green, visible, glabrous & marginal, ending at apex. Stamens 3; anther 1.0 mm long, yellow, oblong, lanceolate, glabrous; filament 0.5 mm long, slender, glabrous, longer than anther. Lodicules 2, 0.2 mm long, 0.1 mm broad, hyaline, glabrous, nerveless, oblanceolate, slightly wavy at apex. Pistil 1.5 mm long; ovary 0.2 mm long, ovate, glabrous; style 0.8 mm long, slender, glabrous, slightly connate at base, bifid above; stigma 0.5 mm long, brown, hairy; hairs acute, base broad, in 2–rows. Caryopsis 1.0 mm long, 0.5 mm broad, reddish-brown, spindle shaped.

Flowering & Fruiting

August–December; Naugarh, Chandauli, Allahabad, Ramnagar, Basti.

Habitat

Rare, along the rocky soil.

Distribution

India, Ethiopia, South Africa.

18. *E. tenella*

E. tenella (L.) P. Beauv. ex Roem. & Schult., Syst. Veg. 2: 576. 1817; Hook. f. Fl. Brit. Ind. 7: 315. 1896; Cooke, Fl. Bomb. 2: 1023. 1908; Haines, Bot. Bih. & Or. 956. 1924; Bor, Grass. Bur. Cey. Ind. Pak. 513. 1960; Bhattacharya, Bull. Bot. Surv. Ind. 6: 209. 1964; Raizada & Jain, Ind. For. 92: 640. 1966; Rajagopal & Panigrahi, J. Ind. Bot. Soc. 47: 237. 1968; Singh, Bull. Bot. Surv. Ind. 11: 21. 1969; Panigrahi *et al.*, Bull. Bot. Surv. Ind. 11: 112. 1969; Gilliland in Fl. Malaya 3: 71. 1971; Cope in Nasir & Ali, Fl. W. Pak. 143: 88. 1982; Manilal & Sivaraj, Fl. Calic. 348. 1982; Raizada *et al.*, Grass. Up. Gang. Pl. pt. 3:

89. 1983; Roy, Grass. M.P. 92. 1984; Blatter & McCann, Bomb. Grass. 232. 1984; Sreekumar & Nair, Fl. Kera. Grass. 390. 1991; Misra & Verma, Fl. Allah. 442. 1992; Moulik, Grass. Bamb. Ind. 2: 608. 1997. *E. tenella* var. *tenella* (Linn.) Hook. f. in Trimen, Fl. Ceyl. 5: 291. 1900. *E. tenella* var. *plumosa* (Retz.) Stapf. in Hook. f. Fl. Brit. Ind 7: 315. 1896. *E. tenella* var. *insularis* Hubb. in Kew Bull. 1939. 654. 1939. *E. amabilis* (Linn.) Wight et. Arn. ex Hook. Bot. Beechey Voy. 251. 1838. *E. plumosa* Link. Hort. Berol. 1: 192. 1827; Duthie, Grass. N.W. Ind. 38. 1883; Duthie, Fodd. Grass. N. Ind. 64. t. 33. 1888; Aitchis, Cat. Punjab Pl. 170. 1869; Lisboa, J. Bomb. Nat. Hist. Soc. 6: 385. 1893; Fischer in Gamble, Fl. Pres. Madr. 1826. 1934 (Repr. ed. 3: 1264. 1957). *E. despiciens* Schult. Mant. 2: 318. 1824. *Poa amabilis* Linn. Sp. Pl. 68. 1753. *P. tenella* Linn. Sp. Pl. 69. 1753; Retz. Obs. Bot. 5: 19. 1789. *P. plumosa* Retz. Obs. 4: 20. 1786; Kunth. Enum. Pl. 1: 338. 1838; Roxb. Fl. Ind. 1: 337. 1820. *P. despiciens* Link. Enum. Hort. Berol. 1: 88. 1827. *Megastachya tenella* Linn. Bojer Hort. Maurit. 369. 1837. (PLATE–142).

Annual. Rhizome absent; root fibrous. Culm 60 cm tall, 1.0 mm thick, green, glabrous, erect, tufted, geniculately–ascending, very elegant & slender, simple, leafy, rooting at base; nodes glabrous, brown; internodes 6.0 cm long. Leaf–blade 10 cm long, 3.0 mm broad, green, convolute, narrow, linear, lanceolate, acuminate, rounded at base, surface membranous, glabrous, margin hyaline, scabrid, mid–vein distinct on lower surface; 2–lateral veins on each side of mid–vein; lateral veins distinct on both surface; leaf–sheath 4.0 cm long, 2.0 mm broad, green, membranous, glabrous, erect, loose, compressed, shorter than internodes, hairy at mouth; hairs 6.0 mm long, white, acute, base tubercled; ligule 0.3 mm long, a ring of white hairs; hairs acute, base broad; auricle absent. Inflorescence made up of panicle; panicle 20 cm long, 2.5 cm broad, open, erect, elliptical, pyramidal, lax, oblong, ovate, decumpound, delicate, flexuous, pale green at young, turnning into purplish with age, excessively branched; rachis of panicle filiform, minutely scabrid, hairy at nodes; hairs 2.0 mm long, white, acute, base tubercled. Branches 1.5 cm long, suberect, capillary, alternate, minutely scabrid, distant, decreasing in length upwards. Spikelets 4.0 mm long, 2.0 mm broad, innumerable, 9–flowered, ovate, oblong, biseriate, pedicelled; pedicel 5.0 mm long, filiform, glabrous; rachis of spikelet subarticulate. Lower glume 0.8 mm long, 0.4 mm broad, membranous, ovate, oblong, obtuse, lanceolate, glabrous, 1–nerved; nerve green, slightly visible, glabrous, ending at apex. Upper glume 1.0 mm long, 0.5 mm broad, membranous, ovate, oblong, lanceolate, obtuse, glabrous, 1–nerved; nerve green, visible, glabrous, ending at apex. Lemma 1.5 mm long, 1.0 mm broad, membranous, ovate, oblong, acuminate, laterally compressed, surface glabrous between the nerves, also with margin, 3–nerved; nerves green, visible, ending at apex, central one scabrid, lateral nerves glabrous, close to the margin. Palea 1.0 mm long, 0.5 mm broad, hyaline, delicate, oblong, elliptic, lanceolate, glabrous, margin, sharply folded inwards, transversely hairy; hairs 1.0 mm long, acute, base tubercled; 2–nerved; nerves green, slightly visible, marginal & glabrous, ending at apex. Stamens 3; anther 0.3 mm long, deep pink, oblong, glabrous; filament 0.2 mm long, slender, glabrous, shorter than anther. Lodicules absent. Pistil 1.0 mm long; ovary 0.2 mm long, elliptic, glabrous; style 0.3 mm long, slender, glabrous, bifid from base; stigma 0.5 mm long, cream coloured, hairy; hairs acute, base broad. Caryopsis 0.6 mm long, 0.3 mm broad pale brown, polished, ovoid, elliptic.

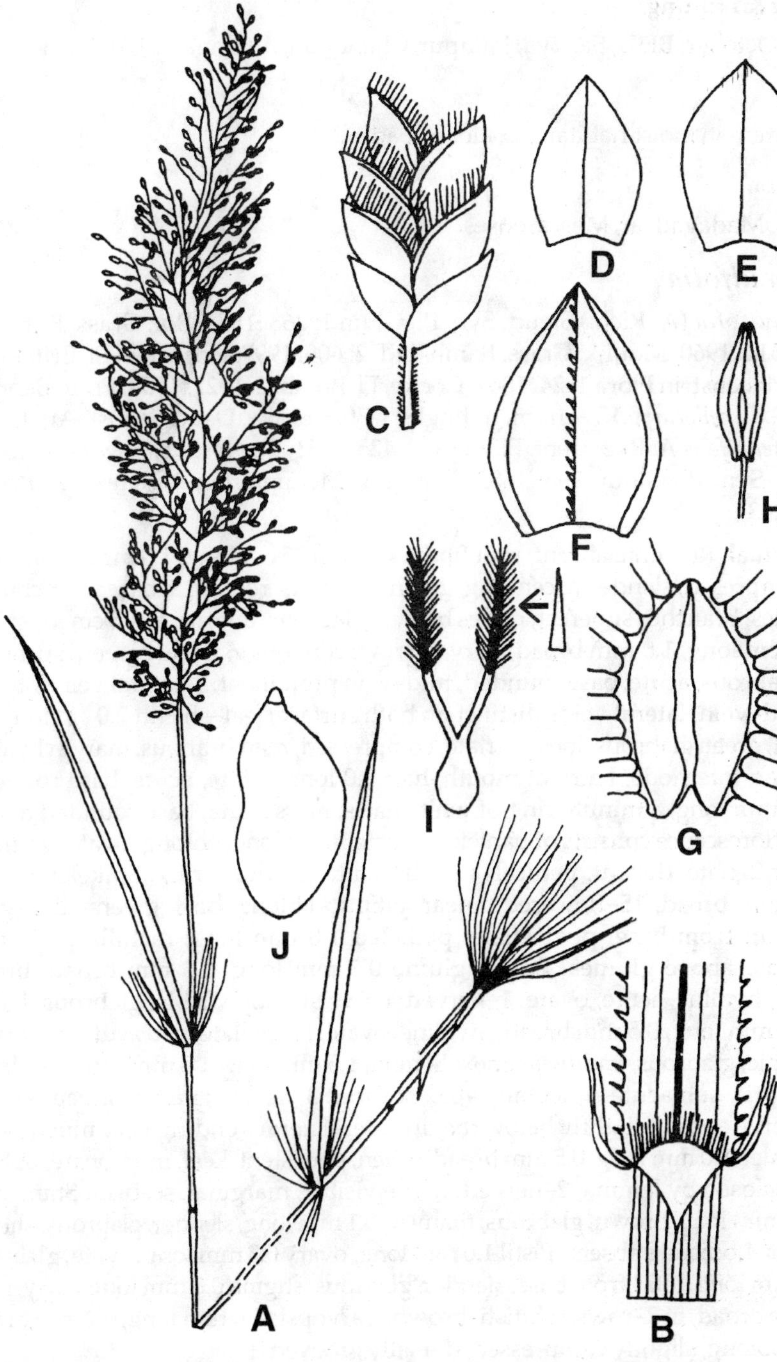

PLATE 142: *Eragrostis tenella* (L.) P. Beauv. Roem. ex Schult.
A. Habit, B. Ligule, C. Spikelet, D. Lower glume, E. Upper glume, F. Lemma, G. Palea, H. Stamen, I. Pistil, J. Caryopsis

Flowering & Fruiting

July–October; BHU, Sarriya, Jaunpur, Chandauli, Azamgarh, Lar, Basti.

Habitat

Common in moist habitates & along road sides.

Distribution

India, Madagadcar, Mascarenses.

19. *E. tenuifolia*

E. tenuifolia (A. Rich.) Steud. Syn. Pl. Glum 1: 268. 1854; Bor, Grass. Bur. Ceyl. Ind. Pak. 514. 1960; Moulik, Grass. Bamb. Ind. 2: 609. 1997; Hochst. in Fl. Brit. Ind. 7: 322. 1896; Hochst. in Flora 1: 24. 1841; Cooke, Fl. Bomb. 2: 1027. 1908; Prain Beng. Pl. 1221. 1903. *E. cellocarpa* K. Schum. in Engler's Pflavizenw. Deutsch–Ost. Afr. C. 114. 1845. *Poa tenuifolia* A. Rich. Tent. Fl. Abys. 2: 425. 1851. *Eragrostis parviglumis* Hochst. ex. Steud. Syn. Pl. Glum. 268. 1855; Blatter & McCann, Bomb. Grass. 240. 1984. (PLATE–143).

Perennial. Rhizome absent; root fibrous. Culm 25 cm long, 1.0 mm thick, erect, tufted, compresed, slender, ascending, green, glabrous; rooting at base; branching at lower nodes; branches suberect, nodes brown, glabrous; internode 4.5 cm long. Leaf blade 5.0 cm long, 1.0 mm broad, very narrow, convolute, linear, green, surface & margin glabrous, acute, base rounded, mid–vein prominent, 3–lateral veins on each side of mid–vein; lateral veins distinct on both surface; leaf–sheath 2.0 cm long, 2.0 mm broad, green glabrous, loose, striate, compressed, membranous, margin hyaline, shorter than internode, hairy at mouth; hair 3.0 long, white, acute, base rounded; ligule 0.1 mm long, a minute ring of white hairs; hairs acute, base rounded auricle absent. Inflorescence consists of panicle; panicle 10 cm long, oblong–ovate, branches solitary or 2–nate, distant, pedicels usually with glandular ring. Spikelet 3.0 mm long, 1.5 mm broad, 15–flowered, linear, elliptic–oblong, dark green–olive–gray, breaking up from base, pedicellate; pedicled 1.0 mm long; rachilla persistent, disarticulate above glumes. Lower glume 0.5 mm long, 0.3 mm broad, broad, lanceolate, hyaline, acute, ovate, 1–nerved; nerve slightly visible, glabrous. Upper glume 1.0 mm long, 0.5 mm broad, hyaline, ovate, lanceolate, sub acute, 1–nerved; nerve visible, glabrous, ending at apex. Lemma 1.5 mm long, 1.0 mm broad, oblong–ovate, elliptic, sub acute–apiculate, apex scabrid, 3–nerved; nerves green, visible, central. nerve ending slightly below the tip, lateral nerves ending very much below the tips. Palea 1.0 mm long, 0.5 mm broad, green, persistent, keel, mucronate, oblong, folded, enclosed by lemma, 2–nerved, nerve visible, marginal, scabrid. Stamens 3; anther 1.0 mm long, brown, glabrous; filament 0.1 mm long, slender, glabrous, shorter than anther. Lodicules absent. Pistil 1.0 mm long; ovary 0.3 mm long, ovate, glabrous; style 0.4 mm long, bifid from base, slender, glabrous; stigma 0.3 mm long hairy; hairs acute, base broad, in 2–rows, whitish–brown. Caryopsis 1.0 mm long, 0.2 mm broad, obovoid, oblong, slightly compressed, dorsally grooved.

Flowering & Fruiting

August–December; Naugarh, Allahabad, Naryanpur, Deoria.

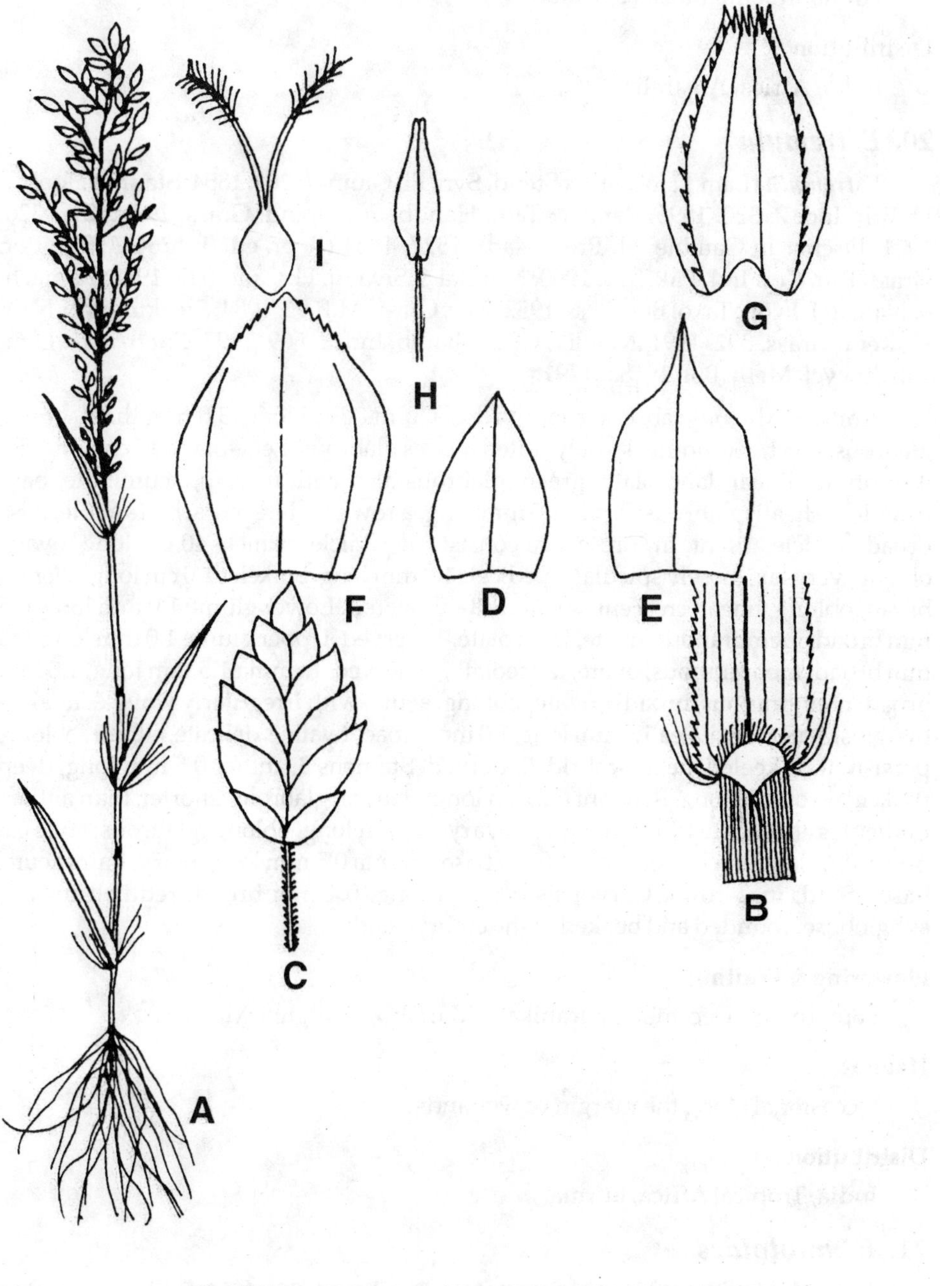

PLATE 143: ***Eragrostis tenuifolia*** **(A. Rich.) Steud.**
A. Habit, B. Ligule, C. Spikelet, D. Lower glume, E. Upper glume, F. Lemma, G. Palea, H. Stamen, I. Pistil

Habitat

Found in semi–moist condition.

Distribution

India, Africa, Australia.

20. *E. tremula*

E. tremula (Lam.) Hochst. ex Steud. Syn. Pl. Glum. 1: 269. 1854; Stapf in Hook. f. Fl. Brit. Ind. 7: 320. 1896; Rang. & Tad., Handb. South. Ind. Grass. 297. t. 219–220. 1921; Fischer in Gamble, Fl. Pres. Madr. 1827. 1934 (Repr. ed. 3: 1264. 1957); Bor, Grass. Bur. Cey. Ind. Pak. 514. 1960; Manilal & Sivaraj, Fl. Calic. 347. 1982; Ramach. & Nair, in J. Econ. Taxo. Bot. 3: 66. 1982; Roy, Grass. M.P. 93. 1984; Sreekumar & Nair, Fl. Kera. Grass. 392. 1991; Moulik, Grass. Bamb. Ind. 2: 609. 1997. *Poa tremula* Lam. Tab. Encycl. Meth. Bot. 1: 185. 1791.

Annual. Rhizome absent; root fibrous. Culm 80 cm long, 3.0 mm thick, green, glabrous, erect, ascending, loosely tufted; nodes glabrous. Leaf–blade 15 cm long, 5.0 mm broad, linear, lanceolate, green, glabrous also with margin, acuminate, base rounded; sheath glabrous; ligule 1.0 mm long, a row of white hairs; hairs acute, base broad; auricle absent. Inflorescence consists of panicle; panicle 40 cm long, ovate, oblong, very lax, loosely spiculate, pedicels 1.0 mm long. Spikelet 1.0 cm long, 2.0 mm broad, oblong, linear, crescent–shaped, 40–flowered. Lower glume 1.0 mm long, 0.5 mm broad, membranous, ovate, lanceolate, 1–nerved. Upper glume 1.0 mm long, 0.5 mm broad, membranous, ovate, lanceolate, 1–nerved. Lemma 1.5 mm long, 1.0 mm broad, membranous, broadly ovate, oblong, acute, with irregularly wavy & hyaline margins, 3–nerved. Palea 1.5 mm long, 1.0 mm broad, hyaline, delicate, elliptic, oblong, persistent. 2–keeled; keels scabrid, 2–nerved. Stamens 3; anther 0.5 mm long, deep pink, glabrous, oblong; filament 0.2 mm long, slender, glabrous, shorter, than anther. Lodicules absent. Pistil 1.0 mm long; ovary 0.2 mm long, oblong, glabrous; style 0.3 mm long, slender, glabrous, bifid from base; stigma 0.5 mm long, hairy; hairs acute, base broad, in 2–rows. Caryopsis 0.6 mm long, 0.4 mm broad, reddish brown, subglobose, rounded and beaked at the embryo end.

Flowering & Fruiting

Septermber–December; Barahikala, Mariahu, Shahgunj, Mairwa.

Habitat

Occasional along the margin of wetlands.

Distribution

India, Tropical Africa, Burma.

21. *E. unioloides*

E. unioloides (Retz.) Nees. ex Steud., Syn. Pl. Glum. 1: 264. 1854; Fischer in Gamble, Fl. Pres. Madr. 1826. 1934 (Repr. ed. 3: 1264. 1957); Bor, Grass. Bur. Cey. Ind. Pak. 515. 1960; Bhattacharya, Bull. Bot. Surv. Ind. 6: 209. 1964; Raizada & Jain, Ind. For. 92: 640. 1966; Panigrahi *et al.*, Bull. Bot. Sur. Ind. 11: 112. 1969; Gilliland in Fl. Malaya 3:

66.1971; Chien–Chang, Hsu in Taiwan Grass. 421. 1975. Cope in Nasir & Ali, Fl. W. Pak. 143: 90. 1982; Raizada *et al.*, Grass. Upp. Gang. Pl. pt. 3: 91. 1983; Roy, Grass. M.P. 93. 1984; Blatter & McCann, Bomb. Grass 235. t. 156. 1984; Sreekumar & Nair, Fl. Kera. Grass. 392. 1991; Misra & Verma, Fl. Allaha. 443. 1992; Moulik, Grass. Bamb. Ind. 2: 609. 1997. *Eragrostis rubens* Hochst ex Miq., Analect. Bot. 2: 26. 1851. *Poa unioloides* Retz. Obs. Bot. 5: 19. 1789. (PLATE–144).

Annual. Rhizome absent; root fibrous. Culm 50 cm tall, 1.5 mm thick, green, glabrous, tufted, erect, geniculately–ascending, slender, leafy, decumbent, simple, rooting at base; nodes glabrous; internodes 6.5 cm long. Leaf–blade 10 cm long, 5.0 mm broad, green, flat, narrow, linear, lanceolate, acuminate, base rounded, surface sub–coriaceous, glabrous, margin hyaline, scabrid, mid–vein distinct on lower surface, 3–lateral veins on each side of mid–vein; lateral veins distinct on upper surface; leaf sheath 4.0 cm long, 4.0 mm broad, green, subcoriaceous, glabrous, margin sub membranous, slightly keeled, erect, loose, striate, slightly compressed, very much shorter than internodes, hairy at mouth; hairs 5.0 mm long, white, acute, base tubercled; veins distinct on outer surface; ligule 0.1 mm long, thinly membranous, narrow rim, lacerate; auricle absent. Inflorescence consist of panicle; panicle 10 cm long, 5.0 cm broad, broadly ovate, narrowly oblong, lax, erect, open, consist of several branches; rachis of panicle scabrid, slender. Branches 3.0 cm long, numerous, ascending, filiform, minutely scabrid, alternate, distant. Spikelets 12 mm long, 5.0 mm broad, ovate, oblong, obtuse, strongly compressed, closely imbricate, biseriate, whitish–green when young, turning into purplish–red with age, 15–flowered; falling downward to upwards at maturaty; rachis of spikelet angular, smooth. Lower glume 1.0 mm long, 0.5 mm broad, membranous, oblong, lanceolate, acute, surface glabrous except the nerves, also with margin, 1–nerved; nerves green, slightly visible, scabrid, ending at apex. Upper glume 1.5 mm long, 1.0 mm broad, membranous, ovate, lanceolate, acute, surface glabrous except the nerves, also with margin, 1–nerved; nerve green, visible, scabrid, ending at apex. Lemma 2.0 mm long, 1.5 mm broad, membranous, broadly ovate, lanceolate, acute, glabrous, laterally compressed, 3–nerved; nerves green, visible, glabrous, ending at apex. Palea 2.0 mm long, 1.0 mm broad, hyaline, delicate, elliptic, ovate, lanceolate, subacute, glabrous, margin sharply folded inwards, 2–keeled, transversely scabrid, 2–nerved; nerves green, visible, glabrous & marginal, ending at apex. Stamens 3; anther 0.5 mm long, deep pink, oblong, glabrous; filament 0.2 mm long, slender, glabrous, shorter than anther. Lodicules 2, 0.2 mm long, 0.1 mm broad, hyaline, nerveless, glabrous, oblanceolate. Pistil 1.5 mm long; ovary, 0.3 mm long, ovate, glabrous; style 0.5 mm long, glabrous, slender, bifid from base; stigma 0.7 mm long, purplish, hairy; hairs acuminate, base broad, in 2–rows. Caryopsis 1.0 mm long, 0.5 mm broad, reddish brown, obovoid, ellipsoid, embryonic end narrow, pointed, lower end rounded & broad.

Vernacular Names

Todha, Asara, Kaluargi.

Flowering & Fruiting

September–March; Navalpur, BHU, Chakiya, Harraua, Bajrangnagar, Lar.

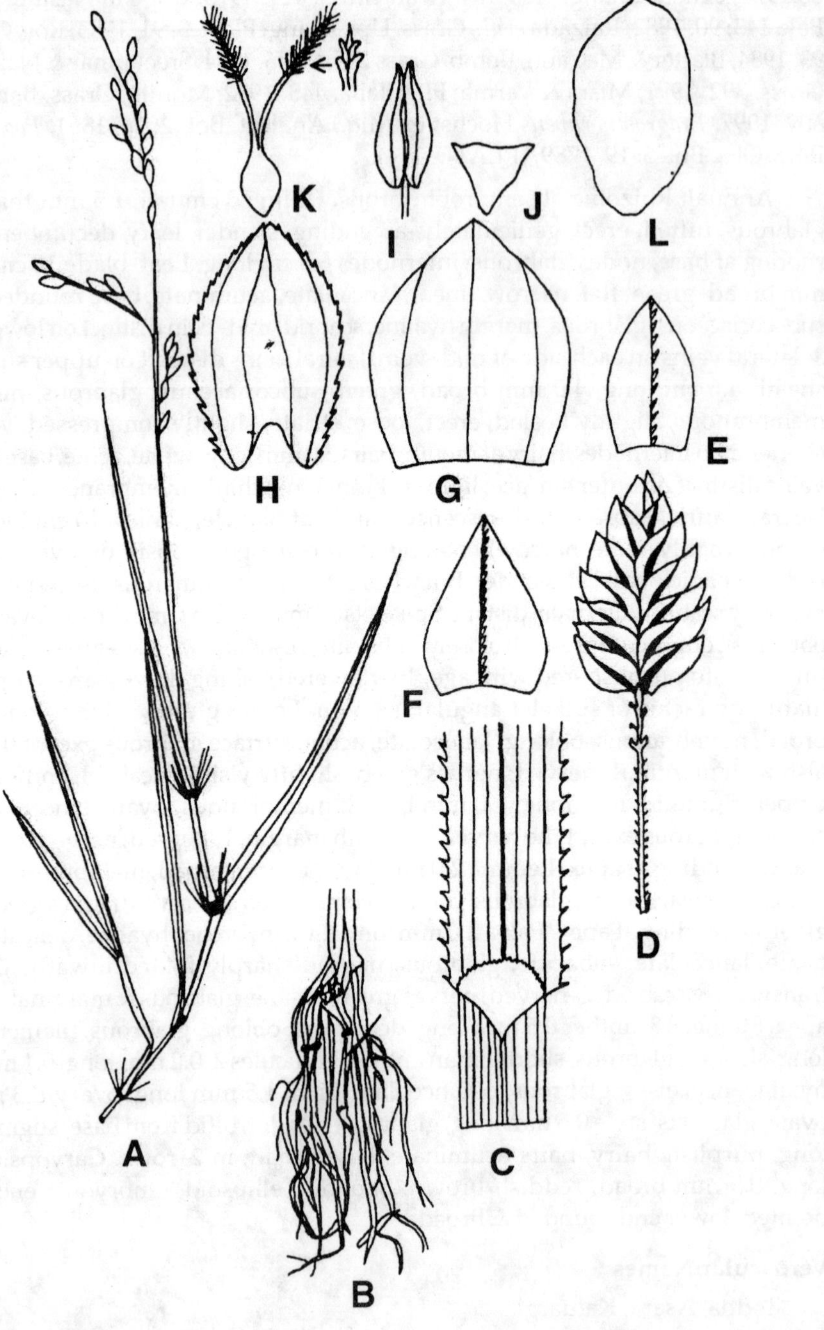

PLATE 144: ***Eragrostis unioloides*** **(Retz.) Nees ex Steud.**
A. Habit, B. Rhizome, C. Ligule, D. Spikelet, E. Lower glume, F. Upper glume, G. Lemma, H. Palea, I. Stamen, J. Lodicule, K. Pistil, L. Caryopsis

Habitat

Common along the ditches, in waterlogged habitat, muddy sandy soil, banks of rivers.

Distribution

India, Tropical Asia, Africa.

22. *E. viscosa*

E. viscosa (Retz.) Trin. in Mem. Acad. Sci. Petersb. 1 (6): 397. 1830; Dalz. & Gibs. Bomb. Fl. 298. 1861; Lisboa, J. Bomb. Nat. Hist. Soc. 7: 386. 1893; Haines, Bot. Bih. &. Or. 957. 1924; Fischer in Gamble, Fl. Pres. Madr. 1826. 1934 (Repr. ed. 3: 1264. 1957); Bor, Grass. Bur. Cey. Ind. Pak. 515. 1960; Roy, Grass. M.P. 94. 1984; Blatter & McCann, Bomb., Grass. 233. 1984; Sreekumar & Nair, Fl. Kera. Grass. 393. 1991. *E. tenella* var. *viscosa* Stapf. in Hook. f. Fl. Brit. Ind. 7: 315. 1896; Cooke, Fl. Bomb. 2: 1024. 1908. *E. mangalorica* Hochst. ex Steud., Syn. Pl. Glum. 1: 265. 1854. *Poa viscosa* Retz. Obs. Bot. 4: 20. 1786. *P. glutinosa* Roxb. ex Stapf in Hook. f. Fl. Brit. Ind. 7: 316. 1896. (PLATE–145).

Annual. Rhizome absent; root fibrous. Culm 45 cm tall, 1.0 mm thick, green, glabrous, erect, densely tufted, very slender, geniculately–ascending, decumbent, leafy, simple, rooting at base, panicle occupying the greater part of the plant; nodes brown, glabrous; internodes 5.0 cm long. Leaf–blade 6.0 cm long, 3.0 mm broad, green, convolute, erecto–patent, narrow, linear, lanceolate, acuminate, base rounded, surface membranous, glabrous, margin hyaline, scabrid, mid–vein distincton lower surface, 2–lateral veins on each side of mid vein; lateral veins distinct on both surface; leaf–sheath 3.0 cm long, 3 mm broad, green, surface membranous, glabrous, margin hyaline, keeled shorter than internode, loose, compressed, hairy at mouth; hairs 5.0 mm long, white, acute base rounded; ligule 0.4 mm long, a line of hairs; hairs acute, base broad; auricle absent. Inflorescence a large panicle; panicle 9.0 cm long, 3.5 cm broad, cylindric, oblong, erect, open, glume & younger parts of stem are sticky, owing to presence of glands, which are usually brownish–black, to which sand particles adhere; rachis fragile. Branches 2.0 cm long, suberect, whorded, corwded more or less spreading, equal, sometimes opposite, binate, decreasing in length upwards, hairs on axils; hairs 3.0 mm long, acute, base rounded; usually glabrous, rarely obscurely bearded at the nodes. Spikelet 3.0 mm long, 1.0 mm broad, lanceolate, oblong, ovate, 5–20 flowered, often purplish. Lower glume 0.8 mm long, 0.4 mm broad, membranous acute, ovate, 1–nerved; nerves green, visible, glabrous, ending at apex. Upper glume 1.0 mm long, 0.5 mm broad, acuminate, membranous, ovate 1–nerved, nerve scabrid, visible, ending at apex. Lemma 1.5 mm long, 1.0 mm broad, rounded at apex, membranous, oblong–elliptic, 3–nerved, nerve green, visible, ending at apex, central one scabrid, marginal & glabrous. Palea 1.0 mm long, 0.5 mm broad, hyaline, keeled; keel hairy; hairs acute, base tubercled, elliptic, folded, enclosed by lemma. Stamen 2; anther 0.5 mm long, yellowish, glabrous, oblong; filament 0.1 mm long slender, glabrous, shorter than anther. Lodicules absent. Pistil 1.0 mm long; ovary 0.2 mm long, ovate, glabrous; style 0.5 mm long, slender, glabrous, bifid from base; stigma 0.3 mm long, purple, hairy; hairs acute, base broad, in 2–rows. Caryopsis 1.0 mm long, 0.5 mm broad, brown, polished, ovoid–sub–globose, enclosed by lemma and palea.

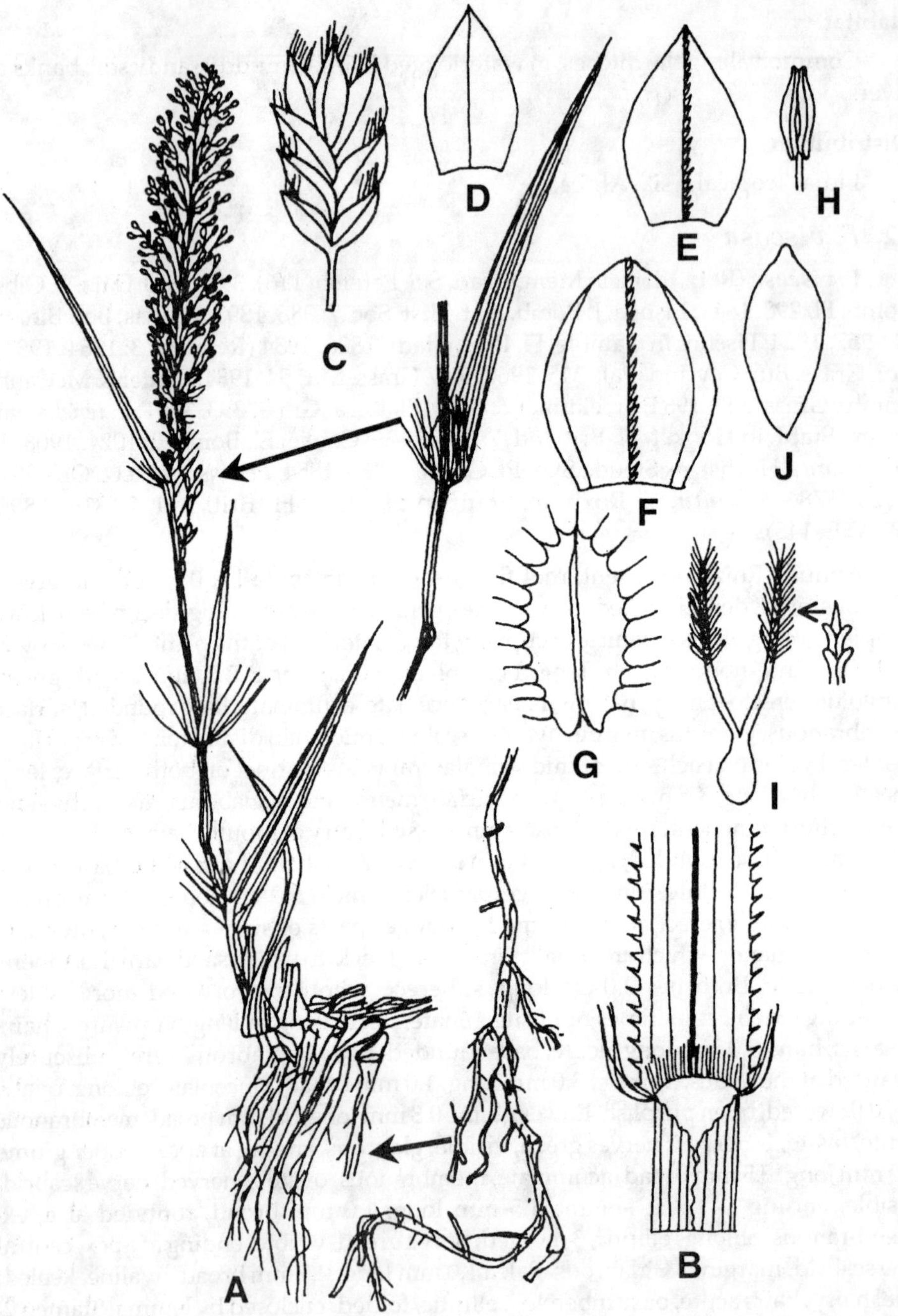

PLATE 145: *Eragrostis viscosa* (Retz.) Trin.
A. Habit, B. Ligule, C. Spikelet, D. Lower glume, E. Upper glume, F. Lemma, G. Palea, H. Stamen, I. Pistil, J. Caryopsis

Flowering & Fruiting

February–April; Tarana, Sindhora, Lohta, Rameshwar, Jaunpur, Mirzapur.

Habitat

Common along road sides & moist places.

Distribution

India, Tropical & South Africa.

23. *E. zeylanica*

E. zeylanica Nees & Mey. in Nov. Act. Nat. Cur. 19. Suppl. 1: 204. 1843; Moulik, Grass. Bamb. Ind. 2: 611. 1997. *E. elongata* sensu. Hook. f. Fl. Brit. Ind. 7: 319. 1896. (PLATE–146).

Perennial. Rhizome absent; root fibrous. Culm 40 cm tall, 2.0 mm thick, green, glabrous, erect, tufted, stiff, slender, leafy, ascending, decumbent, simple, rooting at base; nodes glabrous; internodes 8.0 cm long. Leaf–blade 10 cm long, 4.0 mm broad, green, flat, narrow, glabrous, linear–lanceolate, acuminate, base rounded, margin hyaline, scabrid, mid–vein distinct on lower surface, 2–lateral veins on each side of mid–vein; lateral veins distinct on both surface, slightly distant; leaf–sheath 7.0 cm long, 2.5 mm thick, green, glabrous, membranous, margin hyaline, tightly compressed, striate, shorter than internodes, vein distinct on both surface, hairy at mouth; hairs 3.0 mm long, white, acute, base tubercled; ligule 1.0 mm long, a ring of white, dense hairs; hairs acute, base broad; auricle absent. Inflorescence consists of panicle; panicle 7.0 cm long, 1.0 cm broad, erect, oblong, ovate, pyramidal, falling upwards from downwards with age; rachis of panicle slender, glabrous. Branches 1.0 cm long, simple, very distant, suberect, spreading, decreasing in length upwards, lower in whorl, upper alternate. Spikelets 5.0 mm long, 2.0 mm broad, crowded, deep–brown, linear, ovate, distichous, secund, 15–flowered, glabrous, pedicelled; pedicle 1.0 mm long, slender, scabrid; rachis of spikelets angular, glabrous. Lower glume 0.5 mm long, 0.2 mm broad, membranous, ovate, lanceolate, acute, glabrous, 1–nerved; nerves colourless, glabrous, ending at apex. Upper glume 1.0 mm long, 0.5 mm broad, ovate, lanceolate, acute, glabrous, 1–nerved; nerves green, visible, glabrous, ending at apex. Lemma 1.5 mm long, 1.0 mm broad, membranous, broadly ovate, lanceolate, acuminate, glabrous, laterally compressed, 3–nerved; nerve glabrous, central one deep green, thick at base, continuously thin towards the apex, visible, ending at apex, lateral nerves light green, very slightly visible, ending very below at the apex. Palea 1.0 mm long, 0.5 mm broad, hyaline, delicate, curved, oblong, elliptic, round at apex, surface glabrous, margin sharply folded inwards, glabrous, transversely keeled; keel antrorsely scabrid, 2–nerved; nerves green, slightly visible, marginal & glabrous, ending at apex. Stamens 3; anther 0.8 mm long, yellow, oblong, glabrous; filament 0.2 mm long, slender glabrous, shorter than anther. Lodicules absent. Pistil 1.0 mm long; ovary 0.2 mm broad, elliptic, glabrous; style 0.4 mm long, slender, glabrous, bifid from base; stigma 0.4 mm long, brown, hairy; hairs acute, base broad, in 2–rows. Caryopsis 1.0 mm long, 0.5 mm broad, ovoid, sub–globose, brown, wavy on margin, surface smooth.

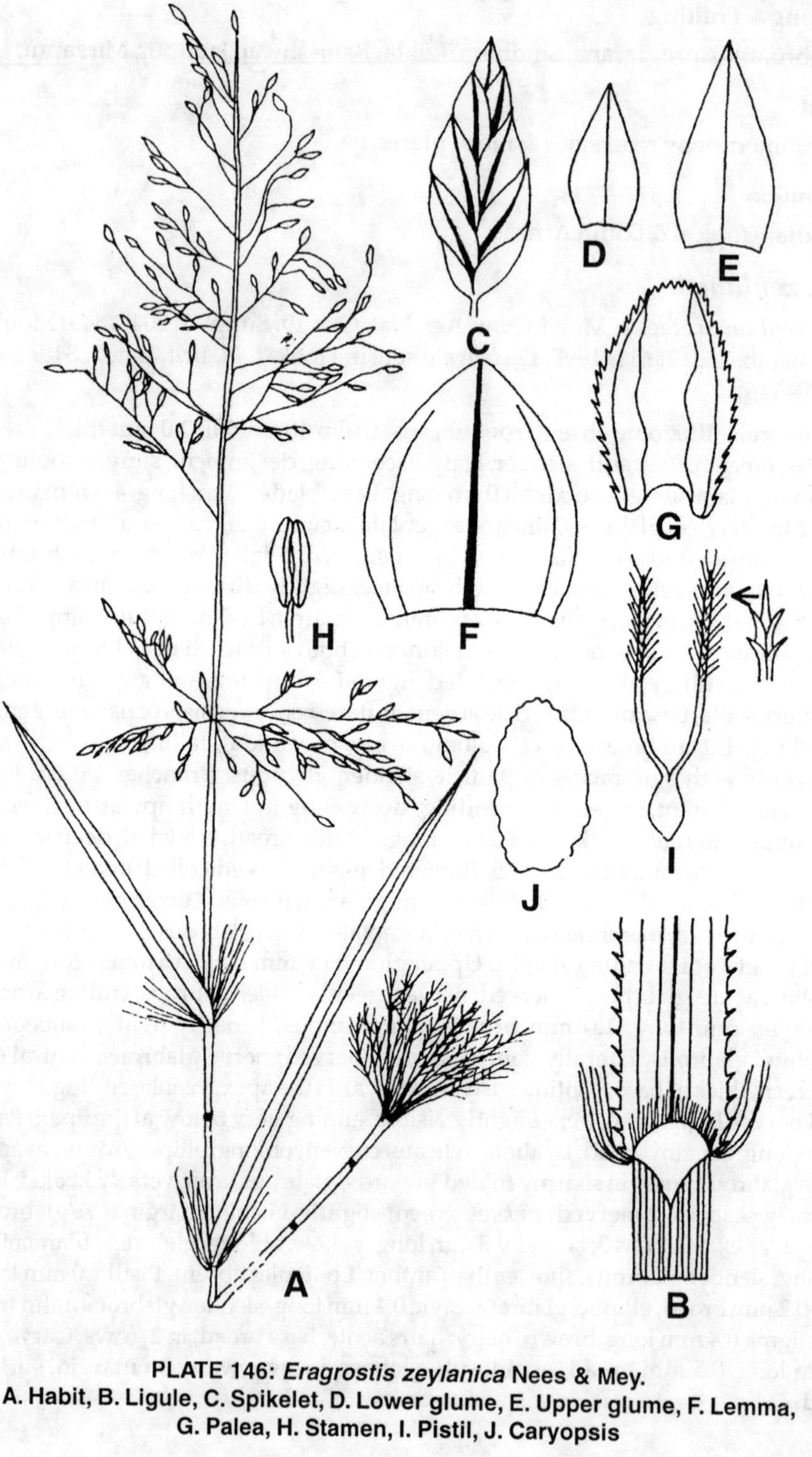

PLATE 146: ***Eragrostis zeylanica*** **Nees & Mey.**
A. Habit, B. Ligule, C. Spikelet, D. Lower glume, E. Upper glume, F. Lemma, G. Palea, H. Stamen, I. Pistil, J. Caryopsis

Flowering & Fruiting

February–April; Chakiya, Chandauli, Naugarh, Ghazipur, Gorakhpur, Ballia.

Habitat

Common along the banks of rivers.

Distribution

India, Burma, Bangladesh, Sri Lanka.

10. *Leptochola* P. Beauv.

KEY TO THE SPECIES

1a. Spikelets 3–7 flowered; lodicules oblanceolate

.... 1. *L. chinensis*

1b. Spikelets 2–3 flowered; lodicules obovate

.... 2. *L. panicea*

1. *Leptochola chinensis*

Leptochola chinensis (Linn.) Nees in Syll. Ratisb. 1: 4. 1824; Nees, Agrost. Bras. 432. 1829; Benth. Fl. Hongk. 430. 1861; Benth. Fl. Austr. 7: 617. 1878; Duthie, Grass. N.W. Ind. 35. 1883; Duthie, Grass. N. Ind. 59. t. 71. 1888; Lisboa, J. Bomb. Nat. Hist. Soc. 7: 372. 1893; Hook. f. Fl. Brit. Ind. 7: 299. 1896; Haines, Bot. Bih. & Or. 972. 1924; Fischer in Gamble, Fl. Pres. Madr. 1820. 1934 (Repr. ed. 3: 1260. 1957); Bor, Grass. Bur. Cey. Ind. Pak. 516. 1960; Chien–Chan Hsu, Taiwan Grass. 423. t. 107. 1975. Roy, Grass. M.P. 108. 1984; Blatter & McCann, Bomb. Grass. 244. 1984; Sreekumar & Nair, Fl. Kera. Grass. 395. 1991. *Poa chinensis* Linn. Sp. Pl. 69. 1753; Roxb. Fl. Ind. 1: 332. 1820. *P. malabarica* Retz. Obs. 5: 19.1786. *P. astenes* Roem. & Schult. Syst. 2: 574. 1817. *P. decipiens* R. Br. Prodr. 181. 1810. (PLATE–147).

Annual. Rhizome absent; root fibrous. Culm 1.0 m tall, 1.0 mm thick, green, glabrous, erect, tufted, geniculately–ascending, slender, leafy, rooting at base, branching at lower nodes; branches suberect, leafy upto inflorescence; nodes glabrous; internodes 11 cm long. Leaf–blade 20 cm long, 6.0 mm broad, green, flat, narrow, linear–lanceolate, acuminate, base rounded, surface scabrid also with hyaline margin, mid–vein distinct on lower surface, 3–lateral veins on each side of mid–veins; lateral veins distinct on both surface; leaf–sheath 8.0 cm long, 4.0 mm broad, green, glabrous, membranous, margin hyaline, keeled, loose, erect, compressed, very much shorter than internodes, hairy on mouth; hairs 5.0 mm long, white, acute, base tubercled, veins distinct on both surface; ligule 1.0 mm long, membranous, lacerate; auricle glabrous. Inflorescence consists of panicle; panicle 20 cm long, erect, open, loose, oblong, ovate, pinkish, consist of several ascending branches; rachis slender, scabrid. Branches 9.0 cm long, slender, simple, opposite, suberect, spreading, slightly distant. Spikelet 5.0 mm long, 2.0 mm broad, 6–flowered, oblong, lanceolate, alternate, biseriate, distant, narrow, pedicelled; pedicel 1.0 mm long, flat, dilated at apex, smooth. Lower glume 1.0 mm long, 0.5 mm broad, membranous, oblong, lanceolate, sub acute, surface glabrous except the nerve, also with margin, 1–nerved; nerves green, visible, scabrid,

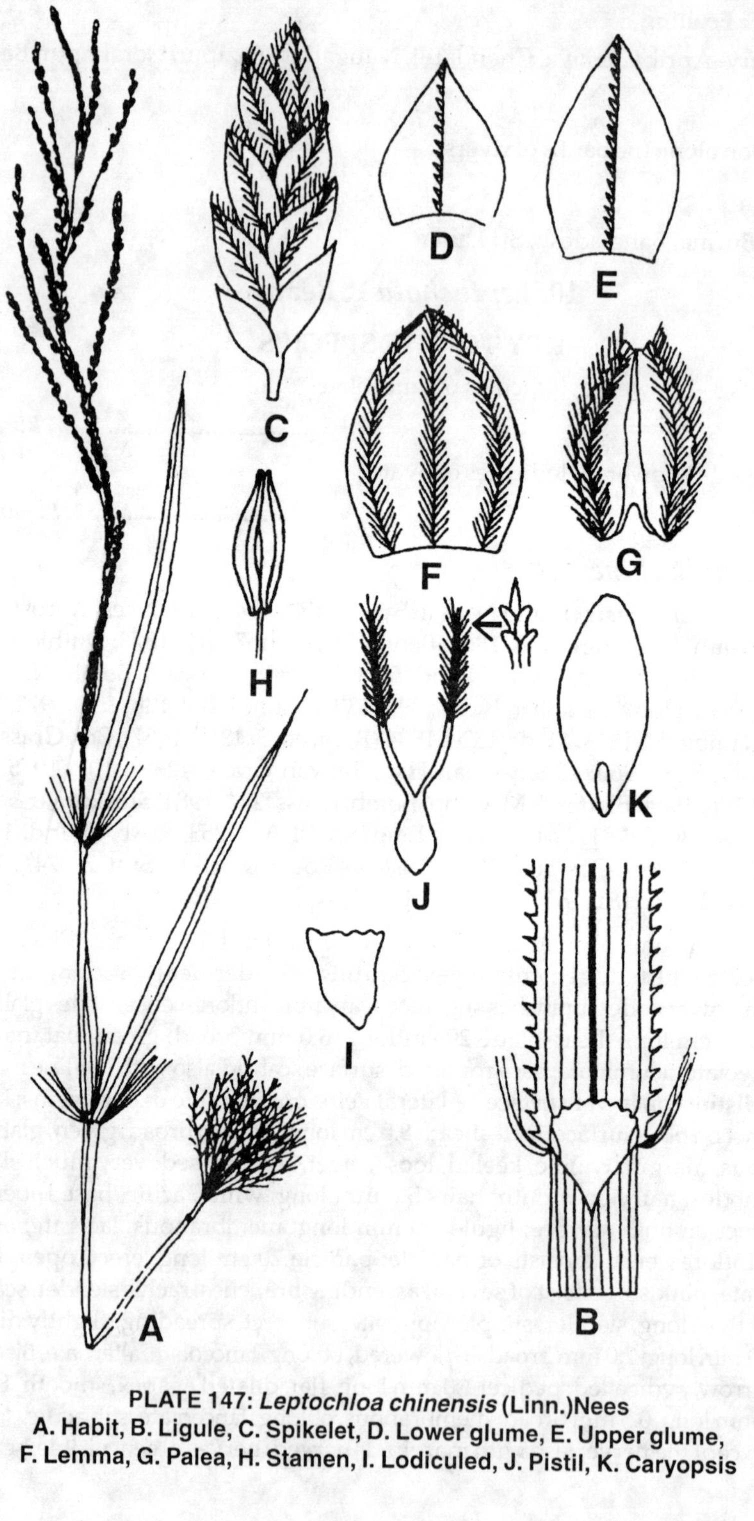

PLATE 147: *Leptochloa chinensis* (Linn.)Nees
A. Habit, B. Ligule, C. Spikelet, D. Lower glume, E. Upper glume, F. Lemma, G. Palea, H. Stamen, I. Lodiculed, J. Pistil, K. Caryopsis

ending at apex. Upper glume 1.5 mm long, 1.0 mm broad, membranous, oblong lanceolate, subacute, surface glabrous except the nerve, also with margin, 1–nerved; nerves green, visible, scabrid, ending at apex. Lemma 2.0 mm long, 1.5 mm broad, membranous, braodly ovate, lanceolate, obtuse, laterally compressed, surface glabrous between the nerves, also with margin, 3–nerved; nerves green, visible, ending at apex, hairy; hairs 0.5 mm long, acute, base tubercled. Palea 1.5 mm long, 1.0 mm broad, hyaline, delicate, enclosed by lemma, oblong, notched at apex, glabrous, margin folded inwards, 2–nerved; nerves green, visible, marginal, ending at apex, hairy; hairs 0.5 mm long, acute, base broad. Stamens 3; anther 0.4 mm long, yellow, oblong, glabrous; filament 0.2 mm long, slender, glabrous, shorter than anther. Lodicules 2, 0.2 mm long, 0.1 mm broad, hyaline, glabrous, lanceolate, nerveless. Pistil 1.0 mm long; ovary 0.2 mm long, obovate, glabrous; style 0.3 mm long, slender, glabrous, bifid from base, stigma 0.5 mm long, brown, hairy; hairs acute, base broad, in 2–rows. Caryopsis 1.0 mm long, 0.5 mm broad, oblong, elliptic, obtusely trigonous, brown, subrugose.

Flowering & Fruiting

September–February; Sarnath, Harraua, Jaunpur Chandauli, Sonbhadra.

Habitat

Common in moist habitat & on old walls.

Distribution

India, Malaya, Japan, China, Australia.

2. *L. panicea*

L. panicea (Retz.) Ohwi, in Bot. Mag. Tokyo 55: 311. 1941; Bor, Grass. Bur. Cey. Ind. Pak. 517. 1960; Raizada & Jain, Ind. For. 92: 641. 1966; Panigrahi & Rajagopal, J. Ind. Bot. Soc. 47: 238. 1968; Gilliland in Fl. Malaya, 3: 75. 1971; Chien–chan, Hus, Taiwan Grass. 425. 1975; Cope in Nasir & Ali, Fl. W. Pak. 143: 74. 1982; Raizada *et al.*, Grass. Upp. Gang. Pl. 93. 1983; Roy, Grass. M.P. 108. 1984; Misra & Verma, Fl. Allah. 452. 1992; Moulik, Grass. Bamb. Ind. 2: 614. 1997. *L. filiformis* Hook. f. Fl. Brit. Ind. 7: 298. 1896; Haines, Bot. Bih. & Or. 972. 1924. *Poa panicea* Retz. Obs. Bot. 3: 11. 1783. *Aira filiformis* Koen ex Roxb. Fl. Ind. 1: 328. 1820. (PLATE–148).

Annual. Rhizome absent; root fibrous. Culm 1.0 m tall, 2.0 mm thick, green, glabrous, erect, tufted, geniculately–ascending, slender, leafy, rooting at base, branching at lower nodes; branches suberect, leafy upto inflorescence; nodes brown, glabrous; internodes 10 cm long. Leaf blade 20 cm long, 8.0 mm broad, green, flat, linear–lanceolate, acute, base rounded, membranous, surface scabrid between the veins & hairy; hairs 2.0 mm long, white, acute, base tubercled; margin hyaline, scabrid, mid–vein distinct on lower surface, 5–lateral veins on each side of mid–vein; lateral veins distinct on both surface; leaf–sheath 8.0 cm long, 5.0 mm broad, green, membranous, scabrid, hairy; hairs 2.0 mm long, white, acute, base tubercled, margin hyaline, keeled & hairy; hairs 1.0 mm long, white, acute, base broad; compressed, loose, erect, shorter than internodes, veins distinct on both surface; ligule 1.5 mm long, membranous, lacerate, ovate; auricle glabrous. Inflorescence consist of panicle;

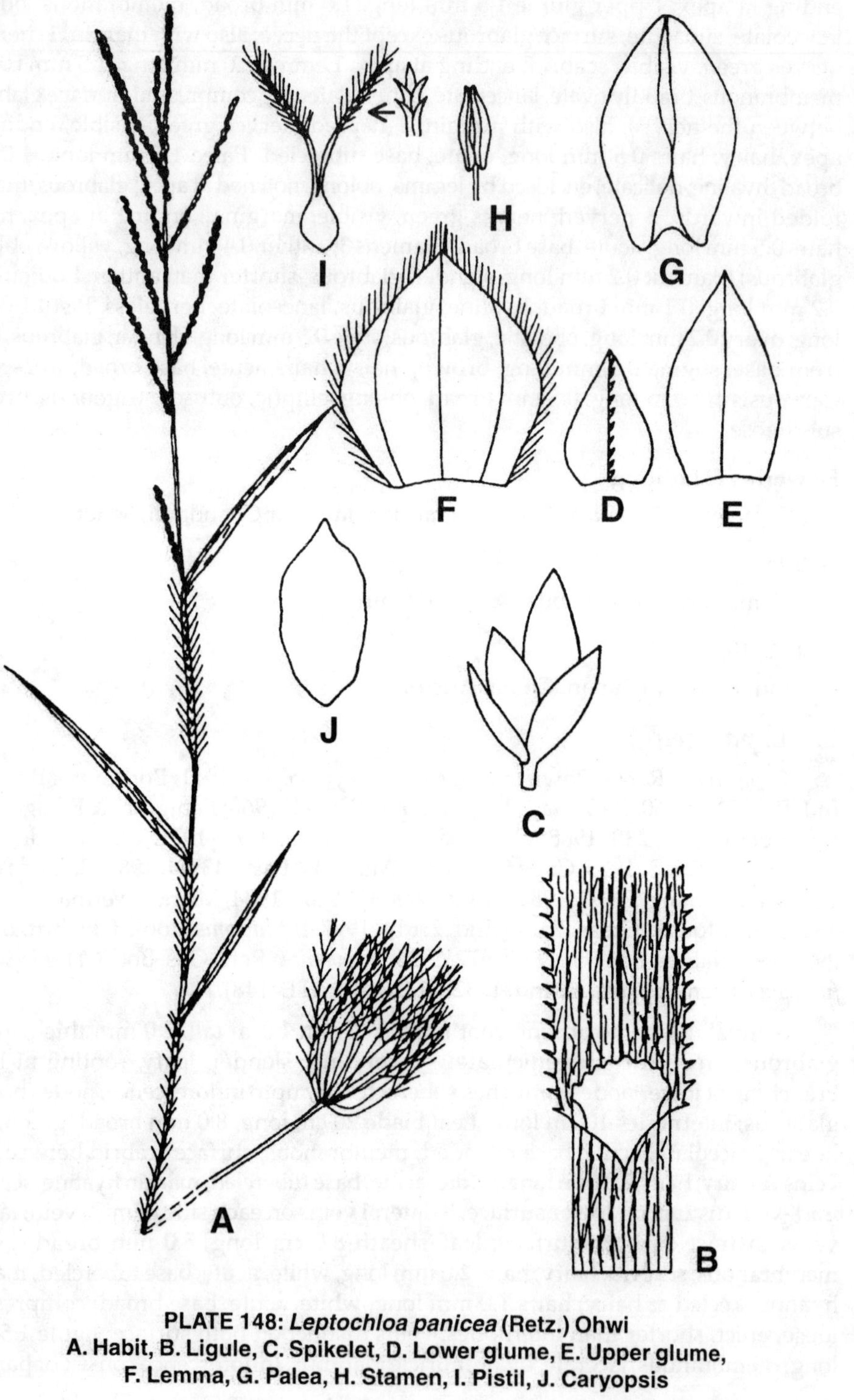

PLATE 148: *Leptochloa panicea* (Retz.) Ohwi
A. Habit, B. Ligule, C. Spikelet, D. Lower glume, E. Upper glume, F. Lemma, G. Palea, H. Stamen, I. Pistil, J. Caryopsis

panicle 30 cm long, elliptic, oblong, erect, terminal, loose, open, consist of several slender ascending branches; rachis filiform, scabrid. Branches 10 cm long, alternate, suberect, rarely spreading, distant, purplish. Spikelets 2.5 mm long, 1.5 mm broad, narrowly elliptic, 3–flowered, overlapping, crowded, subsessile. Lower glume 1.0 mm long, 0.5 mm broad, membranous, ovate, lanceolate, acute, 1–nerved; nerves green, visible, scabrid, ending at apex. Upper glume 1.5 mm long, 1.0 mm broad, membranous, narrowly oblong, obtuse, surface scabrid, margin scabrid at apex, glabrous below, 1–nerved; nerves green, visible, scabrid, ending at apex. Lemma 1.5 mm long, 1.0 mm broad, membranous, elliptic, oblong, apiculate, margin slightly inrolled, hairy; hairs minute, acute, base broad, 3–nerved; nerves green, visible, glabrous, ending at apex. Palea 1.0 mm long, 0.5 mm broad, membranous, elliptic, surface hairy; hairs acute, base broad, margin folded inwards, 2 keeled, nerveless. Stamens 3; anther 0.4 mm long, oblong, yellow, glabrous; filament 0.1 mm long, slender, glabrous, shorter than anther. Lodicules 2, 0.2 mm long, 0.1 mm broad, hyaline, glabrous, nerveless, obovate, cuneate. Pistil 1.0 mm long; ovary 0.2 mm long, ovate, glabrous; style 0.3 mm long, slender glabrous, bifid from base; stigma 0.5 mm long, purple, hairy; hairs acute, base broad, in 2–rows. Caryopsis 1.0 mm long, 0.5 mm broad, broadly elliptic, reticulate on surface, rounded at apex, acute at base, embryo basal, trigonous in cross section.

Flowering & Fruiting

July–November; Sarnath, Cholapur, Sindhora, Navalpur, Jaunpur, Chandauli, Allahabad, Gorakhpur, Kalwari, Padrauna.

Habitat

Very common in gardens, lawns, agricultural fields, banks of rivers & roadsides.

Distribution

India, Tropical Africa, Asia, Australia, Malaya.

11. *Oropetium* Trin.

KEY TO THE SPECIES

1a. Lower glume lacerate at apex; upper glume acute; anther 0.3 mm long; pistil 1.0 mm long

.... **1. *O. thomaeum***

1b. Lower glume truncate at apex; upper glume cuspidate; anther 0.5 mm long; pistil 1.5 mm long

.... **2. *O. villosulum***

1. *Oropetium thomaeum*

Oropetium thomaeum (L.f.) Trin. Fund. Agrost. 98. t. 3. 1820; Kunth Enum. Pl. 1: 464. 1838; Miq. Fl. Ind. Bat. 3: 403. 1837; Duthie, Grass. N.W. Ind. 45. 1883; Duthie, Fodder Grass N. Ind. 69. 1888; Hook. f. Fl. Brit. Ind. 7: 366. 1896; Cooke, Fl. Bomb. 2: 1046. 1908; Haines, Bot. Bih. & Or. 964. 1924; Bor, Grass. Bur. Cey. Ind. Pak. 473. 1960;

Roy, Grass. M.P. 115. 1984; Blatter & McCann, Bomb. Grass. 247. 1984. *Nardus thomacea* L.f. Supple. 105. 1781. *Rottboellia thomacea* Koening in Naturf. 23: 210.1788; Willd. Sp. Pl. 464. 1790; Roxb. Pl. Corm. 2: 17. t. 133. 1832; Roxb. Fl. Ind. 1: 357. 1820; Dalz. & Gibs., Bomb. Fl. 300. 1861. *R. pilosa* Willd. Sp. Pl. 1: 465. 1790. (PLATE–149).

Annual. Rhizome absent; root fibrous. Culm 5.0 cm long, 0.5 mm thick, green, erect, densely tufted, hard, compressed, leafy, flattened, rooting at base, branching at lower nodes; branches suberect; nodes glabrous; internodes 1.0 cm long. Leaf blade 1.5 cm long, 1.0 mm broad, green, setaceous, erect, curved at apex, filiform, acute, base rounded, striate, upper surface hairy; hairs 2.0 mm long, white, acute, base tubercled, lower surface glabrous, margin hyaline, scabrid, 1–veined; vein distinct on both surface; leaf–sheath 9.0 mm long, 1.0 mm broad, green, coriaceous, glabrous margin membranous, keeled, loose, striate, compressed, shorter than internodes, 1–veined; vein distinct on outer surface; ligule 0.1 mm long, membranous, erect, lacerate; auricle glabrous. Inflorescence composed of spike; spike 2.5 cm long, 1.0 mm broad, erect, stright, slightly curved at apex, green at young, brown with age; rachis undulated, tetragonous, angular; angle scabrid. Spikelets 2.0 mm long, 1.0 mm broad, sessile, sunken in the hollow of rachis, 1–flowered, acute. Lower glume 0.5 mm long, 0.2 mm broad, hyaline, ovate, lanceolate, lacerate at apex, glabrous, nerveless. Upper glume 2.0 mm long, 0.5 mm broad, membranous, linear, lanceolate, narrow, acute, surface glabrous, margin scabrid, 3–nerved; nerves green, visible, glabrous, ending at apex, lateral nerves distant to the margin. Lemma 1.2 mm long, 1.0 mm broad, hyaline, boat–shaped at compressed view, semi–circular when spread, notched at apex; notch scabrid, surface glabrous, margin scabrid at apex, glabrous, in middle, a tuft of hairs at base; hairs 1.0 mm long, acute, base tubercled; 3–nerved; nerves green, glabrous, ending at apex, central one visible, lateral 2–nerves slightly visible, very distant to margin. Palea 1.0 mm long, 0.5 mm broad,hyaline, ovate, oblong, glabrous, 2–fid at apex, nerveless. Stamens 3; anther 0.3 mm long, yellow, slightly oblong, glabrous; filament 0.1 mm long, slender, glabrous, shorter than anther. Lodicules absent. Pistil 1.0 mm long; ovary 0.2 mm long, glabrous, oblong; style 0.3 mm long, slender, glabrous, bifid from base; stigma 0.5 mm long, hairy, colourless; hairs acute, base broad, in 2–rows. Caryopsis 0.5 mm long, 0.2 mm broad, linear, oblong, fusiform, reddish–brown.

Flowering & Fruiting

July–September; Naugarh, Sonbhadra, Naini, Nichlaul.

Habitat

Rare, along dry hill slopes.

Distribution

India, Australia, Sri Lanka.

2. *O. villosulum*

O. villosulum Stapf ex Bor in Kew Bull. 1949. 571. 1950; Roy, Grass. M.P. 115. 1984.

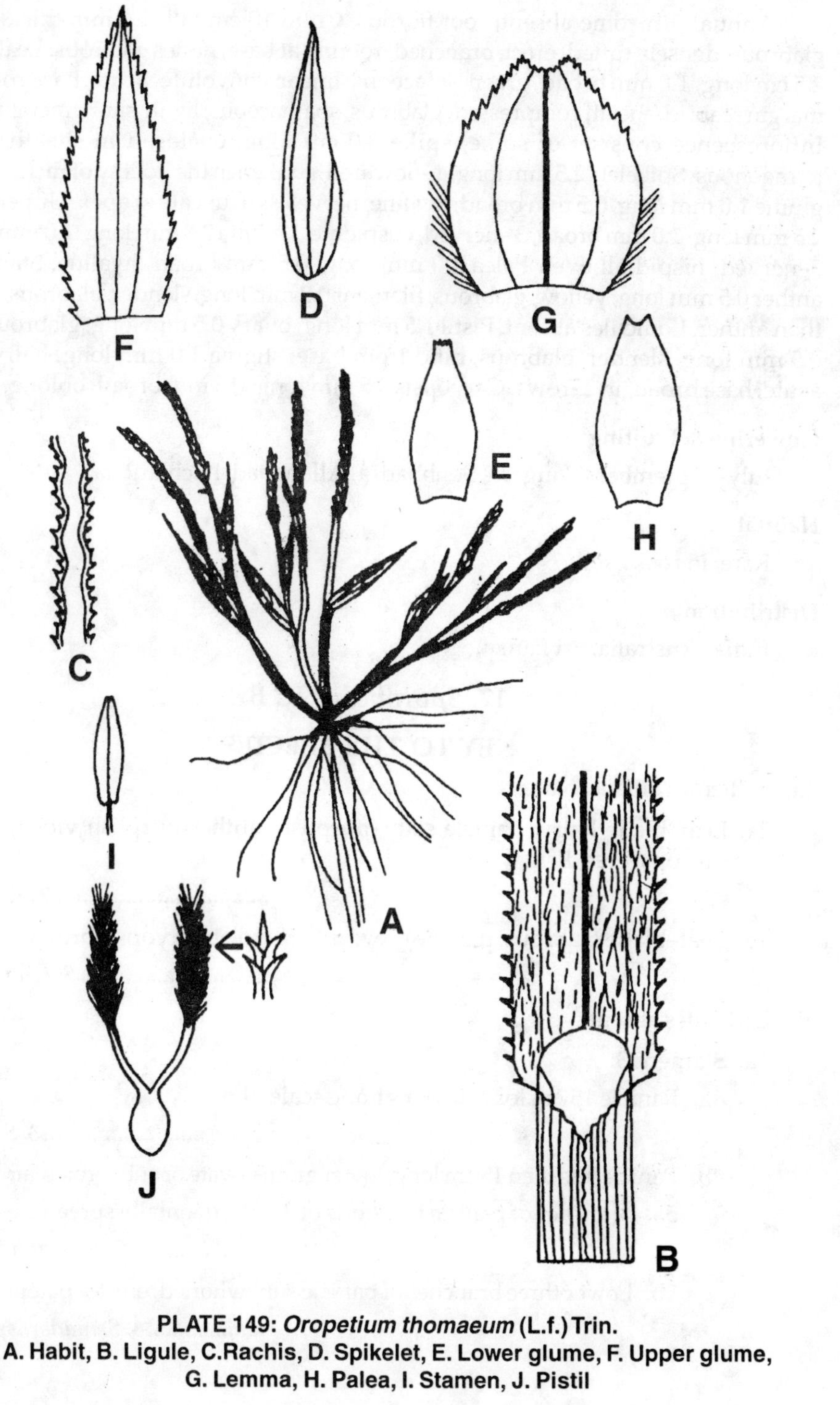

PLATE 149: ***Oropetium thomaeum*** **(L.f.) Trin.**
A. Habit, B. Ligule, C.Rachis, D. Spikelet, E. Lower glume, F. Upper glume, G. Lemma, H. Palea, I. Stamen, J. Pistil

Annual. Rhizome absent; root fibrous. Culm 10 cm tall, 1.0 mm thick, green, glabrous, densely tufted, erect, branched, rooting at base; nodes glabrous. Leaf–blade 2.5 cm long, 1.0 mm broad, green, setaceous, linear, convolute, acute, base rounded, margin scabrid; sheath compressed, glabrous, spathaceous; ligule short, membranous. Inflorescence consists of spike; spike 2.0 cm long, enclosed in sheath; rachis tetragonous. Spikelets 2.5 mm long, 1–flowered, sunken in the hollow of rachis. Lower glume 1.0 mm long, 0.5 mm broad, hyaline, nerveless, truncate at apex. Upper glume 2.5 mm long, 1.0 mm broad, 3– nerved, cuspidate. Lemma 2.0 mm long, 1.0 mm broad, 3–nerved, hispid all over. Palea 1.0 mm long, 0.5 mm broad, hyaline. Stamens 3; anther 0.5 mm long, yellow, glabrous; filament 0.2 mm long, slender, glabrous, shorter than anther. Lodicules absent. Pistil 1.5 mm long; ovary 0.5 mm long, glabrous; style 0.5 mm long, slender, glabrous, bifid from base; stigma 1.0 mm long, hairy; hairs acute, base broad, in 2–rows. Caryopsis 0.5 mm long, 0.3 mm broad, oblong.

Flowering & Fruiting

July–September; Naugarh, Sonbhadra, Allahabad, Nichlaul.

Habitat

Rare, in rocky soil.

Distribution

India, Australia, Sri Lanka.

12. *Sporobolus* R. Br.

KEY TO THE SPECIES

1a. Lodicules oblanceolate:

2a. Leaf–blade hairy; panicle slaty–purplish; anther purplish violet; caryopsis reddish .... **2. *S. diander***

2b. Leaf–blade glabrous; panicle grey; anther violet; caryopsis brown .... **5. *S. tenuissimus***

1b. Lodicules absent:

3a. Stamens 3:

4a. Panicle 15 cm long; lower glume scale like .... **3. *S. fertilis***

4b. Panicle less then 15 cm long; lower glume ovate or sub–orbicular:

5a. Branches of panicle in whorls of 3–6; horizontally spreading .... **1. *S. coromandelianus***

5b. Lower three branches of panicle sub–whorled; erecto–patent .... **4. *S. maderaspatanus***

3b. Stamens 2:

6a. Pedicels 2.0 mm long; lemmas slightly incurved at apex .... **7. *S. wallichii***

6b. Pedicels 0.5 mm long; lemma scabrid at apex .... **6. *S. tetragonus***

1. *Sporobolus coromandelianus*

Sporobolus coromandelianus (Retz.) Kunth. Rev. Gram. 1: 68. 1829; Link. Hort. Berol. 1: 89. 1827; Kunth. Rev. Gram. 1: 68. 1829; Dalz. & Gibs., Bomb. Fl. 296. 1861; Cook, Fl. Bomb. 2: 102. 1908; Haines, Bot. Bih. & Or. 975. 1924; Hook. f. Fl. Brit. Ind. 7: 253. 1896; Bor, Grass. Bur. Cey. Ind. Pak. 627. 1960; Raizada & Jain, Ind. For. 92: 641. 1966; Panigrahi & Rajagopal, J. Ind. Bot. Soc. 47: 241. 1968; Cope in Nasir & Ali, Fl. W. Pak. 143: 142. 1982; Raizada *et al.*, Grass. Upp. Gang. Pl. pt. 3: 109. 1983; Roy, Grass. M.P. 155. 1984; Misra & Verma, Fl. Allah. 468. 1992. Moulik, Grass. Bamb. Ind. 2: 477. 1997. *Agrostis coromandelina* Retz. Obs. Bot. 4: 19. 1786; Roxb. Fl. Ind. 1: 316. 1820. *A indica* Forsk. Fl. Aegy.–Arab. 104. 1775. *Vilfa coromandelina* Beauv. Agrost. 15. 1812; Steud. Syn. Pl. Glum. 153. 1855. *V. discospora* Trin. in Mem. Acad. Petersb. ser. 6, Sc. Nat. 2: 59. 1841. *V. roxburghiana* Nees ex Wight. Cat. no. 1742. *Sporobolus commutatus* Kunth. Enum. 1: 214. 1810; Boiss, Fl. Or. 5: 513. 1884; Aitchis. Cat. Punj. Pl. 165. 1869; Duthie, Grass. N.W. Ind. 29. 1883. *S. discosporus* Nees, Fl. Afr. Austr. 158. 1841. *Triachyrum cordofanum* Hochst ex Steud. Syn. Pl. Glum. 176. 1855.

Annual. Rhizome absent; root fibrous. Culm 30 cm tall, 1.5 mm thick, green, glabrous, erect, densely tufted, ascending, often spreading from the root, copiously leafy at base; nodes glabrous. Leaf–blade 7.5 cm long, 6.0 mm broad, green, glabrous, linear, lanceolate, narrow, acuminate, subcordate at base, spinulosely toothed especially towards the base; lower leaf–sheath glabrous, short, terete, compressed, upper hairy on margin near mouth; ligule 1.0 mm long, membranous, narrow, fringed with long fine hairs. Inflorescence made up of panicle; 10 cm long, 6.0 cm broad, ovate, pyramidal. Branches capillary in rather distant whorls of 3–6, horizontally spreading, naked below, sparingly shortly divided upwards, bearing short, minute red spikelet, pediceles long. Spikelet 1.3 mm long, crowded near branch ends. Lower glume 0.3 mm long, 0.2 mm broad, hyaline, suborbicular, nerveless. Upper glume 1.3 mm long, 0.7 mm broad, hyaline, ovate, oblong, subacute, 1–nerved. Lemma 1.3 mm long, 0.7 mm broad, hyaline, oblong, obtuse, 1–nerved. Palea 1.3 mm long, 0.6 mm broad, hyaline. Stamens 3; anther 0.3 mm long, yellow, glabrous; filament 0.2 mm long, slender, glabrous, shorter than anther. Lodicules absent. Pistil 1.0 mm long; ovary 0.2 mm long, glabrous; style 0.3 mm long, slender, glabrous, bifid from base; stigma 0.5 mm long. Caryopsis 1.0 mm long, 0.5 mm broad, pale reddish–brown broadly ellipsoid, compressed, smooth, endosed in the loose somewhat fleshy, reticulate pericarp.

Flowering & Fruiting

August–February; Babatpur, Jaunpur, Mardah, Lar.

Habitat

Occasional in fallow lands.

Distribution

India, Afghanistan, South Africa.

2. *S. diander*

S. diander (Retz.) P. Beauv. Ess. Agrost. 26: 147, 178. 1812; Hook. f. Fl. Brit. 7: 247. 1896; Fischer in Gamble, Fl. Pres. Madr. 117. 1934 (Repr. ed. 3: 1258. 1957); Bor, Grass. Bur. Cey. Ind. Pak. 629. 1960; Cope in Nasir & Ali, Fl. W. Pak. 143: 136. 1982. Roy, Grass. M.P. 156. 1984; Moulik, Grass. Bamb. Ind. 2: 478. 1978. *Agrostis diandra* Retz. Obs. Bot. 5: 19. 1789. (PLATE–150).

Annual. Rhizome absent; root fibrous. Culm 50 cm tall, 1.0 mm thick, green, glabrous, erect, tufted, very slender, ascending, leafy, simple, rooting at base; nodes glabrous; internodes 6.0 cm long. Leaf–blade 10 cm long, 2.0 mm broad, green, convolute, linear–lanceolate, filiform, acuminate, base rounded, membranous, hairy on upper surface; hairs 1.0 mm long, white, acute, base tubercled; glabrous on lower surface, margin hyaline, scabrid, mid–vein distant on lower surface, 2–lateral nerves on each side of mid–vein; lateral veins distinct on both surface; leaf–sheath 5.0 cm long, 1.5 cm broad, green, membranous, margin hyaline, keeled, tight, compressed, erect, shorter than internodes, hairy at mouth; hairs 3.0 mm long, white, acute, base tubercled; vein distant on both surface; ligule 0.2 mm long, a ring of hairs; hairs white, acute, base broad; acuricle absent. Inflorescence consists of panicle; panicle 10 cm long, 2.0 cm broad, open, erect, oblong, pyramidal, effuse, slaty at young, turnning into purplish with age; rachis minutely scabrid, filiform. Branches 1.0 cm long, suberect, distant, minutely scabrid, decreasing in length upwards. Spikelet 1.5 mm long, 0.7 mm broad, obovoid, glabrous, 1–flowered, sessile. Lower glume 0.3 mm long, 0.2 mm broad, hyaline, delicate, elliptic, ovate, glabrous, apiculate, nerveless. Upper glume 0.8 mm long, 0.4 mm broad, hyaline, elliptic, ovate, lanceolate, acuminate, glabrous, 1–nerved; nerves green, slightly visible, glabrous, ending at apex. Lemma 1.5 mm long, 1.0 mm broad, membranous, ovate, lanceolate, acute, glabrous, 1–nerved; nerves green, visible, glabrous, ending at apex. Palea 1.0 mm long, 0.5 mm broad, hyaline, delicate, oblong, ovate, acuminate, glabrous, nerveless. Stamens 3; anther 0.4 mm long, purplish–violet, oblong, glabrous; filament 0.2 mm long, slender, glabrous, shorter than anther. Lodicules 2, 0.2 mm long, 0.1 mm broad, hyaline, glabrous, nerveless, oblanceolate, wavy at apex. Pistil 1.0 mm long; ovary 0.2 mm long, obovate, glabrous; style 0.3 mm long, slender, glabrous, bifid at base; stigma 0.5 mm long, purple, hairy; hairs acute, base broad, in 2–rows. Caryopsis 1.0 mm long, 0.5 mm broad, reddish–brown, oblong, ovoid, glabrous, shining.

Flowering & Fruiting

September–December; Babatpur, Sindhora, Jaunpur, Rasra, Lar.

Habitat

Common, along moist localities, along roadsides, banks of canals.

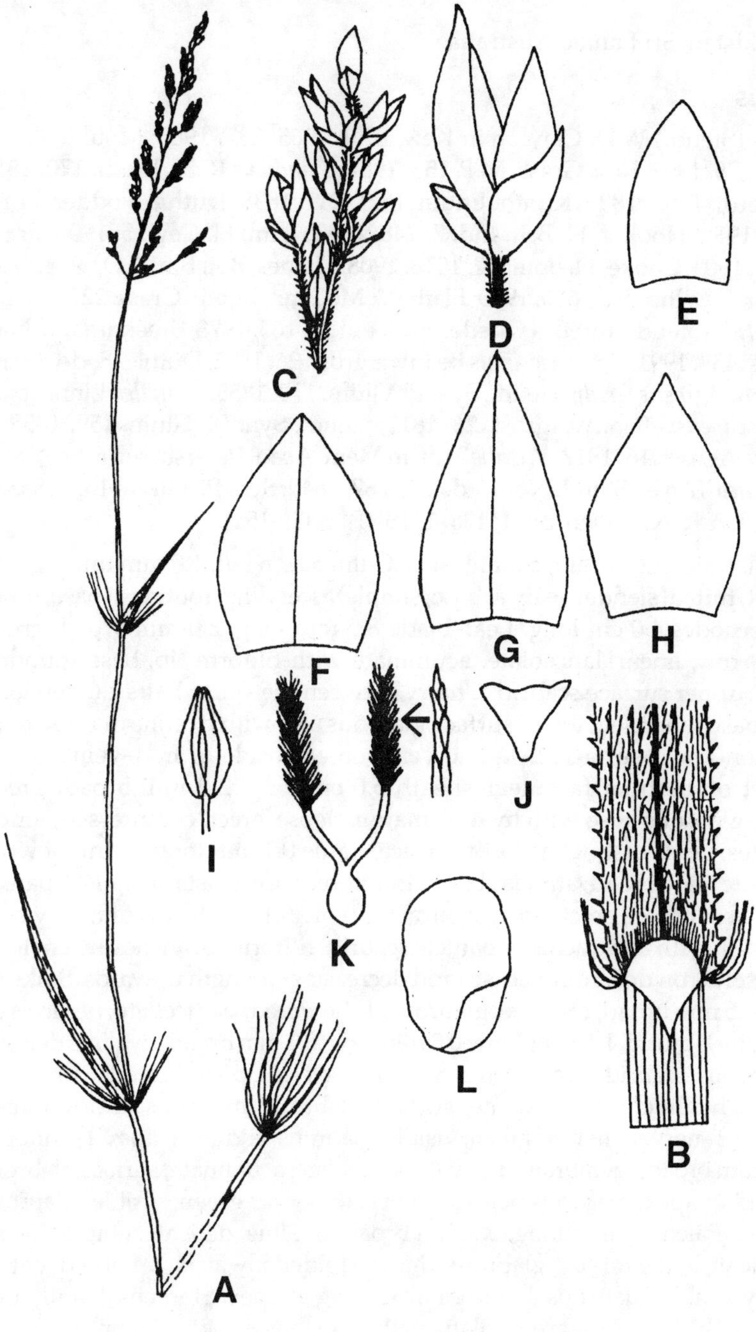

PLATE 150: ***Sporobolus diander*** **(Retz.) P. Beauv.**
A. Habit, B. Ligule, C. Branch of panicle, D. Spikelet, E. Lower glume, F. Upper glume, G. Lemma, H. Palea, I. Stamen, J. Lodicule, K. Pistil, L. Caryopsis

Distribution

India, Pakistan, Sri Lanka, Australia.

3. *S. fertilis*

S. fertilis (Steud.) W.D. Clayton in Kew. Bull. 1965. 287. 1965; Moulik, Grass. Bamb. Ind. 2: 478. 1997. Roy, Grass. M.P. 156. 1984. *S. indicus* R. Br. Prodr. 170. 1810; Link, Hort. Berol. 1: 87. 1827; Kunth. Enum. Pl. 1: 211. 1838; Duthie, Fodder Grass. N.W. Ind. 49. 1988; Hook. f. Fl. Brit. Ind. 7: 248. 1890; Trim, Fl. Ceyl. 5: 1900; Prain, Beng. Pl. 1213. 1903; Cooke, Fl. Bomb. 2: 1018. 1908; Haines, Bot. Bih. & Or. 974. 1924; Bor, Grass. Bur. Cey. Ind. Pak. 630. 1960; Blatter & McCann, Bomb. Grass. 222. 1984. *S. indicus* var. *fertilis* (Steud.) Joret. et Guedes in Taxon 22: 163. 1973; Sreekumar & Nair, Fl. Kera. Grass. 438. 1991. *S. tenacissimus* Beauv. Agrost. 26. 1812; Duthie, Fodd. Grass. N. Ind. 49. 1888. *Agrostis fertilis* Steud., Syn. Pl. Glum. 170. 1855. *A. indica* Linn. Sp. Pl. 63. 1753. *Vilfa capensis* Beauv. Agrost. 26. 1812; Steud., Syn. Pl. Glum. 159. 1855. *V. elongata* Beauv. Agrost. 16. 1812. *V. exilis* Trin. in Mem. Acad. Petersb. ser. 6, Sc. Nat. 2: 89. 1840. *V. indica* Trin ex Steud. Nom. ed. 2, 2: 767. 1840; Syn. Pl. Glum. 162. 1855. *V. tenacissima* H.B. & K. Nov. Gen. Sp. 1: 138. 1816. (PLATE–151).

Perennial. Rhizome absent; root fibrous. Culm 50 cm tall, 1.0 mm thick, green, glabrous, erect, tufted, slender, leafy at base, simple, ascending, rooting at base; nodes glabrous; internodes 4.0 cm long. Leaf–blade 8.0 mm long, 2.0 mm broad, green, convolute, narrow, linear–lanceolate, acuminate with filiform tip, base rounded, membranous, upper surface scabrid & hairy between the veins; hairs 1.0 mm long, white, acute, base tubercled, lower surface glabrous also with hyaline margin, mid–vein distinct on lower surface, 2–lateral veins on each side of mid–veins; lateral veins, distinct on both surface; leaf–sheath 5.0 cm long, 2.0 mm broad, green, membranous, glabrous, also with hyaline margin, loose, erect, compressed, longer than internodes, veins distinct on both surface; ligule 0.1 mm long, a ring of white hairs; hairs acute, base broad; auricle absent. Inflorescence consists of panicle; panicle 15 cm long, 2.0 cm broad, erect, open, contracted, spiciform, interrupted, greyish at young, brownish with age, rachis of panicle scabrid, filiform. Branches 1.0 cm long, sub erect, densely crowded, minutely scabrid decreasing in length upwards. Spikelets 1.5 mm long, 0.5 mm broad, obovate, glabrous, 1–flowered, pedicellate; pedicels 0.7 mm long, slender, scabrid. Lower glume 0.3 mm long, 0.1 mm broad, hyaline, delicate, scale like, glabrous, scabrid & rounded at apex, nerveless. Upper glume 0.6 mm long, 0.3 mm broad, hyaline, oblong, acute, surface glabrous, margin scabrid at apex, glabrous below, 1–nerved; nerves green, visible, glabrous, ending at apex. Lemma 1.0 mm long, 0.5 mm broad, membranous, ovate, lanceolate, acuminate, surface glabrous, margin scabrid at apex, glabrous below, 1–nerved; nerves green, visible, glabrous, ending at apex. Palea 1.0 mm long, 0.5 mm broad, hyaline, delicate, elliptic, ovate, acuminate plicate in the middle glabrous, margin folded inwards, 2–nerved; nerves green, slightly visible, glabrous & marginal, ending at apex. Stamens 3; anther 0.5 mm long, purplish oblong, glabrous; filament 0.1 mm long, slender, glabrous, shorter than anther. Lodicules absent. Pistil 1.0 mm long; ovary 0.2 mm long, ovate, glabrous; style 0.3 mm long, slender, glabrous, bifid from base; stigma 0.5 mm long, purple, hairy; hairs acute, base broad. Caryopsis 0.5 mm long, 0.3 mm broad, red–brown, obovoid, adnate to pericarp.

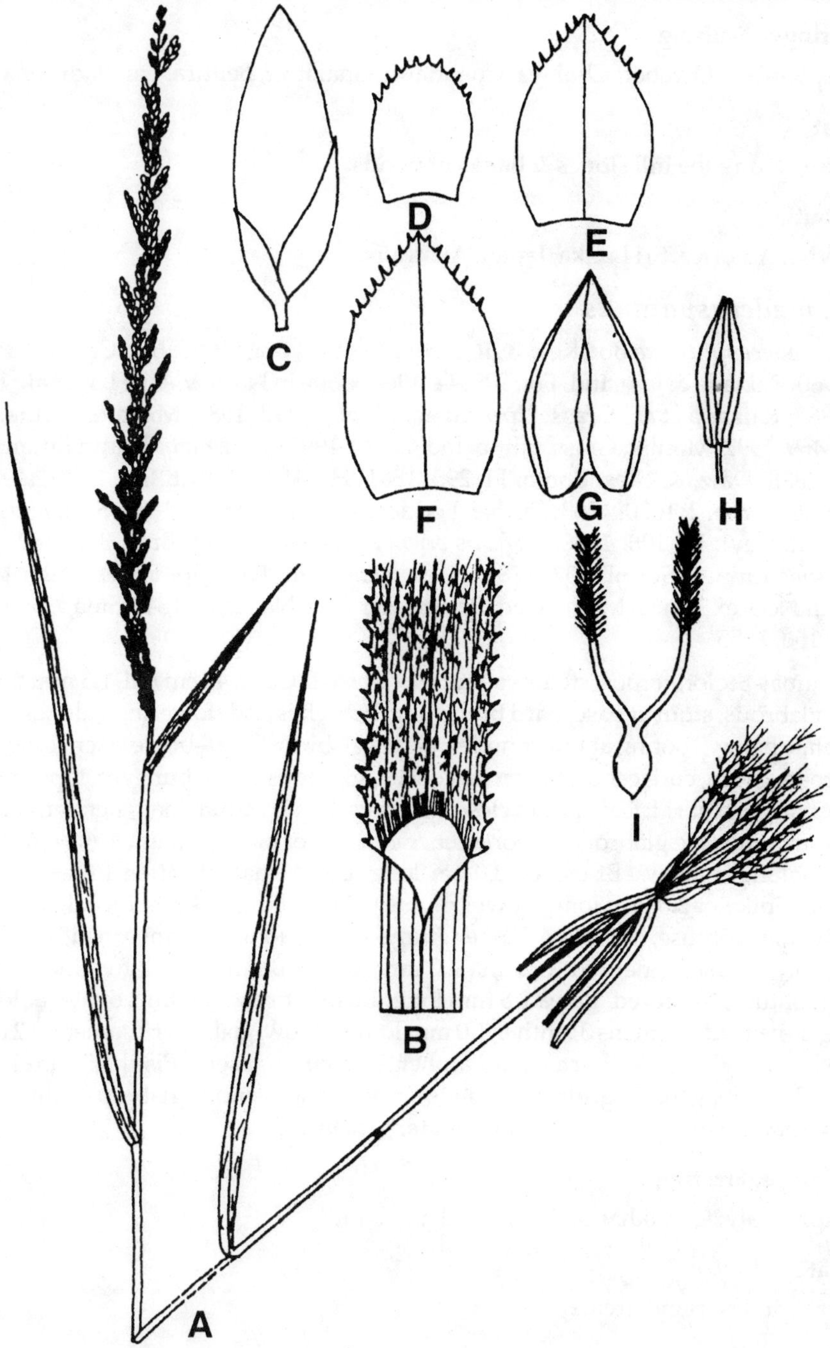

PLATE 151: ***Sporbolus fertilis*** **(Steud.) W.D. Clayton**
A. Habit, B. Ligule, C. Spikelet, D. Lower glume, E. Upper glume, F. Lemma, G. Palea, H. Stamen, I. Pistil

Flowering & Fruiting

September–October; Chakiya, Chandauli, Ghazipur, Belthra, Lar, Padrauna.

Habitat

Rare, along the hill slopes & banks of ponds.

Distribution

India, America, Sri Lanka, Japan, Australia.

4. *S. maderaspatanus*

S. maderaspatanus Bor, Kew Bull. 1957: 234. 1957; Bor, Grass. Bur. Cey. Ind. Pak. 632. 1960; Raizada & Jain, Ind. For. 92: 644. 1966; Cope in Nasir & Ali, Fl. W. Pak. 143: 142. 1982; Raizada *et al.*, Grass. Upp. Gang. Pl. pt. 3: 112. 1983; Misra & Verma, Fl. Allah. 469. 1992; Moulik, Grass. Bamb. Ind. 2: 480. 1997. *S. orientalis* Kunth Enum. Pl. 1: 211. 1838; Dalz. & Gibs. Bomb. Fl. 295. 1861; Hook. f. Fl. Brit. Ind. 7: 251. 1896; Cooke, Fl. Bomb. 2: 1020. 1908; Duthie, Fodder Grass. N. Ind. 49. 1883. *S. humifuscus* Trim. Cat. Ceyl. Pl. 108. 1880. *Agrostis orientalis* Nees. Agrost. Bras. 393. 1829. *A. tenacissima* Linn. f. Suppl. 107. 1781. *A. elongata* Roth. Nov. Sp. Pl. 41. 1821. *Vilfa orientalis* Nees ex Trin. in Mem. Acad. Petersb. ser 7, Sc. Nat. 2: 65. 1840; Steud. Syn. Pl. Glum. 156. 1855.

Annual. Stoloniferous; stolons stout; root fibrous. Culm 40 cm tall, 1.5 mm thick, green, glabrous, stout at base, hard branching; branches and flowering culm usually geniculate below, rooting at lower nodes; nodes glabrous. Leaf–blade 10 cm long, 4.0 mm broad, green, coriaceous, linear, lanceolate, glabrous, more pungant, tapering to the tip, margin scabrid; leaf–sheath close, glabrous; ligule 1.0 mm long, slender, ciliate, membrance; auricle glabrous. Inflorescence consists of panicle; panicle 6.0 cm long, linear, oblong, narrow. Branches 2.0 cm long, erecto–patent, often lowest 3 sub whorled. Spikelet 2.0 mm long. Lower glume 0.7 mm long, 0.4 mm broad, hyaline, ovate, oblong, obtuse, nerveless. Upper glume 2.0 mm long, 1.0 mm broad, hyaline, ovate, sub–obtuse, 1–nerved. Lemma 2.0 mm long, 1.0 mm broad, hyaline, ovate, oblong, obtuse, 1–nerved. Palea 1.8 mm long, 1.0 mm broad hyaline, ovate, oblong, obtuse, 1–nerved. Stamens 3; anther 1.0 mm long, yellow, glabrous; filament 0.2 mm long, slender, glabrous, shorter than anther. Lodicules absent. Pistil 1.5 mm long; ovary 0.3 mm long, oblong, glabrous; style 0.6 mm long, slender, glabrous, bifid from base; stigma 0.6 mm long, hairy; hairs acute, base broad.

Flowering & Fruiting

April–July; Chaukhandi, Vindhyachal, Naini.

Habitat

Rare, in Usarland area.

Distribution

India, Sri Lanka, Diu.

5. *S. tenuissimus*

S. tenuissimus (Schrank) Retz. Rev. Gen. Pl. 3: 396. 1893; Bor, Grass. Bur. Cey. Ind. Pak. 633. 1960; Manilal & Sivaraj, Fl. Calic. 350. 1982; Moulik, Grass Bamb. Ind. 2: 481. 1997; Sreekumar & Nair, Fl. Kera. Grass. 440. 1991. *Panicum tenuissimum* Schrank, in Denkschr. Bot. Ges. Regensb. 2: 26. 1822. *Sporobolus minutiflorus* (Trin.) Link. Hort. Berol. 1: 88. 1827; Hook. f. Fl. Brit. Ind. 7: 248. 1896; Fischer in Gamble, Fl. Pres. Madr. 1816. 1934 (Repr. ed. 3: 1258. 1957). (PLATE–152).

Annual. Rhizome absent; root fibrous. Culms 45 cm tall, 1.0 mm thick, erect, slender, green, glabrous, node brown, glabrous; internodes 5.0 cm long. Leaf–blade 16 cm long, 4.0 mm broad, linear–lanceolate, acuminate, base cordate, green, surface glabrous, margin scabrid, mid–vein prominently, white; leaf–sheath 8.0 cm long, 3.0 mm broad, glabrous, green, loose, longer than internodes; ligule 0.3 mm long, a line of hairs; hairs acute, base rounded; auricle absent. Inflorescence consists of panicle; panicle 19 cm long, diffuse, delicate. Spikelet 2.5 cm long, oblong, lanceolate, terete, gaping, greyish. Lower glume 0.3 mm long, 0.2 mm broad, oblong, membranous, truncate, acute, 1–nerved; nerve green, glabrous, ending at apex. Upper glume 0.7 mm long, 0.3 mm long, ovate, oblong, membranous, sub–acute 1–nerved; nerve green, glabrous, ending at apex. Lemma 1.0 mm long, 0.5 mm broad, ovate, oblong, membranous, shining, greyish 3–nerved; nerve green, glabrous, ending at apex. Palea 1.0 mm long, 0.5 mm broad, deliate, 2–nerved, elliptic–oblong, apex rounded, hyaline. Stamens 2; anther 0.3 mm long, violet, glabrous; filament 0.1 mm long, slender, glabrous, shorter than anther. Lodicules 2, 0.2 mm long, 0.1 mm broad glabrous, hyaline. oblanceolate, broadly cuneate. Pistil 1.2 mm long; ovary 0.2 mm long, glabrous; style 0.5 mm long, glabrous, slender, bifid from base; stigma 0.5 mm long, hairy; hairs acute, base broad, in 2–rows. Caryopsis 0.3 mm long, 0.1 mm broad, brown, glabrous, truncate.

Flowering & Fruiting

August–September; Tarana, Jalalpur, Bhadohi, Ballia.

Habitat

Rare, grow on bunds of rice fields.

Distribution

India, Africa, America.

6. *S. tetragonous*

S. tetragonous Bor in Kew Bull. 1949: 251.1949; Bor, Grass. Bur. Cey. Ind. Pak. 633. 1960; Roy, Grass. M.P. 158. 1984; Moulik, Grass. Bamb. Ind. 2: 481. 1997. *Panicum pulchellus* sensu. Hook. f. Fl. Brit. Ind. 7: 252. 1896 (non R. Br. 1810). (PLATE–153).

Annual. Rhizome absent; root fibrous. Culm 30 cm tall, 1.0 mm thick, green, glabrous, erect, very slender, capillary, tufted, geniculately–ascending, simple, leafy, rooting at base; nodes glabrous; internodes 6.0 cm long. Leaf–blade 6.0 cm long, 3.0 mm broad, green, narrow, convolute, linear–lanceolate, acuminate, base rounded, surface membranous, glabrous, margin hyaline scabrid, mid–vein distinct on lower

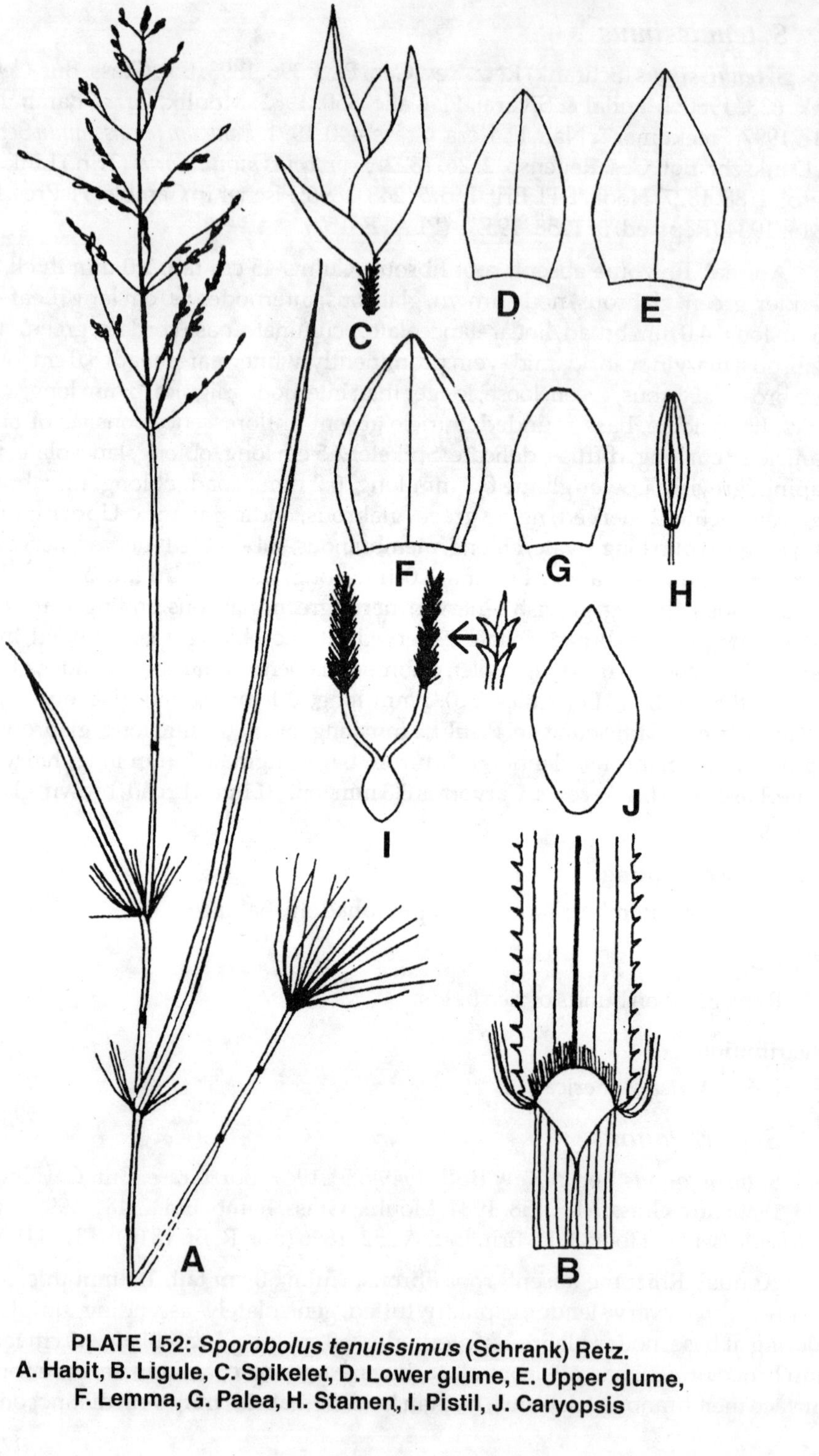

PLATE 152: *Sporobolus tenuissimus* (Schrank) Retz.
A. Habit, **B.** Ligule, **C.** Spikelet, **D.** Lower glume, **E.** Upper glume, **F.** Lemma, **G.** Palea, **H.** Stamen, **I.** Pistil, **J.** Caryopsis

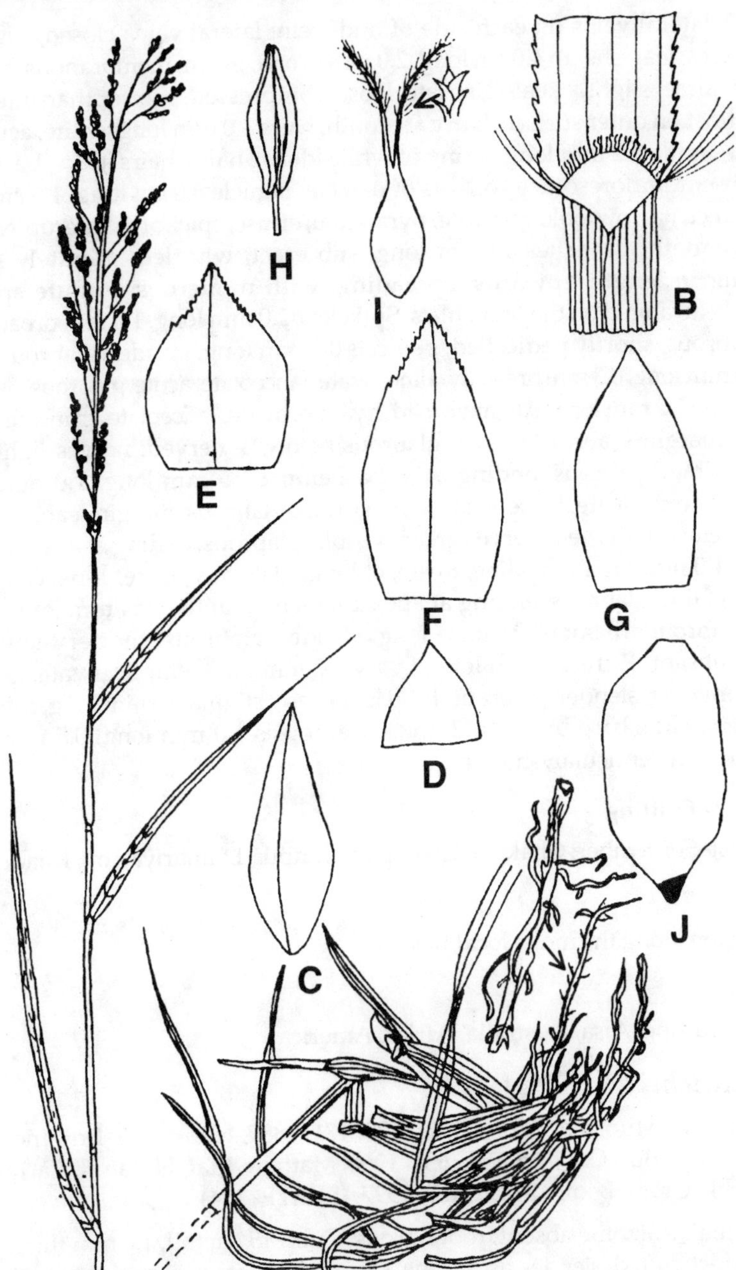

PLATE 153: ***Sporbolus tetragonus*** **Bor**
A. Habit, B. Ligule, C. Spikelet, D. Lower glume, E. Upper glume, F. Lemma, G. Palea, H. Stamen, I. Pistil, J. Caryopsis

surface, 2–lateral veins on each side of mid–vein; lateral veins closed, distinct on both surfaces; leaf–sheath 4.0 cm long, 2.0 mm broad, green, membranous, glabrous, margin hyaline, slightly keeled, striate, loose, compressed, shorter than internodes, veins distinct on outer surface, hairy at mouth; hairs 3.0 mm long, white, acute, base tubercled; ligule 0.2 mm long, a ring of white, dense hairs; hairs acute, base broad; auricle absent. Inflorescence consists of panicle; panicle 8.0 cm long, 1.5 cm broad, greyish, erect, narrow, oblong, ovate, pyramidal, effuse, spiciform, interupted; rachis slender, smooth. Branches 1.0 cm long, sub erect, whorled, minutely scabrid, decreasing in length upwards, spreading, with numerous, minute spikelets, racemosely arranged on the branchlets. Spikelets 2.0 mm long, 1.0 mm broad, ovate, acute, glabrous, shortly pedicelled; pedicels 0.5 mm long, slender, glabrous. Lower glume 0.6 mm long, 0.3 mm broad, hyaline, ovate, lanceolate, acute, glabrous, nerveless. Upper glume 1.2 mm long, 0.6 mm broad, hyaline, ovate, lanceolate, truncate, surface glabrous, margin scabrid at apex, glabrous below, 1–nerved; nerves light green, slightly visible, glabrous, ending at apex. Lemma 1.5 mm long, 1.0 mm broad, submembranous, ovate, lanceolate, acute, surface glabrous, margin scabrid at apex, glabrous below, 1–nerved; nerves green, visible, glabrous, ending at apex. Palea 1.5 mm long, 1.0 mm broad, hyaline, ovate, oblong, glabrous, acute, 1–nerved; nerves colourless, faint, glabrous, ending at apex. Stamens 2; anther 0.4 mm long, yellow, oblong, glabrous; filament 0.1 mm long, slender, glabrous, shorter than anther. Lodicules absent. Pistil 1.0 mm long; ovary 0.2 mm long, elliptic, ovate, glabrous; style 0.3 mm long, slender, glabrous, bifid from base; stigma 0.5 mm long, colourless, hairy; hairs acute, base broad in 2–rows. Caryopsis 1.0 mm long, 0.5 mm broad, reddish–brown, lenticular, smooth.

Flowering & Fruiting

August–September; Chakiya, Ghazipur, Jaunpur, Dumariyaganj, Khadda.

Habitat

Common along the moist localities.

Distribution

India, Europe, Aisa, Australia, Africa, America.

7. *S. wallichii*

S. wallichii Munro & Trin. in J. Bot. 27: 171. 1889; Hook. f. Fl. Brit. Ind. 7: 248. 1896; Bor, Grass. Bur. Cey. Ind. Pak. 634. 1960; Mathew, Mat, Fl. Tamil. Carratic 401. 1981; Moulik, Grass. Bamb. Ind. 2: 482. 1997. (PLATE–154).

Perennial. Rhizome absent; root fibrous. Culm 15 cm tall, 1.0 mm thick, green, glabrous, erect, tufted, slender, ascending, simple, leafy at base, rooting at base; nodes glabrous; internodes 5.0 cm long. Leaf–blade 8.0 cm long, 2.0 mm broad, green, flat, narrow, linear–lanceolate, acuminate, base rounded, surface membranous, margin hyaline, scabrid, mid–vein distinct on lower surface, 2–lateral veins on each side of mid–vein, lateral veins distinct on both surface; leaf–sheath 3.0 cm long, 2.0 mm broad, green, surface membranous, glabrous, margin hyaline, keeled, erect, loose, compressed, shorter than internodes, veins distinct on both surface; ligule 0.3 mm

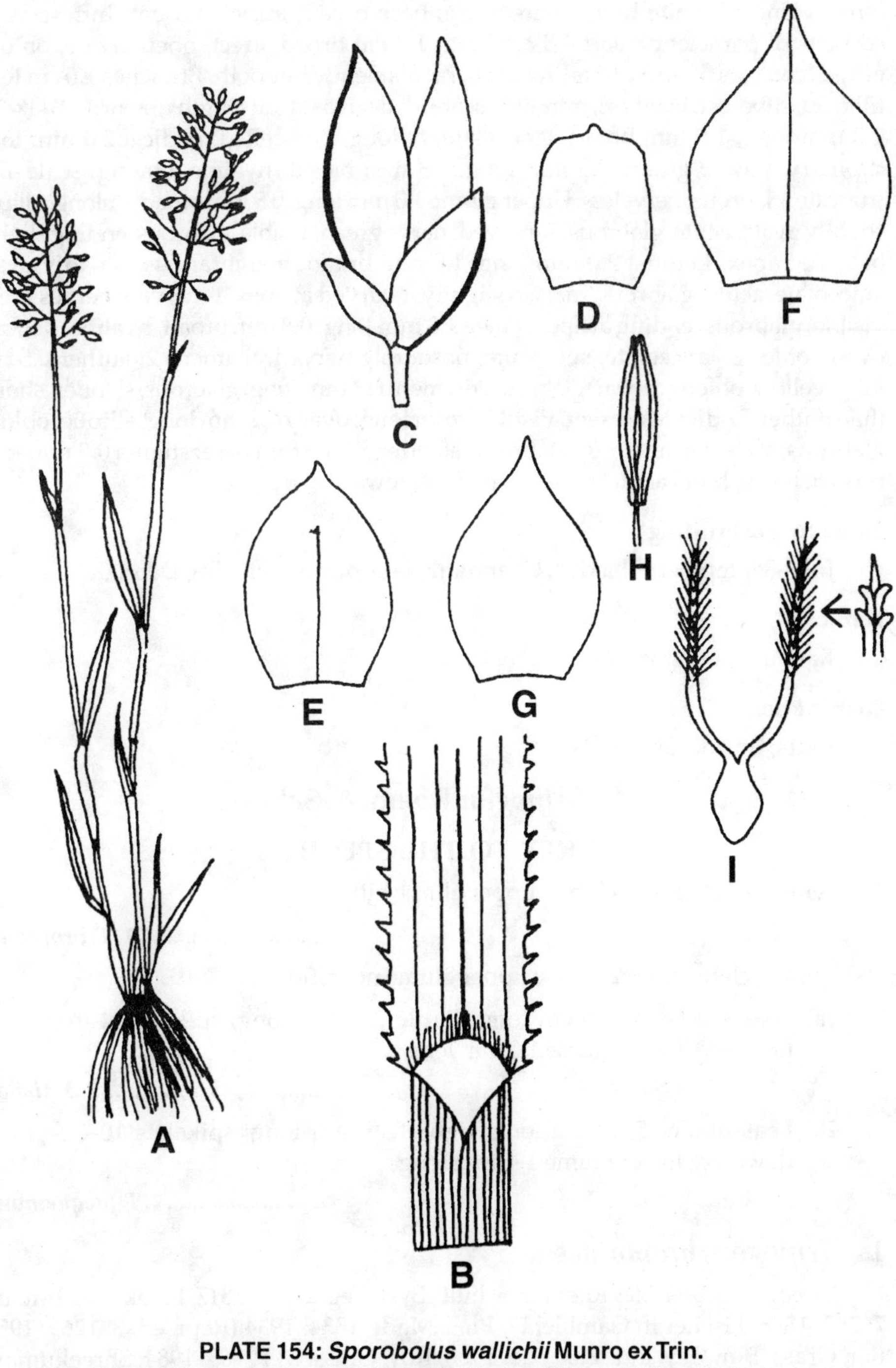

PLATE 154: ***Sporobolus wallichii*** **Munro ex Trin.**
A. Habit, B. Ligule, C. Spikelet, D. Lower glume, E. Upper glume, F. Lemma, G. Palea, H. Stamen, I. Pistil

long, a ring of white hairs; hairs acute, base broad; auricle absent. Inflorescence consists of panicle; panicle 4.5 cm long, 1.5 cm broad, erect, open, ovate, oblong, effuse, consists of numerous branches; rachis slender smooth. Branches 1.0 cm long, filiform, suberect, in whorl, minutely scabrid, decreasing in length upwards. Spikelets 2.0 mm long, 1.0 mm broad, lanceolate, oblong, pedicelled; pedicel 2.0 mm long, capillary. Lower glume 0.5 mm long, 0.3 mm broad, hyaline, oblong, scale–like, truncate, glabrous, nerveless. Upper glume 1.0 mm long, 0.5 mm broad, oblong, elliptic, slightly ovate, acute, glabrous, 1–nerved; nerve green, visible, glabrous, ending slightly below at apex. Lemma 2.0 mm long, 1.0 mm broad, membranous, broadly ovate, lanceolate, acute, glabrous, margin slightly incurved at apex, 1–nerved; nerves green, visible, glabrous, ending at apex. Palea 1.2 mm long, 0.8 mm broad, hyaline, delicate, ovate, oblong, lanceolate, sub acute, obscurely nerved. Stamens 2; anther 0.5 mm long, yellow, oblong, linear, glabrous; filament 0.3 mm long, glabrous, slender, shorter than anther. Lodicules absent. Pistil 1.0 mm long; ovary 0.2 mm long, elliptic, oblong, glabrous; style 0.3 mm long, glabrous, slender, bifid from base; stigma 0.5 mm long, brown, hairy; hairs acute, base broad, in 2–rows.

Flowering & Fruiting

July–September; Chakiya, Chandauli, Jaunpur, Sonbhadra, Deoria.

Habitat

Rare in open grasslands on rocky soils.

Distribution

India, Burma, Sri Lanka.

13. *Tripogon* Roem. & Schult.

KEY TO THE SPECIES

1a. Lemma clefted into 4 lobes; upper glume bifid .... **1. *T. bromoides***

1b. Lemma clefted into 2–lobes; upper glume not bifid:

2a. Leaf–blades 30–60 cm long; ligule 2.0 mm long; spikelets 4–10 flowered; lower glume 1.0 mm long .... **3. *T. lisboae***

2b. Leaf–blades 5–20 cm long; ligule 1.5 mm long; spikelets 10–30 flowered; lower glume 4–5 mm.long .... **2. *T. jacquemontii***

1. *Tripogon bromoides*

Tripogon bromoides Roem. & Schult., Syst. Veg. 2: 600. 1817; Hook. f. Fl. Brit. Ind. 7: 287. 1896; Fischer in Gamble, Fl. Pres. Madr. 1834. 1934 (Repr. ed. 3: 1269. 1957); Bor, Grass. Bur. Cey. Ind. Pak. 521. 1960; Roy, Grass. M.P. 163. 1984; Shreekumar & Nair, Fl. Kera. Grass. 399. 1991; Moulik, Grass. Bamb. Ind. 2: 622. 1997. *Trianthera bromoides* Roth. ex Roem. & Schult., Syst. Veg. 2: 600. 1817.

Perennial. Rhizomatous; root fibrous. Culm 50 cm tall, 4.0 mm thick, green, glabrous, erect, tufted; nodes glabrous. Leaf–blade 30 cm long, 3.0 mm broad, green, glabrous, linear, convolute, margin scabrid, acuminate, base rounded; sheath compressed, glabrous basal ones breaking up into fibres; ligule 2.0 cm long, membranous; auricle glabrous. Inflorescence consists of spikes; rachis stout, slender. Spikelet 10 mm long, 15 flowered. Lower glume 2.5 mm long, 1.0 mm broad, chartaceous, oblong, lanceolate, obtusely toothed on one side, notched at apex, 1–nerved. Upper glume 4.0 mm long, 1.5 mm broad, coriaceous, ovate, oblong, notched at apex, 1–nerved. Lemma 8.0 mm long, 3.5 mm broad, chartaceous, ovate, oblong, bearded at the base, clefted into 4–lobes at apex, 3–nerved, awned. Palea 3.0 mm long, 1.0 mm broad, elliptic, oblong, 2–keeled, 2–nerved; keels scabrid. Stamens 3; anther 1.3 mm long, yellow, glabrous; filament 0.3 mm long, slender, glabrous, shorter than anther. Lodicules absent. Pistil 2.0 mm long; ovary 0.5 mm long, oblong, glabrous; style 0.5 mm long, slender, glabrous, bifid from base; stigma 1.0 mm long, hairy; hairs acute, base broad, in 2–rows.

Flowering & Fruiting

October–December; Natinia Dai, Duddhi, Naini, Belthra.

Habitat

Fairly common along roadside.

Distribution

India, Sri Lanka, Asia.

2. *T. jacquemontii*

T. jacquemontii Stapf in Kew Bull. 1892: 85. 1892; Lisboa in J. Bomb. Nat. Hist. Soc. 7: 370. 1893; Hook. f. Fl. Brit. Ind. 7: 286. 1896; Cooke, Fl. Bomb. 2: 1037. 1908; Haines, Bot. Bih. & Or. 966. 1924; Bor, Grass. Bur. Cey. Ind. Pak. 522. 1960; Roy, Grass. M.P. 163. 1984; Blatter & McCann, Bomb. Grass. 268. t. 181. 1984; Moulik, Grass. Bamb. Ind. 2: 622. 1997. (PLATE–155).

Perennial, semi–aquatic. Rhizomatous; root fibrous. Culms 80 cm tall, 3.0 mm thick, tufted, erect, green, glabrous, ascending, decumbent, rooting at base; nodes brown, glabrous; internodes 11 cm long. Leaf–blade 25 cm long, 4.0 mm broad, mostly basal, convolute, narrow, linear, lanceolate, green, surface rough, margin scabrid, acute, base rounded, mid–vein white, promonent; leaf–sheath 9.0 cm long, 3.0 mm broad, compressed, loose, striate, showing black spot on maturity, surface glabrous, margin scabrid, green, shorter than internode, mid–vein prominent; ligule 1.5 mm long, membranous, lacerate; auricle glabrous. Inflorescence 20 cm long, racemes, alternate; rachis flat, glabrous. Spikelet 9.0 mm long, 10–30 flowered, crowded. Lower glume 2.0 mm long, 1.0 mm broad, membranous, notched at apex, ovate 1–nerved; nerves green, prominent. Upper glume 4.0 mm long, 2.0 mm broad, mucronate, membranous, acute, lanceolate, 1–nerved; nerve green, visiable. Lemma 3.0 mm long, 1.5 mm broad, 2–lobed, apex scabrid, base hairy; hairs acute, base rounded; awn from lobe; awn 0.6 mm long, antrorsely scabrid; lanceolate, membranous, 3–nerved; nerves prominent, green, mid–nerve thick. Palea 3.0 mm long, 1.5 mm broad, hyaline, oblong, mucronate, apex scabrid, hairy at base, hairs acute, base rounded, 2–nerved;

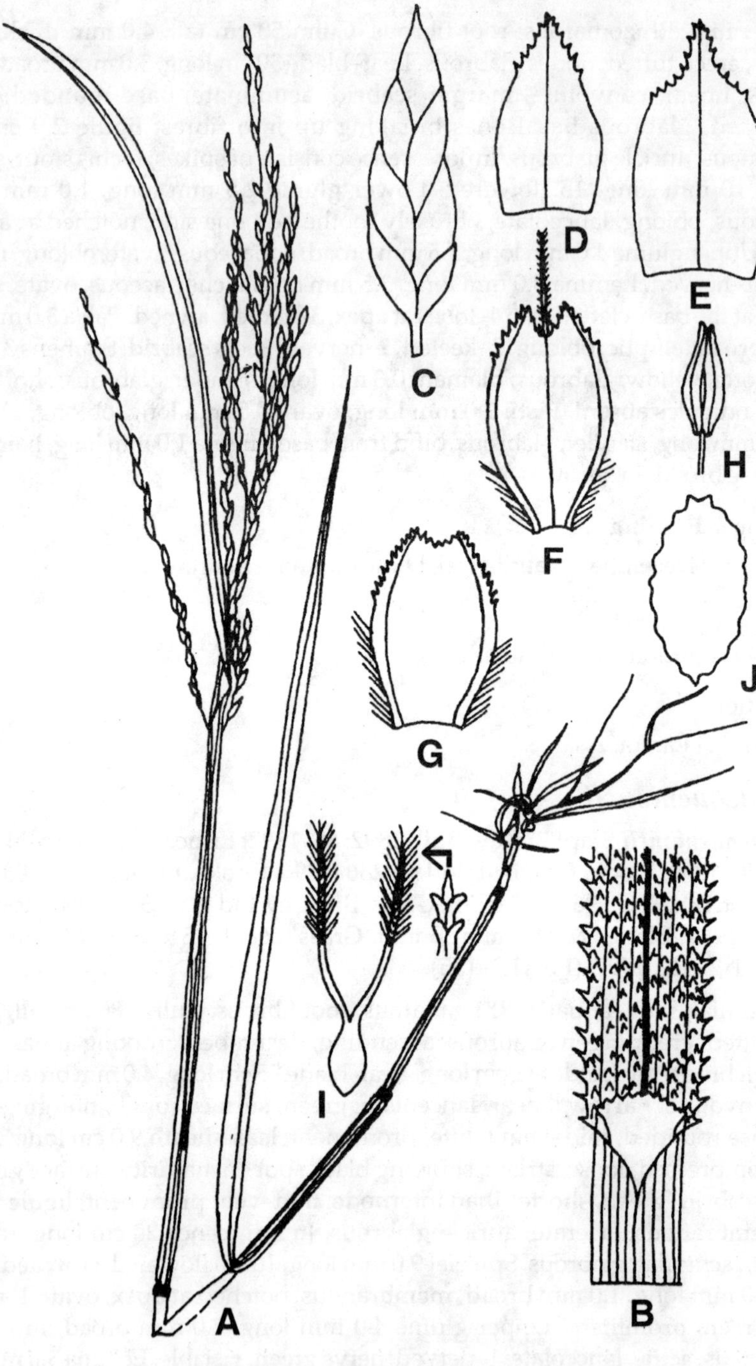

PLATE 155: ***Tripcgon jacquemontii*** **Stapf.**
A. Habit, B. Ligule, C. Spikelet, D. Lower glume, E. Upper glume, F. Lemma, G. Palea, H. Stamen, I. Pistil, J. Caryopsis

nerves green, prominent. Stamens 3; anther 1.0 mm long, yellow, glabrous, keeled; filament 0.5 mm long, slender, glabrous, shorter than anther. Lodicules absent. Pistil 2.0 mm long, ovary, 0.5 mm long, ovate, glabrous; style 0.6 mm long, slender, glabrous, bifid from base; stigma 0.8 mm long, brown, hairy; hairs acute, base broad in 2–rows. Caryopsis 1.5 mm long, 1.0 mm broad, brown, oblong, shining, free between lemma & palea.

Flowering & Fruting

August–December; Navalpur, Bansgaon, Barhaz, Captanganj.

Habitat

Rare, found in semi–aquatic localities.

Distribution

India, Australia, NewZealand.

3. *T. lisboae*

T. lisboae Stapf in Kew Bull. 1892: 84. 1892; Bor, Grass. Bur. Ceyl. Ind. Pak. 522. 1960; Moulik, Grass. Bamb Ind. 2: 622. 1997. (PLATE–156).

Perennial, semi aquatic. Rhizomatous; root fibrous. Culms 60 cm tall, 3.0 mm thick, tufted, erect, decumbend, ascending rooting at base, green, glabrous; nodes brown, glabrous; internodes 12 cm long. Leaf–blade 20 cm long, 4.0 mm broad, narrow, green, linear, lanceolate, acute, base rounded, convolute, surface rough, margin scabrid, mid–vein distinct on lower surface, 3–lateral veins on each side of mid–vein; lateral veins distinct on both surface, closed; leaf–sheath 9.0 cm long, 4.0 mm broad, compressed, loose, glabrous, margin scabrid, mid–vein prominent, membranous, green, shorter than internodes; ligule 2.0 mm long, membranous; auricle glabrous. Inflorescence spikes; spike 19 cm long, 6–10, alternate; rachis stout, lender. Spikelet 1.0 cm long, 5–10 flowered, crowded. Lower glume 2.0 mm long, 1.0 mm broad, margin scabrid, hyaline, lanceolate, obtusely toothed on one side, 1–nerved; nerve green, excurrent. Upper glume 3.0 mm long, 1.4 mm broad, coriaceous, apiculate, lanceolate, margin scabrid, 1–nerved; nerve green, visible. Lemma 4.0 mm long, 2.0 mm broad, membranous, apex scabrid, clefted into 2 lobes, base hairy; hairs acute, base rounded; awned; awn 2.0 mm long, antrorsely scabrid, short, 1–nerved; nerve thick, green, prominent. Palea 3.0 mm long, 1.5 mm broad, hyaline, oblong, mucronate, margin scabrid, keeled, hairy; hairs acute, base rounded; 2–nerved; nerve green, prominent. Stamens 3; anther 0.6 mm long, yellow, glabrous, keeled; filament 0.1 mm long, slender, glabrous, shorter than anther. Lodicules absent. Pistil 1.5 mm long; ovary 0.3 mm long, ovate, glabrous; style 0.5 mm long, slender, glabrous, bifid from base; stigma 0.7 mm long, hairy; hairs acute, base rounded, in 2–rows. Caryopsis 1.0 mm long, 0.5 mm broad, shining, brown, oblong, free, between lemma & palea.

Flowering & Fruiting

October–December; Natinia Dai, Sonbhadra, Dudhi, Basti.

Habitat

Rare, along the wetlands.

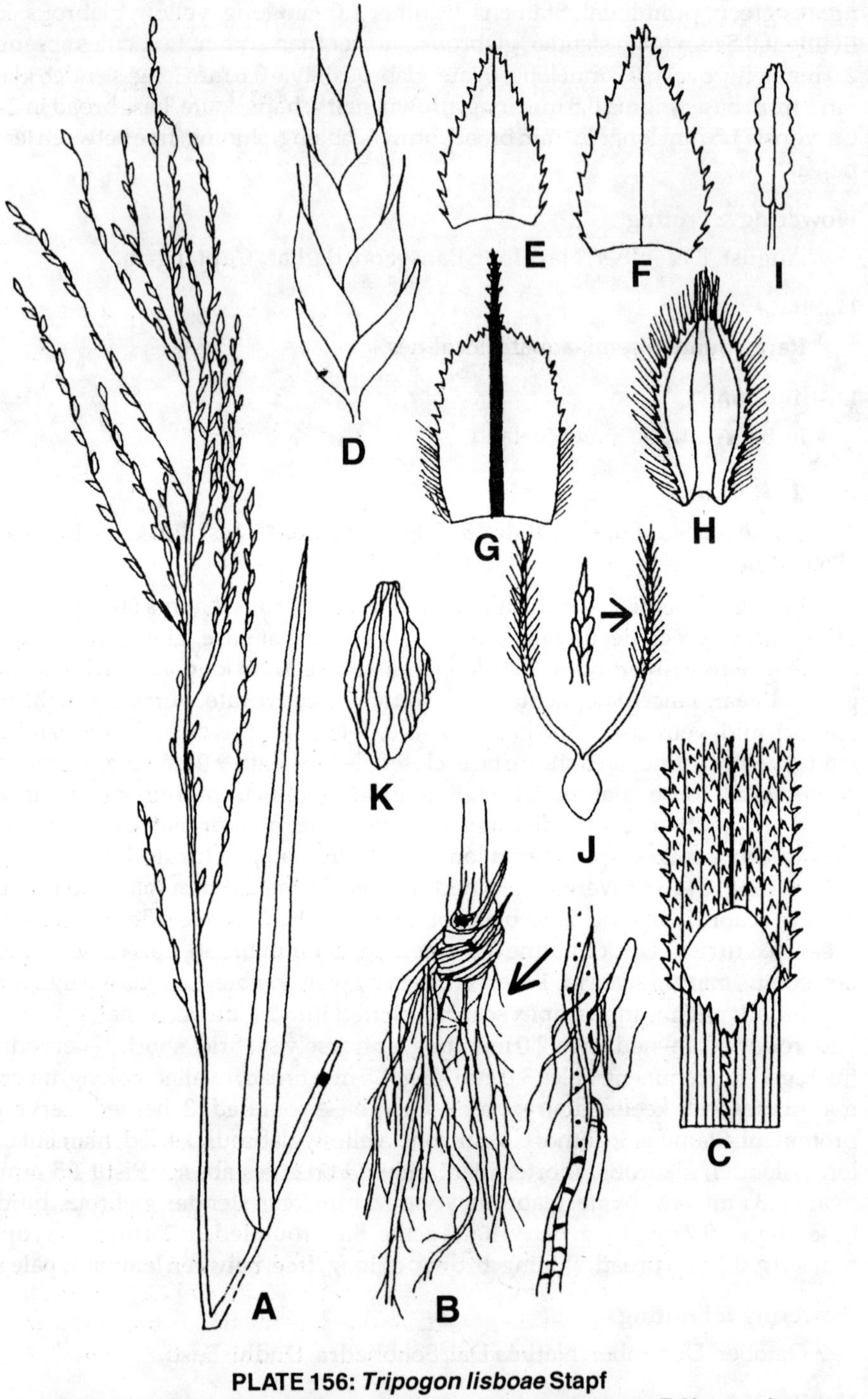

PLATE 156: ***Tripogon lisboae*** **Stapf**
A. Habit, B. Rhizome, C. Ligule, D. Spikelet, E. Lower glume, F. Upper glume, G. Lemma, H. Palea, I. Stamen, J. Pistil, K. Caryopsis

Distribution

India, Asia, Sri Lanka.

4. SUB FAMILY–PANICOIDEAE

KEY TO THE TRIBES

1a. Spikelets paired:

2a. Upper lemma more or less terete, 3–9 nerved; awn twisted & geniculate; hilum linear; starch grains compound

.... **2. Arundinellae**

2b. Upper lemma reduced to geniculate awned, nerveless; hilum sub basal; starch grains simple or compound

.... **1. Andropogoneae**

1b. Spikelets solitary or paired:

3a. Both florets fertile or the lower staminate; lower lemma similar to that of upper floret & indurated

.... **3. Isachneae**

3b. Upper florets alone fertilie, the lower male or barren; lower lemma similar to upper glume & not indurated

.... **4. Paniceae**

1. Tribe – Andropogoneae

KEY TO THE GENERA

1a. Spikelets perfect, if unisexual the male & female spikelet similar:

2a. Spikelets solitary, sessile, laterally compressed

.... **12.** ***Dimeria***

2b. Spikelets paired or rarely in three, both sessile & pedielled spikelets present or rarely the pedicelled spikelet reduced; not laterally compressed:

3a. Inflorescence a true or false panicle:

4a. Panicles interrupted by spathes:

5a. Aromatic grasses; racemes paired, usually one sessile & one pedicelled, collected into much branched panicle

.... **10.** ***Cymbopogon***

5b. Grasses not aromatic; racemes fascicled:

6a. Basal joints of racemes bulbous; spikelets in group of three, one sessile & other two pedicelled; the terminal spikelet reduced to a broad glume like pedicel

.... **2.** ***Apluda***

6b. Basal joints of racemes not bulbous; either sessile or pedicelled & 2 pairs of sessile involucral spikelets at base; the terminal spikelet absent

.... 32. *Themeda*

4b. Panicles not interrupted by spathes:

7a. Panicles densely silky villous:

8a. Panicle upto 20 cm long, contracted, spiciform, both spikelets in a pair pedicelled

.... 19. *Imperata*

8b. Panicle upto 50 cm long or more, effuse; spikelets one sessile & one pedicelled

.... 28. *Saccharum*

7b. Spikelets not silky villous:

9a. Branches of panicles whorled; spikelets awn less; lower glume muricate or spinulose

.... 33. *Vetiveria*

9b. Branches of panicle not whorled; spikelets awned, or awnless; lower glume not as above:

10a. Rachis joints & pedicels with a translucent median line; upper lemma of the sessile spikelet reduced to a hyaline base of awn:

11a. Branches & branchlets of panicle not capillary; glumes pitted

.... 5. *Bothriochloa*

11b. Branches & branchlets of panicle capillary; glumes not pitted

.... 6. *Capillipedium*

10b. Rachis joints & pedicelles not as above; upper lemma of sessile spikelets well developed, often bifid, awn from the sinus:

12a. Spikelets in group of three, one sessile, & other two pedicelled; callus often sharp, golden yellow or rufous brown, hairy; glumes usually aristate, rarely without arista

.... 8. *Chrysopogon*

12b. Spikelets paired; callus not as above; glumes never aristate:

13a. Panicles upto 60 cm or more long; primary branches verticillate or alternate; spikelets dorsally compressed; upper glume of sessile spikelets entire, 2–toothed or notched

.... 31. *Sorghum*

13b. Panicles upto 15 cm long; primary branches often solitary, alternate, spikelets not dorsally compressed; upper glume of sessile spikelets deeply notched

.... **26. *Pseudosorghum***

3b. Inflorescence a solitary, digitate or sub digitate racemes:

14a. Racemes solitary:

15a. Racemes cylindrical; joints & pedicelled fused; upper lemma of sessile spikelets awnless:

16a. Racemes fascicled; lower glume of sessile spikelets orbicular, cartilaginous, pitted

.... **16. *Hackelochloa***

16b. Racemes not fascicled; lower glume of sessile spikelet ovate, oblong, coriaceous, not pitted:

17a. Spikelets & joints reduced to tail like appendage towards the apex; sessile spikelets solitary at each node; pedicelled spikelets present

.... **27. *Rottboellia***

17b. Spikelets & joints uniformly developed; sessile spikelets paired at each node; pedicelled spikelets absent

.... **23. *Mnesithea***

15b. Racemes not cylindrical; joints & pedicels not fused; upper lemma of sessile spikelets awned:

18a. Lower glume of the sessile spikelet grooved with lyriform venation: upper glume awned; joints & pedicels clavate

.... **30. *Sehima***

18b. Lower glume of the sessile spikelet not groved; upper glume with or without awn; joints & pedicels not clavate:

19a. Racemes fragile; joints, pedicels & callus bearded; upper glume awned

.... **25. *Pogonatherum***

19b. Racemes not fragile; joints, pedicels & callus not bearded; upper glume awnless:

20a. Recemes terete; callus pungent; pedicelled spikelets twisted around the sessile spikelets; awn hairy; upto 10 cm long

.... **18. *Heteropogon***

20b. Raceme not terete; callus not pungent; spikelets not as above; awn antrorsely scabrid, upto 5.0 cm long

.... **11. *Dichanthium***

14b. Raceme two to many:

21a. Spikelets closely imbricate; pedicelled spikelets much reduced, usually represented by vestigial pedicels; lower glume of sessile spikelet truncate or emerginate at apex

.... 3. *Apocopis*

21b. Spikelets not closely imbricate; pedicelled spikelets more or less developed; lower glume of sessile spikelets acute or acuminate:

22a. Lower glume of sessile spikelets with a median furrow on dorsal side

.... 22. *Microstegium*

22b. Lower glume of sessile spikelets without median furrow on dorsal side:

23a. Lower glume muicate or spinulose, rarely smooth; upper lemma of sessile spikelet awned from the back, often just above the base

.... 4. *Arthraxon*

23b. Lower glume not as above; upper lemma of sessile spikelets awned from the sinus, often from the middle:

24a. Racemes solitary, frequently collected into a false panicle; joints & pedicels toothed; pedicelled spikelets reduced to small aristate glume

.... 29. *Schizachyrium*

24b. Racemes binate, joints & pedicels not as above; pedicelled spikelets well developed, rarely absent:

25a. Racemes 3–15, rarely solitary; spikelets of each pair similar, bisexual

.... 14. *Eulalia*

25b. Racemes 1–2; spikelets of each pair dissimilar; sessile spikelets with male & bisexual florets; pedicelled spikelets usually male or rarely bisexual:

26a. Lower glume of sessile spikelets usually coriaceous with several nodules & wrinkles, rarely smooth & chartaceous

.... 20. *Ischaemum*

26b. Lower glume of sessile spikelets not as above:

27a. All sessile spikelets on one sides of rachis; lower glume covered with long shaggy hairs

.... 13. *Elionurus*

27b. All sessile spikelets on both sides of rachis; lower glume not hairy:

28a. Racemes spatheate; joints & pedicels slender & solid .... 1. *Andropogon*

28b. Racemes not spatheate; joints & pedicels not as above:

29a. Racemes dorsally compressed; rachilla very tough; joints fused with pedicels .... 17. *Hemarthria*

29b. Racemes not dorsally compressed; rachilla & joints not as above:

30a. Rhizome extensive; spikelets solitary & sessile .... 24. *Ophiuros*

30b. Rhizome not extensive; spikelets paired or fascicled:

31a. Racemes covered with white hairs; lower glume of sessile spikelets 4–toothed .... 15. *Eulaliopsis*

31b. Racemes scabrid or glabrous; lower glumes of sessile spikelets 2–toothed .... 21. *Iseilema*

1b. Spikelets unisexual; male & female conspicuously dissimilar:

32a. Female spikelets completely enclosed in a metamorphosed leaf–sheath, forming spherical or cylindrical bead–like structure .... 9. *Coix*

32b. Female spikelets not enclosed in a false involucre as above:

33a. Male & female spikelets in same inflorescence; female spikelets solitary at base; upper one male .... 7. *Chionachne*

33b. Male & female spikelets in separate inflorescence; female spikelets in longitudinal rows on a very thick axis; male spikelets in a large terminal panicle .... 34. *Zea*

1. *Andropogon* L.

Andropogon pumilus Roxb. Fl. Ind. 1: 273. 1820; Hook. f. Fl. Brit. Ind. 7: 170. 1896; Roy, Grass, M.P. 26. 1984; Moulik, Grass. Bamb. Ind. 1: 209. 1997. *A. pachyarthrus* Hack. in DC. Monogr. Phan. 6: 449. 1889.

Annual. Rhizome absent; root fibrous. Culm 50 cm tall, 2.0 mm thick, green, glabrous, suberect, geniculate; nodes glabrous. Leaf–blade 18 cm long, 5.0 mm broad, green, glabrous, linear, lanceolate, acuminate, base rounded, margin scabrid; leaf–sheath 6.0 cm long, 4.0 mm broad, green, surface glabrous, margin hyaline, keeled; ligule 1.0 mm long, short, truncate, membranous; auricle glabrous. Inflorescence consists of racemes; raceme 2.0 cm long, erect, arising from spathe; spathe 1.2 cm long, flat, glabrous, joints 2.5 mm long, often red on the back, margin ciliate, apex cup shaped, 3–toothed. Sessile, spikelet 5.0 mm long, awned, linear, lanceolate; callus glabrous. Lower glume 5.0 mm long, 2.0 mm broad, membranous, linear, lanceolae, keeled, awned; awn 1.0 mm long. Upper glume 5.0 mm long, 1.5 mm broad, hyaline, narrow, linear, lanceolate. Lower lemma 3.0 mm long, 1.5 mm broad, hyaline, lanceolate, empty, epateate. Upper lemma 2.5 mm long, 1.2 mm broad, hyaline, linear, cleft at apex, bisexual, awned; awn 1.5 mm long, geniculate, lower half brown, upper half white, arising from sinus. Palea 1.0 mm long, 0.5 mm broad, hyaline, oblong. Stamens 3; anther 2.0 mm long, yellow, glabrous, linear; filament 0.5 mm long, slender, glabrous, shorter than anther. Lodicules absent. Pistil 3.0 mm long; ovary 0.5 mm long, oblong, glabrous; style 1.0 mm long, slender, glabrous, bifid from base; stigma 1.5 mm long, brown, hairy; hairs acute, base broad, in 2–rows. Pedicelled spikelet 5.0 mm long, awnless; pedicels 3.0 mm long, margin ciliate.

Flowering & Fruiting

August–December; Chandauli, Mau, Patna Ghat, Salempur.

Habitat

Along the roadside.

Distribution

India, Europe, America, Malaya, Malasian Islands, Australia.

2. *Apluda* L.

Apluda mutica L. Sp. Pl. 82. 1753; Moulik, Grass. Bamb. Ind. 1: 305. 1997; Bor, Grass. Bur. Cey. Ind. Pak. 93. 1960; Panigrahi & Rajagopal, Proc. Nat. Acad. Sci. India Sec. B. 37: 7. 1967; Cope in Fl. W. Pak. 143: 318. 1982; Fischer in Gamble, Fl. Pres. Madras 1750. 1934. (Rep. ed. 3: 1212. 1957); Manilal & Sivaraj, Fl. Calic. 344. 1982; Haines, Bot. Bih. & Or. 1056. 1924; Shreekumar & Nair, Fl. Kera. Grass. 32. 1991. *A. mutica* var. *aristata* (L.) Hack. ex Backer, Handb. Fl. Java 2: 54. 1928; Bor in Fl. Iranica 70: 524. 1970; Pilger in Nat. Pflanzenfam 14e: 130. 1940; Raizada *et al.*, Indian Forester Rec. ser. 2. Bot. 4: 204. 1961. *A. varia* ssp. *aristata* (L.) Hack in DC., Monograph Phan. 6: 196. 1889; Hook. f. Fl. Brit. Ind. 7: 150. 1896. *A. aristata* Linn. Amoen. Acad.4: 303. 1756; Rang. & Tad. Handb. S. Ind. Grass. 177. t. 142–143. 1921. *A. gigantea* Spreng., Syst. Veg. 1: 290. 1825. *A. geniculata* Roxb. Fl. Ind. 1: 327. 1820. *Andropogon glaucus* Retz., Obs. Bot. 5: 20. 1789. *Calamina gigantea* (Spreng.) P. Beauv. Ess. Agrost. 128. 1812. *C. humilis* J.S. Presl ex C.B. Presl, Rel. Haenk. 1: 344. 1831. *C. mutica* P. Beauv., Ess. Agrost. 128. t. 23. f. 1. 1812. (PLATE–157).

Annual. Rhizome absent; root fibrous. Culm 40 cm tall, 1.5 mm thick, erect, ascending from a shortly decumbent base, slender, green, glabrous, leafy, rooting at base, tufted, compressed, unbranched; nodes brown, glabrous; internodes 15 cm long, hollow. Lower leaf–blade 15 cm long, 4.0 mm broad, narrow, flat, elliptic, linear–lanceolate, alternate, usually petiolate, green, glabrous, margin hyaline, scaberulous, acute, base tapering, mid–vein distinct on lower surface, hairy on upper surface; hairs 1.0 mm long, acute, base broad, 3–lateral vein on each sides of mid–vein; lateral veins slender, distinct on ventral surface; upper leaf–blade 1.0 cm long, 1.0 mm broad,

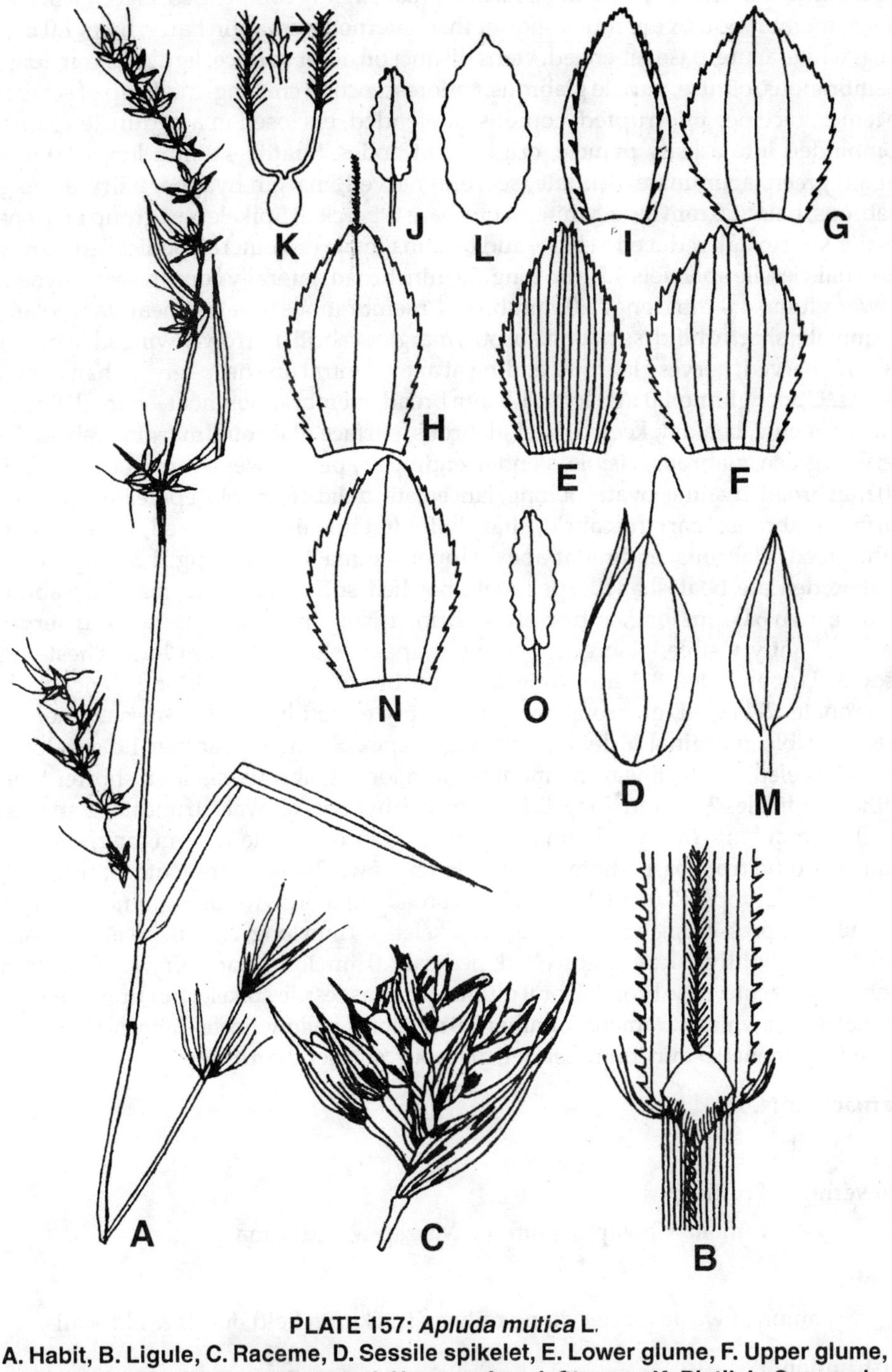

PLATE 157: ***Apluda mutica*** **L.**

A. Habit, B. Ligule, C. Raceme, D. Sessile spikelet, E. Lower glume, F. Upper glume, G. Lower lemma, H. Upper lemma, I. Upper palea, J. Stamen, K. Pistil, L. Caryopsis, M. Pedicelled spikelet, N. Upper lemma, O. Stamen

spathiform; leaf–sheath 4.0 cm long, 1.5 mm broad, tightly compressed, erect, glabrous, green, membranous, very much shorter than internodes, margin hairy; hairs 1.0 mm long, white, acute, base tubercled; veins distinct on outer surface; ligule 0.5 mm long, membranous, obtuse, auricle glabrous. Inflorescence 15 cm long, made up of several racemes; racemes interrupted, copious, subtended, enclosed in a peduncle spathe, combinded into a leafy panicle, originate at nodes. Spathe 4.0 mm long, 2.0 mm broad, green, acuminate, caudate, serveral nerved, margin hyaline, hairy, surface glabrous; hairs 1.0 mm long, white, acute, base tubercled. Spikelets in group of 3, one sessile & bisexual, reduced at base, and terminal male or neuteral, pedicelled, rarely bisexual. Sessile spikelets 4.5 mm long, 2.0 mm broad, laterally compressed, awned. Lower glume 4.0 mm long, 2.0 mm broad, membranous, ovate, linear–lanceolate, acuminate, shortly bifid, surface glabrous, margin scabrid, narrowly winged in upper half, 13–nerved; nerves glabrous, ending at apex, central one deep–green than lateral nerves. Upper glume 4.0 mm long, 1.5 mm broad, membranous, boat–shaped, linear, bilobed, acute, beaked, keeled; keel glabrous, surface glabrous, margin scabrid, 7–nerved, green, glabrous, visible, slender, ending at apex. Lower lemma 4.0 mm long, 2.0 mm broad, hyaline, ovate, oblong, lanceolate, delicate, empty, epateate, awnless, surface glabrous, margin scabrid, sharply folded inwards, 3–nerved; nerves faint, light green, glabrous, ending at apex. Upper lemma 4.0 mm long, 1.5 mm broad, hyaline, delicate, boat–shaped, apex acute, notched, surface glabrous, margin, scabrid, awned; awn 6.0 mm long, antrorsely scabrid, arising from sinus, 3–nerved; nerves green, slightly visible, glabrous, ending at apex; column 3.0 mm long, chest–nut brown. Upper Palea 3.8 mm long, 1.0 mm broad, hyaline, oblong, lanceolate, mucronate, 2–keeled, glabrous, margin sharply folded inwards, 2–nerved; nerves green, visible, marginal & dentate, ending at apex. Stamens 3; anther 1.0 mm long, yellow, keeled, ovate, linear; filament 0.5 mm long, glabrous, slender, shorter than anther. Lodicules 2, 0.5 mm long, 0.2 mm broad, hyaline, obovate, truncate, glabrous. Pistil 2.0 mm long; ovary 0.5 mm long, ovate, glabrous; style 1.0 mm long, slender, glabrous, bifid from base; stigma 0.5 mm long, brown, hairy; hairs acute, base broad, in 2–rows. Caryopsis 2.0 mm long, 1.0 mm broad, light brown, oblong, embryo large, laterally compressed, glabrous. Terminal spikelets 4.0 mm long, 2.0 mm broad, oblong–lanceolate, slightly fulcate, pedicelled; pedical 3.0 mm long, flat, hairy, hairs 1.0 mm long, white, acute, base broad. Floral glumes same as sessile spikelet but upper lemma not notched, awnless. Stamens 3; anther 3.0 mm long, yellow, linear–lanceolate, ovate, keeled; filament 0.5 mm long, slender, glabrous, shorter than anther.

Vernacular Name

Banjuri.

Flowering & Fruiting

July–November; Cholapur, Jaunpur, Mirzapur, Padrauna

Habitat

Common in waste places, along railway tracks, on field bunds & old walls.

Distrbution

India, South–East Asia, Australia.

3. *Apocopis* Nees

Apocopis vaginata Hack. in Ost. Bot. Zeitschr. 41: 8. 1891; Cooke, Fl. Bomb. 2: 967. 1908; Haines, Bot. Bih & Or. 1024. 1924; Bor, Grass. Bur. Cey. Ind. Pak. 96. 1960; Blatter & McCann, Bomb. Grass. 26. f. 17. 1984; Moulik, Grass. Bamb. Ind. 1: 329. 1997. *A. wightii* Nees ex Steud. var. *vaginata* Hook. f. Fl. Brit. Ind. 7: 143. 1896. (PLATE–158).

Annual. Rhizome absent; roots fibrous. Culms 30 cm long, 1.0 mm thick, green, glabrous, erect, tufted, decumbent & branched from base, ascending, slender, rooting at base; nodes hairy; hairs 1.0 mm long, white, acute, base broad; internodes 4.0 cm long, hollow. Leaf blades 5.0 cm long, 2.0 mm broad, flat, narrow, acuminate, base rounded, green, glabrous, linear–lanceolate, mid–veins distinct on ventral surface, 3–lateral veins on each sides of mid–veins, less distinct, both surface hairy; hairs white, acute, base broad, margin hyaline, scaberulous, keeled, hairy; hairs white, acute, base tubercled; leaf–sheath 5.0 cm long, 1.5 mm broad, lax, spathaceous, green, glabrous, longer than internodes, veins distinct on outersurface, margin hyaline, keeled, hairy at mouth; hairs 1.0 mm long, white, acute, base tubercled; ligule 0.2 mm long, membranous, lacerate; acuricle glabrous. Inflorescence of paired racemes; racemes 2.5 cm long, binate, closely appressed together, partly or almost entirely enclosed during flowering in the sheaths of upper leaves supported by a short peduncle, less hairy; hairs brown, acute, base broad; joints much shorter than the sessile spikelets. Spikelets 5.0 mm long, 2.0 mm broad, oblong, brown, nearly uniform in colour, paired, sessile & pedicelled, callus very short, 0.2 mm long, hairy; hairs brown, acute, base broad. Sessile spikelets 3.0 mm long, 1.1 mm broad, bisexual, glabrous, awned, oblong. Lower glume 3.0 mm long, 1.5 mm broad, membranous, cuneately obovate–obcordate, truncate, dark brown below, greenish above, apex hyaline, surface hairy on back; hairs very minute; acute, base broad, margin keeled, hairy; hairs 1.0 mm long, acute, base tubercled; 9–nerved; nerves green, visible, glabrous, abruptly ceasing below the apex. Upper glume 3.0 mm long, 1.0 mm broad, quadrate–oblong, membranous, linear, apex muticous, surface glabrous, margin sharply folded inward, hyaline, truncate, lacerate at apex, hairy from the base to below the apex; hairs 1.0 mm long, acute, base tubercled, 5–nerved; nerves green, visible. Lower lemma 2.5 mm long, 1.0 mm broad, hyaline, linear–oblong, obtuse, acute, empty, epateate, nerveless, surface glabrous, margin hairy at apex; hairs 1.0 mm long, acute, base tubercled. Upper lemma 2.5 mm long, 1.0 mm broad, linear, membranous, narrowly oblong, acutely bilobed, 1–nerved; nerves light green; awned; awn 2.0 cm long, antrorsely scabrid, geniculate. Palea 1.0 mm long, 0.5 mm broad, hyaline, nerveless, slightly ovate, linear–lanceolate, surface glabrous, margin scaberulous. Stamens 3; anther 1.0 mm long, yellow, linear, glabrous; filament 0.5 mm long, glabrous, shorter than anther. Lodicules absent. Pistil 1.5 mm long; ovary 0.4 long, oblong, glabrous; style 0.4 mm long, deeply bifid, glabrous; stigma 0.7 mm long, brown, hairy; hairs acute, base broad. Caryopsis 1.0 mm long, 0.5 mm broad, fusiform. Pedicelled spikelet 5.0 mm long, 1.5 mm broad, male, awnless, same as sessile spikelet, pedicels 2.0 mm long, 0.3 mm broad, flat, hairy; hairs 1.0 mm long, acute, base broad.

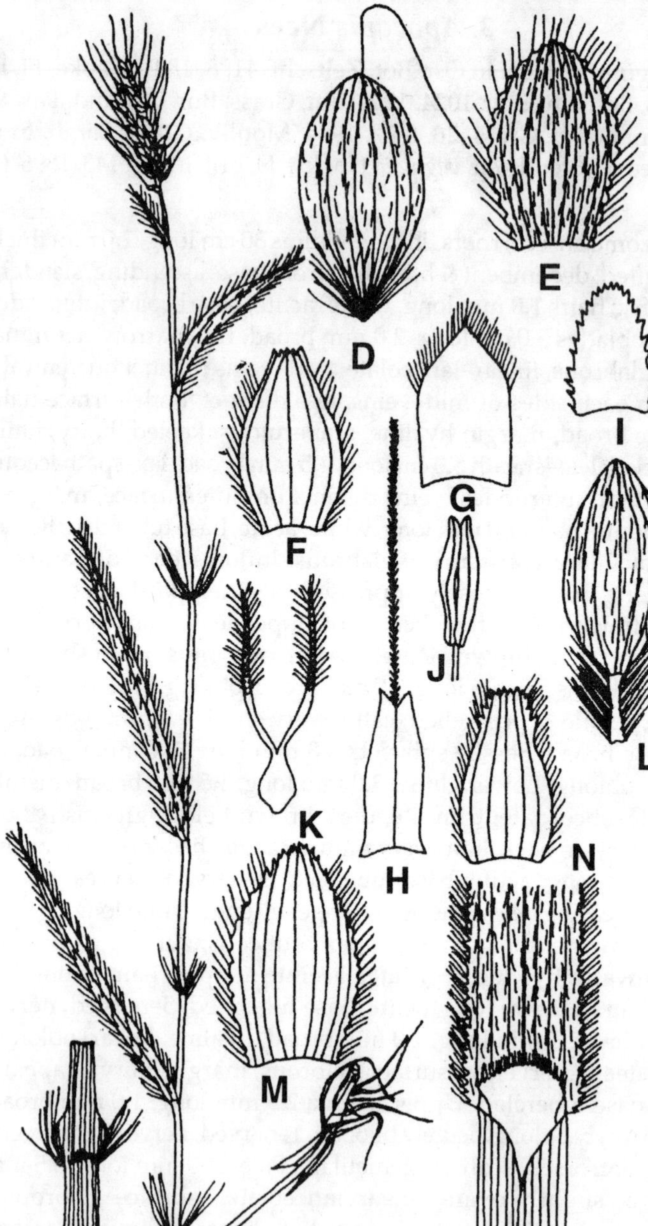

PLATE 158: ***Apocopis vaginata*** **Hack.**

A. Habit, B. Node & internode, C. Ligule, D. Sessile spikelet, E. Lower glume, F. Upper glume, G. Lower lemma, H. Upper lemma, I. Upper palea, J. Stamen, K. Pistil, L. Pedicelled spikelet, M. Lower glume, N. Upper glume

Flowering & Fruiting

September–December; Chandauli, Naugarh, Chakiya, Allahabad, Deoria.

Habitate

Rare, along roadside.

Distribution

Central India, Deccan & West Peninsula, Burma.

4. *Arthraxon* P. Beauv.

KEY TO THE SPECIES

1a. Pedicelled spikelets developed at least in upper half of racemes; caryopsis ± 0.8 mm long, cylindric

.... 2. *A. lancifolius*

1b. Pedicelled spikelets not well developed; caryopsis 2.0 mm long, linear & terete

.... 1. *A. hispidus*

1. *Arthraxon hispidus*

Arthraxon hispidus (Thunb.) Makino. in Bot. Mag. Tokyo 26: 214. 1912. Roy, Grass. M.P. 32. 1984; Moulik Grass. Bamb. Ind. 1: 259. 1997. *Phalaris hispida* Thunb. Fl. Jap. 44. 1784.

Annual. Rhizome absent; root fibrous. Culm 50 cm tall, 1.0 mm thick, green, glabrous, densely tufted, leafy, erect, nodes glabrous. Leaf–blade 2.5 cm long, 1.0 cm broad, ovate, lanceolate, surface hairy also with hyaline, keeled, margin, acuminate, base rounded; leaf–sheath glabrous, striate, ciliate on one margin; ligule 0.5 mm long, membranous, lacerate; auricle glabrous. Inflorescence consists of spikes; spike 3.0 cm long, several; joints of the rachis 2.5 mm long, ciliate, curved. Spikelet 3.0 mm long, solitary, awned, joints 5.0 mm long, suborbicular. Lower glume 3.0 mm long, 1.5 mm broad, membranous, convex on back, scabrid. Upper glume 3.0 mm long, 1.5 mm broad, membranous, convex on back, hispid, muriculate at apex. Lemma 2.0 mm long, 1.0 mm broad, hyaline, bisexual, awned; awn 9.0 mm long, arising from the base of lemma. Palea 1.5 mm long, 1.0 mm broad, hyaline, glabrous. Stamens 2; anther 1.0 mm long, purple, glabrous; filament 0.2 mm long, slender, glabrous, shorter than anther. Lodicules absent. Pistil 1.5 mm long; ovary 0.3 mm long, oblong, glabrous; style 0.5 mm long, slender, glabrous, bifid from base; stigma 0.7 mm long, hairy; hairs acute, base broad, in 2 rows. Caryopsis 2.0 mm long, 0.6 mm broad, linear, terete.

Flowering & Fruiting

September–December; Chakiya, Allahabad, Ballia, Mau.

Habitat

Rare, along hills and on rocky walls.

Distribution

India, Australia, Europe.

2. *A. lancifolius*

A. lancifolius (Trin.) Hochst. in Flora 39: 188. 1856; Moulik, Grass. Bamb. Ind. 260. t. 48. 1997; Jain in J. Ind. Bot. Soc. 51: 176. 1972; Saldhana & Nicolson, Fl. Hasan 709. 1976; Welzen in Blumea 27: 288. 1981; Bor, Grass. Bur. Cey. Ind. Pak. 100. 1960; Fischer in Gamble, Fl. Pres. Madr. 1729. 1934; Hoschst in Flora. 39: 188. 1856. *A. microphyllous* Hack. Monogr. Androp. 351. 1889; inclu. var. *lancifolius* Hook. f. Fl. Brit. Ind. 7: 147. 1896; Cooke, Fl. Bomb. 2: 970. 1908; Haines, Bot. Bih. & Or. 1026. 1924. *A. schimperi* Hochst in Flora 39: 189. 1856. *A. minor* Hochst in Flora 39. 189. 1856. *A. molle* Balf. in Trans. Roy. Soc. Edinb. 31: 315. 1888. *A. ciliaris* Henring in Bolet. Soc. Bot. 13. 1890; Haines Bot. Bih. & Or. 1026. 1924. *Andropogon lancifolius* Trin. in Mem. Acad. Sci. Petersb. ser. 6, 2: 271. 1832. *A. molle* Duthie, Grasses N.W. Ind. 17: 1883, *A. muticaulis* Steud. Syn. Pl. Glum. 1: 383. 1855. *A. microphyllus* Trin. in. Mem. Acad. Sci. Petersb. ser. 6, 2: 275. 1832. *Pleuroplitis lancifolius* (Trin) Regel. in Bull. Acad. Sci. Petersb. 10: 375. 1866. *P. microphyllus* (Trin). Regel in Bull. Acad. Sci. Petersb. 10: 370.1866. *P. schimperi* Regel in Bull. Acad. Sci. Petersb. 10: 369. 1866. *P. ciliata* J. Schmidt. Beiter. Fl. Cap. Verd. 152. 1852. *Psilopogon schimperi* Hochst. ex A. Rich., Tent. Fl. Abyss. 2: 447. 1851. *Bathratherum molle* Nees. et Arn. in Edinb. New Phil. J. 18: 181. 1835. *B. schimperi* Nees ex Hochst in Flora 39: 179. 1856. *Lucaea schimperi* Steud, Syn. Pl. Glum. 1: 414. 1855. (PLATE–159).

Annual. Rhizome absent; root fibrous. Culms 20 cm tall, 2.0 mm thick, green, glabrous, erect, slender, stragling, puberculous below the inflorescence, ascending, leafy, branching from lower nodes; branches sub erect, creeping & geniculate, nodes hairy; hairs 0.5 mm long, white, acute, base broad; internodes 3.0 cm long. Leaf–blade 2.5 cm long, 6.0 mm broad, green, flat, ovate, linear, lanceolate, acuminate, cordate at base, surface glabrous; margin hyaline, scabrid, keeled, hairy; hairs 1.0 mm long, acute, base tubercled; mid vein distinct on upper surface, 4–lateral veins on each side of mid–vein; lateral vein very slender, distinct on both surface; leaf–sheath 2.0 cm long, 2.5 mm broad, green, glabrous, erect, tightly compressed, shorter than internodes, margin hyaline, keeled, hairy; hairs 1.0 mm long, acute, base tubercled, veins distinct on outer surface; ligule 0.5 mm long, hyaline, glabrous, ovate, acuminate; auricle glabrous. Inflorescence racemes; racemes 1.5 cm long, sub–digitate, 2–6 in number, shining; rachilla hairy; hairs 1.0 mm long, acute, base broad. Spikelets 4.5 mm long, 2.5 mm broad, ovate, awned, lanceolate, paired; one sessile & other pedicelled. Sessile spikelet 2.5 mm long, 1.3 mm broad, ovate, oblong, linear–lanceolate, awned, laterally compressed, bisexual. Lower glume 2.5 mm long, 1.5 mm broad, membranous, ovate, oblong, linear–lanceolate, bifid at apex, rounded on back, surface glabrous, margin hairy; hairs minute, acute, base broad, fainty 5–nerved; nerves colourless, glabrous, ending at apex. Upper glume 2.5 mm long, 1.5 mm broad, membranous, ovate, oblong, linear–lanceolate, acuminate, nerveless, surface glabrous, margin keeled & incurved, awned; awn 1.0 mm long, antrorsely scabrid; 5–nerved; nerves colourless, glabrous, ending at apex. Lower lemma 1.5 mm long, 1.0 mm broad, hyaline, delicate, glabrous, oblong, lanceolate, acuminate, empty, epaleate,

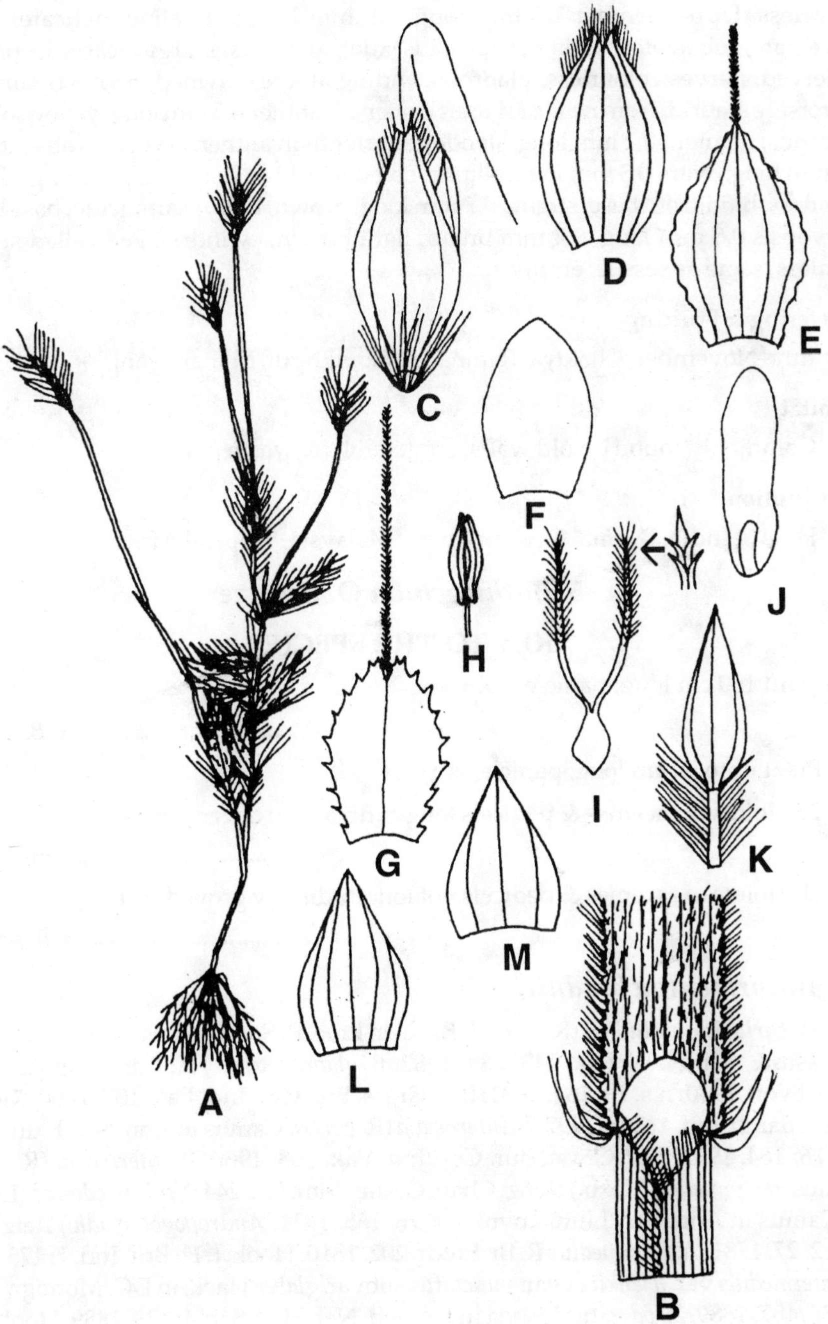

PLATE 159: ***Arthraxon lancifolius*** **(Trin.) Hochst.**
A. Habit, B. Ligule, C. Sessile spikelet, D. Lower glume, E. Upper glume, F. Lower lemma, G. Upper lemma, H. Stamen, I. Pistil, J. Caryopsis, K. Pedicelled spikelet, L. Lower glume, M. Upper glume

nerveless. Upper lemma 1.5 mm long, 1.0 mm broad, hyaline, delicate, ovate, lanceolate, sub acute, biclefet at apex, bisexual, glabrous, margin scabrid, epaleate, 1–nerved; nerves colourless, glabrous, ending at apex; awned; awn 6.0 mm long, antrorsely scabrid, from near the base. Stamens 2; anther 0.5 mm long, yellow, oblong, glabrous; filament 0.2 mm long, slender, shorter than anther. Lodicules absent. Pistil 1.5 mm long; ovary 0.3 mm long, elliptic, oblong, glabrous; style 0.5 mm long, slender, glabrous, bifid from base; stigma 0.7 mm long, brown, hairy; hairs acute, base broad. Caryopsis 0.8 mm long, 0.4 mm broad, light brown, cylindric. Pedicelled spikelet awnless, same as sessile, empty.

Flowering & Fruiting

July–November; Chakiya, Ramnagar, Gorakhpur, Maharajganj.

Habitat

Commonly found on old walls, eroded hills & waste places.

Distribution

Hills of India, Burma, Ceylon, China, Malaysia, Tropical Africa.

5. *Bothriochloa* O. Kuntze

KEY TO THE SPECIES

1a. Plant 150 cm long; panicle glabrous

.... **1. *B. bladhii***

1b. Plant upto 90 cm long; panicle hairy:

2a. Joints of racemes & pedicels longitudinally grooved; anther 2

.... **2. *B. intermedia***

2b. Joints of racemes & pedicels not longitudinally groved; anther 3

.... **3. *B. pertusa***

1. *Bothriochloa bladhii*

Bothriochloa bladhii (Retz.) S.T. Blake in Proc. R. Soc. Queens. 80: 62. 1969; Cope in Nasir & Ali, Fl. W. Pak. 143: 284. 1982. *B. glabra* (Roxb.) A. Camus in Ann. Soc. Linn. Lyon. 1930; n.s. 76. 164. 1931; Bor, Grass. Bur. Cey. Ind. Pak. 107. 1960; Moulik, Grass. Bamb. Ind. 1: 265. 1997. *B. intermedia* (R. Br.) A. Camus in Ann. Soc. Linn. Lyon. n.s. 76: 164: 1931; Bor, Grass. Bur. Cey. Ind. Pak. 108. 1960. *B. intermedia* (R. Br.) A. Camus var. *punctata* (Roxb.) Keng. Chau. Gram. Prim. Sin. 244. 1957. *B. odorata* (Lisboa) A. Camus in Ann. Soc. Linn. Loyn. n. s. 76: 165. 1931. *Andropogon bladhii* Retz. Obs. Bot. 2: 27. 1781. *A. intermedius* R. Br. Prodr. 202. 1810; Hook. f. Fl. Brit. Ind. 7: 175. 1896. *A. intermedius* var. *haenkii* et var. *punctatus* subvar. *glaber* Hack, in DC. Monogr. Phan. 6: 486, 487. 1889. *A. odoratus* Lisboa in J. Bomb. Nat. Hist. Soc. 4: 123. 1889; Hook. f. Fl. Brit. Ind. 7: 177. 1896. *A. caucasicus* Trin. in Mem. Acad. Sci. Petersb. ser. 6, 2: 286. 1832. *A. glaber* Roxb. Fl. Ind. 1: 271. 1820. *A. punctatus* Trin. Sp. Gram. 3: 28. 1836. *A. haenkei* J.S. Presl ex C.B. Rel. Haenk. 1: 340. 1830. *A. vachelli* Hook. et Arn. Bot. Beechey's Voy. 243. 1838. *A. leptanthus* Steud. Syn. Pl. Glum. 1: 391. 1854. *Amphilophis glabra* (Roxb.)

Stapf in Prain. Fl. Trop. Afr. 9: 172. 1917; Fischer in Gamble, Fl. Madr. 1732. 1934 (Repr. ed. 3: 1200. 1957). *Dichanthium intermedium* (R. Br.) Dewet & Harlan in Am. J. Bot. 53: 97. 1966. *D. bladhii* (Retz.) W.D. Clayton in Kew Bull. 32: 3. 1977. *D. glabrum* (Roxb.) Jain & Deshpande in Bull. Bot. Sur. Ind. 20: 134. 1978. *D. odoratum* (Lisboa) Jain & Deshpandey in Bull. Bot. Sur. Ind. 20: 134. 1978. *D. caucasicum* (Trin.) Jain & Deshpande in Bull. Bot. Sur. Ind. 20: 133. 1978. *Bothriochloa caycasica* (Trin) C.E. Hubbard in Kew Bull. 1939: 101. 1939. (PLATE–160).

Perennials. Rhizomatous; rhizome 4.0cm long, horizontal; root fibrous. Culm 37 cm tall, 4.0 mm thick, green, glabrous, erect, tufted, robust, leafy branching at lower nodes; branches suberect; rooting at base; nodes brown, hairy; hairs 1.0 mm long, white, acute, base broad; internodes 6.0 cm long. Leaf–blade 10 cm long, 2.0 mm broad, green, glabrous, glacuous, linear–lanceolate, flat, narrow, acute, base rounded, margin hyaline, scabrid, mid–vein prominent, 3–lateral veins on each side of mid–vein; lateral vein slender, distinct on upper surface; leaf sheath 4.0 cm long, 4.5 mm broad, green, glabrous, membranous, tightly compressed, striate, truncate, margin hyaline, shorter than internodes, veins distinct on outer surface; ligule 0.5 mm long, membranous, ovate, truncate, hairy at tip; hairs white, acute with broad base; auricle glabrous. Inflorescence a false panicle; panicle 10 cm long with a central axis, racemes numerous, in irregular whorls upon simple or branched peduncle, altogether forming an oblong, pyramidal raceme. Racemes 6.0 cm long, whorled or alternate, simple, branched; joints 1.5 mm long, linear, flat, margin ciliate; with a translucent grooved in the centre; callus shortly bearded. Sessile spikelet 4.0 mm long, 2.0 mm broad, elliptic, oblong, bisexual, awned. Lower glume 3.0 mm long, 1.5 mm broad, membranous, oblong, ovate, lanceolate, acuminate, upper surface hairy at apex; hairs minute, acute, base tubercled; glabrous below, lowers surface pitted, margin incurved & hairy; hairs 1.5 mm long, acute, base tubercled; faintly 7–nerved; nerves colourless, glabrous, ending at apex. Upper glume 4.0 mm long, 1.5 mm broad, membranous, oblong, ovate, lanceolate, acuminate, surface glabrous, margin incurved, 3–nerved; nerves colourless, glabrous ending at apex. Lower lemma 3.0 mm long, 1.0 mm broad, hyaline, delicate, ovate, oblong, lanceolate, empty, epaleate, 1–nerved; nerves colourless, glabrous, ending at apex. Upper lemma 2.0 mm long, 0.5 mm broad, reduced, stipitate, paleate, bisexual, acuminate, awned; awn 1.5 cm long, antrorsely scabrid; margin hairy; hairs 1.0 mm long, acute, base broad; column 0.8 mm long, brown. Upper palea 2.0 mm long, 1.0 mm broad, hyaline, oblong, glabrous, 2 keeled, 2 nerved; nerves green, visible, glabrous & marginal, ending at apex. Stamens 3; anther 2.0 mm long, yellow, linear, glabrous; filament 0.5 mm long, slender, shorter than anther. Lodicules 2, 0.5 mm long, 0.2 mm broad, hyaline, obovate, glabrous, nerveless. Pistil. 3.0 mm long; overy 0.5 mm long, oblong, glabrous; style 0.5 mm long, slender, glabrous, bifid from base; stigma 2.0 mm long, brown, hairy; hairs acute, base broad, in 2–rows. Caryopsis 1.5 mm long, 1.0 mm broad, light brown, oblong, glabrous. Pedicelled spikelet 5.5 mm long, 1.5 mm broad, oblong, ovate, lanceolate; pedicel 2.0 mm long, 0.5 mm broad, flat, margin hairy; hairs 1.0 mm long, acute, base broad. Lower glume 3.0 mm long, 1.5 mm broad, membranous, oblong, lanceolate, suriace glabrous, margin infolded, keeled; keel pectinate, pitted at lower view, 11–nerved; nerves colourless scabrid, ending at apex. Upper glume 3.5 mm long, 1.0 mm broad,

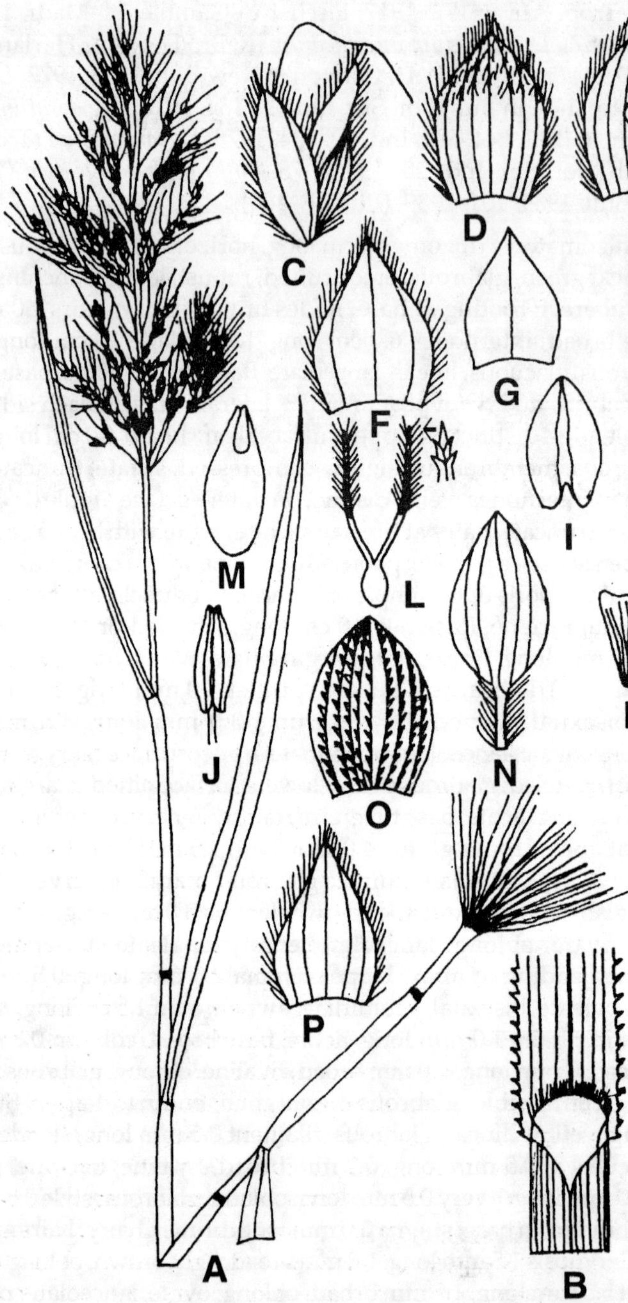

PLATE 160: ***Bothriochloa bladhii*** **(Reyz) C.T. Black**
A. Habit, B. Ligule, C. Sessile spikelet, D & E. Lower glume, F. Upper glume, G. Lower lemma, H. Upper lemma, I. Upper palea, J. Stamen, K. Lodicule, L. Pistil, M. Caryopsis, N. Pedicelled spikelet, O. Lower glume, P. Upper glume

membranous, oblong, lanceolate, surface glabrous, margin hyaline, hairy; hairs 1.0 mm long, acute, base broad; 5–nerved; nerves colourless, glabrous spikelet. Upper lemma 2.5 mm long, 1.0 mm broad, hyaline, delicate, ovate, oblong, glabrous, nerveless, unawned, male, epaleate. Stamens 3; anther 1.0 mm long, yellow, linear, glabrous; filament 0.5 mm long, slender, shorter than anther. Lodicules absent.

Flowering & Fruiting

August–December; Sarnath, Ghazipur, Allahabad, Ballia, Pipiganj.

Habitat

Common along the river banks.

Distribution

India, Sino–Malaya.

2. *B. intermedia*

B. intermedia (R. Br.) A. Camus in Ann. Soc. Lyon. n.s. 76: 164. 1931; Chein–chang, Hsu, in Taiwan Grasses 643–644. 1975. Bor, Grass. Bur Cey. Ind. Pak. 108. 1960. *Andropogon intermedius* R. Br. Prodr. 202, 1810; Hook. f. Fl. Brit. Ind. 7: 175. 1896. *A. intermedius* var. *genuinus* Hackel in DC., Monogr. Phan. 6: 485. 1889. *A. punctatus* Roxb. Fl. Ind. 1: 268. 1820. *A. perfossus* Nees ex Steud. Syn. Pl. Glum. 1: 391. 1854. *Amphilophis intermedia* (R.Br.) Stapf in Agric. News. W. Ind. 15: 179. 1916. (PLATE–161).

Perennial. Rhizomatous; rhizome 1.0 cm long, creeping, woody; root fibrous. Culms 30 cm tall, 3.0 mm thick, erect, tufted, ascending from the base, compressed, green, glabrous, rooting at base, unbranched, stout, decumbent; nodes hairy, brown; hairs white, acute, base broad; internodes 5.0 cm long, glabrous. Leaf–blade 7.0 cm long, 5.0 mm broad, green, surface glabrous, margin scaberulous, acuminate, base rounded, linear–lanceolate, mid–vein distinct on ventral surface, 4–lateral vein each side of mid–vein; lateral vein prominent, hairy at base; hairs white, acute, base broad; leaf–sheath 4.0 cm long, 4.0 mm broad, glabrous, membranous, margin hyaline, shorter than internode, upper sheath compressed, basal sheath loose, veins not distinct, mouth hairy; hairs white, acute, base bulbous; ligule 1.5 mm long, shallow membranous, notched, emerginate; auricle glabrous. Inflorescence panicle; panicle 6.0 cm long, green when young, brown at maturity, hairy, lower branch short. Sessile spikelet 4.0 mm long, 1.5 mm broad, bisexual, awned, acute, ovate, lanceolate; callus 0.5 mm long, hairy; hairs acute, base broad. Lower glume 3.5 mm long, 1.5 mm broad, membranous, ovate, lanceolate, acute, brownish–green, 7–nerved on upper surface; nerves not visible, 5–nerved on lower surface; nerves not visible, marginal two nerves very faint, pitted on lower surface, margin inrolled & scabrid at apex; surface hairy; hairs brownish–green due to surface colour, acute, base bulbous. Upper glume 4.0 mm long, 1.5 mm broad, membranous, ovate–lanceolate, acute, linear, 3–nerved; nerves green, visible, surface glabrous, margin scaberulous, hispid. Lower lemma 4.0 mm long, 1.5 mm broad, membranous, broadly ovate, linear–lanceolate, epaleate, surface & margin glabrous, empty, 3–nerved; nerves green, dendate at apex, visible, central nerves ending at tip, lateral nerves joint to the margins. Upper lemma 1.5 mm long,

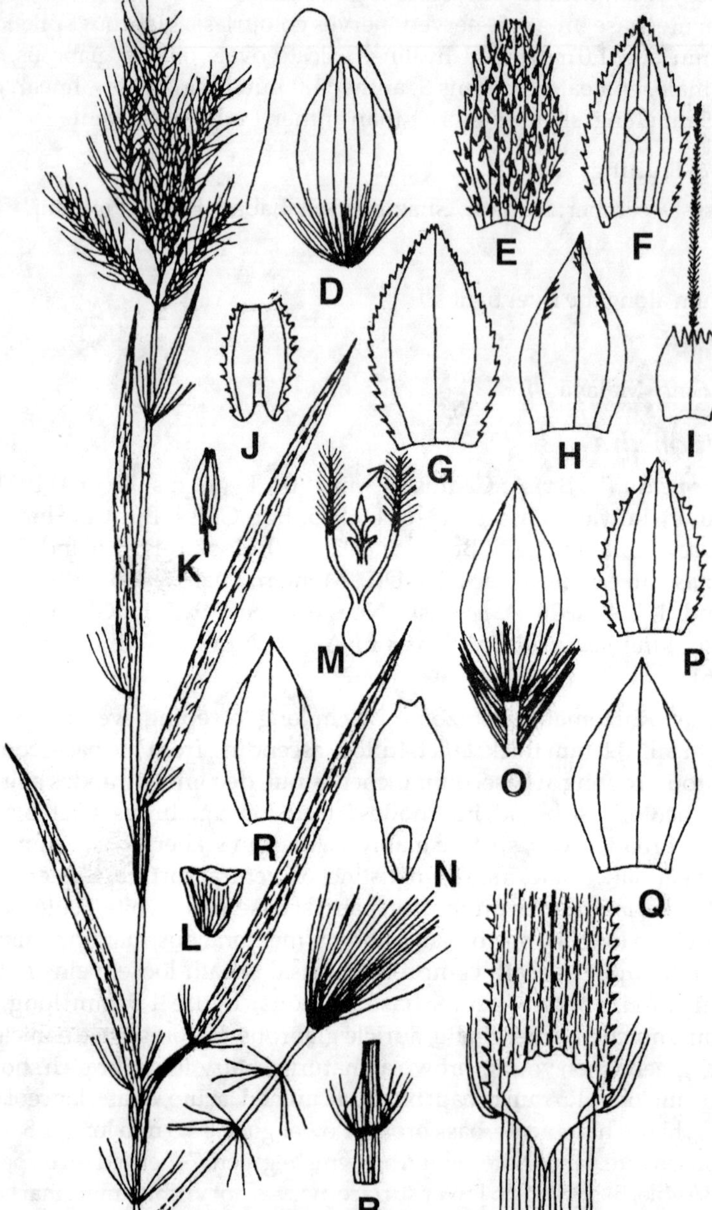

PLATE 161: ***Bothriochloa intermedia*** **(R.Br.) A. Camus**
A. Habit, B. Node & internode C. Ligule, D. Sessile spikelet, E & F. Lower glume, G. Upper glume, H. Lower lemma, I. Upper Lemma, J. Upper Palea, K. Stamen, L. Lodicule, M. Pistil, N. Caryopsis, O. Pedicelld spikelet, P. Lower glume, Q. Upper glume. R. Lower lemma

1.0 broad, hyaline, oblong, linear, notched & hairy at apex; hairs cilia acute, base tubercled; 1–nerved; nerves light green, faint; awned; awn 1.3 cm long, antrorsely scabrid. Upper palea 2.5 mm long, 1.5 mm broad, membranous, lanceolate, surface glabrous, margin scaberulous, apex grooved, 2–nerved; nerves faint, marginal. Stamens 2; anther 1.0 mm long, yellow, ovate–oblong, linear, glabrous; filament; 0.5 mm long, slender, shorter than anther. Lodicule 2; 0.3 mm long, 0.2 mm broad, obovate, waxy at apex, hyaline, surface & margin glabrous, many nerved; nerves faint, divided at apex. Pistil 2.0 mm long; ovary 0.5 mm long, glabrous, light green, ovate, base rounded; style 1.0 mm long, slender, glabrous, bifid from base; stigma 0.5 mm long, brown, hairy; hairs acute, base broad, in 2–rows. Caryopsis 1.2 mm long, 0.6 mm broad, obovate, embryo basal. Pedicelled spikelet 3.5 mm long, 1.3 mm broad, acute, empty, ovate, lanceolate; pedicel 2.0 mm long, 0.5 mm broad, flat, longitudinally with a translucent groove, hairy; hairs white, acute, base broad. Lower glume 3.0 mm long, 1.5 mm broad, membranous, ovate, lanceolate, green, surface glabrous, margin scabrid, 7–nerved, nerves green, visible on upper surface, without pit. Upper glume 3.5 mm long, 1.5 mm broad, membranous, ovate, linear–lanceolate, 3–nerved; nerves faint, light green, acute, surface & margin glabrous. Lower lemma 3.0 mm long, 1.5 mm broad, ovate, lanceolate, membranous, acute, 3–nerved, nerves; green, visible. Upper lemma 1.5 mm long, 0.8 mm broad, oblong, hyaline, glabrous, unawned. Palea absent.

Flowering & Fruiting

August–January; Tarana, Babatpur, Mau, Jaunpur, Basti.

Habitat

Common in open grasslands.

Distribution

India, Australia, Tropical Africa.

3. *B. pertusa*

B. pertusa (L.) A. Camus in Ann. Soc. Linn. Lyon. n.s. 76: 164. 1931; Moulik, Grass. Bamb. Ind. 1: 267. 1997; Bor, Grass. Bur. Cey. Ind. Pak. 109. 1960; Rang. & Tad. Handb. S. Ind. Grass. 191. t. 152–153. 1921. *Amphilophis pertusa* Linn. Nash ex Stapf in Agri. News. Bridge 15: 179. 1916, Stapf in Prain, Fl. Trop. Afr. 8: 175. 1917; Fischer in Gamble, Fl. Pres. Madras 1731. 1934 (rep. ed. 3: 1200. 1957). *Dichanthium pertusum* (Linn.) W.D. Clayton in Kew Bull. 32: 4. 1977. *Holcus pertusus* L. Mant. Alt. 301. 1771. *Andropogon pertusus* L. Wild. Sp. Pl. 4: 922. 1806; Hook. f. Fl. Brit. Ind. 7: 173.1896. *Andropogon angustifolius* Parh. Fl. Palerm 1: 269. 1845. *Lepeocercis pertusa* (Linn.) Hassk. Pl. Jav. Rar. 52. 1848. *Elionurus pertusus* Nees ex Steud. Syn. Pl. Glum. 1: 364. 1854. (PLATE–162).

Perennial. Stoloniferous; root fibrous. Culm 35 cm long, 2.0 mm thick green, erect, ascending, tufted, terete, compressed, decumbent, unbranched, rooting at base, leafy at base, stout, nodes hairy; hairs white, acute, base broad; internodes 4.0 cm long, glabrous. Leaf–blade 6.0 cm long, 5.0 mm broad, green, flat, acuminate, crowded at base, linear–lanceolate, glaucous, mid–vein prominent, 2–lateral veins on each

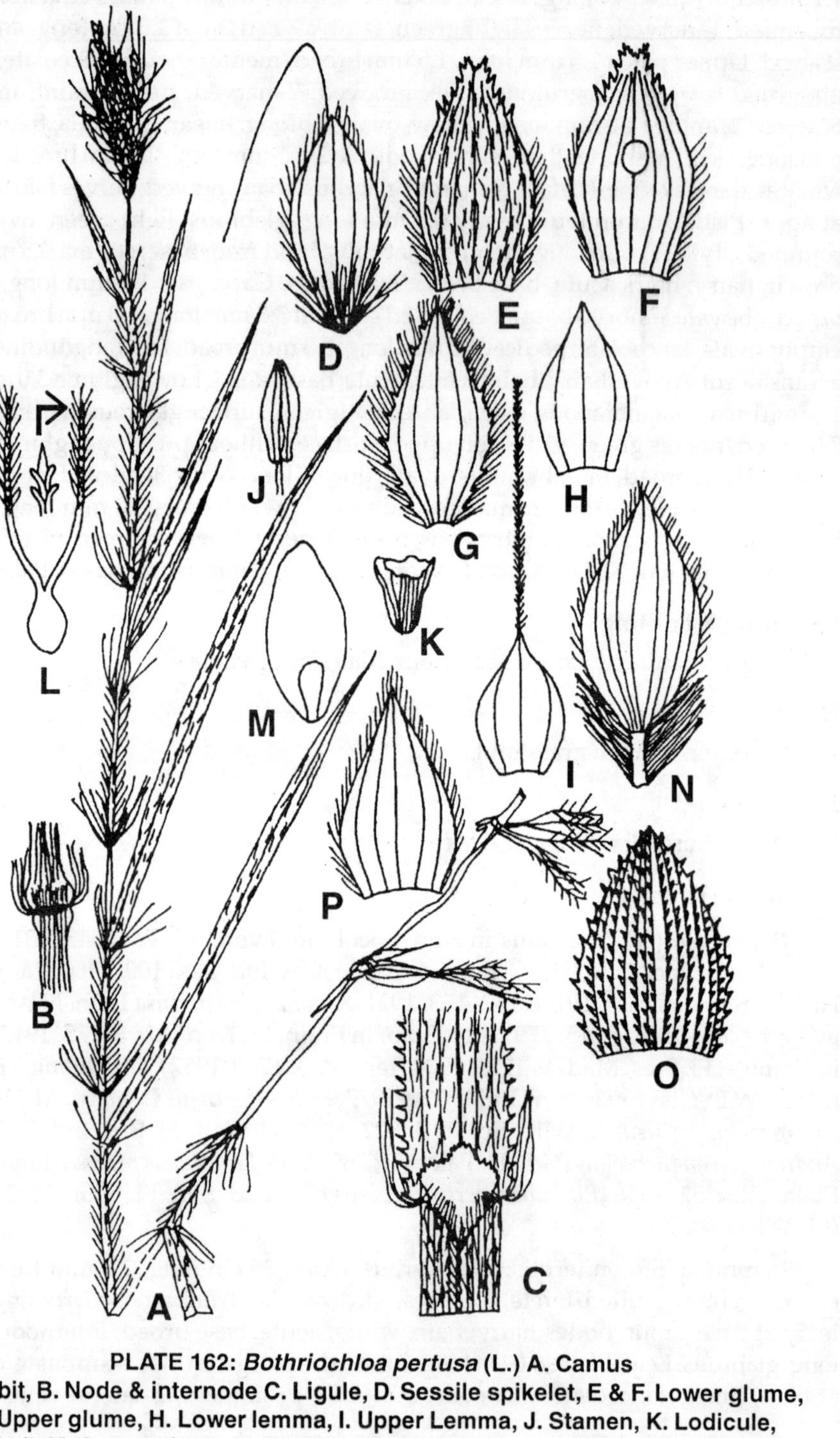

PLATE 162: ***Bothriochloa pertusa*** **(L.) A. Camus**

A. Habit, B. Node & internode C. Ligule, D. Sessile spikelet, E & F. Lower glume, G. Upper glume, H. Lower lemma, I. Upper Lemma, J. Stamen, K. Lodicule, L. Pistil, M. Caryopsis, N. Pedicelld spikelet, O. Lower glume, P. Upper glume

side of mid–vein; surface hairy; hairs acute, base tubercled; margin hyaline, scaberulous; leaf–sheath 3.5 cm long, 2.5 mm broad, green, tightly compressed, membranous, veins not distinct, shorter than internodes, surface hairy, margin keeled, hyaline, mouth hairy; hairs white, acute, base tubercle; ligule 1.0 mm long, membranous, truncate, notched; auricle glabrous. Inflorescence a sub–digitate raceme; raceme, 5.0 cm long, 1.5 mm broad, five on a peduncle, simple, easily breaking up at maturity, silky densely villous, oblong; peduncle short; joints 2.0 mm long, linear, densely villous with long hairs; hairs acute, base rounded. Rachis 0.5 mm broad, filiform, glabrous. Spikelets 4.5 mm long, 2.0 mm broad, paired; upper pedicellate, lower sessile. Sessile spikelets 4.0 mm long, 1.5 mm broad, bisexual, elliptic, lanceolate; callus bearded. Lower glume 4.0 mm long, 1.5 mm broad, membranous, yellow, ovate, elliptic, linear–lanceolate, surface hairy; hairs acute, base rounded, shortly bifid & subobtuse at apex, margin scaberulous at apex, hairy from base to middle; hairs white, acute, base broad, a deep circular pit above the middle on the lower surface, rarely bipitted, 7–nerves; nerves visible, light–green, faintly visible, central one reaching at tip, lateral nerves ending shortly from the apex. Upper glume 4.0 mm long, 1.5 mm broad, acute, boat–shaped, oblong–lanceolate, surface glabrous, margin slightly scaberlous, hairy; hairs acute, base rounded, yellow, membranous, 5–nerved, light green, faint. Lower lemma 3.0 mm long, 1.0 mm broad, ovate–lanceolate, hyaline, acute, delicate, surface glabrous, margin scaberulous, empty, epaleate, 3–nerved; nerves faint, light green. Upper lemma 2.0 mm long, 1.0 mm broad, hyaline, acute, stipitate, kneed, twisted, narrowly linear, epaleate, 3–nerved; nerves faint, central nerve reaching the tip, lateral nerves ending below, awned; awn 1.1 cm long, antrorsely scabrid; column 6.0 mm long, brown. Stamens 3; anther 1.0 mm long, yellow, glabrous; filament 0.2 mm long, slender, shorter than anther. Lodicule 2, 0.2 mm long, 0.1 mm broad, hyaline, obovate, cuneate, wavy at apex, surface & margin glabrous, many nerved; nerves faint. Pistil 2.5 mm long; ovary 0.5 mm long, oblong, glabrous, light green; style 1.0 mm long, slender, glabrous, bifid from base; stigma 1.0 mm long, pink, hairy; hairs acute, base broad, in 2–rows. Caryopsis 1.5 mm long, 1.0 mm broad, oblong, whitish, embryo basal. Pedicellate spikelet 4.0 mm long, 1.0 mm broad, oblong, elliptic, acute, glabrous, empty; pedicels 2.0 mm long, 0.5 mm broad, linear, villous. Lower glume 4.0 mm long, 1.0 mm broad, oblong–elliptic, membranous, yellow, surface glabrous, margin hyaline, scaberulous, acute, 13–nerved; nerved light green, dentate. Upper glume 4.0 mm long, 1.0 mm broad, membranous, margin hyaline, hairy; hairs acute, base broad, 5–nerved; nerves light green. Lower lemma 3.5 mm long, 1.0 mm broad, acute, membranous, nerveless. Upper lemma 3.0 mm long, 1.0 mm broad, lanceolate, hyaline, acute, delicate, unawned, nerveless.

Flowering & Fruiting

August–February; Tarana, Lamhi, Sarnath, Sindhora, Babatpur, Barhikala, Chakiya, Chandauli, Sonbhadra, Ballia, Lar, Padrauna.

Habitat

Found on fallow land, hilly areas, usually along the bunds of fields, waste lands & in open grasslands & forests.

Distribution

India, Kenya, Uganda, Souther & Northern Tropical Africa, Southeast Asia, Caribbean archipelago.

6. *Capillipedium* Stapf.

Capillipedium assimile (Steud.) A. Camus in Lecomte, Fl. Gen. Indo-chine 7: 314. 1922; Bor, Grass. Bur. Cey. Ind. Pak. 110. 1960; Roy,Grass, M.P. 45. 1984; Shreekumar & Nair, Fl. Kera. Grass. 55. 1991; Moulik, Grass. Bamb. Ind. 1: 269. 1997. *Andropogon assimilis* Steud. in Zoll, Syst. Verz. 58. 1854; Syn. Pl. Glum. 1: 397. 1854; Hook. f. Fl. Brit. Ind. 7: 179. 1896. *Capillipedium glaucopsis* (Steud.) Stapf in Hook. Icon. Pl. 31: t.3085. 1922; Ficher in Gamble,Fl. Pres. Madr. 1730. 1934. (Repr. ed. 3: 1199. 1957). *Andropogon glaucopsis* Steud.in Zoll.Syst. Verz. 58.1854. *Capillipedium huegelii* (Hack.) Stapf in Hook. Icon. Pl. 31. t. 3085. 1922; Fischer in Gamble, Fl. Pres. Madr. 1730. 1934 (Repr. ed. 3: 1199. 1957); Bor, Grass. Bur. Cey. Ind. Pak. 110. 1960. *Andropogon huegelii* Hack in DC Monogr. Phan. 6: 492. 1889.

Annual. Rhizome absent; root fibrous. Culm 150 cm tall, erect; nodes glabrous. Leaf–blade 20 cm long, 1.0 cm broad, linear, lanceolate, acuminate; leaf–sheath slightly keeled, glabrous; ligule 1.0 mm long, ovate, acute, membranous; auricle glabrous. Inflorescence consists of panicle; panicle 15 cm long, open, contracted; axils of racemes long, villous; joints 2.5 mm long, hairy. Sessile spikelet 4.0 mm long, ovate, oblong, hairy, callus bearded. Lower glume 3.5mm long,1.5mm broad, subcoriaceous, ovate, acute, hairy, margins pectinately ciliate towards apex, 7-nerved. Upper glune 3.0mm long, 1.0mm broad, chartaceous, boat shaped at compressed view, oblong, lanceolate when spread, keeled. Lower lemma 1.5 mm long, 0.8mm broad, hyaline, delicate, obovate, acute, empty, epaleate. Upper lemma 2.0mm long, 1.0mm broad, hyaline, delicate, bisexual, epaleate, awned; awn 15mm long; column 6.0mm long. Stamens 3; anther 2.0mm long, yellow, linear, glabrous; filament 0.2 mm long, slender, glabrous, shorter than anther. Lodicules 2, 0.2mm long 0.1mm broad, obovate, 2-horned at apex. Pistil 3.0 mm long; ovary 0.5 mm long, oblong; style 1.0mm long, slender, glabrous, bifid from base; stigma 1.5mm long, hairy; hairs acute base broad. Caryopsis 1.0mm long, 0.5mm broad, brownish. Pedicelled spikelets 4.0 mm long, elliptic, lanceolate, unawned; pedicels 2.5mm long. Lower glume 3.0 mm long, 1.0mm broad, chartaceous, oblong, lanceolate, margin long ciliate,13- nerved. Upper glune 3.0 mm long, 1.0mm broad, chartaceous, lanceolate, margin hyaline, ciliate, in upper half. Lower lemma 3.0mm long,1.0mm broad, hyaline, delicate, ovate, oblong.

Flowering & Fruiting

May–January; Naugarh, Sonbhadra, Sonaghati, Mirzapur.

Habitat

Very common along rocky places.

Distribution

India, Asia.

7. *Chionachne* R. Br.

Chionachne koenigii (Spreng.) Thw. Enum. Pl. Zeyl. 357. 1864; Fischer in Gamble, Fl. Pres. Madr. 1706. 1934 (Repr.ed. 3: 1183. 1957); Bor, Grass. Bur. Cey, Ind. Pak. 262. 1960; Roy, Grass. M.P. 49. 1984; Shreekumar & Nair, Fl. Ker. Grass. 205. t. 43. 1991; Moulik, Grass. Bamb. Ind. 1: 172.1997.*C. barbata* (Roxb.) R. Br in Benn., Pl.Jav. Rar. 18: 1838. *Polytoca barbata* (Roxb.) Stapf ex Hook. f. Fl. Brit. Ind. 7: 102.1896; Rang. & Tad., Handb. S. Ind. Grass. 144. t.127. 1921. *Coix koenigii* Spreng. Syst. Veg. 1: 239.1825. *C. barbata* Roxb. Fl. Ind. 1: 569. 1820. *C.arundinacea* Koen ex Willd. Sp. Pl. 4: 203. 1805. *C.crypsoides* C. Mullein Bot. Zeit. 19: 334. 1861. (PLATE–163).

Perennial. Rhizome absent; root fibrous. Culm 45cm tall, 2.0 mm thick, green, glabrous, erect, tufted, robust, decumbent, stout, leafy, simple, rooting at base; nodes glabrous; internodes 8.0 cm long. Leaf–blade 8.0 cm long, 5.0mm broad, green, glabrous, linear, lanceolate, acuminate, base rounded, flat, narrow, margin razor shape, mid-vein distinct on lower surface, 3-lateral veins on each side of mid-vein; lateral viens distinct on upper surface; leaf–sheath 3.0 cm long, 4.0 mm broad, green, glabrous, sub–coriaceous, margin sub membranous, erect, striate, loose, compressed, very much shorter than internodes, veins distinct on outer surface, hairy at mouth; hairs 5.0 mm long, acute, base tubercled; ligule 0.5 mm long, membraneous, truncate; auricle glabrous. Inflorescences consists of raceme; raceme subtended by leaf like spathe, the individual racemes subtended by spatheoles; spathes 2.0 cm long, herbaceous, glabrous. Spikelets monoecious, ovate, elliptic, caudate, acuminate. Male spikelets 10 mm long, linear, lanceolate, elliptic, acuminate, glabrous. Lower glume 10 mm long, 2.0 mm broad, membranous, ovate, acuminate, linear–lanceolate, pectinate, glabrous, margin slightly winged from middle to apex, keels pilose, 11-nerved, nerves green, visible, glabrous, ending at apex. Upper glume 9.5 mm long, 1.5 mm broad, membranous, linear–lanceolate, acuminate, glabrous, 7–nerved; nerves green, visible, glabrous, ending at apex. Lower lemma 5.0mm long, 1.5 mm broad, membranous, elliptic, lanceolate, 2-lobed, apex muticous, surface glabrous, margin scabrid & hairy; hairs 1.0 mm long, acute, base tubercled, empty, paleate, 7- nerved ; nerves green, visible, glabrous, ending below at the apex. Lower palea 4.5mm long, 1.0mm broad, hyaline, delicate, 2-keeled, 2-nerved; nerves green, slightly visible, glabrous, & marginal, ending at the apex. Upper lemma 5.0 mm long, 1.5mm broad, hyaline, delicate, ovate lanceolate, glabrous, bisexual, paleate, 3-nerved; nerves slightly green, faintly visible, glabrous, ending at the apex. Upper palea 4.0mm long, 1.0mm broad, hyaline, linear, glabrous, 2- nerved; nerves green, slightly visible, glabrous & marginal, ending at apex. Stamens 3; anther 6.0mm long, yellow, linear, glabrous; filament 0.5 mm long, slender, glabrous, shorter than anther. Lodicules 2, 0.2 mm long, 0.1 mm broad, hyaline, glabrous, cuneate. Female spikelets 6.0 mm long, sessile, awned, glabrous, surrounded by male spikelets. Lower glume 6.0 mm long, 2.0 mm broad, membranous, ovate, acuminate, linear–lanceolate, pectinate, glabrous, margin slightly winged from middle to apex, keels pilose, 11-nerved, nerves green, visible, glabrous, ending at apex. Upper glume 5.5 mm long, 1.5 mm broad, membranous, linear–lanceolate, acuminate, glabrous, 7–nerved; nerves green, visible, glabrous, ending at apex. Upper lemma 4.0 mm long, 1.5 mm broad, membranous, ovate, linear, glabrous, margin hyaline, 1- nerved; nerves green, glabrous, visible, ending at apex, awned;

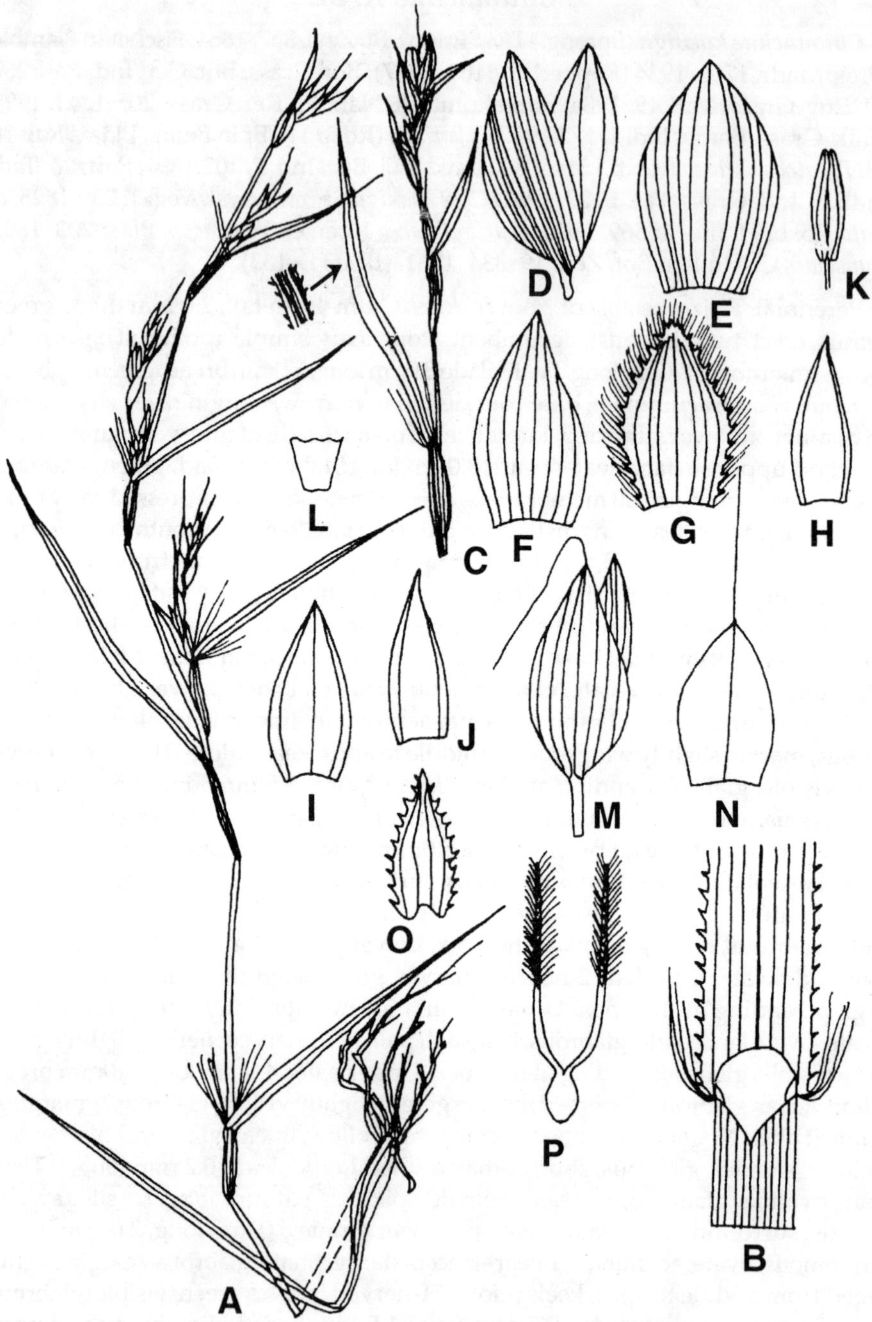

PLATE 163: ***Chionachne koenigii*** **(Spreng.) Thw.**
A. Habit, B. Ligule, C. Raceme, D. Male spikelet, E. Lower glume, F. Upper glume, G. Lower lemma, H. Lower palea, I. Upper lemma, J. Upper palea, K. Stamen, L. Lodicule, M. Female spikelet, N. Lemma. O. Palea, P. Pistil

awn 8.0 mm long, glabrous. Upper palea 3.5mm long, 1.0mm broad, hyaline, surface glabrous,, margin sharply folded inwards, transversely scabrid, 2-nerved; nerves green, slightly visible, glabrous & marginal, ending at apex. Pistil 3.5mm long; ovary 0.5mm long, ovate, glabrous; style 1.0 mm long, slender, glabrous, bifid from base; stigma 2.0 mm long, hairy; hairs acute, base rounded. Lodicules absent. Caropsis 4.0 mm long, 1.0 mm broad, linear, enclosed in hardened lower glume, white, shining, glabrous, oblong, rounded above.

Flowering & Fruiting

September–June; Chaukaghat, Navalpur, Meza road, Naini, Khadda.

Habitat

Common in dry localities, along roadside.

Distribution

India, Burma, Sri–Lanka.

8. *Chrysopogon* Trin.

KEY TO THE SPECIES

1a. Rhizome present; length of the pedicels half of the sessile spikelet .... **1. *C.aciculatus***

1b. Rhizome absent; length of the pedicels less than half of the sessile spikelets:

2a. Ligule membranous; anther yellow .... **3. *C. polyphyllus***

2b. Ligule hairy; anther purple .... **2. *C. fulvus***

1. *Chrysopogon aciculatus*

Chrysopogon aciculatus (Retz.) Trin. Fund. Agrost. 188-1820; Bor, Grass. Bur. Cey. Ind. Pak. 115.1960; Hook. f. Fl. Brit. Ind. 7: 188.1896; Fischer in Gamble, Fl. Pres. Madr. 1738.1934 (Repr.ed. 3: 1205. 1957); Haines.Bot. Bih. & Or. 1035.1924. *Andropogon acicularis* Retz. ex Roem. et.Schult. Syst. Veg. 2: 812.1817. *A. javanicus* Steud. Syn. Pl. Glum.1: 396.1854. *Chrysopogon trivialis* Arn. et Nees. in Nov. Act. Nat. Cur. 19.Suppl.: 171.1843. *Rhaphis javanica* Nees ex Steud. Desn. Opusc. 396.1854. *Andropogon aciculatus* Retz. Obs. Bot. 5: 22.1789. (PLATE–164).

Pernnial. Rhizomatous; rhizome 6.0 cm long, creeping; root fibrous. Culm 35 cm tall, 5.0 mm thick, erect, ascending, green, glabrous, leafy, unbranched, rooting at base, nodes brown, glabrous; internodes 7.0 cm long. Leaf–blade 7.0 cm long, 3.0 mm broad, ovate–lanceolate, narrow, flat, oblong, acute with cordate base, green, surface glabrous, margin scabrid, hairy at mouth; hairs 1.0mm long, white, acute with broad base, mid-vein distinct on lower surface, 4–lateral veins on each side of mid–vein; lateral veins distinct on upper surface; leaf sheath 4.0 cm long, 6.0 mm broad, tightly

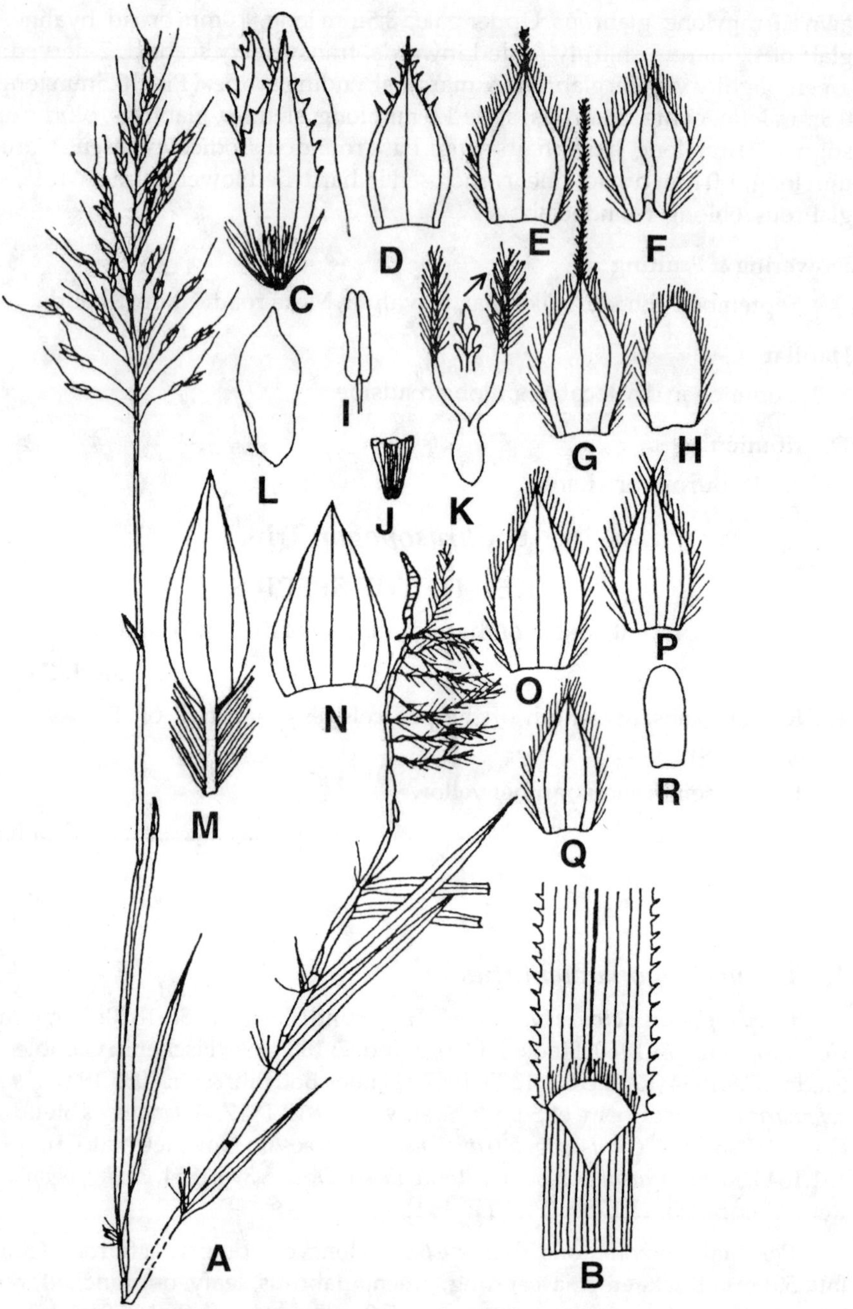

PLATE 164: ***Chrysopogon aciculatus*** **(Retz.) Trin.**
A. Habit, B. Ligule, C. Sessile spikelet, D. Lower glume, E. Upper glume, F. Lower lemma, G. Upper lemma, H. Upper Palea, I. Stamen, J. Lodicule, K. Pistil, L. Caryopsis, M. Pedicelled spikelet, N. Lower glume. O. Upper glume, P. Lower lemma, Q. Upper lemma, R. Upper palea

compressed, green, glabrous, margin slightly keeled, shorter than internodes, flabellate on creeping part, veins distinct on outer surface; ligule 0.5 mm long, a ring of white hairs; hairs acute with broad base; auricle absent. Inflorescence an exserted panicle; panicle 10 cm long, pyramidal, oblong, ovate, purplish dark brown. Branches 3.0 cm long, suberect,in several whorls. Spikelet in group of 3, central one sessile; laterals pedicelled. Sessile spikelet 3.5mm long, 1.5mm broad, lanceolate, awned, bisexual, oblique, acute, callus very sharp with golden brown hairs; hairs 2.0 mm long, acute, base broad. Lower glume 3.0mm long, 1.5mm broad, membranous, ovate, lanceolate, acute, brown, upper part hirsutate, bidentate, keeled; keel pectinate, glabrous, margin folded inwards, 3-nerved; green, visible, ending at apex. Upper glume 3.5 mm long, 1.5 mm broad, membranous, boat–shaped, brown, lanceolate, acute, surface hairy; hairs minute, acute, base broad, margin folded inwards, hyaline, keeled; keel glabrous, 3- nerved; nerves light green, faintly visible, glabrous, ending at apex, awned; awn3.0 mm long, antrorsely scabrid. Lower lemma 3.0 mm long, 1.5mm broad, hyaline, delicate, 2-keeled, ovate, lanceolate, acute, empty, epalate, glabrous, margin slightly folded inwards, 2-nerved; nerves green, visible, marginal & glabrous, ending at apex. Upper lemma 3.0 mm long, 1.0 mm broad, hyaline, delicate, linear, lanceolate, acute, bisexual, paleate, glabrous, 1–nerved; nerves green, visible, glabrous, ending at apex, awned; awn 5.0 mm. long, antrorsely scabrid, margin hairy; hairs 1.0 mm. long, acute, base tubercled. Upper palea 1.0 mm.long, 0.5mm. broad, hyaline, delicate, oblong, ovate, nerveless, hairy; hairs 0.5 mm long, acute, base tubercled. Stamens 3; anther 1.0 mm long, orange–yellow, oblong, lanceolate, glabrous; filament 0.5 mm long, glabrous, slender, shorter than anther. Lodicules 2, 0.5 mm long, 0.2 mm broad, obovate, many nerved, glabrous, hyaline, wavy at apex. Pistil 2.0 mm long; ovary 0.5 mm long, ovate, oblong, glabrous; style 1.0 mm long, slender, bifid from base, glabrous; stigma 0.5mm long, brown, hairy; hairs acute, base broad in 2–rows. Caryopsis 2.0 mm long, 1.0 mm broad, ovate, glabrous. Pedicelled spikelet 4.5 mm long, 1.0mm broad, linear, lanceolate, unawned, male, acute; pedicels 1.5 mm long. Lower glume 3.0 mm long, 1.0 mm broad, linear, lanceolate, membranous, acute, slightly brown, glabrous, 5 nerved; nerves green, slightly visible, glabrous, ending at apex. Upper glume 3.0 mm long, 1.0 mm broad, membranous, linear-lanceolate, acute, brown, glabrous, margin hairy; hairs 1.0mm long, acute, base tubercled. Lower lemma 3.0 mm long, 1.0 mm broad, delicate, hyaline, acute, empty, epaleate, surface glabrous, margin hairy; hairs 1.0mm long, acute, base tubercled, 5 nerved; nerves faint, colourless, glabrous, ending at apex. Upper lamma 3.0 mm long, 1.0 mm broad, hyaline, delicate, male, awnless, mucronate, ovate, lanceolate, margin hairy; hairs 1.0 mm long, acute, base broad, 3-nerved, nerves green, slightly visible, glabrous, reading at apex. Upper palea 1.0 mm.long, 0.5mm. broad, hyaline, delicate, oblong, ovate, nerveless, hairy; hairs 0.5 mm long, acute, base tubercled. Stamens 3; anther 2.5 mm long, orange-yellow, glabrous, linear, oblong; filament0.5 mm long, slender, shorther than anther. Lodicules 2, 0.5 mm long, 0.2 mm broad, obovate, many nerved, glabrous, hyaline, wavy at apex.

Flowering & Fruiting

July–November; Babatpur, Sindhora, Barahikala, Chandauli, Jaunpur, Deoria.

Habitat

Common along road-sides & moist localities.

Distribution

Widely distributed in tropics of Asia, Polynesia, & Australia at low elevations.

2. *C. fulvus*

C. fulvus (Spring) Chiv. Fl. Somala 1: 327.1929; Bor, Grass. Bur. Cey. Ind. Pak. 116. 1960; Moulik. Grass. Bamb. Ind. 1: 240.1997. *Pollinia fulva* Spreng. Pugill. 2: 10.1815. *Andropogon montanus* Koen ex Trin. in Spring. Neue Entdesk 2: 93.1821. *A. monticola* Roem. et Schult. Syst. Veg 2: 665.1827; Hook. f. Fl. Fl. Brit. Ind. 7: 129. 1896. *A. sprengelii* Kunth Rev. Gram. 1: 166.1829. *Chrysopogon montatus* Trin. ex Spreng. Neue Entdeck. 2: 93.1821. *C. monticola* (Roem et schult) Hains in Ind. For. 40: 95.1914. (PLATE–165).

Perennial. Rhizome absent; root fibrous. Culms 50 cm tall, 2.0 mm thick, densely, tufted, erect, ascending, decumbent, green, glabrous, branching at base, nodes glabrous; internodes 10 cm long. Leaf blade 4.0 cm long, 5.0 mm broad, usually basal, acuminate, narrow, surface glabrous, margin scaberulous, linear-lanceolite, oblong, green, mid-vein distinct on lower surface, 3–lateral veins on each side of mid-vein; lateral vein distinct on upper surface; leaf–sheath 6.5 cm long, 3.0 mm broad, tightly compressed, green, surface glabrous, margin keeled, membranous, sub–terete, shorter than internodes, vein distinct on outer surface; ligule 1.0 mm long, a ring of hairs; hairs white, acute, base broad; auricle absent. Inflorescence panicle; panicle 7.0 cm long, narrowly oblong, pyramidal, purplish, open, contracted, erect. Branches 2.5 cm long, filiform, sub–erect, ascending, sharply bearing few spikelet, arranged in whorls. Spikelets in group of three; central one sessile, laterals two pedicelled. Sessile spikelet 3.5 mm long, 1.5 mm broad, dorsally compressed, acuminate, bisexual, lanceolate; callus 0.3 mm long, 0.1 mm broad, densely covered with red hairs; hairs acute, base broad. Lower glume 3.0 mm long, 1.0 mm broad, coriaceous, narrowly oblong, lanceolate, linear, apex tetradendate; central two very small, marginal longer than central, surface glabrous, purplish, margin keeled, spinulous, winged; wing hyaline, glabrous, 3-nerved; nerves very faint. Upper glume 3.0 mm long, 1.5 mm broad, coriaceous, bifid at apex, oblong-lanceolate, surface glabrous, margin hyaline, hairy at apex; hairs acute, base broad, compressed, faintly, 3-nerved; central nerves dentate from apex to middle; denta pectinate, lateral nerves hairy at apex; hairs acute, base tubercle; awned; awn 2.5mm long, antrorsely scabrid. Lower lemma 2.5 mm long, 1.0mm broad, narrow, hyaline, lanceolate, oblong, epaleate, surface glabrous, apex tridendate, empty, margin, inrolled, 2–nerved; nerves faint, hairy; hairs acute, base broad. Upper lemma 3.0 mm long, 1.0 mm broad, narrow, hyaline, acuminate, linear-lanceolate, oblong, epaleate, surface & margin glabrous, awned; awn 5.0 cm long, antrorsely scabrid. Stamens 3; anther 2.0 mm long, yellow, ovate, oblong, linear, glabrous; filament 0.5 mm long, glabrous, slender, shorter than anther. Lodicules 2, 0.4 mm long, 0.2 mm broad, obovate, glabrous, many nerved; nerves faint. Pistil 3.0 mm long; ovary 0.5 mm long, oblong, glabrous; style 1.0 mm long, bifid at base, glabrous; stigma 1.0 mm long, brown, hairy; hairs acute, base broad, in 2–rows. Caryopsis 2.0 mm long, 1.0 mm broad. Pedicelled spikelet 7.0 mm long, 1.5 mm broad,

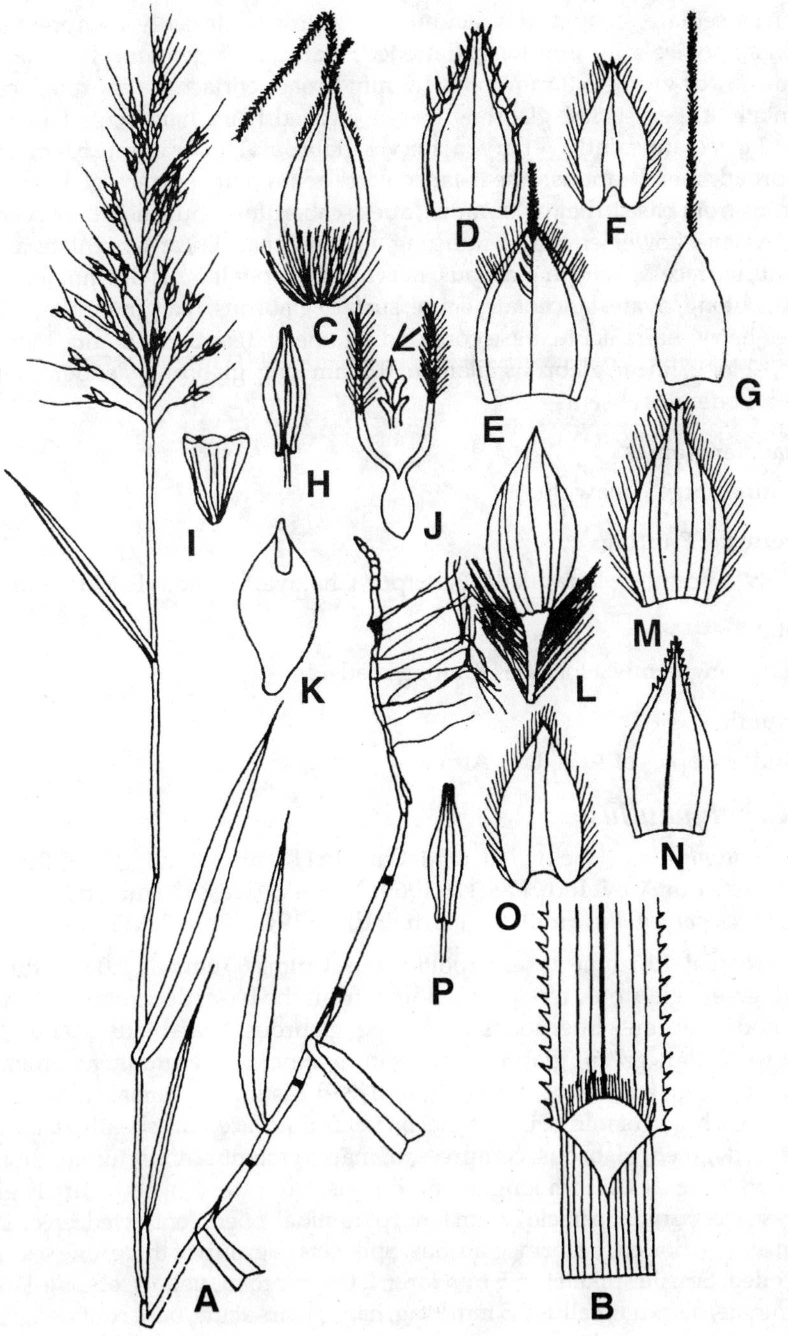

PLATE 165: ***Chrysopogon fulvus*** **(Spreng.) Choiv.**
A. Habit, B. Ligule, C. Sessile spikelet, D. Lower glume, E. Upper glume, F. Lower lemma, G. Upper lemma, H. Stamen, I. Lodicule, J. Pistil, K. Caryopsis, L. Pedicelled spikelet, M. Lower glume, N. Upper glume, O. Lower lemma, P. Stamen

linear-lanceolate, cuspidately acuminate, awnless, dorsally compressed, male, purplish; pedicels 2.0 mm long, flattened, glabrous, base hairy; hairs acute, base broad. Lower glume 5.0 mm long, 1.0 mm broad, coriaceous, purple, acuminate, tridentate, at apex, surface glabrous, margin inrolled, hairy; hairs acute, base tubercled, linear-lanceolate, faintly 7- nerved; nerves glabrous. Upper glume 5.0 mm long, 1.5 mm broad, membranous, linear-lanceolate, acuminate, surface glabrous, margin glabrous from base to below the apex; apex scaberulous, purplish, 3–nerved; nerves green, visible. Lower lemma 3.5 mm long, 1.0 mm broad, linear-lanceolate, acuminate, hyaline, surface & margin glabrous, nerveless. Upper lemma 3.0 mm long, 1.5 mm broad, oblong, ovate, lanceolate, acute, surface glabrous, margin inrolled, 2-nerved, nerves hairy; hairs acute, base rounded. Stamens 3; anther 2.0 mm long, yellow, ovate, oblong, linear, glabrous; filament 0.5 mm long, glabrous, slender, shorter than anther. Lodicules absent.

Vernacular Names

Goria, Jarga, Dhawalu.

Flowering & Fruiting

July–December; Barahikala, Babatpur, Chakiya, Chandauli, Mirzapur, Mau.

Habitat

Growing in moist localties & along roadside.

Distribution

India, tropics of Asia, East Africa.

3. *C. polyphyllus*

C. polyphyllus (Hack) Blatt. & McCann In J Bombay Nat.Hist Soc. 32: 416.1928; Bor, Grass. Bur. Cey. Ind. Pak.118.1960; Moulik, Grass. Bamb Ind. 1: 242.1997. *Andropogon polyphyllus* Hook. f. Fl. Brit. India 7: 194.1896. (PLATE–166).

Perennial. Rhizome absent; root fibrous. Culms 60 cm tall, 2.0 mm thick, erect, tufted, green, glabrous, terete, ascending, from the base, decumbent, stout, leaf at base, node glabrous; internodes 10cm long, glabrous. Leaf–blade 3.5 mm long, 5.0 mm broad, flat, green, glabrous, margin hyaline, scaberulous, acuminate, base rounded, linear, lanceolate, crowded, mid-vein distinct on dorsal surface, 3–lateral veins on each side of mid vein, distinct on ventral surface; leaf-sheath 6.0 cm long, 2.5 mm broad, green, glabrous, compressed, margin membranous, terete, shorter than inter nodes; ligule 0.5 mm long, membranous, truncate, notched, auricle glabrous. Inflorescence panicle; panicle 7.0 cm long, pyramidal, open, contracted, erect. Branches 2.0 cm long, whorled, filiform, glabrous. Spikelets in group of three; one sessile & two pedicelled. Sessile spikelets 5.5 mm long, 2.0 mm broad, yellow, oblong-lanceolate, accuminate, bisexual, callus 0.5 mm long, hairy; hairs acute, base rounded, rusty red. Lowerglume 3.5 mm long, 1.0 mm broad, membranous, ovate, lanceolate, yellow, surface glabrous, margin inrolled, glabrous, strongly compressed, 2 nerved; nerves green, visible, dentate, from apex to middle; denta pectinate, bihaired at apex; hairs acute, base broad. Upper glume 4.0 mm long, 1.0 mm broad, membranous,ovate–

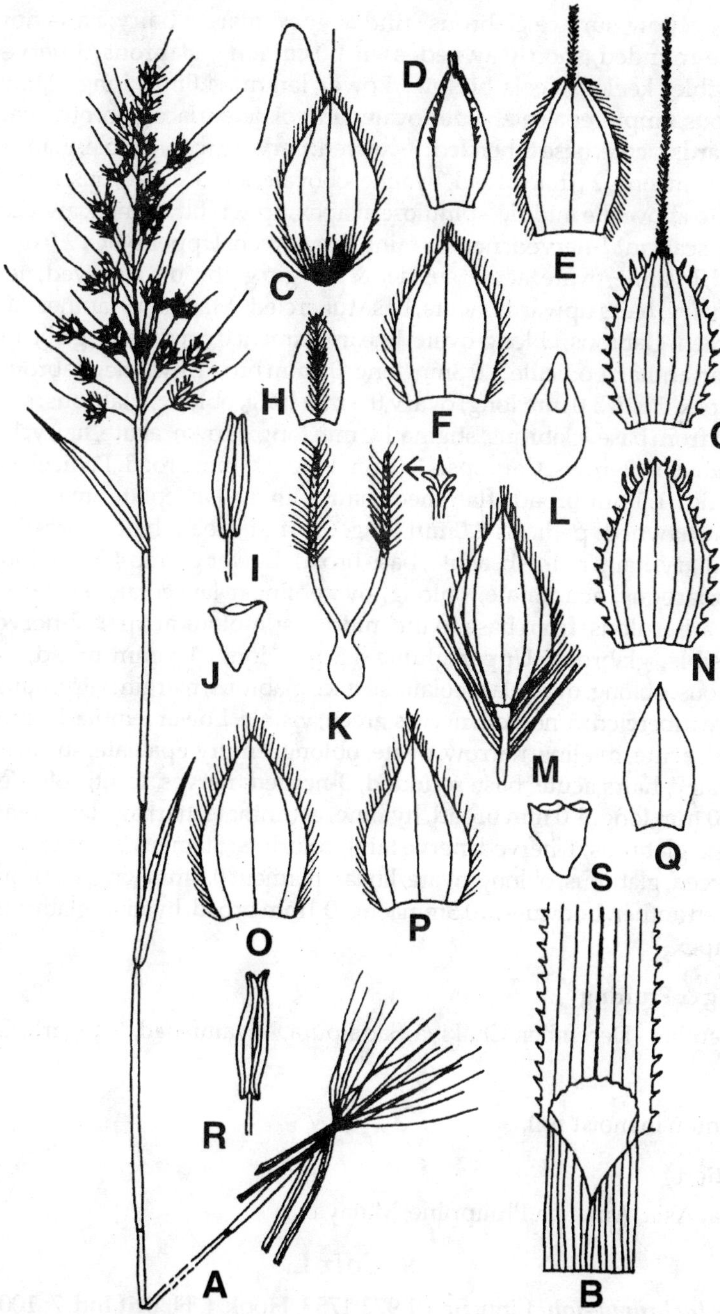

PLATE 166: ***Chrysopogon polyphyllus*** **(Hack.) Blatt. & Mc Cann**
A. Habit, B. Ligule, C. Sessile spikelet, D. Lower glume, E. Upper glume, F. Lower lemma, G. Upper lemma, H.Upper palea, I. Stamen, J. Lodicule, K. Pistil, L. Caryopsis, M. Pedicelled spikelet, N. Lower glume, O. Upper glume, P. Lower lemma, Q. Upper lemma, R. Stamen, S. Lodicule

lanceolate, yellow, surface glabrous, bifid at apex, margin hairy; hairs downwards, acute, base rounded, shortly awned; awn 1.0 cm long, glabrous, 3-nerved; nerves green, visible, keeled; keels hirsute. Lower lemma 4.0mm long, 2.0mm broad, membranous, empty, epaleate, acute, ovate, lanceolate, surface glabrous, margin hairy, hairs upwards, acute, base tubercled, 3–nerved; nerves green, visible, glabrous. Upper lemma 5.0 mm long, 2.0 mm broad, hyaline, obovate, surface glabrous, margin scarbid from base to above the middle, spinulose at apex, apex bifid, awned; awn 2.5 cm long, antrorsely scabrid, 7-nerved; nerves faint, light green. Upper palea 2.0 mm long, 0.5 mm broad, hyaline, ovate, acute, surface & margin glabrous; 1-nerved; nerves faint, colorless hairy; hairs upwards, acute, base tubercled. Stamens 3; anther 2.0mm long, yellow–green, glabrous, oblong-ovate, linear; filament 0.5 mm long, glabrous, slender, shorter than anther. Lodicule 2, 0.3mm long, 0.1mm broad, hyaline, glabrous, obovate, wavy at apex. Pistil 3.0 mm long; ovary 0.5 mm long, oblong, glabrous; style 1.5 mm long, bifid from base, glabrous; stigma 1.0 mm long, brown, acute, hairy; hairs acute, base rounded, in 2–rows. Caryopsis 2.0 mm long, 1.0 mm broad. Pedicelled spikelets 4.5mm long, 1.5mm broad, flattened, hairy, linear-lanceolate, male, acuminate, glabrous, unawned; pedicels 2.0mm long, 0.2 mm broad, half of sessile spikelets, hairy capillary, hairs reddish, acute, base broad. Lower glume 4.5 mm long, 1.5 mm broad, coriaceous, acuminate, oblong, ovate, linear-lanceolate, surface glabrous, margin scarberulous, from base to the middle, spinolous at apex, 7-nerved, nerves green, visible, glabrous. Upper glume 4.5 mm long, 1.5 mm broad, acuminate, membranous, oblong, ovate, lanceolate, surface glabrous, margin hairy; hairs upwards acute, base tubercled; 3-nerved; nerves green, visible. Lower lemma 3.5 mm long, 1.0 mm broad, acute, hyaline, narrow, ovate, oblong, empty, epaleate, surface glabrous, margin hairy; hairs acute, base rounded, 3-nerved; nerves faint, colouless. Upper lemma 3.0 mm long, 1.0 mm broad, hyaline, acuminate, narrow, lanceolate, surface and margin glabrous, 1-nerved; nerve faint, colorless. Stamens 3; anther 2.0mm long, yellow–green, glabrous, oblong-ovate, linear; filament 0.5 mm long, glabrous, slender, shorter than anther. Lodicule 2, 0.3mm long, 0.1mm broad, hyaline, glabrous, obovate, wavy at apex.

Flowering & Fruiting

September –December; Chakiya, Sadalpura, Niyamtabad, Naugarh, Gorakhpur.

Habitat

common in moist soil.

Distribution

India, Asia, America,Philippine, Malaya.

9. *Coix* L.

Coix lachryma–jobi Linn. Sp P1 972.1753; Hook. f. Fl. Brit.Ind 7: 100 1896; Bor, Grass. Bur. Cey. Ind. Pak. 264.1960; Moulik, Grass. Bamb. Ind. 1: 175 1997; Raizada *et al.*, Indian Forest Rec.sec.2.Bot. 4: 176. 1961; Gilliland in F1. Malaya 3: 304 1971; Cope in F1.W. Pak.143: 349.1982; Cooke,F1.Bomb.2: 997.1908; Rang.&Tad.Handb.S.Ind. Grass.141.t.126.1921; Fischer in Gamble, F1. Pres. Madr. 1705.1934 (Repr.ed.3:

1182.1957); Manilal & Sivaraj, F1. Calic. 331.1982; Haines Bot. Bih. & Or. 1063.1924; Saldhana & Nicolson, F1. Hassan 718. 1976, Blatter & McCann, Bomb. Grass 3. t 1.1984. *Coix lachryma* Linn. Syst.ed. 10: 1261. 1759; P. Beauv. Agrost.137, t. 24, fig 5.1812; Roxb. F1. Ind 3: 568.1832; Grah. Cat.Bomb. P1.289.1839; Duthie, Grass. N.W. Ind 10. 1883; Fodder Grasses N. Ind. 300. 1888. *C. arundinacea* Lam. Encycl. 3: 442.1665. *C.ovata* Stockes. Bot. Mat. Med. 5: 342.1812. *C. pendula* Salisb. Prodr. 28.1796. *C. puellarum* Balans in J.deBot 5: 77.1890. *C. stigmatosa* Kock & Bauch Ind Sem. Hort. Berol. 9. 1855. *Lithagrostis lacrymajobi* Gaertn. Fruct 1: 7.t.1 fig. 10. 1788. *Coix exaltata* Jacq. Ecolg. Gram.60. t. 40 1820. (PLATE–167).

Annuals. Rhizomatous; Rhizome 6.0 cm long, erect; root fibrous. Culms 1.5 m tall, 1.0 cm thick, erect, green, polished, robust, densely tufted, glabrous, leafy, stout, rooting at lower nodes, spongy, branched; branches suberect; nodes glabrous, solid; internodes 10 cm long. Leaf–blade 50cm long, 3.5cm broad, green, flat, narrow, acuminate, base cordate, linear-lanceolate, upper surface scabrid from apex to above the base, hairy below; hairs very minute, acute, base tubercled, colourless, lower surface glabrous, margin hyaline, scaberulous, one sided wavy, mid-vein stout, slightly distinct on upper surface, distinct on ventral surface, 6-lateral nerves on both side of mid-vein; lateral veins distinct on both surface, slender; leaf–sheath 15 cm long, 1.5cm broad, erect, green, glabrous, margin hyaline, overlapped, loose, very much longer than internode, mid-vein distinct on outer surface, faint on inner surface; ligule 1.0 mm long, membranous, truncate, lacerate, ovate, narrow, apex 4–dentate; auricle glabrous. Inflorescence made of suberect spikes of racemes; racemes 6.0 cm long, axillary, nodding, drooping from long peduncle. Male spikelets 8.0 mm long, 2.0 mm broad, binate, one sessile & other pedicelled, terminal, imbricating on a slender rachis, located in distal part of inflorescence. Lower glume 8.0 mm long, 4.0 mm broad, papyraceous, ovate, elliptic, lanceolate, bifid at apex, surface glabrous, winged along the inturned margin; wing many nerved, membranous, densely scaberulous; many-nerved; nerves deep green, visible, glabrous, not reaching at apex. Upper glume 7.0 mm long, 3.5 mm broad, membranous, ovate, oblong, lanceolate, acuminate, surface glabrous, margin deep green, scabrid at apex, glabrous below, 9–nerved; nerves light green, faint, glabrous, central one ending at apex, all lateral nerves ending below the apex and meat together with side nerves. Lower lemma 7.0 mm long, 3.5 mm broad, membranous, male, ovate-oblong, lanceolate, acuminate, surface glabrous, margin hyaline & scabrid at apex, glabrous below, 5-nerved; nerves light green, glabrous, faint, reaching at apex. Lower palea 6.0 mm long, 3.0 mm broad, hyaline, oblong, lanceolate, surface glabrous, margin scaberulous at apex, glabrous below, rounded, 3- nerved; nerves green, slightly visible, glabrous, reaching at apex. Stamens 3; anther 4.0 mm long, orange–yellow, linear, glabrous; filament 1.0 mm long, glabrous, slender, shorter, than anther. Lodicules 2; 1.0 mm long, 1.5 mm broad, hyaline, glabrous, broder than length. Upper lemma 6.0 mm long, 3.0 mm broad, membranous, acute, ovate, oblong, lanceolate, surface & margin glabrous, male, 5- nerved; nerves green, visible, glabrous, reaching at tip. Upper palea 5.5mm long, 2.5 mm broad, hyaline, acute, nerveless, ovate-lanceolate, surface & margin glabrous. Stamens 3; anther 3.0 mm long, orange–yellow, linear, glabrous; filament 1.0 mm long, glabrous, slender, shorter than anther. Female spikelets 7.5 mm long, 6.5 mm broad, ovoid, 3 in number,

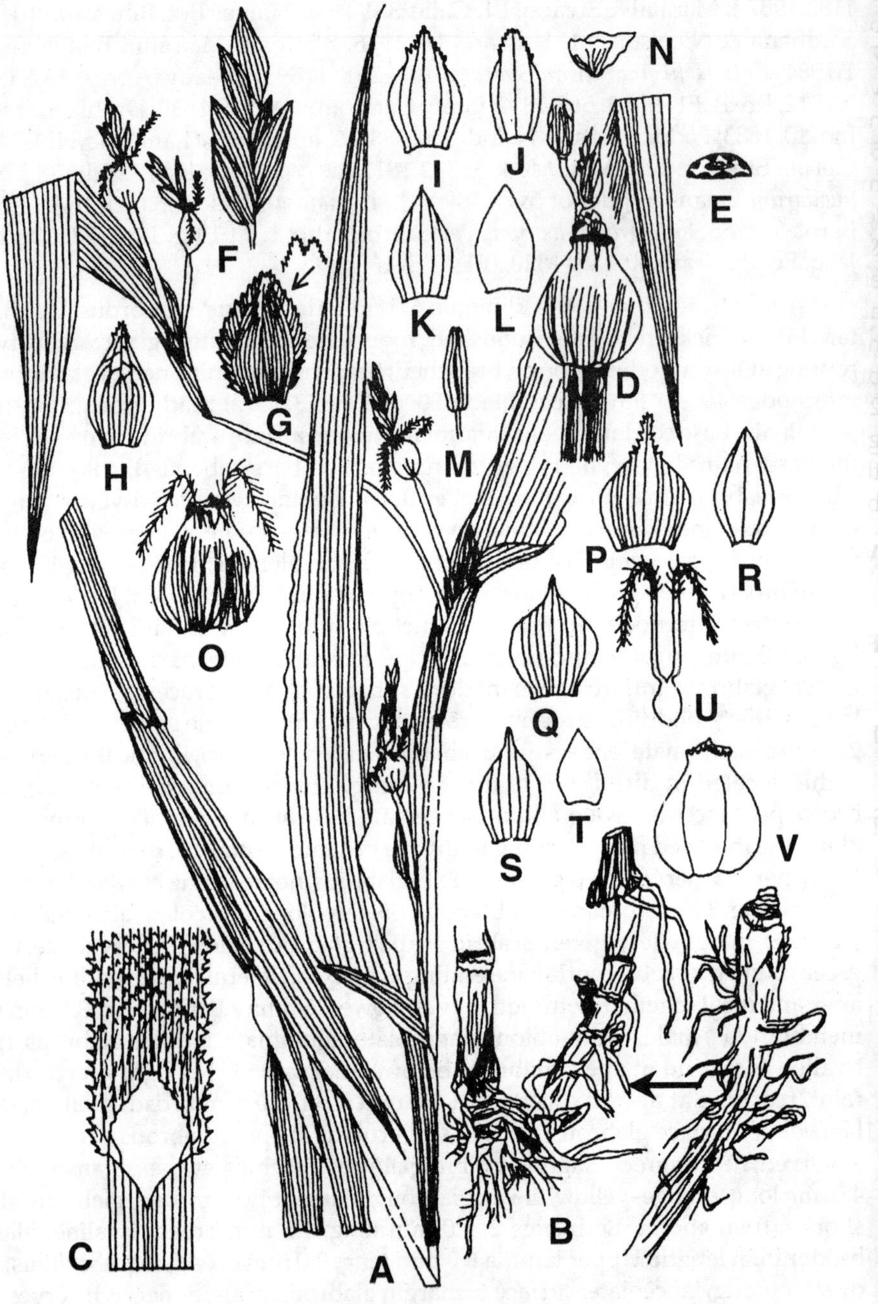

PLATE 167: ***Coix lachryma-jobi*** **L.**

A Habit, B. Rhizome, C. ligule, D. Spike, E. T.S. of axis, F. Male spikelet, G. Lowerglume, H. Upper glume, I. Lower lemma, J. Lower palea, K. Upper lemma, L. Upper palea, M. Stamen, N. Lodicule, O. Female spikelet, P. Lower glume, Q. Upper glume, R. Lower lemma, S. Upper lemma, T. Upper palea, U. Pistil, V. Caryopsis

1–fertile & 2- sterile, entirely surrounded by the hardened bract forming a bead–like bony involucre, bluish–grey at young, yellowish-white when mature. Lower glume 7.0 mm long, 6.0 mm broad, papyraceous, ovate, oblong, acute, dorsally concave, ventrally convex, surface glabrous, margin scabrid at apex, glabrous below, 11- nerved; nerve light green, very faint, glabrous, reaching at apex. Upper glume 6.5 mm long, 5.5 mm broad, membranous, ovate, oblong, acuminate, attenuate at apex, surface & margin glabrous, 9- nerved; nerves faint, colourless, glabrous, ending at apex. Lower lemma 6.0 mm long, 3.0 mm broad, membranous, ovate, lanceolate, acuminate, surface & margin glabrous, empty, epaleate, 3–nerved; nerves faint, colourless, glabrous, ending at apex. Upper lemma 6.0 mm long, 3.0 mm broad, membranous, ovate, oblong, acuminate, attenuate at apex, surface & margin glabrous, paleate, female, 5-nerved; nerves faint, colourless, glabrous, ending at apex. Upper palea 3.0 mm long, 1.5 mm broad, hyaline, ovate-lanceolate, surface & margin glabrous, acute nerveless. Pistil 2.3 cm long; ovary 2.0 mm long, oblong, glabrous, colourless; style 6.0 mm long, glabrous, slender, bifid from base; stigma 1.5 cm long, purple, extend outwards through involucre, hairy; hairs acute, base tubercled. Caryopsis 9.0 mm long, 7.0 mm broad, ovoid, sub globose, slaty-black.

Vernacular Names

Gorlu, Sankru, Vaijanti.

Flowering & Fruiting

September- December; Jansa, Lohta, Naini, Chunar, Ballia, Khadda.

Habitat

Found on wet grounds or in marshy localities.

Distributation

India, Sri Lanka, Africa, South America.

10. *Cymbopogon* Spreng

KEY TO THE SPECIES

1a. Leaf -blade filiform; ligule notched; hairs of pedicels & joints conceal the sessile spikelets

.... 3. *C. schoenanthus*

1b.

Leaf-blade flat; ligule truncate; hairs of pedicels & joints not conceal the sessile spikelets:

2a. Leaf-blade spatheolate at base; lower glume of sessile spikelets with 2 shallow pit at the base

.... 2. *C. pendulus*

2b. Leaf-blade cordate & amplexicaul at base; lower glume of sessile spikelets grooved in lower half

.... 1. *C. martinii*

1. *Cymbopogon martinii*

Cymbopogon martinii (Roxb.) Wats. in Atkins. Gaz. N.W. Pror. Ind. 392.1882; Fischer in Gamble, Fl. Pres. Madr. 1756.1934 (Repr. ed. 3: 1217.1957); Bor, Grasses, Bur. Cey.Ind. Pak. 129.1960; Moulik, Grass. Bamb. Ind. 1: 215.1997; Soenarko in Reinwartia 9: 330.1977; Stapf in Kew Bull. 359.1906; Haines, Bot. Bih. & Or. 1046. 1924. *C. martinianus* Schult. Mant 2: 459.1824. *C. pachnodes* (Trin.) Wats in Atkins, Gaz.N.W. Pror. Ind. 392.1882. *Andropogon martinii* Roxb. Fl. Ind. 1: 280.1820. *A. pachnodes* Trin. in Mem. Acad. Petersb. ser. 6, 2: 234. 1833; Spec. Gram. in Mem. Acad. Petersb. tab. 327.1836. *A. calamus* var. *aromaticus* Royte, Essay, Antiq. Hind. Med.33.1827. *A. nardoides* Nees. Fl. Afr. Austr. 116.1841. *A. schoenanthus* var. *genuinus* Hack. Monogr. Androp. 609.1889. *A. schoenanthus* var. *martinii* Hook. f. Fl. Brit. Ind. 7: 204.1896. (PLATE–168).

Perennial. Rhizomatous ; rhizome 5.0 cm long, root fibrous. Culm 1.2 m tall, 6.0 mm thick, tufted, erect, decumbent, green, glabrous, leafy at base, unbranched, rooting at base, compressed; nodes brown, glabrous; internodes 10 cm long. Leaf–blade 20 cm long, 1.5 cm broad, green, linear-lanceolate, acuminate, semi–amplexicaul at base, glaucous, flat, narrow, surface rough upwards, glabrous below; margin hyaline, scabrid, mid-vein thick, distinct on upper surface, 6- lateral veins on each side of mid-vein; lateral vein slender, distinct on lower surface; leaf-sheath 12 cm long, 8.0 mm broad, green, glabrous, erect, terete, tightly compressed, membranous, margin hyaline, longer than internodes, mid-vein thick; ligule 1.5 mm long, membranous, ovate, acute, lacerate, emerginate; auricle glabrous. Inflourescence an open panicle; panicle 35 cm long, pyramidal, compound. Spathe 3.0 cm long. Spatheole 2.0 cm long, enclosing 3-6 florets. Spikelets 1.0 cm long, green, glabrous, lanceolate, ovate, paired, one sessile & one pedicelled. Sessile spikelet 6.0 mm long, 2.0 mm broad, elliptic, ovate, oblong, bisexual, awned; callus hairy; hairs white, acute, base broad. Lower glume 5.0 mm long, 2.0 mm broad, membranous, ovate, oblong, obtuse, lanceolate, center becoming oblanceolate, upper surface glabrous, slit–like, grooved on the back, margin folded inwards, keeled, narrowly winged in upper half, 4- nerved; nerves green, visible, glabrous, ending at apex. Upper glume 6.0 mm long, 3.0 mm broad, membranous, ovate, lanceolate, acute, nerveless, keeled, winged on upper half; wings glabrous, margin hyaline, hairy, serrulate; hairs 1.0 mm long, acute, base rounded. Lower lemma 4.0 mm long, 2.0 mm broad, hyaline, delicate, ovate, oblong, lanceolate, obtuse, empty, epaleate, glabrous, margin folded inwards, hairy; hairs 1.0 mm long, acute, base tubercled, 1-nerved; nerves green, visible, glabrous, ending at apex. Upper lemma 4.0 mm long, 2.0 mm broad, hyaline, delicate, ovate, oblong, lanceolate, bisexual, surface glabrous, margin hairy; hairs 1.0 mm long, acute,base tubercled; 3–nerved; nerves green, slightly visible, glabrous, ending at apex; awned; awn 1.0 cm long, antrorsely scabid, arising from column. Upper palea 4.0 mm long, 2.0 mm broad, hyaline, elliptic, oblong, lanceolate, nerveless, glabrous. Stamens 3; anther 2.0 mm long, yellow-brown, oblong, lanceolate, glabrous; filament 0.5 mm long, glabrous, slender, shorter than anther. Pistil 3.0 mm long; ovary 0.5 mm long, ovate, oblong, glabrous; style 1.0 mm long, flat, slender, glabrous, bifid from the base; stigma 1.5 mm long, brown, hairy; hairs acute, base broad in 2 rows. Caryopsis 2.0 mm long, 1.0 mm broad, ovate, glabrous. Pedicelled spikelet 5.0 mm long, 2.0 mm broad, oblong, lanceolate, acute,

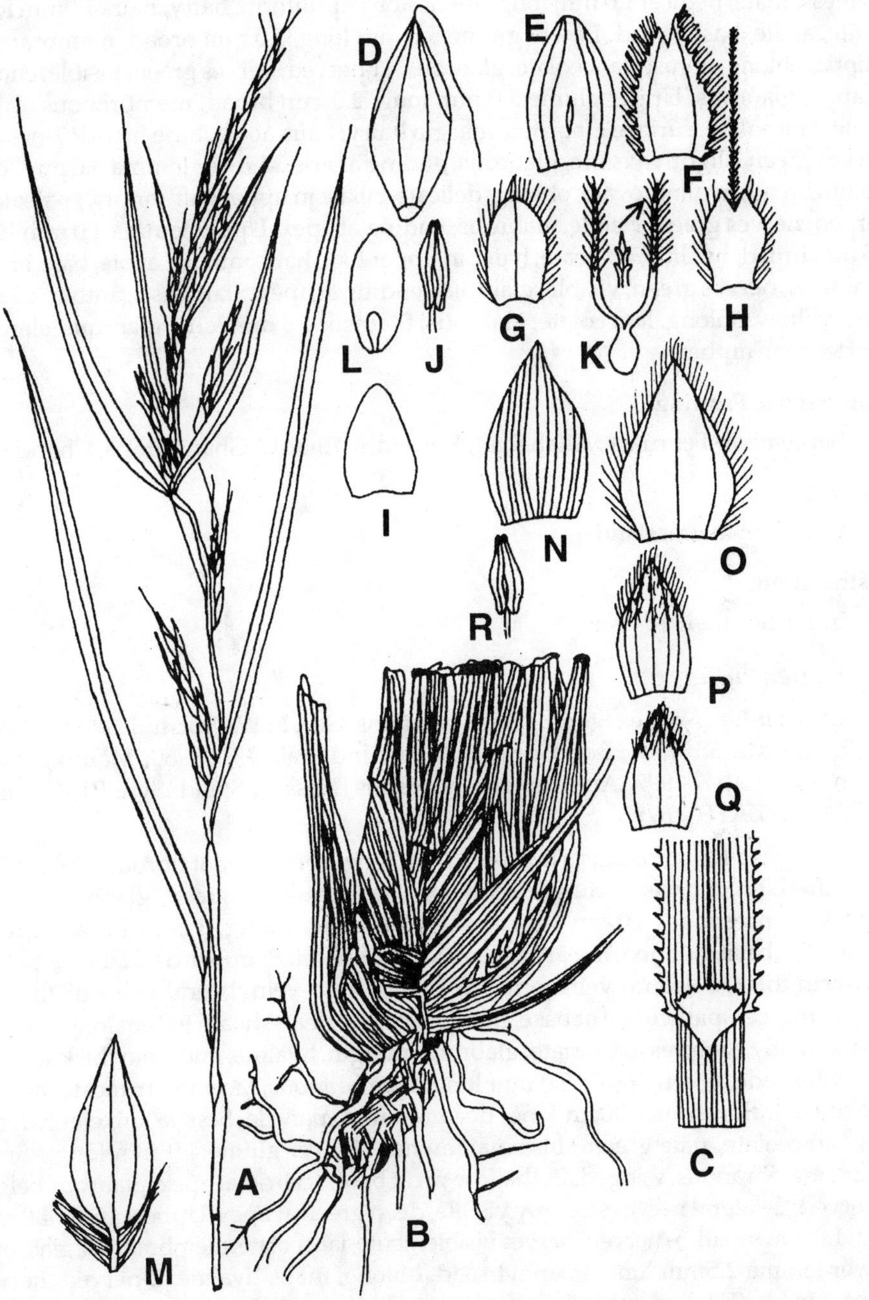

PLATE 168: *Cymbopogon martinii* **(Roxb.) Wats.**
A. Habit, B. Rhizome, C. Ligule, D. Sessile spikelet, E. Lower glume, F. Upper glume, G. Lower lemma, H. Upper lemma, I. Upper palea, J. Stamen, K. Pistil, L. Caryopsis, M. Pedicelled spikelet, N. Lower glume, O. Upper glume, P. Lower lemma, Q. Upper lemma, R. Stamen

awnless, male; pedicel 1.0 mm long, abscission cupuliform, hairy; hairs 1.0mm long, white, acute, base broad. Lower glume 5.0 mm long, 2.0 mm broad, membranous, elliptic, oblong, obtuse, lanceolate, glabrous, 11 nerved; nerves green, visible, ending at apex, glabrous. Upper glume 5.0 mm long, 2.0 mm broad, membranous, ovate, acute, lanceolate, surface glabrous, margin hairy; hairs acute, base broad; 3-nerved; nerves green, slightly visible, glabrous, ending at apex. Lower lemma 5.0 mm long, 2.0mm broad, hyaline, ovate, oblong, delicate, ciliate in upper half, empty, epaleate, 5-nerved; nerves green, visible, glabrous, ending at apex. Upper lemma 4.0 mm long, 1.5 mm broad, hyaline, epaleate, hairy in upper half; hairs minute, acute, base broad, 3- nerved; nerves green, visible, glabrous, ending at apex. Stamens 3; anther 2.5 mm long, yellow, oblong, lanceolate, glabrous; filament 0.5 mm long, glabrous, slender, shorter than anther.

Flowering & Fruiting

September- February; Allahabad, Karondhi, Chauka Ghat, Tarana, Chandauli.

Habitat

Common in grasslands.

Distribution

India, South–East Asia.

2. *C. pendulus*

C. pendulus (Nees ex Steud.) Waits in Atkins. Gaz. N.W. Pror. Ind. 392.1882; Roy in Grasses M.P.58. 1984; Bor, Grass. Bur. Cey Ind. Pak. 131. 1960; Moulik, Grass. Bamb. Ind. 1: 217. 1997. Andropogon pendulus Ness ex Steud. Syn. P1. Glum. 1: 388.1854. (PLATE–169).

Perennial. Rhizomatous; rhizome 6.0 cm long, vertical; root fibrous. Culm 1.5 m tall, tufted, erect or ascending, decumbent, leafy at base, green, glabrous, nodes glabrous, internodes 20 cm long. Leaf–blade 15cm long, 1.5cm broad, green, acuminate, base rounded, linear-lanceolate, surface rough upwards, glabrous below, mid-vein thick, 8–lateral veins on each side of mid-vein; lateral veins distinct on upper surface, spatheolate at base, margin scabrid; leaf–sheath 9.0 cm long, 7.0 mm broad, tightly compressed, striate, glabrous, margin, hyaline, mid-vein thick, shorter than internode, green; ligule 5.0 mm long, membranous, lacerate, truncate; auricle glabrous. Inflorescence 40cm long, decompound, panicle. Sessile spikelet 5.0 mm long, lanceolate, pale green, bisexual, awned. Lower glume 4.0 mm long,1.5mm broad, membranous, winged, 2–shallow pit at base, scabrid at apex, glabrous below, 4–nerved; 2–lateral nerves strongly visible, deep green at apex. Upper glume 4.0 mm long, 1.0 mm broad, 5- nerved, nerves visible, acute, lanceolate, membranous, glabrous. Lower lemma 2.5mm long, 0.5mm broad, oblong, male, hyaline, 2-nerved; nerves green, visible, 2- lobed, folded, surface hairy; hairs acute, base rounded. Stamens 3; anther 1.0 mm long, glabrous, oblong, yellow; filament 0.5 mm long. Upper lemma 4.0 mm long, 1.0 mm broad, 3-nerved ; nerves green, central one dentate, marginal hairy, hairs acute, base rounded, epaleate, membranous, lanceolate, acute, awned; awn 7.0 mm long, antrorsely scabrid. Stamens 3; anther 1.5 mm long, linear lanceolate

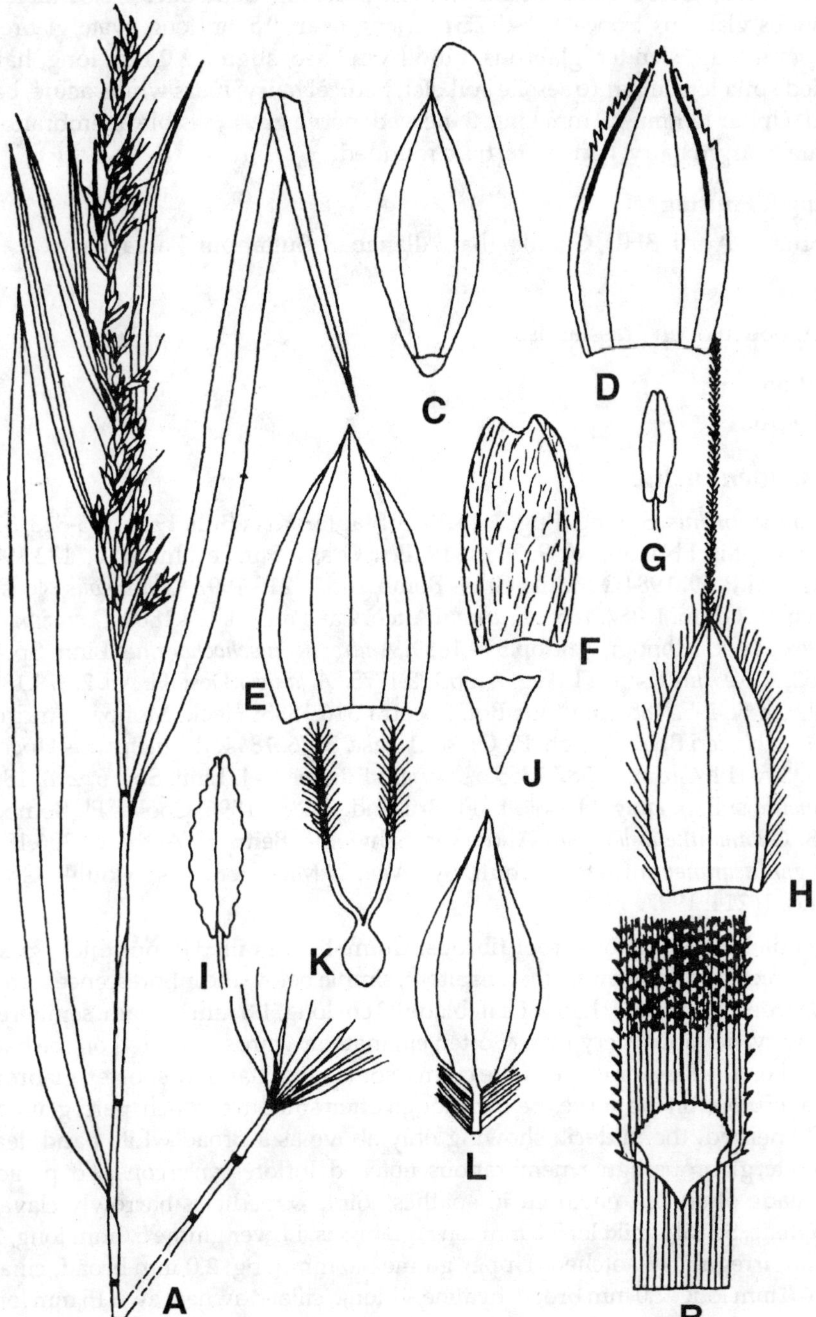

PLATE 169: *Cymbopogon pendulus* (Nees ex Steud.) Wats.
A. Habit, B. Ligule, C. Sessile spikelet, D. Lower glume, E. Upper glume, F. Lower lemma, G. Stamen of lower lemma, H. Upper lemma, I. Stamen, J. Lodicule, K. Pistil, L. Pedicelled spikelet

yellowish-green, keeled, glabrous; filament 1.0 mm long. Lodicuiles 2, 0.3 mm long, membranous, glabrous, truncate. Pistil 2.5 mm long; ovary 0.5 mm long, ovate, glabrous; style 1.0 mm long, slender, glabrous, bifid from base; stigma 1.0 mm long, hairy. Pedicelled spikelet similar to sessile spikelet; pedicel hairy; hairs white, acute, base rounded. Upper lemma 4.0 mm long, 3–nerved; nerve green, visible, membranous, lanceolate, margin hairy; hair acute, base rounded.

Flowering & Fruiting

February–April; BHU, Chaukaghat, Allahabad, Sultanpur, Ballia, Basti.

Habitat

Common in open grasslands.

Distribution

India, Asia.

3. *C. schoenanthus*

C. schoenanthus Spreng. Pug. 2: 15.1815; Stapf in Kew Bull. 1906. 303-313, 312-353.1906; Stapf.in Fl. Trop. Afr 9: 268.1919; Bor, Grass. Bur.Cey. Ind. Pak. 132 1960; Roy, Grass. M.P. 59. 1984; Moulik, Grass Bomb. Ind.1: 218.1997. *C. arabious* Nees ex Steud. Syn. Pl. Glum. 1: 387. 1855. *C. arriani* Aitch. Cat. Panj. Pl. 174.1869. *C. circinnatus* Hochst. ex Hack. Monogr. Andop.599.1889. *Andropogon schoenanthus* Linn. Sp. Pl. 1046. 1753. *A. bicornis* Forsk, Fl. Aeg–Arab.173.1775. *A. laniger* Desf. Fl. Atl. 2; 379.1800; Boiss. Fl. Or. 5: 465. 1881; Balfour Bot. Socotra 316.1888; Hack. Monogr. Androp. 598.1889. *A. olivieri* Boiss. Diagh. Pl. Or. ser.1, fasc. 5: 76. 1844. *A. circinnatus* Hochst. ex Steud. Syn. Pl. Glum. 1: 387.1855. *A. arriani* Edgew. In J. Linn. Soc. 6: 208. 1862. *A. iwaraneusa* subsp. *laniger* Hook. f. Fl. Brit. Ind. 7: 203. 1896; Cooke, Fl. Bomb. 2: 976.1908. *Gymnanthelia lanigera* Anders in Schweinf. Beitr. Fl. Aethiop. 306.1867. *Cymbopogon iwaraneusa* (Jones) Schult, Syst.Veg. 2. Mant.458.1824; Moulik, Grass. Bamb. Ind. 1: 214. 1997.

Perennial. Rhizomatous; root fibrous. Culm 45 cm tall, 3.0 mm thick, green, glabrous, erect, slender, compactly caespitose, simple below the inflorescence, terete; nodes glabrous, very rarely hairy. Leaf–blade 23 cm long, 1.0 mm in diam. semiterete, filiform, wiry, flexuous, very firm & often circinate upwards, rounded on the back, channelled on the face, or those of the culms, some what flatler & shorter, glabrous, finely scaberulous on the nerves below, though often smooth to touch, pale, glaucous evenly 7– nerved, the mid–rib showing only above as a broad white band; leaf–sheath terete, glabrous; ligule membranous, notched. Inflorescence consist of panical; panicle made of spikes, enclosed in spathes; joints & pedicles narrowly clavate, margin ciliate. Sessile spikelet 5.0 mm long, glabrous. Lower glume 5.0mm long, 2.0 mm broad, irregularly notched. Upper glume 5.0 mm long, 2.0 mm broad, ciliate. Lemma 4.0 mm long, 2.0 mm broad, hyaline, oblong, ciliate, awned; awn 15 mm long. Stamens 3; anther 2.0 mm long, pinkish, glabrous; filament 0.5 mm long, slender, glabrous, shorter than anther,. Pistil 4.0 mm long; ovary 1.0 mm long, glabrous; style 1.5 mm long, slender, glabrous, bifid from base; stigma 1.5 mm long, plumose. Pedicelled spikelet similar to sessile but male.

Flowering & Fruiting

September–November; Chaukaghat, Ghazipur, Deoria.

Habitat

Common in open fields.

Distribution

India, Asia.

11. *Dichanthium* Willemet

KEY TO THE SPECIES

1a. Racemes solitary; lower glume pitted

.... 4. *D. foveolatum*

1b. Racemes 2-9; lower glume not pitted:

2a. Ligule hairy; upper glume bifid

.... 2. *D. aristatum*

2b. Ligule membranous, upper glume not as above:

3a. Raceme green; lower glume of sessile spikelets ovate, winged to wards the apex

.... 3. *D. caricosum*

3b. Raceme purplish, lower glume of sessile spikelets boat shaped, not winged

.... 1. *D. annulatum*

1. *Dichanthium annulatum*

Dichanthium annulatum (Forssk.) Stapf. in Prain. F1 Trop. Afr. 9: 178. 1917; Moulik, Grass. Bamb. Ind. 1: 271. 1997; Bor, Grass. Bur. Cey. Ind. Pak. 133. 1960; Raizada *et al.,* Indian Forest. Rec. ser. 2. Bot. 4: 233. 1961; Rajagopal & Panigrahi, Proc. Nat. Acad. Sci. India Sec. B. 37: 9. 1967; Gilliland in F1. Malaya 3: 204. 1971; Cope in Fl. W. Pak. 143: 280. 1982; Deshpande in Fasc. F1. Ind. 15: 5 1964, Fischer in Gamble, F1. Pres. Madr. 1714. 1934 (Repred. 3: 1206. 1957); Rang. & Tad. Handb. S. Ind. Grass. 204. t. 160–161. 1921; Haines, Bot. Bih. & Or. 1039. 1924; Blatter & McCann, J. Bomb. nat. Hist. Soc. 32: 426. 1928. *Andropogon annulatus* Forssk. F1. Aeg-Arab. 173. 1775; Duthie, Fodder Grasses N. Ind. 33. t. 20. 1888; Hack. Monogr. Androp. 570. 1889; Lisboa, Bomb. Grasses. 82. 1896; Cooke, F1. Bomb. 2: 988. 1908, Boiss, F1. Or. 5: 463, 1884. *A. bladhii* Retz. Obs. 2: 27 1781; Roxb. F1. Ind. 1: 259. 1820; Duthie, Grasses N.W..Ind. 19. 1883. *A. comosus* Link. Enum. Hort. Berol. 1: 239 1827. *A. obtusus* Nees in Hook. & Arn. Bot. Beech. Voy. 243. 1838. *A. scandens* Roxb. F1. Ind. 1: 58. 1820; Grah. Cat. Bomb. P1. 238. 1839; Dalz. & Gibs. Bomb. F1. 301 1861. *A. garipensis* Stued. Syn. P1. Glum. 1: 379. 1854. *A. ischaemum* Roxb. ex Wight. cat. 1834. *Lepeoceris annulata* Nees F1. Afr. Austr. 98. 1841. (PLATE–170).

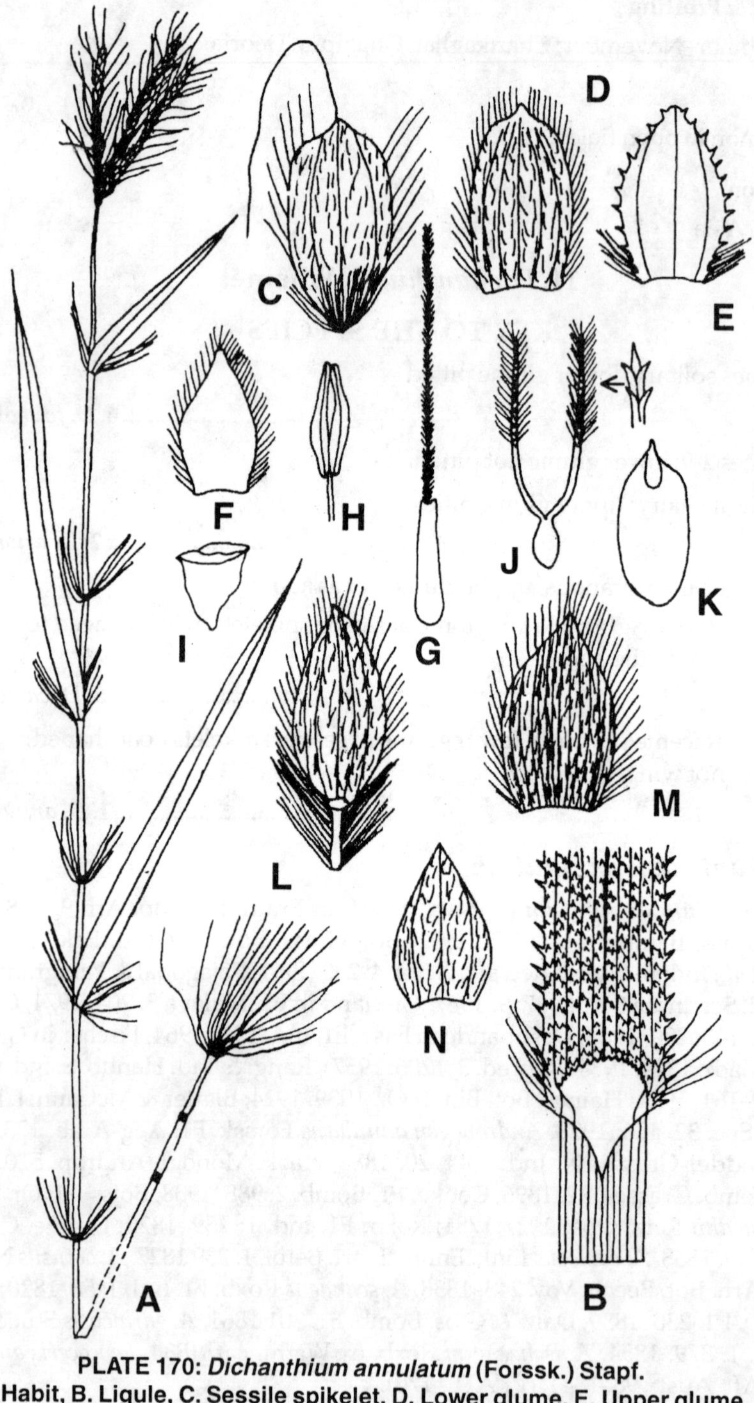

PLATE 170: ***Dichanthium annulatum*** **(Forssk.) Stapf.**
A. Habit, B. Ligule, C. Sessile spikelet, D. Lower glume, E. Upper glume, F. Lower lemma, G. Upper lemma, H. Stamen, I. Lodicule, J. Pistil. K. Caryopsis, L. Pedicelled spikelet, M. Lower glume, N. Upper glume

Perennials. Rhizomatous; rhizome 6.0 cm tall, 5.0 mm thick, green glabrous, erect, densely tufted, compressed, geniculate, ascending, leafy, branching at lower nodes, rooting at base; nodes hairy; hairs 1.0mm long, white, acute, base broad; internodes 7.0 cm long, green, under the sheath, purplish above the sheath. Leaf-blade 12 cm long, 5.0 mm broad, green, glaucous, flat, acuminate, base rounded, linear- lanceolate, surface hairy below; hairs 2.0 mm long, white, acute, base tubercled, rough above, margin scabrid, mid-vein distinct on lower surface; 4–lateral veins on each sides of mid-vein; lateral veins distinct on upper surface; leaf–sheath 5.0 cm long, 7.0 mm broad, green, glabrous, tightly compressed, membranous margin hyaline, shorter than internodes, erect, vein distinet on outer surface; ligule 1.0 mm long, membranous, oblong, ovate, obtuse, lacerate; auricle present. Inflorescence consists of racemes; raceme 5.0 cm long, 2–9 in number, sub–digitate purplish, peduncled; peduncle 4.0 mm long, glabrous. Sessile spikelet 4.0 mm long, 2.0 mm broad, elliptic, oblong, ovate, dorsally compressed, acute, closely imbricating; callus densely hairy; hairs 1.0 mm long, acute, base tubercled, white. Lower glume 4.0 long, 2.0 mm broad, membranous, purple, boat shaped, elliptic, oblong, obtuse, acuminate, margin narrowly incurved, keeled, hairy; hairs 1.0 mm long, acute, base tubercled; surface hairy; hairs minute, acute, base tubercled, 5–nerved; nerves greenish-brown, visible, glabrous, ending at apex. Upper glume 4.0 mm long, 1.5 mm broad, membranous, narrow, lanceolate, subacute, purplish, surface glabrous, margin scabrid & hairy at base; hairs 1.5 mm long, acute, with tubercled base; 3 nerved; nerves purplish–green, glabrous, reaching at apex. Lower lemma 2.5 mm long, 1.0 mm broad, hyaline, delicate, empty, epaleate, glabrous, linear, oblong, obtuse, nerveless, very narrow, keeled, margin hairy; hairs 1.0 mm long, acute, base tubercled. Upper lemma 2.0 mm long, 0.5 mm broad, reduced, stipitate, oblong, awned; awn 1.5 cm long, antrorsely scabrid; column 5.0 mm long, bisexual, epaleate. Stamens 3; anther 1.5 mm long, reddish, linear, glabrous; filament 0.5 mm long, glabrous, slender, shoter than anther. Lodicules 2, 0.6 mm long, 0.3 mm broad, hyaline, glabrous, obovate, nerveless. Pistil 2.0 mm long; ovary 0.5 mm long, oblong, glabrous; style 1.0 mm long, slender, glabrous, bifid from base; stigma 0.5 mm long, hairy; hairs acute, base broad, in 2–rows. Caryopsis 1.5 mm long, 1.0 mm broad, ovate, elliptic, plano-convex, brown. Pedicelled spikelet 4.5 mm long, 2.0 mm broad, oblong, flattened, empty, unawned, lanceolate, acute; pedicels 1.5 mm long, hairy; hairs acute, base broad. Lower glume 4.0 mm long, 2.0 mm broad, membranous, ovate-oblong, obtuse, elliptic, densely bristly, hairy especially along the margin; hairs 1.0 mm long, acute, base tubercled, 7–nerved; nerves green, visible, glabrous, ending at apex. Upper glume 4.0 mm long, 2.0 mm broad, membranous, narrow, ovate, lanceolate, obtuse, acuminate, surface glubrous, margin hairy; hairs 1.0 mm long, acute, base broad; 3- nerved; nerves green, visible, glabrous, ending at apex. Lower lemma 3.0 mm long, 1.5 mm broad, hyaline, delicate, empty, epaleate, ovate, obtuse, lanceolate, hairy; hairs 1.0 mm long, acute, base broad. Upper lemma 2.0 mm long, 1.0 mm broad, hyaline, empty, awnless, ovate, obsolete, epaleate.

Vernacular Names

Jaunera, Jarga, Barlu, Apang.

Flowering & Fruiting

December–February; UPC, BHU, Tarana, Chandmari, Chandauli Chakiya, Sakaldiha, Allahabad, Sultanpur, Pharenda, Khadda, Rudrapur.

Habitat

Very common along road sides.

Distribution

India, Burma, Tropical & Northern Africa.

2. *D. aristatum*

D. aristatum (Poir.) C. E. Hubb. in Kew. Bull. 1939: 654. 1940; Bor, Grass. Bur. Cey. Ind. Pak 134 1960; Chien-chang, Hus, Taiwan Grass 667. t. 229. 1975; Roy, Grass. M. P. 65. 1984; Moulik, Grass. Bamb. Ind. 1: 272. 1997. *D. nodosum* Willem. in Usteri, Ann. Bot. 18: 11. 1796. *Andropogon aristatus* Poir in Lamk., Encycl. Meth. Bot. Suppl. 1: 585. 1810; *A. mollicum* Kunth. Rev. Gram. 1: 365. t. 96. 1830. *A. caricosus* Linn. ssp. *mollicomus* Hack. var. *mollicomus* (Kunth.) Hack. in Dc Monogr. Phan. 6: 569. 1889. *A. nodosus* (Willem.) Nash. in N. Amer. F1. 17: 122 1912. *Diplasanthum lanosum* Desv. Opuso. 67. t. 5. f. 1.1831. *Lepeocercis mollicoma* Kunth. Nees in Edinb. New. Phil. J. 18: 185. 1835. (PLATE–171).

Perennial. Rhizomatous; rhizome 3.0 cm long, creeping; root fibrous. Culm 80 cm tall, 2.0 mm thick, green, glabrous, erect, tufted, geniculately– ascending, leafy slender, stout at base, branched; branches subcrect, rooting at lower nodes; nodes brown, hairy; hairs 1.0 mm long, white, acute, base broad, internodes 10 cm long. Leaf–blade 10 cm long, 6.0 mm broad, green, coriaceous, flat, linear, lanceolate, acuminate, base rounded, upper surface hairy; hairs 3.0 mm long, white, acute, base tubercled; lower surface glabrous, margin hyaline, scabrid; mid-vein distinct on lower surface; 4–lateral veins on each side of mid- vein; lateral veins distinct on upper surface, closed; leaf–sheath 8.0 cm long, 4.0 mm broad, green, surface membranous, glabrous, margin hyaline, keeled, loose, compressed, striate, erect, shorter than internodes, hairy at mouth; hair 5.0 mm long, white, acute, base tubercled; veins distinct on outer surface; ligule. 1.0 mm long, a ring of white, dense hairs; hairs acute, base broad; acuricle glabrous. Inflorescence consists of spike–like racemes, raceme 5.0 cm long, 1.5 mm broad, 5-in number, sub–digitate, erect, binate, fascicled, peduncled; peduncle 3.0 mm long, green; joints 2.0 mm long. Sessile spikelets 5.0 mm long, 2.5 mm broad, obovate, compressed, awned, hairy, bisexual; callus 0.5 mm long, hairy; hairs 1.0 mm long, white, acute, base broad. Lower glume 5.0 mm long, 2.5 mm broad, sub–coriaceous, obovate, lanceolate, elliptic, oblong, obtuse, acute, surface glabrous, margin slightly inrolled, keeled, scabrid, hairy; hairs, 2.0 mm long, acute, base tubercled, 9- nerved; nerves green, visible, glabrous, ending at apex. Upper glume 4.0 mm long, 2.0 mm broad, sub–coriaceous, ovate, oblong, ellipitic, lanceolate, bifid at apex, surface glabrous, margin, slightly inrolled, scabrid, keeled, hairy; hairs 2.0 mm long, acute, base tubercled; 3- nerved; nerves green, visible, glabrous, ending at apex. Lower lemma 3.5 mm long, 1.5 mm broad, hyaline, delicate, oblong, ovate, lanceolate, truncate at apex, surface glabrous, margin hairy; hairs 1.0 mm long, acute

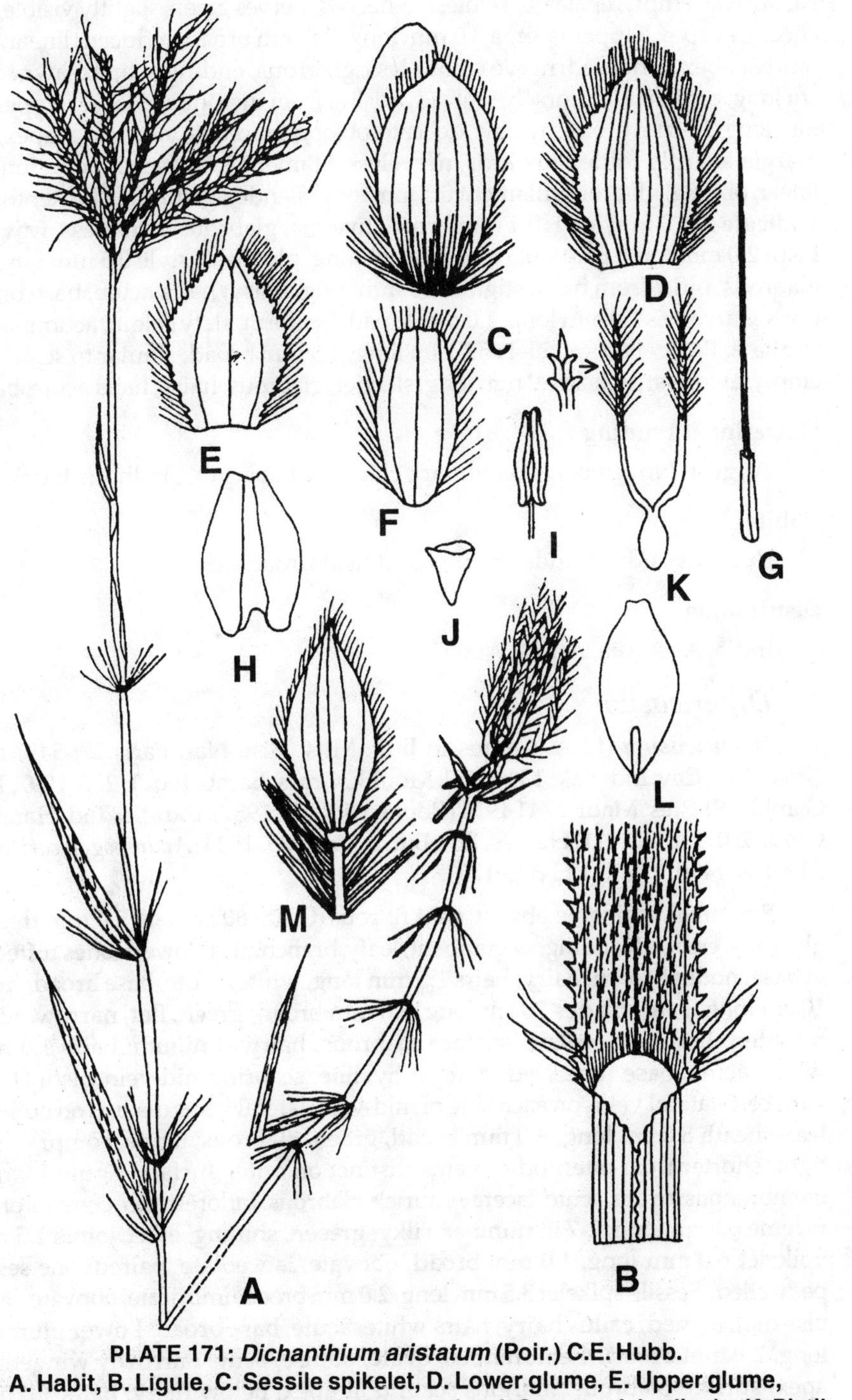

PLATE 171: ***Dichanthium aristatum*** **(Poir.) C.E. Hubb.**
A. Habit, B. Ligule, C. Sessile spikelet, D. Lower glume, E. Upper glume, F. Lower lemma, G. Upper lemma, H. Upper palea, I. Stamen, J. Lodicule, K. Pistil, L. Caryopsis, M. Pedicelled spikelet

base broad, empty, epaleate, awnless, 3-nerved; nerves green, slightly visible, glabrous, ending at apex. Upper lemma 3.0 mm long, 0.5 mm broad, reduced, linear, glabrous, bifid at apex, 1-nerved; nerve colour less, glabrous, ending at apex, awned; awn 2.0 cm long, capillary, antrorsely scabrid, arising from the sinus of lobes. Upper palea 3.0 mm long, 1.5 mm broad, hyaline, delicate, oblong, lanceoate, notched at apex, glabrous, margin sharply folded inwards, nerveless. Stamens 3; anther 1.0 mm long, yellow, linear, oblong, glabrous; filament 0.5 mm long, slender, glabrous, shorter than anther. Lodicule 2, 0.3 mm long, 0.1 mm broad, hyaline, glabrous, nerveless, wavy at apex. Pistil 2.0 mm long; ovary 0.5 mm long, oblong, glabrous; style 0.5 mm long, slender, glabrous, bifid from base; stigma 1.0 mm long, hairy; hairs acute, base broad, in 2-rows. Caryopsis 2.0 mm long, 1.0 mm broad, fusiform, slaty colour, acuminate, shortly stipitate. Pedicelled spikelets 5.0 mm long, 1.5 mm broad, similar to sessile, narrow, empty, awnless; pedicle 1.0 mm long, slender, glabrous, hairy; hairs acute, base broad.

Flowering & Fruiting

August- November; Chandmari, Harrava, Babatpur, Sindhora, Basti, Mau.

Habitat

Common along bunds of cultivated fields, road sides.

Distribution

India, Australia, Africa, America.

3. *D. caricosum*

D. caricosum (L) A. Camus in Bull. Mus. Hist. Nat. Paris 27: 549. 1921; Bor, Grass. Bur. Cey. Ind. Pak. 134. 1960; Moulik, Grass. Bamb. Ind. 1: 273. 1997; Fischer in Gamble, F1 Pres. Madr. 1741 1934 (Repr.ed 3: 1206.·1957); Rang. & Tad. Handb. S. Ind. Grass. 201. t. 159. 1921; Haines, Bot. Bih. & Or. 1039. 1924. *Andropogon caricosus* L. Sp. P1 ed. 2: 1480. 1763. (PLATE–172).

Perennial. Rhizome absent; root fibrous. Culm 60 cm tall, 5.0 mm thick, green, glabrous, erect, ascending, decumbent, leafy, branching at lower nodes tufted, rooting at base, nodes brown, hairy; hairs 1.5 mm long, white, acute, base broad; internodes 10 cm long. Leaf- blades 20 cm long, 8.0 mm broad, green, flat, narrow, acute, base rounded, linear lanceolate, surface glabrous, hairy at mouth; hairs 1.0 mm long, white, acute, base tubercled, margin hyaline, scabrid, mid-vein distinct on lower surface, 5-lateral veins on each side of mid-vein; lateral veins distinct on both surfaces; leaf–sheath 8.0 cm long, 8.0 mm broad, green, glabrous, terete, compressed, erect, tight, shorter than internodes, veins distinct on outer surface; ligule 1.0 mm long, membranous, ovate, acute, lacerate; auricle glabrous. Inflorescence consists of racemes; raceme 6.0 cm long; 2-7 in number, silky–greeen, shining, erect; joints 1.5 mm long. Spikelet 6.0 mm long, 4.0 mm broad, obovate, lanceolate, paired; one sessile, one pedicelled. Sessile spikelet 3.5 mm long, 2.0 mm broad, imbricate, obovate, lanceolate, bisexual, awned, callus hairy; hairs white, acute, base broad. Lower glume 3.5 mm long, 2.0 mm broad, membranous, ovate, obtuse, acute narrowly winged towards apex, surface glabrous, margins scabrid at apex, below hairy; hairs 1.0 mm long, acute, base tubercled; 7- nerved; nerves green, visible, glabrous, ending at apex. Upper

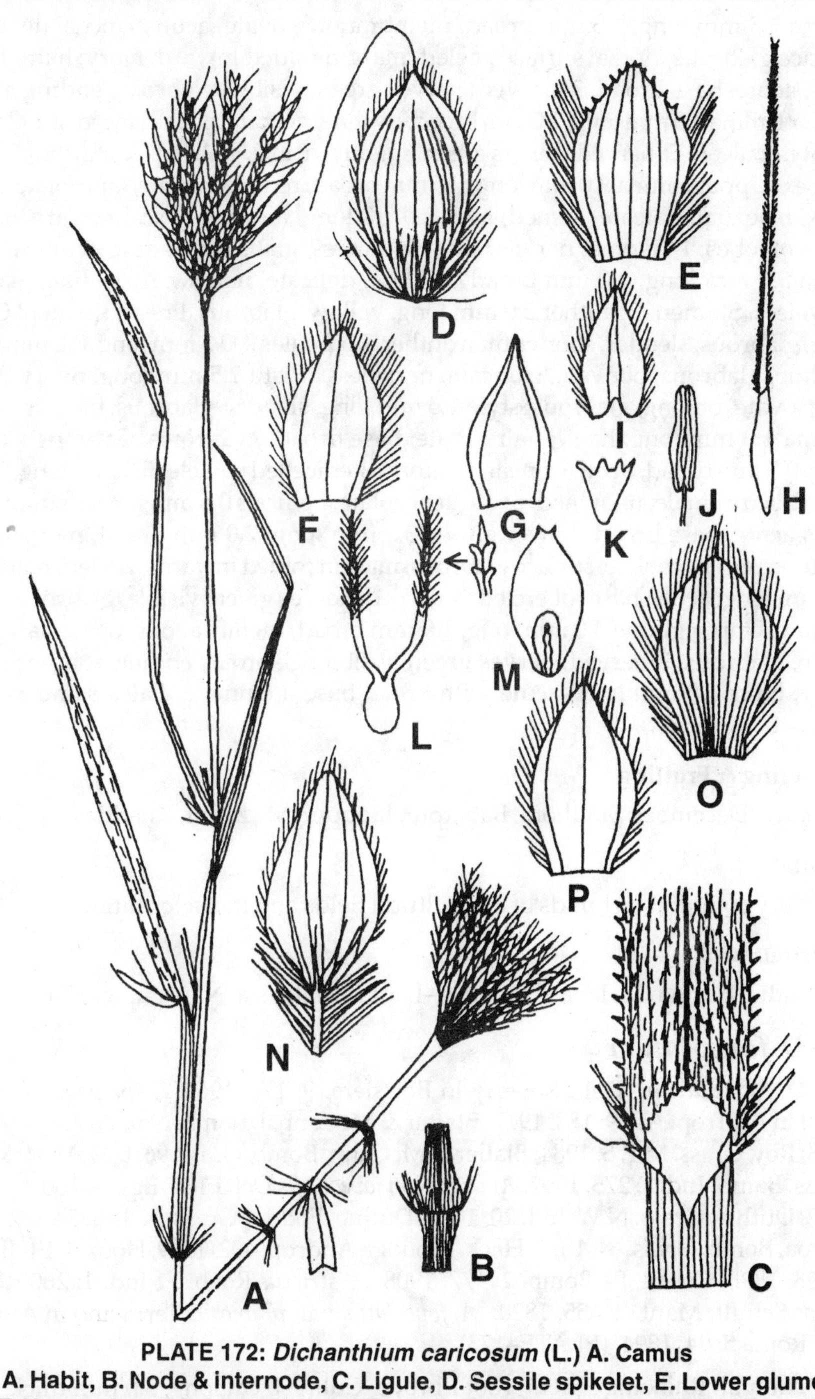

PLATE 172: *Dichanthium caricosum* (L.) A. Camus
A. Habit, B. Node & internode, C. Ligule, D. Sessile spikelet, E. Lower glume, F. Upper glume, G. Lower lemma, H. Upper lemma, I. Upper palea, J. Stamen, K. Lodicule, L. Pistil, M. Caryopsis, N. Pedicelled spikelet, O. Lower glume, P. Upper glume

glume 3.5 mm long, 1.5 mm broad, membranous, ovate, acute, lanceolate, ventral surface glabrous, dorsal surface keeled, margin folded inward, hairy; hairs 1.0 mm long, acute, base broad; 3–nerved; nerves green, visible, glabrous, ending at apex. Lower lemma 2.0 mm long, 1.0 mm broad, hyaline, delicate, ovate, lanceolate, glabrous, empty, epaleate, 2-nerved; nerves green, slightly visible, glabrous & marginal, ending at apex. Upper lemma 3.0 mm long, 0.5 mm broad, reduced, ovate, lanceolate, bifid at apex, bisexual, paleate, awned; awn 2.0 cm long, antrorsely scabrid, arising from sinus of lobes; 1–nerved; nerves faint, colourless, glabrous, ending at apex. Upper palea 3.0 mm long, 1.0 mm broad, hyaline, delicate, narrow, scale–like, glabrous, nerveless. Stamens 3; anther 2.0 mm long, yellow, glabrous, linear; filament 0.5 mm long, glabrous, slender, shorter than anther. Lodicules 2, 0.3 mm long, 0.2 mm broad, hyaline, glabrous, obovate, truncate, nerveless. Pistil 2.5 mm long; ovary 0.5 mm long, ovate, oblong, glabrous; style 1.0 mm long, slender, glabrous, bifid from base; stigma 1.0 mm long, hairy; hairs acute, base broad, in 2–rows. Caryopsis 1.5 mm long, 0.5 mm broad, brown, linear, fusiform. Pedicelled spikelet 5.0 mm long, 2.0 mm broad, acute, male, unawned, ovate, lanceolate; pedicels 0.5 mm long, densely hairy; hairs acute, base broad. Lower glume 3.5 mm long, 2.0 mm broad, membranous, ovate, acute, lanceolate, surface glabrous, margin folded inwards, keeled, hairy; hairs 1.0 mm long, acute, base tubercled; 3-nerved; nerves green, visible, glabrous, ending at apex. Upper glume 3.0 mm long, 1.0 mm broad, membranous, ovate, lanceolate, acute, glabrous, 3-nerved; nerves green, visible, glabrous, ending at apex; margin hairy; hairs 1.0 mm long, acute with broad base. Lemma & Palea same as sessile spikelet.

Flowering & Fruiting

July- December; Sindhora, Babatpur, Jaunpur, Mirzapur, Khadda.

Habitat

Very common on bunds of agricultural fields and moist localities.

Distribution

India, Burma, Malaya, China, Sri–Lanka. Malaysia, New guinea, Fiji.

4. *D. foveolatum*

D. foveolatum (Del.) Roberty in Boissiera. 9: 170. 1960. *Eremopogon foveolatus* Stapf in F1. Trop. Afr. 9: 183. 1917; Blatter & McCann, J. Bomb. Nat. Hist. Soc. 32: 427. 1928; Roy, Grass. M.P. 5. 1984; Blatler & McCann, Bomb. Grass. 96. t. 64. 1984; Moulik, Grass. Bamb. Ind. 1: 275. 1997. *Andropogon foveolatus* Del. F1. d' Egypt. 160. t. 8. fig. 2. 1912; Duthie, Grass. N.W. Ind. 20. 1883; Duthie, Fodder Grass. N. Ind. 34. t. 21. 1888; Lisboa, Bomb. Grass. 64. 1896; Hack. Monogr. Androp. 402. 1889; Hook. f. F1. Brit Ind. 7: 168. 1896; Cooke, F1 Bomb. 2: 977. 1908. *A. strictus* Roxb. F1 Ind. 1: 260. 1820. *A. orthos* Schult. Mant. 2: 455. 1820. *A. foveolatus* var. *plumosus* Terracino in Ann. Istit. Bot. Roma 5: 94. 1894. (PLATE–173).

Annual. Rhizome absent; root fibrous. Culm 60 cm long, 3.0 mm thick, green, globrous, densely tufted, usually deumbent at base, then geniculately ascending, very slender, leafy, nodes hairy. Leaf–blade 15 cm long, 2.5 mm broad, linear, lanceolate,

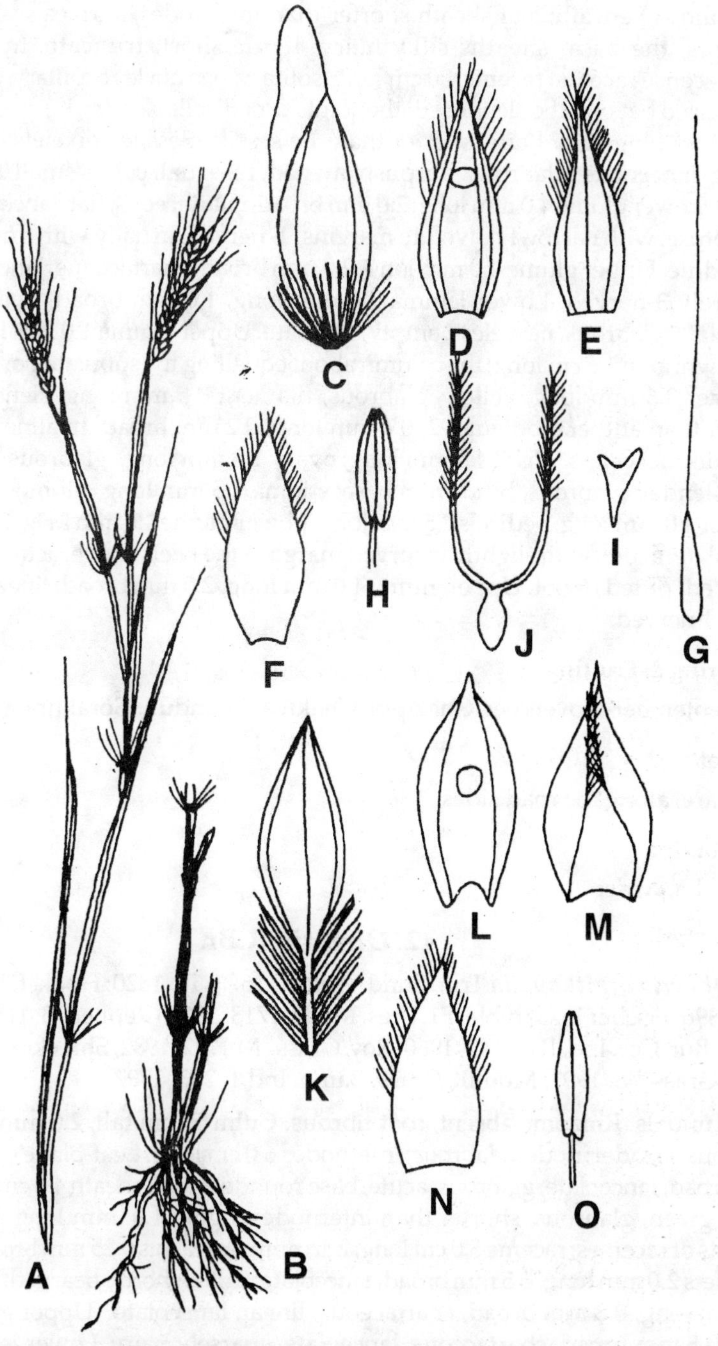

PLATE 173: ***Dichanthium foveolatum*** **(Del.) Roberty**
A. Habit, B. Rhizome, C. Sessile spikelet, D. Lower glume, E. Upper glume, F. Lower lemma, G. Upper lemma, H. Stamen, I. Lodicule, J. Pistil, K. Pedicelled spikelet, L. Lower glume, M. Upper glume, N. Lower lemma, O. Stamen

scaberulous beneath; leaf sheath shorter than internodes, surface scabrid, margin glabrous, the basal sheaths silky hairy; ligule short, truncate, hyaline, hairy. Inflorescence raceme; raceme 4.0 cm long, solitary, peduncles capillary, erect, usually far exserted beyond the slender spathe; joints & pedicells slender, hairy on 2–opposite sides with long silky hairs, shorter than the sessile. Sessile spikelets 4.0 mm long, oblong, lanceolate, glabrous, purplish, awned, bisexual, callus small, shortly hairy at base. Lower glume 4.0 mm long, 2.0 mm broad, chartaceous, flat, lanceolate, oblong, sub-obtuse, with narrowly involute margins, 4- nerved, usually with a deep pit above the middle. Upper glume 4.0 mm long, 2.0 mm broad, chartaceous, lanceolate, acute, obscurely 3-nerved. Lower lemma 2.0 mm long, 1.0 mm broad, hyaline, oblong, lanceolate, glabrous, nerveless, empty, epaleate. Upper lemma 1.0 mm long, reduced to an awn; awn 1.6 cm long; the column about equalling the subulate portion. Stamens 3; anther 2.5 mm long, yellow, glabrous; filament 0.5 mm long, slender, glabrous shorter than anther. Lodicules 2, 0.4 mm long, 0.2 mm broad, hyaline, oblancolate, glabrous, nerveless. Pistil 4.5 mm long; ovary 1.0 mm long, glabrous; style 1.5 mm long, slender, glabrous, bifid from base; stigma 1.5 mm long, plumose. Pedicelled spikelet 4.0 mm long; pedicels 2.5 mm long. Lower glume 4.0 mm long, 2.0 mm broad, lanceolate, acute, with slightly incurved margins, the keels above, aculeately scabrid, 5-nerved, pitted or not. Upper glume 4.0 mm long, 2.0 mm broad, linear, lanceolate, acute, 3-nerved.

Flowering & Fruiting

September-November, Ghazipur, Chakiya, Chanduli, Gorakhpur, Padrauna.

Habitat

Rare, along the road sides

Distribution

India, Africa.

12. *Dimeria* R.Br.

Dimeria ornithopoda Trin. Fund. Agrost. 167.t. 14. 1820; Hook. f. F1 Brit. Ind. 7: 104. 1896; Fischer in Gamble, F1. Pres. Madr. 1713. 1934 (Repr.ed. 3: 1188. 1957); Bor, Grass. Bur. Cey. Ind. Pak. 142, 1960; Roy, Grass. M.P. 72. 1982; Shreekumar & Nair, Fl. Kera. Grass. 98. 1991; Moulik, Grass. Bamb. Ind 1: 283. 1997.

Annuals. Rhizome absent; root fibrous. Culm 50 cm tall, 2.0 mm thick, green, glabrous, slender; nodes glabrous; internodes 6.0 cm long. Leaf-blade 6.0 cm long, 4.0 mm broad, lanceolate, glabrous, acute, base rounded; leaf- sheath 4.0 cm long, 4.0 mm thick, green, glabrous, shorter than internodes; ligule 1.0 mm long. Inflorescence consists of racemes; raceme 5.0 cm long, 2 in number; rachis 0.25 mm broad, trigonous. Spikelets 2.0 mm long, 0.5 mm broad, lanceolate; callus shortly bearded. Lower glume 2.0 mm long, 0.3 mm broad, chartaceous, linear, lanceolate. Upper glume 2.0 mm long, 0.5 mm broad, chartaccous, lanceolate, sparsely hairy. Lower lemma 1.5 mm long, 0.3 mm broad, hyaline, delicate, oblanceolate, ciliate at apex, empty, epaleate. Upper lemma 2.0 mm long, 0.5 mm broad, notched, hyaline, bisexual, epaleate, awned; awn 12 mm long, column 3.0 mm long, brown. Stamens 2; anther 0.5 mm long, yellow, glabrous, filament 0.2 mm long, slender, glabrous, shorter than anther. Pistil 1.0 mm long; ovary 0.3mm long, oblong; style 0.3mm long, slender, glabrous, bifid from base; stigma 0.4 mm long. hairy; hairs acute, base broad. Caryopsis 1.6 mm long, 0.8 mm broad, cylindric, 1/3 of grain is embryo.

Flowering & Fruiting

August-March; Rajghat, Ghazipur, Ballia, Sonbhadra, Mau.

Habitat

Very common along road side, hill slopes.

Distribution

India, Europe, North America, Sri–Lanka, Malaya.

13. *Elionurus* Willd.

Elionurus royleanus Nees ex A. Rich. Tent. F1. Abyss. 2: 471. 1851; Roy, Grass. M.P. 77. fig. 6. 1984; Moulik, Grass. Bamb. Ind. 1: 180. 1997. *Ratzeburgia schimperi* Steud., Nom. Bot. ed. 2. 2: 439. 1841. *Rottboellia degantissima* Hochst. et Steud. ex Steud. Nom. Bot. ed. 2. 2: 474. 1841. *Elionurus grisebachii* Schmidt, Beitr. Zur. F1. Capverd. 154. 1852 *Andropogon elegantissimus* Steud., Syn. F1. Glum. 1: 364. 1854. *A. grisebachii* Schmidt ex Steud. Syn. P1. Glum. 1: 365. 1854.

Annual. Rhizome absent; root fibrous. Culm 25 cm tall, 1.5 mm thick, green, glabrous, tufted, erect, decumbent, slender, leafy, branched; branches suberect; rooting at base, nodes glabrous. Leaf–blade 6.0 cm long, 4.0 mm broad, green, glabrous, linear, lanceolate, acute, base rounded, margin hyaline, scabrid, flat, involute, mid vein distinct on lower surface, 9- lateral veins on each side of mid-vein; lateral veins distinct on upper surface, leaf-sheath compressed, glabrous, erect, striate, hairy at mouth; hairs 3.0 mm long, acute, base tubercled; veins distinct on outer surface; ligule 1.0 mm long, membranous, lacerate; auricle glabrous. Inflorescence consists of raceme; raceme solitary, often tinged with purple, partly enclosed in a flattened sheath; joints 2.5 cm long, densely bearded at apex with long white hairs, callus short, bearded. Sessile spikelet 12 mm long, linear, awned. Lower glume 12 mm long, 4.0 mm broad, coriaceous, linear, lanceolate, apex beaked; beak 6.0 mm long, bifid, violet coloured, 2–keeled; keels hairy, margins below the beak with violet coloured tubercles each provided with a pencil of whitish glistending hair; 9–nerved. Upper glume 5.5 mm long, 2.5 mm broad, membranous, boat–shaped, acuminate, aristate, 1–nerved. Lower lemma 4.5 mm long, 2.5 broad, hyaline, empty, epaleate. Upper lemma 3.0 mm long, 1.5 mm broad, hyline, obovate, lanceolate, bisexual, 1-nerved, epaleate. Stamens 3; anther 1.5 mm long, yellow, linear, glabrous; filament 0.5 mm long, slender glabrous, shorter than anther. Lodicules absent. Pistil 3.0 mm long; ovary 0.5 mm long, oblong, glabrous; style 1.0 mm long, slender, glabrous, bifid from base; stigma 1.5 mm long, hairy; hairs acute, base broad, in 2-rows. Caryopsis 3.0 mm long, 1.5 broad, linear. Pedicelled spikelet 9.0 mm long, 3.0 mm broad; pedicels 2.5 mm long. Lower glume 9.0 mm long, 3.0 mm broad, herbaceous, obscurely nerved, keeles winged. Upper glume 7.0 mm long, 2.5 mm broad, membranous, concave, awns scabrid. Lemma hyaline. Stament 3; anther 1.5 mm long, yellow, glabrous, linear; filament 0.5 mm long, slender, glabrous, shorter than anther.

Flowering & Fruiting

August-November; Pindra, Mirzapur, Jaunpur, Ballia.

Habitat

Rare, along the road side.

Distribution

India, Africa, Egypt & Arabica.

14. *Eulalia* Kunth

Eulalia trispicata (Schult.) Henr. in Blumea 3: 453. 1940; Bor, Grass. Bur. Cey. Ind. Pak. 157. 1960; Manilal & Sivaraj, F1 Calic 334. 1982; Roy, Grass. M.P. 97. 1984; Sur in J. Econ. Tax. Bot. 6 (1): 187. 1985; Shreekumar & Nair, F1. Kera. Grass 104. 1991; Moulik, Grass. Bamb. Ind. 1: 335. 1997. *Andropogon tristaschyus* Roxb. Hort. Beng. 6.1814. *A. trispicatus* Schult. Syst. Veg 2: 452. 1824. *Eulalia tristachya* O. Kuntze, Rev. Gen. P1. 2: 775. 1891; Fischer in Gamble, F1. Pres. Madr. 1715. 1934 (Repr.ed. 3: 1189. 1957). *Pollinia argentea* Trin in Bull. Acad. Sci. Petersb. 1: 71. 1836; Hook. f. Fl. Brit. Ind 7: 11. 1896.

Perennial. Root fibrous. Culm 100 cm tall, 3.0 mm thick, erect, glabrous, tufted, green at young, turning reddisn with age. Leaf-blade 40 cm long, 8.0 mm broad, green glabrous, linear–lanceolate, acuminate, base rounded; leaf–sheath compressed, glabrous, erect; ligule membranous, thin with a tuft of long hairs around. Inflorescence consists of receme; raceme 20 cm long, 15 in number, digitate; rachis 2.0 mm broad, densely villous. Sessile spikelet 3.5 mm long, elliptic, lanceolate, awned. Lower glume 3.0 mm long, 1.0 mm broad, chartaceous, elliptic, 2-keeled, 2-nerved, densely villous. Upper glume 3.0 mm long, 1.0 mm broad, chartaceous, elliptic, margin hyaline, hairy, 3–nerved. Lower lemma 3.0 mm long, 1.0 mm broad, hyaline, elliptic, lanceolate, empty, epaleate, hairy towards the apex. Upper lemma 2.5 mm long, 1.0 mm broad, hyaline, deeply notched, bisexual, epaleate, awned; awn 12 mm long; column 5.0 mm long. Stamens 3; anther 2.0 mm long, yellow, oblong, linear, glabrous; filament 0.5 mm long, slender, glabrous, shorter than anther. Lodicules absent. Pistil 3.0 mm long; ovary 0.5 mm long, elliptic; style 1.0 mm long, slender, glabrous, bifid from base; stigma 1.5 mm long, plumose. Caryopsis oblong. Pedicelled spikelet 3.5 mm long, elliptic, lanceolate, awned or not; pedicels 2.0 mm long, densely villous; glumes & florets similar to those of sessile spikelets.

Flowering & Fruiting

September–May; Jaunpur, Naugarh, Naini, Khadda.

Habitat

Vary common along the dry rocky places & bunds of fields.

Distribution

India, SriLanka, Malaysia, Australia.

15. *Eulaliopsis* Honda

Eulaliopsis binata (Retz) Hubb. in Hook. Icon. Pl. t. 3262. 1935; Bor in Assam Grass 5: 334. 1940; Bor, U.P. Grass. 140. 1941; Rhind, Grass. Burma 64. 1945; Raizada in Ind. For. Rec. 4: 99. 1954; Raizada, Bhard & Jain, Grass. Upp. Gang. P1.4 (7): 191. 1954; Bor, Grass. Bur. Cey Ind. Pak. 158. 1960; Moulik, Grass. Bamb. Ind. 1: 335. 1997. *Andropogon binatus* Retz. Obs. Bot. 5: 21 1789. *Spodiopogon angustifolius* Trin. in Mem. Acad. Sci. Petersb. ser. 6.2: 300. 1932. *Andropogon notopogon* Steud. Syn. P1. Glum. 1. 373. 1854. *A. involutus* Steud. et *A. obvallatus* Steud. Syn. P1. Glum. 1: 373. 1854. *Ischaemum angustifolium* (Trin) Hack in DC., Monogr. Phan. 6: 241. 1889; Hook. f. F1. Brit. Ind. 7: 129. 1896. *Eulaliopsis angustifolium* (Trin.) Honda in Bot. Mag. Tokyo. 38: 56. 1924. *Pollinidium angustifolium* (Trin.) Haines, Bot. Bih & Or. 1020. 1924. *P. binatum* (Retz.) C.E. Hubb. in Kew Bull. 72. 1932; Blatter & McCann, Bomb. Grass. 24 1935. (PLATE–174).

Perennials. Rhizomatus; rhizome horizontal, root fibrous. Culms 40 cm tall, 5.0 mm broad, erect, densely tufted, ascending, branching at lower nodes, green, glabrous, compressed, base slightly swollen, woody, rooting at base; nodes brown, hairy; hairs white, acute, base broad; internodes 6.0 cm long, solid, glabrous. Leaf-blade 11 cm long, 7.0 mm broad, flat, linear-lanceolate, acute, rounded at base, brown, upper surface glabrous, lower surface rough, hairy; hairs acute, base tubercled; margin scaberulous, mid-vein distinct on lower surface, 3–lateral veins on each side mid–vein; lateral veins distinct on upper surface; leaf-sheath 6.5 cm long, 5.0 mm broad, outer surface yellowish-green, hairy; hairs acute, base broad; inner surface purplish, glabrous, margin hyline, glabrous, mid–vein distinct on outer surface, lateral vein not distinct, shorter than internodes, basal sheath longer than internodes; ligule 1.0 mm long, hairy; hairs white, acute, base broad; auricle glabrous. Inflorescence consists of racemes; raceme, 6.0 cm long, 6.0 mm broad, 2-4 in number, glabrous, axillry & terminal, filiform, peduncles 2.0 mm long, erect. Spikelets 7.5 mm long, 2.0 mm broad oblong, lanceolate, paired; lower sessil & upper pedicelled. Sessile spikelets 5.0 mm long, 1.5 mm broad, hairy, acute, lanceolate, glabrous, bisexual, awned, callus 0.2 mm long, 0.1 mm broad, hairy; hairs rusty, acute, base broad. Lower glume 4.5 mm long, 1.5 mm broad, membranous, acute, elliptic, oblong, 4– toothed, broadly ovate when spered, linear- lanceolate, surface glabrous, margin sharply folded inwards & hairy; hairs 1.0 mm long, acute, base rounded; 5–nerved; nerves light green, glabrous, all nerves reaching at the apex. Upper glume 5.0 mm long, 1.0 mm broad, membranous, slightly oblique, boat–shaped, acute, apex bidentate, ovate-lanceolate when spread, surface glabrous, margin sharply folded inwards, hairy; hairs 1.0 mm long, acute, base rounded; 5–nerved; nerves light brown, glabrous, all nerves reaching at tip. Lower lemma 3.0 mm long, 0.5 mm broad, hyaline, oblong-obtuse, acute, male, epaleate, surface glabrous, margin scabrid, hairy; hairs 1.0 mm long, acute, base tubercled; 3–nerved; nerves colourless, slightly visible, glabrous. Upper lemma 3.0 mm long, 0.6 mm broad, reduced, nerveless, bisexual, paleate, bidentate, surface & margin glabrous, lanceolate, linear, awned; awn 6.0 mm long, antrorsely scabrid. Upper palea 2.0 mm long, 1.0 mm broad, hyaline, glabrous, margin scaberulous, neverless, oblong, apex rounded. Stamens 3; anther 1.0 mm long, yellow, linear, glabrous; filament 0.5 mm long, glabrous, slender, shorter than anther. Lodicules 2, 0.3 mm long, 0.1 mm broad,

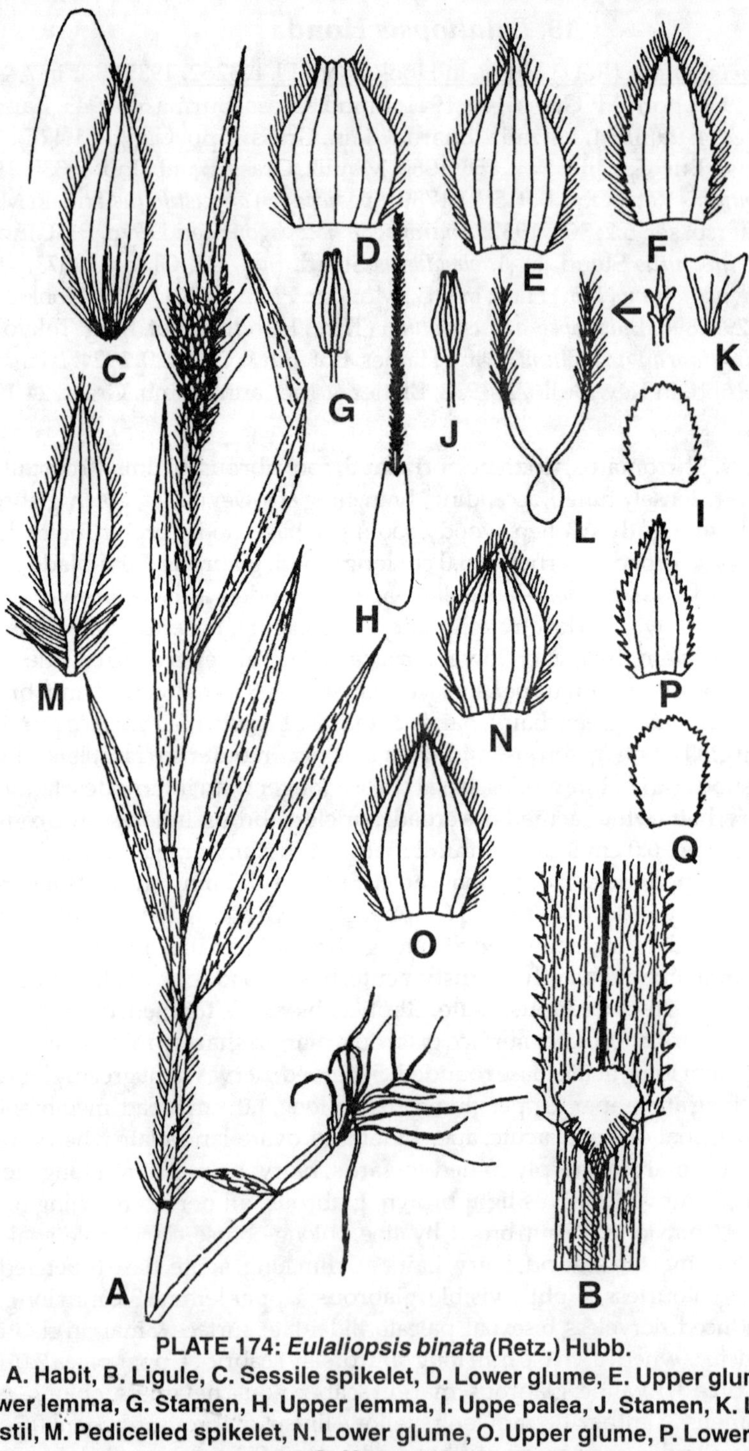

PLATE 174: ***Eulaliopsis binata*** **(Retz.) Hubb.**
A. Habit, B. Ligule, C. Sessile spikelet, D. Lower glume, E. Upper glume, F. Lower lemma, G. Stamen, H. Upper lemma, I. Uppe palea, J. Stamen, K. Lodicule, L. Pistil, M. Pedicelled spikelet, N. Lower glume, O. Upper glume, P. Lower lemma, Q. Upper palea

hyaline, glabrous, obovate, faintly 3– nerved. Pistil 1.5 mm long; ovary 0.5 mm long, glabrous, oblong; style 0.3 mm long, slender, glabrous, bifid from base; stigma 1.0 mm long, acuminate much longer than style, brown, hairy; hairs acute, base broad, in 2–rows. Caryopsis 1.5 mm long. 1.0 mm broad, black. Pedicelled spikelets 6.5 mm long, 1.5 mm broad, empty, awnless glabrous, hairy; pedicels 3.0 mm long, 0.5 mm broad; hairy; hairs 1.0 long, acute, base tubercled. Lower glume 3.5 mm long, 1.5 mm broad, membranous, acute, ovate, linear-lanceolate, surface glabrous, margin sharply folded inwards, hairy; hairs acute, base tubercled; 7–nerved; nerves light green, glabrous, ending at apex. All florets same as sessile spikelet.

Flowering & Fruiting

June-August; Chandauli, Chakiya, Sonaghatti, Duddhi, Narayanpur.

Habitat

Common along moist hill rocks and banks of hill canals.

Distribution

India, Myanmar, Afghanistan, China, Philippines, Siam.

16. *Hackelochloa* O. Ktze.

Hackelochloa granularis (L.) O. Ktze. Rev. Gen. P1. 2: 776. 1891; Moulik, Grass. Bamb. Ind. 1: 193. 1997; Fischer in Gamble, F1. Pres. Madr. 1758. 1934 (Repr.ed 3: 1217. 1957); Bor, Grass. Bur. Cey. Ind. Pak. 159. 1960. *Cenchrus granularis* Linn. Mant. P1. 2: 575. 1771. *Manisuris granularis* (Linn.) Linn. f. Nov. Gram. Gen. 40. 1779, Sw. Prodr. Veg. Ind. Occ. 25. 1788; P. Beauv. Agrost 21. fig 10. 1812; Roxb. P1. Corom. 2: 11. t. 118. 1830; Roxb. F1. Ind. 1: 352. 1820; Grah. Cat. Bomb. P1. 234. 1839; Dalz. & Gibs. Bomb. F1. 300. 1861; Hack. Monogr. Androp. 314. 1889; Duthie, Grasses N.W. Ind. 18. 1883; Duthie, Fodder Grasses N. Ind. 29. t. 46. 1888; Lisboa, Bomb. Grass. 62. 1896; Hook. f. F1. Brit. Ind. 7: 150. 1896; Cooke, F1. Bomb. l2: 955. 1908; Stapf. F1. Trop. Afr. 9: 57. 1917; Haines, Bot. Bih. & Or. 1057. 1924; Blatter & McCann, Bomb. Grass. 32. t. 22. 1984. *Rytilix granularis* Skeels in U.S. Dept. Ag. Bur. P1. Indust. Bull. 2: 282. 1913. *Manisuris polystachya* P. Beauv. F1. Owar. et Ben. 14. 1855. Raizada *et al.*, Indian Forest. Rec. ser. 2. Bot. 4: 205. 1961. Gilliland in F1. Malaya 3: 277. 1971; Cope in F1. W. Pak. 143: 345. 1982. (PLATE–175).

Annuals. Rhizome creeping; root fibrous. Culm 30 cm tall, 2.0 mm thick, erect, tufted, slender, compressed, green, glabrous, diffusely branched; branches erect, or ascending from creeping rhizome, rooting at base, nodes brown, hairy; hairs white, acute, base broad; internodes 7.0 cm long, glabrous. Leaf-blade 7.0 cm long, 4.0 mm broad, oblong, linear-lanceolate, membranous, acute, base cordate, flat, green, mid vein distinct on lower surface, 5–lateral veins on each side of mid-vein; lateral vein distinct on upper surface, both surface hairy; hairs white, acute, base tubercled; margin hyaline, scaberulous, keeled, hairy; hairs white, acute, base broad; leaf-sheath 2.0 cm long, 3.5 mm broad, loose, glabrous, membranous, veins distinct on outer surface, much shorter than internodes, surface hairy; hairs white, acute, base tubercled; margin hyaline, keeled, hairy; hairs acute, base tubercled; ligule 1.0 mm long, membranous, ovate, hairy at tip; hairs white, acute, base broad; auricle glabrous. Inflorescence a

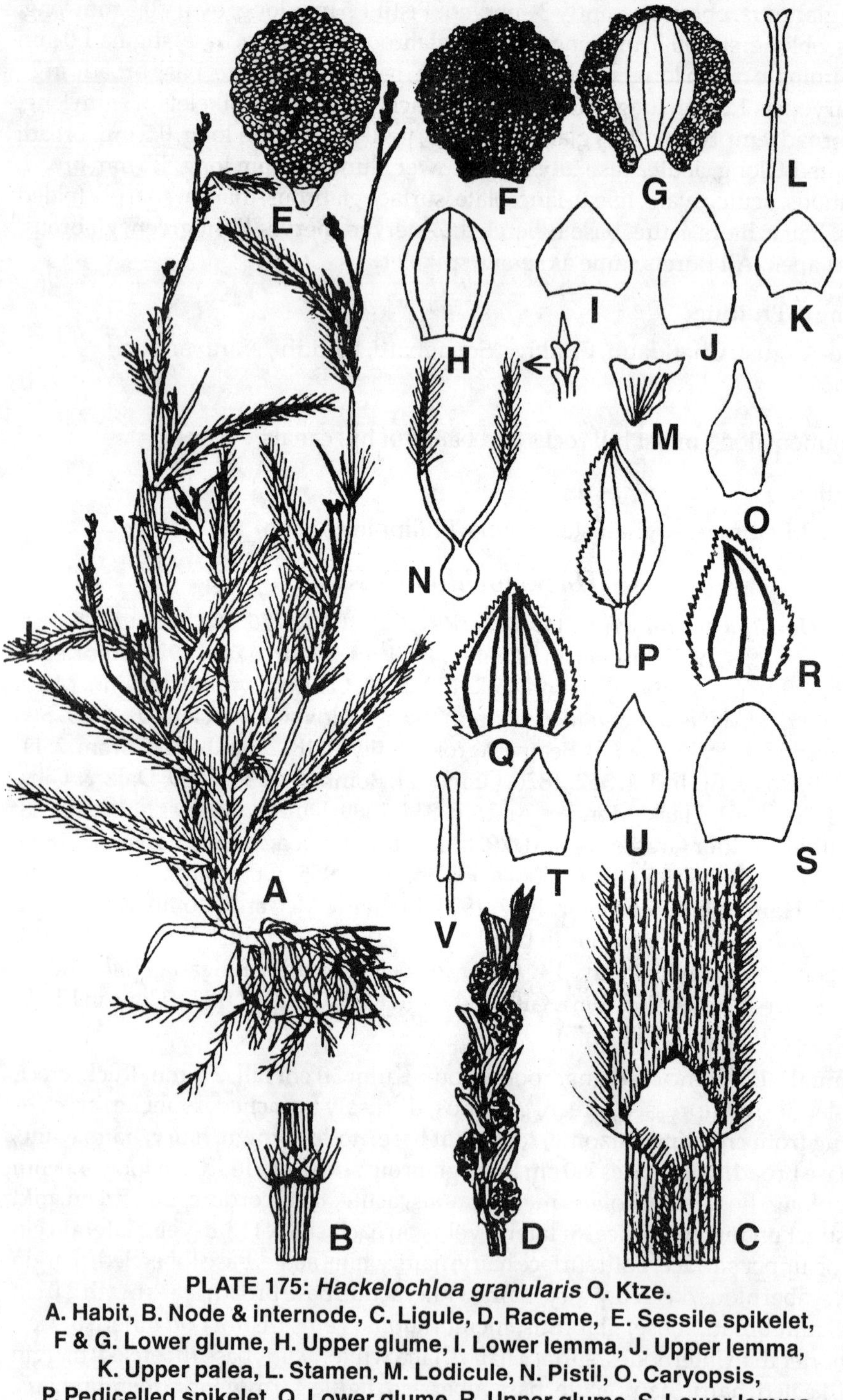

PLATE 175: *Hackelochloa granularis* **O. Ktze.**
A. Habit, B. Node & internode, C. Ligule, D. Raceme, E. Sessile spikelet, F & G. Lower glume, H. Upper glume, I. Lower lemma, J. Upper lemma, K. Upper palea, L. Stamen, M. Lodicule, N. Pistil, O. Caryopsis, P. Pedicelled spikelet, Q. Lower glume, R. Upper glume, S. Lower lemma, T. Upper lemma, U. Upper palea, V. Stamen

terminal & axillary racemes; racemes 1.5 cm long, erect, resembling a string of minute beeds, solitary, or seemingly fascicled in the axils of the leaves, & form spatheate false panicle. Spikelets 3.0 mm long, 1.5 mm broad, paired, lower pedicellate, upper sessile. Sessile spikelets 1.5 mm long, 1.0 mm broad, globose, awnless, bisexual. Lower glume 1.5 mm long, 1.0 mm broad, rounded–orbicular, pitted & tubercled, cub-shaped, coriaceous, margin very strongly folded inwards, 7–nerved; nerves green, very faint on upper surface, but visible on lower surface. Upper glume 1.0 mm long, 0.5 mm broad, subcoriaceous in upper half, membranous in lower half, oblong, obtuse, elliptic, inclosed & adhering below in the cavity of lower glume, surface & margin glabrous, faintly 3- nerved; nerves light green. Lower lemma 1.0 mm long, 0.5 mm broad, broadly ovate, delicate, hyaline, lanceolate, empty, epaleate, nerveless, surface & margin glabrous, apiculate. Upper lemma 1.0 mm long, 0.5 mm broad, ovate, obtuse, hyaline, lanceolate, bisexual, nerveless, apiculate, surface & margin glabrous. Upper palea 0.5 mm long, 0.3 mm broad, about half of the lemmas, hyaline, nerveless, ovate, obtuse, lanceolate, surface & margin glabrous. Stamens 3; anther 0.3 mm long, orange-yellow, linear, glabrous, lobes divaricate; filament 0.6 mm long, glabrous, slender, much longer than anther. Lodicules 2; 0.3 mm long. 0.1 mm broad, obovate, hyaline, surface & margin glabrous, many nerved. Pistil 1.0 mm long; ovary 0.2 mm long, glabrous, light green, oblong; style 0.3 mm long, slender, glabrous, bifid from base; stigma 0.5 mm long, acuminate, hairy; hairs acute, base broad, in 2–rows. Caryopsis 0.6 mm long, 0.3 mm broad, ovoid, shining. Pedicelled spikelets 2.0 mm long, 0.5 mm braod, ovate, acute, lanceolate, male; pedicels 1.0 mm long, fused with rachis, flat, glabrous. Lower glume 2.0 mm long, 1.0 mm broad, broadly ovate, lanceolate, chartaceous, acuminate, laterally compressed, flat on back, 1–winged, surface glabrous, margin scaberulous, hyaline, 7–nerved; nerves deep green, visible, central one ending below the apex. Upper glume 2.0 mm long, 1.0 mm broad, obliquely ovate, lanceolate, chartaceous, acuminate, surface glabrous, margin hyaline, scaberulous, strongly keeled, 3-nerved; nerves deep green, visible, ending below the apex & join to each below the apex. Lower lemma 1.5 mm long, 1.0 mm broad, broadly ovate, lanceolate, apiculate, delicate, hyaline, empty, epaleate, nerveless, surface & margin glabrous. Upper lemma 1.0 mm long, 0.8 mm broad, broadly ovate, lanceolate, hyaline, male, acute, surface & margin glabrous. Upper palea 1.0 mm long, 0.5 mm broad, ovate, lanceolate, acute, hyaline, surface & margin glabrous. Stamens 3; anther 0.3 mm long, orange-yellow, linear, glabrous, lobes divaricate; filament 0.6 mm long, glabrous, slender, much longer than anther. Lodicules absent.

Vernacular Name

Trinpali.

Flowering & Fruiting

July-November; BHU, Chaukaghat, Naini, Padrauna, Turtipar.

Habitat

Frequent along open grassland & bunds of cultivated fields.

Distribution

India, Tropical part of the world.

17. *Hemarthria* R. Br.

Hemarthria compressa (L.f.) R. Br Prodr. Nov. Holl. 207. 1810; Kunth. Enum. 1: 465. 1833; Grah. Cat. Bomb. P1. 233. 1839; Duthie, Grasses N. Ind. 18. 1883; Duthie, Fodder Grasses N. Ind. 30. 1888; Gamble in Fischer, F1. Pres. Madr. 1762. 1934 (Repr.ed 3: 1221. 1957); Bor, Grass. Bur. Cey. Ind. Pak. 161. 1960; Moulik, Grass. Bamb. Ind. 1: 195. 1997; Raizada *et al.*, Indian Forest. Rec. ser. 2. Bot. 4: 21: 207. 1961; Gilliland in F1. Malaya 3: 266. 1971; Cope in F1.W. Pak. 143: 346. 1982; Haines Bot. Bih & Or. 1061. 1924; Cook, Aqu. Wet. Pl. Ind. 294. f. 302 c–d. 1996. *Rottboellia compressa* L.f. Suppl. P1. 114. 1781; Hook.f. F1. Brit. Ind. 7: 153. 1896; Lisboa, Bomb. Grass. 56. 1896; Cooke. F1. Bomb. 952. 1908. *R. compressa* var. *genuina* Hack. Androp. 286. 1889. *R. glabra* Roxb. F1. Ind. 1: 353. 1820. *Hemarthria laxa* Nees ex Steud. Syn. P1. Glume. 358. 1854. *Hemarthria coromandelina* Steud. Syn. P1. Glume. 1: 358. 1855. (PLATE–176).

Perennial. Rhizomatous; rhizome 12 cm long, creeping; root fibrous. Culms 90 cm tall, 3.0 mm thick, glabrous, erect, tufted, compressed, striate, leafy, branched, ascending from base, rooting at lower nodes; nodes glabrous; internodes 7.0 cm long. Leaf–blade 6.0 cm long, 2.0 mm broad, green, linear, lanceolate, flat, narrow, acuminate, base sub–cordate, surface glabrous, margin scabrid, hyaline, mid–vein distinct on lower surface, 4–lateral veins on each side of mid–vein; lateral veins distinct on upper surface; leaf–sheath 4.0 cm long, 4.0 mm broad, loose, green, glabrous, erect, shorter than internodes, membranous, margin hyaline, veins distinct on outer surface, hairy at throat; hairs 2.0 mm long, white, acute, base tubercled; ligule 0.2 mm long, membranous. Inflorescence a solitary racemes; raceme 3.5 cm long, spiciform, curved, or straight, herbaceous, deep green; rachis almost in articulate, internodes broad, striate, glabrous. Spikelets 6.5 mm long, 2.0 mm broad, acuminate, paired; one sessile & one pedicelled. Sessile spikelets 4.0 mm long, 1.5 mm broad, acuminate, glabrous, unawned, bisexual; callus small, naked. Lower glume 3.0 mm long, 1.0 mm broad, coriaceous, tip shortly membranous, bluntly acuminate, obtuse, narrow, linear, lanceolate, winged; wings glabrous; margin hairy; hairs 1.0 mm long, acute, base broad, 5- nerved; nerves green, slightly visible, glabrous, ending at apex. Upper glume 4.0 mm long, 1.5 mm broad, membranous, oblong, linear– lanceolate, acute, boat-shaped, apex mucronate, adhering the cavity of internodes, 3–nerved; nerves green, visible, glabrous, ending at apex. Lower lemma 3.0 mm long, 1.5 mm broad, hyaline, oblong, ovate, lanceolate, acute, empty, epaleate, nerveless. Upper lemma 2.0 mm long, 1.0 mm broad, hyaline, delicate, oblong, lanceolate, acute, nerveless, bisexual. Upper Palea 2.0 mm long, 1.0 mm broad, hyaline, linear-lanceolate, nerveless. Stamens 3; anther 1.1 mm long, yellow, oblong, glabrous; filament 0.5 mm long, glabrous, slender, shorter than anther. Lodicules absent. Pistil 2.0 mm long; ovary 0.5 mm long, oblong, glabrous; style 0.5 mm long, glabrous, slender, bifid from base; stigma 1.0 mm long, brown, hairy; hairs acute, base broad. Caryopsis 1.0 mm long, 0.5 mm broad, brownish, oblong. Pedicelled spikelet 4.5 mm long, 1.0 mm broad, narrow than sessile, empty. Lower glume 3.0 mm long, 1.0 mm broad, membranous. Upper glume 4.5 mm long, 1.3 mm broad, chartaceous, 5–nerved. Floral glumes same as sessile spikelet.

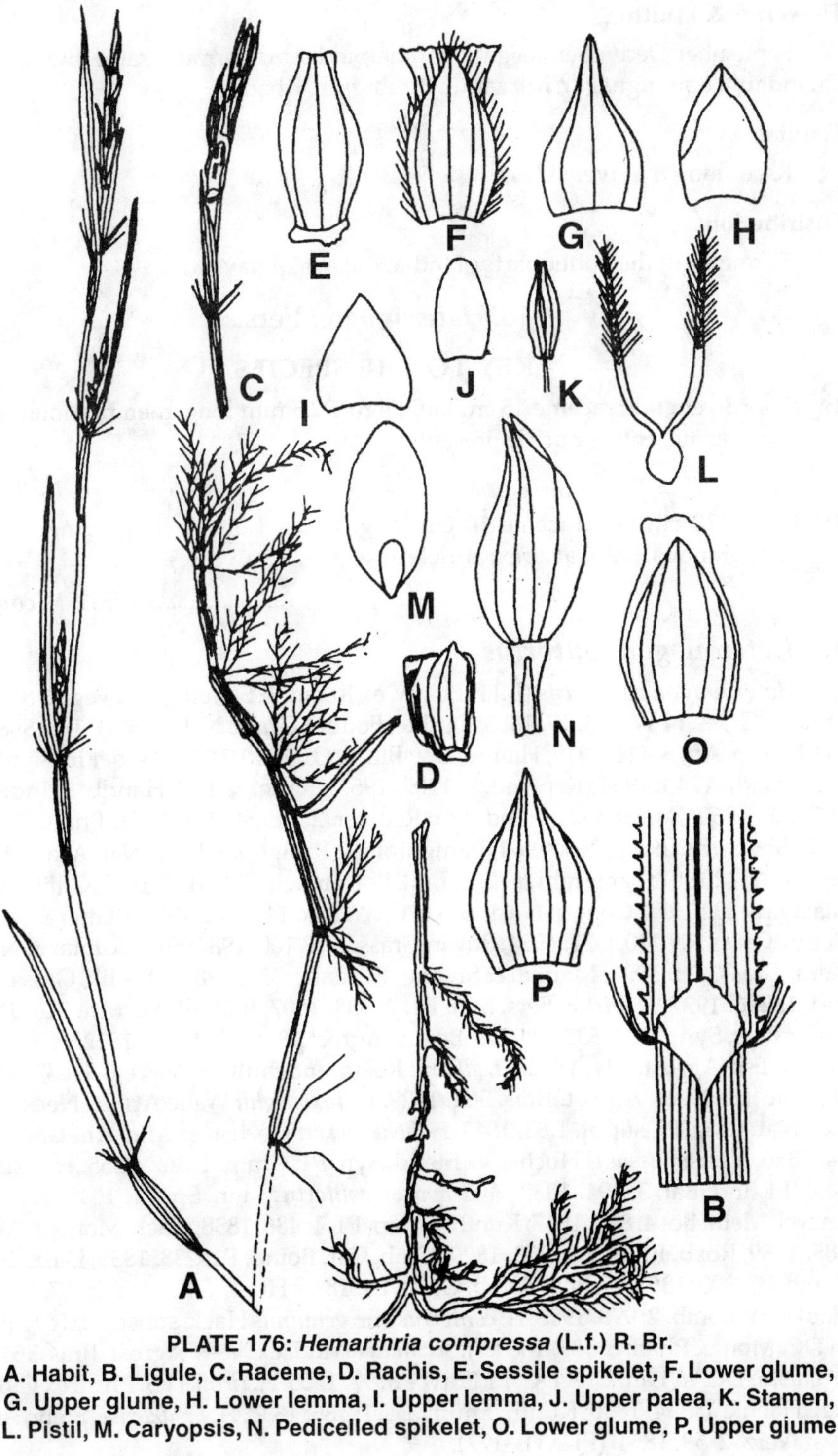

PLATE 176: *Hemarthria compressa* (L.f.) R. Br.
A. Habit, B. Ligule, C. Raceme, D. Rachis, E. Sessile spikelet, F. Lower glume, G. Upper glume, H. Lower lemma, I. Upper lemma, J. Upper palea, K. Stamen, L. Pistil, M. Caryopsis, N. Pedicelled spikelet, O. Lower glume, P. Upper giume

Flowering & Fruiting

September-December; Rajghat, Tarana, Sindhora, Harraua, Rameshwar, Jaunpur, Chandauli, Bajrangnagar, Mirzapur, Azamgarh, Khadda.

Habitat

Rare along the river sides.

Distribution

Throughout the hotter parts of India, Burma, Malaysia.

18. *Heteropogon* Pers.

KEY TO THE SPECIES

1a. Plant 30 cm tall; raceme2.5 cm long; ligule 2.5 mm long, membranous, ovate, acute, entire; auricle present

.... **2.** ***H. fischerianus***

1b. Plant 120 cm tall; raceme 10 cm long; ligule 1.0 mm long, fringed membranous, hairy at apex; auricle absent

.... **1.** ***H. contortus***

1. *Heteropogon contortus*

Heteropogon contortus (L.) P. Beauv. ex Roem. & Schult. Syst. Veg. 2: 836. 1817; Duthie, Grass. N.W. Ind. 19. 1883; Duthie, Fodder Grass. N. Ind. 32. t. 19. 1888; Stapf in F1. Trop. Afr. 9: 411. 1919; Haines, Bot. Bih. & Or. 1040. 1924; Fischer in Gamble, F1. Pres. Madr. 1743. 1934 (Repr. ed. 3: 1207. 1957); Rang. & Tad. Handb. S. Ind. Grass. 207. t. 6. 1960; Raizada *et al.*, Ind. For. Rec. ser. 2. Bot. 4: 247. 1961; Bhattacharya in Bull. Bot. Surv. Ind. 6: 209. 1964; Panighrahi & Rajagopal, Proc. Nat. Acad. Sci. Ind. Sec. B. 37: 12 1907; Panigrahi *et al.*, in Bull. Bot. Surv. Ind. 11: 112. 1969; Gilliland in F1. Malaya 3: 292.1971; Cope in Nasir & Ali, F1.W. Pak. 143: 312. 1982; Blatter & McCann, Bomb. Grass 109.110. t. 71. 1984; Roy in Grass. M.P. 100. 1984; Shreekumar & Nair, F1. Kera. Grass. 111. 1991; Manilal & Sivaraj, F1. Calic. 332. 1982; Moulik, Grass. Bamb. Ind. 1: 226. 1997. *H. hirtus* Pers. Syn. P1. 2: 533. 1807; Balf. F1. Socotra 316. 1888. *H. glaber* Pers. Syn. P1. 2: 533. 1807; P. Beauv. Agrost.134. t. 23 fig. 8. 1812. *H. hirsutus* P. Beauv. Ess. Agrost. 134. 1812. *H. allionii* Roem. & Schult., Syst. Veg. 2: 835. 1817. *H. polystachyus* Nees. Agrost. Bras. 364. 1829. *H. roxburghii* Walk. Arn ex Nees. in Nov. Act. Nat. Curr. 19, suppl. 1.83. 1843. *H. hohenackeri* Hochst. ex Miq. Anal. Bot. Ind. 2: 24. 1856. *H. hispidissmus* Hochst. ex Steud. Syn. P1. Glum. 1: 367. 1855. *H. besukiensis* Miq. F1. Ind. Bat. 3: 494. 1837. *Andropogon contortus* Linn. Sp. P1. 1045. 1753; Lam. Encycl. Meth. Bot 4: 840. 1817; Kunth. Enum. P1. 1: 486. 1838; Hack. Monogr. Androp. 585. 1889; Roxb. F1. Ind. 1: 253. 1820; Grah. Cat. Bomb. P1. 238. 1839; Dalz. & Gibs., Bomb. F1. 300. 1861; Lisboa, Bomb. Grass. 84. 1896; Hook. f. F1. Brit. Ind 7: 199. 1896; Cooke F1. Bomb. 2: 990. 1908. *A. contortus* var. *genuinus* Hack. subsp. *roxburghii* Hack in DC. Mongr. Phan. 6: 586. 1889. *A. secundus* Willd. ex Nees. Agrost. Bras. 364. 1829. *A. allionii* DC. in Lamk., et DC. F1 Fravce 3: 97. 1805. *A. firmus* J.S. Presl. ex C.B. Presl Rel. Haenk. 1: 334. 1830; Kunth. Enum. P1. 1: 486. 1838. *A. besukiensis* Steud. in Zoll. Syst. Verz. 2: 59. 1854. (PLATE–177).

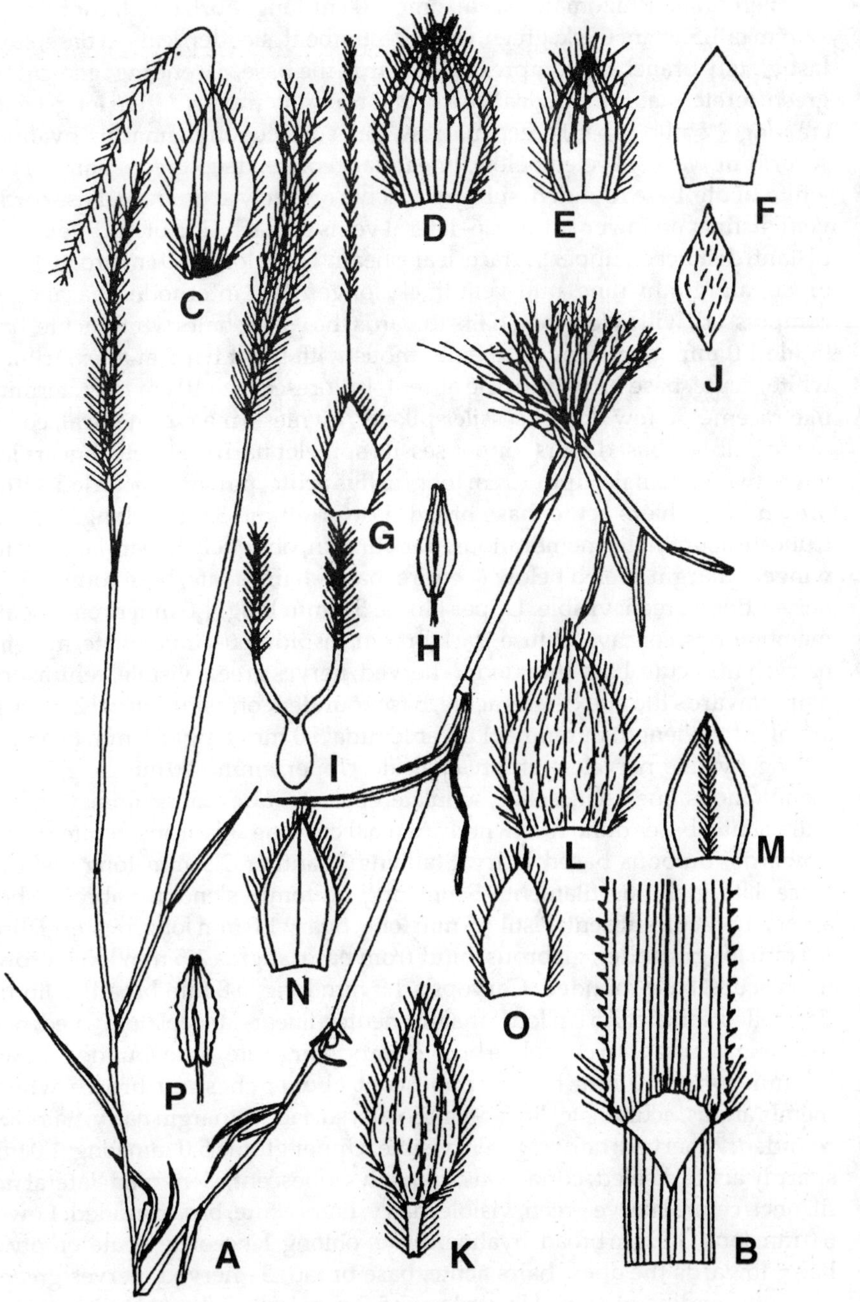

PLATE 177: *Heteropogon contortus* (L.) P. Beauv. ex Roem. & Schult.
A. Habit, B. Ligule, C. Sessile spikelet, D. Lower glume, E. Upper glume, F. Lower lemma, G. Upper lemma, H. Stamen, I. Pistil, J. Caryopsis, K. Pedicelled spikelet, L. lower glume, M. Upper glume, N. Lower lemma, O. Upper lemma, P. Stamen

Perennial. Rhizomatous; rhizome 6.0 cm long, horizontal; root fibrous. Culm 120 cm tall, 5.0 mm thick, green, erect, decumbent, slender, leafy at base, simple, sub–fastigiately branched, compressed towards the base, ascending, geniculate, tufted, green, terete, glabrous; nodes brown, glabrous; internodes 7.0 cm long. Leaf-blade 25 cm long, 1.5 mm broad, green, narrow, flat, surface rough, margin hyaline, sharply scabrid upwards, keeled below, linear–lanceolate, deep green, hairy at base; hairs white, acute, base rounded; sub–erect, acute or rarely acuminate, base cordate, mid–vein distinct on lower surface, 6–lateral veins on each side of mid-vein; lateral vein distant, distinct on upper surface; leaf–sheath 11 cm long, 1.0 cm broad, loose, striate, green, margin hyaline, mid-vein thick, longer than internodes, basal one laterally compressed, with a few long hairs towards the collar; hairs white, acute, base broad; ligule 1.0 mm long, fringed, membranous with short hairs at apex, truncate; hairs white, acute, base broad; auricle absent. Infloresecence 10 cm long, terminal spike-like racemes, a few lower sessile spikelet, awnless, male or neutral, covered with white balbous based hairs, upper sessile spikelet narrow, terete, linear- lanceolate, long–awned, female, upto 10 cm long; callus acute, pungent, bearded with reddish-brown hairs; hairs acute, base broad. Lower glume 5.0 mm long, 1.0 mm broad, truncate acuminate, membranous, dark brown, obliquely linear- lanceolate, obtuse, winged, margin keeled below & entire, hairy; hairs acute, base rounded; 7-nerved; nerves deep green, visible. Upper glume 5.0 mm long, 1.0 mm broad, boat-shaped, membranous, concave, obtuse, dark-brown, hispidulous, mucronate, margin hyaline, hairy; hairs acute, base rounded; 7- nerved; nerves green, visible, central one keeled, hairy towards the apex; hairs acute, base rounded; one side lateral 2- nerves ending at half of the length of glume. Lower lemma 5.0 mm long, 1.5 mm broad, truncate, oblong, hyaline, nerveless, empty, epaleate. Upper lemma 4.0 mm long, 1.0 mm broad, membranous, linear-lanceolate, subulate, epaleate, bisexual, acuminate, margin hairy; hairs acute, base rounded; awned; awn 6.0 cm long, glabrous, white below, brown upwards, bulbous based hairy. Stamens 3; anther 2.5 mm long, yellow, linear-lanceolate, glabrous; filament 0.5 mm long, unifomly slender, glabrous, shorter than anther. Lodicules absent. Pistil 3.5 mm long; ovary 1.0 mm long, ovate, glabrous; style 1.0 mm long, slender, glabrous, bifid from base; stigma 1.5 mm long, brown, hairy; hairs acute, base rounded. Caryopsis 1.5 mm long, 0.8 mm broad, cylindric, hairy. Pedicelled spikelet 7.5 cm long, male or neuter, linear–lanceolate, awned; pedicels 2.0 mm long, acute, with chocolate brown hairs; hair acute, base rounded. Lower glume 7.5 mm long, 1.0 mm broad, pubescent above, chestnut brown when young, membranous, acuminate, linear–lanceolate, surface & margin hairy; hairs acute, base rounded; 7–nerved; nerves green, visible. Upper glume 5.0 mm long, 1.0 mm broad, sharply acute, keeled, coriaceous, sparsely pubescent, 3–nerved; lateral nerves not distinct, central nerve green, visible, hairy; hairs acute, base rounded. Lower lemma 6.0 mm long, 3.0 mm broad, hyaline, ovate, oblong, lanceolate, acute, empty, epaleate, hairy towards the apex; hairs acute, base broad; 3– nerved, nerves green, visible, glabrous, ending at apex. Upper lemm 5.0 mm long, 2.0 mm broad, ovate, oblong, acute, epaleate, membranous, lanceolate, hairy; hairs acute, base broad. Stamens 3; anther 3.0 mm long, yellow, linear-lanceolate, keeled; filament 0.5 mm long, uniformly glabrous, slender, shorter than anther.

Vernacular Names

Dabsuli, Parua, Surbala.

F1owering & Fruiting

August–January; B.H.U, Cholapur, Jaunpur, Chandauli, Mirzapur, Deoria.

Habitat

Common in open fields, along railway tracks & river beds.

Distribution

Tropical and Subtropical countries of the world.

2. *H. fischerianus*

H. fischerianus Bor in Kew Bull. 170. 1957; Bor, Grass. Bur. Cey. Ind. Pak. 165. 1960; Moulik, Grass. Bamb. Ind. 1: 227. 1997. *H. contortus* (Linn.) P. Beauv. ex Roem. et Schult., Syst. Veg. 2. 836. 1817. *H contortus* var. *distichus* C.E.C. Fischer in Gamble, F1. Pres. Madr. 1743. 1934. (PLATE–178).

Perennial. Rhizomatous; root fibrous. Culms 30 cm tall, 1.5 mm thick, green, glabrous, erect, tufted, compressed, ascending, geniculate at base, leafy at base, rooting at base, unbranched, slender, minutely hairy below the inflorescence; hairs acute, base rounded; nodes glabrous; internodes 5.0 cm long, glabrous. Leaf–blade 5.0 cm long, 2.0 mm broad, narrow, acute, base rounded, linear-lanceolate, green, surface glbrous, margin hyaline, scaberulous, midvein distinct on both surface, 3–lateral veins on each side of mid vein; lateral veins distinct on ventral surface; leaf-sheath 2.0 cm long, 1.5 mm broad, green, glabrous, tightly compressed, much shorter than internode, veins distinct on outer surface; ligule 2.5 mm long, membranous, ovate, entire, acute; auricle glabrous. Inflorescence a solitary terminal spike like raceme; raceme 2.5 cm long, 4.0 mm broad, oblong. Spikelets 1.0 cm long, paired, upper sessile, lower pedicellate. Sessile spikelet 7.0 mm long, 2.0 mm broad, hairy, acute, lanceolate, bisexual, callus hairy; hair acute, base broad. Lower glume 7.0 mm long, 1.5 mm broad, chartaceous, linear-lanceolate, acute, oblique, obtuse, surface glabrous, margin scabrid, hairy; hairs 1.0 mm long, acute, base broad; 11–nerved; nerves green, visible, central one ending at apex, lateral nerves ending below the apex; awned; awn 4.0 mm long, antrorsely scabrid. Upper glume 3.5 mm long, 2.0 mm broad, chartaceous, ovate-oblong, lanceolate, obtuse & bifid at apex, surface glabrous, margin inrolled, scaberulous, 3–nerved; nerves green, visible, central one dentate, hairy, denta pectinate; hairs 1.0 mm long upwards, acute, base broad, lateral nerves hairy at apex; hairs acute, base broad; awned; awn 1.1 cm long, arising from sinus, antrorsely scabrid. Lower lemma 3.5 mm long, 1.5 mm broad, hyaline, margin inrolled, ovate, linear-lanceolate, surface glabrous, empty, paleate, acute, margin glabrous from base to slightly below the middle, hairy above; hairs 1.0 mm long, acute, base tubercled; 2–nerved; nerves green, marginal slightly visible. Lower palea 3.5 mm long, 1.5 mm broad, hyaline, broadly lanceolate, slightly ovate, acute, surface glabrous also with inrolled margin, 3–nerved; nerves green, visible, central one glabrous, lateral nerves scabrid at apex. Upper lemma 2.0 mm long, 1.0 mm broad, broadly ovate, lanceolate,

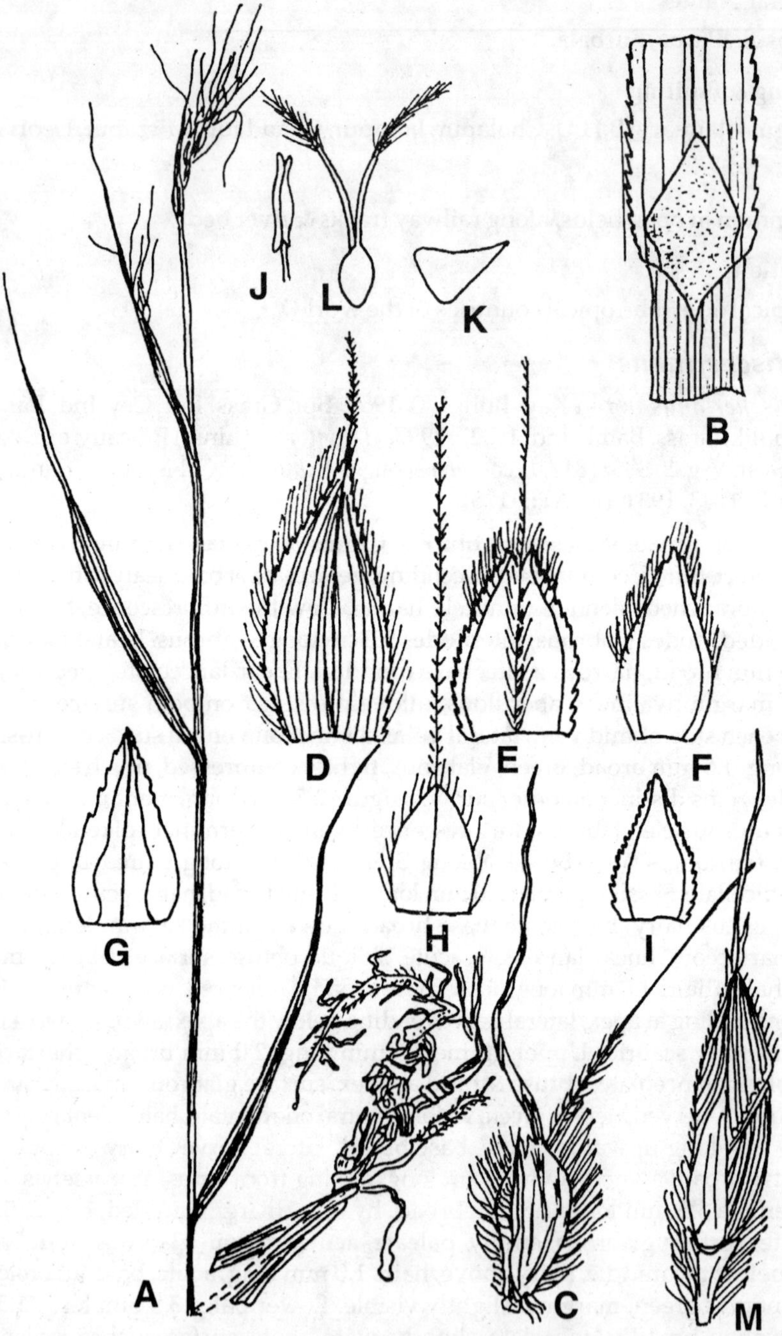

PLATE 178: ***Heteropogon fischerianus*** **Bor**
A. Habit, B. Ligule, C. Sessile spikelet. D. Lower glume, E. Upper glume, F. Lower lemma, G. Lower palea, H. Upper lemma, I. Upper palea, J. stamen, K. Lodicule, L. Pistil, M. Pedicelled spikelet

acute, membranous, bisexual, surface glabrous, margin hairy; hairs 1.2 mm long upward, acute, base tubercled, 1–nerved; nerves green, visible, glabrous; awned; awn 3.0 cm long, antrorsely scabrid, arising from base. Upper palea 2.0 mm long, 1.0 mm broad, hyaline, ovate-lanceolate, surface glabrous, margin scabrid, 2-nerved; nerves slightly visible, light brown, glabrous & marginal. Stamens 3; anther 1.0 mm long, linear, ovate, yellow, glabrous, lobes divaricate; filament 0.5 mm long, slender shorter than anther. Lodicules 2; 0.2 mm long, 0.1 mm broad, hyaline, glabrous, nerveless. Pistil 2.0 mm long; ovary 0.4 mm long, oblong, glabrous, light green; style 0.8 mm broad, glabrous, bifid from base, slender; stigma 0.8 mm long, acuminate brown, equal to style in lengh, hairy; hairs acute, base broad. Pedicelled spikelet 9.0 mm long, 2.0 mm broad, lanceolate, elliptic, hairy, male; pedicel 3.0 mm long 0.3 mm broad, flat, hairy. Lower glume 6.0 mm long, 1.5 mm broad, chartaceous, linear–lanceolate, acute, obtuse, surface glabrous, margin scabrid, hairy; hairs acute, base tubercled; 9 nerved; nerves gereen, visible, central one ending at apex, lateral nerves not reaching at apex, unawned. Upper glume 3.0 mm long, 1.5 mm broad, chartaceous, bifid at apex, ovate-lanceolate, surface & margin glabrous, 3–nerved; nerves green, visible, cental nerves scabrid, lateral nerves glabrous. Lower lemma 3.0 mm long, 1.5 mm broad, broadly ovate, lanceolate, acute, membranous, empty, epaleate, 2–nerved; nerves green, visible, marginal, hairy at apex; hairs 1.0 mm long, acute, base tubercled. Upper lemma 2.0 mm long, 1.0 mm broad, acute, hyaline, male, epaleate, surface & margin glabrous, 1-nerved; nerves green, visible, glabrous. Stamens 3; anther 1.0 mm long, linear, ovate, yellow, glabrous, lobes divaricate; filament 0.5 mm long, slender shorter than anther. Lodicules absent.

Flowering & Fruiting

September–December; Chakiya, Naygarh, Meja, Vindham.

Habitat

Common in moist places.

Distribution

Tropical & sub–tropical countries of the world.

19. *Imperata* Cyr.

Imperata cylindrica (Linn.) P. Beauv. Agrost. 165. t. 5. fig. 1. 1812; Stapf. in F1. Trop. Afr. 9: 187. 1917; Blatter & McCann, in Jaurn. Bomb. Nat. Hist. Soc. 32: 281. 1927; Paeusch, Nom. Bot. ed. 3: 10. 1797; Bor, Grass. Bur. Cey. Ind. Pak. 170. 1960; Raizada *et al.*, Ind. For. Rec. ser. 2. Bot. 4: 182. 1961; Panigrahi & Rajagopal, Proc. Nat. Acad. Sci. Ind. Sec. B. 37: 13. 1967; Singh in Bull. Bot. Sur. Ind. 11: 21. 1969; Panigrahi et at., in Bull. Bot. Sur. Ind 11: 112. 1969; Cope in Nasir & Ali, F1.W. Pak. 143: 152. 1982; Gilliland in F1. Malaya 3: 220. 1971; Roy in Grass. M.P. 101. 1984; Blatter & McCann, Bomb. Grass.42. t. 30. 1984; Shreekumar & Nair, F1. Kera. Grass. 114. 1991; Moulik, Grass. Bamb. Ind. 1: 311. 1997. *I. cylindrica* (L) P. Beauv var. *major* (Nees.) C.E. Hubb. ex Hubb. & Vaughon, Grass. Maur. 96. 1940. *I. cylindrica* var. *koenigii* (Retz.) T.A. Durand & Schinz. Consp F1. Afr. 694. 1894; Fischer in Gamble, F1. Pres. Madr 1707. 1934 (Repr.ed.3: 1184. 1957). *I. koenigii* (Retz.) P. Beauv var. *major* Nees. F1. Afr. Austr. 90.

1941. *I. arundinacea* Cyrill. P1. Rar. Neap. Fasc 2: 26. t. 11. 1792; Hack. Monogr. Androp. 92. 1889; Boiss. F1 Or. 5: 452. 1881; Duthie, Grasses N.W. Ind. 14. 1883; Duthie, Indig Fodder Grass. t. 15. 1886; Duthie, Fodder Grass N. Ind. 22. 1888; Hook. f. F1. Brit. Ind 7: 106. 1896; Lisboa, Bomb. Grass. 45. 1896; Cooke, F1. Bomb. 2: 246. 1908; Hole in Ind. For. Mem. 1: 91. 1911; Rang. & Tad. Handb. South. Ind. Grass 147. t.128. 1921. *I. arundinacea* var. *latifolia* Haines, Bot. Bih & Or. 1015. 1924. *I. alang* Jungh. in Tifdschr. Nat. Gesch. 7: 295. 1840. *I. condensata* Steud. Syn. P1. Glum. 1: 431. 1855. *I. filifolia* Nees ex Steud. Syn. P1. Glum. 1: 432. 1855. *I. koenigii* P. Beauv. Ess. Agrost.65. 1812. *I. pedicellata* Steud. in Flora 24: 22 1846. *Saccharum cylindricum* Lam. Encycl. Meth. Bot. 1: 594. 1791; Roxb. F1. Ind. 1: 234. 1820; Grah. Cat. Bomb. P1. 239. 1839. *S. diandrum* Koen. ex Retz. Obs. Bot. 5: 16. 1789. *S. koenigii* Retz. Obs. Bot. 5: 17. 1789. *S. laguroides* Pourr. in Mem. Acad. Toul. 3: 326. 1788. *Lagurus cylindricus* Linn. Syst. Nat. ed. 10. 878. 1759. *Calamagrostis lagurus* Koel. Descr. Gram. 112. 1802. (PLATE–179).

Strongly perennial. Rhizomatous; rhizome 15 cm long, creeping, white, root stock; root fibrous. Clum 60 cm tall, 2.0 mm thick erect, slender, filiform, tufted, brownish–green, glabrous; nodes hairy; hairs white, acute, base broad; internodes 8.0 cm long, glabrous. Upper most leaf–blade 8.0 cm long, 4.0 mm wide, mucroniform; lower leaf–blade 35 cm long, 12 mm wide, erect, linear- lanceolate, acute, base narrow & tapering, mostly collected at base, glaucous, surface glabrous, margin scabrid, green, mid vein white, distinct on lower surface, 4–lateral veins on each side of mid vein, lateral veins distinct on upper surface; leaf–sheath 6.0 cm long, 4.0 mm broad, loose, glabrous, shorter than internodes, basal sheath disintegrated into fibres, mouth villous; ligule 2.0 mm long, lacerate, membranous, ovate, thin, truncate, dorsaly silky; auricle glabrous. Inflorescence 6.0 cm long, compact, cylindrical panicle, very dense; branches & branchlets numerous, crowded, appressd. Spikelets 3.0 mm long, paired, each 1–flowered, bisexual, linear-lanceolate, with long white silky hairs from the callus & glumes; hairs 1.0 cm long, acute, base rounded; purple stigma hanging at matunity, glabrous, pubescent, pedicel 1.0 mm long with fine clavate tips. Lower glume 3.0 mm long, 1.0 mm broad, acute, lanceolate, apex hyaline, membranous below, slightly thickend toward the base, margin incurved, 5–nerved, hairy in lower half; hairs white, acute, base rounded, overtopping the glume. Upper glume 2.5 mm long, 1.0 mm broad, acute, lanceolate, membranous, margin hairy from base to below the apex; hairs 5.0 mm long, white, acute, base rounded; glabrous at apex, 5–nerved; nerves faintly visible, ending at apex, glabrous. Lower lemma 1.5 mm long, 1.0 mm broad, hyaline, epaleate, slightly 3-lobed at apex, sub–truncate, oblong, lacinitate, denticulate, lanceolate, hairy at apex; hairs acute, base rounded. Upper lemma 1.0 mm long, 0.5 mm broad, elliptic, ovate, lanceolate, slightly each lobe acuminate, lacinitate, denticulate, 3–lobed at apex, margin hairy at apex; hairs acute. base broad; hyaline, nerveless. Upper palea 1.0 mm long, 0.5 mm broad, obovate, quadrate, hyaline, nerveless, apex denticulate or sometimes lacinitate. Stamens 2; anther 1.5 mm long, apiculate, base rounded, glabrous, yellow at young, purple at maturity; filament 0.5 mm long, glabrous, slender, shorter than anther. Pistil 2.5 mm long; ovary 0.5 mm long, ovate, glabrous; style 1.0 mm long, glabrous, slender, bifid from base; stigma 1.0 mm long, purple, hairy; hairs acute with broad base. Caryopsis 1.0 mm long, oblong. Pedicelleded spikelets similar to sessile spikelets.

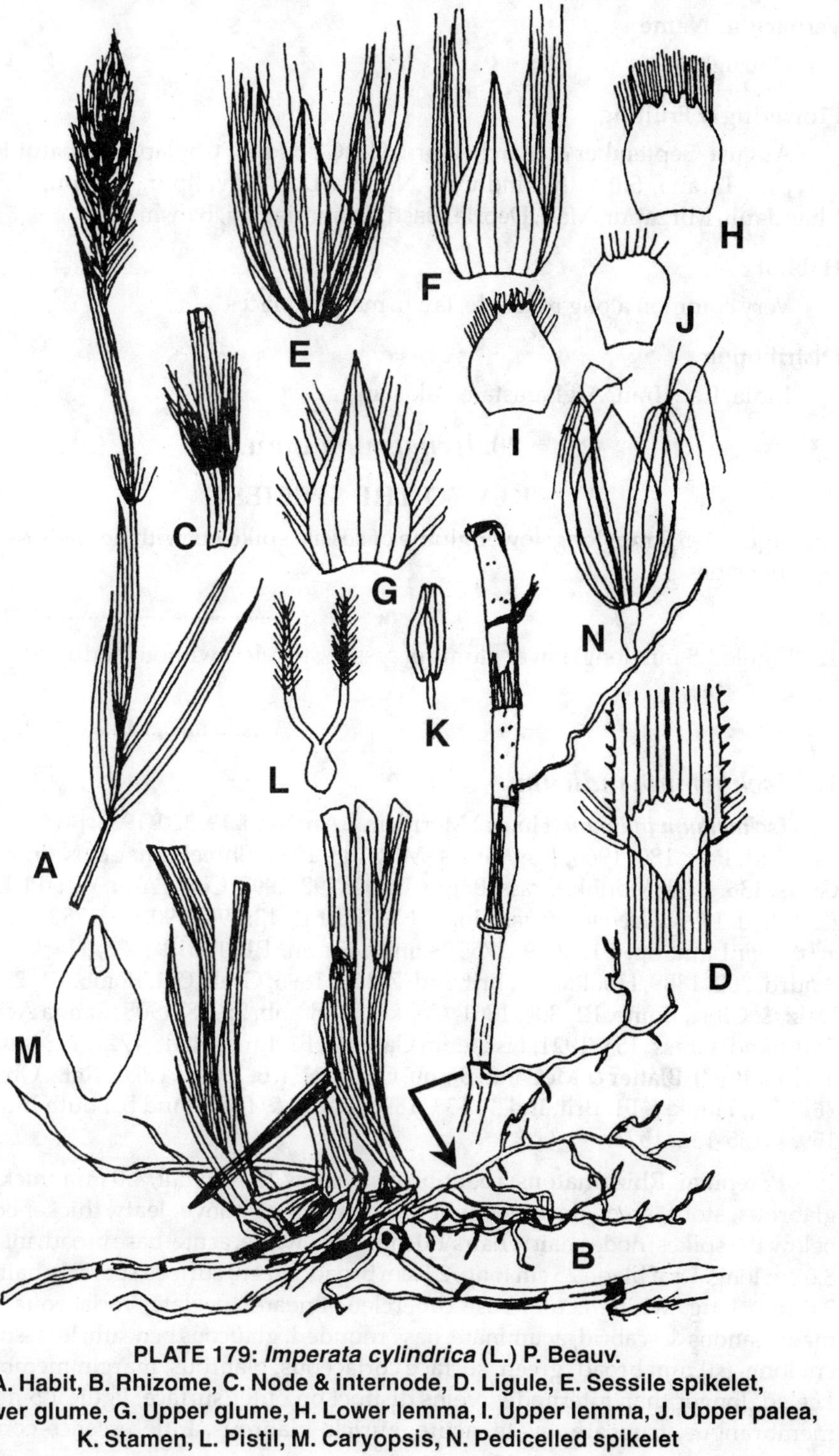

PLATE 179: *Imperata cylindrica* (L.) P. Beauv.
A. Habit, B. Rhizome, C. Node & internode, D. Ligule, E. Sessile spikelet, F. Lower glume, G. Upper glume, H. Lower lemma, I. Upper lemma, J. Upper palea, K. Stamen, L. Pistil, M. Caryopsis, N. Pedicelled spikelet

Vernacular Name

Dhirughas.

Flowering & Fruiting

August–September & March–April; U.P.C., B.H.U., Cholapur, Babatur Harrua, Shivpur, Tarana, Sarnath, Sindhora, Natinia Dai, Navallpur, Jaunpur, Chakiya, Chandauli, Mirzapur, Mau, Deoria, Basti, Dumariaganj, Banshi.

Habitat

Very common along roadside, lawn, marshy places.

Distribution

India, Iraq, Iran, Afghanistan, Pakistan.

20. *Ischaemum* Linn.

KEY TO THE SPECIES

1a. Ligule 4–5.0 mm long; lower glume of sessile spikelets with noduels & rugosities

.... **2. *I. rugosum***

1b. Ligule 1.5 mm long; lower glume of sessile spikelets without nodules & rugosities

.... **1. *I. indicum***

1. *Ischaemum indicum*

Ischaemum indicum (Houtt.) Merril. in J. Arn. Arb. 19: 320. 1938; Bor, Grass. Bur. Cey. Ind. Pak. 180. 1960; Roy, Grass. M.P. 103. 1984; Shreekumar & Nair, F1. Kera. Grass. 136. 1991; Moulik, Grass. Bamb. Ind. 1: 292. 1997; Cook, Aqu. Wet. Pl. Ind. 299. f. 307 a–d. 1996. *Phleum indicum* Houtt. Nat. Hist 11, 13: 198. t. 90. t. 2. 1782. *Ischaemum aristatum* Linn. Sp. P1. 1049. 1753; Kunth. Enum. P1. 1: 512 1833; Hack. Monogr. Andro. 202. 1889; Hook. f. F1. Brit. Ind. 7: 126. 1896; Grah. Cat. Bomb. P1. 239. 1839; Dalz. & Gibs., Bomb. P1. 306. 1861; Cooke, F1. Bomb. 2: 958. 1908; Ranga Achariyar, South Ind. Grass. 151. 1921; Fischer in Gamble, F1. Pres. Madr. 1721. 1934 (Repr. ed. 3: 1194. 1957); Blatter & McCann, Bomb. Grass. 11. t. 6. 1984. *I. ciliare* Retz. Obs. Bot. 6: 36. 1791; Hook. f. F1. Brit. Ind. 7: 133. 1896; Rang. & Tad. Hand b. South Ind. Grass. 159. t. 135-136. 1921.

Perennial. Rhizomatous; root fibrous. Culms 120 cm tall, 4.0 mm thick, green, glabrous, stout, erect, decumbent at base, branched above, leafy, thickened at top below the spikes; nodes hairy; hairs 1.0 mm long, white, acute, base broad; internodes 8.0 cm long. Leaf blade 25 cm long, 2.5 cm broad, green, surface sparsely hairy; hairs 2.0 mm long, white, acute, base tubercled; linear–lancelate, coriaceous, margin membranous & scabrid, acuminate, base rounded, glaucous beneath; leaf–sheath 9.0 cm long. 6.0 mm broad, green, surface coriaceous, glabrous, margin membranous, keeled, longer than internodes, veins distinct on outer surface; ligule 1.5 mm long, membranous, truncate, ovate, acute; auricle glabrous. Inflorescence consists of racemes; raceme 5.0 cm long, usually 2 in number, stout, slender; joints 2.0 mm long, turbinate, densely villous. Sessile spikelet 5.0 mm long, 2.0 mm broad, ovate, oblong, awned; callus shortly bearded. Lower glume 5.0 mm long, 1.5 mm broad, sub–coriaceous below, chartaceous above, oblong, ovate, often purplish below the middle, with 2–4 marginal nodules on each edge, transverselly connected with shallow ridges, green, anastomosing 9– nerves above the middle, tip obtuse, bidentate, broadly winged on either side in the upper half; wings ear shaped, margin long ciliolate below the wings. Upper glume 5.0 mm long, 20 mm broad, chartaceous, boat–shaped, dorsaly keeled, minutely winged & ciliate along the keeles, aristate; arista 3.0 mm long, 5-nerved. Lower lemma 4.0 mm long, 1.5 mm broad, chartaceous, ovate, lanceolate, margin delicate, hyaline, male, paleate, faintly 3–nerved. Lower palea 3.5 mm long, 1.0 mm broad, chartaceous, elliptic, lanceolate, margin infolded, hyaline, 2–keeled, 2–nerved. Stames 3; anther 2.0 mm long, yellow, glabrouos; filament 0.2 mm long, slender, glabrous, shorter than anther. Upper lemma 4.0 mm long, 2.0 mm broad, hyaline, delicate, notched, bisexual, paleate, faintly 3–nerved, awned; awn 15 mm long, antrorsely scabrid; column 6.0 mm long. Upper palea 3.5 mm long, 1.0 mm broad, hyaline, delicate, oblong, 2–keeled, 2–nerved. Stamens 3; anther 2.0 mm long, yellow, linear, glabrous; filament 0.5 mm long, slender, glabrous, shorter than anther. Lodicules absent. Pistil 3.5 mm long; ovary 0.5 mm long, oblong, glabrous; style 1.5 mm long, slender, glabrous, bifid from base; stigma 1.5 mm long, hairy; hairs acute, base broad in 2- rows. Caryopsis 1.5 mm long, 1.0 mm broad, with a persistent style. Pedicelled spikelets 5.0 mm long, ovate, acute, awned; pedicels 2.0 mm long, densely villous. Lower glume 4.0 mm long, 1.5 mm broad, chartaceous, ovate, lanceolate, acuminate, winged on one margin, densely villous in the upper half, 11-nerved. Upper glume & florets similar to those of sessile spikelts.

Flowering & Fruiting

August-December; Jaunpur, Naugarh, Rajdari, Kushinagar.

Habitat

Very common along the bunds of paddy fields and roadsides, usually in moist places.

Distribution

India, South-East Asia, Sri Lanka.

2. *I. rugosum*

I. rugosum Salisb. Icon. Strip. Rar. 1: t. 1. 1791; Moulik, Grass. Bamb. Ind. 1: 226. 1997; Hook. f. F1. Brit. Ind. 7: 127. 1896; Raizada *et al.*, Indian Forester Rec. ser. 2. Bot. 4: 200. 1961; Gilliland in F1. Malaya 3: 259. 1971; Cope in F1. W. Pak. 143: 321. 1982; Bor, Grass. Bur. Cey. Ind. Pak. 184. 1960; Rang. & Tad. Handb. S. Ind. Grass. 153. t. 132. 1921; Fischer in Gamble, F1. Pres. Madr. 1722. 1934 (Repr.ed. 3: 1194. 1957); Blatter & McCann, Bomb. Grass. 12. t. 7.1984; Roy, Grass. M.P. 104. 1984; Shreekumar & Nair, F1. Kera. Grass. 159. 1991; Cook, Aqu. Wet. Pl. Ind. 301. f. 309 a–b. 1996; Roxb. F1 Ind. 1: 320. 1820. Hack. Monogr. Androp. 206. 1889; Duthie, Grass. N.W. Ind. 18. 1883; Duthie, Fodder Grasses N. Ind. 31. 1888; Grah, Cat. Bomb. P1. 239. 1839; Dalz.

& Gibs. Bomb. F1. 305. 1861; Watt. Dict. Econ. Prodr. 4: 531. 1828; Cooke, F1. Bomb. 2: 959. 1908; Ranga Achariyar, South Ind. Grass. 31. 1921; Haines, Bot. Bih. & Or. 1021. 1924. *I. muticum* L. Sp. P1. 1049. 1753. *I. segetum* Trin. in Mem. Acad. Petrsb. ser. 6, 2: 294. 1833. *Meoschium griffithii* Nees & Arn. in Ann. & Mag. Nat. Hist. 1: 284. 1838. *M. arnottianum* Nees in Nov. Act. Nat. Curr. 19, Suppl. 1: 199. 1843. *M. elegans* Arn. & Nees in Nov. Act. Nat. Curr. 19. Suppl. 1: 199. 1843; Nees Syn. P1. Glum. 1: 210. 1854. *M. rugosum* Nees in Wall, Cat. no. 8864. 1828. *M. wightianum* Nees in Nov. Acat. Nat. Cur. 19: Supp. 1: 199. 1843. *M. royleanum* Nees ex Steud. Syn. Gram. 375. 1854. *Calladoa distachya* Cav. Ic. 37. t. 460. 1799. *Meoschium rugosum* Nees in Nov. Act. Nat. Curr. 19. Suppl. 1: 199. 1843. *Andropogon arnottianus* Steud. Syn. P1. Glum. 375. 1854. *A. griffithsiae* Steud. Syn. P1. Glum. 375. 1854. *A. segetum* Steud. Syn. P1. Glum. 376. 1854. *A. tong-dong* Steud. Syn. P1. Glum. 376. l854. (PLATE–180).

Annual. Rhizome absent; root fibrous. Clums 50 cm tall, 5.0 mm thick, erect or ascending, tufted, stoloniferous, geniculate, slightly thickend beneath the inflorescence, leafy, compressed, decumbent, green, glabrous, rooting at base; upper nodes hairy; hairs 1.0 mm long, white, acute, base broad; lower nodes brown, glabrous; internodes 8.0 cm long. Leaf–blade 20 cm long, 1.0 cm broad, narrow, flat, acuminate, base cordate, linear-lanceolate, green, surface glabrous, margin acabrid, hyaline, hairy; hairs acute, base rounded; midvein thick & distinct on both surface; 5–lateral veins on each side of mid–vein; lateral veins distinct on lower surface; upper most leaves often reduced to spathiform, lanceolate, sheath which partially enclosed the racemes; leaf-sheath 10 cm long, 8.0 mm broad, loose, compressed, green, coricaceous, margin hyaline, mid-vein thick, longer than internodes; ligule 5.0 mm long, membranous, truncate, ovate, subacute, entire; auricle glabrous. Inflorescence raceme; racemes 10 cm long, usually 2, erect, pale yellow, fragile; rachis trigonous, ciliate on dorsal angle. Spikelet linear, oblong, paired, geminate, remain closed set & erect, for long time peduncle long, slender; joints trigonous. Sessile spikelet 5.5 mm long, 2.0 mm broad, ovate, oblong, obtuse, lanceolate, awned; callus bearded. Lower glume 5.5 mm long, 2.5 mm broad, lower 2/3 part chartaceous below with prominent side nodules connected by deeper shallow, ridges & furrows, pale yellow, shining, concave, margin keeled, upper 1/3 part membranous, flattened, thinner, obtuse, 9–nerved; nerves green, visible, glabrous, slightly oblique, acuminate, oblong, ovate. Upper glume 5.5 mm long, 1.5 mm broad, membranous, boat-shaped, oblong- lanceolate when spread, slightly 2-dentate, acuminate, keeled; keel minutely winged & scabrid, hairy; hairs acute, base rounded. Lower lemma 4.5 mm long, 1.0 mm broad, delicate, lanceolate, acute, nerveless, paleate, empty. Lower palea 4.5 mm long, 1.0 mm broad, hyaline, delicate, oblong, elliptic, lanceolate, 2 keeled, margin infolded, 2- nerved; nerves green, visible, glabrous & marginal ending at apex. Upper lemma 4.0 mm long, 1.5 mm broad, delicate, a few wart like growth towards the apex, scabrid below, bisexual, oblong, deeply cleft into 2–acute lobes, hyaline, margin scabrid, awned; awn 1.2 cm long, geniculate, glabrous. Upper palea 3.0 mm long, 1.0 mm broad, hyaline, delicate, acute, bifid, oblong, linear–lanceolate, elliptic, margin infolded. Stamens 3; anther 2.0 mm long, yellow, glabrous, linear–lanceolate; filament 0.5 mm long, uniformly slender, glabrous, shorter than anther. Pistil 3.5 mm long; ovary 0.5 mm long, ovate, glabrous; style 1.0 mm long, glabrous, slender, bifid from base; stigma 2.0 mm long,

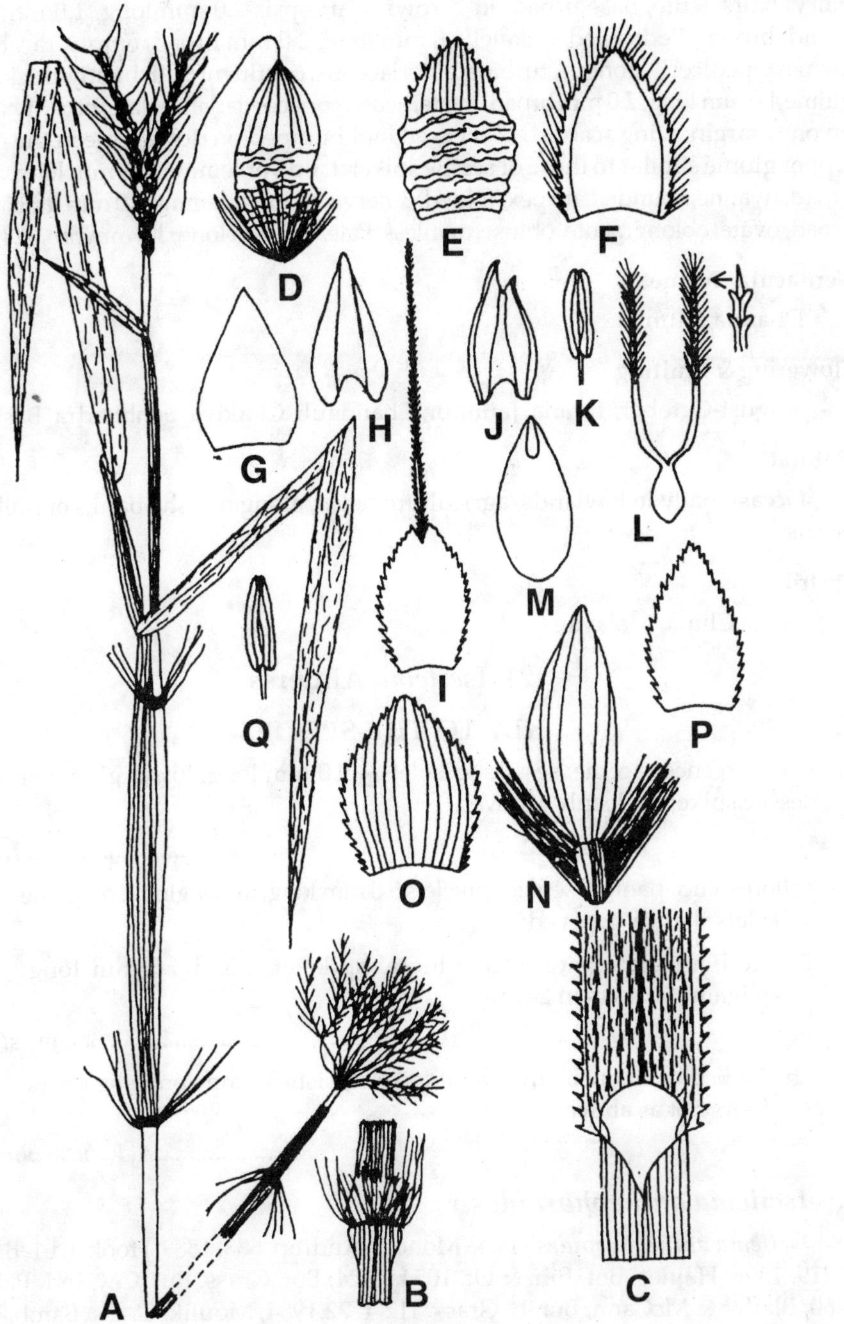

PLATE 180: ***Ischaemum rugosum*** **Salisb.**

A. Habit, B. Node & internode, C. Ligule, D. Sessile spikelet, E. Lower glume, F. Upper glume, G. Lower lemma, H. Lower palea, I. Upper lemma, J. Upper palea, K. Stamen, L. Pistil, M. Caryopsis, N. Pedicelled spikelet, O. Lower glume, P. Upper glume, Q. Stamen

hairy; hairs acute, base broad, in 2–rows. Caryopsis 2.0 mm long, 1.0 mm broad, ovoid, brown. Pedicelled spikelet 4.0 mm long, 2.0 mm broad, empty, rarely male, awnless; pedicel 3–gonous, turbinate, coriaceous, confluent with bristly, thick. Lower glume 4.0 mm long, 2.0 mm broad, ovate, acute, coriaceous, lanceolate, broadly winged, on one margin; wing scabrid, 11–nerved, not humped on dorsal side in the middle. Upper glume similar to those of sessile spikelet. Lower lemma, 2.0 mm long, 1.0 mm broad, hyaline, rounded at apex, folded, 3-nerved. Upper lemma 2.0 mm long, 1.2 mm broad, ovate, oblong, acute, obtuse, awnless. Palea 1.5 mm long, 1.0 mm broad, hyaline.

Vernacular Names

Dhanua, Murchi.

Flowering & Fruiting

August-October; Tarana, Jaunpur, Chandauli, Chakiya, Sonbhadra, Basti.

Habitat

Occasionally in lowlands, agriculture fields, along marshy banks of nullas and ponds.

Distribution

India, China, Malaya.

21. *Iseilema* Anderss

KEY TO THE SPECIES

1a. Inflorescence raceme; sessile spikelets ± 9.0 mm long; lower glume of sessile spikelets deeply grooved

.... **2. *I. laxum***

1b. Inflorescence panicle; sessile spikelets 5.0 mm long; lower glume of sessile spikelets slightly grooved:

2a. Lodicules obovate in involucral spikelet; pistil 1.5 mm long; stigamatic hairs in 2–rows

.... **3. *I. prostratum***

2b. Lodicules absent in involucral spikelet; pistil 3.0 mm long, stigamatic hairs not as above

.... **1. *I. anthephoroides***

1. *Iseilema anthephoroides*

Iseilema anthephoroides Hack. Monogr. Androp. 683. 1889; Hook. f. F1. Brit. Ind 7: 219. 1896; Haines, Bot. Bih. & Or. 1054. 1924; Bor, Grass. Bur. Cey. Ind. Pak. 187. 1960; Blatter & McCann, Bomb. Grass. 112. t. 72.1984; Moulik, Grass. Bamb. Ind. 1: 235. 1997. (PLATE–181).

Annual. Rhizome absent; root fibrous. Culms 45 cm tall, 2.0 mm thick, tufted, glabrous, base creeping, decumbent, green, rarely pink, very leafy below with branch;

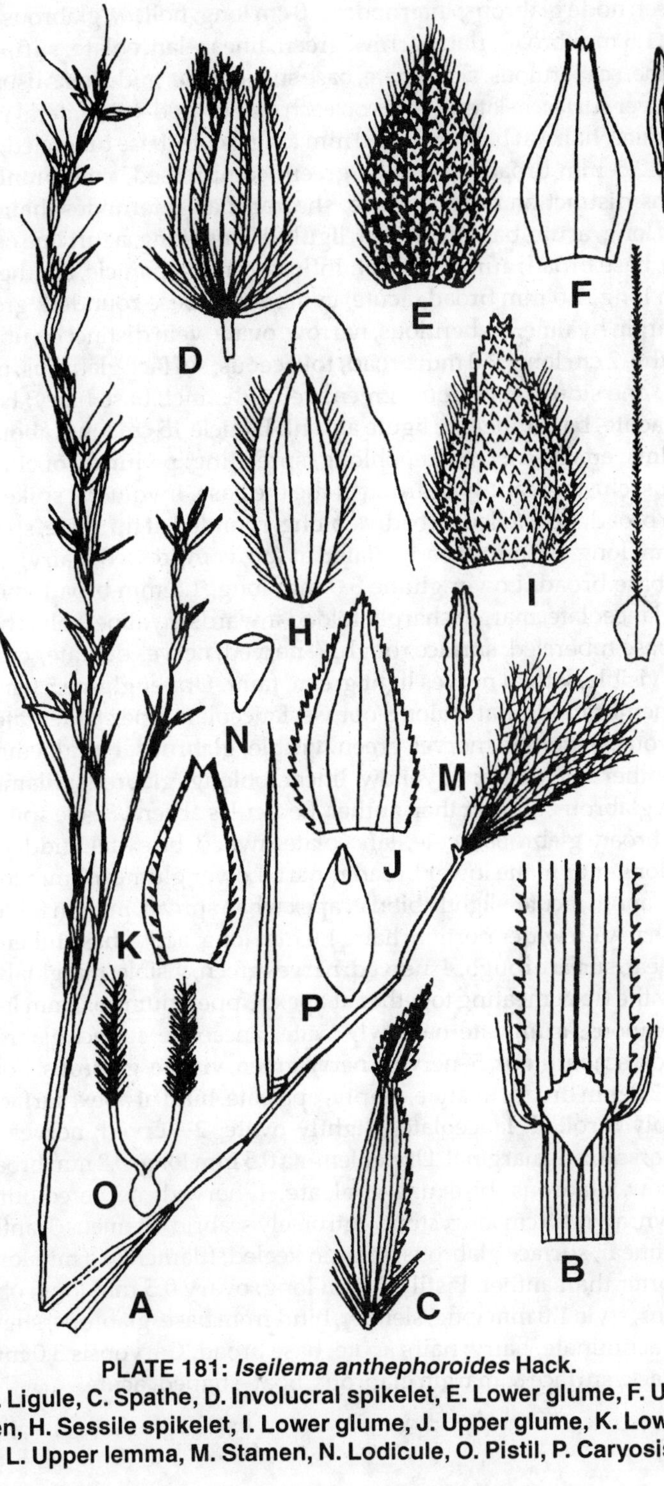

PLATE 181: *Iseilema anthephoroides* Hack.
A. Habit, B. Ligule, C. Spathe, D. Involucral spikelet, E. Lower glume, F. Upper glume, G. Stamen, H. Sessile spikelet, I. Lower glume, J. Upper glume, K. Lower lemma, L. Upper lemma, M. Stamen, N. Lodicule, O. Pistil, P. Caryosis

branches erect; node glabrous; internodes 7.0 cm long, hollow, glabrous. Leaf–blade 3.5 cm long, 1.5 mm broad, flat, narrow, green, linear–lanceolate, surface glabrous, margin hyaline, scaberulous, acuminate, base sub–obtuse, mid–veins usually uniform distinct on lower surfac, 3–lateral veins on each side of mid-vein; lateral veins distinct on upper surface, hairy at base; hairs 1.0 mm long, acute, base tubercled; leaf–sheath 4.0 cm long, 2.5 mm broad, glabrous, green, compressed, membranous, margin hyaline, veins distinct on outer surface, shorter than internodes, hairy at mouth; hairs 1.0 mm long, acute, base tubercled; ligule 0.5 mm long, a tuft line of hairs; hairs white, acute, base broad; auricle absent. Inflorescence a panicle, spatheolate; spath blade 1.0 cm long, 1.0 mm broad, acute, cymbiform, base rounded, green, surface glabrous, margin hyaline, scaberulous, narrow, ovate, vein distinct on upper surface; spathe-sheath 1.2 cm long, 2.0 mm broad, foliaceous, surface glabrous, margin one–side hyaline, one side scaberulous, green, veins distinct; base hairy; hairs 1.0 mm long, white, acute, base tubercle; ligule absent. Panicle 15 cm long, about 1/3 to the lenght of culms, erect, strict, linear, oblong, spreading; peduncle of cluster 2.5 mm long, glabrous; cluster scantily bearded at the the base. Involucral spikelets 5.0 mm long, 1.5 mm broad, pedicelled, broadly oblong, rounded at tip, male, sparsely hairy; pedicel 1.0 mm long, 0.5 mm broad, dilated at top, compressed, hairy; hairs 1.0 mm long, acute, base broad. Lower glume 5.0 mm long, 1.5 mm broad, membranous, acute, ovate, lanceolate, margin sharply folded inwards, hyaline, hairy; hairs 1.0 mm long. acute, base tubercled; surface rough, 7–nerved; nerves dentate, central nerves deep, green, visible, lateral nerves light green, faint. Upper glume 5.0 mm long, 1.5 mm broad, membranous, flat, oblong-obtuse, lanceolate, apex tridentate, surface & margin glabrous, 3–nerved; nerves green, visible, glabrous. Floral lemmas absent. Stamens 3; anther 2.5 mm long, yellow, linear, oblong, glabrous; filament 0.5 mm long, slender, glabrous, shorter than anther. Lodicules absent. Sesile spikelet 6.0 mm long, 1.5 mm broad, glabrous, acute, lanceolate, awned, bisexual, suddenly tapering part or beak longer than the lower broader part. Lower glume 6.0 mm long, 1.5 mm broad, membranous, acute, slightly bifid at apex when spread, margin folded inwards, scaberulous, hairy on wider portion; hairs 1.0 mm long, acute, base tubercled, ovate-lanceolate, linear, surface rough, 4–nerved; nerves green, visible, marginal & glabrous, ending below the tip & meating together at apex. Upper glume 6.0 mm long, 1.5 mm broad, membranous, acuminate, narrowly ovate, lanceolate, surface glabrous, margin folded inwards, scaberulous, 5–nerved; nerve green, visible, glabrous. Lower lemma 4.0 mm long, 2.0 mm broad, hyaline, empty, epaleate, bifid at apex, surface glabrous, margin sharply inrolled, lanceolate, slightly ovate, 2–nerved; nerves colourless, slightly visible, scabrid, marginal. Upper lemma 0.5 mm long, 0.2 mm broad, hyaline, slender, narrow, glabrous, bisexual, epaleate, 1–nerved; nerve colourless, faint, reduced to awn; awn 1.5 cm long, stout, antrorsely scabrid. Stamens 3; anther 2.5 mm long, yellow, linear, surface glabrous, margin keeled; filament 0.5 mm long, slender, glabrous, shorter than anther. Pistil 3.0 mm long, ovary 0.5 mm long, oblong, light green, glabrous; style 1.0 mm long, slender, bifid from base, glabrous; stigma 1.5 mm long, brown, acuminate, hairy; hairs acute, base broad. Caryopsis 3.0 mm long, 1.0 mm broad, black, surface & margin glabrous, boat–shaped, acute.

Flowering & Fruiting

July–December, Narayanpur, Chakiya, Chandauli, Mirzapur, Basti.

Habitat

In damp places and open grasslands.

Distribution

Western India, Asia, Australia.

2. *I. laxum*

I. laxum R. Br. Prodr. 205. 1810; Hack. Monogr. Androp. 682. 1889; Lisboa, Bomb. Grass. 88. 1896; Duthie, Fodder Grass. N. Ind. 43. 1888; Hook. f. F1 Brit. Ind. 7: 218. 1896; Cooke, F1. Bomb. 2: 996. 1908, Roy, Grass. M.P. 106. 1984; Moulik, Grass. Bamb. Ind. 1: 226. 1997.

Perennial. Rhizomatous; root fibrous wiry. Culm 50 cm tall, 2.0 mm thick, green, ascending from a stout, hard sometimes shortly creeping root stock base, very slender, simple, glabrous, leafy. Leaf–blade 15 cm long, 3.0 mm broad, all cauline, the lower radical with equitant sheath, linear, obtuse, hairy near the base, scaberulous above, margin scabrid; sheath lax, quite glabrous, ligule short, truncate, membranous, ciliate. Inflorescence consists of panicle; panicle 20 cm long, racemiform, narrow, distant, axillary, pedunclate fascicles; fascicles 13 mm long, outer spathes 2.5 cm long, narrowly lanceolate, surface glabrous, margin with few hairs, proper spathes 13 mm long, elliptic, lanceolate. Involucral spikelet 4.0 mm long, truly whorled, pedicellate, oblong, lanceolate, male, acute, pedicels shorts, flattened, bearded at base. Lower glume 3.5 mm long, 2.0 mm broad, membranous, acute, sparsely hairy, 5–nerved. Upper glume 3.5 mm long, 2.0 mm broad, oblong, lanceolate, acuminate,sparsely hairy, 3–nerved. Lower lemma 3.0 mm long, 1.0 mm broad, hyaline,linear, glabrous. Stamens 3; anther 3.0 mm long, yellow, linear, narrow. Lodicules 2; 0.3 mm long, 0.2 mm broad, hyaline, cuneate, retuse. Pedicellate spikelets on long hairy pedicels, lanceolate, glabrous. Sessile spikelets 5.0 mm long, narrowly lanceolate, narrowed into a slender stipe, glabrous, bisexual, awned. Lower glume 5.0 mm long, 2.5 mm broad, chartaceous, bifid, truncate, margin scabrid at the apex, glabrous below, thin, faintly 5–nerved. Upper glume 4.5 mm long, 2.0 mm broad, chartaceous, lanceolate, acuminate, scabrid at the apex, glabrous below, 1- nerved. Lower lemma 4.0 mm long, 2.0 mm broad, hyaline, linear, glabrous, empty, epaleate. Upper lemma 4.0 mm long, 2.0 mm broad, hyaline, bisexual, epaleate, 2–lobed, awned; awn 3.5 cm long, antrorsely scabrid. Upper palea 3.0 mm long, 1.5 mm broad, hyaline, ciliate. Stamens 3; anther 3.0 mm long, yellow, linear, glabrous; filament 1.0 mm long, slender, glabrous, shorter than anther. Lodicules absent. Pistil 4.5 mm long; ovary 0.5 mm long, glabrous; style 1.5 mm long, slender, glabrous, bifid from base; stigma 2.5 mm long, plumose. Caryopsis 3.5 mm long, 1.5 mm broad, boat–shaped.

Flowering & Fruiting

Octover–November; Sindhora, Mariyahun, Mardah, Ballia, Khadda.

Habitat

Common along the roadsides.

Distribution

India, Sri Lanka.

3. *I. prostratum*

I. prostratum (L) Anderss. Nova. Acta. Regiae Soc. Sci: Upsal. ser.3.2: 251.1856; Bor, Grass. Bur. Cey. Ind. Pak. 188.1960; Moulik, Grass. Bamb Ind. 1: 236.1997;.Raizada *et al.*, India Forest. Rec. Sec.2. Bot. 4: 257.1961; Cope in Fl.W. Pak.143: 315.1982. *Andropogon prostratum* L., Mant. Alt. 304.1771. *Iseilema wightii* Anderss. in Nov. Acta. Soc. Regiae Sci. Upsal. ser. 3,2: 251. 1856; Hack. Monogr. Androp. 679.1889; Lisboa, Bomb. Grasses 87. 1896; Hook. f. F1. Brit. Ind. 7: 218.1896; Cooke, F1. Bomb.2: 996.1908; Haines, Bot.Bih.& Or. 1055. 1924; Blatter & McCann, Bomb. Grass.113.1984. *Anthistiria prostata* Willd. Sp. Pl. 5: 901.1760; Roxb. F1. Ind. 1: 250. 1820; *A. cimicina* Edgew. in Journ. Soc. Beng. 182.1852; Duthie, Grasses N.W. Ind. 24. 1883. *A. wightii* Nees ex Steud. Syn. Pl. Glum. 400.1855. *Cymbopogon glandulosus* Spreng.Pug.2: 14.1858. (PLATE–182).

Perennial. Rhizomatous; root fibrous. Culms 30 cm tall, 2.0 mm thick, cylindrical, tufted, compressed, prostrate, ascending, decumbent, green, glabrous, rooting at base, branched; branches slender; nodes hairy; hairs white, acute, base broad.; internodes 5.0 cm long, hollow. Leaf–blade 8.0 cm long, 2.0 mm broad, narrow, flat, linear, acute, base rounded, glaucescent or sometimes as well as the whole plant reddish, surface glabrous, margin hyaline, glabrous from base to middle, scaberulous above, mid–vein distinct on lower surface, 3-lateral veins on each side of mid-vein; lateral veins distinct on upper surface; leaf–sheath 4.0 cm long, 2.5 mm broad, lax, compressed, green, glabrous, margin hyaline, shorter than internodes, veins distinct on outer surface, membranous; ligule 1.5mm long, membranous, truncate, apex hairy; hairs 0.5 mm long, white, acute, base broad. Inforescence a panicle, spatheolate; spathe-blade 1.0 cm long, 1.5 mm broad, acute, base cordate, green, linear- lanceolate, surface glabrous, margin keeled, scaberulous, hyaline, faintly veined; spathe–sheath 1.8 cm long, 3.0mm broad, green, surface glabrous, one side margin hyline, glabrous, one side tubercled, veins distinct, more or rarely less compound. Involucral spikelets 4.5mm long, 1.5 mm broad, oblong- lanceolate, male. Lower glume 4.5mm long, 1.5mm broad, acute- oblong, linear–lanceolate, membranous, surface scabrid, margin scaberulous, sharply folded inwards, hairy; hairs 1.5 mm long, acute, base tubercled, apex tridentate, 5–nerved; central nerve deep green, thicker than lateral nerves, all nerves visible on lower surface. Upper glume 3.0 mm long, 1.0mm broad, membranous, inflexed on margins, surface & margin glabrous, apex muticous, winged; wing hyaline, bicleft, ovate, lanceolate, 1–nerved; nerves green, visible. Lemma 3.0 long, 0.5 mm broad, quadrate, hyaline, male, nerveless, surface & margin glabrous, scale–like, epaleate, linear–lanceolate. Stamens 3; anther 2.0 mm long, light green, glabrous, linear; filament 0.5 mm long, glabrous, slender, shorter than anther. Lodicules 2; 0.5 mm long, 0.1 mm broad, obovate, cuneate, glabrous nerveless. Pedicelled spikelets same as involucral spikelet, but more delicate; callus 1.0 mm long, 0.5mm broad,

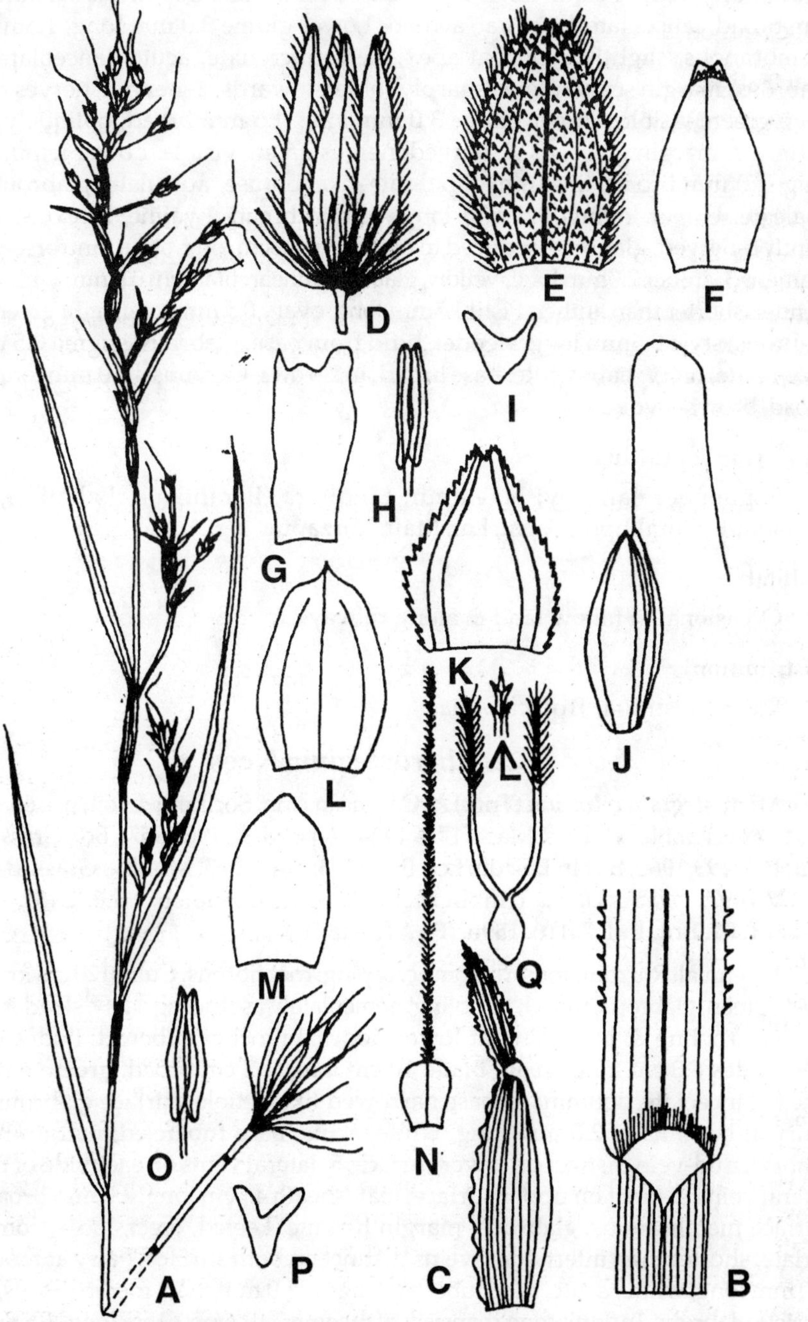

PLATE 182: *Iseilema prostratum* (L.) Anderss.

A. Habit, B. Ligule, C. Spathe, D. Involucral spikelet, E. Lower glume, F. Upper glume, G. Lemma, H. Stamen, I. Lodicule, J. Sessile spikelet, K. Lower glume, L. Upper glume, M. Lower lemma, N. Upper lemma, O. Stamen, P. Lodicule, Q. Pistil

white, hairy; hairs 1.0 mm long, acute, base broad. Sessile spikelets 3.0 mm long, 1.5 mm broad, lanceolate, bisexual, awned. Lower glume 3.0 mm long, 1.5mm broad, membranous, slightly bicleft at apex; cleft mucronate, acute, lanceolate, surface glabrous, margin scaberulous, sharply folded inwards, 4–nerved; nerves marginal, deep green, visible. Upper glume 3.0 mm long, 1.5 mm broad, obliquely carniate, surface & margin glabrous, 3–nerved; nerves green, visible. Lower lemma 2.0 mm long, 1.0 mm broad, linear, hyaline, sale–like, obtuse, apiculate, glabrous, empty, epaleate. Upper lemma 0.5 mm long, 0.1mm broad, hyaline, bisexual, epaleate, faintly1–nerved, glabrous, reduced to awn; awn 1.6cm long, stout, antrorsely scabrid. Stamens 3; anther 2.0mm long, yellow, glabrous, linear; filament 0.5 mm long, glabrous, slender, shorter than anther. Pistil 1.5mm long; ovary 0.5 mm long, light green, oblong, glabrous; style 0.6mm long, slender, bifid from base, glabrous; stigma 0.5 mm long, acuminate, hairy; hairs acute, base broad, in 2–rows. Caryopsis 2.0 mm long, 1.0 mm broad, brown, ovoid.

Flowering & Fruiting

September–January; Navalpur, Sindhora, Barahikala, Babatbur, Tarana, Chandauli, Gorakhpur, Bansi, Lar, Mau, Mirzapur.

Habitat

Occasional in fallow land & along raiways.

Distribution

India, Australia, Tropical Asia

22. *Microstegium* Nees

Microstegium ciliatum (Trin.) A. Camus in Ann. Soc.Linn. Lyon, n.s. 68: 201.1921; Fischer in Gamble, F1. Pres. Madr. 1716 1934 (Repr.ed.3: 1190.1957); Bor, Grass. Bur.Cey. Ind. Pak.193.1960; Sur in J. Econ.Tax. Bot. 6 (1): 169. 1985. Moulik, Grass. Bamb. Ind. 1: 337.1997. *Pollinia ciliata* Trin. in Mem. Acad. Sci. Petersb. ser.6. 2 (3): 306. 1832; Hook. f. F1. Brit. Ind. 7: 116. 1896. (PLATE–183).

Perennial. Rhizomatous; rhizome creeping; root fibrous. Culm 120 cm long, 2.0mm thick, green, glabrous, tufted, rembling, geniculately ascending, leafy, slender, creeping at base, rooting & branching at lower nodes; branches suberect; nodes glabrous; internodes 6.0 cm long. Leaf– blade 10 cm long, 1.2 cm broad, green, flat, elliptic, linear, lanceolate, acuminate, base narrowed like petiole, surface glabrous at apex, hairy at base; hairs 2.0 mm long, white, acute, base tubercled; margin hyaline & scabrid, mid-vein distinct on lower surface, 5–lateral veins on each side of mid-vein; lateral veins distinct on upper surface; leaf- sheath 4.0cm long, 4.0 mm broad, green, surface membranous, glabrous, margin hyaline, keeled, erect, loose, compressed, striate, shorter than internodes, veins distinct on both surface, hairy at mouth; hairs 5.0 mm long, white, acute, base tubercled; ligule 1.0 mm long, membranous, obscure; auricle glabrous. Inflorescence consists of racemes; raceme 10 cm long, 1.5 mm broad, sub–digitately arranged, 4–in number, flexuous, alternate; joints 4.0mm long, linear, clavate, densely hairy in 2–rows; hairs acute, base broad. Spikelets in alternate pairs. Sessile spikelet 5.0mm long, 1.0mm broad, oblong, linear, lanceolate, narrow, hairy,

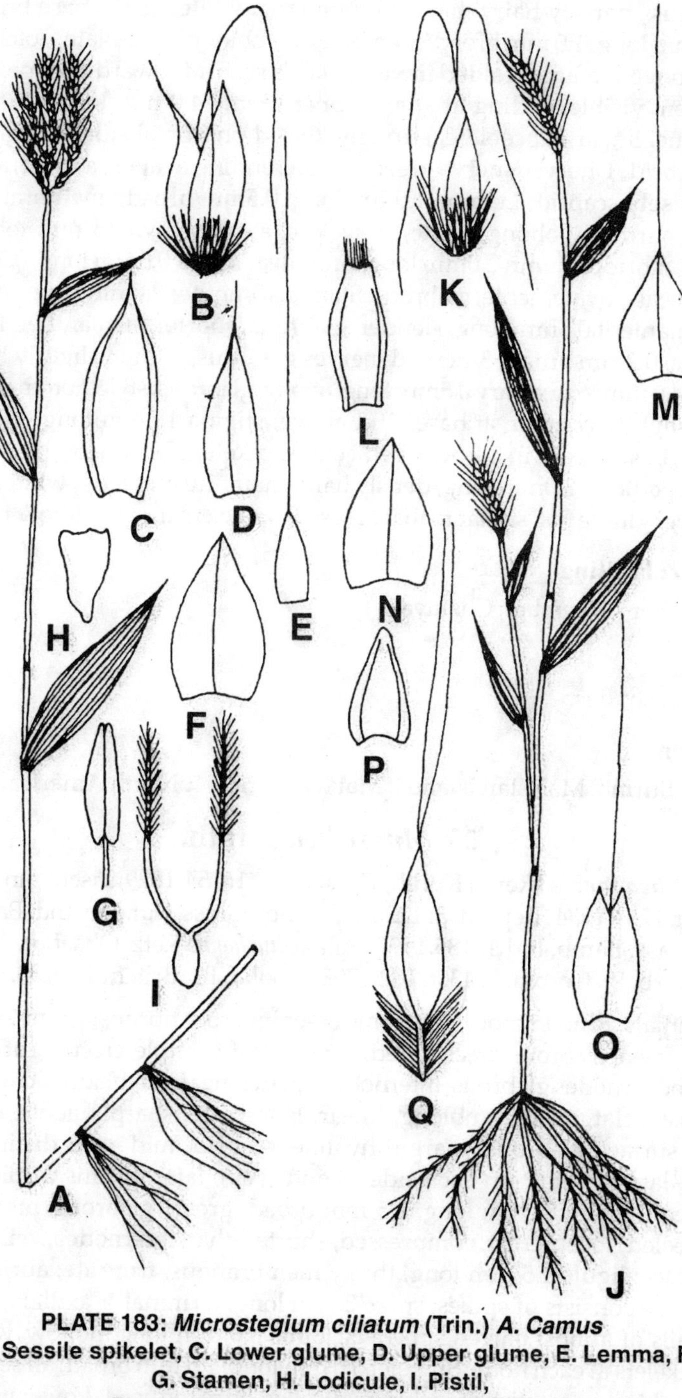

PLATE 183: ***Microstegium ciliatum*** **(Trin.)** ***A. Camus***
A. Habit, B. Sessile spikelet, C. Lower glume, D. Upper glume, E. Lemma, F. Palea, G. Stamen, H. Lodicule, I. Pistil,
Pogonatherum crinitum **(Thunb.) Kunth**
J. Habit, K. Sessile spikelet, L. Lower glume, M. Upper glume, N. Lower lemma, M. Upper lemma, P. Upper palea, Q. Pedicelled spikelet

awned; callus sparsely hairy; hairs 1.0 mm long, white, acute, base broad. Lower glume 5.0 mm long, 1.0 mm broad, membranous, oblong, lanceolate, notched at apex, dorsally grooved, margin folded inwards, keels scabrid toward the apex, 5-nerved; nerves green, visible, ending at apex. Upper glume 4.0 mm long, 0.5 mm broad, membranous, linear, lanceolate, strongly keeled on dorsal side, margin infolded, hyaline, scabrid, 1-nerved; nerve green, visible, ending at apex, awned; awn 2.0 mm long, antrorsely scabrid. Lemma 1.0mm long, 0.5mm broad, membranous, ovate, lanceolate, narrowly obong, acute, nerveless, awned; awn 15 mm long, slender, antrorsely scabrid; column 2.0mm long, brownish. Palea 0.5mm long, 0.2mm broad, hyaline, delicate, ovate, acute, glabrous. Stamens 3; anther 2.0mm long, yellow, linear, glabrous; filament 0.5mm long, slender, glabrous, shorter than anther. Lodicules 2, 0.5mm long, 0.2mm broad, 3-nerved; nerves glabrous, ending slightly below at the apex. Pistil 4.0 mm long; ovary 0.5mm long, obong, glabrous; style 2.0mm long, slender, glabrous, slightly connate at base,bifid above; stigma 1.5mm long, purple, hairy; hairs acute, base broad, in 2–rows. Pedicelled spikelets 5.0mm long, 1.0 mm broad, lanceolate; pedicels 3.0mm long, densly hairy; hairs 1.0 mm long, white, acute, base broad. Lower glume flat, simiar to that of sessile spikelet; upper glume & floret similar.

Flowering & Fruiting

September–December; Chakiya.

Habitat

Rare, in shades.

Distribution

India, Burma, Malesian Islands, Malaya, Africa, & North America.

23. *Mnesithea* Kunth.

Mnesithea laevis (Retz.) Kunth, Rev. Gram. 1: 154.1829; Fischer in Gamble, F1. Pres. Madr. 1761.1934 (Repred. 3: 1220.1957); Bor, Grass. Bur. Cey.Ind. Pak. 344. 1960; Moulik, Grass. Bamb. Ind.1: 185.1997. *Rottboellia laevis* Retz. Obs. Bot. 3: 11. 1783. *R. perforata* Roxb. Pl. Corom. 2: 43. t. 182.1798; Hook.f. F1. Brit. Ind.7: 158.1896.

Perennials. Rhizomatours; rhizome creeping; root fibrous. Culm 150cm tall, 2.0 mm thick, green, glabrous, erect, tufted, slender, leafy, simple, creeping at base, rooting at lower nodes; nodes glabrous; internodes 10 cm long. Leaf- blade 50cm long, 8.0mm broad, green, flat, narrow, oblong, linear, lanceolate, sharply acute at apex, base rounded, surface glabrous, margin hyaline, scabrid, mid–vein distinct on lower surface, 4–lateral veins on each side of mid- vein; lateral veins distinct on upper surface; leaf–sheath 8.0 cm long, 3.5 mm broad, green, glabrous, margin hyaline, slightly keeled, loose, erect, compressed, shorter–than internodes, veins distinct on outer surface; ligule 1.5 mm long, thinly membranous, truncate; auricle glabrous. Inflorescence consists of spikes; spike 20 cm long, terminal & axillary, ending with narrow tails of almost barren spikelets; joints 6.0 mm long, hollow, with a pair of sessile spikelets at each node. Spikelet 5.0 mm long, 1.5 mm broad, all sessile, oblong, awnless, sunken in cavities of rachis, pedicelled ones reduced. Lower glume 5.0 mm long, 1.5mm broad, coriaceous, oblong, 7-nerved; nerves visible only on inner side.

Upper glume 5.0 mm long, 2.0 mm broad, hyaline, delicate, oblong, faintly 3–nerved. Lower lemma 4.5 mm long, 1.5 mm broad, hyaline, delicate, oblong, empty, epaleate, 3–nerved. Upper lemma 4.0mm long, 1.5mm broad, hyaline, delicate, oblong, faintly 3–nerved. Upper palea 3.5 mm long, 1.0 mm broad, hyaline, delicate, oblong. Stamens 3; anther 3.0 mm long, pinkish, linear, glabrous; filament 0.5mm long, slender, glabrous, shorter than anther. Lodicules absent. Pistil 3.0 mm long; ovary 0.5mm long, oblanceolate, glabrous; style 1.0 mm long, slender, glabrous, bifid from base; stigma 1.5 mm long, plumose. Caryopsis 2.0mm long, 0.6mm broad.

Flowering & Fruiting

September- November; Naugarh, Sonbhadra, Bherafarm, Duddhi.

Habitat

Rare, usually in dry sandy soil & along hill slopes.

Distribution

India, Burma, Sri Lanka, Malaya.

24. *Ophiuros* Gaertn.f

Ophiuros exaltatus (L) O. Ktze. Rev. Gen. Pl.2: 780.1891. Bor, Grass.Bur. Cey. Ind. Pak. 199.1960; Moulik,Grass.Bamb. Ind. 1: 198.1997. *Aegilops exaltata* L. Mant. Pl. 2: 575.1771. (PLATE–184).

Strongly perennial. Rhizomatous; rhizome 10 cm long, creeping; root fibrous. Culms 95cm tall, 2.0mm thick, green,glabrous, erect, tufted, compressed, ascending, bulbous at base, rooting at nodes; nodes brown, glabrous; internodes 11 cm long, hollow. Leaf–blade 8.0 cm long, 2.0 mm broad, green, glabrous, margin scaberulous, hyaline, acute, base rounded, narrow, flat, linear, hairy at base; hairs very minute, acute, base tubercled, mid–vein distinct on both surface; 3–lateral veins at each side of mid–vein; lateral veins distinct on lower surface; leaf–sheath 2.0 cm long, 4.0 mm broad, green, glabrous, margin hyaline, loose, very much shorter than internodes, veins distinct on outer surface; ligule 0.5 mm long, membranous, truncate, hairy at apex; hairs acute, base tuberceled; auricle glabrous. Inflorescence conists of spike–like racemes; racemes 7.0 cm long, 2.0 mm broad, 6 in number, slender, corymbosely fascicled, curved, erect, originated at nodes. Spikelets 3.0 mm long, 1.2 mm broad, solitary, sessile, bisexual, acuminate, glabrous, awnless, lanceolate. Lower glume 3.0 mm long, 1.0 mm broad, membranous, slightly oblique, ovate, linear, lanceolate, acuminate, surface glabrous, margin winged; wing hyaline, scaberulous at apex, 9–nerved; nerves green, visible, glabrous. Upper glume 3.0 mm long, 1.5 mm broad, boat shaped, obtuse, cuspidate, surface glabrous, margin glabrous below, scaberulous at apex, membranous, 5–nerved; nerves green, visible, nerves ending at the apex. Lower lemma 2.0 mm long, 1.0 mm broad, hyaline, empty, ovate, lanceolate, oblong, apiculate, surface glabrous, margin scaberulous, faintly 2–nerved; nerves colourless, glabrous, ending at apex. Lower palea 2.0 mm long, 1.0 mm broad, hyaline, acuminate, surface glabrous, margin sharply inrolled, glabrous, 2–nerved; nerves light green, marginal, scaberulous. Upper lemma 2.0 mm long, 1.0 mm broad, hyaline, nerveless, bisexual, linear, apex rounded, surface & margin glabrous. Upper palea 1.5 mm long, 0.5 mm

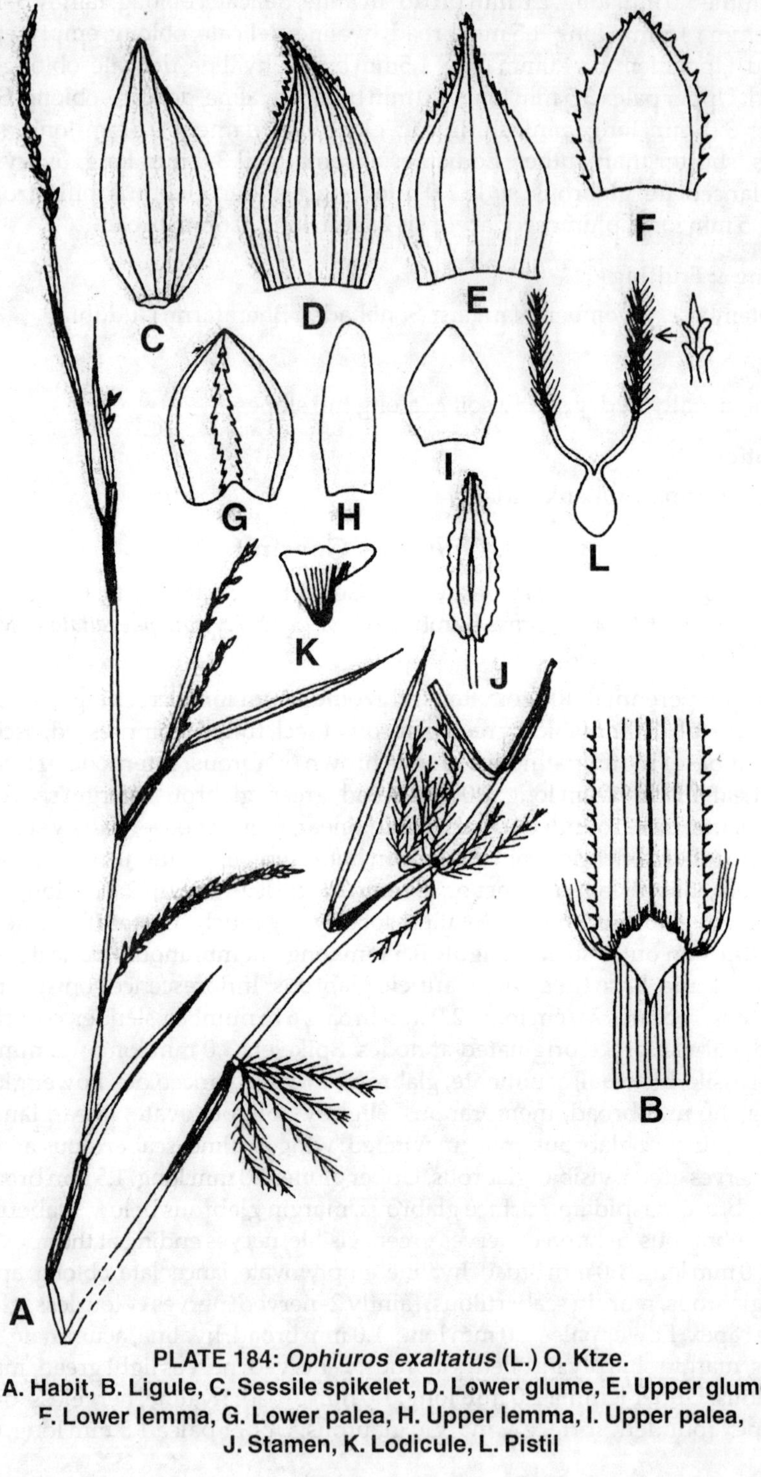

PLATE 184: *Ophiuros exaltatus* (L.) O. Ktze.
A. Habit, B. Ligule, C. Sessile spikelet, D. Lower glume, E. Upper glume, F. Lower lemma, G. Lower palea, H. Upper lemma, I. Upper palea, J. Stamen, K. Lodicule, L. Pistil

broad, acuminate, nerveless, hyaline, lanceolate, glabrous. Stamens 3; anther 1.5 mm long, yellow, linear, glabrous; filament 0.5 mm long, glabrous, slender, shorter than anther. Lodicules 2; 0.3 mm long, 0.1 mm broad, obovate, cuneate, hyaline, many nerved, glabrous. Pistil 2.0 mm long; ovary 0.5 mm long, oblong, ovate, glabrous; style 0.5 mm long, slender, glabrous, bifid from base; stigma 1.0 mm long, brown, acuminate hairy; hairs acuminate, base broad, in 2–rows. Caryopsis 1.5 mm long, 1.0 mm broad, slightly dorsally compressed.

Flowering & Fruiting

September–December; Rajghat, banks of Varuna rivers, Chandauli, Sadalpura, Babatpur, Jaunpur, Gorakhpur, Bakhira, Padrauna.

Habitat

Common in moist localities, banks of rivers.

Distribution

India, Australia, China.

25. *Pogonatherum* P. Beavu.

KEY TO THE SPECIES

1a. Spikelets 2.5-3.0 mm long; callus hairs of pedicelled spikelets upto 2.0 mm long; lower floret male; upper floret with 2 stamens; awns 10-15 mm long

.... 2. *P. panicum*

1b. Spikelets upto 2.0 mm long; callus hairs of pediclled spikelets 2–4.0 mm long; lower floret empty; upper floret with 1 stamen; awns 20-30 mm long

.... 1. *P. crinitum*

1. *Pogonatherum crinitum*

Pogonatherum crinitum (Thunb.) Kunth. Enum. Pl.1: 478. 1833; Hook. f. Fl. Brit. Ind. 7: 141. 1896; Bor. Grass, Bur. Cey. Ind. Pak.200.1960; Sur in Journ. Econ.Tax. Bot. 6(3): 665. 1985; Roy, Grass. M.P.132.1984; Shreekumar & Nair, Fl. Kera. Grass. 176.1991; Moulik, Grass. Bamb.Ind. 1: 342. 1997. *Andropogon crinitus* Thunb.Fl. Jap. 40.t. 7.1784. *Pogonatherum paniceum* acut. non (Lam.) Hack. 1906; Alst. Suppl. Fl. Ceyl. 330. 1931; Fischer in Gamble, Fl. Pres. Madr. 1713.1934 (Repr. ed.3: 1188. 1957). (PLATE–183).

Perennial. Rhizomatous; rhizome woody with fibrous root stock. Culms 40 cm tall, 2.0 mm thick, green, glabrous, densely tufted, erect, ascending, leafy throughout, branched; branches suberect, rooting at base, nodes hairy; hairs 1.0mm long, white,acute, base broad; internodes 6.0 cm long.Leaf–blade 8.0 cm long, 4.0 mm broad, green, soft, flat, elliptic, linear, lanceolate, acuminate, base narrowed, glabrous, margin hyaline, scabrid, mid–vein distinct on lower surface, 3–lateral veins on each side of mid–vein; lateral veins distinct on both surface; leaf–sheath 3.5 cm long, 3.0 mm broad, green, surface membranous, glabrous, margin hyaline, keeled, hairy along one side; hairs 1.0 mm long, white, acute, base tubercled, loose, erect, compressed, shorter than internodes, hairy at mouth; hairs 3.0 mm long, white, acute, base tubercled; veins distinct on both surface; ligule 1.0 mm long, thinly membranous, hairy at apex; hairs minute, acute, base broad; auricle glabrous. Inflorescence consists of racemes; racemes 3.5 cm long, 1.0 mm broad, solitary, terminal, slender; joints 1.0 mm long, turbinate, densely hairy along the angles; hairs 2.0 mm long, white, acute, base tubercled. Sessile spikelets 2.0 mm long, 1.0 mm broad, lanceolate, awned, cream–yellow. Lower glume 1.5 mm long, 0.5 mm broad, membranous, oblong, truncate & hairy at apex; hairs 1.0 mm long, acute, base broad, slightly 2–keeled, 2–nerved; nerves yellow, visible, glabrous & marginal, ending below at the apex. Upper glume 2.0 mm long, 1.0 mm broad, membranous, boat shaped at compressed veiw, ovate, lanceolate when spread, glabrous, 3–nerved; nerves yellow, glabrous, visible, ending at apex; awned; awn 25 mm long, slender, antrorsely scabrid. Lower lemma 1.0 mm long, 0.5 mm broad, hyaline, delicate, ovate, oblong, obtuse at apex, empty, epaleate, 1–nerved; nerves colourless, glabrous, ending below at the apex. Upper lemma 1.5 mm long, 0.5 mm broad, hyaline, delicate, linear, notched at apex, bisexual, paleate, 1–nerved; nerve colourless, glabrous, ending at apex; awned; awn 30 mm long, slender, antrorsely scabrid, arising from the sinus of notch. Upper palea 1.0 mm long, 0.5 mm broad, hyaline, delicate, ovate, oblong, rounded at apex, glabrous, 2–keeled, 2–nerved; nerves slightly visible, glabrous, & marginal, ending slightly below at the apex. Stamens 1; anther 1.0 mm long, pale yellow, linear, glabrous; filament 0.5 mm long, slender, glabrous, shorter than anther. Lodicules absent. Pistil 2.2 mm long; ovary 0.2 mm long, elliptic, glabrous; style 1.0 mm long, glabrous, slender, bifid from base; stigma 1.0 mm long, cream coloured, hairy; hairs acute, base broad, in 2–rows. Caryopsis 1.5 mm long, 0.5 mm broad, linear. Pedicelled spikelet similar; callus hairy; hairs 4.0 mm long, white, acute, base tubercled.

Flowering & Fruiting

July–October; Rajdhari, Duddhi, Sonbhadra.

Habitat

Rare, along the hill slopes and moist rocky places.

Distribution

India, China, Malaya.

2. *P. paniceum*

P. paniceum (Lam.) Hack, in Allg. Bot. Z. 12: 178. 1906. Bor, Grass. Bur. Cey. Ind. Pak. 202. 1960; Cope in Nasir & Ali, Fl. W. Pak. 143: 271, 272. t.29 (1–2). 1982; Sur, Journ. Econ. Tax. Bot. 6 (3): 666.1985; Roy, Grass. M.P. 132.1984; Shreekumar & Nair, Fl. Kera. Grass. 178. 1991; Moulik, Grass. Bamb. Ind. 1: 342.1997. *Saccharum paniceum* Lam. Encycl. Meth. Bot. 1: 595. 1785. *Pogonatherum saccharoideum* P.Beauv. Ess. Agrost. 177. 1812; Hook. f. Fl. Brit. Ind. 7: 141. 1896.

Perennial. Rhizomatous; rhizome woody, with fibrous root stock. Culms 50cm tall, 1.5mm thick, green, glabrous, erect, densely tufted, ascending, leafy throughout, slender, branched; branches suberect, rooting at base; nodes hairy; hairs 1.0 mm long, white, acute, base broad; internodes 7.0 cm long. Leaf–blade 8.0 cm long, 3.0 mm broad, green, soft, narrow, linear, lanceolate, acuminate, base rounded, margin hyaline & scabrid, mid–vein distinct on lower surface, 2–lateral veins on each side of mid–vein; lateral veins distinct on upper surface; leaf–sheath 4.0 cm long, 2.5 mm broad, green, membranous, margin hyaline, keeled, hairy towards the mouth; hairs 2.0 mm long, white, acute, base tuberded, glabrous below, erect, shorter than internodes, loose compressed; ligule 1.0 mm long, thinly membranous, hairy at apex; hair minute, actue, base broad; acuricle glabrous. Inflorescence consists of racemes; raceme 4.0 cm long, solitary, terminal, yellowish–cream; joints 1.5 mm long, turbinate, densely hairy along the margins; hairs 2.0 mm long, white, actue, base broad. Sessile spikelet 3.0 mm long, 1.0 mm broad, oblong, lanceloate, awned; callus 2.0 mm long, hairy; hairs acute, base tuberecled. Lower glume 2.5mm long,1.0 mm broad, membranous, oblong, 2–lobed, denticulate & densely hairy at apex; hairs 1.0 mm long, acute, base tubercled; slightly 2–keeled, 2–nerved; nerves visible, glabrous & marginal, ending at apex. Upper glume 3.0 mm long,1.0 mm broad, membranous, boat shaped at compressed view, ovate, lanceolate when spread, keeled on dorsal sides, margin hyaline, awned; awn 15 mm long, slender, antrorsely scabrid. Lower lemma 2.0 mm long, 0.5 mm broad, hyaline, oblong, truncate, hairy at apex; hairs 0.5 mm long, acute, base tubercled; male, paleate, slightly keeled, 2–nerved; nerves slightly visible, glabrous & marginal, ending at apex. Lower palea 2.0 mm long, 0.5 mm broad, hyaline, delilate, linear, oblong, nerveless. Stamens 2; anther 1.0 mm long, glabrous; filament 0.5 mm long, slender, glabrous, shorter than anther. Upper lemma 2.0 mm long, 0.5 mm broad, hyaline, notched, bisexual, paleate, nerveless, awned; awn 15 mm long slender, antrorsely scabrid. Upper palea 2.0 mm long, 1.0 mm broad, hyaline, acute, oblong, truncate, hairy at apex; hairs acute, base broad, slightly 2–keeled, 2–nerved; nerves slightly visible, glabrous & marginal, ending at apex. Stamens 2; anther 1.5 mm long, oblong, linear, glabrous; filament 0.5 mm long, slender, glabrous, shorter than anther. Lodicules absent. Pistil 3.0 mm long; ovary 0.2 mm long, oblong; style 1.0 mm long, slender, glabrous, bifid from base; stigma 1.8 mm long, hairy; hairs acute, base broad, in 2– rows. Caryopsis 1.5 mm long, 1.0 mm broad, linear. Pedicelled spikelets 2.5mm long, 1.0 mm broad, lanceolate, awned; pedicels 2.0 mm long, linear, turbinate, hairy along the angles; hairs 2.0 mm long, acute, base tubercled. Glumes & florest similar to those of sessile spikelets.

Flowering & Fruiting

May–June; Chakiya, Chandauli, Sonbhadra, Mirzapur, Chunar.

Habitat

Rare, in dry rocky places and hill slopes.

Distribution

India, Burma, SriLanka.

26. *Pseudosorghum* A. Camus

Pseudosorghum fasciculare (Roxb.) A. Camus in Bull. Mus. Hist. Nat. Paris 26: 662. 1920; Fischer in Gamble, Fl. Pres. Madr. 1734.1934 (Repr. ed. 3: 1202.1957); Bor,

Grass. Bur. Cey. Ind. Pak. 205. 1960; Roy, Grass. M.P. 135. 1984; Shreekumar & Nair, Fl. Kera. Grass.182.1991; Moulik, Grass. Bamb. Ind. 1: 247. 1997. *Andropogon fascicularis* Roxb. Fl. Ind. 1: 269. 1920; Hook. f. Fl. Brit. Ind. 7: 117. 1896. *A. gangeticus* Hack. In DC. Monogr. Phan. 6: 539. 1889. *A. tonkinensis* Balansa in Morot. J. de. Bot. 4: 112. 1890. *A. nitidulus* Hook. f. Fl. Brit. Ind. 7: 199. 1896. *Sorghum fasciculare* Haines, Bot. Bih. & Or. 1034.1924.

Perennial. Rhizomatous; root fibrous. Culm 150 cm tall, green, glabrous, erect, stout, leafy simple; nodes hairy; hairs acute, base tubercled; internode 15 cm long. Leaf–blade 40 cm long, 8.0 mm broad, flat, linear–lanceolate, acuminate, base rounded, glabrous, margin hyaline, & scabrid; leaf sheath 18 cm long, erect, compressed, striate, longer than internodes & densely hairy; hairs acute, base tubercled; veins distinct on outer surface; ligule 5.0 mm long, ovate, truncate, acute, membranous; auricle glabrous. Inflorescence consists of panicle; panicle 15 cm long, open; joints 3.0 mm long, dorsally rounded, ventrally flat, densely bristly. Sessile spikelets 6.0mm long, 1.0 mm broad, lanceolate, awned; callus densly hairy; hairs 2.0 mm long, acute, base broad. Lower glume 5.0 mm long, 2.0 mm broad, subcoriaceous, elliptic, lanceolate, truncate, slightly notched at apex, keeled, pectinate towards the apex, 7–nerved; nerves colourless, faintly glabrous, ending at apex. Upper glume 5.0 mm long, 1.5 mm broad, subcoriaceous, glabrous, boat–shaped, dorsal side slightly keeled towards the apex, 7–nerved; nerves colourless, glabrous, ending at apex. Lower lemma 5.0 mm long, 1.0mm broad, elliptic, lanceolate, hyaline, delicate, surface glabrous, margin hairy; hairs 1.0 mm long, white, acute, base broad, empty, epaleate, awnless, 5- nerved; nerves green, visible, glabrous, ending at apex. Upper lemma 4.0 mm long, 1.0 mm broad, hyaline, delicate, bicleft into two acute lobes, bisexual, epleate, awned; awn 25 mm long, slender, antrorsely scabrid, arising from the sinus of lobes; column 12 mm long, brownish. Stamens 3; anther 1.0 mm long, yellow, glabrous; filament 0.5 mm long, slender, glabrous, shorter than anther. Lodicules absent. Pistil 2.5 mm long; ovary 0.5 mm long, ovate; style 1.0 mm long, slender, glabrous, bifid, from base; stigma 1.0 mm long, hairy; hairs acute, base broad, in 2–rows. Caryopsis 4.0 mm long, 1.5 mm broad, linear, lanceolate. Pedicelled spikelet 5.0 mm long, 1.5 mm broad, linear, lanceolate, unawned, pedicels 4.0 mm long. Lower glume 5.0 mm long, 1.5 mm broad, subcoriaceous, elliptic, oblong, linear, lanceolate, keels pectinate towards the apex, 13–nerved; nerves colourless, glabrous, ending at apex. Upper glume 5.0 mm long, 1.0 mm broad, subcoriaceous, surface glabrous, margin minutely ciliate, 9-nerved; nerves colorless, glabrous, ending at apex. Lower floret empty. Upper floret male or empty.

Flowering & Fruiting

September–December; Jaunpur, Gorakhpur, Basti.

Habitat

Rare, along the banks of streams.

Distributation

India, South–East Asia, Burma.